AF737351

© Woodhead Publishing Limited, 2013

© Woodhead Publishing Limited, 2013

© Woodhead Publishing Limited, 2013

Contents

© Woodhead Publishing Limited, 2013

Published by Woodhead Publishing Limited,
80 High Street, Sawston, Cambridge CB22 3HJ, UK
www.woodheadpublishing.com
www.woodheadpublishingonline.com

Woodhead Publishing, 1518 Walnut Street, Suite 1100, Philadelphia, PA 19102-3406, USA

Woodhead Publishing India Private Limited, G-2, Vardaan House, 7/28 Ansari Road, Daryaganj, New Delhi – 110002, India
www.woodheadpublishingindia.com

First published 2013, Woodhead Publishing Limited
© Woodhead Publishing Limited, 2013. The publisher has made every effort to ensure that permission for copyright material has been obtained by authors wishing to use such material. The authors and the publisher will be glad to hear from any copyright holder it has not been possible to contact.
The authors have asserted their moral rights.

This book contains information obtained from authentic and highly regarded sources. Reprinted material is quoted with permission, and sources are indicated. Reasonable efforts have been made to publish reliable data and information, but the authors and the publishers cannot assume responsibility for the validity of all materials. Neither the authors nor the publishers, nor anyone else associated with this publication, shall be liable for any loss, damage or liability directly or indirectly caused or alleged to be caused by this book.

Neither this book nor any part may be reproduced or transmitted in any form or by any means, electronic or mechanical, including photocopying, microfilming and recording, or by any information storage or retrieval system, without permission in writing from Woodhead Publishing Limited.

The consent of Woodhead Publishing Limited does not extend to copying for general distribution, for promotion, for creating new works, or for resale. Specific permission must be obtained in writing from Woodhead Publishing Limited for such copying.

Trademark notice: Product or corporate names may be trademarks or registered trademarks, and are used only for identification and explanation, without intent to infringe.

British Library Cataloguing in Publication Data
A catalogue record for this book is available from the British Library.

Library of Congress Control Number: 2012953436

ISBN 978-0-85709-119-2 (print)
ISBN 978-0-85709-746-0 (online)
ISSN 2042-8049 Woodhead Publishing Series in Food Science, Technology and Nutrition (print)
ISSN 2042-8057 Woodhead Publishing Series in Food Science, Technology and Nutrition (online)

The publisher's policy is to use permanent paper from mills that operate a sustainable forestry policy, and which has been manufactured from pulp which is processed using acid-free and elemental chlorine-free practices. Furthermore, the publisher ensures that the text paper and cover board used have met acceptable environmental accreditation standards.

Typeset by Toppan Best-set Premedia Limited
Printed and bound in the UK by the MPG Books Group

Woodhead Publishing Series in Food Science, Technology and Nutrition:
Number 242

Advances in aquaculture hatchery technology

Edited by
Geoff Allan and Gavin Burnell

WOODHEAD PUBLISHING

Oxford Cambridge Philadelphia New Delhi

© Woodhead Publishing Limited, 2013

Related titles:

Infectious disease in aquaculture: Prevention and control
(ISBN 978-0-85709-016-4)

New technologies in aquaculture: Improving production efficiency, quality and environmental management (ISBN 978-1-84569-384-8)

Shellfish safety and quality (ISBN 978-1-84569-152-3)

Details of these books and a complete list of titles from Woodhead Publishing can be obtained by:

- visiting our web site at www.woodheadpublishing.com
- contacting Customer Services (e-mail: sales@woodheadpublishing.com; fax: +44 (0) 1223 832819; tel.: +44 (0) 1223 499140 ext. 130; address: Woodhead Publishing Limited, 80 High Street, Sawston, Cambridge CB22 3HJ, UK)
- in North America, contacting our US office (e-mail: usmarketing@woodheadpublishing.com; tel.: (215) 928 9112; address: Woodhead Publishing, 1518 Walnut Street, Suite 1100, Philadelphia, PA 19102-3406, USA)

If you would like e-versions of our content, please visit our online platform: www.woodheadpublishingonline.com. Please recommend it to your librarian so that everyone in your institution can benefit from the wealth of content on the site.

We are always happy to receive suggestions for new books from potential editors. To enquire about contributing to our Food Science, Technology and Nutrition series, please send your name, contact address and details of the topic/s you are interested in to nell.holden@woodheadpublishing.com. We look forward to hearing from you.

The Woodhead team responsible for publishing this book:
Commissioning Editor: Sarah Hughes
Publications Coordinator: Anneka Hess
Project Editor: Rachel Cox
Editorial and Production Manager: Mary Campbell
Production Editor: Adam Hooper
Copyeditor: Helen MacFadyen
Proofreader: George Moore
Cover Designer: Terry Callanan

© Woodhead Publishing Limited, 2013

Advances in aquaculture hatchery technology

© Woodhead Publishing Limited, 2013

© Woodhead Publishing Limited, 2013

© Woodhead Publishing Limited, 2013

© Woodhead Publishing Limited, 2013

Contributor contact details

(* = main contact)

Editors

Geoff Allan
Port Stephens Fisheries Institute
Locked bag 1
Nelson Bay
NSW 2315
Australia

E-mail: geoff.allan@dpi.nsw.gov.au

Gavin Burnell
School of Biological, Earth & Environmental Sciences
North Mall Campus
University College Cork
Distillery Fields
Cork
Ireland

E-mail: g.burnell@ucc.ie

Chapter 1

Odd-Ivar Lekang
Department of mathematical science and technology
Norwegian University of Life Sciences
1430 Aas
Norway

E-mail: odd-ivar.lekang@umb.no

Chapter 2

Neil Duncan*
IRTA Sant Carles de la Rapita
P.O. Box 200
E-43540 Sant Carles de la Rapita
Tarragona
Spain

E-mail: Neil.duncan@irta.es

© Woodhead Publishing Limited, 2013

Anna K. Sonesson and Hervé Chavanne
Nofima AS
P.O. Box 210
1431 Ås
Norway

E-mail: anna.sonesson@nofima.no; herve.chavanne@nofima.no

Chapter 3

Dr Catherine Labbé
INRA
UR1037 Fish Physiology and Genomic (LPGP)
Campus de Beaulieu
F-35000 Rennes
France

E-mail: Catherine.Labbe@rennes.inra.fr

Chapter 4

Malcolm R. Brown* and Susan I. Blackburn
Energy Transformed and Food Futures Flagships
CSIRO Marine and Atmospheric Research
G.P.O. Box 1538
Hobart 7001
Tasmania
Australia

E-mail: malcolm.brown@csiro.au

Chapter 5

Patrick Sorgeloos
Laboratory of Aquaculture & Artemia Reference Center
Ghent University
Rozier 44
Ghent
Belgium

E-mail: patrick.sorgeloos@ugent.be

Chapter 6

Sagiv Kolkovski
Department of Fisheries
Western Australia
P.O. Box 20
North Beach
WA 6920
Australia

E-mail: skolkovski@fish.wa.gov.au

Chapter 7

T. J. Bowden
Aquaculture Research Institute
Department of Animal & Veterinary Sciences
University of Maine
5735 Hitchner Hall
Orono
ME 04469-5735
USA

E-mail: timothy.bowden@umit.maine.edu

© Woodhead Publishing Limited, 2013

Chapter 8

E. F. Goulden, L. Høj and M. R. Hall*
Australian Institute of Marine Science (AIMS)
PMB No. 3
Townsville Mail Centre
Townsville
Qld 4810
Australia

E-mail: m.hall@aims.gov.au

Chapter 9

M. R. Hall*, M. Kenway, M. Salmon, D. Francis, E. F. Goulden and L. Høj
Australian Institute of Marine Science (AIMS)
PMB No. 3
Townsville Mail Centre
Townsville
Qld 4810
Australia

E-mail: m.hall@aims.gov.au

Chapter 10

Dr Jim Wyban
High Health Aquaculture Inc.
Kona Kailua
HI 96740
USA

E-mail: jim.wyban@gmail.com

Chapter 11

Pauline Kamermans
IMARES
Wageningen UR
P.O. Box 77
4400 AB Yerseke
The Netherlands

E-mail: pauline.kamermans@wur.nl

Chapter 12

Dr J. Iglesias
Oceanographic Center of Vigo
Spanish Institute of Oceanography (IEO)
Subida a Radio Faro 50
36390 – Vigo
Spain

E-mail: jose.iglesias@vi.ieo.es

Chapter 13

Dr J. E. Purcell*
Shannon Point Marine Center
Western Washington University
1900 Shannon Point Road
Anacortes
Washington 98221
USA

E-mail: Jennifer.Purcell@wwu.edu

E. J. Baxter
Vet-Aqua International
Unit 7B
Oranmore Business Park
Oranmore
Co. Galway
Ireland

© Woodhead Publishing Limited, 2013

Dr V. L. Fuentes
Marine Biology and Oceanography Department
Instituto de Ciencias del Mar (CSIC)
Paseo Marítimo de la Barceloneta 37–49
08003 Barcelona
Spain

Chapter 14

Annie Mercier*
Ocean Sciences Centre (OSC)
Memorial University of Newfoundland
St John's
NL A1C 5S7
Canada

E-mail: amercier@mun.ca

Jean-François Hamel
Society for the Exploration and Valuing of the Environment (SEVE)
21 Phils Hill Road
St Philips
NL A1M 2B7
Canada

E-mail: jfhamel.seve@gmail.com

Chapter 15

Dr Gavin J. Partridge
Australian Centre for Applied Aquaculture Research
Challenger Institute of Technology
1 Fleet St Fremantle
WA 6160
Australia

E-mail: gavin.partridge@challenger.wa.edu.au

Chapter 16

Nguyen Thanh Phuong
College of Aquaculture and Fisheries
Can Tho University
Vietnam

E-mail: ntphuong@ctu.edu.vn

Chapter 17

Neil Duncan* and Alicia Estévez
IRTA Sant Carles de la Rapita
P.O. Box 200
E-43540 Sant Carles de la Rapita
Tarragona
Spain

E-mail: Neil.duncan@irta.es; alicia.estevez@irta.cat

Hipólito Fernández-Palacios
Grupo de Investigación en Acuicultura
Instituto Canario de Ciencias Marinas (ICCM) & Universidad de las Palmas de Gran Canaria
P.O. Box 56
E-35200 Telde
Las Palmas
Canary Islands
Spain

E-mail: pipo@iccm.rcanaria.es

© Woodhead Publishing Limited, 2013

Chapter 18

D. Stewart Fielder
Port Stephens Fisheries Institute
NSW Department of Primary Industries
Locked Bag 1
Nelson Bay
NSW 2315
Australia

E-mail: stewart.fielder@industry.nsw.gov.au

Chapter 19

Stuart J. Rowland
NSW Fisheries
Grafton Aquaculture Centre
PMB 2
Grafton
NSW 2460
Australia

E-mail: ikei@westnet.com.au

Chapter 20

D. Meritt
University of Maryland
Center for Environmental Science
Horn Point Laboratory
P.O. Box 13
Cambridge
Maryland 21613
USA

E-mail: meritt@hpl.umces.edu

© Woodhead Publishing Limited, 2013

Woodhead Publishing Series in Food Science, Technology and Nutrition

1 **Chilled foods: a comprehensive guide** *Edited by C. Dennis and M. Stringer*
2 **Yoghurt: science and technology** *A. Y. Tamime and R. K. Robinson*
3 **Food processing technology: principles and practice** *P. J. Fellows*
4 **Bender's dictionary of nutrition and food technology Sixth edition** *D. A. Bender*
5 **Determination of veterinary residues in food** *Edited by N. T. Crosby*
6 **Food contaminants: sources and surveillance** *Edited by C. Creaser and R. Purchase*
7 **Nitrates and nitrites in food and water** *Edited by M. J. Hill*
8 **Pesticide chemistry and bioscience: the food-environment challenge** *Edited by G. T. Brooks and T. Roberts*
9 **Pesticides: developments, impacts and controls** *Edited by G. A. Best and A. D. Ruthven*
10 **Dietary fibre: chemical and biological aspects** *Edited by D. A. T. Southgate, K. W. Waldron, I. T. Johnson and G. R. Fenwick*
11 **Vitamins and minerals in health and nutrition** *M. Tolonen*
12 **Technology of biscuits, crackers and cookies Second edition** *D. Manley*
13 **Instrumentation and sensors for the food industry** *Edited by E. Kress-Rogers*
14 **Food and cancer prevention: chemical and biological aspects** *Edited by K. W. Waldron, I. T. Johnson and G. R. Fenwick*
15 **Food colloids: proteins, lipids and polysaccharides** *Edited by E. Dickinson and B. Bergenstahl*
16 **Food emulsions and foams** *Edited by E. Dickinson*
17 **Maillard reactions in chemistry, food and health** *Edited by T. P. Labuza, V. Monnier, J. Baynes and J. O'Brien*
18 **The Maillard reaction in foods and medicine** *Edited by J. O'Brien, H. E. Nursten, M. J. Crabbe and J. M. Ames*

© Woodhead Publishing Limited, 2013

19 **Encapsulation and controlled release** *Edited by D. R. Karsa and R. A. Stephenson*
20 **Flavours and fragrances** *Edited by A. D. Swift*
21 **Feta and related cheeses** *Edited by A. Y. Tamime and R. K. Robinson*
22 **Biochemistry of milk products** *Edited by A. T. Andrews and J. R. Varley*
23 **Physical properties of foods and food processing systems** *M. J. Lewis*
24 **Food irradiation: a reference guide** *V. M. Wilkinson and G. Gould*
25 **Kent's technology of cereals: an introduction for students of food science and agriculture Fourth edition** *N. L. Kent and A. D. Evers*
26 **Biosensors for food analysis** *Edited by A. O. Scott*
27 **Separation processes in the food and biotechnology industries: principles and applications** *Edited by A. S. Grandison and M. J. Lewis*
28 **Handbook of indices of food quality and authenticity** *R. S. Singhal, P. K. Kulkarni and D. V. Rege*
29 **Principles and practices for the safe processing of foods** *D. A. Shapton and N. F. Shapton*
30 **Biscuit, cookie and cracker manufacturing manuals Volume 1: ingredients** *D. Manley*
31 **Biscuit, cookie and cracker manufacturing manuals Volume 2: biscuit doughs** *D. Manley*
32 **Biscuit, cookie and cracker manufacturing manuals Volume 3: biscuit dough piece forming** *D. Manley*
33 **Biscuit, cookie and cracker manufacturing manuals Volume 4: baking and cooling of biscuits** *D. Manley*
34 **Biscuit, cookie and cracker manufacturing manuals Volume 5: secondary processing in biscuit manufacturing** *D. Manley*
35 **Biscuit, cookie and cracker manufacturing manuals Volume 6: biscuit packaging and storage** *D. Manley*
36 **Practical dehydration Second edition** *M. Greensmith*
37 **Lawrie's meat science Sixth edition** *R. A. Lawrie*
38 **Yoghurt: science and technology Second edition** *A. Y. Tamime and R. K. Robinson*
39 **New ingredients in food processing: biochemistry and agriculture** *G. Linden and D. Lorient*
40 **Benders' dictionary of nutrition and food technology Seventh edition** *D. A. Bender and A. E. Bender*
41 **Technology of biscuits, crackers and cookies Third edition** *D. Manley*
42 **Food processing technology: principles and practice Second edition** *P. J. Fellows*
43 **Managing frozen foods** *Edited by C. J. Kennedy*
44 **Handbook of hydrocolloids** *Edited by G. O. Phillips and P. A. Williams*
45 **Food labelling** *Edited by J. R. Blanchfield*
46 **Cereal biotechnology** *Edited by P. C. Morris and J. H. Bryce*
47 **Food intolerance and the food industry** *Edited by T. Dean*
48 **The stability and shelf-life of food** *Edited by D. Kilcast and P. Subramaniam*
49 **Functional foods: concept to product** *Edited by G. R. Gibson and C. M. Williams*
50 **Chilled foods: a comprehensive guide Second edition** *Edited by M. Stringer and C. Dennis*

© Woodhead Publishing Limited, 2013

51 **HACCP in the meat industry** *Edited by M. Brown*
52 **Biscuit, cracker and cookie recipes for the food industry** *D. Manley*
53 **Cereals processing technology** *Edited by G. Owens*
54 **Baking problems solved** *S. P. Cauvain and L. S. Young*
55 **Thermal technologies in food processing** *Edited by P. Richardson*
56 **Frying: improving quality** *Edited by J. B. Rossell*
57 **Food chemical safety Volume 1: contaminants** *Edited by D. Watson*
58 **Making the most of HACCP: learning from others' experience** *Edited by T. Mayes and S. Mortimore*
59 **Food process modelling** *Edited by L. M. M. Tijskens, M. L. A. T. M. Hertog and B. M. Nicolaï*
60 **EU food law: a practical guide** *Edited by K. Goodburn*
61 **Extrusion cooking: technologies and applications** *Edited by R. Guy*
62 **Auditing in the food industry: from safety and quality to environmental and other audits** *Edited by M. Dillon and C. Griffith*
63 **Handbook of herbs and spices Volume 1** *Edited by K. V. Peter*
64 **Food product development: maximising success** *M. Earle, R. Earle and A. Anderson*
65 **Instrumentation and sensors for the food industry Second edition** *Edited by E. Kress-Rogers and C. J. B. Brimelow*
66 **Food chemical safety Volume 2: additives** *Edited by D. Watson*
67 **Fruit and vegetable biotechnology** *Edited by V. Valpuesta*
68 **Foodborne pathogens: hazards, risk analysis and control** *Edited by C. de W. Blackburn and P. J. McClure*
69 **Meat refrigeration** *S. J. James and C. James*
70 **Lockhart and Wiseman's crop husbandry Eighth edition** *H. J. S. Finch, A. M. Samuel and G. P. F. Lane*
71 **Safety and quality issues in fish processing** *Edited by H. A. Bremner*
72 **Minimal processing technologies in the food industries** *Edited by T. Ohlsson and N. Bengtsson*
73 **Fruit and vegetable processing: improving quality** *Edited by W. Jongen*
74 **The nutrition handbook for food processors** *Edited by C. J. K. Henry and C. Chapman*
75 **Colour in food: improving quality** *Edited by D. MacDougall*
76 **Meat processing: improving quality** *Edited by J. P. Kerry, J. F. Kerry and D. A. Ledward*
77 **Microbiological risk assessment in food processing** *Edited by M. Brown and M. Stringer*
78 **Performance functional foods** *Edited by D. Watson*
79 **Functional dairy products Volume 1** *Edited by T. Mattila-Sandholm and M. Saarela*
80 **Taints and off-flavours in foods** *Edited by B. Baigrie*
81 **Yeasts in food** *Edited by T. Boekhout and V. Robert*
82 **Phytochemical functional foods** *Edited by I. T. Johnson and G. Williamson*
83 **Novel food packaging techniques** *Edited by R. Ahvenainen*
84 **Detecting pathogens in food** *Edited by T. A. McMeekin*
85 **Natural antimicrobials for the minimal processing of foods** *Edited by S. Roller*
86 **Texture in food Volume 1: semi-solid foods** *Edited by B. M. McKenna*

© Woodhead Publishing Limited, 2013

87 **Dairy processing: improving quality** *Edited by G. Smit*
88 **Hygiene in food processing: principles and practice** *Edited by H. L. M. Lelieveld, M. A. Mostert, B. White and J. Holah*
89 **Rapid and on-line instrumentation for food quality assurance** *Edited by I. Tothill*
90 **Sausage manufacture: principles and practice** *E. Essien*
91 **Environmentally-friendly food processing** *Edited by B. Mattsson and U. Sonesson*
92 **Bread making: improving quality** *Edited by S. P. Cauvain*
93 **Food preservation techniques** *Edited by P. Zeuthen and L. Bøgh-Sørensen*
94 **Food authenticity and traceability** *Edited by M. Lees*
95 **Analytical methods for food additives** *R. Wood, L. Foster, A. Damant and P. Key*
96 **Handbook of herbs and spices Volume 2** *Edited by K. V. Peter*
97 **Texture in food Volume 2: solid foods** *Edited by D. Kilcast*
98 **Proteins in food processing** *Edited by R. Yada*
99 **Detecting foreign bodies in food** *Edited by M. Edwards*
100 **Understanding and measuring the shelf-life of food** *Edited by R. Steele*
101 **Poultry meat processing and quality** *Edited by G. Mead*
102 **Functional foods, ageing and degenerative disease** *Edited by C. Remacle and B. Reusens*
103 **Mycotoxins in food: detection and control** *Edited by N. Magan and M. Olsen*
104 **Improving the thermal processing of foods** *Edited by P. Richardson*
105 **Pesticide, veterinary and other residues in food** *Edited by D. Watson*
106 **Starch in food: structure, functions and applications** *Edited by A.-C. Eliasson*
107 **Functional foods, cardiovascular disease and diabetes** *Edited by A. Arnoldi*
108 **Brewing: science and practice** *D. E. Briggs, P. A. Brookes, R. Stevens and C. A. Boulton*
109 **Using cereal science and technology for the benefit of consumers: proceedings of the 12th International ICC Cereal and Bread Congress, 24–26th May, 2004, Harrogate, UK** *Edited by S. P. Cauvain, L. S. Young and S. Salmon*
110 **Improving the safety of fresh meat** *Edited by J. Sofos*
111 **Understanding pathogen behaviour: virulence, stress response and resistance** *Edited by M. Griffiths*
112 **The microwave processing of foods** *Edited by H. Schubert and M. Regier*
113 **Food safety control in the poultry industry** *Edited by G. Mead*
114 **Improving the safety of fresh fruit and vegetables** *Edited by W. Jongen*
115 **Food, diet and obesity** *Edited by D. Mela*
116 **Handbook of hygiene control in the food industry** *Edited by H. L. M. Lelieveld, M. A. Mostert and J. Holah*
117 **Detecting allergens in food** *Edited by S. Koppelman and S. Hefle*
118 **Improving the fat content of foods** *Edited by C. Williams and J. Buttriss*
119 **Improving traceability in food processing and distribution** *Edited by I. Smith and A. Furness*
120 **Flavour in food** *Edited by A. Voilley and P. Etievant*
121 **The Chorleywood bread process** *S. P. Cauvain and L. S. Young*
122 **Food spoilage microorganisms** *Edited by C. de W. Blackburn*
123 **Emerging foodborne pathogens** *Edited by Y. Motarjemi and M. Adams*

© Woodhead Publishing Limited, 2013

124 **Benders' dictionary of nutrition and food technology Eighth edition** *D. A. Bender*
125 **Optimising sweet taste in foods** *Edited by W. J. Spillane*
126 **Brewing: new technologies** *Edited by C. Bamforth*
127 **Handbook of herbs and spices Volume 3** *Edited by K. V. Peter*
128 **Lawrie's meat science Seventh edition** *R. A. Lawrie in collaboration with D. A. Ledward*
129 **Modifying lipids for use in food** *Edited by F. Gunstone*
130 **Meat products handbook: practical science and technology** *G. Feiner*
131 **Food consumption and disease risk: consumer-pathogen interactions** *Edited by M. Potter*
132 **Acrylamide and other hazardous compounds in heat-treated foods** *Edited by K. Skog and J. Alexander*
133 **Managing allergens in food** *Edited by C. Mills, H. Wichers and K. Hoffman-Sommergruber*
134 **Microbiological analysis of red meat, poultry and eggs** *Edited by G. Mead*
135 **Maximising the value of marine by-products** *Edited by F. Shahidi*
136 **Chemical migration and food contact materials** *Edited by K. Barnes, R. Sinclair and D. Watson*
137 **Understanding consumers of food products** *Edited by L. Frewer and H. van Trijp*
138 **Reducing salt in foods: practical strategies** *Edited by D. Kilcast and F. Angus*
139 **Modelling microorganisms in food** *Edited by S. Brul, S. Van Gerwen and M. Zwietering*
140 **Tamime and Robinson's Yoghurt: science and technology Third edition** *A. Y. Tamime and R. K. Robinson*
141 **Handbook of waste management and co-product recovery in food processing Volume 1** *Edited by K. W. Waldron*
142 **Improving the flavour of cheese** *Edited by B. Weimer*
143 **Novel food ingredients for weight control** *Edited by C. J. K. Henry*
144 **Consumer-led food product development** *Edited by H. MacFie*
145 **Functional dairy products Volume 2** *Edited by M. Saarela*
146 **Modifying flavour in food** *Edited by A. J. Taylor and J. Hort*
147 **Cheese problems solved** *Edited by P. L. H. McSweeney*
148 **Handbook of organic food safety and quality** *Edited by J. Cooper, C. Leifert and U. Niggli*
149 **Understanding and controlling the microstructure of complex foods** *Edited by D. J. McClements*
150 **Novel enzyme technology for food applications** *Edited by R. Rastall*
151 **Food preservation by pulsed electric fields: from research to application** *Edited by H. L. M. Lelieveld and S. W. H. de Haan*
152 **Technology of functional cereal products** *Edited by B. R. Hamaker*
153 **Case studies in food product development** *Edited by M. Earle and R. Earle*
154 **Delivery and controlled release of bioactives in foods and nutraceuticals** *Edited by N. Garti*
155 **Fruit and vegetable flavour: recent advances and future prospects** *Edited by B. Brückner and S. G. Wyllie*
156 **Food fortification and supplementation: technological, safety and regulatory aspects** *Edited by P. Berry Ottaway*

© Woodhead Publishing Limited, 2013

157 **Improving the health-promoting properties of fruit and vegetable products** *Edited by F. A. Tomás-Barberán and M. I. Gil*
158 **Improving seafood products for the consumer** *Edited by T. Børresen*
159 **In-pack processed foods: improving quality** *Edited by P. Richardson*
160 **Handbook of water and energy management in food processing** *Edited by J. Klemeš, R. Smith and J.-K. Kim*
161 **Environmentally compatible food packaging** *Edited by E. Chiellini*
162 **Improving farmed fish quality and safety** *Edited by Ø. Lie*
163 **Carbohydrate-active enzymes** *Edited by K.-H. Park*
164 **Chilled foods: a comprehensive guide Third edition** *Edited by M. Brown*
165 **Food for the ageing population** *Edited by M. M. Raats, C. P. G. M. de Groot and W. A Van Staveren*
166 **Improving the sensory and nutritional quality of fresh meat** *Edited by J. P. Kerry and D. A. Ledward*
167 **Shellfish safety and quality** *Edited by S. E. Shumway and G. E. Rodrick*
168 **Functional and speciality beverage technology** *Edited by P. Paquin*
169 **Functional foods: principles and technology** *M. Guo*
170 **Endocrine-disrupting chemicals in food** *Edited by I. Shaw*
171 **Meals in science and practice: interdisciplinary research and business applications** *Edited by H. L. Meiselman*
172 **Food constituents and oral health: current status and future prospects** *Edited by M. Wilson*
173 **Handbook of hydrocolloids Second edition** *Edited by G. O. Phillips and P. A. Williams*
174 **Food processing technology: principles and practice Third edition** *P. J. Fellows*
175 **Science and technology of enrobed and filled chocolate, confectionery and bakery products** *Edited by G. Talbot*
176 **Foodborne pathogens: hazards, risk analysis and control Second edition** *Edited by C. de W. Blackburn and P. J. McClure*
177 **Designing functional foods: measuring and controlling food structure breakdown and absorption** *Edited by D. J. McClements and E. A. Decker*
178 **New technologies in aquaculture: improving production efficiency, quality and environmental management** *Edited by G. Burnell and G. Allan*
179 **More baking problems solved** *S. P. Cauvain and L. S. Young*
180 **Soft drink and fruit juice problems solved** *P. Ashurst and R. Hargitt*
181 **Biofilms in the food and beverage industries** *Edited by P. M. Fratamico, B. A. Annous and N. W. Gunther*
182 **Dairy-derived ingredients: food and neutraceutical uses** *Edited by M. Corredig*
183 **Handbook of waste management and co-product recovery in food processing Volume 2** *Edited by K. W. Waldron*
184 **Innovations in food labelling** *Edited by J. Albert*
185 **Delivering performance in food supply chains** *Edited by C. Mena and G. Stevens*
186 **Chemical deterioration and physical instability of food and beverages** *Edited by L. H. Skibsted, J. Risbo and M. L. Andersen*
187 **Managing wine quality Volume 1: viticulture and wine quality** *Edited by A. G. Reynolds*

© Woodhead Publishing Limited, 2013

188 **Improving the safety and quality of milk Volume 1: milk production and processing** *Edited by M. Griffiths*
189 **Improving the safety and quality of milk Volume 2: improving quality in milk products** *Edited by M. Griffiths*
190 **Cereal grains: assessing and managing quality** *Edited by C. Wrigley and I. Batey*
191 **Sensory analysis for food and beverage quality control: a practical guide** *Edited by D. Kilcast*
192 **Managing wine quality Volume 2: oenology and wine quality** *Edited by A. G. Reynolds*
193 **Winemaking problems solved** *Edited by C. E. Butzke*
194 **Environmental assessment and management in the food industry** *Edited by U. Sonesson, J. Berlin and F. Ziegler*
195 **Consumer-driven innovation in food and personal care products** *Edited by S. R. Jaeger and H. MacFie*
196 **Tracing pathogens in the food chain** *Edited by S. Brul, P. M. Fratamico and T. A. McMeekin*
197 **Case studies in novel food processing technologies: innovations in processing, packaging, and predictive modelling** *Edited by C. J. Doona, K. Kustin and F. E. Feeherry*
198 **Freeze-drying of pharmaceutical and food products** *T.-C. Hua, B.-L. Liu and H. Zhang*
199 **Oxidation in foods and beverages and antioxidant applications Volume 1: understanding mechanisms of oxidation and antioxidant activity** *Edited by E. A. Decker, R. J. Elias and D. J. McClements*
200 **Oxidation in foods and beverages and antioxidant applications Volume 2: management in different industry sectors** *Edited by E. A. Decker, R. J. Elias and D. J. McClements*
201 **Protective cultures, antimicrobial metabolites and bacteriophages for food and beverage biopreservation** *Edited by C. Lacroix*
202 **Separation, extraction and concentration processes in the food, beverage and nutraceutical industries** *Edited by S. S. H. Rizvi*
203 **Determining mycotoxins and mycotoxigenic fungi in food and feed** *Edited by S. De Saeger*
204 **Developing children's food products** *Edited by D. Kilcast and F. Angus*
205 **Functional foods: concept to product Second edition** *Edited by M. Saarela*
206 **Postharvest biology and technology of tropical and subtropical fruits Volume 1: fundamental issues** *Edited by E. M. Yahia*
207 **Postharvest biology and technology of tropical and subtropical fruits Volume 2: açai to citrus** *Edited by E. M. Yahia*
208 **Postharvest biology and technology of tropical and subtropical fruits Volume 3: cocona to mango** *Edited by E. M. Yahia*
209 **Postharvest biology and technology of tropical and subtropical fruits Volume 4: mangosteen to white sapote** *Edited by E. M. Yahia*
210 **Food and beverage stability and shelf life** *Edited by D. Kilcast and P. Subramaniam*
211 **Processed Meats: improving safety, nutrition and quality** *Edited by J. P. Kerry and J. F. Kerry*

© Woodhead Publishing Limited, 2013

212 **Food chain integrity: a holistic approach to food traceability, safety, quality and authenticity** *Edited by J. Hoorfar, K. Jordan, F. Butler and R. Prugger*
213 **Improving the safety and quality of eggs and egg products Volume 1** *Edited by Y. Nys, M. Bain and F. Van Immerseel*
214 **Improving the safety and quality of eggs and egg products Volume 2** *Edited by F. Van Immerseel, Y. Nys and M. Bain*
215 **Animal feed contamination: effects on livestock and food safety** *Edited by J. Fink-Gremmels*
216 **Hygienic design of food factories** *Edited by J. Holah and H. L. M. Lelieveld*
217 **Manley's technology of biscuits, crackers and cookies Fourth edition** *Edited by D. Manley*
218 **Nanotechnology in the food, beverage and nutraceutical industries** *Edited by Q. Huang*
219 **Rice quality: a guide to rice properties and analysis** *K. R. Bhattacharya*
220 **Advances in meat, poultry and seafood packaging** *Edited by J. P. Kerry*
221 **Reducing saturated fats in foods** *Edited by G. Talbot*
222 **Handbook of food proteins** *Edited by G. O. Phillips and P. A. Williams*
223 **Lifetime nutritional influences on cognition, behaviour and psychiatric illness** *Edited by D. Benton*
224 **Food machinery for the production of cereal foods, snack foods and confectionery** *L.-M. Cheng*
225 **Alcoholic beverages: sensory evaluation and consumer research** *Edited by J. Piggott*
226 **Extrusion problems solved: food, pet food and feed** *M. N. Riaz and G. J. Rokey*
227 **Handbook of herbs and spices Second edition Volume 1** *Edited by K. V. Peter*
228 **Handbook of herbs and spices Second edition Volume 2** *Edited by K. V. Peter*
229 **Breadmaking: improving quality Second edition** *Edited by S. P. Cauvain*
230 **Emerging food packaging technologies: principles and practice** *Edited by K. L. Yam and D. S. Lee*
231 **Infectious disease in aquaculture: prevention and control** *Edited by B. Austin*
232 **Diet, immunity and inflammation** *Edited by P. C. Calder and P. Yaqoob*
233 **Natural food additives, ingredients and flavourings** *Edited by D. Baines and R. Seal*
234 **Microbial decontamination in the food industry: novel methods and applications** *Edited by A. Demirci and M.O. Ngadi*
235 **Chemical contaminants and residues in foods** *Edited by D. Schrenk*
236 **Robotics and automation in the food industry: current and future technologies** *Edited by D. G. Caldwell*
237 **Fibre-rich and wholegrain foods: improving quality** *Edited by J. A. Delcour and K. Poutanen*
238 **Computer vision technology in the food and beverage industries** *Edited by D.-W. Sun*
239 **Encapsulation technologies and delivery systems for food ingredients and nutraceuticals** *Edited by N. Garti and D. J. McClements*
240 **Case studies in food safety and authenticity** *Edited by J. Hoorfar*

© Woodhead Publishing Limited, 2013

241 **Heat treatment for insect control: developments and applications** *D. Hammond*
242 **Advances in aquaculture hatchery technology** *Edited by G. Allan and G. Burnell*
243 **Open innovation in the food and beverage industry: concepts and case studies** *Edited by M. Garcia Martinez*
244 **Trends in packaging of food, beverages and other fast-moving consumer goods (FMCG)** *Edited by N. Farmer*
245 **New analytical approaches for verifying the origin of food** *Edited by P. Brereton*
246 **Microbial production of food ingredients, enzymes and nutraceuticals** *Edited by B. McNeil, D. Archer, I. Giavasis and L. Harvey*
247 **Persistent organic pollutants and toxic metals in foods** *Edited by M. Rose and A. Fernandes*
248 **Cereal grains for the food and beverage industries** *E. Arendt and E. Zannini*
249 **Viruses in food and water: risks, surveillance and control** *Edited by N. Cook*
250 **Improving the safety and quality of nuts** *Edited by L. J. Harris*
251 **Metabolomics in food and nutrition** *Edited by B. Weimer and C. Slupsky*
252 **Food enrichment with omega-3 fatty acids** *Edited by C. Jacobsen, N. Skall Nielsen, A. Frisenfeldt Horn and A.-D. Moltke Sørensen*
253 **Instrumental assessment of food sensory quality: a practical guide** *Edited by D. Kilcast*
254 **Food microstructures: microscopy, measurement and modelling** *Edited by V. J. Morris and K. Groves*
255 **Handbook of food powders: processes and properties** *Edited by B. R. Bhandari, N. Bansal, M. Zhang and P. Schuck*
256 **Functional ingredients from algae for foods and nutraceuticals** *Edited by H. Domínguez*
257 **Satiation, satiety and the control of food intake: theory and practice** *Edited by J. E. Blundell and F. Bellisle*
258 **Hygiene in food processing: principles and practice Second edition** *Edited by H. L. M. Lelieveld, J. Holah and D. Napper*

© Woodhead Publishing Limited, 2013

Foreword

It is clear that global aquaculture production needs to expand significantly, even to maintain average per capita consumption of foodfish, let alone to increase it (as is desirable to enhance human nutrition and food supply). Capture fisheries cannot meet these demands. This fact, and the growing realisation that responsible, sustainable aquaculture is potentially beneficial for the environment, is becoming (somewhat grudgingly, perhaps) accepted by those who once strongly opposed its expansion.

This book, edited by Geoff Allan and Gavin Burnell, is timely – since the necessity for efficient hatchery production for aquaculture and release has never been more important. These distinguished scientists have combined their own complementary practical and theoretical experience on the hatchery rearing of a wide variety of finfish and molluscs with that of a team of other authors working with other species and in other disciplines.

Since I first became involved in aquaculture in the 1960s, basic hatchery technology for a much wider range of species has been developed. However, satisfactory data on the quantity, species and characteristics of hatcheries has not yet been received by FAO; thus no global statistics exist. As the editors comment in their preface, there are now only a few farmed aquatic species for which juveniles need to be sourced from the wild. However, although appropriate hatchery technology may exist or be emerging, there remain some types of aquaculture for which wild seeds are still primarily or partially used for various reasons. These include eels (*Anguilla* spp.); flathead grey mullets (*Mugil cephalus*) in Egypt, SE Asia and southern China; milkfish (*Chanos chanos*) in the Philippines, Indonesia and Taiwan; mud crabs (*Scylla serrata*) in Asia; and certain species of groupers. In some countries and for some species, including freshwater prawns (*Macrobrachium*

© Woodhead Publishing Limited, 2013

spp.), wild seed capture remains an important source of rural income, despite the availability of hatchery technology. For other important farmed species, including southern bluefin tuna (*Thunnus maccoyii*), yellowfin tuna (*Thunnus albacares*) and spiny lobsters (*Panulirus* spp.), hatchery technology is still novel. In addition to the production of juveniles for aquaculture, the hatchery production of many species of finfish (and some crustaceans, notably *Homarus* spp.) for fisheries enhancement or recreational fishing has long been commonplace.

Although the stocking requirements of so many species can be supplied through hatcheries, the refinement of existing hatchery technology is essential. The fact that the health and performance of farmed foodfish depends on the quality of the juveniles stocked is *sine qua non*. Furthermore, the health and genetic characteristics of broodstock are of fundamental importance in aquaculture. This book, by reviewing some outstanding successes in hatchery technological development, highlighting current research and pointing out the needs for the future, is an outstanding contribution to aquaculture, fisheries enhancement and conservation.

Michael New, OBE
Past-President, World Aquaculture Society
Past-President, European Aquaculture Society

© Woodhead Publishing Limited, 2013

Preface

Aquaculture is the fastest growing food producing sector in the world, growing at an annual rate of nearly 9 %. This growth is driven by demand for seafood that is increasing as global population and per capita consumption increase. The growth in demand is particularly strong in Asia. If seafood production is to meet demand, an additional 42 Mt of seafood will need to be produced by 2020 if global per capita consumption remains constant and an additional 116 Mt if consumption continues to increase at the current rate. Given limited expectations for any increase in capture fisheries, this increase in production will have to come from aquaculture.

One of the primary constraints to continuing growth of aquaculture is the supply of juveniles from hatcheries. There are now only a few species where juveniles are sourced from the wild, and, increasingly, advances in hatchery technology and genetic improvement programs are being adopted in order to ensure sufficient quantities of high quality juveniles are available when demanded to maximise production from growout facilities. However, advances in hatchery technology are often incremental, driven by technical managers of hatcheries and are poorly captured in the primary scientific literature. These advances are not well shared because of real or perceived concerns about losing commercial advantage, but this tends to slow industry development and will reduce the chance of meeting global targets for production.

This book reviews current and emerging technologies in key areas of aquaculture hatchery technology. It is hoped that this book will present a convenient, comprehensive and systematic review of recent advances in hatchery technology that will be of interest to hatchery operators, scientists and educators. The book is divided into four parts: Reproduction and larval

© Woodhead Publishing Limited, 2013

rearing; Closing the life cycle and overcoming challenges in hatchery production of selected invertebrate species; Closing the life cycle and overcoming challenges in hatchery production for selected fish species; and Emerging issues and future trends.

Part I. Reproduction and larval rearing: The first section of the book deals with new techniques and technologies in hatchery husbandry. In Chapter 1, Odd-Ivar Lekang emphasises the importance of hatchery location in relation to water quality. He gives valuable information on how to measure and maintain suitable water quality with very practical information on the type of equipment required for both delivery and treatment. He predicts that in the future good sites for both freshwater and marine aquaculture will be at a premium and this will force operators to become more and more reliant on recirculation technology.

The topic of broodstock management is large and complex and is expertly reviewed by Neil Duncan and his colleagues in Chapter 2. Genetic improvement through selection can yield from 5–20 % improvement in selected traits per generation, but only if you get it right from the start. This comprehensive chapter gives all the latest advances along with the caveats and cautions accumulated by the authors' considerable experience in this field. One of the most exciting and significant additions to the modern hatchery manager's toolbox is the ability to store gametes by cryopreservation. Chapter 3 in this section, by Catherine Labbé, complements the previous chapter and gives an insight into the challenges posed by this innovative (to aquaculture) technology. In addition, she introduces cutting edge science by considering the challenges posed by the use of embryonic and somatic cells to reconstruct fish.

Various aspects of larval feeding are addressed in the following three chapters. The first one by Malcolm Brown and Sue Blackburn (Chapter 4) explains the production and use of live algae for the hatchery production of bivalves, abalone, shrimp and fish. In a comprehensive review, they cover the species used, the key nutrients, biosecurity and the types of production systems. Of particular interest is the section on heterotrophic production which could reduce costs significantly. Chapter 5 by Patrick Sorgeloos and his colleagues (with the contribution on copepods by Josianne Stottrup) in the University of Ghent covers live feeds. Each of the three main cultured organisms (rotifers, *Artemia* and copepods) is dealt with in turn with detailed husbandry protocols and many useful tips that that have been acquired from many years of experimental and industrial experience. However, many hatchery operators would like to be able to dispense with the unpredictability and expense of live diets. Formulated microdiets are one possible solution and in Chapter 6, Sagiv Kolkovski describes how these diets are manufactured and reviews their performance so far. Although there have been some encouraging results published, there is still a lot of research required and the best we can hope for at the moment are feeding regimes that integrate live feeds with microdiets.

© Woodhead Publishing Limited, 2013

After nutrition the most important component of good husbandry practice in the hatchery is maintaining a healthy environment. In Chapter 7, Timothy Bowden and Ian Bricknell combine their renowned expertise in the area of disease management. They review ways of improving health issues in invertebrate hatcheries and explain how we can promote the health of larvae through the use of immunostimulants and vaccines. '*Although a hatchery is operated to produce larvae of a target aquaculture species, it is actually a complete ecosystem*'. With this quote, Evan Goulden and his co-authors (Chapter 8) challenge us to think of the hatchery environment in a more holistic way. They review the latest advances in microbiological management from prebiotics to probiotics and from bacteriophages to quorum sensing and advise us that as our understanding of the microbial ecology and virulence mechanisms of important hatchery pathogens increases, a new world of specific biocontrol strategies will also emerge.

As the demand for aquatic products continues to increase and the supply of wild stocks is diminishing, it is important that aquaculture embraces the ecosystem approach to management just as the fisheries sector has done. In Parts 2 and 3 of the book we describe the culture of species that either currently rely upon wild juveniles for their production or have done so until recently.

Part II. Closing the life-cycle and overcoming challenges in hatchery production of selected invertebrate species: Chapter 9 covers a group of invertebrate species that are proving very difficult to breed, the Palinurid lobsters. They have one of the longest planktonic larval stages recorded for marine invertebrates varying from 4 to 22 months depending upon the species. In this chapter, Mike Hall and his co-authors explain that although the life-cycle has been closed for several species the numbers produced are still not economic. According to them, health and nutrition must be optimised in order to produce a commercial rate of larval survival from phyllosoma to puerulus to juvenile. Through a focus on disease control and biosecurity, there have been significant improvements in white shrimp larval production and Jim Wyban (Chapter 10) documents the production of High Health postlarvae (PL) from Specific Pathogen Free (SPF) broodstock. It is perhaps surprising that the blue mussel is included in this section but, until recently, both the seabed dredging and longline culture industries were supplied by wild juveniles (seed). Pauline Kamermans and her co-authors (Chapter 11) collaborated in an EU project to investigate its hatchery production. Despite the fact that other mussel species are commercially produced in New Zealand, North America and China, it is not yet economic in Europe.

Cuttlefish, squid and octopus have also been successfully reared from egg to adult but, as with the previous two species, the percentage survival in the larval stages is still very poor. José Iglesias and his co-authors review the literature in Chapter 12 and give details of their own experience with the common octopus. Once again, understanding the nutrition of the

© Woodhead Publishing Limited, 2013

paralarvae is going to be the key to unlocking this life-cycle. In Chapter 13 on jellyfish, Jenny Purcell and her co-authors not only review their fisheries and aquaculture but also consider the negative impacts of these coelenterates on fish farming. There is some evidence that various anthropogenic activities are contributing to the increasing prevalence of their blooms. Last in this category is the sea cucumber where a collapse in wild stocks stimulated aquaculture initiatives resulting in successful hatchery production. Annie Mercier and Jean-François Hamel give a fascinating insight in Chapter 14 into these benthic invertebrates and describe how aquaculture is now even contributing to restocking and enhancement in some countries.

Part III. Closing the life-cycle and overcoming challenges in hatchery production for selected fish species: One of the most valuable groups of farmed fish are the tuna and in Chapter 15 Gavin Partridge comprehensively reviews the current state of this industry. Although several hatcheries have closed the cycle, the day when significant numbers of juvenile tuna will be routinely cultured from egg to adult is still some way off. Next in this section (Chapter 16) Nguyen Phuong and his co-authors document the explosive growth in the production of the striped catfish in Vietnam. As with the previous species, the early growth of this industry was fuelled by wild-caught juveniles until overfishing and conservation issues changed the economics in favour of hatchery production. Despite the success with this species there is room for improvement and research is concentrating on all the usual topics of nutrition, genetic improvement and health. Meagre (Chapter 17) are a fish with enormous potential. Their hatchery production is relatively easy and they have remarkable growth rates. But, as Neil Duncan and his co-authors point out, the markets are still very niche, and this is probably the area most in need of attention. Finally in this section, we have the yellow kingfish (Chapter 18). Despite successful commercial hatchery production, there are still challenges with this species, particularly with larval deformities. Stewart Fielder reviews progress to date.

Part IV. Emerging issues and future trends: Endangered freshwater fish are a serious issue in many countries, and in Chapter 19 Stuart Rowland describes how hatcheries can be used in conservation programs by supplying juveniles for restocking. However, prevention is better than cure and in the final chapter (Chapter 20), Don Meritt considers how we can augment the teaching and learning process by bringing the classroom into the hatchery. He argues that both public and private hatcheries offer the opportunity to create exciting educational experiences for training new generations of environmentally aware and technically competent students.

Geoff Allan and Gavin Burnell

© Woodhead Publishing Limited, 2013

Part I
Reproduction and larval rearing

1

Aquaculture hatchery water supply and treatment systems

O.-I. Lekang, Norwegian University of Life Sciences, Norway

DOI: 10.1533/9780857097460.1.3

Abstract: Good water quality is of major importance to production results in a hatchery, particularly in those which work more intensively. A low cost water transfer system is also important economically, affecting both investment and running costs. The low cost, however, should not be allowed to compromise the efficacy of the system, leading to operating problems or a lower water quality, for example. This chapter will take a brief look at some important factors in the location of hatcheries, before progressing to assess different water supply systems, encompassing both fresh water and sea water. Before the water enters the production unit, it has to be treated to fulfill the requirement of the aquatic organism as closely as possible. A description of different water treatment equipment/methods is also included in the chapter.

Key words: hatchery, water supply, site selection, water treatment.

1.1 Introduction

Water quality is integral to the production results of any hatchery. Larvae/fry have higher water quality requirements than larger fish – the larger the fish, the less significant the water quality – and it is therefore crucial to find suitable sites for the establishment of hatcheries and to ensure that the water is treated in the optimal way before it is used in the hatchery.

The water quality requirements will, to a certain extent, be dependent on the given species. Larvae or fry may be divided into two different categories: fish such as cod which, in their natural environment, will drift passively around the water, and 'stationary' species such as salmonids, which remain still in a continuous water flow. For the former, the water volume around the larvae is quite constant, while for the latter, the larvae must adapt to the quality of the given water.

Hatcheries can be categorized, based on the intensity of production (in other words, the number of produced larvae/fry per m^3 water and growth

© Woodhead Publishing Limited, 2013

rate), into extensive, semi-intensive and intensive. They may also be divided into freshwater or seawater hatcheries depending on the species grown and if sea water or fresh water is used. Hatcheries that use fresh water may, furthermore, be separated into hatcheries that utilize gravity for water transport into the farm and ones which use pumped water which must be lifted into the hatchery. The former is preferable due to lower running costs. In this chapter, the general focus will be on land-based hatcheries with intensive production. Hatcheries may, however, also float in the sea, which reduces the pumping cost but makes it difficult to control water quality and renders them more exposed to weather.

1.1.1 Site selection

The selection of a suitable site for a hatchery is crucial to the cost of establishing and running the hatchery, and thus to the hatchery's economic efficiency. The better and more suitable the raw water quality is, the less water treatment is necessary, leading to a reduction in investment and running costs. Theoretically, hatcheries may also be established where the raw water sources are less than optimal, but these will require more water treatment, resulting in higher production costs. For sub-optimal sites, recirculation aquaculture systems (RAS) are an alternative because the amount of raw water that has to be treated will be reduced.

Normally, inlet pipe(s) represent an important cost in hatchery installation. The pipes can be up to several kilometers long, in order to get enough water of satisfactory quality. The same might be the case with the outlet pipes, in order to reach a depth and an area where contaminated outlet water can be sufficiently diluted and local pollution avoided. The ideal is, of course, to find a site where both a short inlet and outlet pipe can be used while still achieving satisfactory water quality. In addition, for optimum efficiency to be maintained, the raw water quality ought to be as stable as possible throughout the year in order to avoid both expensive systems for water quality regulation and fluctuation in water quality. If it is not possible to utilize gravity water and pumps have to be used, sites with lower lifting heads are favorable. Large tide differences offer no advantage regarding the pumping cost, unless the hatchery is established in the tidal zone and the tide is utilized for water exchange. The location of the inlet pipes when they meet the shore and that of the pumping station must not be too exposed to the elements. Waves can be particularly dangerous in this regard.

1.1.2 Brief characterization of water

Small, newly hatched larvae or fry in particular need a good water quality to achieve optimal growth and survival rate. Natural water may have a sub-optimal temperature – typically it is too cold but it may also be too warm – and water temperature regulation may be necessary. The gas saturation

© Woodhead Publishing Limited, 2013

in the incoming water may also be sub-optimal – generally, this is due to a high content of nitrogen gas, or super-saturation. The oxygen content may also be too low, leaving the water under-saturated, resulting in a higher water supply than necessary. The CO_2 content may also be too high on some sites, such as those using groundwater. A high content of foreign particles in the water – typically a result of sand or clay – is also detrimental, as it may clog the gills. Water sources with too high content of particles should therefore be avoided. A high concentration of micro-organisms, such as bacteria, virus and fungi, is, of course, also unfavorable. The pH in the water, either alone or in association with dissolved metallic ions, may also be significant for the survival and growth rate.

For newly hatched fry it is of major importance that the water quality is stable; there should be no fluctuation. However, even if the raw water quality is stable, it might become unstable after water treatment. During treating, there will be reactions in the water of a chemical, microbiological or physical nature. Such reactions may take some time to be completed, meaning that the water composition may still be changing when the water reaches the production units, leading to instability. For this reason, it may be advisable to mature the water in large tanks with a long retention time to ensure that water quality is stable prior to its entry into the larva/fry tanks.

Another notable feature of water treatment in hatcheries is that water velocity in the production units must be very low. The water exchange rate will thus be low, as will the water supply to the production unit. Long water retention time in the production unit, together with high concentrations of organic substances/nutrients – such as dead larvae, feed loss, eventually live feed (rotifers and artemia) and microalgae – provide good conditions for the growth of micro-organisms and of biofilm. This often results in sub-optimal water quality in the rearing unit. Opportunistic, pathogenic bacteria may follow the raw water to the production unit or follow the added live feed when this is used in the production (Vadstein *et al.*, 2004; Brunvold *et al.*, 2007). Microbial maturation of the water prior to the production unit may thus be necessary to control growth of useful bacteria. Water quality inside the production unit is currently one of the major challenges in all hatcheries growing species with small larvae size or species using live feed.

1.1.3 The components in a system

The main components in any water supply system are the inlet pipe, the pumping station and the water treatment. The suitability of the last two components to the given water treatment depends on the raw water quality, the species grown and the intensity (number of kg of aquatic organisms per kg water supplied). For intensive hatcheries, water treatment typically includes units for particle removal, micro-organism control, heating/cooling, pH adjustment (freshwater sites) and degassing before the water reaches the production units.

© Woodhead Publishing Limited, 2013

1.2 The water supply and its main components

1.2.1 Inlet pipes

Inlet pipes are conventionally used to bring the water from the desired depth in the water source to the hatchery if gravity water is used, or to the pumping station if the water has to be lifted to the hatchery (Huguenin and Colt, 2002; Colt *et al.*, 2008; Pulido-Calvo *et al.*, 2008). It is also possible to utilize the tide to 'pump' the water into the hatchery or to place the hatchery in the tidal zone, but water quality control is more difficult in such cases. Materials commonly used in the inlet pipe are polyethylene plastic (PE) and PVC. PE is a low cost material that is easy to handle and weld and does not release toxic substances. When dimensioning the inlet pipes, normal water velocity in the pipe is 1–1.5 m/s, but it is important to find a specific, suitable dimension.

To stabilize the inlet pipe, it must either be put in ditches or moored to the ground, usually achieved with concrete lump weights clamped to the pipe (Fig. 1.1). The appropriate distance between the lumps and weight of the lumps depends on the diameter of the pipe and whether it is in the water or on shore. It is important that there is a continuous slope on the inlet pipes if the water is transferred with gravity, as this avoids creating points vulnerable to degassing. It is also advisable to avoid using siphon construction on the inlet pipe. The top of the siphon will always be a critical point.

It is advantageous to be able to clean the inlet pipe easily as it may be susceptible to fouling. Cleaning can, to give one example, be achieved by sending a cleaning plug through the pipe. Fouling will result in a large reduction in the amount of water that could flow through the pipe. Usually, the

Fig. 1.1 Water transfer pipes from the water source to the hatchery. The concrete block is used for stabilization of the pipes.

© Woodhead Publishing Limited, 2013

closer the water is to the surface of the source, the more susceptible it will be to fouling, although this also depends on the degree of eutrophication in the source. It is therefore advisable to take the inlet water from as depths below 20–30 m if possible, thereby avoiding the layers with the highest concentration of fouling organisms. Taking the raw water from deeper water layers will also facilitate a stable water temperature.

The water may also be transferred in open channels for the whole or parts of the distance from the water source to the hatchery. However, this reduces the control one has over the water quality. There has been a recent increase in the use of PE pipes, PE parts and pre-manufacturing of combined PE parts that are difficult to get properly welded on site.

1.2.2 Pumps and pump stations

The pump station can be placed either on shore using dry placed pumps or below the low tide water level using submerged pumps (Rishel, 2002; Pulido-Calvo *et al.*, 2006; Lekang, 2007; Jones *et al.*, 2008; Ratnayaka *et al.*, 2009). Both systems are in use. Dry placed pumps have the advantages that they are easy to install, control and maintain, while submerged pumps pose no risk regarding the suck of air on the sucking site of the pump and require no special equipment to fill the suck part of the pipe prior to starting the pump.

Before the water enters the pump station it ought to be passed through a grid or filter to prevent unwanted objects from finding their way into the pump. This may be a grid at the entrance to the inlet pipe. With submerged pumps, it may be a grid inside the pumping station, leaving the inlet pipes open with no inlet restriction.

Centrifugal or propeller type pumps are the best to fit for water supply to hatcheries (Fig. 1.2). Propeller pumps will typically have the lowest pumping cost for larger water flows and a lifting head below 10 m. Increase of the total lifting head to above 15 m will result in a considerable rise in the pumping costs; in such a case, a higher degree of oxygenation or RAS could be considered.

An important issue to consider is whether to pump more water or to add pure oxygen gas. Depending on the costs of electricity and liquid oxygen, respectively, the amount of pumped water can be reduced and more pure oxygen can be added to fulfill the oxygen requirements of the fry. The amount of pumped water can then be reduced to only what is necessary for dilution of the concentration of CO_2, suspended solids (SS) or total ammonium nitrogen (TAN) to acceptable levels for the grown organism (see later description).

One significant advance in this area is the application of more and better electronic devices in the pump stations, such as improved frequency converters and computer programs which optimize the running costs of the pumps. Sensors can also be added to monitor the conditions inside the

© Woodhead Publishing Limited, 2013

Fig. 1.2 Dry placed centrifugal pumps for water supply.

pumps, such as moisture content in submerged pumps and temperatures in bearings.

1.3 Water treatment systems

1.3.1 Particle removal

There is a variety of equipment available catering for the removal of different sizes of particle (Tchobanoglous *et al.*, 2003; Timmons and Ebeling, 2007; Spellman, 2009). A typical method is to install a sieve with a rotating filter cloth able to catch particles smaller than the mesh (Fig. 1.3a, b). The filter cloth should be cleaned, or back flushed, regularly to avoid clogging and to ensure that the particles exit through the particle outlet (Lekang, 2007). This is an area effective treatment method that typically removes particles down to 20 μm. Some parasites will also be removed with such filter mesh, and this is particularly useful for hatcheries. Depth filters may also be included to remove smaller sized particles, but these require more intensive operation, due to the necessity of regular back flushing of the filter mass. Filter masses of sand or crushed glass are common in depth filters. In-line filters are also used to an extent, but they are prone to head loss due to their dimensions.

Protein skimmers, or foam fractionators, may also be used for water treatment, especially when growing larvae/small aquatic organisms (Chen *et al.*, 1993a, b). These are normally used in tandem with another particle filter such as a micro-sieve or a depth filter. The removal of small sized particles in foam fractionators is achieved by gas bubbles, usually of air, which are released at the bottom of a water column. When the bubbles rise to the surface of the water column, particles and some dissolved organic

© Woodhead Publishing Limited, 2013

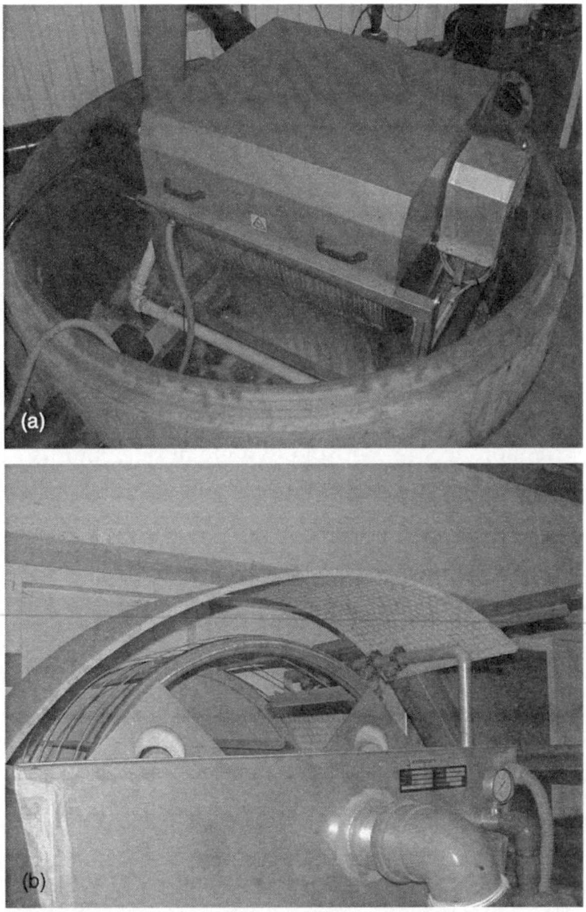

Fig. 1.3 (a, b) Drum filter for particle removal.

matter will attach to the bubble surface due to chemical and physical interaction forces. When the bubbles hit the surface they may create foam, depending on whether the bubbles are stabilized (particles and dissolved organic matter on the bubble surface will stabilize the bubble preventing it from bursting). The bubbles will burst immediately after entering the water surface if they are not stabilized. The collected particles can be removed easily with the foam. Protein skimmers also allow smaller particles (1–50 μm), and even some colloidal or dissolved substances (below 1 μm), to be removed (surface-active substances). Since protein skimmers remove small particles, they will also remove some micro-organisms such as parasites and bacteria, meaning that the skimmer will, to some extent, double up as a disinfectant.

© Woodhead Publishing Limited, 2013

To improve the efficiency of protein skimmers, ozone gas can be used instead of air for bubble creation (ozone splits the organic matter so that the amount of surface-active components increases). The amount of ozone supplied is lower than that required to achieve full disinfection. It is important to be aware that foam will not be created in all water qualities; some pollution in the water is required. If no foam is created by using air, the use of ozone can improve the foam creation.

Advances in particle filtration include the removal of smaller particles (colloids), meaning that protein skimming becomes more important. Membrane separation or filtration is another technology that can lead to improved water quality (Szmukala and Szaniawska, 2009; Peinemann and Nunes, 2010). Compared to a traditional macrofilter, such as a drum filter, membrane filtration facilitates the removal of smaller particles, and it is possible to control the size of the impurities passing the membrane by choosing the appropriate membrane size. Put simply, a membrane filter is a filter with a very fine mesh size or pore size. Indeed, the smallest size is so compact that it is not even described as a mesh. By pressurizing the water before it passes through the membrane, pure water will be pressed through while impurities will be stopped according to the pore size in the membrane. Membrane filtration may be categorized into microfiltration (>100 nm–0.1 μm), ultrafiltration 2–50 nm, nanofiltration (2–5 nm) and reversed osmosis (<1 nm) based on the size of impurity that is stopped. In nanofiltration and reversed osmosis, for example, ions are inhibited from passing the membrane. Due to the small pore size, membrane filters may also be used for the removal of parasites, bacteria and viruss, thereby acting as a disinfectant. Although development in this field is accelerating, membranes are still to a certain extent exposed to fouling, representing the major drawback with their use. The sizes actually used in hatcheries are micro- and ultrafiltration membranes. The method remains, however, quite expensive and, if there are high concentrations of particles, fouling may become an issue. Pretreatment with macrofiltration in the form of a drum filter or protein skimmers is thus advisable.

1.3.2 Microorganism control

In order to reduce the chances of disease in the hatchery, disinfection of the inlet water is crucial. Disinfection can reduce the amount of fish pathogens typically as much as 99.9 % (log 3), although complete sterilization is practically impossible. The most common system is ultraviolet (UV), but it is also possible to use ozone (Summerfelt, 2003; Masters *et al.*, 2008). Disinfection should be carried out after particle removal in order to avoid any negative influence from the particles during disinfection, UV shadowing, or unnecessary reduction of the ozone concentration.

A UV lamp will send out UV radiation, with a sufficient dosage to inactivate the micro-organism. The UV dose that the micro-organism is exposed

© Woodhead Publishing Limited, 2013

to depends on the UV light intensity (power of lamp), the distance from the micro-organism to the UV light source, the ease with which the UV light passes through water (the UV transmission of the water) and the length of time for which micro-organisms are exposed (the exposure time). The dose needed to induce inactivity will vary with the type of micro-organisms; fungi and viruses, for example, need a high dose, while bacteria require a slightly lower dose.

UV lights are usually placed within the water flow, but they may also be placed above the surface. In the first case, the lights are placed within glass pipes for protection while, in the second case, the UV radiation has to cross the water surface and continue into the water. UV lights placed in the water flow are exposed to fouling and are thus usually equipped with cleaning mechanisms.

Ozone is a very strong oxidizing agent, and a high enough dose will oxidize organic matter in the water, including micro-organisms. Use of ozone will therefore have additional effects which UV light does not have, such as reducing water turbidity, water color, the amount of organic carbon, the number of metal connections and the amount of algae. There is, therefore, an increased interest in the use of ozone. Ozone gas (O_3) is produced on location with an ozone generator; the gas is then mixed into the water with a venturi. It then needs a certain retention time in the water in order to have an effect on the micro-organisms. However, ozone gas is also lethal to fry and larvae in the hatcheries. Before the water enters the hatcheries, the retention time has to be long enough to ensure that the residual ozone level is well below critical values for the eggs or fry. Special de-ozonation efforts may also be included before sending the water into the production units, such as sending the water through an active coal filter or adding de-ozonation chemicals.

The amount of ozone that must be added to achieve disinfection depends on the water quality. Water containing a high amount of organic matter requires a higher ozone dose, because ozone is utilized to decompose the organic matter. Normally the amount of ozone added is quantified by the amount of ozone remaining after a given time: for example, above 0.1 mg/L after 3 min contact time. This can be measured with an ozone meter. Most pathogens are killed with a dose of 0.1–1 mg/L and a contact time between 1 and 10 min. There is also a correlation between the ozone concentration and the red-ox potential; thus a red-ox measurement can help to deduce ozone levels, although this not as accurate as measuring ozone directly. It can also be difficult to control red-ox potential; the measurements are not as reliable and the sensor is exposed to fouling, especially in sea water. Toxic connections, like bromide, may be created when using ozone in sea water and care should be taken.

Ozone can be used in combination with protein skimmers. Here, a small doze of ozone, instead of air, is added to the water in the skimmer, improving the skimming. However, this is not a full disinfection and less ozone is

© Woodhead Publishing Limited, 2013

added compared to full disinfection. For sea water, typical red-ox values are in the area 250–350 mW, while a 'full' disinfection requires red-ox values of 700–800 mW.

It is very important to be aware that the toxicity of ozone applies to humans as well as fish. Breathing in ozone gas will injure the lungs and a high concentration is directly toxic. An alarm system that monitors the ozone concentration in the air should therefore be used, preferably with an automatic close down mechanism on the ozone generator.

A significant new group of methods to improve disinfection has recently emerged under the banner of advanced oxidation technology (AOT) or advanced oxidation processes (AOP) (Comninellis *et al.*, 2008; Joseph *et al.*, 2009; Oller *et al.*, 2011; Sievers, 2011). This may be divided into chemical AOP and photochemical AOP (Tunay *et al.*, 2010).

AOP utilizes the high oxidation potential of free radicals to oxidize organic substances. The major free radical created is the hydroxyl radical, which has higher oxidation potential than ozone, resulting in a very fast reaction. There are a number of methods of creating the free radicals but all involve a combination of several oxidation/disinfection methods. The most commonly used combinations are ozone + hydrogen peroxide and UV light + hydrogen peroxide. A combination of all three is also possible, with a catalyst to enhance reaction. Other examples include photocatalytic processes combining UV and titanium dioxide (TiO_2) and the photo-Fenton reaction combining iron and hydrogen peroxide and light (Chong *et al.*, 2010). Such methods are, for example, used for ballast water treatment (Tsolaki and Diamadopoulos, 2010). Due to its oxidation potential it may also be used for reducing total organic carbon (TOC) and nitrate (Mook *et al.*, 2012; Virkutyte and Jegatheesan, 2009). These methods are also advantageous as they tend not to release toxic bromate substances, which can pose problems with ozone and sea water when used for aquaculture (Brookman *et al.*, 2011).

1.3.3 Heating/cooling

It is very common to install heating or cooling plants or, indeed, both in a combined system to control the temperature in hatcheries (Huguenin and Colt, 2002; Lekang, 2007). The reason for this is that the natural raw water temperature is sub-optimal for growth and survival for the different life stages. One solution can be to take water from different water depths in the water source, collecting water with varying temperatures, but this will only give limited possibilities and, if the inlet pipe in the water source is close to the surface, it will be exposed to fouling organisms. If possible, temperate/hot industrial wastewater or geothermal water may also be used as a source, either directly or in combination with heat exchangers to transfer the energy, if the water quality is too bad for direct use in the hatchery.

© Woodhead Publishing Limited, 2013

The natural water temperature also changes throughout the year, and a stable water temperature all year around is preferable in the hatchery, especially for temperature-sensitive larvae and fry. A water heating/cooling system allows a more stable and consistent water temperature all year round. Any cooling or heating mechanism should also include a water temperature regulation system.

When dealing with small amounts of water or small temperature differences, the energy requirement for heating of the water is low. In such cases, the most common heating plant is simply a direct heater. This can be run either on electricity (resistance heater), or oil or gas and the transferal of combustion energy into the water. When using direct heating, it is commonplace to use a heat exchanger that recovers a part of the supplied energy from the outlet water (Fig. 1.4). In freshwater hatcheries lying by the coast, heat exchangers may be used for transferring heat from the sea water to the incoming fresh water if the seawater temperature is higher than the freshwater, or vice versa.

In large hatcheries, there may be a vast amount of energy that has to be transferred to the water. In this case, heat pumps, including heat exchangers, are normally used (Fig. 1.5). Heat pumps need a low temperature source where the energy is 'removed' and compressor energy is added and finally

Fig. 1.4 Heat exchangers for heating of water.

© Woodhead Publishing Limited, 2013

Fig. 1.5 Heat pumps (3) for heating of water to a hatchery.

translated into a small amount of high temperature energy. The heat conversion rate is typically around 3–5, meaning that for each kW of electricity added to the compressor 3–5 kW of energy is transferred to the water as heat. When using heat pumps, the system is normally combined with heat exchangers in the outlet water, allowing the heat conversion rate to be significantly higher (20–30).

If the water must be chilled, a refrigeration plant must be installed and the water has to pass through this before entering the production unit. The working principle of a refrigeration plant is the same as that of the heat pump but it utilizes the opposite side of the circuit: energy is taken from the water, not added to it.

Heating or chilling of water, however, inevitably means both investment and running cost, and when the water flow is large the running cost can be considerable. It is therefore advantageous to find alternative sources for heating or chilling of water. When selecting a site for a new hatchery, such factors must be considered. One solution, for example, could be to locate the hatchery close to geothermal water sources or an industry that releases temperate wastewater.

It is important to remember that heating and cooling may change the gas composition of the water. After heating, the amount of gas that can be diluted in water is reduced. If the water is then sent directly to the hatchery without degassing, it may be super-saturated with gas and thus toxic for the organism grown.

Advances in this sector include heat pumps/chilling plants with more parts and valves which are electronically controlled, resulting in an improved coefficient of performance. Improvement in the design of heat exchangers in the heat pumps is also notable.

© Woodhead Publishing Limited, 2013

1.3.4 Aeration, oxygenation and degassing

Water in equilibrium with air may contain a certain amount of gases, as dictated by Henry's law. The major gases are nitrogen, oxygen and CO_2 (the last normally in smaller amounts), with approximately 60 % nitrogen and 40 % oxygen. If the total amount of gases in the water is too high compared to the amount that can be dissolved, the total gas pressure (TGP) will be above 100 % and free gas bubbles will be released, leading to super-saturation. There are many possible reasons for super-saturation of gases in the water in a hatchery, such as the heating of water, biological reactions in the water source, the sucking of air into the water (under vacuum) or the pressurization of the air/water combination.

Larvae and fry are sensitive to high gas saturation in the water, and can quickly become diseased as a result. The real problems occur when the total gas pressure is above 100 %, and free bubbles are released. Gas bubble disease is created if the nitrogen gas concentration is above 101–102 %, which indicates super-saturation (Colt, 1986; Gunnarsli *et al.*, 2008; Noga, 2010; Plumb and Hanson, 2010). Super-saturation of oxygen is less critical and the fish are generally able to tolerate it (Espmark *et al.*, 2010). CO_2 super-saturation is not common because, although the water may contain much dissolved CO_2, its solubility is higher than that of oxygen and nitrogen. The concentration of CO_2 in water is also related to the alkalinity and pH. The tolerance of aquatic organisms for CO_2 in the water is, however, limited (Gil Martens *et al.*, 2006; Good *et al.*, 2010), and it must still be controlled.

To avoid problems with the super-saturation of gases and to ensure that the water contains the maximum amount of oxygen, aerators are used (Fig. 1.6). If the water is in contact with the air with atmospheric pressure for a

Fig. 1.6 Aerators for equalizing the gas composition in the water and in the air under atmospheric pressure.

© Woodhead Publishing Limited, 2013

sufficient amount of time, there will be no super-saturation, and an equilibrium of the gas concentration between the air and water will occur due to gas exchange over the water surface. This is, however, a lengthy process; thus aerators are used as a catalyst. Aerators aim to create a large water–air surface in relation to the water volume so as to facilitate a faster gas exchange, resulting in a more swiftly achieved equilibrium between the gases in the water and in the air. There are a number of possible aerator designs, the most prevalent of which are gravity aerators and surface aerators. Due to the consequences of super-saturation in hatcheries, there is a trend towards more installations of vacuum aerators, in which the total pressure inside the aerator is reduced. The gas exchange is therefore done in an atmosphere with less gas, where pressure is lower. The amount of gases in the water will therefore be below equilibrium when the water comes out of the vacuum aerator and into normal pressure. The gas pressure is typically reduced by 5 % inside the vacuum aerator. However, it is important to be aware that, when both the nitrogen and the oxygen concentration are reduced, if no pure oxygen gas is added to the water, the water supply to the fish tanks must be increased to cover the oxygen consumption.

Some water sources may contain too much dissolved CO_2 gas, a problem that has become ever more highlighted in recent years. Specially designed aerators must be used for CO_2 removal, due to its high solubility (Moran, 2010). Either a large amount of air is blown through the column or the column aerator is built into a tower of 4–6 m, with separated elements at 1 m intervals to ensure that enough time is allowed for gas exchange and that a new equilibrium occurs between the elements. Specially designed CO_2 aerators may also be used.

To reduce the amount of water that has to flow through the hatchery, pure oxygen gas can be added. Each liter of water supplied to the production unit will now contain more oxygen, and the necessary water flow can be reduced. The water flow must only be large enough to ensure a satisfactory dilution of the metabolic excess products such as suspended solids, CO_2 and ammonia (TAN).

If enough water is available then there will be an evaluation as to whether it is most economical to invest in oxygen and reduce the amount of water that should be pumped into the hatchery or to add no extra oxygen and simply increase the amount of water that has to be pumped.

It is, however, important to be aware that if the inlet water is supersaturated with gas, oxygen gas bubbles may be released in critical places in the water inlet system, in the water distribution system or in the production system and, in the worst case scenario, could block the water flow in the system. Oxygen bubbles may also be released and cause damage, for example by moving eggs at critical development stages when the bubbles float to the surface. This will, of course, depend on how and where the oxygen gas is injected.

© Woodhead Publishing Limited, 2013

Fig. 1.7 Cones for transferring oxygen gas into the inlet water.

Typical methods of injecting of oxygen gas into water include the use of cones (Fig. 1.7) and direct injection into the tank inlet pipe. What it is important here is the efficiency of the equipment. As much as possible of the supplied oxygen has to be available for the fish, and this figure can vary wildly from a few percent up to above 90 %, depending on the equipment and the design of the internal water distribution system.

When adding pure oxygen to the water in a hatchery the amount needed is normally quite large. Pressurized oxygen in standard gas bottles is therefore impractical, although this can be used as an emergency backup. Under normal conditions, the best method is either to transform liquid oxygen into gas (Fig. 1.8) or, alternatively, to produce oxygen gas on location by sieving of air (Lekang 2007). The optimum solution depends on electricity costs, as well as the availability and the price of liquid oxygen in the area.

One current trend is to install regulation systems that can regulate the oxygen concentration in each individual production unit (tank). This is usually achieved either by separate oxygenation of each unit or by central oxygenation of the main flow to a certain level and subsequent modification for each unit. There have also been developments in the means by which the oxygen gas is supplied to the water, for instance by installing a device on top of the oxygen cone to avoid release of undissolved oxygen from the cone, which again prevents bubbles and gas pockets from forming in the

© Woodhead Publishing Limited, 2013

Fig. 1.8 Tank for storing liquid oxygen including evaporators for transferring oxygen from liquid to gas.

pipe system. Reuse of this oxygen can also render a more efficient utilization of the supplied oxygen.

1.3.5 pH adjustment

In some cases, freshwater sources may have a pH that is sub-optimal for the survival and growth of the organism. Sources may have pH which is too low, due to exposure to acid rain, for instance. Even if the pH is not dangerous, it may end up creating toxic metal compounds, such as aluminum and iron, meaning that the pH has to be increased before sending the water to the production units (Kristensen *et al.*, 2009).

Regulation of pH in the water can be done with different chemicals, some of which simply increase pH, while others actively increase alkalinity. If the alkalinity is increased, the water will have some buffering capacity against later drops in pH. A typical chemical of the first group is lye, while $CaCO_3$ is a chemical included in the second group. Moreover, if there are problems with metal ions, a chemical such as silica lye can be used as treatment. Before choosing which pH regulating agent to use, it is therefore crucial to check if the problem is rooted solely in the pH, or if it extends to problems with the metal.

© Woodhead Publishing Limited, 2013

After a pH regulation chemical has been added, the water quality may become unstable, meaning that some retention time is required. Different pH regulating chemicals will have different reaction time before the water quality is stable again.

The pH regulation chemical is used either in fixed or dissolved form. If it is dissolved, it is easy to mix into the water, either by mixers or by adding air through bubbles in the mixing basin before the pumping stage. Fixed substances like $CaCO_3$ have to be dissolved prior to mixing. This can be achieved by first making a slurry and then mixing it with the water. It is important to leave some time between adding the pH adjuster and measuring the pH. The measured pH is then used as a signal for how much pH adjusting chemicals to add. With some chemicals, such as lye, it is quite easy to overdose. Due to our improved knowledge of metal–pH interactions, the use of silica lye has increased in pH regulation.

1.3.6 Stabilization or maturation of the water

Some organisms, such as marine larvae, are very sensitive to water quality and changes therein. Unstable water quality and temporary changes may result in increased mortality rate and reduced growth. Several of the methods used for water treatment result in changes in water quality. Such changes, however, may take some time, meaning there is a time window where water quality or water composition is unstable. Examples of this include heating of the water, disinfecting the water and regulating pH. When heating, the excess gases have to be released from the water before the water enters the fry; this is usually achieved through an aerator and occurs quite quickly. A simple means for general stabilization is to store the water in a tank with a long retention time before it is sent into the larvae and fry. A significant amount of time, however, is required to achieve a totally stable water quality.

Disinfecting water removes large amounts of micro-organisms, but it does not sterilize the water completely. After disinfection, depending on method and agent, the water will therefore be full of dead and inactivated bacteria. This decomposed organic matter is a good food source for the remaining bacteria. Disinfection agents such as ozone may 'cut' the long-chain organic molecules into smaller ones, thus making it more available for the remaining bacteria.

This is an issue which is yet to be properly analyzed. It could lead to a resurgence in bacteria levels, and it is important that it is not the opportunistic photogenic bacteria which lead that resurgence. In such a transient phase, water quality will also be unstable, and it would be inadvisable to utilize this water for the larvae. Ozone may also have effect on ions in the water making them unstable. Microbiological stabilization, microbiological water maturation or water ripening may therefore be a suitable water treatment for disinfected water or raw water, to help control the bacterial

© Woodhead Publishing Limited, 2013

development (Olafsen, 2001; Brunvold *et al.*, 2007; Brunvold, 2010; van der Meeren *et al.,* 2011). One method is to send the water through a filter where biological organisms are growing naturally, and to let these organisms stabilize the water quality by inhibiting growth of photogenic bacteria. This is also known as microbial maturation. Another theoretical method is to reduce the water retention time between treatment and disinfection and disposal from the fish tank. This would have to be short enough that the water quality would not have enough time to change and so that any unwanted growth would occur after the larvae tank. This is, however, a problem as too short retention time in the fish tank will create too large a flow of water flow inside the production units, with larva mortality increasing as a result.

1.4 Future trends

Due to the growth in the aquaculture industry and the increased demand for aquatic food, there will not only be an increase in the number of aquaculture production sites but also in the production per site. Knowledge of optimal water supply systems and water treatment systems will therefore be of great importance in the future. The evidence covered here, however, shows that the freshwater resources available for aquaculture purposes are limited, and that the aquaculture industry has to compete with other users for the available freshwater. This means that efficient utilization of the fresh water will become ever more important in aquaculture. There is also likely to be an increase in use of sub-optimal sites, with lower raw water quality, which will increase the need for knowledge in optimal water treatment. For seawater aquaculture, the available resources are theoretically vast, but the hatcheries are usually established in the coastal zone, where there are already a number of sites, and thus a degree of conflict. Good sites will therefore not be easy to find in the future and the best sites will thus be utilized for more intensive fish farms. Due to lack of fresh water and lack of good land-based seawater sites for hatcheries, knowledge of good water reuse systems will increase in importance in the future. The production of aquatic biomass per kg supplied raw water will increase, but this will not decrease the importance of good water supply and water treatment systems. Stable water quality throughout the year is also becoming ever more important, in order to get the maximum growth out of the organism.

1.5 References

BROOKMAN R M, LAMSAL R *et al.* (2011) 'Comparing the formation of bromate and bromoform due to ozonation and UV-TiO2 oxidation in seawater.' *Journal of Advanced Oxidation Technologies* **14**(1): 23–30.

© Woodhead Publishing Limited, 2013

BRUNVOLD L (2010) *Microbial ecology in the early life stages of intensively reared marine species,* PhD Thesis, University of Bergen.
BRUNVOLD L, SANDAA R-A *et al.* (2007) 'Characterisation of bacterial communities associated with early stages of intensively reared cod (*Gadus morhua*) using Denaturing Gradient Gel Electrophoresis (DGGE).' *Aquaculture* **272**(1–4): 319–327.
CHEN S T, BISOGNI M B, TIMMONS J J and ANESHANSLEY D J (1993a) 'Protein and its removal by foam fractionation.' *The Progressive Fish-Culturist* **55**(2): 76–82.
CHEN S T, BISOGNI M B, TIMMONS J J and ANESHANSLEY D J (1993b) 'Suspended-solids removal by foam fractionation.' *The Progressive Fish-Culturist* **55**(2): 69–75.
CHONG M N, JIN B *et al.* (2010) 'Recent developments in photocatalytic water treatment technology: A review.' *Water Research* **44**(10): 2997–3027.
COLT J (1986) 'Gas supersaturation – impact on the design and operation of aquatic systems.' *Aquacultural Engineering* **5**(1): 49–85.
COLT J, SUMMERFELT S *et al.* (2008) 'Energy and resource consumption of land-based Atlantic salmon smolt hatcheries in the Pacific Northwest (USA).' *Aquaculture* **280**(1–4): 94–108.
COMNINELLIS C, KAPALKA A *et al.* (2008) 'Advanced oxidation processes for water treatment: advances and trends for R&D.' *Journal of Chemical Technology & Biotechnology* **83**(6): 769–776.
ESPMARK Å M, HJELDE K *et al.* (2010) 'Development of gas bubble disease in juvenile Atlantic salmon exposed to water supersaturated with oxygen.' *Aquaculture* **306**(1–4): 198–204.
GIL MARTENS L, WITTEN P E *et al.* (2006) 'Impact of high water carbon dioxide levels on Atlantic salmon smolts (*Salmo salar* L.): Effects on fish performance, vertebrae composition and structure.' *Aquaculture* **261**(1): 80–88.
GOOD C, DAVIDSON J *et al.* (2010) 'The effects of carbon dioxide on performance and histopathology of rainbow trout *Oncorhynchus mykiss* in water recirculation aquaculture systems.' *Aquacultural Engineering* **42**(2): 51–56.
GUNNARSLI K S, TOFTEN H *et al.* (2008) 'Effects of nitrogen gas supersaturation on growth and survival in juvenile Atlantic cod (*Gadus morhua* L.).' *Aquaculture* **283**(1–4): 175–179.
HUGUENIN J E and COLT J (2002) *Design and Operating Guide for Aquaculture Seawater Systems*. Amsterdam: Elsevier Science.
JONES G M, SANKS R L *et al.* (2008) *Pumping Station Design*. Oxford/Burlington, MA: Elsevier/Butterworth-Heinemann.
JOSEPH C G, LI PUMA G *et al.* (2009) 'Sonophotocatalysis in advanced oxidation process: A short review.' *Ultrasonics Sonochemistry* **16**(5): 583–589.
KRISTENSEN T, ÅTLAND Å *et al.* (2009) 'Important influent-water quality parameters at freshwater production sites in two salmon producing countries.' *Aquacultural Engineering* **41**(2): 53–59.
LEKANG O-I (2007) *Aquaculture Engineering*. Oxford: Blackwell.
MASTERS A L, VINCI B J *et al.* (2008) 'Performance characterization of influent and effluent treatment systems: A case study at Craig Brook National Fish Hatchery.' *Aquacultural Engineering* **38**(1): 66–76.
MOOK W T, CHAKRABARTI M H *et al.* (2012) 'Removal of total ammonia nitrogen (TAN), nitrate and total organic carbon (TOC) from aquaculture wastewater using electrochemical technology: A review.' *Desalination* **285**: 1–13.
MORAN D (2010) 'Carbon dioxide degassing in fresh and saline water. I: Degassing performance of a cascade column.' *Aquacultural Engineering* **43**(1): 29–36.
NOGA E J (2010) *Fish Disease: Diagnosis and Treatment*. Ames, IA: Wiley.
OLAFSEN J A (2001) 'Interactions between fish larvae and bacteria in marine aquaculture.' *Aquaculture* **200**(1–2): 223–247.

© Woodhead Publishing Limited, 2013

OLLER I, MALATO S *et al.* (2011) 'Combination of Advanced Oxidation Processes and biological treatments for wastewater decontamination – A review.' *Science of The Total Environment* **409**(20): 4141–4166.

PEINEMANN K V and NUNES S P (2010) *Membranes for Water Treatment*. Weinheim: Wiley-VCH.

PLUMB J A and HANSON L A (2010) *Health Maintenance and Principal Microbial Diseases of Cultured Fishes*. Ames, IA: Wiley.

PULIDO-CALVO I, GUTIÉRREZ-ESTRADA J C *et al.* (2006) 'Optimal design of pumping stations of inland intensive fishfarms.' *Aquacultural Engineering* **35**(3): 283–291.

PULIDO-CALVO I, GUTIÉRREZ-ESTRADA J C *et al.* (2008) 'Pipes size selection of water distribution systems of fishfarms.' *Aquacultural Engineering* **39**(1): 43–52.

RATNAYAKA D D, BRANDT M J *et al.* (2009) *Water Supply*. Oxford: Elsevier.

RISHEL J B (2002) *Water Pumps and Pumping Systems*. New York: McGraw-Hill.

SIEVERS M (2011) Advanced oxidation processes, in Wilderer P (ed.), *Treatise on Water Science*. Oxford: Elsevier, 377–408.

SPELLMAN F (2009) *Handbook of Water and Wastewater Treatment Plant Operations*. Boca Raton, FL: CRC Press/Taylor & Francis.

SUMMERFELT S T (2003) 'Ozonation and UV irradiation – an introduction and examples of current applications.' *Aquacultural Engineering* **28**(1–2): 21–36.

SZMUKALA M and SZANIAWSKA D (2009) 'Application of ceramic membranes in water treatment for fish hatchery supplying purposes.' *Desalination* **240**(1–3): 117–126.

TCHOBANOGLOUS G, BURTON F L *et al.* (2003) *Wastewater Engineering: Treatment and Reuse*. Boston, MA: McGraw-Hill.

TIMMONS M B and EBELING J M (2007) *Recirculating Aquaculture*. Ithaca, NY: Cayuga Aqua Ventures.

TSOLAKI E and DIAMADOPOULOS E (2010) 'Technologies for ballast water treatment: a review.' *Journal of Chemical Technology & Biotechnology* **85**(1): 19–32.

TUNAY O, KABDASLI I *et al.* (2010) *Chemical Oxidation Applications for Industrial Wastewaters*. London: IWA Publishing.

VADSTEIN O, MO T A and BERGH Ø (2004) 'Microbial interactions, prophylaxis and diseases,' in Moksness E, Kjorsvik E and Olsen Y (eds), *Culture of Cold-Water Marine Fish*. Oxford: Blackwell, 28–72.

VAN DER MEEREN T, BRUNVOLD L, SANDAA R, BERGH Ø, CASTBENG T, THYRHAUG R and MANGOR JENSEN A (2011) 'Water quality and microbial community structure in juvenile Atlantic cod (*Gadus morhua* L.) cultures.' *Aquaculture* **316**(1–4): 111–120.

VIRKUTYTE J and JEGATHEESAN V (2009) 'Electro-Fenton, hydrogenotrophic and Fe2+ ions mediated TOC and nitrate removal from aquaculture system: Different experimental strategies.' *Bioresource Technology* **100**(7): 2189–2197.

© Woodhead Publishing Limited, 2013

2

Principles of finfish broodstock management in aquaculture: control of reproduction and genetic improvement

N. J. Duncan, IRTA, Spain and A. K. Sonesson and H. Chavanne, Nofima, Norway

DOI: 10.15333/9780857097460.1.23

Abstract: Broodstock management, the control of reproduction and genetic improvement are central parts of an aquaculture business that allow a hatchery to continually improve efficiency and productivity of the entire business. Control of reproduction, has been classified into critical points and management points. The chapter identifies and explains how critical points must be controlled to obtain successful spawning of good quality eggs and how management points can be controlled to maximize production. Genetic improvement gives the potential to shape an organism to meet human needs, improving productivity and product quality. The chapter explains the bases of setting up a genetic improvement program and identifies the associated benefits and risks of this long term investment.

Key words: finfish, broodstock management, genetic improvement, reproduction, gamete quality.

2.1 Introduction

Broodstock management requires that biological (particularly reproductive) characteristics are understood and used to create a culture environment to enable the organism to reach advanced stages of maturation, vitellogenesis and spermiation from which the spawning can be obtained with adequate egg quality and quantity for commercial hatchery production. Once protocols to achieve adequate spawning have been established, aspects of reproduction and genetic traits of the broodstock or population need to be managed to ensure the progeny have improved characteristics for ongrowing and the market. This chapter aims to describe broodstock management, the principles of control of reproduction and genetic

© Woodhead Publishing Limited, 2013

improvement applied to finfish species, with emphasis on the principal aquaculture species in Europe.

Control of reproduction (Section 2.2) has been classified into critical points and management points. Critical points or limiting factors are those that need to be considered when setting up and managing broodstock to obtain adequate egg quality and quantity for commercial hatchery production. Management points are those related to the logistics of egg and juvenile production or how aspects of the control of reproduction can be managed to improve production or provide particular characterises in the progeny that are needed for production.

Perhaps the most important aspect of broodstock management, after spawning of the required quality and quantity has been achieved, is genetic improvement, the domestication process by which genetic selection improves desired traits over generations. Genetic improvement is the cornerstone to domestication as it gives the potential to shape an organism to meet human needs. It is a cumulative process of improvement and, as such, is a long-term activity which has been very successful for aquaculture species. Genetic improvement is a field that can contribute to sustainable aquaculture by improving traits like growth, product quality, disease resistance, general robustness of individuals and digestion of sustainable feeds. The principles of genetic improvement (Section 2.9) are described: how to start up a genetic improvement program, the important concepts of selected traits, breeding values and inbreeding, the main design of genetic improvement programs and the importance of genomic resources for genetic improvement application. Genetic improvement represents a long-term investment that brings long-term results; the main risks for broodstock subject to genetic improvement and ways to mitigate such risks are detailed (Section 2.9).

2.2 Control of reproduction

All fish species have evolved to occupy a niche in the aquatic environment. Evolution has shaped the reproductive strategy, ultimately giving the species some advantages necessary to survive. The reproductive strategy encompasses the following aspects: sexual differentiation, size at first maturity, reproductive nutritional requirements, development of maturation in relation to environmental changes, spawning in relation to the environment, spawning behaviour, parental care and egg parameters. A wide variation in reproductive strategies exists amongst the fish species:

- *Sexual differentiation* is controlled genetically, environmentally, socially or through a combination of these factors. The period of time between hatching and sexual differentiation varies (can be days, months or years) and sex can be fixed for life (gonochoristic) or can change (hermaphrodites). Most aquaculture species are gonochoristic such as in Atlantic

© Woodhead Publishing Limited, 2013

salmon (*Salmo salar*), but a number of species are either protandrous hermaphrodites that change sex from males to females such as the gilthead seabream (*Sparus aurata*) (Zohar *et al.*, 1995) or protogynous hermaphrodites that change sex from females to males such as the greasy grouper (*Epinephelus tauvina*) (Abu-Hakima, 1987). This information can be used to indicate sex ratios for spawning and when fish of a desired sex can be identified and selected.

- *First maturation* can be early at a small size, late at a large size or plastic with many maturational episodes or many opportunities for which the timing is genetically and/or environmentally controlled. Aquaculture species exhibit a wild range of strategies, Nile tilapia (*Oreochromis niloticus*) first mature below 40 g within six months after fertilisation (MacIntosh and Little, 1995), while at the other end of the scale, Pacific bluefin tuna (*Thunnus Orientalis*) mature at >30 kg (Sawada *et al.*, 2005). Atlantic salmon exhibit plasticity in age of maturation and can mature at 50 g in fresh water and again at 3+ kg in sea water, and different genetic strains of Atlantic salmon present different strategies (Meerburg, 1986). This information indicates the minimum size and optimal size for spawning. The size of mature fish also indicates the tank size that may be needed.
- *Nutrition* plays a critical role in the provision of the nutrients required for reproduction. Diets vary between species and nutrition is involved in determining the age of first maturity, the period between maturation episodes and fecundity. Most aquaculture species are omnivorous or carnivorous and diets range from those typical of species low in the food web to those for top predators. Low in the food web species include flat head mullet (*Mugil cephalus*) and common carp (*Cyprinus carpio carpio*) with dietary items dominated by plants, zooplankton and detritus to give respective trophic values of 2.13 ± 0.18 and 2.96 ± 0.32 (Froese and Pauly, 2011). Senegalese sole (*Solea senegalensis*) is a species intermediate in the food web with diet items from the zoobenthos, such as polychaetes, small crustaceans and molluscs with trophic values of 3.13 ± 0.33 (Froese and Pauly, 2011). Examples of top predators are Atlantic salmon and Atlantic bluefin tuna (*Thunnus thynnus*) that predominantly eat other fish and have trophic values of 4.43 ± 0.06 and 4.43 ± 0.78, respectively (Froese and Pauly, 2011). This information helps ensure that the broodstock diets provide the nutritional requirements for spawning.
- *Gametogenesis in relation to the environment*: species have evolved to use distinct environmental cues or proximal factors to entrain the progress of maturational development to ensure that spawning and critical stages of larval and juvenile development coincide with optimal environmental conditions for survival. Photoperiod and temperature are the most common proximal factors, but food availability, lunar or tidal cycles, weather cycles (e.g. rainfall), ocean currents and pressure have also been implicated. Most aquaculture species appear to use

© Woodhead Publishing Limited, 2013

photoperiod and temperature to entrain the progress of maturation. It has been demonstrated that photoperiod with constant optimal temperatures entrains the maturational process in rainbow trout (*Oncorhynchus mykiss*), including the initiation of vitellogenesis, the timing of progress of vitellogenesis, final maturation, ovulation and spermiation (Bromage *et al.*, 1993, 2001). Entrainment appears to be similar in other temperate species and some sub-tropical species. However, different aspects of the photoperiod and thermocycle have different importance in different species; for example, in rainbow trout vitellogenesis lasts 8–10 months (Bromage *et al.*, 1993, 2001). Trout initiate vitellogenesis during the winter under a short but increasing photoperiod and cold winter temperatures; vitellogenesis then continues through the spring and summer with the photoperiod increasing to long days and temperature rising. In the autumn, with the decrease in photoperiod and the decline of temperature, the trout enters final maturation before spawning. The gilthead seabream also spawns in the winter, but vitellogenesis takes approximately four months (Zohar *et al.*, 1995). Seabream complete sexual determination in September and initiate vitellogenesis with the photoperiod and temperatures declining. Vitellogenesis in bream is completed under the declining photoperiod and temperature. The arrival of winter temperatures and the gradual increase from the short winter day entrains that the bream complete final maturation and initiate spawning. This information enables the broodstock manager to provide the optimal environment required for gametogenesis to advance to final oocyte maturation and spermiation.

- *Spawning in relation to the environment*: spawning often requires a quite precise environment, particularly temperature, space and substrates. For example, Atlantic salmon require cold (>8 °C) fresh water and a gravel substrate to make a redd (gravel nest in which the eggs can be buried) if spontaneous spawning is required (Mañanós *et al.*, 2009). Common carp require a certain number of degree days (1000–2000 degree days depending on age) of spring temperatures increasing from 15 °C, a specially prepared pond and spawning mats or vegetation (FAO, 1985). Red spotted grouper (*Epinephelus akaara*) require large tanks with a water column of 3.5 m for the spawning behaviour in which the fish swim up through the water column to spawn at the water's surface (Okumura *et al.*, 2003). Grouper in smaller tanks with 1.7 m water depth were observed to jump out of the water at the peak of the ascent through the water column; fertilisation rates were significantly lower than spawning pairs that did not jump during spawning behaviour in a tank with 3.5 m water depth. This kind of information enables the broodstock manger to provide the optimal environment required for spontaneous spawning.
- *Spawning frequency*: ovarian development has been classified into asynchronous, batch synchronous and synchronous. Asynchronous development is associated with daily or close to daily spawning during the

© Woodhead Publishing Limited, 2013

spawning season. Batch synchronous is associated with repeated spawns with a short interval of between a week and a month between spawns and synchronous development is when a single batch of oocytes develops for a single spawn per year or in a lifetime. For example, female rainbow trout that have passed puberty spawn once a year (Bromage *et al.*, 1993), female European seabass (*Dicentrarchus labrax*) spawn three or four times during the three to four month spawning season (Mylonas *et al.*, 2003) and female gilthead seabream can spawn daily or close to daily during the three to four month spawning season (Zohar *et al.*, 1995). This information can be used to estimate numbers of broodstock required and logistics of stocking larval facilities.

- *Spawning behaviour* can be in pairs (monogamy), a female with many males, a male with many females or mass spawning involving males and females. Senegalese sole spawn in pairs and five out of six randomly selected spawns were assigned with microsatellites to a single spawning pair from a Senegalese sole broodstock of 19 fish (Porta *et al.*, 2006). Nile tilapia males control a territory with a spawning nest, 'lek', and females visit the area to spawn with the males (MacIntosh and Little, 1995). Sparidae spawning behaviour was generally described as the spawning female swims up through the water column liberating ova that are fertilised by the numerous following males (Mylonas *et al.*, 2011). This information can be used to indicate sex ratios and tank design.
- *Egg care*: eggs can be scattered uncared for into the environment, hidden, cared for in different ways such as in nests or in the mouth of the female or male and incubated internally and the juvenile 'born' (ovoviviparity). Many marine aquaculture species, such as gilthead seabream, European seabass and Senegalese sole, scatter floating eggs into the water and these can be collected with surface water collectors. Common carp have sticky eggs that are scattered into vegetation to which the eggs stick. Salmonids, such as Atlantic salmon, hide eggs in gravel nests, 'redd', when a gravel substrate is available (Mañanós *et al.*, 2009). Channel catfish (*Ictalurus puntatus*) incubate and care for eggs in a nest or spawning containers provided by the broodstock manager (Dupree, 1995). Female Nile tilapia care for and incubate eggs and newly hatched larvae in the mouth (MacIntosh and Little, 1995). This information indicates the type of egg collection required or if stripping should be considered, how eggs may be incubated and care required for the initial stages of larval culture.
- *Egg size* varies from large to small and egg size has a strong influence on fecundity with the spawning of a few large eggs or many small eggs. These opposing strategies have been termed K and r strategies, where large eggs produce an offspring with higher rates of survival (K strategy) while many small eggs give more opportunities for survival (r strategy). Gilthead seabream have annual fecundities of 2–3 million eggs kg^{-1} of small eggs 0.94–0.99 mm (Zohar *et al.*, 1995) compared to Atlantic

© Woodhead Publishing Limited, 2013

salmon that have annual fecundities of 2000–4000 eggs kg^{-1} of large 5–6 mm eggs (Mañanós *et al.*, 2009). This information can be used to estimate numbers of broodstock, incubation systems and care required for the initial stages of larval culture.

Each species exhibits a different combination of these aspects to give a large multitude of different reproductive strategies amongst fish species. However, despite this complexity, the environment is the common factor that directly and indirectly controls the progress of the reproductive strategy and the success of spawning. Therefore, most species share common points in the reproductive strategy or maturational development for which the environmental requirements are species specific. Maturational development can be considered to be a continual process that starts from hatching and is completed each time spawning is achieved. When the correct environmental signals are not present, maturational development does not proceed and development is arrested or, in extreme cases, maturational development may be suspended and developing oocytes may enter atresia. These points where maturational development may be arrested or suspended can be considered to be critical points in the control of reproduction.

2.3 Critical points in the control of reproduction: forming a broodstock, culture environment and nutrition

The critical points in the control of reproduction are periods in the reproductive development of a species that require control of the culture environment to provide the optimal conditions that allow maturation to proceed (Table 2.1). The captive environment can be very different from the natural environment and the broodstock manager must define what the optimal captive environmental conditions are that address the critical points for the species in question. The principal critical points can be ordered, each factor needing to be surpassed and maintained at optimal levels for maturation to progress: forming or selecting a broodstock > the captive environment and fish welfare > nutrition > environment during gametogenesis > environment during final maturation and ovulation/spermiation (Table 2.2). This classification into critical points helps understand the captive environmental conditions that are required to complete maturation and spawning, but in reality maturational development is progressive with subtle changes in the environment entraining the gradual progress of maturation.

2.3.1 Forming a broodstock

Broodstock can be formed from two sources, wild caught fish or fish reared in captivity. Wild fish are best caught with professional fishermen using passive arts of fishing in the following preference; almadraba or traps, surrounding or seine type nets, hook and line and lastly gill nets which will in

© Woodhead Publishing Limited, 2013

Table 2.1 The stages of maturational development for which different reproductive strategies and the control can be identified and how these stages relate to different critical and management points

Stages of maturational development	Sexual differentiation		Primary gonadal development		First maturation or puberty		Secondary gonadal development		FOM, ovulation and spermiation		Spawning
Different strategies	Gonochoristic, hermaphrodite	→		→	Maturation at a small size/ large size	→	Short to long timeframe	→	Timing and frequency, social cues	→	Timing and frequency, social cues
Control of strategies	Genetic and environment	→		→	Genetic growth, nutrition and environment	→	Environment and nutrition	→	Environment, social environment	→	Environment, social environment
Critical points in captivity		→	Captive environment and welfare	→	Captive environment, welfare and nutrition	→	Optimal nutritional components and captive environment	→	Optimal captive environment or use of hormonal induction protocols	→	Optimal captive and social environment or use of strip spawning protocols
Management points	Sex can be controlled to enhance growth or genetic selection	→		→	Nutrition and photoperiod can control percentage maturation and fecundity	→	Environmental control to obtain out-of-season spawning	→	Environmental control to advance or elongate spawning period	→	Strip spawning provides many opportunities

© Woodhead Publishing Limited, 2013

Table 2.2 Brief description of requirements for optimal points for the five top species produced in Europe

Species	Critical points in captivity			
	Forming a broodstock	Culture environment during vitellogenesis/spermatogenesis	Culture environment during FOM ovulation and spermiation	Spawning
Atlantic salmon	3+ years old, sex ratios 1:1 or biased to females, 2:1	Seawater, cages or large tanks with <5 kg/m^3, environment natural spring – summer – autumn, photoperiod (LD12:12→LD16:8→LD8:16) and temperature (7→13→8 °C)	Transfer to freshwater large tanks with <5 kg/m^3, decreasing autumn photoperiod (→LD8:16) and temperature (<8 °C).	Strip spawning, fresh water temperature <8 °C
Rainbow trout	2+ years old, sex ratios 1:1 or biased to females, 2:1	Ponds or large tanks with <5 kg/m^3, environment natural spring–summer–autumn photoperiod (LD12:12→LD16:8→LD8:16) and temperature (8→16→<8 °C)	Ponds or large tanks with <5 kg/m^3, decreasing autumn photoperiod (→LD8:16) and temperature (<8 °C).	Strip spawning, temperature <8 °C
Gilthead Seabream	Males, 2–3 years old, 300–500 g, females 4–6 years old, 1–1.5 kg, sex ratio 1:1 or biased to females, 2:1	Large tanks with <5 kg/m^3, environment natural autumn photoperiod (LD12:12→LD9:15) and temperature (25→16 °C)	Large tanks with <5 kg/m^3, environment natural winter–spring photoperiod (LD9:15→LD12:12) and temperature (16→18 °C)	Spontaneous natural tank spawning
European seabass	Males, 3–4 years old, 700 g, females 6–8 years old, 1–1.5 kg, sex ration 1:1 or biased to males, 1:5:1	Large tanks with <5 kg/m^3, environment natural autumn photoperiod (LD12:12→LD9:15) and temperature (20→10 °C)	Large tanks with <5 kg/m^3, environment natural winter–spring photoperiod (LD9:15→LD12:12) and temperature (13→15 °C). Hormone induction may be needed for captivity bred broodstock.	Spontaneous natural tank spawning
Common carp	2–5 years old, 1.5 kg, sex ratios vary from 1.5:1 to 1:2, males:females	Ponds, environment natural summer–autumn–winter–spring, photoperiod (LD14:10→LD10:14→LD12:12) and temperature (30→13→20 °C)	Ponds, spring, increasing photoperiod (→LD14:10) and temperature (15→20 °C: 1–2000 degree days). Hormone induction also routinely used for broodstock transferred to the hatchery	Natural pond spawning on spawning mats. Strip spawning of hormonally induced fish.

© Woodhead Publishing Limited, 2013

some cases severely damage the fish. Cultured fish can be selected from the stock produced by the farm or from other farms. At this point, it is important to consider that the rearing performance and long-term genetic management of a hatchery broodstock are conditioned by the origin of the fish forming the base population. For the genetic aspects of forming the broodstock, see further details below in Section 2.9 on Genetic improvement.

The size (weight or length) of the fish and sex ratios within the new stock must also be considered and measures taken to ensure that the stock has the desired characteristics. A clear understanding of size at first maturity or optimal size for gamete production and mechanisms of sexual determination and especially sex-specific growth differences will enable the broodstock manager to select the organisms required. Ideally, the stock will have the sex ratio desired in the spawning stock and a size that will ensure that the broodstock have an optimal size in the year spawning is required. Errors in these basic aspects can result in a delay in the spawning of the stock or inappropriate sex ratios. For example, European seabass stocks exhibit variable sex ratios resulting from environmental influences in sex determination and differential growth rates in sexes, which can mean that the routine practice of size grading can bias a size grade to a particular sex (Piferrer *et al.*, 2005). Selection of a potential European seabass broodstock must, therefore, first consider the sex ratio of the stock. In another example, gilthead seabream are protandrous hermaphrodites that change sex from males to females (Zohar *et al.*, 1995). The sex change appears to be controlled socially and younger males were observed to change to females when few large females were present in the stock. Under culture conditions, optimal spawning sizes for males are 300–500 g and females 1–1.5 kg (Moretti *et al.*, 1999). The broodstock can, therefore, be renewed with potential males of the correct size, which are held with older males that change to females. The timing and size of newly entered gilthead seabream stock can be critical to obtaining the correct size and proportion of sexes in the spawning broodstock. An interesting consideration in the families produced for genetic improvement of a stock of hermaphrodite fish such as the gilthead seabream is that fish may participate in early year classes as males and in later year classes as females.

On arrival of new fish at the hatchery, care needs to be taken with quarantine and acclimatisation to the new holding conditions. The entrance of new broodstock is a path through which diseases can enter into a hatchery. Therefore, a quarantine period should be managed before entering the fish into the broodstock unit and the broodstock unit should be managed as an area quarantined from the other hatchery facilities. Ideally, the quarantine should be in tanks with an environment similar or identical to that in the broodstock unit. Three classes of disease should be considered: parasites, bacteria and virus. All fish should be examined for external parasites and if the fish have wounds from capture, the wounds should be treated with an

© Woodhead Publishing Limited, 2013

antiseptic (iodine or hydrogen peroxide H_2O_2-based) and the fish should be given an antibiotic injection. Five percent of the fish can be sacrificed and an autopsy made by a fish veterinarian/qualified biologist to determine if internal parasites, bacteria and virus exist.

All the fish that enter the broodstock facility should pass through the predetermined steps for acclimatisation/quarantine during which procedures and prophylactic treatments should be varied depending on the pathogens that were found in the stock, the species and the previous environment of the fish, i.e. wild fish, fish from cages that have an interaction with wild populations, fish from a facility with a known disease history. The protocol should be developed and applied with a fish veterinarian/qualified biologist. For example, the following protocol has been successfully used for wild meagre (*Argyrosomus regius*) that were observed to have skin parasites (monogenean worms and skin copepods commonly referred to as sea lice) and small wounds from capture and transport (Duncan *et al.* in this volume): Week 1: leave the fish with no disturbance including no feeding. Week 2: A formalin bath, an H_2O_2 bath (Table 2.3) and offer feed from the natural diet of the fish, i.e. squid and sardines. Week 3: A formalin bath, an H_2O_2 bath and offer feed from the natural diet of the fish. Week 4: A formalin bath, an H_2O_2 bath and offer feed from the natural fish diet. Week 5 or when fish are feeding: an oral treatment for internal and external parasites such as prazinquantel (Table 2.3). A similar protocol would work well for most marine species and can be adapted for freshwater species. For euryhaline species that originate from the marine environment, a freshwater bath will kill many skin parasites and should be applied at weekly intervals as described above for formalin. At the end of the acclimatisation period the fish should be feeding and as close to disease free as is possible and can be moved into the broodstock facility.

2.3.2 Culture environment and fish welfare

The culture environment for broodstock and potential broodstock should promote a high level of welfare in the fish population. In most fish species good welfare promotes the progress of maturation, whilst poor welfare can postpone maturational development until conditions and welfare improve. Fish welfare is a new area of study which has recently attracted public concern in Northern Europe, prompting the EU parliament to recommend that the welfare of fish is safeguarded (Council of Europe, 2006) in the same way that the welfare of farm animals is protected. Defining fish welfare and how to measure or determine the welfare status of a fish is complex (Huntingford *et al.*, 2006). Huntingford *et al.* (2006) promotes a definition that fish should be free from suffering and discusses the five freedoms (promoted by UK Farm Animal Welfare Council, 2005) that form an accepted framework for terrestrial animals to determine the areas in which welfare may be compromised. However, despite the complexity of the issues and absence

© Woodhead Publishing Limited, 2013

Table 2.3 Chemical treatments used during quarantine and to disinfect eggs

Pathogen to be controlled	Chemical*	Spacies	Dosing	Administration
External parasites, protozoa, monogenea, copepods. Also external bacteria	Formalin	Fish	100 μL/L	1 hour bath
External bacterial and protozoa (Amylodinium)	H_2O_2	Fish	150 μL/L	1 hour bath
Internal and external parasites	Praziquantel	Fish	5 mg/kg	In feed or orally with a canula
Internal bacteria	Enrofloxicina	Fish	10 mg/kg	Injection every other day for minimum period of 10 days
Internal bacteria	Oxytetracycline	Fish	10 mg/kg	Injection every other day for minimum period of 10 days
External bacteria, fungus and protozoa	H_2O_2	Fish eggs	500 μL/L	15 minute bath
External bacteria and protozoa	Active iodine	Fish eggs	50 mg/L	10–60 minute bath

* The authors do not recommend the use of unapproved chemicals to treat fish. All treatments should be previously tested on a small number of organisms that will not cause financial loss to the farm.
Sources: US Fish and Wildlife Service's Aquatic Animal Drug Approval Partnership Program, 2011; personal observation.

of agreed definitions, fish welfare schemes have been developed and successfully applied to aid consumers in purchasing Atlantic salmon that did not suffer from adverse welfare (RSPCA, 2010) and which are, to a large extent, based on describing the environment and environmental limits under which a species exhibits good welfare. The RSPCA (2010) have adapted the five freedoms to give the following definitions as the framework to maintaining a high level of fish welfare:

1) Freedom from thirst, hunger and malnutrition, by access to an appropriate high quality diet and an environment in which fluid and electrolyte balance can be maintained. 2) Freedom from discomfort, by maintaining the water at an appropriate temperature and chemical composition and providing well

© Woodhead Publishing Limited, 2013

designed enclosures or tanks, with shading if necessary. 3) Freedom from pain, injury and disease, by avoiding situations which are likely to cause pain, injury or disease, by rapid diagnosis and treatment of disease and humane killing. 4) Freedom to exercise normal behaviour, by providing the appropriate space and environment for the species. 5) Freedom from fear and distress, by minimising stressful situations such as handling or predator attack as far as possible, by making gradual changes to husbandry and water quality, and by humane slaughter.

In practice, the framework indicates that water quality, temperature, tank environment and feeding should be optimal, while invasive manipulation of the stock should be minimal. Therefore, for a broodstock or potential broodstock, the water quality should be the best available, oxygen levels, pH and contaminants (ammonia, nitrates and pollutants such as fertilisers, herbicides and pesticides) need to be in optimal ranges for the species.

Knowledge of the species requirements and common sense need to be applied to determine the environment that promotes optimal welfare. A clean water supply with flowthrough water exchange rates of five to six tank volumes per day or recirculation with 50 % renovation of water per day can be recommended and should provide optimal conditions for many species. Temperature should be optimal for the species in relation to the season. The tank should be self-cleaning with no sharp edges that may injure the fish or complicate tank cleaning. Stocking densities should be low, even for species that have a schooling behaviour. Below 5 kg/m^3 is often considered optimal for broodstock. Tank size should provide the space required that enables normal swimming, schooling and social behaviour. High quality diets that meet all the nutritional requirements should be offered to the fish regularly to avoid both over- and under-feeding and to ensure that at the start of feeding the fish have a good appetite and feeding behaviour. Husbandry operations, feeding, tank cleaning and regulation/monitoring of water quality should be conducted in a way that does not startle or stress the fish with minimal noise and without rapid movements that cast a shadow over the fish. More invasive operations, such as moving fish to a new tank, weighing fish, sedation, biological sampling and assessment of maturity, that require that the fish be crowded and/or removed from the water should be reduced to a minimum and conducted quickly and efficiently. Campbell *et al.* (1992) demonstrated that rainbow trout that were repeatedly briefly left without water by draining the tank at random points during a nine month period before the spawning season had delayed ovulation, reduced egg quality, lower sperm counts and lower progeny survival compared to rainbow trout that were not disturbed.

2.3.3 Nutrition

After fish welfare, nutrition can be considered to be the next limiting factor that must be surpassed to maintain the progress of maturational

© Woodhead Publishing Limited, 2013

development. Nutrition affects the decision to proceed with maturation and the availability of the nutritional components that are necessary to form the gametes and, therefore, is an aspect that controls gamete quality and quantity. Food availability and growth have been demonstrated to determine when a fish proceeds to the completion of maturation. Reducing the ration fed reduced the percentage of fish that completed maturation (Thorpe *et al.*, 1990; Cerda *et al.*, 1994; Bromage 1995). Thorpe (1986) and Thorpe *et al.* (1990) suggested that Atlantic salmon have a biochemical threshold to signal the size and/or age when conditions were optimal to proceed with maturation. In addition to growth and bio-energetics, an adequate nutrition must be provided to avoid detrimental effects on egg quality and quantity (see review by Izquierdo *et al.*, 2001). From before and during gametogenesis until spawning, attention needs to be placed on providing adequate quality of protein, lipid/fatty acid composition, vitamins and minerals (Izquierdo *et al.*, 2001). Unfortunately, broodstock nutrition is poorly studied. The importance is recognised, but prohibitive costs of working with large expensive fish compared to small markets for broodstock diets have resulted in few comprehensive studies. Izquierdo *et al.* (2001) reviewed how optimal levels of essential fatty acids, particularly arachidonic acid (ARA), docosahexaenoic acid (DHA) and eicosapentaenoic acid (EPA), and vitamin E levels improved fecundity, fertilisation, embryo development and larval survival. Protein level and amino acid profile, carotenoids and vitamins C, A and B_6 were also related to improved gamete quality, but less convincingly, indicating the need for further studies. The relationship between essential nutrients and gamete quality was often that increasing dietary levels increased gamete quality up to a point, beyond which an excess of the nutrient had a negative effect, indicating that optimal levels should be administered. For example, increased dietary ARA (up to 2–3% of total fatty acid content with an EPA:ARA ratio of 2–4) has increased aspects such as fertilisation and hatching success, sperm quality and larval survival in European seabass (Bell *et al.*, 1997; Bruce *et al.*, 1999), gilthead seabream (Harel *et al.*, 1992; Fernandez-Palacios *et al.*, 1995) and Atlantic halibut (*Hippoglossus hippoglossus*) (Mazorra *et al.*, 2003). However, higher levels of ARA in broodstock diets may have negative effects as has been observed in larval nutrition (Estevez *et al.*, 1999; Koven *et al.*, 2001).

Generally, fish broodstock appear to have evolved to conserve nutrients destined for the gametes and only produce gametes when the broodstock nutrition is adequate. When nutrition is limiting, the first observed effect is a reduction in fecundity as maximum numbers of oocytes develop with the nutritional resources available. The reduced fecundity appears to be often, but not always, associated with a reduction in egg or sperm quality. There is no quick solution to poor spawning caused by an inadequate diet and the only solution may be to improve the nutrition for the following maturation episode, which is often the following year. Some species, such as gilthead

© Woodhead Publishing Limited, 2013

seabream, can quickly assimilate into the eggs the dietary components supplied and egg quality has been observed to improve over the spawning season as the diet was improved.

The easiest way to avoid nutritional problems is to offer a varied diet that includes dietary components that are or were derived from organisms that are naturally found in the species diet. For well-established aquaculture species, such as Atlantic salmon and rainbow trout, proven commercial broodstock diets exist. The broodstock diets that are available tend to be very conservative, utilising high levels of fish meal and oils of good quality with boosted levels of essential fatty acids, vitamins and minerals. However, some farming companies working with well-established species such as the European seabass and the gilthead seabream will supplement commercial diets with previously frozen natural diet items such as squid, mussels, cuttlefish, polychaetes and krill or small shrimp (Izquierdo *et al.*, 2001, personal observation). For species new to aquaculture it can be recommended to use a commercial broodstock diet formulated for a closely related species and diet items from the natural diet. Good egg quality was obtained from wild meagre that were fed with squid, sardines and a commercial broodstock diet designed for European seabass or gilthead seabream. (Vitalis REPRO and Vitalis CAL, Skretting, Netherlands) (Duncan *et al.* this volume). When there are difficulties feeding an extruded diet to wild broodstock, a semi-humid paste may be more readily eaten or natural diet items can be used with a small supplement to ensure that essential lipids, vitamins and minerals are delivered. Delivery of essential dietary components can be achieved by feeding small squid filled with a commercial marine fish diet or lipid, vitamin and mineral supplements, or injecting supplements into fish fillet or shrimp. However, the use of natural diet items and pastes also has an associated sanitary risk. Natural diet items and pastes will reduce the cleanliness of the tank environment and increased cleaning may be required. Diet items should also be of good hygiene quality, free from decomposition and bacterial contamination, i.e. as good or better than human food requirements. Items should also be frozen (–20 °C) for a minimum of 48 h before feeding to ensure that most parasites are killed.

2.4 Environment during gametogenesis and spawning: the optimal environment

The captive environment during gametogenesis and spawning can be a limiting factor once adequate welfare and nutrition have been provided. The environment entrains the maturational process and particularly vitellogenesis in females and the latter stages of spermatogenesis in males. Different species use different environmental cues or proximal factors to entrain the process of maturational development. Photoperiod and temperature are the most common proximal factors for reproduction, but food

© Woodhead Publishing Limited, 2013

availability, lunar or tidal cycles, weather cycles (e.g. rainfall), ocean currents and pressure have also been implicated. The proximal factors must entrain stages of gametogenesis during months or even years to ensure that the timing of spawning or an early stage of juvenile development coincide with optimal environmental parameters for offspring survival. The captive environment must, therefore, provide these environmental cues or proximal factors to ensure the completion of gametogenesis and spawning in captivity.

The periods of maturational development for which the correct proximal factors must be provided as part of the captive environment can be considered as (i) the later stages of gametogenesis and (ii) the spawning period. In females this is (i) the secondary growth phase from the initiation of vitellogenesis until the final stages of vitellogenesis and the start of final oocyte maturation (FOM) when the germinal vesicle begins to migrate and (ii) the spawning period with the complete process of FOM, ovulation and spawning. In males this is (i) from when spermatocytes are formed until the start of spermiation and (ii) the spawning period when the males have free-flowing sperm. Generally the same proximal factors will control maturation in both males and females. When it is necessary to identify these maturational periods, the period of gametogenesis is easier to identify in females through observation of the size of the ovary, histological determination of oocyte development or endocrine changes. During vitellogenesis the ovary of many species undergoes considerable growth and, for example, an increase in GSI (gonado-somatic index) from less that 1 % to 12–15 % in rainbow trout with oocytes undergoing a 100-fold increase in volume (Bromage and Cumaranatunga, 1988). In females the inclusion of vitellogenin, which stains pink (HE staining) in a histological section, is easily identified in the oocyte cytoplasm and has been termed stage 5 of vitellogensis until stage 7 when the germinal vesicle is observed to be off centre (Bromage and Cumaranatunga, 1988; Mañanós *et al.*, 2009). This period can also be identified by the endocrine changes with increasing presence of oestradiol and vitellogenin in the blood, and the start of the spawning period is associated with an increase in luteinising hormone (LH) which controls final maturation and ovulation. The spawning period can be identified as the period in which males have free-flowing sperm when abdominal pressure is applied. Females can also be canulated to obtain an ovarian biopsy with large vitellogenic oocytes which can be examined to determine the stage of ovary development. Canulation is conducted by inserting a flexible polyethylene tube (the authors use a cut feeding tube made for feeding babies born prematurely) with a 1 mm internal diameter approximately a quarter to a maximum of half the length from the ovary duct entrance to the pelvic fin and applying a slight suction with the mouth. Higher suction can be applied with a syringe, but this should be avoided and a suction of 1 ml with a syringe is more than sufficient. The pore of the ovary duct is often recognisable as a faint line perpendicular to the lateral

© Woodhead Publishing Limited, 2013

line and is generally between the anal pore (closest to the head) and the urinary pore (closest to the tail).

The environmental requirements for gametogenesis are generally more flexible or have a wider optimal range than the requirements for spawning, which can have a narrow range of precise environmental requirements. This has been inadvertently demonstrated when environmental manipulations have changed the timing of spawning and consequently altered the timing of late gametogenesis to coincide with environmental parameters very different to these naturally experienced, but with little effect on egg quality in rainbow trout (Bromage *et al.*, 2001; Davies and Bromage, 2002), Atlantic salmon (Taranger *et al.*, 1999) and European seabass (Carrillo *et al.*, 1995). Therefore, for gametogenesis the combination of good welfare, natural photoperiod, either with natural daylight or simulated with artificial light, and temperatures close to those experienced in the species' natural environment should ensure that vitellogenesis proceeds to late stages and that spermatozoa are produced. The environmental requirements for spawning can be quite precise and attention must be paid to all the aspects addressed above to give optimal welfare and, in particular, to final cues that indicate that conditions are optimal for spawning. Some final cues are difficult or impossible to control or manipulate, such as moon phases or tides and weather patterns such as rains, but others, such as temperature, social environment, salinity or presence of a particular substrate, can and should be manipulated to control spawning. Most species have a narrow range of optimal temperature in which they will spawn and this can be used to make quite precise calculations that predict spawning. For example, degree days after the temperature rises above 15 °C predict when common carp can be spawned (Rothbard and Yaron, 1995) and the start of common sole (*Solea solea*) spawning was positively correlated ($r = 0.9$) with winter temperatures (Baynes *et al.*, 1993).

Behaviour can change considerably during the spawning period and spawning behaviour may require particular sex ratios or stocks may require more space than outside of the spawning period. Generally fish that appear to spawn in a group will need to be held in a group, preferably more than 10 fish, that enables the formation of an aggregation, and space should be provided for the behaviour of the aggregation. For example, gilthead seabream broodstock do not spawn well when kept in low numbers in smaller tanks and the spawning response is better when 10+ fish are kept at densities below 5 kg/m^3 (personal observation). Commercial hatcheries often use groups of more than 30 gilthead seabream broodstock. Some species exhibit increased aggression during the spawning season and this may be related to confinement limiting expression of normal behaviour or because the species spawn in pairs or a single male with many females and the fish are competing for partners. The Pacific cubera snapper (*Lutjanus novemfasciatus*) was observed to become aggressive during the spawning season resulting in large open wounds (up to a quarter of the body area) or fish jumping

© Woodhead Publishing Limited, 2013

from the tanks, presumably in an attempt to escape from an aggressive fish, and this situation led to the recommendation to work with single spawning pairs isolated in a spawning tank (Duncan *et al.*, 2011).

Diadromus species migrate to water bodies with different salinities to spawn and it may be necessary to change the salinity to obtain spawning. Catadromous species such as eels of the genus *Anguilla* migrate to the marine environment to spawn. Anadromous species such as the salmonids of the genera *Salmo* and *Oncorhynchus* migrate to fresh water to spawn. For example, Atlantic salmon had higher percentage ovulation and egg quality when broodstock were transferred to fresh water with a naturally decreasing temperature (Magwood *et al.*, 2000) compared to broodstock left in sea water. Spawning substrates may be needed for egg release and spawning; for example, common carp are given spawning mats and channel catfish spawn naturally in ponds, prepared with 20–40 L spawning containers (Dupree, 1995).

2.5 Sub-optimal environment: reproductive dysfunctions

When incorrect proximal factors are provided or the broodstock are stressed and suffering with a poor level of welfare, maturation will be arrested until conditions improve. In extreme conditions maturation may be aborted and the gonads regressed to early stages of development. This situation has been termed a reproductive dysfunction, and three types of reproductive dysfunction have been described (Zohar and Mylonas, 2001). The first dysfunction is arrest in the early stages of maturational development, the second dysfunction is arrest in late stages of gametogenesis and the third dysfunction is arrest before gamete release.

The first type of dysfunction is when maturation is arrested in the early stages of development, before or in the early stages of vitellogenesis in females. This indicates that captive conditions were not appropriate for the later stages of gametogenesis. The classic example of the first type of dysfunction is the eel. The European eel (*Anguilla anguilla*) and Japanese eel (*Anguilla japonica*) do not mature in captivity. Long-term hormone therapies have been used to induce gametogenesis and the production of gametes in the eel, but the protocols have not produced sufficient quality and quantity for commercial production (Ohta *et al.*, 1997; Pedersen, 1997; Van Ginneken and Maes, 2005).

The second type of dysfunction is when maturation is arrested in late stages of gametogenesis, late vitellogenesis in females and reduced sperm production in males. This indicates that the captive environment was optimal for all stages until the spawning period. When welfare cannot be improved or the correct cues cannot be provided, hormone therapies can be used to produce the quality and quantity of gametes required for commercial production. Zohar *et al.* (1995) described the second type of dysfunction in

© Woodhead Publishing Limited, 2013

gilthead seabream females where the hypothalamus did not produce gonadotropin releasing hormone (GnRH) to release the LH present in the pituitary, thereby, blocking the brain–pituitary–gonad axis that controls final oocyte maturation and ovulation.

Two types of hormone therapies have been used to control this dysfunction, the application of exogenous gonadotropins (GtH) and the application of exogenous GnRHa, an agonist of GnRH. The exogenous GtH acts directly on the gonads, while the exogenous GnRHa liberates the LH from the pituitary of the treated fish which in turn acts on the gonads (Mañanós *et al.*, 2009). Both treatments stimulate FOM and ovulation in the ovaries and spermiation in the testis. Some species, particularly freshwater species, have a dual control for the release of LH from the pituitary, GnRH control release and dopamine inhibits release. For these species, a dopamine antagonist (such as pimozide or metaclopramide) must be used in combination with the GnRHa to induce FOM or spermiation. Exogenous GtH are often favoured for species that have a strong dopamine activity and which have a long tradition for the use of GtH, for example the cyprinids. The sources of GtH most commonly used are purified carp pituitary extract (cPE) and human chorionic gonadotropin (hCG). The GnRHa-induced spawning therapies are commonly used for other species. The release of LH stimulated by an acute GnRHa injection lasts approximately 24 h and returns to pre-injection levels after 48 h (Zohar *et al.*, 1995). The short duration is because GnRHa is quickly broken down and repeated injections or implants must be used if a longer period of circulating LH is required. The form of applying GnRHa should be tested to determine the most suitable method, a single injection, multiple injections or a slow release implant. The type of ovarian development is an important aspect in this decision; multiple spawning fish with an asynchronous ovarian development are best treated with implants to induce a multiple spawning response. Injections may be favoured for fish that spawn once in a season or that have an interval of more than a week between spawns. These species generally have a synchronous or batch synchronous ovarian development. The exogenous agonist most commonly used is des-Gly10, [D-Ala6]-GnRH which is the same as and often sold under the name Gly10, [D-Ala6]- luteinising hormone releasing hormone (LHRHa).

The two most important aspects of a hormone therapy are dose and stage of ovarian development measured as oocyte size (Ibarra-Castro and Duncan, 2007; Mañanós *et al.*, 2009). If a dose is not known, the best approach is to carry out a dose response study (Garcia, 1989; Ibarra-Castro and Duncan, 2007). However, dose can also be estimated from doses successfully used in closely related species. Mañanós *et al.* (2009) showed that a selection of species from a wide range of families (Percichthyidae, Moronidae, Centropomidae, Sparidae, Sciaenidae, Carangidae, Scombridae, Paralichthyidae, Scophthalmidae, Tetraodontidae, Chanidae, Salmonidae, Acipenseridae) all required a similar dose of GnRHa, an injection of

© Woodhead Publishing Limited, 2013

1–40 μg kg^{-1} or an implant of 25–100 μg kg^{-1}, and a selection of species from a narrower range of families (Lutjanidae, Ictaluridae, Clariidae, Mugilidae, Cyprinidae) required higher doses of GnRHa (injection of +100 μg kg^{-1}), GnRHa combined with a dopamine antagonist (20–50 μg kg^{-1} of GnRHa with an antagonist pimozide or metaclopramide) or a source of GtH (hCG: 44–4000 IU kg^{-1}, carpGtH: 300–600 μg kg^{-1}). Lists of optimal doses can also be found in Tucker (1998), Mylonas and Zohar (2001) and Zohar and Mylonas (2001).

The minimum oocyte diameter can be based on the morphology of the oocytes, particularly the size of oocytes that have an off-centre migrating germinal vesicle. Oocytes with a central germinal vesicle and a similar diameter to an oocyte with an off-centre migrating germinal vesicle are close to FOM and should respond to a hormone induction therapy. The minimum oocyte diameter of successful spawning induction was also correlated ($r^2 = 0.62$) to egg size and, if the egg size of a species is known, an estimate of the minimum oocyte size can be calculated with the equation, minimum oocyte size = −95.9872 + 0.623725 x egg size (Mañanós *et al.*, 2009). Although it is temperature and species dependent, a hormone therapy with an optimal dose and oocyte size should give spawns of good quality and quantity starting 24–72 h after the hormone application (latency period). Poor egg quality or quantity from a hormone therapy may indicate that the optimal dose or oocyte size has not been selected. Generally, a higher than optimal dose gave poor egg quality (Garcia, 1989; Ibarra-Castro and Duncan 2007) and a lower than optimal dose a low egg quantity (Garcia, 1989; Barbaro *et al.*, 1997; Ibarra-Castro and Duncan, 2007). Smaller than optimal oocyte size gave no response or low quality and quantity of eggs (Lee *et al.*, 1986; Berlinsky *et al.*, 1996, 1997; Mugnier *et al.*, 2000; Ibarra-Castro and Duncan, 2007) and too large an oocyte size has given a short latency period and poor egg quality (personal observation). When these poor responses are obtained from a hormone therapy the protocol should be revised and adjusted.

The third dysfunction is arrest before gamete release; the oocytes are ovulated, but the ova are not released by the female to be fertilised by the male. The cause is generally environmental or social and results in the ova being reabsorbed or liberated without being fertilised. Environmental factors are commonly no spawning substrate such as for salmonids and carp. Not enough space or incorrect social stimulus are factors that appear to affect spontaneous spawning in flat fish species. This third type of dysfunction is generally solved by providing substrate as for carp or stripping the gametes from the broodstock by applying pressure to the abdomen, as for salmonids, carp, turbot (*Psetta maxima*) and Atlantic halibut. An aspect that needs to be considered with strip spawning is that once ovulated the ova left in the peritoneal cavity over-ripen or age and, with time, lose viability (see review in Bromage, 1995). Some species have been observed to go through a period of ripening after ovulation during which viability improves

© Woodhead Publishing Limited, 2013

before over-ripening begins. Both processes are species and temperature dependent. Optimal stripping times can be determined by stripping and fertilising ova in a time course from the moment of ovulation. Optimal stripping time varies from four to 20 days in salmonids to an hour for Nile tilapia (Bromage, 1995; Tucker, 1998). Strip spawning is a labour-intensive operation, but it does also provide important management options (see Section 2.8) that are not available if fish allowed to spawn voluntarily.

2.6 Egg quality and incubation

Egg quality indicates if optimal conditions were provided and egg production of the quantity and quality needed for the hatchery indicates that critical points were addressed with optimal environmental conditions and diet. The quantity of eggs is simply measured by estimating the number of eggs spawned, usually by (i) measuring the volume of eggs and calculating the number of eggs with a measure of egg diameter or counting eggs in a sub-sample of small volume, (ii) placing the eggs in a known volume of water and counting the eggs in a small volume sub-sample of the total volume of water or (iii) by weighing the eggs and counting the eggs in a weighed sub-sample. The number of eggs produced by a female is referred to as the fecundity, and a more useful measure for comparing fish or stocks is the relative fecundity, which is the number of eggs per kilo of female body weight. Fecundities can be expressed in many ways, for example per spawn or per spawning season.

Measures of quality are more complex or not exact, but are an essential aspect of hatchery production. The most reliable indicator of good egg quality is high survival, through egg hatching and the yolk sac stage of larval development before exogenous feeding is required. Relatively simple measures have been developed and successfully used, where small batches of eggs were incubated separately to hatching and survival of starved larvae assessed until the yolk sac was fully absorbed, but before the larvae began to die due to starvation (Fernandez-Palacios *et al.*, 1995, 2009) or, similarly, individual eggs have been placed in the wells of 96 well plates and the survival of eggs and hatched larvae assessed on a daily basis (Shields *et al.*, 1997; Gimenez *et al.*, 2006). These measures are particularly good for species with short incubation; for cold water species with long incubation periods, external factors (such as fungal infections) unrelated to the intrinsic qualities of the eggs may begin to have strong influences, and the incubation environment between egg batches needs to be identical if good comparisons are to be made. However, these 'end-point' measures of quality are not especially useful for the hatchery which requires methods that can predict good quality before production tanks are stocked.

Generally, the measures used in hatcheries are reliable in identifying poor quality eggs but are not reliable for identifying good quality eggs.

© Woodhead Publishing Limited, 2013

This is because the methods focus on identifying eggs that will not or are not developing; higher proportions of these eggs clearly indicate poor egg quality, but high proportions of developing eggs do not necessarily indicate good egg quality as the eggs or larvae may die before exogenous feeding begins. For example, dead eggs are often, but not always, quite distinct in appearance (opaque) compared to developing eggs, and the proportion of dead eggs can be estimated to indicate a poor batch of eggs. The quality of pelagic marine fish eggs is often assessed by separating the eggs into the proportion that floats and the proportion that sinks in a large measuring cylinder. With full strength sea water (35 %) the eggs that sink are all dead and, therefore, a high proportion of sinking eggs is an indication of a poor quality batch of eggs, but again floating eggs are not necessarily of good quality. In transparent eggs, fertilisation success (early cell divisions) can be assessed, low fertilisation indicating poor quality. In salmonids, stripped ova that are over-ripe are easily recognised as the individual over-ripe ova are transparent, glassy with a single bright orange spot inside the ova; again, high proportions of over-ripe eggs indicate poor egg quality, but the absence of over-ripening does not necessarily indicate good egg quality. For rainbow trout eggs a process called shocking is often used to identify dead or poor quality eggs (Bromage, 1988). The eyed eggs are poured or siphoned over a 50–60 cm drop, which kills poor quality eggs that turn white and opaque and are easily identified and removed. Early morphology of dividing cells (blastula) can be assessed in transparent eggs, and studies have found that a high proportion of asymmetrical divisions indicates poor quality, while symmetrical divisions indicate good quality; however, other studies did not agree that early morphology was an indicator of egg quality (see review by Lahnsteiner *et al.*, 2009). Hatching success is clearly an indicator that the eggs successfully produced larvae, but again the larvae may be weak and die before exogenous feeding. Many other criteria have been investigated, such as changes in egg size during water hardening, microbial load, biochemical parameters of ovarian fluid, biochemical parameters of the eggs, hormone levels and gene expression (Lahnsteiner *et al.*, 2009). However, many of these measures involve time-consuming laboratory procedures that may require specialised analysing equipment. A practical measure of egg quality should be easily measured to provide results in a short time period to indicate if eggs should be used for production.

Lahnsteiner *et al.* (2009) concluded that the most promising measures of egg quality were percentage weight increase during hardening and pH of ovarian fluid for freshwater fish and shape of lipid vesicle and cleavage pattern of early embryos for pelagic marine fish eggs. As previously mentioned, the criteria in use tend to give an early indication of poor egg quality enabling egg batches that will not perform well to be discarded; egg batches that fulfil criteria of not being a poor can be taken forward and progress assessed through hatching success and early larval survival. When a sudden

© Woodhead Publishing Limited, 2013

deterioration in survival is observed, the production may be discarded at an early stage.

Batches of eggs destined for incubation are disinfected and placed in incubators. Iodine-based disinfectants are generally used for the disinfection of freshwater eggs; for example, rainbow trout eggs can be disinfected after water hardening with a 10–60 min immersion in solution with 50–100 mg/L of active iodine (US Fish & Wildlife Service, 2012). For marine fish eggs, gluteraldehyde, H_2O_2, UV treated water and sodium hypochlorite have been used (Lahnsteiner *et al.*, 2009). Moretti *et al.* (1999) recommended using in a marine fish hatchery 8 L of a solution of 50 mg/L of active iodine for 10 min to disinfect 1×10^6 European seabass or 1.5×10^6 gilthead seabream eggs.

Once disinfected, the eggs are passed to the incubation system. Four general systems exist that are utilised for different types of eggs: for dermersal or sinking eggs, incubation is often in trays (system 1, used for salmonids) or at high densities in hatchery jars or vertical columns (system 2) with a conical bottom and an upwelling water flow (salmonids and carp); pelagic eggs are held at lower densities in small tanks with a conical bottom (system 3) or stocked directly into larval rearing tanks (system 4). For example, rainbow trout and Atlantic salmon eggs are placed in trays in a layer of one or two eggs and a baffle system is used to ensure that the water upwells through the tray and the eggs (Bromage, 1988). Generally, 1 L of eggs can be placed in each tray and flows of 3–5 L/min per litre of eggs are required. The second method is a vertical column that holds 5–100 L of eggs and has a perforated base over a conical bottom – the water enters at the bottom and upwells through the eggs (Bromage, 1988). Water flows of 3 L/min per 4–5 L of eggs are required. A vertical incubator that is transparent enables the entire column of eggs to be visually checked for fungal infections. Pelagic eggs, such as gilthead seabream and European seabass, can be incubated in tanks of 100–250 L with a conical bottom, stocked with 10000–15000 eggs/L with gentle aeration and water exchange (Moretti *et al.*, 1999). Water exchange is increased before and during hatching. Debris from the hatching is purged before the larvae are transferred by gravity and water flow to the larval rearing tanks. Alternatively, pelagic eggs can be stocked directly into larval tanks where the larvae are reared after hatching. The drawback with this method is that the debris and fluids from the hatching eggs provide a good substrate for bacterial growth and the tank bottom must be carefully siphoned after hatching (Moretti *et al.*, 1999).

2.7 Management points: fecundity, out-of-season spawning and sexual differentiation

Once the spawning of good quality eggs has been achieved, management points need to be considered to ensure the number and quality of juveniles

© Woodhead Publishing Limited, 2013

are produced when required to ultimately provide the production needed for the market and to sustain the business. These management points can control the number of eggs produced, the timing of the production and aspects of progeny quality such as sex and sterility that can have significant impacts on the growth rates of the production stock (Table 2.4).

2.7.1 Fecundity

A good prediction of the fecundity is required in order to estimate the number or biomass of fish that will be needed in the spawning broodstock. The number of potential broodstock must be decided between one and five years (species-dependent) before the spawning period, and the broodstock manager must have a clear idea of the maximum number of eggs that will be required to calculate the number of fish to grow through to maturity. The fecundity can be influenced by broodstock management, and different species have very different fecundities that require different forms of broodstock management. Nutrition, fish size, genetic strain and environmental control can affect fecundity (see review by Bromage, 1995). In salmonids, nutrition and growth which will affect the size of the fish are the most influential factors in fecundity. Larger fish produce more eggs and, therefore, the growth achieved in a potential broodstock has significant effects on the eggs produced by the broodstock. The availability of essential nutrients will also affect the fecundity. Studies on ARA, vitamins A and E have shown that varying the amounts of these essential nutrients available from the diet significantly affects the fecundity but not the size of the fish. For example, increased vitamin A levels in the diet of rainbow trout (Fontagné-Dicharry *et al.*, 2010) and Japanese flounder (*Paralichthys olivaceus*) (Furuita *et al.*, 2001, 2003) broodstock were observed to increase fecundity over the spawning season, but high levels in rainbow trout were also observed to have negative effects on progeny survival. This highlights the need for optimal nutrition made above in Section 2.3.3 on Nutrition. In addition to optimal nutrition, the ration fed has been shown to have profound effects in rainbow trout, and Bromage (1995) described how a period of time during early vitellogenesis or window of recruitment existed when varying the ration could significantly affect fecundity resulting in the need to have twice as many fish or twice as much biomass. In some groups this was despite achieving similar mean weights from the feeding regime. The window of opportunity for recruitment can also be affected by environmental manipulation. Advancing the timing of spawning and reducing the window of recruitment can both reduce the percentage of broodstock maturing and the fecundity (Bromage *et al.*, 2001). Fecundity also has a genetic component and this should be considered when implementing a genetic improvement program. Bromage (1995) showed that different strains of rainbow trout gave very different fecundities with some strains having fecundities twice as high as other strains.

© Woodhead Publishing Limited, 2013

Table 2.4 Brief details of management points commonly used for the five top species produced in Europe

Species	Management points			
	Fecundity	Out-of-season spawning	Sexual differentiation	Spawning
Atlantic salmon	2–4000 eggs/kg per spawn. 1 spawn per year	Generally out-of-season smolt production is preferred	May be controlled for all female triploid production	Strip spawning, short- and long-term gamete storage used Triploid production
Rainbow trout	2–4000 eggs/kg per spawn. 1 spawn per year	Photoperiod control used to provide eggs all year round	XX males used to produce all female stocks	Strip spawning, short- and long-term gamete storage used. XX males produced
Gilthead seabream	2–3 million eggs/kg over spawning season, 3–4 months of daily spawning.	Photoperiod and temperature control used to provide eggs all year round		Spontaneous natural tank spawning protocols required to avoid inbreeding
European bass	300–600 000 eggs/kg over spawning season (3–4 spawns)	Photoperiod and temperature control used to provide eggs all year round	Temperature manipulation can reduce the number of males, high proportion of males is a problem	Spontaneous natural tank spawning protocols required to avoid inbreeding
Common carp	100–300 000 eggs/kg per spawn. 1–5 spawns per year	Warm tropical temperatures can give all-year-round spawning		Strip spawning or natural spawning in pools

© Woodhead Publishing Limited, 2013

As first mentioned, different species have very different fecundities depending on the reproductive strategy. In currently established food finfish species produced for human consumption, fecundities range from 2000–4000 eggs/kg for salmonids and Nile tilapia to 2–3 000 000 for marine species such gilthead seabream. To obtain a given number of eggs 1000 times more biomass will be required for salmonids and Nile tilapia compared to some marine species. Thus, the fecundity strongly influences the architecture and size of genetic improvement programs. For poorly fecund species, a multiplication stage must be included comprised between the breeding nucleus and the on-growers. The genetic improvement programs for Atlantic salmon typically illustrate this incremental organisation, also dictated by the necessity to rapidly propagate the genetic progress to the whole on-growing sector. On the other hand, when species are characterised by a high fecundity, the fries derived from the breeding nucleus can be directly used by the on-growers. In this case, the genetic progress is immediately available for the production, thus avoiding the delay caused by the multiplication step.

There may also be the necessity to synchronise the spawning of the fish in the broodstock to obtain a large batch of eggs that can facilitate hatchery management. In salmonids, hormone therapies have been used to synchronise spawning. For example, in Atlantic salmon an implant of 50 $\mu g\ kg^{-1}$ synchronised spawning (Taranger *et al.*, 2003) and for rainbow trout doses of 20–50 $\mu g\ kg^{-1}$ synchronised spawning (Breton *et al.*, 1990; Arabaci, *et al.*, 2004). For Nile tilapia, the spawning of females was synchronised to obtain larger batches of eggs for the hatchery and fecundity per kilo of female was increased through a broodstock management program (MacIntosh and Little, 1995). Essentially, eggs and larvae were removed from mouth brooding Nile tilapia females to increase the frequency of spawning and female spawning was synchronised with reconditioning females in single sex groups and broodstock replacement of females in mixed sex spawning groups. For these species, salmonids and Nile tilapia, the holding of large numbers of broodstock can facilitate maintaining a genetic diversity and a range of different families. The opposite concerns are relevant to highly fecund marine species. A relatively small number of fish can provide the eggs required and good or large batches of eggs can be selected and poorer or smaller batches discarded. However, maintaining the genetic variability and a genetic improvement program is more complicated when a small number of fish form the broodstock.

2.7.2 Out-of-season spawning

As described above the changes in the environment entrain the maturational development and care is required to ensure that the captive environment provides the full ambient cycle of proximal factors to entrain particularly gametogenesis and spawning. This environmental entrainment ensures the seasonality in the spawning of many if not most fish species,

© Woodhead Publishing Limited, 2013

which under farming conditions results in a seasonality of production that may not be related to market demand. An important management tool is to manipulate entraining proximal factors to obtain out-of-season spawning (see reviews by Bromage *et al.*, 1993; Bromage *et al.*, 2001; Mañanós, *et al.*, 2009). The aim of out-of-season spawning is to give control over the timing of production of market-sized fish. For example, a single natural spawning period will result in a single large harvesting period which may not coincide with market demand. Out-of-season spawning can be used to ensure that market-sized fish are produced in relation to market demand or that all-year-round spawning gives a company a constant year round production of similar size and quality. All-year-round production also enables a company to optimise the use of facilities, avoiding periods when facilities are not used. Although many proximal factors have been implicated, the two principal factors that have been successfully used to manipulate the timing of spawning are photoperiod and temperature. Other factors, such as food availability, lunar or tidal cycle, rainfall, currents and pressure, that have been implicated as proximal factors are probably final cues of maturation and spawning after photoperiod and temperature have entrained vitellogenesis/spermatogenesis. In addition, these factors are difficult or impossible to control.

Photoperiod manipulation, often combined with temperature, has been shown to change the timing of maturation of many species from both temperate latitudes – rainbow trout (Bromage *et al.*, 1993; Bromage *et al.*, 2001), Atlantic salmon (Tarranger *et al.*, 2003), European seabass (Carrillo *et al.*, 1995), gilthead seabream (Zohar *et al.*, 1995), red drum (*Sciaenops ocellatus*) (Thomas *et al.*, 1995), Atlantic cod (*Gadus morhua*) (Davie *et al.*, 2007), Atlantic halibut (Smith *et al.*, 1991), common sole, turbot (Girin and Devauchelle 1978) – and tropical latitudes – Nile tilapia (Campos-Mendoza *et al.*, 2004), flat head grey mullet (Kelly *et al.*, 1991), catfish (*heteropneusfes fossilis*) (Sundararaj and Sehgal, 1970) and common carp (Davies and Hanyu, 1986; Davies *et al.*, 1986a,b). The simplest way to adjust the timing of spawning is to phase shift photoperiod and thermocycles by the number of months it is necessary to advance or delay the timing of spawning. For example, a phase advance of three months would give the winter solstice on 21 September and this should coincide with cold winter temperatures.

However, a number of important considerations must be made to ensure the timing of spawning is correctly entrained to coincide with the new photoperiod:

1. The maturational processes need time to adjust to the new cycle and the bigger the change the more time required to adjust. Ideally, cycles would be adjusted two years before spawning is required to give the fish a full 12-month phase shifted cycle before the second phase shifted cycle under which out-of-season spawning will be expected.

© Woodhead Publishing Limited, 2013

2. If adequate time is not given, spawning can be expected at a time intermediate to the spawning date of the ambient cycle and the new expected spawning date.
3. The initial change from the ambient cycle to a new cycle will also affect the timing of maturation. For example, applying a decreasing autumn cycle (decreasing from Light Dark, LD 12:12 and decreasing temperature) in the winter (LD 6:18) would be the equivalent of applying a direct increase (spring) to a short summer followed by an autumn decrease. For this reason, ideally the change should be made in the same direction as the required manipulation, for example an advance in the cycle would make a sharp increase from winter to the spring part of the cycle and follow with the continued cycle through to the summer.

From a more technical point of view it is important that no natural daylight enters the photoperiod facility, especially if the night illumination being used is dim (below 1000 lux) compared to natural daylight and this includes leaving a door open during working hours. It is also recommended to use a 1 lux or lower night light during the night and to have twilight periods at the start and finish of the day period as the sudden switch on of lights startles the fish. As indicted above, a phase shifted 12 months photoperiod and thermocycle is the simplest and perhaps the most effective way to change the spawning period. However, in species for which environmental control has been thoroughly investigated and tested, compressed photoperiod (a full 12 month photoperiod compressed to less than 12 months) and square wave photoperiod (direct increases or decreases in day length) regimes can be used (see reviews by Bromage *et al.*, 1993; Bromage *et al.*, 2001; Mañanós *et al.*, 2009): for example, rainbow trout (Bromage *et al.*, 1993, 2001), European seabass (Carrillo *et al.*, 1995), Sciaenidaes (Thomas *et al.*, 1995) and gilthead seabream (Zohar *et al.*, 1995).

Lastly, small modifications can be made to the natural spawning period either advancing or delaying by advancing, delaying or extending optimal spawning conditions, for example temperature was used to advance and delay the spawning period of the channel catfish. Heating pond water during February to April advanced spawning by two months (Hall *et al.*, 2002). Maintaining the temperature at 18 °C from April for 109 days before increasing to 26 °C gave delayed spawning in August and November (Brauhn, 1971). Taking these practices further, a few tropical species, Nile tilapia (MacIntosh and Little, 1995), common carp (Beverage and Haylor, 1998) and one sub-tropical species, red drum (Thomas *et al.*, 1995), have been maintained spawning continuously for long periods extending to over a year by maintaining optimal spawning conditions, the most impressive being two male and two female red drum that spawned continually for seven years under extended optimal spawning conditions of LD 12:12 and temperatures close to 24 °C (Arnold, 1988; Thomas *et al.*, 1995).

© Woodhead Publishing Limited, 2013

2.7.3 Sexual differentiation

Sexual differentiation in fish exhibits an enormous variation in controlling mechanisms and consequently in strategies and timing of sexual determination (see review by Devlin and Nagahama, 2002). The control and understanding of sexual differentiation enables the production of single sex populations which in some species, like the European seabass and the Nile tilapia, offer growth advantages over mixed sex populations due to different growth rates between sexes. Sexual differentiation is also an important consideration for broodstock management and a genetic improvement program as the culture environment in some species can result in high proportions of a particular sex (for example European seabass populations reared in captivity can be predominantly males) (Piferrer *et al.*, 2005) and this strongly influences which fish can be selected as potential broodstock and, therefore, the range of crosses and families that could be made for genetic improvement.

Most gonochoristic species important to aquaculture determine sex during the first year after fertilisation and protocols to produce single sex populations have been developed and are routinely applied in species such as rainbow trout and Nile tilapia that exhibit differential growth between sexes. Generally, sex reversal is achieved through the administration of steroids in the feed during the period of sex determination; feminisation can be achieved by feeding 17B-oestradiol and masculinisation by feeding 17a-methyl testosterone. Genetic sex reversal can also be used for sex reversal in species with XX and XY sex determining chromosomes. Genetic sex reversal is achieved in two steps and is common practice in production of rainbow trout. All female populations are used in the rainbow trout industry to avoid loss of growth due to precocious maturation in the males. In the first step, a population of mixed sex rainbow trout is masculinised by feeding 3 mg/kg of 17a-methyl testosterone during the first 70 days after first feeding (Bromage, 1988). These fish are then reared to maturity and the second step is to select a phenotypic male that has a female genotype (XX). These XX males can often be recognised because the sperm duct does not develop correctly and, therefore, a male from which sperm cannot be obtained is an XX male and males with flowing sperm are XY males. The XX males are sacrificed and the testis and sperm used to fertilise the ova from normal XX females to produce all female offspring. Nile tilapia males grow faster than females and all male populations are preferred for ongrowing. The most common method to produce all male populations is to feed 17a-methyl testosterone at a concentration of 30–60 mg/kg of feed during the first 25–60 days after first feeding (MacIntosh and Little, 1995). In a similar approach to that described for trout, 'super' YY Nile tilapia males have been produced through androgenesis and back crosses to confirm the YY chromosomes, and these 'super' males can be used to produce all male populations (Myers *et al.*, 1995; Mair *et al.*, 1997; Ezaz *et al.*, 2004).

© Woodhead Publishing Limited, 2013

2.8 Gamete stripping and spawning

Gamete stripping or spawning provides opportunities and problems for broodstock management. Gamete stripping, although labour intensive, provides opportunities for gamete selection (see Section 2.9 on genetic improvement) and gamete manipulation, such as ploidy manipulation at the point of fertilisation to obtain sterile stock, long-term cryopreservation and short-term gamete preservation.

Strip spawning is a straightforward process, which is similar for most species that have been strip spawned, although some species such as carp and sturgeon have some specific differences (Urbanyi *et al.*, 2009). The ova or sperm are extracted by applying gentle abdominal pressure to the anesthetised or dead (female salmon) broodstock. Before stripping, the fish is washed in clean water to remove residues of anaesthesia and dried around the genital pore to avoid contamination of the gametes with water or mucus. Pressure is applied to the abdomen from the head to the tail. The ova are usually collected in a clean, dry bowl. Sperm can also be collected in a bowl or a syringe or into a flask using a catheter inserted into the sperm duct. Once the gametes have been obtained, gamete quality can be assessed and ova fertilised with the sperm. The most common fertilisation method is dry fertilisation where the sperm is added to the ova. In some species the sperm will be activated by the ovarian fluid, whilst for other species, with little ovarian fluid, clean hatchery water should be added to activate the sperm. The eggs are left to complete fertilisation and hydration before being washed, disinfected and placed into the incubation systems.

Ploidy manipulation offers the opportunity to obtain stock with a desired type or number of chromosomes (Thorgaard, 1986; Ihssen *et al.*, 1990; Rothband, 2006) which can give the advantages of sterility and improved growth. A technique that has been applied commercially in aquaculture is the production of polyploid progeny, often triploid, that have more than two sets of chromosomes (diploid) and that were consequently sterile offering improved growth during ongrowing. For example, in the 1990s triploid all-female Atlantic salmon were commercially produced using a pressure shock soon after fertilisation to retain the polar body and a third set of chromosomes. The female triploid stock produced did not mature, avoiding grading and forced undersize harvesting before the maturing fish reached market size. However, during the same period, improved diets and genetic improvement for growth resulted in the achievement of harvest size before seawater maturation and the methods were not required commercially. Presently, the methods are being re-examined to reduce genetic contamination of wild stocks by escapees and again for growth advantages. Gynogenesis and androgenesis offer the possibility to produce single sex stocks and clones (Komen and Thorgaard, 2007). However, survival rates are generally low frustrating commercial application of the techniques.

© Woodhead Publishing Limited, 2013

Cryopresevation offers the opportunity for the long-term storage of sperm to assist in production and storage of genetic material to improve future aquaculture production or for conservation purposes (see Chapter 3 on cryopreservation of gametes for aquaculture). Cryopreserved sperm can be used when sperm volume may be limiting and to store the genetic material for use in later generations, for example to cross year classes that do not mature at the same time or preserve sperm with desired genetic traits after the male or family line has died. Fish eggs, embryos or oocytes have not been successfully cryopreserved for commercial use, but techniques are under development and these would offer full control to store all genetic material and produce larvae when required.

Short-term chilled storage of sperm and eggs offers a procedure that can facilitate the logistics of strip spawning large numbers of broodstock or the crossing of genetic strains held in different geographical locations (Bobe and Labbé, 2009). Undiluted, sperm from many species has been stored for hours to days at 0–4 °C in a thin layer of 2–4 mm in bags or tubes. For example, in large farming operations of Atlantic salmon when hundreds of salmon must be strip spawned at the same time, stripping was conducted over two days (personal observation). The males were stripped on the first day and the sperm stored in a normal 4 °C fridge. The sperm was stored in tissue culture flasks with a 2–4 mm layer of sperm when the flask was laid on its side. The following day, females were stripped and the ova fertilised with the sperm that had been stored overnight. Rainbow trout sperm has been stored in this way for as long as 34 days (Stoss and Holtz, 1983). Increasing the oxygen level in the storage container and dilution of the sperm with extenders may increase the length of time for which the sperm can be stored (Bobe and Labbé, 2009). For example, turbot sperm fertilisation capacity was maintained unchanged for three days when stored diluted with an artificial seminal liquid (ASL2) compared to undiluted stored sperm that significantly lost fertilisation capacity after 4 h (Chereguini *et al.*, 1997). Ova, without dilution or the use of oxygen, can also be stored in a similar way at low temperatures and this can be useful to transport gametes between production centres to make crosses between stocks in different geographical locations. Fertilisation and survival rates can be similar to ova and sperm that were not transported and can offer an alternative to transporting recently fertilised eggs that can be delicate to transport procedures. Ova and sperm from Atlantic salmon that were transported gave similar survival to hatching compared to newly fertilised eggs that were transported (personal observation).

Lastly, strip spawning and artificial fertilisation has advantages for genetic improvement programs as it offers full control over which male is crossed with which female and, therefore, the families produced. This advantage is lost when groups of fish are left to spawn spontaneously in mass spawnings as in the production of many marine fish. Mass spawning has clear advantages from a labour point of view as large quantities of good quality eggs

© Woodhead Publishing Limited, 2013

can be collected from spawning tanks on a daily basis, with no requirement of manpower to manipulate the broodstock. However, this gives a situation where the broodstock manager is not in full control of fertilisation or the families produced for genetic improvement. For example, Brown *et al.* (2005) found that the effective spawning population of gilthead seabream was considerably lower than the population size of the entire broodstock. Four groups of gilthead bream broodstock were studied, the groups comprising 48–58 mature fish, but the effective spawning population size ranged from 14–18.3 which indicated 26.6–32.9 % of the population contributed to the offspring analysed. Brown *et al.* (2005) conclude that this situation may lead to higher levels of inbreeding than would be expected. Attempts to strip spawn or individually spawn species such as gilthead seabream and European seabass have been complicated as spawning is poor in small groups, the fish can be sensitive to excessive manipulations and volumes of ova and sperm can be low, and generally attempts have not been successful enough to establish the bases of a genetic improvement program. The development of protocols to describe and control the reproduction of mass spawning species becomes then crucial for the development of genetic improvement programs for such species, also because these important aquaculture species currently have the least developed genetic improvement programs. This control can be achieved through the understanding of spawning kinetics with molecular markers combined with the accurate use of hormonal therapies. In this context, the role of molecular markers for kinship reconstruction becomes central.

The effective control of reproduction in cultured species is crucial to successively implementing genetic improvement programs. One fundamental regards the way brood fish are involved in the mating and the resulting structure of the sibling groups submitted to genetic improvement. Thus, both the understanding of the sex differentiation and the mastering of reproduction will be necessary to obtain the required high number of parents contributing to the progeny and balance between the families produced. Another basic principle is to ensure a short generation interval in order to maximise the annual genetic gain. Here, the control of the reproductive cycle through the optimisation of the maturation process will be crucial in reducing the generation interval and accelerating the genetic progress.

2.9 Genetic improvement: traits, breeding values and application of genomic resources

Starting up a genetic improvement program means taking the decision to make a long-term investment in order to adapt the biological features of a species to the market demand. Critical choices must be made regarding the traits to improve, the genetic improvement program to implement and, as

© Woodhead Publishing Limited, 2013

a corollary, the type of facilities required and the supporting tools to develop. The first step includes the setting-up of a synthetic founder population made up of different strains collected from the natural geographical area of the species or from commercial strains. As a general rule of thumb, Holtsmark *et al.* (2006) recommend the use of at least four strains and cross individuals from these either at random or in a diallel cross. The first Norwegian Atlantic salmon genetic improvement program (NLA) was founded by crossing strains originating from 40 Norwegian river systems (Gjedrem, 2000). Previous knowledge about the history of the strains is valuable to ensure that there is enough genetic variation within and between the strains. Note that these performance tests are regrettably often performed only for one growth trait and maybe one or two other traits although, as the genetic improvement program develops, many more traits that are economically important may be included. In addition to strain performance tests as for the Nile tilapia or the European seabass (Eknath *et al.*, 1993; Vandeputte *et al.*, 2009 and others), neutral markers represent a straightforward tool to assess the level of variability (Hayes *et al.*, 2006). This strategy was followed in particular for the brown trout (*Salmo trutta*) (Chevassus *et al.*, 2004), the rainbow trout (Kause *et al.*, 2005) and the Nile tilapia (Eknath *et al.*, 2007).

2.9.1 Traits

Breeding goals, the desired outcomes of breeding programs, have a long-term perspective. Some of the traits in the breeding goal are pure production traits, others are more closely related to the quality of the product or disease resistance traits. The breeding goal should fit market needs, although the traits and their relative weights in the breeding goal may be altered due to changes in market requirements and/or the general development of the genetic improvement program. Every additional trait included in the breeding goal represents an additional cost; therefore the financial capacities of the organisation implementing the program will also determine the type and number of improved traits that are included in the breeding goal. For example, AquaGen Atlantic salmon genetic improvement program (www.aquagen.no) measures 22 traits in their population (2011), although not all traits may be included in the breeding goal.

The selected traits in aquaculture are divided into two categories, depending on the parentage relationship of fish on which the measurements are performed. The highest accuracy of selection can be made for traits measured on the candidates, but, as can be seen in Table 2.5, other important traits cannot be measured on the selection candidates, so the information for those traits is recorded on sibs of the candidates. The heritability, reported in Table 2.5, is the parameter which calculates for each trait the ratio of the genetic variation over the total phenotypic variation (genetic and environmental) in a population. For example, for a trait like growth,

© Woodhead Publishing Limited, 2013

Table 2.5 Ranges of heritability estimates (%) for aquaculture species traits selected on the candidates or on their sibs

Candidates		Sibs of candidates	
Traits	Heritability	Traits	Heritability
Growth	30–60	Disease resistance	10–50
Morphology/appearance	10–50	Carcass quality	10–30
Age at maturation	10–40	Processing yield	20–50

the variation in the growth rate attributed to genetic effects (30–60 %) would be divided by the whole variation observed in the population. Therefore, heritability ranges between 0 and 100 % and directly affects the genetic response.

The genetic correlation is another genetic parameter which deserves attention as it gives insights into how selection for one trait impacts on the other traits. As an example, strong positive genetic correlation (0.78–0.83) was found between IPN (infectious pancreatic necrosis) challenge tests and IPN field outbreaks, suggesting that the performed challenge tests are good predictors for the disease field mortality (Wetten *et al.*, 2007). Conversely, undesirable genetic correlations require a cautious monitoring as selection causes certain traits to be altered. Hence, selection for growth in rainbow trout will produce more rotund fish due to an unfavourable correlation with the condition factor (Kause *et al.*, 2003). For the European seabass, this unfavourable relation is expressed only in certain environmental conditions (Dupont-Nivet *et al.*, 2008). In a more general way, it is highly advisable to monitor the potential effects of selection on non-selected traits.

The interaction between the genetic variation of traits and the environment where they are expressed, the so-called GxE interaction, will also impact on the genetic improvement strategy. In general, the rearing environment must be optimised for the species, but Dupont-Nivet *et al.* (2010) showed in the European seabass the necessity of selecting the strains for several environments. These considerations may have significant effects on the trait when the organism is in production, for example a strain (genotype) selected to give high performance of a desired trait (phenotype) in a particular environment may not perform well in another environment. Practically put, a disease-resistant strain of Atlantic salmon from cold waters (Northern Europe) may not exhibit the same resistance in warmer water (Chile). This highlights the fact that evidence of GxE effects may also become useful when choosing the testing environment in programs where tests on the sibs of candidates are conducted. A problem with most GxE studies in aquaculture is that they mainly focus on production traits, whereas other traits may be more important for GxE tests, such as disease resistance traits.

© Woodhead Publishing Limited, 2013

2.9.2 Breeding values

Although not strictly a breeding value, the individual record of the trait for each candidate can be used to select only for traits that can be measured in the candidates. In aquaculture, these individual records are used to select for easy to measure production traits (growth and often age at sexual maturation). For genetic improvement programs incorporating several traits, economic weights are derived representing the relative importance of the traits. Altogether, the weights associated to the corresponding trait breeding values form the selection index used to rank the different tested candidates. The selection index should correspond as much as possible to the breeding goal.

Best linear unbiased prediction (BLUP) breeding values are calculated using statistics that incorporate information on the relationship between animals in the population included as random effects in the model in addition to fixed effects such as year effects (Henderson, 1984). In aquaculture, BLUP breeding values are used efficiently in family-based genetic improvement programs, as they require that individuals are individually tagged to be able to set up the pedigree. Information on one or few genetic markers linked to a quantitative trait loci (QTL) can be included for marker-assisted BLUP breeding values as in Fernando and Grossman (1989). See also Sonesson (2007) for marker-assisted selection in aquaculture.

Genomic breeding values as in Meuwissen *et al.* (2001) use genetic markers covered over the whole genome (e.g. SNP markers) to calculate the breeding values of the candidates. The genetic marker effects are estimated in one part of the population (typically sibs of the candidates) that has both genotypic and phenotypic records using, for example, the BLUP method. Each candidate has genotypic records only, and their breeding values are simply the sum of the effect of each of the marker variants that they have.

There are several statistical packages that can be used to calculate traditional and genomic BLUP breeding values and the variance components of the traits that are needed for the breeding values. The most commonly used are ASREML (Gilmour *et al.*, 2006), DMU (Madsen and Jensen, 2007) and VCE (Groeneveld *et al.*, 2008).

2.9.3 Restriction of rates of inbreeding in practice

Inbreeding is inevitable in closed genetic improvement programs, because at a certain moment it is impossible to avoid mating of unrelated individuals. However, inbreeding should increase at a controlled rate. Inbreeding leads to inbreeding depression and reduced genetic variation of traits, as well as fluctuation in genetic gain through increased risk of random fixation of alleles due to genetic drift.

The number of breeders effectively contributing to the next generation, the so-called effective population size N_e, is of particular importance as it

© Woodhead Publishing Limited, 2013

is directly linked to the rate of inbreeding ΔF by the relation (Falconer and MacKay, 1996):

$$\Delta F = (1/N_e)/2$$

So, prime attention should be given when setting up and managing the breeding population to obtain N_e sufficiently large (>50) to maintain ΔF below the generally accepted limit of 1 % per generation (Meuwissen and Woolliams, 1994; Pante *et al.*, 2001). Other factors influence positively the N_e value, beyond the census number of sires and dams contributing to the new generation, like the equal contribution between male and female parents or the reduced variance of family size. One better understands here the role assumed by reproduction in genetic improvement and the importance of fully mastering artificial insemination.

In practical genetic improvement programs for aquaculture, the rate of inbreeding is reduced by first ensuring that the number of candidates from each family is the same and then by merging families after the survival has become stable. At selection, the number of selected sibs per family is restricted and sometimes within-family selection is done. Within-family selection minimises rates of inbreeding if performed strictly (Falconer and Mackay, 1996), but it also severely reduces the genetic gain, because only 50 % of the total genetic variation is utilised for selection.

Optimum contribution selection was first presented by Meuwissen (1997) and later by Hinrichs *et al.*, (2006) for aquaculture populations. It uses as input breeding value and pedigree files, and calculates optimum contribution of each candidate, while controlling the rates of inbreeding in the population. This method is very efficient in combination with a preselected group of individuals that have been genotyped for parentage testing, because it can handle different family sizes well as in Sonesson (2005).

In the AquaBreeding survey conducted over 37 genetic improvement programs (see Section 2.10.3 on On-going genetic improvement programs), the number of parents per generation ranged between 200 and 500 for half of the programs; eight relied on more than 800 parents while the others counted less than 200 parents (AquaBreeding Survey, 2009). This gives an idea of the size of broodstock being used by the industry in genetic improvement programs; however, it should be noted that these values differ from N_e as the latter depends on the number of families produced and the number of offspring per family. Ideally, an efficient genetic improvement program will have N_e close to the total number of the broodstock, but this is not necessarily the actual situation that is encountered.

2.9.4 Design of genetic improvement programs

The design of a genetic improvement program has the purpose of structuring groups of (test) individuals in the population such that accurate breeding values for the traits under selection can be obtained and ensuring

© Woodhead Publishing Limited, 2013

the intended selection intensity and generation intervals, i.e. the components that contribute to genetic gain. Control of the reproductive cycle is one of the main limitations of the different designs together with investment and running costs of the genetic improvement program.

In general, the genetic improvement programs rely on 50–400 families. This number should ensure that the rate of inbreeding is not too high, and that there is high genetic variation to select from in the population. Each family produces around 15 selection candidates and 15–100 test individuals. Mating ratio of males:females often differs from 1:1, especially in family-based genetic improvement programs where there are tank effects that need to be separated from the parents' effects. A male:female ratio of 1:2 is often used, and sometimes 2:2 in order to better estimate the female parent effect, but then the number of parents selected may be too low to keep an acceptable rate of inbreeding and ensure the production of offspring. For mass spawning populations, the number of full-sib families is often not a large limitation, and a higher degree of factorial matings can be used. Minimum co-ancestry matings and different degrees of factorial matings can reduce rates of inbreeding and/or increase genetic gain and should be considered (Sonesson and Meuwissen, 2000; Sonesson and Ødegård, 2012).

The choice of design of a genetic improvement program differs between the types of system that it is used in:

- The least costly design of aquaculture genetic improvement programs is where all candidates are measured for a trait and truncation selected for that trait, i.e. the candidates with trait values above a certain threshold point are selected as parents. This *mass*, or *individual selection scheme*, is, however, a method that may generate high rates of inbreeding due to the possible very high selection intensity resulting in the aquaculture genetic improvement program. It can also only be used to select for one (or very few) traits that can be measured in the selection candidates. It assumes in the best case that the survival of all families is similar, such that the selection pressure for all families is the same. However, for species with very skewed reproductive contribution or survival, this will not be the case, resulting in increased rates of inbreeding for this scheme.
- The *PROSPER method* (Chevassus *et al.*, 2004) was developed to overcome possible factors like low genetic variability, inbreeding, maternal effects or competitiveness, that may contribute to the low success of individual selection. It uses a high number of breeders ($N_e > 100$) and is characterised by fertilisation groups and recurrent growth challenges to contain non-genetic effects. The limitation with this scheme is that few traits can be selected on the candidates only.
- The *walk-back selection scheme* (Doyle and Herbinger, 1994) uses genetic markers to assign the candidates to the most probable parent pair. In this way, a pedigree file is set up and within-family selection

© Woodhead Publishing Limited, 2013

becomes a way to control inbreeding. This scheme was further developed by Sonesson (2005) to be combined with optimum contribution selection, a method that restricts rates of inbreeding in the population. BLUP breeding values are calculated and the selection is performed both within and between families. This scheme can also only be used to select for one (or very few) trait that can be measured on the selection candidates. It is, however, less sensitive to different number of candidates per family than the mass selection scheme. In addition to trait recording costs, one part of the group of candidates must be genotyped and individually tagged, which increases costs of the scheme compared to the mass selection scheme.

- The *traditional family-based genetic improvement scheme* (Gjedrem, 1985) keeps each family separately until they can be tagged individually with e.g. a PIT-TAG. Then the families are split up into one group of candidates and several groups of test individuals (for product quality and disease resistance traits). Multi-trait BLUP breeding values that combine both between- and within-family breeding values are calculated, and parents are either truncated selected for these or the optimum contribution selection method is applied (Hinrichs *et al.*, 2006). This scheme is very flexible regarding numbers and types of traits in the breeding goal. The main disadvantages are the investment and running costs.
- *Genomic selection* for genomic breeding values (Meuwissen *et al.*, 2001) can either be done in the same types of design as the traditional family-based genetic improvement program as presented in Nielsen *et al.* (2009) for truncation selection or Nielsen *et al.*, (2011) for optimum contribution selection. Then, the marker effects are estimated in the group of sibs of the candidates that are genotyped. Candidates are also genotyped to be able to estimate their genomic breeding value. This is a very costly scheme, as long as the genotyping costs are still high. Sonesson *et al.* (2010) proposed to pool a number of families and estimate the marker effects in these pools instead of on individuals. The candidates are preselected for growth, and only these preselected individuals are genotyped. That scheme does not need to keep families separately until tagging, and is therefore a cost-efficient way to select for genomic breeding values that are generally more accurate than the BLUP breeding values (Nielsen *et al.*, 2009). For situations where the whole production is affected by one disease that cannot be vertically transmitted or where the eggs can be treated such that the breeding nucleus is not at risk, Sonesson *et al.* (2011) developed a method where the population is firstly challenged for the disease, and part of the survivors are genotyped for parentage testing, and selected for BLUP breeding values for growth using optimum contribution selection (see Section 2.9.3 on restriction of rates inbreeding in practice). This scheme was very efficient when the genetic correlation between the traits was zero or positive, but when the

© Woodhead Publishing Limited, 2013

genetic correlation was negative, the heritability of the disease trait should not be too low for the scheme to be efficient. This scheme has low costs, while selecting for at least two traits.

2.9.5 Genomic resources for application in genetic improvement

Today, the markers of choice are the biallelic single-nucleotide polymorphisms (SNPs), which are progressively supplanting microsatellites markers. With the ever more flexible solutions proposed by the genotyping platforms, they can be included in chips, offering a wide range of applications. One application concerns the parentage inference, allowing pedigree reconstruction for individuals reared together from the egg stage onward. The challenge is to unambiguously assign the parenthood to a high percentage of the progeny (above 95 %) derived from complex mating designs. This is achieved with specific parentage allocation software (see technical review of Jones *et al.*, 2010) combining both the exclusion and the likelihood-based approaches (Herlin *et al.*, 2007). When the number of markers increases as with SNPs (Hauser *et al.*, 2011), such software must account for eventual mutations or genotyping errors, lowering the threshold level for parentage inclusion (Anderson and Garza, 2006). SNP chips represent a prime solution as they associate a high discriminative power (Anderson and Garza, 2006; Hauser *et al.*, 2011) to rapidly evolving genotyping solutions expected to decrease the genotyping costs. These conditions are highly favourable for a wider diffusion of kinship reconstruction in aquaculture genetic improvement.

SNP chips or genotype by sequencing techniques are also essential for the calculation of genome-wide breeding values (Meuwissen *et al.*, 2001), and for the mapping of quantitative trait loci (QTL) at a fine scale. It is important to notice that today there are no SNP chips with high enough numbers of SNPs to estimate the genomic breeding values with high accuracy for any aquaculture species. With the advent of cheaper high-throughput genome sequencing technology, which can produce hundreds of thousands of SNPs, this will hopefully change soon such that tailor-made SNP chips for the aquaculture genetic improvement programs can be made.

2.10 Genetic improvement: risks, evidence of genetic response and current research programs in Europe

2.10.1 Risks and mitigating solutions

Some of the risks pertaining to a genetic improvement program are of genetic origin, and the mitigating solutions will depend on the genetic improvement strategy implemented. The low genetic variability of a breeding population represents a first genetic risk causing a limited capacity to adapt to new selection pressures. It is often a direct consequence of the

© Woodhead Publishing Limited, 2013

scarce attention given to the genetic make-up of the founder population. Thus, a founder effect will be observed in the breeding population if only few founders are included in the base population. The breeding population may also undergo a genetic bottleneck (drastic decrease in population size), resulting in a rapid increase of both rates of inbreeding and genetic drift with severe and long-lasting repercussions on the animals' phenotypic characteristics (Tave, 1999). Increased rates of inbreeding as a result of high selection pressures and small effective population size N_e will accelerate the erosion of the genetic variability observed during the selection process. As already mentioned, increased rates of inbreeding also lead to increased inbreeding depression and genetic drift, which reduces the response to selection.

Other genetic risks can be assessed through the genetic parameters. As mentioned above, the presence of unfavourable genetic correlations will lead to negative genetic trends on selected and unselected traits, the negative indirect selection being more important when only one trait is selected. Also, the G × E interactions, which express the capacity of different genotypes to exhibit variable performance according to the environment, may result in a variation of the genetic progress according to the ongrowing location and production system. Lastly, the genetic progress relies mainly on additive genetic variation on fish, while the phenotype depends also on non-genetic and genetic non-additive effects (environmental, maternal, dominance and epistasis). Such effects bias the true genetic value of candidates and may be responsible for poor or negative selection response.

The mitigation of the above genetic risks will be achieved by putting into practice key principles of genetic improvement. One regards the necessity to keep the effective population size N_e large to avoid genetic bottlenecks and limit inbreeding and genetic drift. Well-designed breeding strategy and analytical procedures are necessary to distinguish between effects influencing the animals' phenotype. In this context, rearing conditions that magnify non-genetic effects (overcrowding, food shortage) should be avoided. The advancements in artificial fertilisation (stripping of gametes, cryopreservation) and the genetic markers are used to better control the family contribution. Having such tools available, controlled selection and mating based on pedigree information is preferred in order to avoid mating between highly related individuals. Long-term and sustainable genetic progress will also depend on other factors like the application of an equilibrated selection pressure, the monitoring of more characters during the selection process and a good knowledge of genetic parameters.

Still today, there is a gap between short-term economic view and long-term sustainability in breeding goals. This difference leads to the attribution of a market or non-market value to the breeding traits (Olesen *et al.*, 2000). With the increasing societal expectations of sustainable development, more efforts on non-market value traits (animal welfare, reduced environmental impact, safe and healthy food) are required. With an

© Woodhead Publishing Limited, 2013

increasing aquaculture production, growing concerns have been expressed about the effects that escaped cultured fish may have on wild populations, and little known what the genetic effects on the wild populations are. Also, different types of incidents may put at risk the breeding nucleus and jeopardise the genetic improvement program itself. As genetic improvement represents a long-term investment, securing the facilities holding the breeding nucleus must become a constant priority, ensuring the effective control of diseases and maintaining a continuous and high level technical supervision. The breeding population can also be divided into cohorts rather than maintained as a unique line or even entirely duplicated, but the latter may become economically unsustainable.

2.10.2 Evidence of genetic response

Different approaches are available to estimate the response to selection. The most common method is the BLUP animal model, where the complete pedigree from the start of the genetic improvement program is used to plot the average BLUP breeding values each year or generation. However, in order to get these genetic trends unbiased by the year effect, there is a need to ensure genetic ties between the generations, by reusing parents in several generations. Alternatively, control lines can also be used in fish (Moav *et al.*, 1976). Such a line, possibly built from the same base population, is carried alongside the selected line. With this approach, the deviation between the two lines permits the genetic gain to be estimated without interference due to the environmental fluctuations. Still, the method is little used because the recorded differences between both lines could be due to G × E interaction effects and it is difficult to guarantee the genetic stability of the control, and due to the running costs associated with its propagation (Hill, 1972).

The first evidence of genetic gain resulting from selective breeding experiments in fish has been reported on the rainbow trout for growth and spawning date traits (Lewis, 1944; Donaldson and Olson, 1955). Since then, the response to selection has been recurrently confirmed for most species subject to selection, demonstrating the relevance of genetic selection to improve aquaculture production:

- In the rainbow trout, the Norwegian National genetic improvement program produced a 4.3 % increase in growth rate per year (Gjerde, 1986), while the Finnish National genetic improvement program produced an increase of 5–12 % (Kause *et al.*, 2005)
- In Atlantic salmon, a response of 2 g or 7 % per year was estimated for body weight at 190 days (Gjedrem, 1983). A considerable gain was observed for age at sexual maturation, equivalent to 1.03 years, between offspring of parents aged 5 and parents aged 4 (Gjerde, 1984). Neira *et al.* (2006) reported for a four-generation genetic improvement program

© Woodhead Publishing Limited, 2013

conducted on the Coho salmon (*Oncorhynchus kisutch*) a mean selection response of 10.2 % per generation on harvest weight.

- In the brown trout, after four generations of an optimised mass selection procedure, the increase of weight at one year of age reached 21.5 % per generation (Chevassus *et al.*, 2004).
- In the European seabass, Vandeputte *et al.* (2009) report high genetic gains on the first selected generation, with an increase of weight at 20 months of 42 % and 23 % in two different testing environments (mixed and separate tanks). The difference is attributed to reduced competition and sub-optimal rearing conditions in the separate tanks.
- In the gilthead seabream, industry records report a selection response ranging from 5–10 % per generation (Knibb, 2000). Thorland *et al.* (2007) predicted a response to selection for weight of 22 % for the first generation. Similar results were obtained in a field test comparing the selected F1 progeny with a control line.
- The case of the common carp is peculiar, the species being cultivated since Roman times (Balon, 1995) and thus empirically selected over centuries. This long domestication history probably explains why little evidence of growth improvement has been reported in carp (Vandeputte, 2003; Vandeputte *et al.*, 2008). Response to selection has been obtained for height/length ratio (Ankorion *et al.*, 1992) and resistance to a bacterial disease, the carp dropsy (Kirpichnikov *et al.*, 1993).

2.10.3 On-going genetic improvement programs in Europe

A survey conducted in the European aquaculture industry (AquaBreeding survey, 2009) reported 34 breeding organisations and 37 genetic improvement programs for 14 selected species (Table 2.6). The programs deeply differ on the selection strategy implemented, the number (1 to 8) and type of improved traits, the number of selected generations performed (1–26), the diffusion on the market of the genetic material produced (1–15 countries) and the operative costs supported.

Most of the genetic improvement programs are owned and run by private organisations. As the breeding sector is knowledge intensive, most of the breeding organisations rely on consultancy services to run their program. Consultancy services are provided by specialised companies, producers associations, universities or freelance to run the program. However, companies that have reached a critical size or have their core business centred on breeding may have inhouse expertise to run the breeding program.

The type of facilities required to run a breeding program depends on the breeding scheme implemented. When families are reared separately, 100–400 larval tanks will typically be used. In contrast, only a few tanks will be used when the families are reared in common, like in the individual selection or when the pedigree can be traced with molecular markers in

© Woodhead Publishing Limited, 2013

Table 2.6 Number of genetic improvement programs run on European aquaculture species in 2009

Species		Nb	Country
	*Oncorhynchus mykiss*	1	Denmark
		1	Finland
		4	France
		2	Norway
	*Salmo salar*	1	Ireland
		4	Norway
		1	UK
	*Sparus aurata*	1	France
		3	Greece
		1	Israel
	*Cyprinus carpio*	1	Bulgaria
		1	Czech Republic
		1	Hungary
		1	Poland
	*Dicentrarchus labrax*	1	France
		1	Greece
	*Gadus morhua*	2	Norway
	*Scophtalmus maximus*	1	France
		1	Spain
	*Crassostrea gigas*	2	France
	*Salmo trutta*	1	France
	*Acipenser baeri*	1	France
	*Solea solea*	1	The Netherlands
	*Oreochromis niloticus*	1	The Netherlands
	*Tinca tinca*	1	Czech Republic
	*Argyrosomus regius*	1	France

Source: AquaBreeding survey 2009.

© Woodhead Publishing Limited, 2013

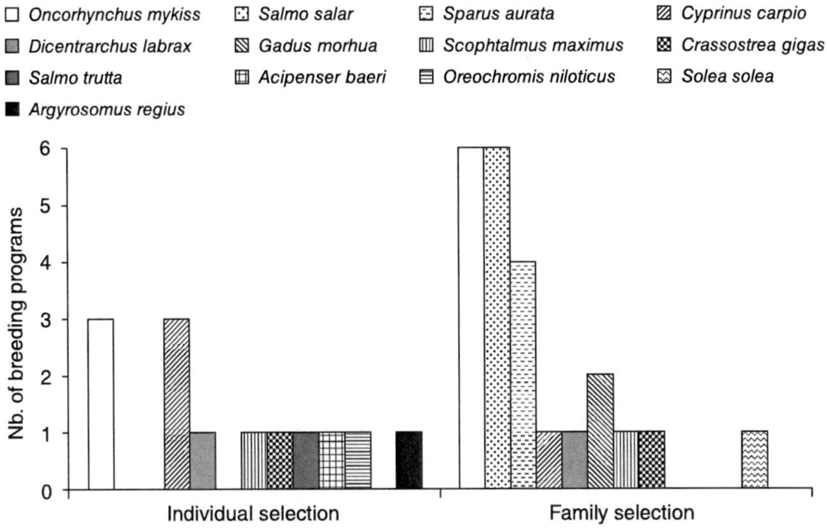

Fig. 2.1 Types of genetic improvement programs implemented on European aquaculture species in 2009.

more complex schemes. At later stages, when the tagging size allows the common rearing of families, the number and the dimension of the tanks/cages will differ significantly depending on the fish size at performance tests, the number of families produced and the tested environments (field tests, disease challenges tests).

Artificial fertilisation is systematically used on salmonids genetic improvement programs, even if some also rely on mass spawning. Instead, the latter technique is the most popular for species characterised by difficulties in fully mastering the artificial fertilisation like the Atlantic cod, the European seabass, the gilthead seabream and the common sole.

Regarding the genetic improvement programs implemented (Fig. 2.1), the results show a clear preponderance of the family selection (23 out of 37) over individual based genetic improvement programs (13).

2.11 Conclusion

Broodstock management, control of reproduction and genetic improvement allow a hatchery to continually improve efficiency and productivity of the whole farming process. Control of reproduction enables the production of juveniles of the quality and quantity required while genetic selection can provide a continual improvement in production efficiency that can form an essential part in maintaining and improving the ability of a business to compete. The two are inseparably linked; genetic improvement maintains

© Woodhead Publishing Limited, 2013

and improves the stock avoiding inbreeding while control of reproduction is essential to capitalise on advances in genetic improvement. Examples of potential problems in the control of reproduction that would limit selection include limited spawning from the entire stock due to a low percentage of individuals achieving maturation, a limited number of individuals participating in spawning or unbalanced levels in the sexes leading to unequal contribution to progeny. The principles to achieve complete broodstock management are outlined. However, information to achieve this is lacking for many if not all species of finfish in aquaculture. Areas for future research are many and include improved understanding and control of sex determination, a fuller understanding of gamete quality both to link quality to critical rearing factors and to enable the prediction of good quality through to juvenile stages or even market size, improved understanding of puberty, improved sterilisation methods such as ploidy manipulation or alternatives, genetic selection programs based on natural mass spawning (Reprosel – http://www.reprosel.eu/SitePages/Home.aspx), improved selection for traits and knowledge of interaction between traits such as adaptation to alternative feed sources, disease resistance, feed efficiency, fillet yield, flesh quality, nutritional profile and human health factors, improved tools, particularly genomic tools to improve genetic improvement programs (EATiP – http://www.eatip.eu).

2.12 Acknowledgements

The authors gratefully acknowledge the following projects: REPROSEL, project FP7-SME-2010-1-262523. REPROduction protocols and molecular tools for mass spawning and communal rearing based SELective breeding schemes applied to multiple-spawning marine fish and AquaBreeding, project FP6-2005-SSP-044424. Towards enhanced and sustainable use of genetics and breeding in the European aquaculture industry. Both coordinated by Hervé Chavanne, in Reprosel Anna Sonesson coordinated WP5 (work package) and Neil Duncan WP2. Warm thanks are given to Carlos Mazorra and Lynne Overton for reviewing the chapter.

2.13 References

ABU-HAKIMA R (1987) 'Aspects of the reproductive biology of the grouper, *Epinephelus tauvina* (Forskål), in Kuwaiti waters', *Journal of Fish Biology*, vol. 30, no. 2, pp. 213–222.

ANDERSON E C and GARZA J C (2006) 'The power of single-nucleotide polymorphisms for large-scale parentage inference', *Genetics*, vol. 172, pp. 2567–2582.

ANKORION Y, MOAV R and WOHLFARTH G W (1992) 'Bidirectionnal mass selection for body shape in common carp', *Genetic Selection Evolution*, vol. 24, pp. 43–52.

AQUABREEDING SURVEY (2009) Survey on the breeding practices in the European aquaculture industry, in *Towards enhanced and sustainable use of genetics and*

© Woodhead Publishing Limited, 2013

breeding in the European aquaculture industry, EU project no. 044424, available at: http://www.aquabreeding.eu/LinkClick.aspx?fileticket=Ilx9ZR26NYw%3d&tabid=98&mid=438 (accessed September 2012).

ARABACI M, DILER I and SARI M (2004) Induction and synchronisation of ovulation in rainbow trout, *Oncorhynchus mykiss*, by administration of emulsified buserelin (GnRHa) and its effects on egg quality. *Aquaculture*, vol. 237, pp. 475–484.

ARNOLD C R (1988) Controlled year-round spawning of red drum *Sciaenops ocellatus* in captivity. *Contributions in Marine Science*, vol. 30 (supplement), pp. 65–70.

BALON E K (1995) 'Origin and domestication of the wild carp, *Cyprinus carpio*: from Roman gourmets to the swimming flowers', *Aquaculture*, vol. 129, pp. 3–48.

BARBARO A, FRANCESCON A, BOZZATO G, MERLIN A, BELVEDERE P and COLOMBO L (1997) 'Induction of spawning in gilthead seabream, *Sparus aurata* L., by a long-acting GnRH agonist and its effects on egg quality and daily timing of spawning', *Aquaculture*, vol. 154, pp. 349–359.

BAYNES S M, HOWELL B R and BEARD T W (1993) 'A review of egg production by captive sole, *Solea solea* (L.)', *Aquaculture Research*, vol. 24, pp. 171–180.

BELL J G, FARNDALE B M, BRUCE M P, NAVAS J and CARILLO M (1997) 'Effects of broodstock dietary lipid on fatty acid compositions of eggs from sea bass (*Dicentrarchus labrax*)', *Aquaculture*, vol. 149, pp. 107–119.

BERLINSKY D L, KING V W, SMITH T I J, HAMILTON RDII, HOLLOWAY J JR and SULLIVAN C V (1996) 'Induced ovulation of southern flounder *Paralichthys lethostigma* using gonadotropin releasing hormone analogue implants', *Journal of World Aquaculture Society*, vol. 27, pp. 143–152.

BERLINSKY D L, KING W, HODSON R G and SULLIVAN C V (1997) 'Hormone induced spawning of summer flounder *Paralichthys dentatus*', *Journal of World Aquaculture Society*, vol. 28, pp. 79–86.

BEVERIDGE M C M and HAYLOR G S (1998) 'Warm-water farmed species', in Black K D and Pickering A D (eds), *Biology of Farmed Fish*. Sheffield: Academic Press, pp. 838–406.

BOBE J and LABBE C (2009) Chilled storage of sperm and eggs. in Cabrita E, Robles V and Herraez P (eds), *Methods in reproductive Aquaculture Marine and Freshwater Species*. CRC Press, Boca Ration, FL, pp. 219–235.

BRAUHN J L (1971) Fall spawning of channel catfish. *Progressive Fish Culturist*, vol. 33, pp. 150–152.

BRETON B, WEIL C, SAMBRONI E and ZOHAR Y (1990) Effects of acute versus sustained administration of GnRHa on GtH release and ovulation in the rainbow trout, *Oncorhynchus mykiss*. *Aquaculture*, vol. 91, pp. 373–383.

BROMAGE N (1995) 'Broodstock management and seed quality – general considerations', in Bromage N R and Roberts R J (eds), *Broodstock Management and Egg and Larval Quality*. Oxford: Blackwell Science, pp. 1–26.

BROMAGE N (1988) Propagation and stock improvement, in Shepherd J and Bromage N (eds), *Intensive Fish Farming*. Oxford: Blackwell Science, pp. 103–153.

BROMAGE N and CUMARANATUNGA R (1988) 'Egg production in the rainbow trout', in Muir J F and Roberts R J (eds), *Recent Advances in Aquaculture*, vol. 3. Oxford: Blackwell, pp. 64–138.

BROMAGE N R, RANDALL C R, THRUSH M and DUSTON J (1993) 'The control of spawning in salmonids', in Muir J F and Roberts R J (eds), *Recent Advances in Aquaculture*, Vol. 4. Oxford: Blackwell, pp. 55–65.

BROMAGE N R, PORTER M J R and RANDALL C F (2001) 'The environmental regulation of maturation in farmed finfish with special reference to the role of photoperiod and melatonin', *Aquaculture*, vol. 197, pp. 63–98.

BROWN R C, WOOLLIAMS J A and MCANDREW B J (2005) ('Factors influencing effective population size in commercial populations of gilthead seabream, *Sparus aurata*', *Aquaculture*, vol. 247, pp. 219–225.

© Woodhead Publishing Limited, 2013

BRUCE M, OYEN F, BELL G, ASTURIANO J F, FARNDALE B, CARRILLO M, ZANUY S, RAMOS J and BROMAGE R (1999) 'Development of broodstock diets for the European Sea Bass (*Dicentrarchus labrax*) with special emphasis on the importance of n-3 and n-6 highly unsaturated fatty acid to reproductive performance', *Aquaculture*, vol. 177, pp. 85–97.

CAMPBELL P M, POTTINGER T G and SUMPTER J P (1992) 'Stress reduces the quality of gametes produced by rainbow trout', *Biology of Reproduction*, vol. 47, pp. 1140–1150.

CAMPOS-MENDOZA A, MCANDREW B J, COWARD K and BROMAGE N (2004) 'Reproductive response of Nile tilapia (*Oreochromis niloticus*) to photoperiodic manipulation; effects on spawning periodicity, fecundity and egg size', *Aquaculture*, vol. 231, pp. 299–314.

CARRILLO M, ZANUY S, PRAT F, CERDA J, RAMOS J, MAÑANÓS E *et al.* (1995) 'Sea bass (*Dicentrarchus labrax*)', in Bromage N R and Roberts R J (eds), *Broodstock Management and Egg and Larval Quality*. Oxford: Blackwell Science, pp. 138–168.

CERDA J, CARRILLO M, ZANUY S and RAMOS J (1994) 'Effect of food ration on estrogen and vitellogenin plasma levels, fecundity and larval survival in captive sea bass, *Dicentrarchus labrax*: preliminary observations', *Aquatic Living Resources.*, vol. 7, pp. 255–256.

CHEREGUINI O, CAL R, DREANNO C, OGIER DE BAULNY B, SUQUET M and MAISSE G (1997) (Short-term storage and cryopreservation of turbot (*Scophthalmus maximus*) sperm, *Aquatic Living Resources*, vol. 10, pp. 251–255.

CHEVASSUS B, QUILLET E, KRIEG F, HOLLEBECQ M G, MAMBRINI M, FAURE A *et al.* (2004) 'Enhanced individual selection for selecting fast growing fish: the "PROSPER" method, with application on brown trout (*Salmo trutta fario*)', *Genetic Selection Evolution*, vol. 36, pp. 643–661.

COUNCIL OF EUROPE (2006) *Standing Committee of the European Convention for the Protection of Animals Kept for Farming Purposes (T-AP) Recommendation Concerning Farmed Fish*. Strasbourg: Council of Europe.

DAVIE A, PORTER M J R, BROMAGE N R and MIGAUD H (2007) 'The role of seasonally altering photoperiod in regulating physiology in Atlantic cod (*Gadus morhua*). Part I. Sexual maturation', *Canadian Journal of Fisheries and Aquatic Sciences*, vol. 64, pp. 98–112.

DAVIES B and BROMAGE N (2002) 'The effects of fluctuating seasonal and constant water temperatures on the photoperiod advancement of reproduction in female rainbow trout, *Oncorhynchus mykiss*', *Aquaculture*, vol. 205, pp. 183–200.

DAVIES P R and HANYU I (1986) 'Effect of temperature and photoperiod on sexual maturation and spawning of the common carp. 1. Under conditions of high temperature', *Aquaculture*, vol. 51, pp. 277–288.

DAVIES P R, HANYU I, FURUKAWA K and NOMURA M (1986a) 'Effect of temperature and photoperiod on sexual maturation and spawning of the common carp. 2. Under conditions of low temperature', *Aquaculture*, vol. 52, pp. 51–58.

DAVIES P R, HANYU I, FURUKAWA K and NOMURA M (1986b) 'Effect of temperature and photoperiod on sexual maturation and spawning of the common carp. 3. Induction of spawning by manipulating photoperiod and temperature', *Aquaculture*, vol. 52, pp. 137–144.

DEVLIN R H and NAGAHAMA Y (2002) 'Sex determination and sex differentiation in fish: an overview of genetic, physiological, and environmental influences', *Aquaculture*, vol. 208, pp. 191–364.

DONALDSON L R and OLSON P R (1955) 'Development of rainbow trout broodstock by selective breeding', *Transactions of the American Fisheries Society*, vol. 85, pp. 93–101.

© Woodhead Publishing Limited, 2013

DOYLE R W and HERBINGER C H (1994) 'The use of DNA fingerprinting for high-intensity within-family selection in fish breeding', *Proceedings of the 5th World Congress on Genetics Applied to Livestock Production*, 19, 7–12 August, Guelph, pp. 364–371.

DUNCAN N J, IBARRA-ZATARAIN Z, HERNÁNDEZ C, GARCÍA N, VELASCO-BLANCO G, RODRÍGUEZ-IBARRA E *et al.* (2011) 'Maduración del pargo prieto (*Lutjanus novemfasciatus*) en cautiverio', in Ruiz Luna A, Berlanga Robles C A and Betancourt Lozano M (eds), *Avances en Acuicultura y Manejo Ambiental*. Mexico: Trillas, pp. 39–57.

DUPONT-NIVET M, VANDEPUTTE M, VERGNET A, MERDY O, HAFFRAY P, CHAVANNE H *et al.* (2008) 'Heritabilities and GxE interactions for growth in the European sea bass (*Dicentrarchus labrax* L.) using a marker-based pedigree', *Aquaculture*, vol. 275, pp. 81–87.

DUPONT-NIVET M, KARAHAN-NOMM B, VERGNET A, MERDY O, HAFFRAY P, CHAVANNE H *et al.* (2010) 'Genotype by environment interactions for growth in European seabass (*Dicentrarchus labrax*) are large when growth rate rather than weight is considered', *Aquaculture*, vol. 306, pp. 365–368.

DUPREE H (1995) 'Channel catfish (*Ictalurus punctatus*)', in Bromage N R and Roberts R J (eds), *Broodstock Management and Egg and Larval Quality*. Oxford: Blackwell Science, pp. 220–241.

EKNATH A E, TAYAMEN M M, PALADA-DE VERA M S, DANTING J C, REYES R A, DIONISIO E E *et al.* (1993) 'Genetic improvement of farmed tilapias: the growth performance of eight strains of *Oreochromis niloticus* tested in different farm environments', *Aquaculture*, vol. 111, pp. 171–188.

EKNATH A E, BENTSEN H B, PONZONI R W, RYE M, NGUYEN N H, THODESEN *et al.* (2007) 'Genetic improvement of farmed tilapias: Composition and genetic parameters of a synthetic base population of *Oreochromis niloticus* for selective breeding', *Aquaculture*, vol. 273, pp. 1–14.

ESTEVEZ A, MCEVOY L A, BELL G and SARGENT J R (1999) Growth, survival, lipid composition and pigmentation of turbot (*Scophthalmus maximus*) larvae fed live prey enriched in AA and EPA. *Aquaculture*, vol. 180, pp. 321–343.

EZAZ M, MYERS J M, POWELL S F, MCANDREW B J and PENMAN D J (2004) Sex ratios in the progeny of androgenetic and gynogenetic YY male Nile tilapia, *Oreochromis niloticus* L., *Aquaculture*, vol. 232, pp. 205–214.

FALCONER D S and MACKAY T F C (1996) *Introduction to Quantitative Genetics* (4th edn). Harlow: Longmans Green.

FAO (1985) *Mass production of eggs and early fry. Part 1. Common Carp, Collection: Capacitation No. 8*. Rome: FAO.

FERNANDEZ-PALACIOS H, IZQUIERDO M S, ROBAINA L, VALENCIA A, SALHI M and VERGARA J M (1995) Effect of n-3HUFA level in broodstock diets on egg quality of gilthead sea bream (*Sparus aurata* L.), *Aquaculture*, vol. 32, pp. 325–337.

FERNÁNDEZ-PALACIOS H, SCHUCHARDT D, ROO J, HERNÁNDEZ-CRUZ C M and DUNCAN N (2009) Efecto de distintas dosis de GnRHa sobre la calidad de la puesta de corvina (*Argyrosomus regius*), In Beaz Paleo, D, Villarroel Robinson, M and Cardenas Rojas S (eds), *Libro de Resúmenes, XII Congreso Nacional de Acuicultura*, 24–26 Noviembre, Madrid. Madrid: CICEGRAF Artes Graficas, pp. 554–555 (in Spanish, abstract in English).

FERNANDO R L and GROSSMAN M (1989) 'Marker assisted selection using best linear unbiased prediction,' *Genetic Selection Evolution*, vol. 21, pp. 467–477.

FONTAGNÉ-DICHARRY S, LATAILLADE E, SURGET A, BRÈQUE J, ZAMBONINO-INFANTE J L and KAUSHIK S J (2010) 'Effects of dietary vitamin A on broodstock performance, egg quality, early growth and retinoid nuclear receptor expression in rainbow trout (*Oncorhynchus mykiss*)', *Aquaculture*, vol. 303, pp. 40–49.

© Woodhead Publishing Limited, 2013

FROESE R and PAULY P (EDS) (2011) *FishBase*, World Wide Web electronic publication, www.fishbase.org, version, viewed 21 October 2011.

FURUITA H, TANAKA H, YAMAMOTO T, SHIRAISHI M and TAKEUCHI T (2001) 'Effects of high dose of vitamin A on reproduction and egg quality of Japanese flounder *Paralichthys olivaceus*', *Fisheries Science*, vol. 67, pp. 606–613.

FURUITA H, TANAKA H, YAMAMOTO T, SUZUKI N, TAKEUCHI T (2003) 'Supplemental effect of vitamin A in diet on the reproductive performance and egg quality of the Japanese flounder Paralichthys olivaceus (T & S)', *Aquaculture Research*, vol. 34, pp. 461–468.

GARCIA L M B (1989) 'Dose-dependent spawning response of mature female sea bass, *Lates calcarifer* (Bloch), to pelleted luteinizing hormone-releasing hormone analogue (LHRHa)', *Aquaculture*, vol. 77, pp. 85–96.

GILMOUR, AR, GOGEL, BJ, CULLIS, BR and THOMPSON, R (2006) *ASReml User Guide Release 2.0*, VSN International Ltd, Hemel Hempstead, HP1 1ES, UK.

GIMÉNEZ G, ESTÉVEZ A, LAHNSTEINER F, ZECEVIC B, BELL J G, HENDERSON R J, PIÑERA J A and SANCHEZ-PRADO J A (2006) 'Egg quality criteria in common dentex (*Dentex dentex*)', *Aquaculture*, vol. 260, pp. 232–243

GIRIN M and DEVAUCHELLE N (1978) 'Décalage de la période de reproduction par raccourcissement des cycles photopériodique et thermique chez des poissons marins', *Annales de Biologie Animale, Biochimie, Biophysique*, vol. 18, pp. 1059–1065.

GJEDREM T (1983) 'Genetic variation in quantitative traits and selective breeding in fish and shellfish', *Aquaculture*, vol. 33, pp. 51–72.

GJEDREM T (1985) 'Improvement of productivity through breeding scheme', *Geo Journal*, vol. 10, pp. 233–241.

GJEDREM T (2000) 'Genetic improvement of cold-water fish species', *Aquaculture Research*, vol. 31, pp. 25–33.

GJERDE B (1984) 'Response to individual selection for age at sexual maturity in Atlantic salmon', *Aquaculture*, vol. 38, pp. 229–240.

GJERDE B (1986) 'Growth and reproduction in fish and shellfish', *Aquaculture*, vol. 57, pp. 37–55.

GROENEVELD E, KOVAC M and MIELENZ N (2008) *VCE user's guide and reference manual, version 6.0.* Neustadt: Institute of Farm Animal Genetics.

HALL S G, FINNEY J, LANG R P and TIERSCH T R (2002) 'Design and development of a geothermal temperature control system for broodstock management of channel catfish *Ictalurus punctatus*', *Aquacultural Engineering*, vol. 26, pp. 277–289.

HAREL M, TANDLER A, KISSIL G W and APPLEBAUM S (1992) The kinetics of nutrient incorporation into body tissues of gilthead seabream *S. aurata* females and the subsequent effects on egg composition and egg quality. *Israeli Journal of Aquaculture*, vol. 44, p. 127.

HAUSER L, BAIRD M, HILBORN R, SEEB L W and SEEB J E (2011) 'An empirical comparison of SNPs and microsatellites for parentage and kinship assignment in a wild sockeye salmon (*Oncorhynchus nerka*) population', *Molecular Ecology Resources*, vol. 11 (Suppl. 1), pp. 150–161.

HAYES B, HE J, MOEN T and BENNEWITZ J (2006) 'Use of molecular markers to maximise diversity of founder populations for aquaculture breeding programs', *Aquaculture*, vol. 255, pp. 573–578.

HENDERSON C R (1984) *Applications of Linear Models in Animal Breeding*. Guelph, Ont: University of Guelph.

HERLIN H, TAGGART J B, MCANDREW B J and PENMAN D J (2007) 'Parentage allocation in a complex situation: A large commercial Atlantic cod (*Gadus morhua*) mass spawning tank', *Aquaculture*, vol. 272S1, pp. S195–S203.

HILL W G (1972) 'Estimation of genetic change, I: general theory and design of control populations', *Animal Breeding Abstracts*, vol. 40, pp. 1–15.

© Woodhead Publishing Limited, 2013

HINRICHS D, WETTEN M and MEUWISSEN T H E (2006) 'An algorithm to compute optimal genetic contributions in selection programs with large number of candidates', *Journal of Animal Science*, vol. 84, pp. 3212–3218.

HOLTSMARK M, SONESSON A K, GJERDE B and KLEMETSDAL G (2006) 'Number of contributing subpopulations and mating design in the base population when establishing a selective breeding program for fish', *Aquaculture*, vol. 258, pp. 241–249.

HUNTINGFORD F A, ADAMS C, BRAITHWAITE V A, KADRI S, POTTINGER T G, SANDØE P *et al.* (2006) 'Current issues in fish welfare', *Journal of Fish Biology*, vol. 68, pp. 332–372.

IBARRA-CASTRO L and DUNCAN N J (2007) 'GnRHa-induced spawning of wild-caught spotted rose snapper *Lutjanus guttatus*', *Aquaculture*, vol. 272, pp. 737–746.

IHSSEN P E, MCKAY L R, MCMILLAN B I and PHILLIPS R B (1990) 'Ploidy manipulation and gynogenesis in fishes: Cytogenetic and fisheries applications', *Transactions of the American Fisheries Society*, vol. 119(4), pp. 698–717.

IZQUIERDO M S, FERNÁNDEZ-PALACIOS H and TACON A G T (2001) 'Effect of broodstock nutrition on reproductive performance of fish', *Aquaculture*, vol. 197, pp. 25–42.

JONES A G, SMALL C M, PACZOLT K A and RATTERMAN N L (2010) 'A practical guide to methods of parentage analysis', *Molecular Ecology Resources*, vol. 10, pp. 6–30.

KAUSE A, RITOLA O, PAANANEN T, ESKELINEN U and MANTYSAARI M (2003) 'Big and beautiful? Quantitative genetic parameters for appearance of large rainbow trout', *Journal of Fish Biology*, vol. 62, no 3, pp. 610–622.

KAUSE A, RITOLA O, PAANANEN T, WAHLROOS H and MANTYSAARI E A (2005) 'Genetic trends in growth, sexual maturity and skeletal deformations, and rate of inbreeding in a breeding program for rainbow trout (*Oncorhynchus mykiss*)', *Aquaculture*, vol. 247, no. 1–4, pp. 177–187.

KELLY C D, TAMARU C S, LEE C S, MORIWAKE A and MIYAMOTA G (1991) 'Effects of photoperiod and temperature on the annual ovarian cycle of the striped mullet, *Mugil cephalus*', in Scott A P, Sumpter J P, Kime D E and Rolfe M S, (eds), *Reproductive Physiology of Fish*. Sheffield: FishSymp, pp. 142–144.

KIRPICHNIKOV V S, ILYASOV I, SHART L A, VIKHMAN A A, GANCHENKO M V, OSTAHEVSKY A L *et al.* (1993) 'Selection of Krasnodar common carp (*Cyprinus carpio*) for resistance to dropsy: principal results and prospects', *Aquaculture*, vol. 111, pp. 7–20.

KNIBB W (2000) 'Genetic improvement of marine fish – which method for the industry ?', *Aquaculture International*, vol. 31, pp. 11–23.

KOMEN H and THORGAARD G (2007) Androgenesis, gynogenesis and the production of clones in fishes: A review, *Aquaculture*, vol. 269, pp. 150–173.

KOVEN W, BARR Y, LUTZKY S, BEN-ATIA I, WEISS R, HAREL M, BEHRENS P and TANDLER A (2001) The effect of dietary arachidonic acid (20:4n-6) on growth, survival and resistance to handling stress in gilthead seabream (*Sparus aurata*) larvae, *Aquaculture*, vol. 193, no. 1–2, pp. 107–122.

LAHNSTEINER F, SOARES F, RIBEIRO L and DINIS M T (2009) Egg quality determination in teleost fish, in Cabrita E, Robles V and Herraez P (eds), *Methods in Reproductive Aquaculture*. Boca Raton, FL: CRC Press, pp. 149–180.

LEE C-S, TAMARU C S, KELLEY C D and BANNO J E (1986) Induced spawning of Milkfish, *Chanos chanos*, by a single application of LHRH-analogue. *Aquaculture*, vol. 58, pp. 87–98.

LEWIS R C (1944) 'Selective breeding of rainbow trout at Hot Creek hatchery', *California Fish Game*, vol. 30, pp. 95–97.

MACINTOSH D J and LITTLE D C (1995) 'Nile tilapia (*Oreochromis niloticus*)', in Bromage N R and Roberts R J (eds), *Broodstock Management and Egg and Larval Quality*. Oxford: Blackwell Science, pp. 277–320.

MADSEN P and JENSEN J (2007) *A User's Guide to DMU A package for analyzing multivariate mixed models, Version 6, release 47*, available at: http://dmu.agrsci.dk/dmuv6_guide-R4-6-7pdf (accessed September 2012).

© Woodhead Publishing Limited, 2013

MAGWOOD S J, BROMAGE N, DUNCAN N J and PORTER M (2000) 'The influence of salinity on reproductive success in female Atlantic salmon (*Salmo salar*) grilse', in Taranger G L, Norberg B, Stefansson S, Hansen T, Kjesbu O and Andersson E, (eds), *Reproductive Physiology of Fish*. Bergen: Department of Fisheries and Marine Biology, University of Bergen, p. 346.

MAIR G C, ABUCAY J S, SKIBINSKI D O F, ABELLA T A and BEARDMORE J A (1997) Genetic manipulation of sex ratio for the large-scale production of all-male tilapia, *Oreochromis niloticus. Canadian Journal of Fisheries and Aquatic Science*, vol. 54, pp. 396–404.

MAÑANÓS E, DUNCAN N and MYLONAS C (2009) 'Reproduction and control of ovulation, spermiation and spawning in cultured fish', in Cabrita E, Robles V and Harraez P (eds), *Methods in Reproductive Aquaculture*. Boca Raton, FL: CRC Press Taylor and Francis Group, pp. 3–80.

MAZORRA C, BRUCE M, BELL J G, DAVIE A, ALOREND E, JORDAN N, REES J F, PAPANIKOS N, PORTER M and BROMAGE N (2003) Dietary lipid enhancement of broodstock reproductive performance and egg and larval quality in Atlantic halibut (*Hippoglossus hippoglossus*). *Aquaculture*, vol. 227, pp. 21–33.

MEERBURG D J (1986) *Salmonid age at maturity*, Canadian Special Publication of Fisheries and Aquatic Sciences, no. 89. Ottawa: Department of Fisheries and Oceans.

MEUWISSEN T H E (1997) 'Maximizing the response of selection with a predefined rate of inbreeding', *Journal of Animal Science*, vol. 75, pp. 934–940.

MEUWISSEN T H E and WOOLLIAMS J A (1994) 'Effective sizes of livestock populations to prevent a decline in fitness', *Theoretical Applied Genetics*, vol. 89, pp. 1019–1026.

MEUWISSEN T H E, HAYES B J and GODDARD M E (2001) 'Prediction of total genetic value using genome wide dense marker maps', *Genetics*, vol. 157, pp. 1819–1829.

MOAV R, BRODY T, WOHLFARTH G and HULATA G (1976) 'Application of electrophoretic genetic markers to fish breeding I Advantages and methods', *Aquaculture*, vol. 9, pp. 217–228.

MORETTI A, FERNANDEZ-CRIADO M P, CITTOLIN G and GUIDASTRI R (1999) *Manual on Hatchery Production of Seabass and Gilthead Seabream,* Volume 1. Rome: FAO, (available at: http://www.fao.org/docrep/005/x3980e/x3980e00.htm).

MUGNIER C, GUENNOC M, LEBEGUE E, FOSTIER A and BRETON B (2000) 'Induction and synchronisation of spawning in cultivated turbot (*Scophthalmus maximus* L) broodstock by implantation of a sustained-release GnRH-a pellet', *Aquaculture*, vol. 181, pp. 241–255.

MYERS J M, PENMAN D J, BASAVARAJU Y, POWELL S F, BAOPRASERTKUL P, RANA K J, BROMAGE N and MCANDREW B J (1995) Induction of diploid androgenetic and mitotic gynogenetic Nile tilapia (*Oreochromis niloticus* L.). *Theoretical and Applied Genetics*, vol. 90, pp. 205–215.

MYLONAS C C and ZOHAR Y (2001) 'Use of GnRHa-delivery systems for the control of reproduction in fish', *Reviews in Fish Biology and Fisheries*, vol. 10, pp. 463–491.

MYLONAS C C, SIGELAKI I, DIVANACH P, MANANOS E, CARRILLO M and AFONSO-POLYVIOU A (2003) 'Multiple spawning and egg quality of individual European sea bass (*Dicentrarchus labrax*) females after repeated injections of GnRHa', *Aquaculture*, vol. 221, pp. 605–620.

MYLONAS C, ZOHAR Y, PANKHURST N and KAGAWA H (2011) 'Reproduction and broodstock management', in Pavlidis M A and Mylonas C (eds), *Sparidae: Biology and aquaculture of Gilthead Seabream and others species*. Oxford: Blackwell Science, pp. 95–131.

NEIRA R, DÍAZ N F, GALL G A E, GALLARDO J A, LHORENTE J P and MANTEROLA R (2006) 'Genetic improvement in Coho salmon (*Oncorhynchus kisutch*) I: Selection

© Woodhead Publishing Limited, 2013

response and inbreeding depression on harvest weight', *Aquaculture*, vol. 257, pp. 9–17.

NIELSEN H-M, SONESSON A K, YAZDI H and MEUWISSEN T H E (2009) 'Comparison of accuracy of genome-wide and BLUP breeding value estimates in sib based aquaculture breeding schemes', *Aquaculture*, vol. 289, pp. 259–264.

NIELSEN H-M, SONESSON A K and MEUWISSEN T H E (2011) 'Optimum contribution selection for traditional Best Linear Unbiased Prediction and genome-wide breeding values in aquaculture', *Journal of Animal Science*, vol. 89, pp. 630–638.

OHTA H, KAGAWA H, TANAKA H, OKUZAWA K, IINUMA N and HIROSE K (1997) 'Artificial induction of maturation and fertilization in the Japanese eel, *Anguilla japonica*', *Fish Physiology and Biochemistry*, vol. 17, pp. 163–169.

OKUMURA S, OKAMOTO K, OOMORI R, SATO H and NAKAZONO A (2003) 'Imporved fertilization rates by using a large volume tank in red spotted grouper (*Epinephelus akaara*)', *Fish Physiology and Biochemistry*, vol. 28, pp 515–516.

OLESEN I, GROEN A F and GJERDE B (2000) 'Definition of animal breeding goals for sustainable production systems', *Journal of Animal Science*, vol. 78, pp. 570–582.

PANTE M J R, GJERDE B and MCMILLAN I (2001) 'Inbreeding levels in selected populations of rainbow trout, *Oncorhynchus mykiss*', *Aquaculture*, vol. 192, pp. 213–224.

PEDERSEN B H (1997) 'Induced sexual maturation of the European eel *Anguilla anguilla* and fertilisation of the eggs', *Aquaculture*, vol. 224, pp. 323–338.

PIFERRER F, BLÁZQUEZ M, NAVARRO-MARTÍN L and GONZÁLEZ A (2005) Genetic, endocrine, and environmental components of sex determination and differentiation in the European sea bass (*Dicentrarchus labrax* L.). *General and Comparative Endocrinology*, vol. 142, pp. 102–110.

PORTA J, PORTA J M, MARTÍNEZ-RODRÍGUEZ G and CARMEN ALVAREZ M C (2006) 'Development of a microsatellite multiplex PCR for Senegalese sole (*Solea senegalensis*) and its application to broodstock management', *Aquaculture*, vol. 256, pp.159–166.

ROTHBARD S (2006) 'A review of ploidy manipulations in aquaculture: The Israeli experience', *The Israeli Journal of Aquaculture – Bamidgeh*, vol. 58(4), pp. 266–279.

ROTHBARD S and YARON Z (1995) 'Carps (Cyprinidae)', in Bromage N R and Roberts R J (eds), *Broodstock Management and Egg and Larval Quality*. Oxford: Blackwell Science, pp. 321–352.

RSPCA (2010) *Welfare standards for farmed Atlantic salmon*, available at: http://www.rspca.org.uk/ImageLocator/LocateAsset?asset=documentandassetId=1232712364659andmode=prd (accessed September 2012).

SAWADA Y, OKADA T, MIYASHITA S, MURATA O and KUMAI H (2005) Completion of the Pacific bluefin tuna *Thunnus orientalis* (Temminck et Schlegel) life cycle, *Aquaculture Research*, vol. 36, pp. 413–421.

SHIELDS R J, BROWN N P and BROMAGE N R (1997) 'Blastomere morphology as a predictive measure of fish egg viability', *Aquaculture*, vol. 155, pp. 1–12.

SMITH P, BROMAGE N R, SHIELDS R, GAMBLE J, GILLESPIE M, DYE J *et al.* (1991) 'Photoperiod controls spawning time in the Atlantic halibut (*Hippoglossus hippoglossus*)', in Scott A P, Sumpter J P, Kime D E and Rolfe M S (eds), *Reproductive Physiology of Fish*. Sheffield: FishSymp, p. 172.

SONESSON A K (2005) 'A combination of walk-back and optimum contribution selection for fish', *Genetic Selection Evolution*, vol. 37, pp. 587–599.

SONESSON A K (2007) 'Within-family marker-assisted selection in breeding schemes for aquacultural species', *Genetic Selection Evolution*, vol. 39, pp. 301–317.

SONESSON A K and MEUWISSEN T H E (2000) 'Mating schemes for optimum contribution selection with constrained rates of inbreeding', *Genetic Selection Evolution*, vol. 32, pp. 231–248.

© Woodhead Publishing Limited, 2013

SONESSON A K and ØDEGÅRD J (2012) Mating structures for genomic selection breeding programs in aquaculture (Submitted).

SONESSON A K, GODDARD M E and MEUWISSEN T H E (2010) 'The use of communal rearing of families and DNA pooling in aquaculture genomic selection schemes', *Genetic Selection Evolution*, vol. 42, no. 1, p. 41.

SONESSON A K, GJERDE B and ROBINSON N (2011) 'A simple selection scheme to improve disease resistance and growth', *Aquaculture*, vol. 319, pp. 337–341.

STOSS J and HOLTZ W (1983) Successful storage of chilled rainbow trout (*Salmo gairdneri*) spermatozoa for up to 34 days, *Aquaculture*, vol. 31, pp. 269–274.

SUNDARARAJ B I and SEHGAL A (1970) 'Effects of a long or an increasing photoperiod on the initiation of ovarian recrudescence during the preparatory period in the catfish *heteropneusfes fossilis* (bloch)', *Biology of Reproduction*, vol. 2, pp. 413–424.

TARANGER G L, STEFANSSON S O, OPPEDAL F, ANDERSON E, HANSEN T and NORBERG B (1999) 'Photoperiod and temperature affects gonadal development and spawning time in Atlantic salmon (*Salmo salar*)', in Taranger G L, Norberg B, Stefansson S, Hansen T, Kjesbu O and Andersson E (eds), *Reproductive Physiology of Fish*. Bergen: Department of Fisheries and Marine Biology, University of Bergen, p. 345.

TARANGER G L, VIKINGSTAD E, KLENKE U, MAYER I, STEFANSSON S O, NORBERG B, HANSEN T, ZOHAR Y and ANDERSSON E (2003) Effects of photoperiod, temperature and GnRHa treatment on the reproductive physiology of Atlantic salmon (*Salmo salar* L.) broodstock, *Fish Physiology and Biochemistry*, vol. 28, pp. 403–406.

TAVE D (1999) *Inbreeding and brood stock management*, Fisheries Technical Paper No 392. Rome: FAO.

THOMAS P, ARNOLD C R and HOLT G J (1995) 'Red drum and other Sciaenids', in Bromage N R and Roberts R J (eds), *Broodstock Management and Egg and Larval Quality*. Oxford: Blackwell Science, pp. 118–137.

THORGAARD G H (1986) 'Ploidy manipulation and performance', *Aquaculture*, 57, pp. 57–64.

THORLAND I, PAPAIOANNOU N, KOTTARAS L, REFSTIE T, PAPASOLOMONTOS S and RYE M (2007) 'Family based selection for production traits in gilthead sea bream (*Sparus aurata*) and European sea bass (*Dicentrarchus labrax*) in Greece', *Aquaculture*, vol. 272, S314.

THORPE J E (1986) 'Age at first maturity in Atlantic salmon, *Salmo salar*: Freshwater period influences and conflicts with smelting, in Salmonid age at maturity', *Canadian Journal of Fisheries and Aquatic Sciences*, vol. 89, pp. 7–14.

THORPE J E, TALBOT C, MILES M S and KEAY D S (1990) 'Control of maturation in cultured Atlantic salmon, *Salmo salar*, in pumped seawater tanks, by restricting food intake', *Aquaculture*, vol. 86, pp. 315–326.

TUCKER J W (1998) *Marine Fish Culture*. Norwell, MA: Kluwer Academic Publishers.

URBANYI B, HORVATH A and BOKOR Z (2009) Artificial fertilization in aquaculture species: from normal practice to chromosome manipulation, fish', in Cabrita E, Robles V and Harraez P (eds), *Methods in Reproductive Aquaculture*. Boca Raton, FL: CRC Press Taylor and Francis Group, pp. 183–218.

UK FARM ANIMAL WELFARE COUNCIL (2005) *The Five Freedoms*, available at: http://wwwfawc.org.uk/freedoms.htm (accessed September 2012).

US FISH and WILDLIFE SERVICE (2012) Iodopor disinfection of eggs, in *Handbook of Aquatic Animal Health Procedures and Protocols, Volume 3: Disinfection, Isolation, and Quarantine Guidelines*, available at: http://www.fws.gov/policy/AquaticHB.html (accessed September 2012).

© Woodhead Publishing Limited, 2013

US FISH and WILDLIFE SERVICE'S AQUATIC ANIMAL DRUG APPROVAL PARTNERSHIP PROGRAM (2011) *Quick Desk Reference Guide to: Approved Drugs for Use in Aquaculture*. American Fisheries Society Fish Culture and Fish Health Sections and Association of Fish & Wildlife Agencies – Fisheries and Water Resources Policy Committee's Drug Approval Working Group, available at: http://www.fws.gov/fisheries/aadap/desk-reference_introduction.htm (accessed September 2012).

VAN GINNEKEN V J T and MAES G E (2005) 'The European eel (*Anguilla anguilla*, Linnaeus), its lifecycle, evolution and reproduction: a literature review', *Reviews in Fish Biology and Fisheries*, vol. 15, pp. 367–398.

VANDEPUTTE M (2003) 'Selective breeding of quantitative traits in the common carp (*Cyprinus carpio*): a review', *Aquatic Living Resources*, vol. 16, pp. 399–407.

VANDEPUTTE M, KOCOUR M, MAUGER S, RODINA M, LAUNAY A, GELA D *et al.* (2008) 'Genetic variation for growth at one and two summers of age in the common carp (*Cyprinus carpio* L): Heritability estimates and response to selection', *Aquaculture*, vol. 277, no. 1–2, pp. 7–13.

VANDEPUTTE M, GAROUSTE R, DUPONT-NIVET M, HAFFRAY P, CHAVANNE H and CHATAIN B (2009) 'Multi-local evaluation of the rearing performances of 5 natural populations of European sea bass (*Dicentrarchus labrax*)', *Proceedings of the 10th International Symposium for Genetics in Aquaculture*, 23–26 June, Bangkok.

WETTEN M, AASMUNDSTAD T, KJØGLUM S and STORSET A (2007) 'Genetic analysis of resistance to infectious pancreatic necrosis in Atlantic salmon (*Salmo salar* L)', *Aquaculture*, vol. 272, pp. 111–117.

ZOHARY and MYLONAS C C (2001) 'Endocrine manipulations of spawning in cultured fish: from hormones to genes', *Aquaculture*, vol. 197, pp. 99–136

ZOHAR Y, HAREL M, HASSIN S and TANDLER A (1995) 'Gilt-head sea bream (*Sparus aurata*)', in Bromage N R and Roberts R J (eds), *Broodstock Management and Egg and Larval Quality*. Oxford: Blackwell Science, pp. 94–117.

© Woodhead Publishing Limited, 2013

3

Cryopreservation of gametes for aquaculture and alternative cell sources for genome preservation

C. Labbé, INRA, France and V. Robles and M. P. Herraez, University of León, Spain

DOI: 10.1533/9780857097460.1.76

Abstract: In this chapter the benefits of short- to long-term storage of genetic material from aquatic species in liquid nitrogen are introduced. The biophysical and chemical challenges that arise during freeze-thawing are presented from the perspective of the large degree of structural and biological diversity found in fish, mollusc and crustacean gametes. Sperm cryopreservation procedures and the challenges raised by cryopreservation of oocytes and embryos or larvae are then discussed with a special focus on the importance of the genetic integrity of the thawed cells. Alternatives to the cryobanking of gametes and embryos rely on the cryopreservation of cells of somatic, embryonic and germinal origin. The challenges posed by the use of such cells to reconstruct fish are then described.

Key words: crystallization, cryoprotectant, chromatin structure, nuclear transfer, cell transplantation.

3.1 Introduction

Cell cryopreservation and cryobanking are unique tools that can be used in the aquaculture industry to preserve the genomes of domesticated species over the many steps of genetic selection, diffuse genetic progress and facilitate broodstock management by extending or delaying offspring production. Gamete morphology and biology vary among species; therefore the cryopreservation strategy also varies accordingly. Compacted spermatozoa with a reduced cytoplasmic compartment are much easier to cryopreserve than oocytes and whole embryos, which are larger and have more complex compartmentalization. As a result, methods for sperm cryopreservation have been described for most domesticated species, whereas oocyte

© Woodhead Publishing Limited, 2013

and whole embryo cryopreservation has been successful so far only in invertebrates.

One growing concern with regard to cryopreservation of gametes and embryos is that the genetic information borne by the cellular genome might be altered by the freezing, cryobanking and thawing processes, thus leading to development abnormalities and losses of genetic information. Alternative cell sources are being considered additionally, especially in fish where progress with the preservation of a diploid genome is stalling. Somatic cells and tissue pieces such as fins are easy to cryopreserve, but the reconstruction method involves nuclear transfer, a technology which is still far from applicable to aquaculture. Diploid cells collected from early embryos as well as primordial germ cells are also valuable sources of the diploid genome, as cryopreservation of such cells is possible. The transplantation of the thawed cells into recipient/surrogate fish is a promising reconstruction method, although it is still technically and biologically challenging. Ongoing research into these technologies will be outlined and their promise, pitfalls and practicability in aquaculture will also be discussed.

3.2 Gamete cryopreservation in aquacultured species

3.2.1 The benefits of gamete cryopreservation

Gamete cryopreservation in aquatic species has been developing for almost 60 years since Blaxter (1953). The aim of storing cells in liquid nitrogen (–196 °C) is to maintain cell viability and functionality over long durations. Such storage requires the ready availability of liquid nitrogen, and many reports of dramatic collection loss underline the importance of a ready supply. Cryobanking (i.e. storage in facilities dedicated to cryopreservation and equipped with a liquid nitrogen supply and liquid nitrogen level control) is thus the most reliable method of storage at present. National or supranational management of such cryobanks should ensure the safety of irreplaceable genetic resources, even in the event of political or economic upheaval.

The most represented type of cell in cryobanks is sperm, fertilization with cryopreserved sperm being relatively simple. Several advantages of sperm cryopreservation are similar to those of chilled short-term storage (at and above 0 °C) (Bobe and Labbé, 2008). Among these advantages, sperm cryopreservation facilitates management of males during the breeding season; allows milt to be shipped all over the world; gives time for veterinary diagnosis or quarantine before sperm is utilized; and allows sperm to be optimized when the male is lost in the process of extraction, as is the case in fishing, or when sperm is obtained from testes. To date, cryopreserved sperm is mainly used to support breeding programs for genetic selection. It is high risk and costly to keep original strains, including wild strains and the fish produced every few selected generations, in fish farms, so sperm from

© Woodhead Publishing Limited, 2013

representative individuals is cryopreserved. Cryopreserved sperm allows genetic progress to be assessed, and helps preserve selection steps.

Embryo or diploid cell cryopreservation has an advantage over sperm and oocytes in that both parental genomes are represented in one sample. In species where embryos can be cryopreserved, cryopreservation is a unique means to extend the fry production period. In oyster hatcheries, for example, facilities are overwhelmed during the reproductive season, and fry diffusion cannot be extended throughout the favourable rearing period. As will be discussed in Section 3.6, the necessary technologies to reconstruct a fish from diploid cells, other than a whole embryo, remain highly complex. Thus, although diploid resources are indeed being cryopreserved, the reconstruction technologies are not yet ready.

3.2.2 The physical limits of cryopreservation

Theoretically, cell and organism storage at ultra-low temperature (–130 °C and below) should not damage cells (Mazur, 1984) because all water is either in a crystalline or glassy state; viscosity is higher than 10^{13} poises; and thermal energy is too low to allow molecular motion. Alteration by cosmic rays and ionizing radiation is possible, but it would take several thousand years to kill the stored cells (Mazur, 1984). The main challenge of cryopreservation is therefore in the cooling, freezing, and thawing processes.

To cope with transition from living temperature to sub-zero temperature and back to living temperature, cells must survive damage from chilling. This phenomenon is well described in mammals at temperatures between 37 °C and 0 °C, where it is referred to as 'cold-shock'. Cold-shock induces alteration of membrane organization (Ilieva *et al.*, 1992) and reduces sperm motility (Wales and White, 1959). Cold-shock, which depends on membrane lipid composition, is caused by transient heterogeneity of the membrane surface at low temperatures, when lipids coexist in the liquid and gel phase. In consequence, lateral phase separation of membrane components occurs upon rewarming (Quinn, 1985; Drobnis *et al.*, 1993). Aquatic organisms, however, do not regulate their body temperature, and therefore the cellular membranes are relatively better adapted to temperature changes, with lipid composition providing a more fluid environment for membrane components (Labbé *et al.*, 1997). Due to this fact, the chilling damage at 0°C is of little concern for spermatozoa of most species, and to oocytes of cold living species (Bobe and Labbe, 2008), although chilling damage at 0 °C is more problematic in oocytes from warm water species. Besides, temperatures much below 0 °C can be reached before any crystallization has occurred yet. It is in this supercooling area that overall chilling damage can affect the more complex structures present in oocytes and embryos, whatever the chilling sensitivity at 0 °C (see Section 3.4.2). Once crystallization is triggered, separating chilling effect from damage caused by crystal growth is impossible.

© Woodhead Publishing Limited, 2013

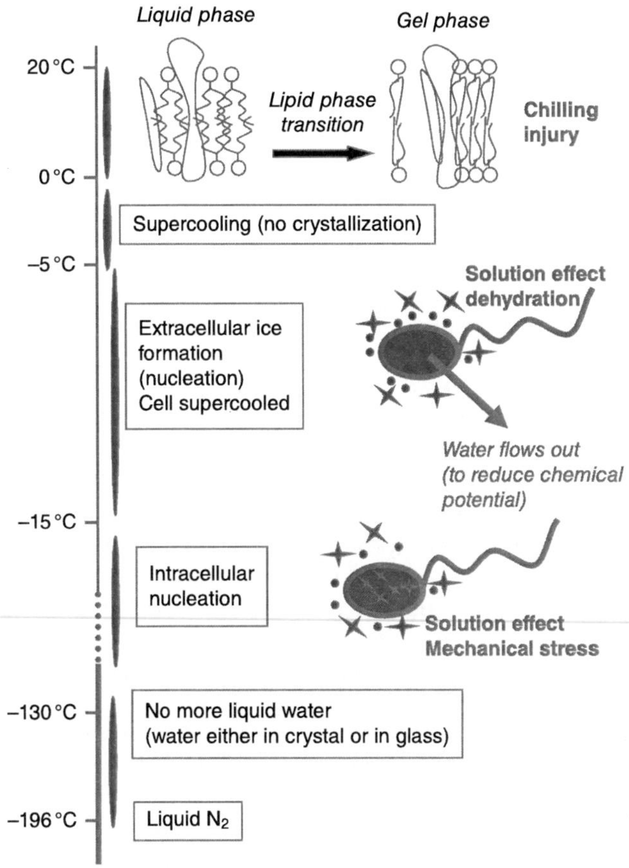

Fig. 3.1 Challenges to cells during temperature drop upon freezing.

As temperature decreases below 0°C (Fig. 3.1), intra- and extracellular media will supercool. This metastable state where water crystallization has not yet started will depend on media composition (salts, cryoprotectants, other additives) and on the concentration in ice-nucleating agents (dust, bacteria) in relation to temperature. Provided that chilling damage is controlled, supercooling may be a good way to cold-store most cells unable to withstand freezing (Haga, 1982; Harvey and Ashwoodsmith, 1982). However, supercooling state is highly unstable and prone to stochastic ice nucleation, leading to serious water crystallization. Ice nucleation is induced by ice-nucleating agents present in the media. This heterogeneous ice nucleation differs from self-induced homogeneous crystallization of the media, which occurs when the required temperature is reached. Heterogeneous ice nucleation is therefore triggered at higher temperatures than homogeneous ice nucleation (Chandrasekaran and Pitt, 1992). Ice nucleation always begins in the external medium which contains the nucleating agents (Fahy, 1995).

© Woodhead Publishing Limited, 2013

As a consequence, solutes excluded from ice will increase osmolality sensed by cell membranes, and supercooled water from the cellular compartment will flow out of the cell. As temperature decreases further, intracellular water will eventually freeze, most likely because of heterogeneous ice nucleation induced by external ice through membrane pores, or through surface contact with the membrane. The lower the temperature upon crystallization, the smaller and less damaging the internal ice crystals. Upon thawing, the smallest ice crystals will melt first, at a temperature still favourable to formation of larger ice crystals. Should the thawing rate be too slow, the melted ice will then re-crystallize into those larger ice crystals. These crystals are damaging to cells if too much internal water was present upon freezing (Mazur, 1984).

In summary, frozen–thawed cells are exposed to more or less extensive dehydration; volume shrinking; solution effect when external and internal medium osmolality rises; and to physical limits induced by ice crystals. Given these factors, freezing rates are carefully assessed in cell cryopreservation protocols in order to find a compromise. Overly gradual cooling will prevent internal ice crystals forming, but dehydration will be more severe. Conversely, overly sudden cooling will induce formation of very small internal ice crystals, doing little cell damage, but being very unstable upon thawing. As a general rule, thawing should be as fast as possible to minimize recrystallization, although no thawing procedure faster than hot water-bathing has yet been executed successfully. Cells with very little cytoplasm, in common with almost all spermatozoa from aquatic species, are *de facto* less sensitive to fast freezing rates. Cells with complex compartmentalization and variable water permeability within compartments, such as fish oocytes and embryos, will be more difficult to cryopreserve (see Section 3.4.2.). In cells with high water (cytoplasmic) content, ice nucleation may be induced manually, for example by touching samples with the tips of supracooled forceps. This seeding procedure is to prevent the external medium supercooling to temperatures at which the intracellular ice would form readily without having allowed sufficient intracellular dehydration. Serious crystallization can also be prevented by adding a mix of cryoprotectant at very high concentration and by application of very high freezing rates. This method, known as vitrification (Fahy *et al.*, 1984) because external and internal media will adopt a glassy state without any dramatic crystallization, should strike a balance between the benefit of absent ice crystals, and the toxicity of concentrated cryoprotective solution. Achieving very high freezing rates with large samples is another difficulty associated with vitrification.

3.2.3 Specificity of gametes from aquatic species

It is difficult to speak of gametes from aquatic species as a whole, even when restricted to aquacultured species. This is due to extreme dissimilarities in

© Woodhead Publishing Limited, 2013

gamete biology and structure among species associated with evolutionary adaptation. Indeed, through the pressure of different living conditions, including marine and freshwater environments, vertebrate and invertebrate species have developed extremely variable reproductive strategies and modes of fertilization (namely, internal and external).

The diversity of sperm ultrastructure in aquatic species reflects species classification to some extent (Jamieson, 1991). Sperm morphology was recently reviewed for fish spermatozoa (Lahnsteiner and Patzner, 2008). Spermatozoa of the Neopterygii, the infraclass to which more than 20000 Teleostei species belong, have no acrosome. After sperm release in water, egg fertilization is performed through the micropyle, a cone-shaped aperture in the external protective layer (chorion) of an egg. The sperm head bears the packed nucleus where the chromatin proteins (protamines, histones, and histone-like proteins) vary from one species to another (Saperas *et al.*, 1993, 1994, Frehlick *et al.*, 2006). The flagellar axoneme has a 9+2 arrangement of microtubules. It is more or less deeply inserted within a nucleus depression, where the two centrioles are located. The deeper the depression, the stronger the head-to-tail attachment will be. The mitochondria vary in number, from one single ring-shaped mitochondria in Salmonidae, to the six found in Labridae (Lahnsteiner and Patzner, 2008). The Chondrostei infraclass fish, to which Acipenseridae (including sturgeons) belong, do have a complex acrosome (Psenicka *et al.*, 2007), although the eggs bear several micropyles at the animal pole. Except that they have a small acrosome, mollusc sperm organization, as in oysters and abalone, is similar to that of Teleosteii, with little mitochondria at the base of the head (Bozzo *et al.*, 1993; Dong *et al.*, 2005). Crustacea spermatozoa are different again. They are aflagellated, and display a spiked cap, unistellate in natantian decapods (shrimps), and multistellate in the reptantians (lobsters, crabs, crayfish). The nucleus is composed of diffuse unpacked chromatin filaments with unique basic proteins (Tudge, 2009). Peneaoid shrimp and prawn spermatozoa are glued together in spermatophores attached externally to the male or female, or stored in internal receptacles in the female after release by the male (Lynn and Clark, 1983; Bauer and Min, 1993). These differences add up to diversity in sperm biology where the energy metabolism, motility activation signal, and duration of motility for those species having a flagella have little in common between species (Bobe and Labbé, 2010). Such dissimilarities in sperm structure and biology lead to differences among species in sperm fitness for cryopreservation.

Diversity extends to oocytes and embryos whose organization (mero- or holoblastic segmentation) and development (duration, pelagophil or benthophil) is connected to complex structures. Most aquatic species will release eggs in water where the embryo develops on its own, with little metabolic contribution of the surrounding medium except for the temperature and chemical properties (osmolality) of the water. In this context, eggs contain yolk and other nutriments (including proteins, nucleotides, and

© Woodhead Publishing Limited, 2013

vitamins) necessary for the embryo to sustain itself up to the first feeding stage (Brooks *et al.*, 1997). Protective envelopes (chorion in fish) will buffer the mechanical and chemical influences of the external media. Depending on the species, the yolk will partition inside each sister embryonic cell, in the case of holoblastic cleavage of the embryos, or will remain located as a single giant cell at the vegetal pole of meroblastic eggs. When cleavage takes place over the entire egg volume (holoblastic cleavage), as in sturgeon, oysters, or shrimp, embryo structure is relatively more homogeneous with regard to cryoprotectant permeation (see Section 3.4.2). Conversely, embryo development at the top and around the giant yolk cell (meroblastic cleavage), as in Teleostei, induces a compartmentalization in embryo structure which makes cryoprotection more difficult to achieve. Another source of diversity among species which influences egg and embryo behaviour during cryopreservation is egg size. Egg size varies, from the tiny oyster (*Crassostrea gigas*) egg (about 40 μm diameter), the bigger Cyprinidae or Acipensaridae eggs (0.5–1 mm diameter), up to the large Salmonidae egg (5–6 mm diameter). Egg and embryo envelopes will also have variable permeability. This may affect cryoprotectant effectiveness, particularly in oocytes where they cannot be removed, in contrast to embryos after the cortical reaction.

3.3 Sperm cryopreservation methods and adaptation to hatcheries

3.3.1 Methods of cryopreservation for the main aquacultured species

Hundreds of references to sperm cryopreservation in aquatic species are available today, and this number grows each month. Many authors summarized published methods in critical reviews (see Table 3.1). Anyone attempting sperm cryopreservation for the first time is referred to these reviews as a starting point. They will find ample information on the procedure and the factors which should be considered in a given species. In this section, we will focus on the items most common in all aquatic species, and describe the steps of a cryopreservation procedure (Fig. 3.2).

Initial sperm quality: Evidently, if the initial quality of sperm is high, this will result in high fertilizing ability after freeze–thawing. In many cases, the need to extend the reproductive season encourages broodstock manipulation by temperature (cyprinids, oysters); feeding (oysters); and photoperiod (salmonids), and broodstock sampling very early or very late in the breeding season. The fertilization rates obtained with an excess of fresh sperm are hardly ever affected by such manipulation. As a consequence, breeders are not always aware of reduced sperm quality. This becomes a problem when seeking good fertilization rates after the more challenging cryopreservation process. Cryopreservation can even become a test for global sperm quality assessment, as it will reveal weakness that would not be detectable otherwise in standard fertilization tests.

© Woodhead Publishing Limited, 2013

Table 3.1 Some reviews on sperm cryopreservation in aquatic species, other than in books

Title	Species	Additional focus	Reference
Germplasm cryobanking in zebrafish and other aquarium model species	Zebrafish, medaka	Sperm analysis	Robles *et al.*, 2011
Cryopreservation of fish sperm: applications and perspectives	Aquacultured species, aquarium model species	DNA integrity and progeny identity	Cabrita *et al.*, 2010
Review: Current status of sperm cryopreservation in biomedical research fish models: Zebrafish, medaka, and Xiphophorus	Aquarium model species	High-throughput processing, standardization, repositories	Yang and Tiersch, 2009
Sperm quality and cryopreservation of Brazilian freshwater fish species: a review	Characiformes, Siluriformes	Gene banking in Brazil	Viveiros and Godinho, 2009
Sperm cryopreservation in fish and shellfish	Fish (Xiphophorus) and shellfish (oyster)	Need for standardization	Tiersch *et al.*, 2007
Cryopreservation and short-term storage of sturgeon sperm, a review	Acipenseridae	Motility activation solutions	Billard *et al.*, 2004
Cryopreservation of finfish and Shellfish gametes and embryos	Fish, abalone, oyster, crab	Advanced research in Taiwan	Chao and Liao, 2001
Cryopreservation of aquatic invertebrate semen	Crustacea, molluscs	Spermatophores and sperm viability assay in crustacea	Gwo, 2000
Cryopreservation of sperm in marine fish	Marine finfish	Initial sperm quality	Suquet *et al.*, 2000

© Woodhead Publishing Limited, 2013

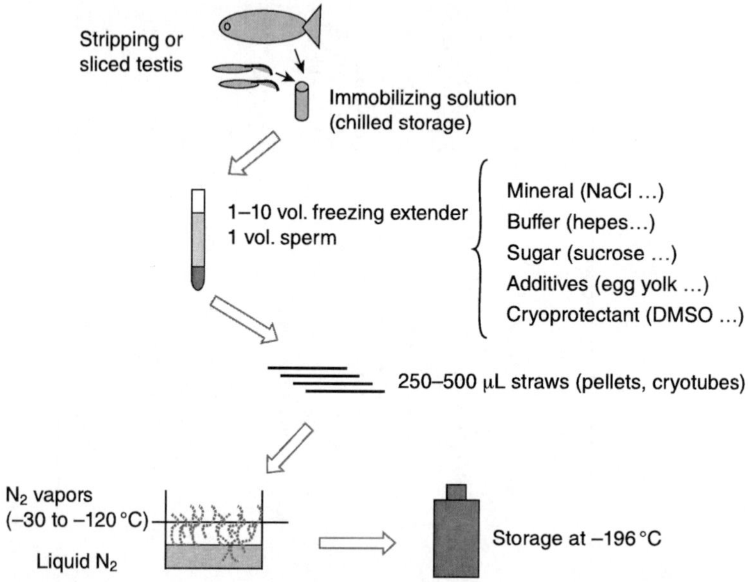

Fig. 3.2 Main steps of the freezing method. Gamete collection, immobilizing solution, freezing extender, and height above liquid nitrogen are to be determined according to the species.

In species with low cryopreservation ability, attempts to improve intrinsic sperm fitness for cryopreservation by way of broodstock rearing conditions (such as temperature and diet) have shown little success (Labbé *et al.*, 1993; Labbé and Maisse, 1996, 2001). Fresh sperm quality measured according to cryopreservation fitness results from a complex combination of factors including: plasma membrane quality and permeability; mitochondria efficiency and energy resource level; the ability to activate and sustain motility (for species bearing a flagellum); and the ability for acrosomal reaction (for species bearing one). Unfortunately, none of these parameters taken individually can accurately and reliably predict sperm fitness for cryopreservation. We believe that a cryopreservation challenge to a number of sperm samples in hatcheries is the best way to characterize sperm fitness for cryopreservation, on the basis of fertilization and development rates obtained after thawing. However, this is only possible when eggs are easily available.

- *Sperm collection and storage*: Depending on the species and context, sperm and spermatophores are collected either by stripping the male, or by shredding dissected testes. Urine (in fish) and water contamination should be avoided as both can induce motility and cell alteration. In the case of molluscs, sperm can also be obtained after induction (i.e. temperature and hormonal treatment) of sperm release in water,

© Woodhead Publishing Limited, 2013

although this procedure also induces extensive sperm dilution and motility activation. In some species (i.e. oysters), motility activation is compensated for by the long duration of motility (several hours (Suquet *et al.*, 2010)). Sperm obtained by stripping is already diluted in seminal plasma, but a suitable dilution medium must be used when sperm is obtained from testes. Collected sperm can be stored successfully from a few hours to several days or weeks (Bobe and Labbe, 2008) in extenders formulated to: (i) prevent sperm motility activation; (ii) fit sperm needs for osmolality, pH, nutrients; and (iii) prevent bacterial and fungal growth. More often, extenders mimic seminal fluid composition (in fish) or the external medium (sea water). Most spermatozoa can be stored without any additional protection at 0–4 °C prior to cryopreservation.

- *Cryoprotectant*: Dimethyl sulfoxide (DMSO) is by far the most widely used cryoprotectant for sperm in aquatic species, although methanol (MeOH), dimethyl acetamide (DMA), ethylene glycol (EG), 1,2-propanediol (PDiol), and even glycerol are, in fact, more efficient in some species. Concentration ranges between 5 % (volume ratio) and 20 %. These molecules permeate the plasma membrane and enter the spermatozoa, meaning that internal structures will be cryoprotected. Permeating cryoprotectant is toxic to spermatozoa, especially at high temperature; the choice of cryoprotectant is again a compromise between toxicity and cryoprotection. Impermeant molecules such as sugars (sucrose, trehalose, glucose), proteins and lipids (BSA, egg yolk, skimmed milk), are often added. The mechanism by which such additives protect spermatozoa is still a matter of debate. All these cryoprotective molecules are diluted in sea or fresh water or, more frequently, in extenders whose ionic composition is highly variable, even intraspecies. The dilution ratio of sperm to cryoprotective extender is again highly variable, from 1:1 to 1:10. We believe that the highest sperm concentration should be sought; the cryoprotectant concentration being adapted accordingly to reach the same final concentration. In this way, one straw will fertilize more eggs. An overly high sperm concentration might induce sperm agglutination (Dong *et al.*, 2007). However, this phenomenon appeared non-deleterious to fertilization rates in this study. Equilibration time of the cell in the cryoprotectant is not critical, and the common range is 3–30 min at refrigerated temperature.
- *Packaging*: Packaging should be chosen according to the purpose of the frozen sperm. Straw is the best packaging for long-term storage of sperm samples. The easiest to use are 0.25–0.5 mL French straws, on which sample name and reference can be printed. Larger straws (1.5 mL) also give good results. Although they are more difficult to handle in liquid nitrogen tanks, larger straws simplify the handling procedure during thawing and fertilization. Cryovials, which allow storage of glass capillary tubes, or that of frozen sperm pellets, are less common.

© Woodhead Publishing Limited, 2013

- *Freezing device*: Sperm of aquatic species is usually frozen in a very simple device consisting of an insulated box in which straws are layered on a polystyrene frame floating above a surface of liquid nitrogen. The distance of straw to liquid nitrogen, depending on frame thickness (3–6 cm), will to some extent determine the freezing rate. Such a device does not allow for a constant freezing rate. The rate will be rapid before crystallization. Crystallization being exothermic, the rate will decrease up to the end of sample crystallization, and then increase again (Haffray *et al.*, 2008). Except for crustaceans, many spermatozoa are not particularly sensitive to freezing rates between 5 and 50 °C/min. Direct plunging into liquid nitrogen, however, is deleterious whatever the species. The use of programmable freezers has also been reported. Their use allows more control over the plateau duration during crystallization. Although more expensive, programmable freezers may be considered in order to improve standardization of the freezing procedure, and when large numbers of straws are to be cryopreserved at once.

Thawing is always performed in a water-bath, where temperature exchange is more advantageous than in air. The temperature of the water-bath ranges from 25 to 40 °C with little influence on sperm quality. It is important to remove samples from the water-bath as soon as they are melted, to prevent sample temperatures climbing above 4 °C.

- *Quality of thawed spermatozoa*: The ultimate quality parameter for thawed spermatozoa is the ability to fertilize an egg and produce an embryo which will develop as well as one fertilized with fresh sperm. Beyond the assessment of fertilization rates and of development quality, research describes the sub-cellular alterations endured by spermatozoa, including damage to plasma membrane (viability), mitochondria, acrosome, and to nucleus, together with a highly integrative parameter known as motility. Plasma membrane quality is almost the only test available for crustacean sperm, by way of eosin-nigrosin staining of spermatozoa (Nimrat *et al.*, 2006). These quality tests are very important when establishing a new method for cryopreservation, because they help us understand where the fragility of a cell stands. However, it is very difficult to use the results of these tests to predict fertilization and development outcomes. Therefore, for routine use of cryopreserved sperm, we believe the best way to estimate the quality of cryopreserved sperm is a procedure where a few straws from the cryopreserved batch are used in sample fertility tests. Although fertilization tests are difficult to perform in crustaceans, preliminary attempts at artificial insemination are being made by placing thawed spermatophores into the female thelycum (Vuthiphandchai *et al.*, 2007).

In extreme conditions, when no fertility of thawed sperm is maintained, intracytoplasmic sperm injection (ICSI) into a recipient oocyte can be

© Woodhead Publishing Limited, 2013

considered (Poleo *et al.*, 2001; Liu *et al.*, 2011). This requires facilities equipped for micromanipulation and micro-injection, and these should be connected to facilities capable of nuclear transfer and cell transplantation (see Section 3.6).

3.3.2 Standardization, high throughput procedures and biosafety

As Tiersch *et al.* (2007) correctly note, the diversity of published cryopreservation methods plagues their evaluation. This diversity inhibits the development of a standardized method for hatchery purposes, in which all inconsequential variables should be expunged. It also limits development of multi-species procedures where only a limited number of parameters would vary. Since testing all protocols published in order to determine the most effective is impossible, we propose that procedures between groups mainly vary along what are largely inconsequential parameters. At the French National Cryobank, an attempt was made to use multi-species extenders for fish sperm collection and freezing (Haffray *et al.*, 2008). In a centralized facility, the same staff attain fresh sperm from various species collected and shipped from all over the country, and process it according to the same freezing procedure, with the same success as with species-specific established methods.

That a cryopreservation procedure is successful in laboratory conditions does not necessarily mean it can be used in hatcheries where mass production is required. Procedures established in laboratories must be scaled up, so that more straws can be processed and handled by inventory systems. Automated apparatus must be tested and set up for straw printing and straw filling, and streamlined procedures and quality guarantee enacted (Haffray *et al.*, 2008; Hu *et al.*, 2011; Tiersch *et al.*, 2011). In this context, the expertise built up by livestock centres is of great value for sample management and equipment provision. Scaling up often leads to longer delays between sperm dilution in cryoprotectant and freezing, and between sperm thawing and fertilization. Another example of potential difficulties is temperature stress due to manipulation of large samples in non-refrigerated rooms. Lastly, concerns also arise from the use of animal proteins and lipids in several cryopreservation media. The improvement brought by egg yolk, BSA, or skimmed milk should bring a clear advantage over extenders containing only sugars and permeating chemicals as cryoprotectant. Some groups propose alternative protein and lipid sources, such as coconut powdered water (Viveiros *et al.*, 2010), or chemically defined liposomes.

Due to high diversity in aquacultured species, no unique directive concerning disease control exists. For example, rules set up for fish species may not apply to molluscs, and little is known on how disease can be transmitted from one species to another. Establishment of cryobanks where several aquatic species will be stored together is therefore a new challenge for sanitary regulation (Tiersch and Jenkins, 2003). Beyond existing work to

© Woodhead Publishing Limited, 2013

minimize the occurrence and spread of disease in hatcheries, specific biosecurity rules are necessary for cryobanking. Possible regulation includes the setting up of a quarantine procedure for straws before they can enter the main tank room. It is essential to ensure that a straw will not spread any of its content into liquid nitrogen and the surroundings, and that no contaminant in the liquid nitrogen will enter the straws. The French National Cryobank recommends that all sperm from aquatic species be stored in high security CBS™ straws.

3.4 Trials on egg and embryo cryopreservation

3.4.1 Problems in egg and embryo cryopreservation

The fact that cryopreservation of both embryos and oocytes remains elusive in fish is due to three factors: (i) the low surface-area-to-volume ratio of embryos and oocytes, resulting from their large size, which leads to a delay in the flux of water and cryoprotectants; (ii) the high yolk content; and (iii) the high chilling sensitivity (Robles *et al.*, 2009). However, there are some differences between embryos and oocytes regarding cryopreservation. Embryos have a multi-compartmental structure, with differing permeability to water and cryoprotectant, whereas oocytes constitute a single compartment, do not have a fully formed (hardened) chorion (Zhang *et al.*, 2008), and lack some important barriers to cryoprotectants (Seki *et al.*, 2007) such as the yolk syncytial layer (YSL) and the enveloping layer (EVL) in fish. These characteristics explain why some reports find oocyte cryopreservation more feasible than embryo cryopreservation.

Cryobanking of embryos, even if success remains distant, may be more advantageous than oocyte preservation, in that both the maternal and paternal genomes are preserved. Cryoinjuries occurring in the embryo cellular compartment and in the yolk sac, and preventing normal embryo development are poorly characterized. Morphological injuries, including mechanical breakdown of the yolk sac, have been frequently reported after freezing–thawing (Robles *et al.*, 2003; Fornari *et al.*, 2010). Ice crystal formation, and specifically intracellular ice formation (IIF), is assumed to be a lethal factor during cooling (Li *et al.*, 2009). However, insufficient cryoprotection of the yolk compartment due to low permeability of the YSL could be responsible for modifications to their content, rendering it unsuitable for embryo utilization (Zhang and Rawson, 1998; Cabrita *et al.*, 2003). Indeed, the YSL, a thin multi-nucleate acellular layer formed between the blastoderm and the yolk around the 10th mitosis (Kimmel and Law, 1985) and enveloping the yolk sac after gastrulation, is considered the most impermeable and sensitive (Hagedorn *et al.*, 1997a, 1998; Robles *et al.*, 2005). The embryo compartment seems easier to protect, but the ratio of cells preserving signs of normal activity in terms of enzymatic activity, cell viability, or ability to survive under culture conditions has been considered

© Woodhead Publishing Limited, 2013

too low to support further embryo development (Robles *et al.*, 2004; Martinez-Paramo *et al.*, 2009a). Both the yolk sac and cellular compartment are essential to continued development, and different approaches have been taken to implement cryopreservation strategies.

3.4.2 Strategies for egg and embryo cryopreservation

Choosing the appropriate developmental stage could be crucial for oocyte cryopreservation. In order for cryopreservation to be deemed successful, immature oocytes must remain viable for *in vitro* maturation and fertilization after thawing. Cryopreservation of oocytes prior to ovulation, when the yolk content is reduced or when meiotic maturation is halted, has been studied mainly on warm water fish species where chilling injury is a major outcome. Zebrafish (*Danio rerio*) oocytes at late developmental stages (III and IV) are reported to be very sensitive to chilling (Isayeva *et al.*, 2004). Stage II follicle survival is significantly reduced after 4 h at –1 °C whereas stages I and II can be stored at this temperature for up to 24 h (Tsai *et al.*, 2009). Apart from sensitivity to chilling, permeability to water and cryoprotectants also varies with developmental stage. In medaka (*Oryzias latipes*), the permeability of immature oocytes is higher than in mature (Valdez *et al.*, 2005) and similar observations have been made in zebrafish, indicating that immature oocytes may be more suited to cryopreservation (Seki *et al.*, 2007). In order to improve water and cryoprotectant permeability, the aquaporin AQP3 was artificially expressed in immature medaka oocytes, increasing permeability to water and to the cryoprotectant, propylene glycol (Valdez *et al.*, 2006). Although the ability of oocytes to mature and be fertilized was maintained, this promising strategy has not yet been successfully applied to cryopreservation.

Oocyte viability after cryopreservation has been reported. However, most tests are not based on subsequent maturation and fertilization ability, but rather on membrane integrity. Such tests include: trypan blue (TB) exclusion, and fluorescein diacetate (FDA)/propidium iodide (PI) staining. Tests on mitochondria metabolism include: thiazol methyl thiazol tetrazolium (MTT) staining and ATP assay. Using the TB exclusion test, Guan and his group obtained 88 % viability after cryopreserving stage III zebrafish oocytes (Guan *et al.*, 2008) and oocyte vitrification gave promising viability just after thawing (Guan *et al.*, 2010). Tsai and his colleagues reported that FDA/PI staining, ADP/ATP assay and *in vitro* maturation leading to germinal vesicle breakdown (GVBD) were more sensitive than the TB exclusion test (Tsai *et al.*, 2009, 2010). Other methods such as cathepsin activity assays have also been explored for evaluating zebrafish oocyte quality (Zhang *et al.*, 2008). In all, *in vitro* maturation is considered the most reliable method for assessing ovarian follicle viability (Tsai *et al.*, 2010), but post-thaw culturing conditions still need to be optimized (Guan *et al.*, 2010).

© Woodhead Publishing Limited, 2013

In molluscs, exciting success was reported in the cryopreservation of the very small mature oocytes of Pacific oysters (Tervit *et al.*, 2005). Using very low cooling rates in EG obtained fertilization rates above 50 %. Greenshell mussel (*Perma canaliculus*) oocytes, however, did not handle cryopreservation very well, most likely due to chilling sensitivity (Adams *et al.*, 2009).

The consensus regarding embryos is that late stage embryos are more resistant to cryopreservation: permeability to cryoprotective agents and resistance to chilling seem lower at cleavage or blastula stages, so gastrulation or somitogenesis stages are usually considered for cryopreservation (reviewed by Robles *et al.*, 2009). A range of species have been studied, from model species, such as zebrafish or medaka (Robles *et al.*, 2009), to farmed species, such as turbot (*Scophtalmus maximus*), gilthead seabream (*Sparus aurata*), red seabream (*Pagrus major*), pacu (*Piaractus mesopotamicus*), Japanese whiting (*Sillago japonica*), etc. (Robles *et al.*, 2003, 2007a, b; Cabrita *et al.*, 2006; Fornari *et al.*, 2010). Most effort is directed at overcoming permeability barriers and increasing freezing resistance. Chorion, the external envelope, can be eliminated or partially digested using proteases or chemical treatments (Robles *et al.*, 2007b), but permeabilization of the YSL is a more significant challenge. Various strategies have been employed upon zebrafish embryos to this end. Among these, ultrasound treatments increased methanol penetration; artificial AQ3 expression was achieved in embryo membranes without affecting survival or developmental ability (Hagedorn *et al.*, 2002; Lance *et al.*, 2004); and femtolaser permeabilization was developed to deliver exogenous material in the same species. Nevertheless, no reports on the freezing ability of these embryos have been published so far. Micro-injection of cryoprotectants has also been performed in seabream, turbot and zebrafish, increasing low temperature resistance (Robles *et al.*, 2006; Beirao *et al.*, 2006). However, some stages are extremely sensitive to the process, and puncture seems to increase mechanical cryodamage with the growth of ice crystals (Martinez-Paramo *et al.*, 2008).

The challenge posed by extremely high chilling sensitivity in warm water species, mainly attributed to the yolk content, was undertaken by Zhang's group, who removed the yolk prior to freezing (Liu *et al.*, 2001), a time-consuming method with limited benefits for cryopreservation. Yolk removal also involves the potential loss of maternal transcripts present during early development in the yolk (Pelegri, 2003), necessary for embryo development (Dekens *et al.*, 2003). Vitrification (an ultrafast procedure requiring a much higher concentration of cryoprotective agents, which reduces the exposure to the cold shock conditions) instead of slow freezing, is often preferred in order to reduce damage at temperatures between 4 and –30 °C (Hagedorn *et al.*, 2004). Nevertheless the required degree of cryoprotection has not been reached. Winter flounder (*Pseudopleuronectes americanus*) embryos, naturally expressing antifreeze proteins (AFPs) (Young and Fletcher, 2008), preserved some developmental ability when vitrified using a protocol designed for turbot embryos (Robles *et al.*, 2005). The effectiveness of

© Woodhead Publishing Limited, 2013

incorporating these AFPs from Arctic species into other embryos before vitrification needed confirming. A widespread distribution of AFPs between compartments has not yet been accomplished, either by micro-injection or by other methods. However, their entrance into some derivatives of the endoderm was reported by Martinez-Paramo *et al.* (2008) when embryos were incubated at the appropriate stage with these proteins. These beneficial effects of AFPs on preserving activity in those cells were also demonstrated by Martinez-Paramo *et al.* (2009a). Successful embryo cryopreservation will probably require the synergy of two approaches: (i) increasing the cryoprotection of the cellular compartment, and (ii) increasing the permeabilization and dehydration of the yolk sac. Both could benefit from the artificial expression of exogenous proteins, namely aquaporins and AFPs.

3.4.3 Cryopreservation of marine invertebrate embryos and larvae

Unlike in teleosts, cryopreservation of both embryos and larvae from marine invertebrates was achieved some time ago, as reviewed by Lin *et al.* (1999), making this technology more easily transferrable to the aquaculture industry. Most of the studies have focused on the Pacific oyster, due to its commercial importance globally. The first successful attempts were reported by Chao *et al.* (1994, 1997) and Gwo (1995). Chao and his colleagues developed freezing protocols with the aim of establishing long-term preservation techniques for application in hatcheries. The authors indicate that initial quality of embryos is an important factor for success, whatever the freezing procedure used (stepwise freezing or vitrification). Regardless of conditions, a high percentage of ciliated blastula preserve active rotary motion 1–2 h after thawing. Studies performed by the same group (Lin *et al.*, 1999) at the same stage showed better results when a two-step freezing procedure was developed and optimized: cooling to −10 °C, seeding by touch with forceps to trigger ice nucleation, and freezing by immersion in LN2. With late stage oyster embryos (trochophores) with a broad range of cooling rates (from 20.5 to 216 °C/min), 78 % and 83 % of embryos with motion were achieved using 2 M DMSO or glycerol, respectively. Nevertheless, survival at hatching could be compromised, as was later observed in other species. It was apparent from the results obtained by different researchers (Gwo, 1995; *Usuki et al.*, 2002) that a wide range of optimal parameters supporting freezing of late embryos exists, including the development stage. Indeed, Gwo (1995) showed that trochophore larvae are more resistant than morula and gastrula embryos, and that two to eight cells stages were unsuccessful.

Protocols have been developed for the cryopreservation of oyster at early to late trochophore stages. It should be noted that, frequently, the trochophore stage refers to trochophore embryos as well as trochophore larvae, depending on time after fertilization. Successful cryopreservation of Eastern oyster (*Crassostrea virginicaI*) larvae was first reported by

© Woodhead Publishing Limited, 2013

Paniagua-Chavez *et al.* (Paniagua-Chavez *et al.*, 1998; Paniagua-Chavez and Tiersch, 2001), who obtained almost 100 % motile trocophores immediately post-thawing when larvae were frozen at low concentration (125 larvae per macrotube) using 10–15 % propylene glycol as cryoprotectant. Large-scale cryopreservation of larvae was attempted by these authors, but the increasing concentration of larvae caused a decrease in survival (less than 10 % survival when 50 000 larvae were loaded in the same macrotube) (Paniagua-Chavez and Tiersch, 2001). For production purposes, higher survival rates must be achieved and survival at further larval stages should be analysed. Other species such as hard clam (*Meretrix lusoria*) (Chao *et al.*, 1997), mangrove oyster (*Crassostrea rhizophorae*), and pearl oyster (*Pinctada fucata martensii*) have also received attention, demonstrating good freezability: 43.1 % and 91 % survival rates for trochophores and D-shape larvae, respectively, in pearl oyster (Choi and Chang, 2003).

Larvae of other invertebrate species such as sea urchins (*Evechinus chloroticus* or *Paracentrotus lividus*), have also shown strong ability to resist the cryopreservation process. Adams and her group (Adams *et al.*, 2003, 2006) reported some four-armed pluteus larvae from the first species that survived to slow freezing in 1.5 M DMSO, that developed through to metamorphosis and settled. More recently, Bellas and Paredes (2011) published an optimized protocol for *P. lividus* blastula, providing 98 % larval growth. Motivation for the cryopreservation of sea urchins derives from their use as biomarkers in ecotoxicology. Conservationists have made attempts to cryopreserve lace coral (*Pocillopora damicornis*) and mushroom coral (*Fungia scutaria*) which face severe environmental pressures. Preliminary studies on cryoprotectant toxicity, water and cryoprotectant permeability, ice nucleation temperatures, chilling sensitivity, and settlement of coral larvae have been performed (Hagedorn *et al.*, 2006). Both species appeared suitable for cryopreservation, apart from their extreme sensitivity to chilling, which indicated that vitrification could be the choice in place of slow freezing.

Attempts to cryopreserve penaeid embryos and larvae have also shown promising results, although further efforts should provide higher success rates. Gwo and Lin (1998) evaluated the most suitable developmental stage for freezing penaid shrimp (*Penaeus japonicus*). The survival rate of nauplii and zoea treated with 10 % methanol in sea water and frozen to -15 °C was 85 %, and some nauplii and zoea survived freezing to –25 and –196 °C. However, no treatment yielded normal nauplii or zoea after freezing, demonstrating that further optimization of the protocol is required. Other penaeids are also being studied and basic information about resistance to cooling or toxicity of different cryoprotectants has been developed in species such as Pacific blue shrimp (*P. stylirostris*), Carabali shrimp (*Trachypenaeus Byrdi* and Asian tiger shrimp (*P. monodon*) (Alfaro *et al.*, 2001; Vuthiphandchai *et al.*, 2005).

All these studies reflect that embryo and larvae cryopreservation is, or could be, possible in most of the invertebrate species analysed.

© Woodhead Publishing Limited, 2013

Nevertheless, an optimization of freezing protocols and standardization of evaluation procedures is required, because survival after thawing does not imply developmental ability. Invertebrate development is a complex process and the outcome of thawed individuals should be checked before scaling up protocols.

3.5 Genetic integrity and epigenetic perspective

3.5.1 Maintenance of genetic integrity and variability after cell banking

Cell cryopreservation is carried out to preserve valuable genetic material. Cryoinjuries at any cellular level are not important in themselves, but because they compromise the ability of a cell to carry genetic code. Genetic traceability is of utmost importance when frozen material is to be used in genetic selection programs, or to preserve high value strains. Nevertheless, the control of genetic integrity is still in its infancy.

Any cell subjected to cryopreservation will experience several physico-chemical aggressions (see Section 3.2.2). Freezing acts as a selection mechanism for a subset of cells. In the case of sperm, it is clear that milt is a heterogeneous suspension of cells, with spermatozoa carrying different genetic information, originating from different germ cells after meiosis, and often showing different degrees of maturation. Different spermatozoa sub-populations have been clearly identified in the sperm of Senegalese sole (*Solea senegalensis*) or gilthead seabream according to their motility pattern (Martinez-Pastor *et al.*, 2008; Beirao *et al.*, 2011), the structure of sub-populations changing after freeze–thawing (Beirao *et al.*, 2011). The factors responsible for cryoresistance have not been comprehensively enumerated. Membrane resistance to osmotic shock or DNA integrity have been linked to differential resistance to freezing in some species (Cabrita *et al.*, 2008a; Perez-Cerezales *et al.*, 2010a). Whatever the cause, modification to the profile of viable cells after thawing could bias the characteristics of progeny obtained. To our knowledge, few groups have explored this important issue (Van der Walt *et al.*, 1993; Martinez-Paramo *et al.*, 2009b). So far, no such genetic drift has been observed in offspring sired with cryopreserved sperm in the two species studied: the African catfish (*Clarias gariepinus*) and the brown trout (*Salmo trutta*). However, several groups have reported embryonic deformation after fertilization with thawed sperm (Tiersch *et al.*, 1994; Suquet *et al.*, 1998; Horvath and Urbanyi, 2000). Therefore, more extensive work is needed to identify the relevance of cryopreservation's genetic impact, taking into account the biological roots of the question: DNA structure and function. Detailed study of genetic integrity is desirable both before and after cryopreservation in order to identify risks during the process and guarantee the gene banking accuracy and conformity of frozen samples with regard to the genetic material stored.

© Woodhead Publishing Limited, 2013

The impact in future offspring of fertilization with an injured DNA genome needs careful study. Many cells, and particularly oocytes, have the ability to repair DNA failures using specific pathways according to the nature of the insult (for example, simple or double strand breaks, base modifications, and base mismatch). The activation of these mechanisms has been noted in oocytes fertilized with frozen DNA cryodamaged sperm (Kopeika *et al.*, 2004; Perez-Cerezales *et al.*, 2010b), but this ability is limited and could introduce changes to the code, for which potential consequences should be evaluated.

3.5.2 Chromatin damage during cryopreservation

Chromatin condensation with protamines, by reducing access to DNA, should provide some protection against damaging agents. However, some chromosomal elements remain unprotamined, retaining histones. Recent data for zebrafish indicate that these less packed regions, much more sensitive to damage, are necessary for proper embryogenesis after egg fertilization (Wu *et al.*, 2011). These important regions are at risk of being altered by cryopreservation. In mammals, it has been suggested that the strong selective process which sperm undergo in the female tract is probably unrelated to a DNA-scanning mechanism, but some kind of link exists between sperm phenotype and genotype which positively discriminates those cells with higher levels of chromatin integrity (Hourcade *et al.*, 2010). Selection is much weaker in teleosts; spermatozoa carrying at least 10 % fragmented chromatin kept the ability to fertilize in rainbow trout (*Oncorhyncus mykiss*) (Perez-Cerezales *et al.*, 2010b). Moreover, chromatin proteins in sperm from aquatic species may not provide an utterly compacted structure (see Section 3.2.3), thereby enhancing DNA susceptibility to cryodamage (Surralles *et al.*, 1998) in a species-dependent manner. Research is needed in fish to identify chromatine regions with different structure and cryosensitivity due to inter-species variation. Some interesting data have already been published. A higher susceptibility of telomeres to fragmentation was detected in rainbow trout, leading to a shortening of telomeres in frozen spermatozoa (Perez-Cerezales *et al.*, 2011). Similar effects on telomeres have also been reported recently by Moskovtsev *et al.* (2010) in humans, and this fact may have important implications.

DNA fragmentation in fish has been described after UV radiation or H_2O_2 treatment of brown trout, common carp (*Cyprinus carpio*) and lamprey (*Petromyzon marinus*) spermatozoa (Ciereszko *et al.*, 2005; Dietrich *et al.*, 2005); treatment with duroquinone of common carp sperm; cold storage at 4 °C with rainbow trout (Perez-Cerezales *et al.*, 2009); and cryopreservation of sperm from various species (Labbé *et al.*, 2001; Zilli *et al.*, 2003; Cabrita *et al.*, 2005; Perez-Cerezales *et al.*, 2009, 2010a; Martinez-Paramo *et al.*, 2009b). DNA fragmentation has also been observed in zebrafish primordial germ cells submitted to some vitrification protocols, even if

© Woodhead Publishing Limited, 2013

cell viability seemed unaffected, demonstrating both: (i) the cryosensitivity of DNA; and (ii) the importance of these criteria in choosing the most adequate cryopreservation protocol for a specific cell type. All results reported showed a difference in susceptibility to strand breaks between species. For example, the sperm DNA fragmentation index after cryopreservation is significantly lower in brown trout than rainbow trout (Cabrita *et al.*, 2005; Perez-Cerezales *et al.*, 2009; Martinez-Paramo *et al.*, 2009b). Moreover cold storage at 4 °C for four days promoted a higher degree of DNA bases oxidization than cryopreservation, due to a higher production of reactive oxygen species (ROS) in rainbow trout (Perez-Cerezales *et al.*, 2009). Progression throughout the breeding period has also been shown as a contributing factor to chromatin instability and susceptibility to damage. A certain degree of strand breakage at the end of the breeding season, and a significantly higher level of breakage after cryopreservation of this milt, has been reported in rainbow trout at the end of breeding period (Perez-Cerezales *et al.*, 2010a), even if none of the standard parameters of quality or fertilization ability of fresh sperm revealed any change with respect to the sperm collected in preceding weeks.

Oxidative stress is always perceived to be the most harmful factor for DNA. In fact, the increase in ROS production during cryopreservation, mainly due to damaged mitochondria and acrosomes, has been observed by several authors in mammals. In fish, the presence of oxidized bases (oxo-cytosine and oxo-guanosine) has been reported in rainbow trout sperm subjected to cold storage (4 °C) and cryostorage (Perez-Cerezales *et al.*, 2009). The use of antioxidants in the extender is being optimized to reduce cryoinjuries; chromatin integrity was very much improved when the low-density lipoprotein fraction of egg yolk (LDL) was added to the freezing extender (Perez-Cerezales *et al.*, 2010a).

Beyond DNA in itself, chromatin proteins deserve attention, as it is now known that in many species modification of the tail end of histones, together with the addition of methyl groups to DNA cytosine bases, will affect genome transcription (Collas *et al.*, 2007). These so-called epigenetic marks will sequentially change chromatin organization from a permissive to a repressive state, depending on the developmental stage and tissue. Modulation of epigenetic marks during early development plays a major role in differentiation, and such marks are increasingly perceived as important variables which should be controlled during cryopreservation (Hammoud *et al.*, 2009; Carrell and Hammoud, 2010). Inventory of the histone tail profile in zebrafish spermatozoa and embryos was recently reported (Wu *et al.*, 2011; Lindeman *et al.*, 2010; Andersen *et al.*, 2012), suggesting that the tools are now available for exploration in non-mammalian species. At the DNA methylation level, care should be taken when considering the widespread use of DMSO as a cryoprotectant. DMSO can become a strong methylating agent in specifically oxydative environments (Kawai *et al.*, 2010). Another challenge is to understand how DMSO could affect DNA

© Woodhead Publishing Limited, 2013

methylation in cryopreserved cells, thereby affecting the demethylation triggered during early development (Mhanni and McGowan, 2004; MacKay *et al.*, 2007). To our knowledge, no exploration of the epigenetic changes in cryopreserved cells has been reported in aquatic species, although such changes may offer another explanation for the development abnormalities observed after cryopreservation.

3.5.3 Methods for DNA integrity evaluation

Many different and complementary methods have been developed for the analysis of chromatin structural integrity. The most commonly used in fish gametes is the comet assay, SCGE (single-cell gel electrophoresis), reported in cryopreservation research (Labbé *et al.*, 2001; Zilli *et al.*, 2003; Cabrita *et al.*, 2005) and toxicological studies (Ciereszko *et al.*, 2005; Dietrich *et al.*, 2007, 2010; Devaux *et al.*, 2011). This technique measures the migration of fragmented DNA from the previously demembranated cell nucleus, submitted to acidic or basic challenge, and placed in an electrophoresis gel. The smaller the DNA strands the faster they migrate during electrophoresis, leading to the occurrence of a comet-like tail preceding the nucleus DNA (head of the comet) when stained with ethidium bromide. Different parameters might be obtained by analyzing the comet images using specific software. The percentage of Tail DNA (% DNAt) and the Olive Tail Moment (Mt), which includes both length and intensity of DNA in the tail, are the most common. This method shows a high versatility, because some modifications in the procedure allow detection of single- and/or double-strand breaks. Moreover, the combination of previous digestion with specific endonucleases demonstrates the presence of sensitive regions in the tested enzymes, according to the increase in fragmentation rate (i.e. digestion with ENDIII cut the DNA in oxidized cytosines, revealing their presence in non-fragmented positions of the strand). This method can also be applied to detect the fragmentation of specific genes when combined with fluorescence *in situ* hybridization (FISH)

The SCSA (sperm chromatin structure assay) is based on the same principle as the comet assay. It is another measure of the susceptibility of sperm nuclear DNA to acid-induced *in situ* denaturation (Evenson and Darzynkiewicz, 1990), which is dependent on the degree of packaging between DNA and proteins and is also related to strand breaks in the DNA. DNA integrity is measured by changes in the fluorescence pattern of acridine orange when associated with loosely or densely packaged chromatin. SCSA is a simple test which can be assessed by flow cytometry, thereby allowing the analysis of a high number of cells per sample. This method is frequently used with mammalian sperm, but not with fish samples, due to a different pattern of DNA/proteins packaging, which leads to difficulties in setting up the required parameters.

A very sensitive method is the TUNEL assay (terminal deoxynucleotidyl transferase mediated dUTP-biotin end-labelling), that allows the direct

© Woodhead Publishing Limited, 2013

detection of DNA breaks through the addition of a fluorescent labelled nucleotide to the 3′OH end of the strand, and the measurement of fluorescence emitted. Fluorescence increases with fragmentation due to the appearance of free 3′OH ends (Fatehi *et al.*, 2006; Fernandez-Gonzalez *et al.*, 2008). Analysis by flow cytometry is also compatible with this method, rendering it accurate. Its relatively high cost should be mentioned as a negative.

The use of quantitative PCR (QPCR), based in the blockage of amplification caused by some DNA insults, is being assayed for a more precise identification of damage to nuclear (nDNA) and mitochondrial (mtDNA) mammalian spermatozoa genomes (Bennets and Aitken, 2005). This is particularly interesting in species whose genome is sequenced (such as zebrafish; and rainbow trout and sea bass, very soon), as results can be analysed on ships themselves.

3.5.4 Consequences of the use of frozen gametes with non-tested genetic integrity for the progeny

Achievements in fish sperm cryopreservation are expanding the use of cryopreserved sperm, sometimes without quality control, because success is attained with fertility. In the previous paragraphs we have shown how the process could affect the sperm genome, generally viewed as an 'inert' chromatin whose role begins when bulk transcription starts in the embryo (at gastrulation). Recent studies in mammals describe how sperm chromatin damage affects early embryo development more than fertilization (Speyer *et al.*, 2010). This is due to the important role of sperm chromatin in the regulation of gene expression at early stages (Evenson and Wixon, 2006; Carrell and Hammoud, 2010; Delbes *et al.*, 2010; Ward, 2010), including by way of genome imprinting (repression of some genes from one parental allele). This fact changes our perception of sperm chromatin's role, suggesting, as noted regarding epigenetics marks, that some elements in the paternal genome are not reset at fertilization, thereby affecting gene control of the embryo.

This topic deserves the increased interest it is receiving in the field of human and cattle reproduction because, in spite of the sperm selection process taking place before fertilization between mammals, DNA damage does not always obstruct the fertilization ability of sperm (Ahmadi and Ng, 1999) and can go on to cause deleterious effects on embryo development, such as impaired embryo cleavage (Morris *et al.*, 2002), implantation failures (Speyer *et al.*, 2010), or an increase in abortion rate (Lin *et al.*, 2008). Their impact on the reproductive outcome of assisted reproductive technologies (ART) is also a matter of controversy. Fertilization with sperm carrying some kind of chromatin damage seems to be linked to morphopathological alterations, such as a higher incidence of cancer, abnormal body size, early ageing, and increase of imprinting-related syndromes

© Woodhead Publishing Limited, 2013

or altered behaviour, not only in the F1 but also in subsequent generations (reviewed by Delbés *et al.*, 2010).

Among fish, the possible effects of sperm DNA injury on offspring performance are unclear and reported results are scarce, fragmented and sometimes contradictory. In trout, Labbé *et al.* (2001) and Young *et al.* (2009) showed that quality of embryo and early fry was not different when produced with fresh or cryopreserved sperm, although DNA fragmentation was increased in the cryopreserved milt (Labbe *et al.*, 2001). At later stages, Hayes and his colleagues (2005) reported that juveniles produced with fresh or cryopreserved sperm had the same length and body weight, but that a different response to stress, measured by cortisol level, existed in some larvae batches obtained with frozen sperm whose DNA could have been cryodamaged. Kopeika *et al.* (2003) reported failures in the loach (*Misgurnus fossilis*) offspring from batches fertilized with frozen milt, and Horvath *et al.* (2007) detected clastogenic effects in carp obtained with cryostored semen. More recently, our work revealed that telomeres from trout larvae developed from sperm with shortened telomeres were surprisingly longer than those of their respective controls (Perez-Cerezales *et al.*, 2011). The analysis by reverse transcriptase-polymerase chain reaction (RT-PCR) of gene expression of eight genes related to growth and development (*Igf1, Igf2, Igfr1a, Igfr1b, Gh1, Gh2, Ins1* and *Ins2*), revealed the overexpression of five of them after hatching (*Igf1, Igfr1a, Igfr1b, Gh1, Gh2*). These findings point to a failure in gene regulation, supporting the implication of sperm chromatin in the control of embryo development. Generating genetically traceable progenies is essential for commercial aquaculture, and for biodiversity preservation programs. Sperm chromatin seems to play a more important role than expected.

The evidence indicates that control of DNA damage during cryopreservation renders cryobanking a safe, accurate and useful tool in fish reproduction. However, the use of non-optimal procedures could have important consequences in recovering a population from frozen cells. The analysis of chromatin integrity deserves, in our opinion, special attention when frozen samples are employed to obtain androgenetic progenies, where the single paternal contribution should increase the importance of sperm chromatin status. A wider analysis of phenotypical and molecular effects of chromatin damage upon offspring could also be helpful in order to develop molecular markers of later reproductive outcome.

3.6 Alternative cell sources for cryobanking in fish

3.6.1 Cryopreservation of somatic cells and challenges of nuclear transfer technology

When gametes and embryos are not easily available, somatic cells should be considered for fish genome cryobanking. Although especially valuable

© Woodhead Publishing Limited, 2013

with non-domesticated species, somatic cell cryopreservation is also relevant to hatcheries with species whose sexual maturity is reached only after several years, as in the case of some sturgeons, and in tuna. Here, cryopreserving somatic cells from immature fish is a means to preserve the genetics of the population. Somatic cells, in particular fin cells, have been successfully cryopreserved for decades (Wolf and Quimby, 1969; Mauger *et al.*, 2006), but whole fin pieces can also be cryopreserved, giving fin cells in culture after thawing (Moritz and Labbé, 2008). Fin cryopreservation allows significant cryopreservation of somatic resources without the need for specific laboratory equipment; conventional chest freezers are sufficient. Fins are also easy to collect on live fish, with the animal suffering little damage. However, no reproductive cells are present in fin samples. Fish reconstruction therefore requires the use of non-conventional nuclear transfer technology: a diploid nucleus is transferred, either by injection or by electrofusion, into a recipient oocyte which has previously been enucleated. The recipient oocyte can belong to the same species as that of the donor cell, or come from a closely related species. In both cases, whatever its nuclear DNA background, the developing embryo will use the cytoplasmic components of the recipient oocyte (mitochondria and its small DNA, proteins, maternal mRNAs). Fish reconstructed by nuclear transfer are therefore referred to as *nucleo-cytoplasmic hybrids*.

Nuclear transfer can be carried out with a nucleus donor obtained from embryonic cells (whose cryopreservation is possible, see Section 3.6.2), or with a somatic cell nucleus. Both possibilities, which are referred to respectively as embryonic cell nuclear transfer (ECNT) and somatic cell nuclear transfer (SCNT), are still a long way off application in hatcheries. This technique has been developed in fish since the 1960s, mainly in cyprinids with embryonic donor cells, with a generally low efficiency of approximately 1 %. It was several years after the report of successful SCNT in mammals (Wilmut *et al.*, 1997) that viable offspring (2 %) were obtained after SCNT in zebrafish (Lee *et al.*, 2002). Since then, the main methodological improvements achieved have been the use of metaphase II oocytes as recipients in place of activated eggs (Siripattarapravat *et al.*, 2009; Le Bail *et al.*, 2010); control of the location at which the nucleus is injected within the oocyte (Le Bail *et al.*, 2010); and use of laser-ablation of the recipient DNA (Siripattarapravat *et al.*, 2009). Enucleation of the recipient may not always be necessary, as successful diploid developments have been reported using non-enucleated eggs as recipients in medaka, zebrafish (Niwa *et al.*, 1999; Wakamatsu, 2008; Hattori *et al.*, 2011), and in goldfish (*Carassius aurata*) (Labbé *et al.*, personal observation). Nuclear transfer primarily addresses a biological rather than a methodological level. Indeed, most researchers suspect that the DNA reprogramming process taking place on gametic chromatin during normal embryo development (Mhanni and McGowan, 2004) does not operate properly when the chromatin is from somatic origin (Pei *et al.*, 2007a, 2008, 2009; Liu *et al.*, 2008; Le Bail *et al.*, 2010;

© Woodhead Publishing Limited, 2013

Siripattarapravat *et al.*, 2010). Accordingly, Zhu and collaborators in Wuhan, China, have been working extensively on gene expression in clones, and have identified several genes whose expression is altered in cloned embryos (Pei *et al.*, 2007a, 2008, 2009; Liu *et al.*, 2008; Luo *et al.*, 2009, 2011). How the exogenous chromatin can be modified in order to undergo a proper reprogramming upon development initiation is currently the main focus for research in the field.

3.6.2 Pluripotent cell cryopreservation and reconstruction with blastomeres *versus* embryonic stem cells

Pluripotent cells are embryonic or embryonic-derived cells with the ability to differentiate into multiple cell types. Initially, pluripotent cells are blastomeres obtained from early stage embryos. They are then either directly cryopreserved and used for reconstruction, or cultured in conditions which will enable them to maintain pluripotency, together with their self-renewal potential. Such cultured cells are referred to as embryonic stem cells (ES cells).

Blastomeres have been cryopreserved with different degrees of success in several species, including rainbow trout (Leveroni-Calvi and Maisse, 1998), common carp (Leveroni-Calvi and Maisse, 1999), chum salmon (*Oncorhynchus keta*) (Kusuda *et al.*, 2002), goldfish (Kusuda *et al.*, 2004), zebrafish (Kopeika *et al.*, 2005; Cardona-Costa and Garcia-Ximenez, 2007; Lin, 2009), and leopard danio (*Brachydanio frankei*) (Routray *et al.*, 2010). Different cryoprotectants have been used for blastomere cryopreservation, including DMSO, glycerol, EG, and Pdiol. Some studies present DMSO as the best option for either zebrafish blastomere vitrification (Cardona-Costa and Garcia-Ximenez, 2007) or chum salmon blastomere cryopreservation (Kusuda *et al.*, 2002). Interestingly, the addition of an antifreeze protein originating from fish, AFP type I, has significantly improved zebrafish blastomere survival after cryopreservation (Martinez-Paramo *et al.*, 2009a). In zebrafish, both controlled slow cooling and vitrification have been studied for blastomere preservation (Cardona-Costa and Garcia-Ximenez, 2007; Lin *et al.*, 2009b). Using slow cooling, the survival rate was 70.2 % in 1.5 M DMSO and 0.1 M sucrose (Lin *et al.*, 2009b). Blastomere survival rates after vitrification were higher than 90 %.

Once thawed, blastomeres cells can be transplanted into a recipient embryo, leading to the formation of a chimera. The purpose of such transplantation is so injected cells contribute to formation of the germline. Chimera production with cryopreserved cells has been successfully reported in rainbow trout (Takeuchi *et al.*, 2001, 2004), together with successful germline transmission, although transmission rates were very low. Germline transmission rates were improved after the recipient cells were weakened by irradiation (Joly *et al.*, 1999). In zebrafish, only two adult chimaeric specimens were obtained from a total of 47 injected embryos

© Woodhead Publishing Limited, 2013

(Cardona-Costa *et al.*, 2009), and no germline transmission was achieved. It has also been reported that cryopreservation can disrupt normal gene expression patterns in zebrafish embryonic blastomeres; this disruption may have had detrimental effects on embryo development (Lin *et al.*, 2009a).

Studies on ES cells in fish began almost 20 years ago (Wakamatsu *et al.*, 1994; Ma *et al.*, 2001). However, it is likely that blastula cell lines rather than ES cell lines proper have been established (Robles *et al.*, 2011), and the term *ES-like cells*, rather than *ES cells*, should be used for the majority of these cultures. Nowadays, ES-like cells have been derived in several species, including zebrafish, medaka, gilthead sea bream, red sea bream, sea perch (*Lateolabrax japonicas*), Asian sea bass (*Lates calcarifer*), Atlantic cod (*Gadus morhua*) (reviewed by Hong *et al.*, 2011), and major carp (*Catla catla*) (Dash *et al.*, 2010). Some of these cultures remain pluripotent and germline competent for some passages (Ma *et al.*, 2001; Fan *et al.*, 2004a, b). As a consequence, germline transmission of ES cells was reported after short-term culture, although germline transmission of ES cells cultured for long periods has not yet been reported. One reason for this could be that ES-like cells lose pluripotency, or at least germline competence (Hong *et al.*, 2011). Some pluripotency genes such as *oct4/pou5f1*, which is highly expressed in blastomeres (Christen *et al.*, 2010), are not always detected in ES-like cell cultures (Robles *et al.*, 2011).

Considering that both blastomeres and ES-like cells possess germline colonization ability once transplanted into a recipient embryo, we could conclude that both cell types may be optimal for cryobanking. However, we must also point out that cells in culture can suffer modifications after some passages that could affect, for example, chromosome stability and germline transmission ability. With this in mind, blastomeres would be a better option for cryobanking. However, the reduced germline transmission is hampered by the small number of primordial germ cells within the whole blastomere population.

3.6.3 Primordial germ cells and spermatogonial stem cells cryopreservation

Focusing on primorgial germ cells (PGCs), those embryonic cells destined to migrate towards the genital ridges and to develop into spermatogonia or oogonia, is a relatively straightforward means to increase germline transmission. Spermatogonia stem cells (SSCs) constitute a small cell population in mature testes which maintains the ability to re-colonize genital ridges upon transplantation. Both cell types are probably the best alternative cell source for cryobanking in fish. As recently reviewed by Lacerda *et al.* (2013), both can colonize sexually undifferentiated embryonic gonads, and resume gametogenesis (Takeuchi *et al.*, 2003; Yoshizaki *et al.*, 2005; Okutsu *et al.*, 2006).

The development of PGC visualization methods has helped significantly in PGC isolation and manipulation for cryopreservation. These methods

© Woodhead Publishing Limited, 2013

include: the creation of transgenic fish lines with *vasa* regulatory regions controlling green fluorescent protein (GFP) expression (Yoshizaki *et al.*, 2000; Krovel and Olsen, 2002); transgenic fish carrying the Tol-kop-EGFP-F-nos1-3′UTR transgene (Boldajipour *et al.*, 2008); or injecting GFP-nos 1 3′UTR mRNA (Saito *et al.*, 2006). GFP fluorescent PGCs from rainbow trout have been isolated by flow cytometry (Kobayashi *et al.*, 2004), successfully cryopreserved using 1.8M ethylene glylcol, and transplanted into the peritoneal cavities of allogenic trout hatchlings. Fertile F1 fish were obtained by fertilizing eggs derived from those cryopreserved PGCs with cryopreserved spermatozoa (Kobayashi *et al.*, 2007). Moreover, the mechanism of PGC migration is highly conserved. This makes inter-species transplantation and migration of PGCs possible, as has been reported for cyprinid fish (Saito *et al.*, 2010). In addition, using xenogenesis across species (goldfish and loach as donor species and zebrafish as recipients) it is possible to achieve inter-species transplantation by transplanting a single PGC from a donor species (Saito *et al.*, 2008).

In zebrafish, embryos have been vitrified with the aim of recovering their PGCs, which were subsequently transplanted to recipient embryos; fertile animals with germ cells originated from these PGCs were successfully obtained (Higaki *et al.*, 2010). Genital ridges were also successfully cryopreserved in zebrafish using a combination of cryoprotectants (DMSO 5 M, ethylene glycol (EG) 1 M, polyvinyl pyrrolidone (PVP) 4 %) incorporated in a stepwise manner. After thawing, no differences with control (non-cryopreserved) PGCs were detected in viability, pseudopodial emission, or DNA integrity (Riesco *et al.*, 2012a). Therefore, this protocol guarantees high cell survival (around 90 %) and the absence of DNA fragmentation. However, there are several other types of genetic damage which are undetectable by Comet assay. The authors are currently using a new method for DNA evaluation after PGC cryopreservation based on QPCR. This method allows the quantification of lesions per Kb in a specific gene or genome region of interest. Using the above mentioned cryopreservation protocol, the number of DNA lesion per 10 Kb in the nuclear and mitochondrial genome regions analysed was never higher than 4.68 (Riesco *et al.*, 2012b).

3.7 Conclusions

Cryopreservation of gametes from aquatic species, either for routine use in hatcheries or for long-term cryobanking, has reached varied levels of development (Table 3.2) under the expertise of a world-wide scientific community (Fig. 3.3). Sperm is easily cryopreserved in almost all species, although spermatophores in crustaceans are still difficult to handle and to test for fertilization. Cryopreservation of oocytes and larvae in molluscs shows promise, although in fish it is still a field of active research.

© Woodhead Publishing Limited, 2013

Table 3.2 Advantages and pitfalls of the different cell types for cryopreservation and reconstruction

Cell type	Availability	Cryopreservation fitness	Reconstruction method	Genetic conformity to the donor
Spermatozoa	Requires mature broodstock /reproduction control	+++ in fish, mollusc +/– in crustacean	Fertilization	Cross breeding, requires females from close strains
Oocyte	Requires mature broodstock /reproduction control	– in fish +/– in molluscs ? in crustacean	Fertilization	Cross breeding, requires males from close strains
Embryos/larvae	Requires mature broodstock /reproduction control	– in fish + in molluscs +/– in crustacean	Normal development	Total
Somatic cells (fin cells in fish)	Available on every fish	+++ in fish ? in invertebrates	Nuclear transfer	Nucleo-cytoplasmic hybrid (mtDNA from maternal origin) Allows inter-specificity
Blastomeres/ES-like cells	Requires mature broodstock /reproduction control	+++ in fish ? in invertebrates	Grafting on recipient embryos Nuclear transfer	Total if germline transmission Nucleocytoplasmic hybrid
PGCs/SSCs	Requires larvae or adult males whatever the maturational state	+ in fish ? in invertebrates	Cell transplantation in surrogate fish	Total if genital ridge colonization Allows inter-specificity

© Woodhead Publishing Limited, 2013

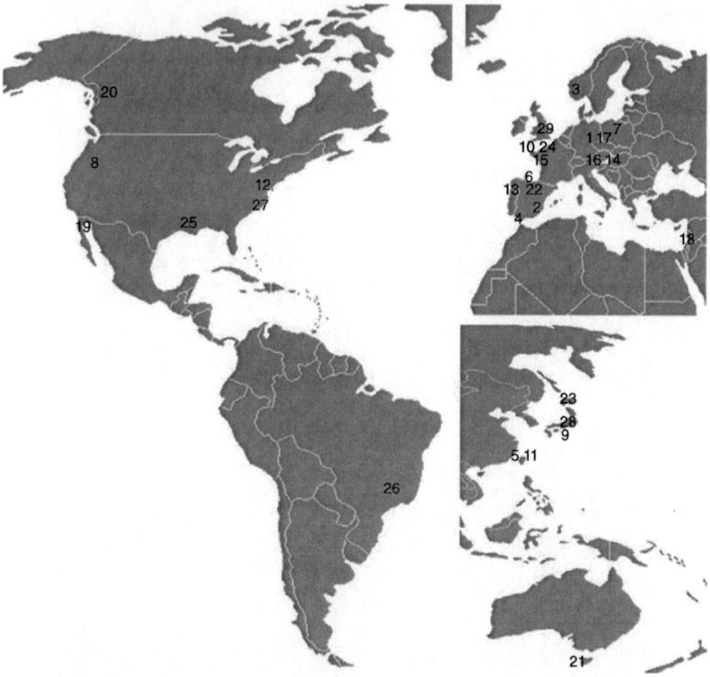

Group/Lab	Affiliation
0 Dr Serean Adams	Cawthron Institute, Nelson, New Zealand
1 Dr Hadi Alavi	University of South Bohemia, Czech Republic.
2 Dr Juan F. Asturiano	University of Valencia, Spain
3 Dr Igor Babiak	Bodo University College, Norway
4 Dr Elsa Cabrita	ICMAN, Cadiz, Spain
5 Dr Nai_Hsien Chao	National Taiwan Ocean University, Taiwan
6 Dr Olvido Chereguini	IEO, Santander, Spain
7 Dr Andrzej Ciereszko	Polish Academy of Sciences, Poland
8 Dr Joseph Cloud	University of Idaho, USA
9 Dr Keisuke Edashige	Kochi University, Japan
10 Dr Christian Fauvel,	IFREMER, Palavas, France
11 Dr Jin-Chywan Gwo	Taiwan National Ocean University, Taiwan
12 Dr Mary Hagedorn	Smithsonian Institution, Washington, USA
13 Dr Paz Henáez	University of León, Spain
14 Dr Ákos Horvath	Szent István University, Hungary
15 Dr Catherine Labbé	INRA, France
16 Dr Franz Lahns teiner	University of Salzburg, Austria
17 Dr Otomar Linhart	University of South Bohemia, Czech Republic
18 Dr Esther Lubzens	National Institute of Oceanography, Haifa, Israel
19 Dr Carmen Paniagua-Chavez	CICESE, Baja California, México
20 Dr Rick M. Rideout	Northwest Atlantic Fisheries Centre, Canada
21 Dr Arthur Jeremy Ritar	University of Tasmania, Australia
22 Dr Vanesa Robles	University of León, Spain
23 Dr Taiju Saito	University of Hokkaido, Japan
24 Dr Marc Suquet	IFREMER, France
25 Dr Terrence R. Tiersch	Louisiana State University, USA
26 Dr Ana Viveiros	Federal University of Lavras, Brazil
27 Dr Curry Woods	University of Maryland, USA
28 Dr Goro Yoshizaki	Tokyo University of Fisheries, Konan, Japan
29 Dr Tiantian Zhang	University of Bedfordshire, UK

Fig. 3.3 Some of the groups working on sperm, oocyte, embryo or PGC cryopreservation throughout the world. (There are many other experts with publications in the field; we apologize to those who could not be included.)

© Woodhead Publishing Limited, 2013

If cryopreservation is now a routine technology transferable for application, at least for sperm, quality tests for cell diffusion are still to be determined. Tests should be favored which aim to assess the genetic conformity of the offspring obtained. When considering alternative cell sources, cryopreservation of somatic cells, cultured cells, blastomeres, PGCs, or SSCs does not appear technically challenging. However, these cell types differ in their accessibility and in the complexity of the reconstruction technology they require (Table 3.2)

The development of cryobanking would benefit from a standardization of protocols used by the scientific community. A scaling up of the method is now being considered for routine hatchery use, making this a more urgent matter. A global strategy must be worked out to allow accurate and standardized identification of samples, follow up sanitary status, and efficiently manage records and data when the collections are incorporated into or transferred out of depositories.

3.8 Sources of further information

- The book by Tiersch and Mazik (2011) gives the basics in sperm, egg and oocyte cryopreservation in fish and shellfish. Above all, this book thoughtfully reviews the relevant factors when dealing with germplasm repositories. The reader is referred to it for more information on databases, sanitary consideration, regulation for transfer, and for economic and ethical considerations in cryobanking.
- The book by Cabrita *et al.* (2008b) gives more detailed information on specific procedures for sperm and egg cryopreservation in marine and freshwater species.
- The book by Alavi *et al.* (2008) develops the biology of fish spermatozoa with extensive reviews of differing cellular organization and function.

3.9 Acknowledgements

The authors would like to thank Ramón y Cajal program RYC-2008-02339, MICINN AGL2009-06994, JCyL LE365A11-2, and Fundación Ramón Areces. Conception and work on the French National Cryobank for aquatic species was funded by CryoAqua IBiSA 2009–2011 in connection with the French National Cryobank, IFREMER and SYSAAF. Pierrick Haffray (SYSAAF, Rennes) is acknowledged for valuable discussion and reliable input on the outcome of sperm and larvae cryopreservation in hatcheries.

3.10 References

ADAMS S L, KLEINHANS F W, MLADENOV P V and HESSIAN P A (2003) Membrane permeability characteristics and osmotic tolerance limits of sea urchin (*Evechinus chloroticus*) eggs. *Cryobiology*, 47, 1–13.

© Woodhead Publishing Limited, 2013

ADAMS S L, HESSIAN P A and MLADENOV P V (2006) The potential for cryopreserving larvae of the sea urchin, *Evechinus chloroticus. Cryobiology*, 52, 139–145.

ADAMS S L, TERVIT H R, MCGOWAN L T, SMITH J F, ROBERTS R D, SALINAS-FLORES L, GALE S L, WEBB S C, MULLEN S F and CRITSER J K (2009) Towards cryopreservation of Greenshell mussel (*Perna canaliculus*) oocytes. *Cryobiology*, 58, 69–74.

AHMADI A and NG SC (1999), Fertilizing ability of DNA-damaged spermatozoa. *J Exp Zool*, 284, 696–704.

ALAVI SMH, LINHART O, COWARD K and RODINA M (2008) *Fish Spermatology: Implications for Aquaculture Management*. Oxford: Alpha Science International Ltd.

ALFARO J, KOMEN J and HUISMAN EA (2001) Cooling, cryoprotectant and hypersaline sensitivity of penaeid shrimp embryos and nauplius larvae. *Aquaculture*, 195, 353–358.

ANDERSEN I S, ØSTRUP O, LINDEMAN L C, AANES H, REINER A H, MATHAVAN S, ALESTRÖM P and COLLAS P (2012) Epigenetic complexity during the zebrafish mid-blastula transition. *Biochem Biophys Res Comm*, 417, 1139–1144.

BAUER R T and MIN L J (1993) Spermatophores and plug substance of the marine shrimp *Trachypenaeus similis* (Crustacea: Decapoda: Penaeidae): Formation in the male reproductive tract and disposition in the inseminated female. *Biol Bull*, 185, 174–185.

BEIRAO J, ROBLES V, HERRAEZ M P, SARASQUETE C, DINIS MT and CABRITA E (2006) Cryoprotectant microinjection toxicity and chilling sensitivity in gilthead seabream (*Sparus aurata*) embryos. *Aquaculture*, 261, 897–903.

BEIRAO J, CABRITA E, PEREZ-CEREZALES S, MARTINEZ-PARAMO S and HERRAEZ M P (2011) Effect of cryopreservation on fish sperm subpopulations. *Cryobiology*, 62, 22–31.

BELLAS J and PAREDES E (2011) Advances in the cryopreservation of sea urchin embryos: Potential application in marine water quality assessment. *Cryobiology*, 62, 174–180.

BENNETS L A AND AITKEN R J (2005) A comparative study of oxidative DNA damage in mammalian spermatozoa. *Mol Reprod Dev*, 71, 77–87.

BILLARD R, COSSON J, NOVEIRI S B and POURKAZEMI M (2004) Cryopreservation and short-term storage of sturgeon sperm, a review. *Aquaculture*, 236, 1–9.

BLAXTER JHS (1953) Sperm storage and cross-fertilization of spring and autumn spawning herring. *Nature*, 26, 1189–1190.

BOBE J and LABBE C (2008) Chilled storage of sperm and eggs, in Cabrita E, Robles V, Herraez MP (eds), *Methods in Reproductive Aquaculture, Marine and Freshwater Species*. Boca Raton, FL: CRC Press Taylor & Francis Group, 219–235.

BOBE J and LABBÉ C (2010) Egg and sperm quality in fish. *Gen Comp Endocrinol*, 165, 535–548.

BOLDAJIPOUR B, MAHABELESHWAR H, KARDASH E, REICHMAN-FRIED M, BLASER H, MININA S, WILSON D, XU Q and RAZ E (2008) Control of chemokine-guided cell migration by ligand sequestration. *Cell*, 132, 463–473.

BOZZO M G, RIBES E, SAGRISTA E, POQUET M and DURFORT M (1993) Fine-structure of the spermatozoa of *Crassostrea-Gigas* (Mollusca, Bivalvia). *Mol Reprod Dev*, 34, 206–211.

BROOKS S, TYLER C R and SUMPTER J P (1997) Quality in fish: what makes a good egg? *Reviews in Fish Biology and Fisheries*, 7, 387–416.

CABRITA E, ROBLES V, CHEREGUINI O, WALLACE J C and HERRAEZ M P (2003) Effect of different cryoprotectants and vitrificant solutions on the hatching rate of turbot embryos (*Scophthalmus maximus*). *Cryobiology*, 47, 204–213.

CABRITA E, ROBLES V, REBORDINOS L, SARASQUETE C and HERRAEZ M P (2005) Evaluation of DNA damage in rainbow trout (*Oncorhynchus mykiss*) and gilthead sea bream (*Sparus aurata*) cryopreserved sperm. *Cryobiology*, 50, 144–153.

© Woodhead Publishing Limited, 2013

CABRITA E, ROBLES V, WALLACE JC, SARASQUETE M C and HERRAEZ M P (2006) Preliminary studies on the cryopreservation of gilthead seabream (*Sparus aurata*) embryos. *Aquaculture*, 251, 245–255.

CABRITA E, ROBLES F and HERRAEZ M P (2008a) Sperm quality assessment, in Cabrita E, Robles V and Herraez MP (eds), *Methods in Reproductive Aquaculture, Marine and Freshwater Species*. Boca Raton, FL: CRC Press Taylor & Francis Group, 93–148.

CABRITA E, ROBLES V and HERRAEZ M P (2008b) *Methods in Reproductive Aquaculture, Marine and Freshwater Species*. Boca Raton, FL: CRC Press Taylor & Francis Group.

CABRITA E, SARASQUETE C, MARTINEZ-PARAMO S, ROBLES V, BEIRAO J, PEREZ-CEREZALES S and HERRAEZ MP (2010) Cryopreservation of fish sperm: applications and perspectives. *Journal of Applied Ichthyology*, 26, 623–635.

CARDONA-COSTA J and GARCIA-XIMENEZ F (2007) Vitrification of zebrafish embryo blastomeres in microvolumes. *CryoLetters*, 28, 303–309.

CARDONA-COSTA J, FRANCISCO-SIMAO M and GARCIA-XIMENEZ F (2009) Can vitrified zebrafish blastomeres be used to obtain germ-line chimaeras? *CryoLetters*, 30, 422–428.

CARRELL D T and HAMMOUD S S (2010) The human sperm epigenome and its potential role in embryonic development. *Mol Hum Reprod*, 16, 37–47.

CHANDRASEKARAN M and PITT RE (1992) On the use of nucleation theory to model intracellular ice formation. *CryoLett*, 13, 261–272.

CHAO N-H and LIAO IC (2001) Cryopreservation of finfish and shellfish gametes and embryos. *Aquaculture*, 197, 161–189.

CHAO N-H, CHIANG C P, HSU H W, TSAI C T and LIN T T (1994) Toxicity tolerance of oyster embryos to selected cryoprotectants. *Aquat Living Resour*, 7, 99–103.

CHAO N-H, LIN T-T., CHEN Y-J, HSU H-W and LIAO I.C (1997) Cryopreservation of late embryos and early larvae in the oyster and hard clam. *Aquaculture*, 155, 31–44.

CHOI Y H and CHANG Y J (2003) The influence of cooling rate, developmental stage, and the addition of sugar on the cryopreservation of larvae of the pearl oyster *Pinctada fucata martensii*. *Cryobiology*, 46, 190–193.

CHRISTEN B, ROBLES V, RAYA M, PARAMONOV I and BELMONTE JCI (2010) Regeneration and reprogramming compared. *BMC Biology*, 8, 5–13.

CIERESZKO A, WOLFE T D and DABROWSKI K (2005) Analysis of DNA damage in sea lamprey (*Petromyzon marinus*) spermatozoa by UV, hydrogen peroxide, and the toxicant bisazir. *Aquat Toxicol*, 73, 128–138.

COLLAS P, NOER A and TIMOSKAINEN S (2007) Programming the genome in embryonic and somatic stem cells. *J Cell Mol Med*, 11, 602–620.

DASH C, ROUTRAY P, TRIPATHY S, VERMA D K, GURU B C, MEHER P K, NANDI S and EKNATH A E (2010) Derivation and characterization of embryonic stem-like cells of Indian major carp *Catla catla*. *J Fish Biol*, 77, 1096–1113.

DEKENS M P, PELEGRI F J, MAISCHEIN HM and NUSSLEIN-VOLHARD C (2003) The maternal-effect gene futile cycle is essential for pronuclear congression and mitotic spindle assembly in the zebrafish zygote. *Development*, 130, 3907–3916.

DELBÉS C, HALES BF and ROBAIRE B (2010) Toxicants and human sperm chromatin integrity. *Mol Hum Reprod*, 16, 14–22.

DEVAUX A, FIAT L, GILLET C and BONY S (2011) Reproduction impairment following paternal genotoxin exposure in brown trout (*Salmo trutta*) and Arctic charr (*Salvelinus alpinus*). *Aquat Toxicol*, 101, 405–411.

DIETRICH G J, SZPYRKA A, WOJTCZAK M, DOBOSZ S, GORYCZKO K, ZAKOWSKI L and CIERESZKO A (2005) Effects of UV irradiation and hydrogen peroxide on DNA fragmentation, motility and fertilizing ability of rainbow trout (*Oncorhynchus mykiss*) spermatozoa. *Theriogenology*, 64, 1809–1822.

© Woodhead Publishing Limited, 2013

DIETRICH G J, ZABOWSKA M, WOJTCZAK M, SLOWINSKA M, KUCHARCZYK D and CIERESZKO A (2007) Effects of different surfactants on motility and DNA integrity of brown trout (*Salmo trutta fario*) and common carp (*Cyprinus carpio*) spermatozoa. *Reprod Biol*, 7, 127–142.

DIETRICH GJ, DIETRICH M, KOWALSKI RK, DOBOSZ S, KAROL H, DEMIANOWICZ W and GLOGOWSKI J (2010) Exposure of rainbow trout milt to mercury and cadmium alters sperm motility parameters and reproductive success. *Aquat Toxicol*, 97, 277–284.

DONG Q X, HUANG C J and TIERSCH T R (2005) Spermatozoal ultrastructure of diploid and tetraploid Pacific oysters. *Aquaculture*, 249, 487–496.

DONG Q X, HUANG C J and TIERSCH T R (2007) Control of sperm concentration is necessary for standardization of sperm cryopreservation in aquatic species: Evidence from sperm agglutination in oysters. *Cryobiology*, 54, 87–98.

DROBNIS E Z, CROWE L M, BERGER T, ANCHORDOGUY T J, OVERSTREET J W and CROWE J H (1993) Cold shock damage is due to lipid phase transitions in cell membranes: a demonstration using sperm as a model. *J Exp Zool*, 265, 432–437.

EVENSON D and DARZYNKIEWICZ Z (1990) Acridine orange-induced precipitation of mouse testicular sperm cell DNA reveals new patterns of chromatin structure. *Exp Cell Res*, 187, 328–334.

EVENSON D P and WIXON R (2006) Clinical aspects of sperm DNA fragmentation detection and male infertility. *Theriogenology*, 65, 979–991.

FAHY GM (1995) The role of nucleation in cryopreservation in Lee RE, Warren GI, Gusta LV eds, *Biological Ice Nucleation and its Applications.* St Paul, MN: APS Press, 315–336.

FAHY G M, MACFARLANE D R, ANGELL C A and MERYMAN H T (1984) Vitrification as an approach to cryopreservation. *Cryobiology*, 21, 407–426.

FAN L C, ALESTROM A, ALESTROM P and COLLODI P (2004a) Development of cell cultures with competency for contributing to the zebrafish germline. *Crit Rev Eukaryot Gene Expr*, 14, 43–51.

FAN L C, CRODIAN J and COLLODI P (2004b) Production of zebrafish germline chimeras by using cultured embryonic stem (ES) cells. *Zebrafish*, 77, 113–119.

FATEHI A N, BEVERS M M, SCHOEVERS E, ROELEN B A, COLENBRANDER B and GADELLA B M (2006) DNA damage in bovine sperm does not block fertilization and early embryonic development but induces apoptosis after the first cleavages. *J Androl*, 27, 176–188.

FERNANDEZ-GONZALEZ R, MOREIRA P N, PEREZ-CRESPO M, SANCHEZ-MARTIN M, RAMIREZ M A, PERICUESTA E, BILBAO A, BERMEJO-ALVAREZ P, DE DIOS HOURCADE J and D E FONSECA F R (2008) Long-term effects of mouse intracytoplasmic sperm injection with DNA-fragmented sperm on health and behavior of adult offspring. *Biol Reprod*, 78, 761–772.

FORNARI D C, RIBEIRO R P, STREIT D P, VARGAS L, BARRERO N M and DE MORAES G V (2010) Freezing injuries in the embryos of *Piaractus mesopotamicus*. *Zygote*, 23, 1–6.

FREHLICK L J, EIRÍN-LÓPEZ J M, PRADO A, SU HW, KASINSKY H E and AUSIÓ J (2006) Sperm nuclear basic proteins of two closely related species of scorpaeniform fish (*Sebastes maliger, Sebastolobus* sp.) with different sexual reproduction and the evolution of fish protamines. *J Exp Zool*, 305A, 277–287.

GUAN M, RAWSON D M and ZHANG T (2008) Cryopreservation of zebrafish (*Danio rerio*) oocytes using improved controlled slow cooling protocols. *Cryobiology*, 56, 204–208.

GUAN M, RAWSON D M and ZHANG T (2010) Cryopreservation of zebrafish (*Danio rerio*) oocytes by vitrification. *CryoLett*, 31, 230–238.

GWO J C (1995) Cryopreservation of oyster (*Crassostrea gigas*) embryos. *Theriogenology*, 43, 1163–1174.

© Woodhead Publishing Limited, 2013

GWO J C (2000) Cryopreservation of aquatic invertebrate semen: a review. *Aquac Res*, 31, 259–271.

GWO J C and LIN C H (1998) Preliminary experiments on the cryopreservation of penaeid shrimp (*Penaeus japonicus*) embryos, nauplii and zoea. *Theriogenology*, 49, 1289–1299.

HAFFRAY P, LABBE C and MAISSE G (2008) Fish sperm cryopreservation in France: from laboratory studies to application in selective breeding programs. *Cybium*, 32, 127–129.

HAGA Y (1982) On the subzero temperature preservation of fertilized-eggs of rainbow-trout. *Bull Jap Soc Sci Fisher*, 48, 1569–1572.

HAGEDORN M, KLEINHANS F W, FREITAS R, LIU J, HSU E W, WILDT D E and RALL W F (1997a) Water distribution and permeability of zebrafish embryos, *Brachydanio rerio*. *J Exp Zool*, 278, 356–371.

HAGEDORN M, KLEINHANS F W, WILDT D E and RALL W F (1997b) Chill sensitivity and cryoprotectant permeability of dechorionated zebrafish embryos, *Brachydanio rerio*. *Cryobiology*, 34, 251–263.

HAGEDORN M, KLEINHANS FW, ARTEMOV D and PILATUS U (1998) Characterization of a major permeability barrier in the zebrafish embryo. *Biol Reprod*, 59, 1240–1250.

HAGEDORN M, LANCE S L, FONSECA D M, KLEINHANS F W, ARTIMOV D, FLEISCHER R, HOQUE A T, HAMILTON M B and PUKAZHENTHI B S (2002) Altering fish embryos with aquaporin-3: an essential step toward successful cryopreservation. *Biol Reprod*, 67, 961–966.

HAGEDORN M, PETERSON A, MAZUR P and KLEINHANS F W (2004) High ice nucleation temperature of zebrafish embryos: slow-freezing is not an option. *Cryobiology*, 49, 181–189.

HAGEDORN M, PAN R, COX E F, HOLLINGSWORTH L, KRUPP D, LEWIS T D, LEONG J C, MAZUR P, RALL W F, MACFARLANE D R, FAHY G M and KLEINHANS F W (2006) Coral larvae conservation: Physiology and reproduction. *Cryobiology*, 52, 33–42.

HAMMOUD SS, NIX DA, ZHANG T, PURWAR J, CARRELL DT and CAIRNS BR (2009) Distinctive chromatin in human sperm packages genes for embryo development. *Nature*, 460, 473–478.

HARVEY B and ASHWOODSMITH M J (1982) Cryoprotectant penetration and supercooling in the eggs of salmonid fishes. *Cryobiology*, 19, 29–40.

HATTORI M, HASHIMOTO H, BUBENSHCHIKOVA E and WAKAMATSU Y (2011) Nuclear transfer of embryonic cell nuclei to non-enucleated eggs in zebrafish, *Danio rerio*. *Int J Biol Sci*, 7, 460–468.

HAYES M C, RUBIN S P, HENSLEIGH J E, REISENBICHLER R R and WETZEL L A (2005) Performance of juvenile steelhead trout (*Oncorhynchus mykiss*) produced from untreated and cryopreserved milt. *Aquaculture*, 249, 291–302.

HIGAKI S, ETO Y, KAWAKAMI Y, YAMAHA E, KAGAWA N, KUWAYAMA M, NAGANO M, KATAGIRI S and TAKAHASHI Y (2010) Production of fertile zebrafish (*Danio rerio*) possessing germ cells (gametes) originated from primordial germ cells recovered from vitrified embryos. *Reproduction*, 139, 733–740.

HONG N, LI ZD and HONG Y H (2011) Fish stem cell cultures. *Int J Biol Sci*, 7, 392–402.

HORVATH A and URBANYI B (2000) The effect of cryoprotectants on the motility and fertilizing capacity of cryopreserved African catfish *Clarias gariepinus* (Burchell 1822) sperm. *Aquac Res*, 31, 317–324.

HORVATH A, MISKOLUI E, MIHALFFY S, OSZ K, SZABO K and URBANYI B (2007) Cryopreservation of common carp (*Cyprinus carpio*) sperm in 1.2 and 5 ml straws and occurrence of haploids among larvae produced with cryopreserved sperm. *Cryobiology*, 54, 251–257.

HOURCADE J, PEREZ-CRESPO M, FERNANDEZ-GONZALEZ R, PINTADO B and GUTIERREZ-ADAN A (2010) Selection against spermatozoa with fragmented DNA after

© Woodhead Publishing Limited, 2013

postovulatory mating depends on the type of damage. *Reprod Biol Endocrinol*, 8, 9.

HU E, YANG H and TIERSCH T R (2011) High-throughput cryopreservation of spermatozoa of blue catfish (*Ictalurus furcatus*): Establishment of an approach for commercial-scale processing. *Cryobiology*, 62, 74–82.

ILIEVA A, IVANOV A G, KOVACHEV K and RICHTER H P (1992) Cryodamage in ram sperm plasma membranes. Energy transfer and freeze-fracture studies. *Bioelectrochem Bioenerg*, 27, 41–44.

ISAYEVA A, ZHANG T and RAWSON D M (2004) Studies on chilling sensitivity of zebrafish (*Danio rerio*) oocytes. *Cryobiology*, 49, 114–122.

JAMIESON BGM (1991) *Fish Evolution and Systematics, Evidence From Spermatozoa*. Cambridge: Cambridge University Press. 1319 p.

JOLY J S, KRESS C, VANDEPUTTE M, BOURRAT F and CHOURROUT D (1999) Irradiation of fish embryos prior to blastomere transfer boosts the colonisation of their gonads by donor-derived gametes. *Mol Reprod Dev*, 53, 394–397.

KAWAI K, LI YS, SONG M F and KASAI H (2010) DNA methylation by dimethyl sulfoxide and methionine sulfoxide triggered by hydroxyl radical and implications for epigenetic modifications. *Bioorg Med Chem Lett*, 20, 260–265.

KIMMEL C B and LAW R D (1985) Cell lineage of zebrafish blastomeres: II. Formation of the yolk syncytial layer. *Dev Biol*, 108, 86–93.

KOBAYASHI T, YOSHIZAKI G, TAKEUCHI Y and TAKEUCHI T (2004) Isolation of highly pure and viable primordial germ cells from rainbow trout by GFP-dependent flow cytometry. *Mol Reprod Dev*, 67, 91–100.

KOBAYASHI T, TAKEUCHI Y, TAKEUCHI T and YOSHIZAKI G (2007) Generation of viable fish from cryopreserved primordial germ cells. *Mol Reprod Dev*, 74, 207–213.

KOPEIKA J, KOPEIKA E, ZHANG T and RAWSON DM (2003) Studies on the toxicity of dimethyl sulfoxide, ethylene glycol, methanol and glycerol to loach (*Misgurnus fossilis*) sperm and the effect on subsequent embryo development. *Cryo-Lett*, 24, 365–374.

KOPEIKA J, KOPEIKA E, ZHANG T, RAWSON DM and HOLT W V (2004) Effect of DNA repair inhibitor (3-aminobenzamide) on genetic stability of loach (*Misgurnus fossilis*) embryos derived from cryopreserved sperm. *Theriogenology*, 61, 1661–1673.

KOPEIKA JE, ZHANG T, RAWSON DM and ELGAR G (2005) Effect of cryopreservation on mitochondrial DNA of zebrafish (*Danio rerio*) blastomere cells. *Mutat Res*, 570, 49–61.

KROVEL A V and OLSEN L C (2002) Expression of a vasa:EGFP transgene in primordial germ cells of the zebrafish. *Mech Dev*, 116, 141–150.

KUSUDA S, TERANISHI T and KOIDE N (2002) Cryopreservation of chum salmon blastomeres by the straw method. *Cryobiology*, 45, 60–67.

KUSUDA S, TERANISHI T, KOIDE N, NAGAI T, ARAI K and YAMAHA E (2004) Pluripotency of cryopreserved blastomeres of the goldfish. *J Exp Zool*, 301, 131–138.

LABBÉ C and MAISSE G (1996) Influence of rainbow trout thermal acclimation on sperm cryopreservation: relation to change in the lipid composition of the plasma membrane. *Aquaculture*, 145, 281–294.

LABBÉ C and MAISSE G (2001), Characteristics and freezing tolerance of brown trout spermatozoa according to rearing water salinity. *Aquaculture*, 201, 287–299.

LABBÉ C, CROWE L M and CROWE J H (1997) Stability of the lipid component of trout sperm plasma membrane during freeze-thawing. *Cryobiology*, 34, 176–182.

LABBÉ C, LOIR M, KAUSHIK S and MAISSE G (1993) The influence of both rearing temperature and dietary lipid origin on fatty acid composition of spermatozoan polar lipids in rainbow trout (*Oncorhynchus mykiss*). Effect on sperm cryopreservation tolerance, in Kaushik S J and Luquet P (eds), *Fish Nutrition in Practise*, Les Colloques No. 61. Paris: INRA, 49–59.

© Woodhead Publishing Limited, 2013

LABBÉ C, MARTORIATI A, DEVAUX A and MAISSE G (2001) Effect of sperm cryopreservation on sperm DNA stability and progeny development in rainbow trout. *Mol Reprod Dev*, 60, 397–404.

LACERDA S, COSTA G, CAMPOS-JUNIOR P, SEGATELLI T, YAZAWA R, TAKEUCHI Y, MORITA T, YOSHIZAKI G and FRANÇA L (2013) Germ cell transplantation as a potential biotechnological approach to fish reproduction, *Fish Physiol Biochem*, in press (DOI 10.1007/s10695-012-9606-4).

LAHNSTEINER F and PATZNER R A (2008) Sperm morphology and ultrastructure in fish, in ALAVI SMH, COSSON JJ, COWARD K, RAFIEE G (eds), *Fish Spermatology*. Oxford: Alpha Science International Ltd, 2–61.

LANCE S L, PETERSON A S and HAGEDORN M (2004) Developmental expression of aquaporin-3 in zebrafish embryos (*Danio rerio*). *Comp Biochem Physiol*, 138, 251–258.

LE BAIL P Y, DEPINCE A, CHENAIS N, MAHE S, MAISSE G and LABBÉ C (2010) Optimization of somatic cell injection in the perspective of nuclear transfer in goldfish. *BMC Dev Biol*, 10, 64.

LEE K Y, HUANG H, J U B, YANG Z and LIN S (2002) Cloned zebrafish by nuclear transfer from long-term-cultured cells. *Nat Biotechnol*, 20, 795–799.

LEVERONI-CALVI S and MAISSE G (1998) Cryopreservation of rainbow trout (*Oncorhynchus mykiss*) blastomeres: Influence of embryo stage on postthaw survival rate. *Cryobiology*, 36, 255–262.

LEVERONI-CALVI S and MAISSE G (1999) Cryopreservation of carp (*Cyprinus carpio*) blastomeres. *Aquat Living Resour*, 12, 71–74.

LI L, ZHANG L L, LIU Q H, XU X Z, XIAO Z Z, MA D Y, XU S H and XUE Q Z (2009) Extra- and intra-cellular ice formation of red seabream (*Pagrus major*) embryos at different cooling rates. *Cryobiology*, 59, 48–53.

LIN T T, CHAO N H and TUNG H T (1999) Factors affecting survival of cryopreserved oyster (*Crassostrea gigas*) embryos. *Cryobiology*, 39, 192–196.

LIN M H, KUO-KUANG L R, LI S H, LU C H, SUN F J and HWU Y M (2008) Sperm chromatin structure assay parameters are not related to fertilization rates, embryo quality, and pregnancy rates in *in vitro* fertilization and intracytoplasmic sperm injection, but might be related to spontaneous abortion rates. *Fertil Steril*, 90, 352–359.

LIN C, SPIKINGS E, ZHANG T and RAWSON D M (2009a) Effect of chilling and cryopreservation on expression of Pax genes in zebrafish (*Danio rerio*) embryos and blastomeres. *Cryobiology*, 59, 42–47.

LIN C, ZHANG T and RAWSON D M (2009b) Cryopreservation of zebrafish (*Danio rerio*) blastomeres by controlled slow cooling. *CryoLett*, 30, 132–141.

LINDEMAN L C, WINATA C L, AANES H, MATHAVAN S, ALESTROM P and COLLAS P (2010) Chromatin states of developmentally-regulated genes revealed by DNA and histone methylamine patterns in zebrafish embryos. *Int J Dev Biol*, 54, 803–813.

LIU X H, ZHANG T and RAWSON D M (2001) Effect of cooling rate and partial removal of yolk on the chilling injury in zebrafish (*Danio rerio*) embryos. *Theriogenology*, 55, 1719–1731.

LIU J, SUN Y H, WANG Y W, WANG N, PEI D S, WANG Y P, HU W and ZHU Z Y (2008) Identification of differential transcript profiles between mutual crossbred embryos of zebrafish (*Danio rerio*) and Chinese rare minnow (*Gobiocypris rarus*) by cDNA-AFLP. *Theriogenology*, 70, 1525–1535.

LIU TM, LIU L, WEI QW and HONG YH (2011) Sperm nuclear transfer and transgenic production in the fish medaka. *Int J Biol Sci*, 7, 469–475.

LUO D, HU W, CHEN S, XIAO Y, SUN Y and ZHU Z (2009) Identification of differentially expressed genes between cloned and zygote-developing zebrafish (*Danio rerio*) embryos at the dome stage using suppression subtractive hybridization. *Biol Reprod*, 80, 674–684.

© Woodhead Publishing Limited, 2013

LUO DJ, HU W, CHEN S P and ZHU Z Y (2011) Critical developmental stages for the efficiency of somatic cell nuclear transfer in zebrafish. *Int J Biol Sci*, 7, 476–486.

LYNN J W and CLARK W H JR (1983) The fine structure of the mature sperm of the freshwater prawn, *Macrobrachium rosenbergii*. *Biol Bull*, 164, 459–470.

MA CG, FAN LC, GANASSIN R, BOLS N and COLLODI P (2001) Production of zebrafish germ-line chimeras from embryo cell cultures. *Proc Nat Acad Sci USA*, 98, 2461–2466.

MACKAY A B, MHANNI A A, MCGOWAN R A and KRONE P H (2007) Immunological detection of changes in genomic DNA methylation during early zebrafish development. *Genome*, 50, 778–785.

MARTINEZ-PARAMO S, PEREZ-CEREZALES S, ROBLES V, ANEL L and HERRAEZ M P (2008) Incorporation of antifreeze proteins into zebrafish embryos by a non-invasive method. *Cryobiology*, 56, 216–222.

MARTINEZ-PARAMO S, BARBOSA V, PEREZ-CEREZALES S, ROBLES V and HERRAEZ M P (2009a) Cryoprotective effects of antifreeze proteins delivered into zebrafish embryos. *Cryobiology*, 58, 128–133.

MARTINEZ-PARAMO S, PEREZ-CEREZALES S, GOMEZ-ROMANO F, BLANCO G, SANCHEZ JA and HERRAEZ M P (2009b) Cryobanking as tool for conservation of biodiversity: Effect of brown trout sperm cryopreservation on the male genetic potential. *Theriogenology*, 71, 594–604.

MARTINEZ-PASTOR F, CABRITA E, SOARES F, ANEL L and DINIS M T (2008) Multivariate cluster analysis to study motility activation of *Solea senegalensis* spermatozoa: a model for marine teleosts. *Reproduction*, 135, 449–459.

MAUGER PE, LE BAIL P Y and LABBÉ C (2006) Cryobanking of fish somatic cells: optimizations of fin explant culture and fin cell cryopreservation. *Comp Biochem Physiol*, 144, 29–37.

MAZUR P (1984) Freezing of living cells: Mechanisms and implications. *Am J Physiol*, 247, C125–C142.

MHANNI A A and MCGOWAN R A (2004) Global changes in genomic methylation levels during early development of the zebrafish embryo. *Dev Genes Evol*, 214, 412–417.

MORITZ C and LABBÉ C (2008) Cryopreservation of goldfish fins and optimization for field scale cryobanking. *Cryobiology*, 56, 181–188.

MORRIS ID, ILOTT S, DIXON L and BRISON DR (2002) The spectrum of DNA damage in human sperm assessed by single cell gel electrophoresis (Comet assay) and its relationship to fertilization and embryo development. *Hum Reprod*, 17, 990–998.

MOSKOVTSEV S I, WILLIS J, WHITE J and MULLEN J B (2010) Disruption of telomere-telomere interactions associated with DNA damage in human spermatozoa. *Biol Reprod Med*, 56, 407–412.

NIMRAT S, SIRIBOONLAMOM S, ZHANG SC, XU YY and VUTHIPHANDCHAI V (2006) Chilled storage of white shrimp (*Litopenaeus vannamei*) spermatophores. *Aquaculture*, 261, 944–951.

NIWA K, LADYGINA T, KINOSHITA M, OZATO K and WAKAMATSU Y (1999) Transplantation of blastula nuclei to non-enucleated eggs in the medaka, *Oryzias latipes*. *Dev Growth Differ*, 41, 163–172.

OKUTSU T, YANO A, NAGASAWA K, SHIKINA S, KOBAYASHI T, TAKEUCHI Y and YOSHIZAKI G (2006) Manipulation of fish germ cell: visualization, cryopreservation and transplantation. *J Reprod Dev*, 52, 685–693.

PANIAGUA-CHAVEZ C G and TIERSCH T R (2001) Laboratory studies of cryopreservation of sperm and trochophore larvae of the eastern oyster. *Cryobiology*, 43, 211–223.

PANIAGUA-CHAVEZ CG, BUCHANAN JT and TIERSCH TR (1998) Effect of extender solutions and dilution on motility and fertilizing ability of eastern oyster sperm. *J Shellfish Res*, 17, 231–237.

© Woodhead Publishing Limited, 2013

PEI D S, SUN Y H, CHEN S P, WANG Y P, HU W and ZHU Z Y (2007a) Identification of differentially expressed genes from the cross-subfamily cloned embryos derived from zebrafish nuclei and rare minnow enucleated eggs. *Theriogenology*, 68, 1282–1291.

PEI D S, SUN Y H, CHEN S P, WANG Y P, HU W and ZHU Z Y (2007b) Zebrafish GAPDH can be used as a reference gene for expression analysis in cross-subfamily cloned embryos. *Anal Biochem*, 363, 291–293.

PEI D S, SUN Y H, CHEN C H, CHEN S P, WANG Y P, HU W and ZHU Z Y (2008) Identification and characterization of a novel gene differentially expressed in zebrafish cross-subfamily cloned embryos. *BMC Dev Biol*, 8, 29.

PEI D S, SUN Y H and ZHU Z Y (2009) Identification of a novel gene K23 over-expressed in fish cross-subfamily cloned embryos. *Mol Biol Rep*, 36, 1375–1380.

PELEGRI F (2003) Maternal factors in zebrafish development. *Dev Dyn*, 228, 535–554.

PEREZ-CEREZALES S, MARTINEZ-PARAMO S, CABRITA E, MARTINEZ-PASTOR F, DE PAZ P and HERRAEZ M P (2009) Evaluation of oxidative DNA damage promoted by storage in sperm from sex-reversed rainbow trout. *Theriogenology*, 71, 605–613.

PEREZ-CEREZALES S, MARTINEZ-PARAMO S, BEIRAO J and HERRAEZ M P (2010a) Evaluation of DNA damage as a quality marker for rainbow trout sperm cryopreservation and use of LDL as cryoprotectant. *Theriogenology*, 74, 282–289.

PEREZ-CEREZALES S, MARTINEZ-PARAMO S, BEIRAO J and HERRAEZ MP (2010b) Fertilization capacity with rainbow trout DNA-damaged sperm and embryo developmental success. *Reproduction*, 139, 989–997.

PEREZ-CEREZALES S, GUTIERREZ-ADAN A, MARTINEZ-PARAMO S, BEIRAO J and HERRAEZ M P (2011) Altered gene transcription and telomere length in trout embryo and larvae obtained with DNA cryodamaged sperm. *Theriogenology*, 76, 1234–1245.

POLEO G A, DENNISTON R S, REGGIO B C, GODKE R A and TIERSCH T R (2001) Fertilization of eggs of zebrafish, *Danio rerio*, by intracytoplasmic sperm injection. *Biol Reprod*, 65, 961–966.

PSENICKA M, ALAVI SMH, RODINA M, GELA D, NEBESAROVA J and LINHART O (2007) Morphology and ultrastructure of Siberian sturgeon (*Acipenser baerii*) spermatozoa using scanning and transmission electron microscopy. *Biol Cell*, 99, 103–115.

QUINN PJ (1985) A lipid-phase separation model of low-temperature damage to biological membranes. *Cryobiology*, 22, 128–146.

RIESCO MF, MARTÍNEZ-PASTOR F, CHEREGUINI O and ROBLES V (2012a) Evaluation of zebrafish (*Danio rerio*) PGCs viability and DNA damage using different cryopreservation protocols. *Theriogenology*, 77, 122–130.

RIESCO M F and ROBLES V (2012b) Quantification of DNA damage by q-PCR in cryopreserved zebrafish Primordial Germ Cells, *J Appl Ichtyol*, 28, 925–929.

ROBLES V, CABRITA E, REAL M, ALVAREZ R and HERRAEZ M P (2003) Vitrification of turbot embryos: preliminary assays. *Cryobiology*, 47, 30–39.

ROBLES V, CABRITA E, DE PAZ P, CUNADO S, ANEL L and HERRAEZ M P (2004) Effect of a vitrification protocol on the lactate dehydrogenase and glucose-6-phosphate dehydrogenase activities and the hatching rates of Zebrafish (*Danio rerio*) and Turbot (*Scophthalmus maximus*) embryos. *Theriogenology*, 61, 1367–1379.

ROBLES V, CABRITA E, FLETCHER GL, SHEARS MA, KING MJ and HERRAEZ MP (2005) Vitrification assays with embryos from a cold tolerant sub-arctic fish species. *Theriogenology*, 64, 1633–1646.

ROBLES V, CABRITA E, ANEL L and HERRAEZ M P (2006) Microinjection of the antifreeze protein type III (AFPIII) in turbot (*Scophthalmus maximus*) embryos: Toxicity and protein distribution. *Aquaculture*, 261, 1299–1306.

ROBLES V, BARBOSA V, HERRAEZ M P, MARTINEZ-PARAMO S and CANCELA M L (2007a) The antifreeze protein type I (AFP I) increases seabream (*Sparus aurata*) embryos tolerance to low temperatures. *Theriogenology*, 68, 284–289.

© Woodhead Publishing Limited, 2013

ROBLES V, CABRITA E, DE PAZ P and HERRAEZ M P (2007b) Studies on chorion hardening inhibition and dechorionization in turbot embryos. *Aquaculture*, 262, 535–540.

ROBLES V, CABRITA E and HERRAEZ M P (2009) Germplasm cryobanking in zebrafish and other aquarium model species. *Zebrafish*, 6, 281–293.

ROBLES V, MARTI M and IZPISÚA BELMONTE J C (2011) Study of pluripotency markers in zebrafish embryos and transient embryonic stem cell cultures. *Zebrafish*, 8, 57–63.

ROUTRAY P, DASH C, DASH S N, TRIPATHY S, VERMA D K, SWAIN S K, SWAIN P and GURU B C (2010) Cryopreservation of isolated blastomeres and embryonic stem-like cells of Leopard danio, *Brachydanio frankei*. *Aquac Res*, 41, 579–589.

SAITO T, FUJIMOTO T, MAEKAWA S, INOUE K, TANAKA M, ARAI K and YAMAHA E (2006) Visualization of primordial germ cells in vivo using GFP-nos1 3'UTR mRNA. *Int J Dev Biol*, 50, 691–699.

SAITO T, GOTOH-KAZETO R, ARAI K and YAMAHA E (2008) Xenogenesis in teleost fish through generation of germ-line chimeras by single primordial germ cell transplantation. *Biol Reprod*, 78, 159–166.

SAITO T, GOTO-KAZETO R, FUJIMOTO T, KAWAKAMI Y, ARAI K and YAMAHA E (2010) Interspecies transplantation and migration of primordial germ cells in cyprinid fish. *Int J Dev Biol*, 54, 1481–1486.

SAPERAS N, LLORIS D and CHIVA M (1993) Sporadic appearance of histonesn histone-like proteins, and protamines in sperm chromatin of bony fish. *J Exp Zool*, 265, 575–586.

SAPERAS N, AUSIO J, LLORIS D and CHIVA M (1994) On the evolution of protamines in bony fish: alternatives to the 'Retroviral horizontal transmission' hypothesis. *J Mol Evol*, 39, 282–295.

SEKI S, KOUYA T, VALDEZ D M, JIN B, HARA T, SAISA N, KASAI M and EDASHIGE K (2007) The permeability to water and cryoprotectants of immature and mature oocytes in the zebrafish (*Danio rerio*). *Cryobiology*, 54, 121–124.

SIRIPATTARAPRAVAT K, PINMEE B, VENTA P J, CHANG C C and CIBELLI J B (2009) Somatic cell nuclear transfer in zebrafish. *Nat Methods*, 6, 733–735.

SIRIPATTARAPRAVAT K, PINMEE B, CHANG E A, MUNOZ J D, KAWAKAMI K and CIBELLI J B (2010) The influence of donor nucleus source on the outcome of zebrafish somatic cell nuclear transfer. *Int J Dev Biol*, 54, 1679–1683.

SPEYER B E, PIZZEY A R, RANIERI M, JOSHI R, DELHANTY J D and SERHAL P (2010) Fall in implantation rates following ICSI with sperm with high DNA fragmentation. *Hum Reprod*, 25, 1609–1618.

SUQUET M, DREANNO C, PETTON B, NORMANT Y, OMNES M H and BILLARD R (1998) Long-term effects of the cryopreservation of turbot (*Psetta maxima*) spermatozoa. *Aquat Living Resour*, 11, 45–48.

SUQUET M, DREANNO C, FAUVEL C, COSSON J and BILLARD R (2000) Cryopreservation of sperm in marine fish. *Aquac Res*, 31, 231–243.

SUQUET M, LABBE C, BRIZARD R, DONVAL A, LE COZ J R, QUERE C and HAFFRAY P (2010) Changes in motility, ATP content, morphology and fertilisation capacity during the movement phase of tetraploid Pacific oyster (Crassostrea gigas) sperm. *Theriogenology*, 74, 111–117.

SURRALLES J, PUERTO S, RAMIREZ MJ, CREUS A, MARCOS R, MULLENDERS L H and NATARAJAN A T (1998) Links between chromatin structure, DNA repair and chromosome fragility. *Mutat Res*, 404, 39–44.

TAKEUCHI Y, YOSHIZAKI G and TAKEUCHI T (2001) Production of germ-line chimeras in rainbow trout by blastomere transplantation. *Mol Reprod Dev*, 59, 380–389.

TAKEUCHI Y, YOSHIZAKI G and TAKEUCHI T (2003) Generation of live fry from intraperitoneally transplanted primordial germ cells in rainbow trout. *Biol Reprod*, 69, 1142–1149.

© Woodhead Publishing Limited, 2013

TAKEUCHI Y, YOSHIZAKI G and TAKEUCHI T (2004) Biotechnology: surrogate broodstock produces salmonids. *Nature*, 430, 629–630.

TERVIT HR, ADAMS SL, ROBERTS RD, MCGOWAN LA, PUGH PA, SMITH JF and JANKE AR (2005) Successful cryopreservation of Pacific oyster (*Crassostrea gigas*) oocytes. *Cryobiology*, 51, 142–151.

TIERSCH T R and JENKINS J A (2003) Biosecurity and regulatory considerations for transfer of cryopreserved sperm and early life stages of aquatic species, in Lee C-S and Bryen O (eds), *Biosecurity in Aquaculture Production Systems: Exclusion of Pathogens and Other Undesirables*. Baton Rouge, LA: World Aquaculture Society, 171–198.

TIERSCH T R and MAZIK P M (2011) *Cryopreservation in Aquatic Species* (2nd edn). Baton Rouge, LA: World Aquaculture Society.

TIERSCH TR, GOUDIE CA and CARMICHEL GJ (1994) Cryopreservation of channel catfish sperm: Storage in cryoprotectants, fertilization trials, and growth of channel catfish produced with cryopreserved sperm. *Transaction of the American Fisheries Society*, 123, 580–586.

TIERSCH T R, YANG H, JENKINS J A and DONG Q (2007) Sperm cryopreservation in fish and shellfish. *Soc Reprod Fertil Suppl*, 65, 493–508.

TIERSCH T R, YANG H and HU E (2011) Outlook for development of high-throughput cryopreservation for small-bodied biomedical model fishes. *Comp Biochem Physiol C*, 154, 76–81.

TSAI S, RAWSON D M and ZHANG T (2009) Studies on chilling sensitivity of early stage zebrafish (*Danio rerio*) ovarian follicles. *Cryobiology*, 58, 279–286.

TSAI S, RAWSON DM and ZHANG T (2010) Development of in vitro culture method for early stage zebrafish (*Danio rerio*) ovarian follicles for use in cryopreservation studies. *Theriogenology*, 74, 290–303.

TUDGE C (2009) Spermatozoal morphology and its bearing on decapod phylogeny, in Martin JW, Crandall KA and Felder DL (eds), *Decapod Crustacean Phylogenetics*. Boca Ratom, FL CRC Press, 101–119.

USUKI H, HAMAGUCHI M and ISHIOKA H (2002) Effects of developmental stage, seawater concentration and rearing temperature on cryopreservation of Pacific oyster *Crassostrea gigas* larvae. *Fisher Sci*, 68, 757–762.

VALDEZ D M, MIYAMOTO A, HARA T, SEKI S, KASAI M and EDASHIGE K (2005) Water- and cryoprotectant-permeability of mature and immature oocytes in the medaka (*Oryzias latipes*). *Cryobiology*, 50, 93–102.

VALDEZ D M, HARA T, MIYAMOTO A, SEKI K, JIN B, KASAI M and EDASHIGE K (2006) Expression of aquaporin-3 improves the permeability to water and cryoprotectants of immature oocytes in the medaka (*Oryzias latipes*). *Cryobiology*, 53, 160–168.

VAN DER WALT L D, VAN DER BANK F H and STEYN G J (1993) The suitability of using cryopreservation of spermatozoa for the conservation of genetic diversity in African catfish (*Clarias garienipus*). *Comp Biochem Physiol A: Mol Integr Physiol*, 106A, 313–318.

VIVEIROS A T M and GODINHO H P (2009) Sperm quality and cryopreservation of Brazilian freshwater fish species: a review. *Fish Physiol Biochem*, 35, 137–150.

VIVEIROS ATM, NASCIMENTO AF, ORFAO LH and ISAU ZA (2010) Motility and fertility of the subtropical freshwater fish streaked prochilod (*Prochilodus lineatus*) sperm cryopreserved in powdered coconut water. *Theriogenology*, 74, 551–556.

VUTHIPHANDCHAI V, PENGPUN B and NIMRA S (2005) Effects of cryoprotectant toxicity and temperature sensitivity on the embryos of black tiger shrimp (*Penaeus monodon*). *Aquaculture*, 246, 275–284.

VUTHIPHANDCHAI V, NIMRAT S, KOTCHARAT S and BART AN (2007) Development of a cryopreservation protocol for long-term storage of black tiger shrimp (*Penaeus monodon*) spermatophores. *Theriogenology*, 68, 1192–1199.

© Woodhead Publishing Limited, 2013

WAKAMATSU Y (2008) Novel method for the nuclear transfer of adult somatic cells in medaka fish (*Oryzias latipes*): use of diploidized eggs as recipients. *Dev Growth Differ*, 50, 427–436.

WAKAMATSU Y, OZATO K and SASADO T (1994) Establishment of a pluripotent cell line derived from a medaka (*Oryzias latipes*) blastula embryo. *Mol Mar Biol Biotechnol*, 3, 185–191.

WALES R G and WHITE I G (1959) The susceptibility of spermatozoa to temperature shock. *J Endocrinal*, 19, 211–220.

WARD S W (2010) Function of sperm chromatin structural elements in fertilization and development. *Mol Hum Reprod*, 16, 30–36.

WILMUT I, SCHNIEKE AE, MCWHIR J, KIND AJ and CAMPBELL KH (1997) Viable offspring derived from fetal and adult mammalian cells. *Nature*, 385, 810–813.

WOLF K and QUIMBY M C (1969) Fish cell and tissue culture, in Hoar WS and Randall DJ (eds), *Fish Physiology*, Vol. III. New York: Academic Press 1, 253–305.

WU S F, ZHANG H and CAIRNS B R (2011) Genes for embryo development are packaged in blocks of multivalent chromatin in zebrafish sperm. *Genome Res*, 21, 578–589.

YANG H P and TIERSCH T R (2009) Current status of sperm cryopreservation in biomedical research fish models: Zebrafish, medaka, and Xiphophorus. *Comp Biochem Physiol C*, 149, 224–232.

YOSHIZAKI G, TAKEUCHI Y, SAKATANI S and TAKEUCHI T (2000) Germ cell-specific expression of green fluorescent protein in transgenic rainbow trout under control of the rainbow trout vasa-like gene promoter. *Int J Dev Biol*, 44, 323–326.

YOSHIZAKI G, TAGO Y, TAKEUCHI Y, SAWATARI E, KOBAYASHI T and TAKEUCHI T (2005) Green fluorescent protein labeling of primordial germ cells using a nontransgenic method and its application for germ cell transplantation in salmonidae. *Biol Reprod*, 73, 88–93.

YOUNG HM and FLETCHER GL (2008) Antifreeze protein gene expression in winter flounder pre-hatch embryos: Implications for cryopreservation. *Cryobiology*, 57, 84–90.

YOUNG F K, WHEELER P and THORGAARD G (2009) No increase in developmental deformities or fluctuating asymmetry in rainbow trout (*Oncorhynchus mykiss*) produced with cryopreserved sperm. *Aquaculture*, 289, 13–18.

ZHANG T and RAWSON D M (1998) Permeability of dechorionated one-cell and six-somite stage zebrafish (*Brachydanio rerio*) embryos to water and methanol. *Cryobiology*, 37, 13–21.

ZHANG T, RAWSON D M, TOSTI D M and CARNEVALI O (2008) Cathepsin activities and membrane integrity of zebrafish (*Danio rerio*) oocytes after freezing to –196 degrees C using controlled slow cooling. *Cryobiology*, 56, 138–143.

ZILLI L, SCHIAVONE R, ZONNO V, STORELLI C and VILELLA S (2003) Evaluation of DNA damage in Dicentrarchus labrax sperm following cryopreservation. *Cryobiology*, 47, 227–235.

© Woodhead Publishing Limited, 2013

4

Live microalgae as feeds in aquaculture hatcheries

M. R. Brown and S. I. Blackburn, CSIRO Marine and Atmospheric Research, Australia

DOI: 10.1533/9780857097460.1.117

Abstract: Microalgae remain an indispensible hatchery food for many aquaculture species, despite some on-going progress in the development of formulated larval feeds. The majority of the microalgae feed production occurs as living cultures on-site within hatcheries. In this review, general characteristics of microalgae are discussed, including chemical profiles, nutritional qualities and methods of mass culture. Key microalgal species are identified that are used as feeds leading to the hatchery production of bivalves, abalone, shrimp and fish. Global demand for microalgae production as hatchery feeds is likely to increase, requiring not only the expansion of existing infrastructure, but technical innovation to develop new production systems that are more intensive and have greater production capacity.

Key words: aquaculture feeds, hatchery, larvae, microalgae, live feeds, photobioreactor.

4.1 Introduction

Microalgae, the microscopic plants present in oceans and waterways, are exploited as an indispensible food source for the commercial production of many aquaculture species. Within the hatchery environment, they are directly eaten by all growth stages of bivalves (broodstock, larvae, juveniles), post-set abalone, the larval stages of some crustacean species, and the very early developmental stages of some fish species. Microalgae are also used as feed to culture zooplankton (e.g. *Artemia*, rotifers, copepods) that are used as food for larval and juvenile stages of many fish and crustacean species.

This chapter reviews the production and use of microalgae within aquaculture hatcheries. The compositional diversity of microalgae is firstly discussed, with emphasis on the concentrations of nutritionally important

© Woodhead Publishing Limited, 2013

polyunsaturated fatty acids (PUFAs) (Section 4.2). The utilisation of microalgae for the hatchery production of bivalves, abalone, shrimp and fish is discussed, with reference to nutritional requirements in terms of quality (species used and key nutrients) and quantity (diet ration) (Section 4.3). Methods for isolating and establishing microalgae stock cultures, the role of culture collections in the provision of starter cultures are described, and the associated biosecurity issues associated with translocation are discussed (Section 4.4). Production systems for scale-up and mass culture are compared, from traditional polythene bags, to tank and pond systems, and to the newer generation photobioreactors (PBRs) with higher surface area to volume ratios (SA:V) (Section 4.5). An overview is provided on the status of preserved microalgae as feeds (Section 4.6). Finally, a perspective is given on the likely trends and applications of microalgae into the future (Section 4.7).

4.2 Compositional diversity of microalgae used in aquaculture

4.2.1 Proximate composition

The chemical composition is a key factor contributing to the nutritional value of microalgae. Microalgae assessed for aquaculture have a broad range of compositional profiles (Enright *et al.*, 1986a; Brown, 1991; Renaud *et al.*, 1999; Volkman and Brown, 2006; Martínez-Fernández *et al.*, 2006; Tzovenis *et al.*, 2009). Under standard growth conditions (i.e. where nutrients are not limited), microalgae typically contain between 25 and 50 % of dry weight (DW) as protein, 5–40 % as carbohydrate, 10–30 % as lipid and 5–40 % as ash (Whyte, 1987; Renaud *et al.*, 1999; Knuckey *et al.*, 2002; Martínez-Fernández *et al.*, 2006). Because of this variability – even between species from the same genera (Wikfors *et al.*, 1996) – it is difficult to categorize algal classes based on proximate composition alone. However, cryptophytes are usually rich in protein (40–60 % DW) (McCausland *et al.*, 1999; Seixas *et al.*, 2009), whereas diatoms have higher amounts of ash (20–40 %; Brown and Jeffrey, 1995; Renaud *et al.*, 1999). When microalgae reach nutrient limitation (stationary phase), the proximate composition can change significantly. For example, when nitrate is limiting, carbohydrate levels can double at the expense of protein (e.g. in *Chaetoceros calcitrans* and *Isochrysis* sp. T.ISO; Harrison *et al.*, 1990; Brown *et al.*, 1993a), whereas when silicate is limiting, carbohydrate and lipid levels can double (e.g. in the diatom *C. gracilis*; Enright *et al.*, 1986b). Other culture parameters such as light (intensity and photoperiod), temperature and the status of other nutrients can also influence the proximate composition (Thompson *et al.*, 1992; Renaud *et al.*, 2002; Guedes *et al.*, 2010).

© Woodhead Publishing Limited, 2013

4.2.2 Amino acids and carbohydrates

The amino acid composition of the protein fraction is very similar between different microalgae and relatively unaffected by light and nutrient conditions (Brown *et al.*, 1997; Daume *et al.*, 2003). On this basis, it is unlikely that protein quality is a major factor contributing to differences in the nutritional value of microalgae.

Most carbohydrate within microalgae (i.e. 80–95 %) occurs as readily-hydrolysable polysaccharide (Whyte, 1987; Brown, 1991), although some microalgae can have significant amounts of fibrous material such as cellulose (e.g. *Tetraselmis* spp.; Parsons *et al.*, 1961) or chitan (e.g. *Thalassiosira* spp.; Falk *et al.*, 1966). Residual carbohydrate comprises the 'simple' sugars, i.e. mono-, di- and oligo-saccharides. Glucose is typically the main sugar of the polysaccharide fraction comprising between 20 and 90 % of total sugars (Brown, 1991), and its percentage increases in most microalgae, except for diatoms, with culture age (Chu *et al.*, 1982; Whyte, 1987). These observations are consistent with glucans (glucose-rich polysaccharides) being the major food reserve in microalgae (Handa and Yanagi, 1969). From a survey of 16 species, Brown (1991) found galactose (1–20 % of polysaccharide sugars) and mannose (2–46 %) were also common, with arabinose, fucose, rhamnose, ribose and xylose found in varying proportions (0–17 %). Haptophytes contained a higher proportion of arabinose than microalgae from other classes, whereas diatoms had higher mannose and galactose than most other species.

4.2.3 Lipids and fatty acids

The lipid profiles of microalgae used in aquaculture have been comprehensively examined (Volkman *et al.*, 1989; Mourente *et al.*, 1990; Renaud *et al.*, 1999; Volkman and Brown, 2006; Patil *et al.*, 2007). The major lipid classes are triacylglycerols (TAG; an energy store), and the polar lipids (structural role); the latter can be further divided into phospholipids and glycolipids. Lipid class composition is dependent on the microalgal species and growth conditions (Volkman and Brown, 2006). In general, microalgae are rich in phospholipids during the logarithmic phase (70–90 % total lipid), but accumulate TAG during the stationary phase when nitrogen is limiting (20–50 % of total lipid; Dunstan *et al.*, 1993, Brown *et al.*, 1996). Fatty acids form a major part of the chemical structure of all the above-mentioned lipids, chemically bonded through ester linkages. The fatty acid compositions of the different lipid classes differ, but higher percentages of saturated and mono-unsaturated fats occur in TAG compared to the polar lipid fraction, whereas for PUFAs the reverse is true (Leblond and Chapman, 2000; Chen *et al.*, 2008). Most of the fatty acid data of microalgae is expressed as percentages of total fatty acid composition. Profiles can change according to light intensity and photoperiod (Brown *et al.*, 1996; Guedes *et al.*, 2010),

© Woodhead Publishing Limited, 2013

culture media (Lin *et al.*, 2007), temperature (Thompson *et al.*, 1992; Durmaz *et al.*, 2009) and pH (James *et al.*, 1988), but many of the changes are species-specific.

Microalgae display a great variety of fatty acids and their distributions are often distinctly different between classes (Volkman *et al.*, 1989). For aquaculture, most of the focus has been on the percentages of the PUFAs docosahexaenoic acid (DHA; 22:6n-3), eicosapentaenoic acid (EPA; 20:5n-3) and arachidonic acid (ARA; 20:4n-6) in feed microalgae, as these are considered to be important, if not essential, for the development of many aquaculture species (Langdon and Waldock, 1981; Sargent *et al.*, 1999). The profiles of DHA, EPA and ARA of selected microalgae commonly used in aquaculture, grown under standard conditions, are displayed in Fig. 4.1. These data, taken from studies at CSIRO's laboratory, should be taken as indicative only as profiles can vary according to growth conditions. Of the diatoms, the benthic *Navicula jeffreyae* and *Nitzschia closterium* and the planktonic *Thalassiosira pseudonana* are rich in EPA (21, 24 and 19 %, respectively, of total fatty acids); the latter also has good percentages of DHA (4 %). *C. calcitrans*, *C. muelleri* and *N. closterium* contained between 4 and 6 % of ARA. Of the prymnesiophytes, *Isochrysis* sp. (T.ISO) is rich in DHA (8 %) but only has trace amounts of EPA, where as *Pavlova lutheri* has a similar percentage of DHA, but is also rich in EPA (20 %). Other species from the same genera have similar profiles (e.g. *I. galbana*, *P. salina*, *P. pinguis*) (Mourente *et al.*, 1990; Volkman *et al.*, 1991; McCausland *et al.*, 1999). *Rhodomonas salina* has a profile similar to diatom species, with moderate levels of all three PUFAs. The prasinophyte *Tetraselmis suecica* has significant EPA (5 %) and a similar profile to *T. chuii* (Dunstan *et al.*, 1992), although another *Tetraselmis* sp. has been reported containing 14 % EPA (Tzovenis *et al.*, 2009). *Dunaliella tertiolecta* is characteristic of most chlorophytes in lacking DHA, EPA and ARA. *Micromonas pusilla* has a relatively high proportion of DHA (8 %). *Nannochloropsis oculata*, like other *Nannochloropsis* species (e.g. *N. salina*, *N. gaditana*) (Mourente *et al.*, 1990; Volkman *et al.*, 1993) has high proportions of EPA (34 %) and ARA (7 %). The dinoflagellate *Heterocapsa niei* has a very high percentage of DHA (17 %) but lacks EPA and ARA (Mansour *et al.*, 2005). Other PUFAs that may be of nutritional significance and occur in variable concentrations in microalgae include the C18 PUFAs 18:2n-6, 18:3n-3 and 18:4n-3 (Volkman and Brown, 2006).

4.2.4 Sterols, alkenones and pigments

All microalgae contain sterols (typically up to 10 % of total lipid); either as the free sterol (most common) or as a fatty acid ester, glycoside or sulphate (Ghosh *et al.*, 1998; Volkman *et al.*, 1998; Milke *et al.*, 2008). Sterol profiles can be characteristic of a particular class, family, genus or even species and so are often applied to chemotaxonomic and phylogenetic studies. Profiles

© Woodhead Publishing Limited, 2013

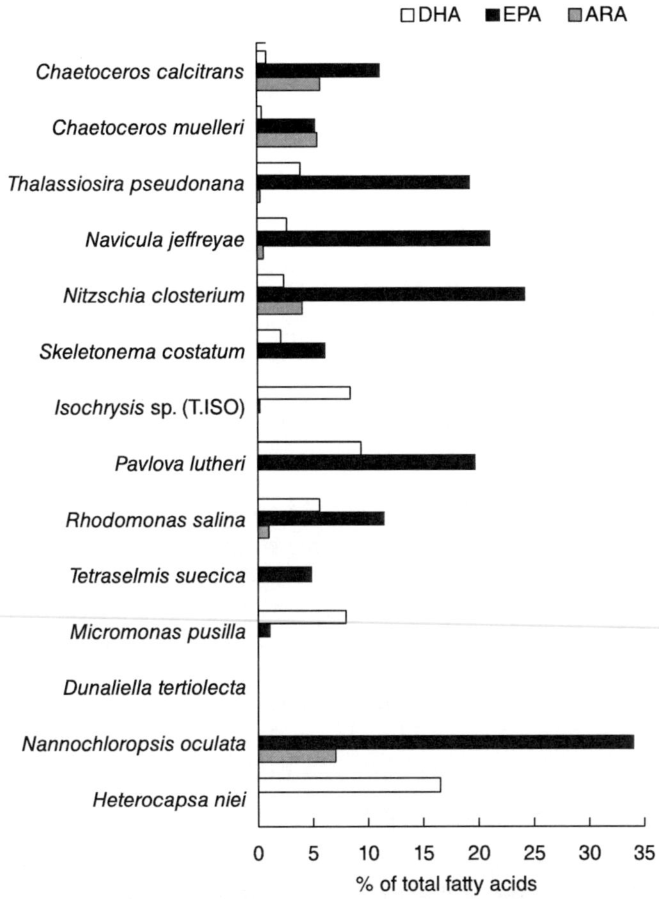

Fig. 4.1 Percentage compositions of the nutritionally-important PUFAs docosahexaenoic acid (DHA; 22:6n-3), eicosapentaenoic acid (EPA; 20:5n-3) and arachidonic acid (ARA; 20:4n-6) of microalgae commonly used in aquaculture hatcheries. Data from various studies undertaken at CSIRO Laboratories (Volkman *et al.*, 1989, 1993; Dunstan *et al.*, 1994, 2005; Mansour *et al.*, 2005).

range from simple (e.g. marine eustigmatophytes usually contain just cholesterol with traces of other sterols) to complex (e.g. dinoflagellates can contain 20 or more 4-desmethyl and 4-methyl sterols) (Volkman, 1986). A few microalgae also contain small amounts of unusual lipids, such as hydrocarbons and alkenones (Volkman and Brown, 2006). Microalgae contain a diverse range of pigments, especially the lipid-soluble carotenoids and chlorophylls, which can often be used in chemotaxonomy (Wright and Jeffrey, 2006). Pigments are also important nutritionally as colouring agents and as antioxidants. Microalgae used in aquaculture usually contain 0.4–1.2 mg g^{-1} β-carotene (Seguineau *et al.*, 1996), whereas under optimal

© Woodhead Publishing Limited, 2013

conditions *D. salina* may contain 10% of its DW as β-carotene (Prieto *et al.*, 2011).

4.2.5 Vitamins and minerals

Of the vitamins, ascorbic acid occurs in highest concentrations and shows the greatest variation (1–16 mg g^{-1} DW) (Brown and Miller, 1992). Other vitamins typically show a two- to four-fold difference between aquaculture species: e.g. β-carotene, 500–1200 μg g^{-1}; niacin, 110–470 μg g^{-1}; α-tocopherol, 70–350 μg g^{-1}; thiamine, 30–110 μg g^{-1}; riboflavin, 25–50 μg g^{-1}; pantothenic acid, 14–38 μg g^{-1}; folates, 7–24 μg g^{-1}; pyridoxine, 4–17 μg g^{-1}; cobalamin 1.7–7.4 μg g^{-1}; biotin, 0.7–1.9 μg g^{-1}; retinol, up to 2.2 μg g^{-1}; and vitamin D, up to 1.0 μg g^{-1} (Seguineau *et al.*, 1996; Brown *et al.*, 1999). Concentrations of α-tocopherol, riboflavin and thiamin can increase by up to two- to three-fold during stationary phase (Brown and Farmer, 1994; Seguineau *et al.*, 1996; Brown *et al.*, 1999).

The main elements within the mineral (or ash) fraction include P, Ca, Na, K, Cl, Fe, Mg and, for diatoms, Si (Fábregas and Herrero, 1986). Species can differ significantly in their composition of these minerals and other trace metals; moreover, their composition is affected by culture environment (Lee and Picard, 1982). For example, *Synechococcus* sp. cells accumulated ions (Co, Zn, Ag, Sn, Hg, Pu and Am) in proportion to the concentration of the same ions in culture media (Fisher, 1985).

4.3 Microalgae applications as aquaculture feeds

4.3.1 Bivalve molluscs

In bivalve hatcheries, microalgae are utilised as feed for conditioning broodstock, and for producing larvae, post-larvae and juveniles. Depending on species, broodstock are fed microalgae for one to three months, at a daily feed rate of 2–6% (DW algae: DW animal meat) (Utting and Millican, 1997; Rico-Villa *et al.*, 2008; Ragg *et al.*, 2010). Typically, mixtures of two or more of species are used (Table 4.1 and Plate I – see colour section between pages 126 and 127), with *C. calcitrans* and *Isochrysis* sp. (T.ISO) being common components.

After spawning and fertilisation, feeding of larvae commences within 24–48 h of fertilisation at the D-larvae stage (Helm and Bourne, 2004). The addition of low concentrations of microalgae (e.g. *C. muelleri* at 1–2 × 10^4 cells ml^{-1}) to tanks prior to feeding may be beneficial, as a source of dissolved organics that may be absorbed and utilised by developing embryos (Helm and Bourne, 2004). Once feeding commences the quality and ration of microalgal feed is critical during the three to four week larval period. Microalgae have been assessed as larval food since the 1930s and much of our knowledge on their quality has been based on their performance as

© Woodhead Publishing Limited, 2013

Table 4.1 Major classes, genera and species of microalgae used as live feeds in aquaculture

Class	Genus/Species	Feeding target and frequency of use
Diatoms	*Chaetoceros* (e.g. *C. calcitrans*, *C. muelleri*)	BL (+++), BP (+++), BB (++), PL (+++), MZ (+)
	Thalassiosira (e.g. *T. pseudonana*, *T. oceanica*)	BL (+), BP (++), BB (++), PL (+)
	Skeletonema (e.g. *S. costatum*, *S. pseudocostatum*)	BL (+++), BP (+++), BB (++), PL (++)
	Phaeodactylum tricornutum	PL (+)
	Nitzschia (e.g. *N. longissima*, *N. closterium*)	AP (+++), PL (+)
	Navicula (e.g. *N. closterium*, *N. jeffreyae*)	AP (+++), PL (+)
	Cocconeis sp.	AP (+++)
Prymnesiophytes	*Isochrysis* (e.g. *I. galbana*, *I.* sp. (T.ISO)	BL (+++), BP (+++), BB (+++), ZM (+++), GW (+)
	Pavlova (e.g. *P. lutheri*, *P. salina*, *P. pinguis*)	BL (++), BP (+++), BB (++), MZ (+++)
Prasinophytes	*Tetraselmis* (e.g. *T. suecica*, *T. chuii*)	BL (+), BP (++), BB (++), PL (+), MZ (+)
	Pyramimonas virginica	BL (+), BP (+)
	Micromonas pusilla	BL (+), MZ (+)
Cryptophytes	*Rhodomonas* (e.g. *R. lens*, *R. salina*)	BL (+), BP (+), MZ (+)
	Proteomonas sulcata	MZ (+)
Chlorophytes	*Dunaliella tertiolecta*	BP (+), BB (+), MZ (+)
	Chlorella (e.g. *C. minutissima*, *C. virginica*)	MZ (+), FZ(+), GW (+)
	Nannochloris (e.g. *N. atomus*, *N. oculata*)	MZ (+), GW (+)
Eustigmatophytes	*Nannochloropsis* (e.g. *N. oculata*, *N. salina*)	MZ (+++), GW (+++)
Dinoflagellates	*Heterocapsa niei*	MZ (+)
Cyanobacteria	*Scenedesmus* (e.g. *S. obliquus*, *S. acuminatus*)	FZ (+)

AP = abalone postlarvae; BL = bivalve larvae; BP = bivalve postlarvae; BB = bivalve broodstock; FZ = freshwater zooplankton; GW = greenwater application; MZ = marine zooplankton; PL = penaeid shrimp larvae.
Source: updated from Tredici *et al.*, 2009.

© Woodhead Publishing Limited, 2013

individual diets (Webb and Chu, 1983). Several species have excellent nutritional value and support growth and development through all larval stages, e.g. *C. calcitrans* forma *pumilum* for Pacific oyster (*Crassostrea gigas*) (Rico-Villa *et al.*, 2006) and *I. galbana* for black-lip oyster (*Pinctada margaritifera*) (Alagarswami *et al.*, 1989). In fact, in an examination of eight independent experiments, Marshall *et al.* (2010) noted that all those producing 88 % larval survival or greater contained *C. calcitrans* either as a mono-specific diet, or in combination with other species. However, mixtures of two or three species (usually a diatom and a haptophyte) generally produce faster growth and better development than individual species (Rico-Villa *et al.*, 2006). Specific diet mixtures have proven successful for the larvae of most of the common bivalves grown in hatcheries (Helm and Bourne, 2004). Microalgae that may be ingested by larvae are in the size range 1.5–15 μm, although smaller larvae, e.g. 90 μm, may only ingest cells < 4–6 μm (Marshall *et al.*, 2010). For these early larval stages, mixtures of *C. calcitrans* with *Isochrysis* sp. (T.ISO) are recommended (Helm and Bourne, 2004; Rico-Villa *et al.*, 2006; Gagné *et al.*, 2010). *C. calcitrans* may be substituted by *T. pseudonana* or *C. muelleri* for larger larvae (e.g. > 55 and > 90 μm respectively), and for larvae > 120 μm, flagellates such as *Tetraselmis* spp. may be included. Some other species belonging to above-mentioned genera, *Pyramimonas virginica*, and other *Skeletonema* and *Nannochoris* spp. have also been successfully used (Webb and Chu, 1983; Brown and Robert, 2002; Martínez-Fernández *et al.*, 2004). *Pavlova* spp. have been successfully fed to various larvae, often in addition to, or substitute for, *Isochrysis* (Nell and O'Connor, 1991; Ponis *et al.*, 2006; Martínez-Fernández and Southgate, 2007), but its value for *C. gigas* larvae has been questioned since it is poorly ingested by this species (Ponis *et al.*, 2003b; Rico-Villa *et al.*, 2006).

Daily feed rations depend on the larval species and growth stage, and culture temperature. Typically, these are based on calculated feed volume, based on a standardised algal size. During larval development at 24 °C, each *C. gigas* larva will consume between 3000–70 000 cells of *Isochrysis* (or equivalent-sized cell), whereas a manila clam (*Ruditapes philippinarum*) larva will consume between 4000 and 15 000 (Helm and Bourne, 2004). Microalgae may be fed in batch mode, where one or several rations are added to the culture tanks each day (ensuring concentrations do not exceed 10^5 cells ml^{-1}; Helm and Bourne, 2004). Alternatively, feed may be added continuously or semi-continuously where part or all of the feed is drip-fed to maintain cell concentration of $3–4 \times 10^4$ cells ml^{-1}; this is a preferred option for high-density larval culture systems (Rico-Villa *et al.*, 2008; Ragg *et al.*, 2010). Recently, a dynamic energy budget growth model was developed to describe the energetic needs of *C. gigas* throughout its larval stages, and concluded that at 27 °C, a food density of 1400 μm μl^{-1} (i.e. equivalent to 3×10^4 cells *Isochrysis* sp. (T.ISO) ml^{-1}) must be maintained to maximise growth and metamorphosis success (Rico-Villa *et al.*, 2010). A further modelling approach combining *C. gigas* larval performance data and microalgal

© Woodhead Publishing Limited, 2013

feed compositional data has taken this a step further to define daily feed rations for protein, carbohydrate, lipid and the essential fatty acids, DHA and EPA (Marshall *et al.*, 2010).

Microalgae feeds that have been successful for larvae have also proven to be successful for post-larvae and juveniles, though for the latter, *Tetraselmis* spp., *C. muelleri*, *T. weissflogii* and *Skeletonema costatum* are common additions (Table 4.1). After metamorphosis, the volume of microalgae required to feed the rapidly growing bivalve biomass increases significantly. Once juveniles reach 2 mm, over 300 l of culture (e.g. based on *Tetraselmis* at 1×10^6 cells ml^{-1}) are required daily to feed each 100 g live weight of spat (45 000 individuals) (Helm and Bourne, 2004). Therefore for economic reasons, the application of mass cultured microalgae as a diet for juveniles is generally restricted to several weeks, and then, often as a supplementary food in conjunction with natural particles in sea water flowing through culture systems (Brown and McCausland, 2000). When juveniles reach 2–3 mm they are usually transferred to local waterways for ongrowing, where they feed on natural food particles, including microalgae (Helm and Bourne, 2004).

Factors that determine the nutritional value of a specific microalga, not only for bivalves but more generally, include: (i) size and shape (ingestibility) (Plate I); (ii) digestibility, which is related to cell wall structure and composition; (iii) biochemical composition (e.g. macro- and micronutrients, enzymes, toxins if present); and (iv) the requirements of the animal feeding on it. Most commonly cultured microalgae appear to be well digested by bivalves, but there are some exceptions, e.g. *Chlorella* spp. which have thick cellulose cell walls (Peirson, 1983). Most of our understanding of the nutritional requirement of bivalves has been based on experiments where animal performance has been directly compared to the composition of the feed microalgae. High levels of carbohydrate produced the best growth for juvenile oysters (*Ostrea edulis*) and larval scallops (*Patinopecten yessoensis*) provided that PUFAs are also present in adequate proportions (Enright *et al.*, 1986b; Whyte *et al.*, 1989). Carbohydrate-rich diets may also be beneficial by increasing the energy content and metamorphosis competence in great scallop (*Pecten maximus*) (Tremblay *et al.*, 2007). In another study utilising mixed diets containing diatoms, the growth of juvenile *C. gigas* was correlated to dietary protein (Knuckey *et al.*, 2002). However, as a cautionary note, a correlation of a component to high growth rates may be a result of that component correlating to a more influential component; also there may be interactions with other nutrients (e.g. vitamins, specific fatty acids) that confound the interpretation (Marshall *et al.*, 2010).

Alterations of the culture environment can change the biochemical composition of microalgae, and hence their nutritional value. For example, *C. gracilis* and *Isochrysis* sp. (T.ISO) when grown under high light produced better growth rates in *C. gigas* larvae than the same microalgae grown under low light (Thompson *et al.*, 1993). The authors attributed this to higher

© Woodhead Publishing Limited, 2013

percentages of the energy-rich saturated fatty acids in high light cultures. It is also feasible to alter cell size of certain microalgae (e.g. *Tetraselmis* spp.) by manipulation of the culture conditions so they can be more readily ingested by larvae (Helm and Bourne, 2004). Most bivalves appear to require the PUFAs, DHA and/or EPA, and the commonly used microalgae typically contain significant proportions of one or both (Fig. 4.1). Specific requirements may depend on species and life stage of the bivalve (Knauer and Southgate, 1999). A requirement was first unequivocally demonstrated in *C. gigas* spat, whereby supplementation of diets deficient in DHA (*T. suecica*) or DHA and EPA (*D. tertiolecta*) with micro-encapsulated oil enriched in DHA improved growth (Langdon and Waldock, 1981). Requirements for EPA have been suggested for the oysters *C. gigas* and *C. rhizophorae* larvae (Helm and Laing, 1987) and *C. virginica* spat (Wikfors *et al.*, 1996), whereas DHA may be important for juvenile *O. edulis*, sea scallop (*Placopecten magellanicus*) and *P. maximus* (Enright *et al.*, 1986b; Coutteau *et al.*, 1996; Gagné *et al.*, 2010) and larvae of *C. gigas*, *P. margaritifera* and *P. maximus* (Thompson *et al.*, 1996; Martínez-Fernández *et al.*, 2006; Gagné *et al.*, 2010).

Quantitative requirements are still poorly understood. For *C. gigas* larvae, growth was not improved by feeding them microalgae containing more than 2% (total fatty acids) of DHA; moreover, the percentage of dietary EPA was negatively correlated to larval growth (Thompson *et al.*, 1993). Clams are less dependent on dietary PUFAs (Helm and Bourne, 2004), and PUFAs are not required by *Tapes philippinarum* larvae (Laing *et al.*, 1990) or grooved carpetshell (*Ruditapes decussatus*) spat (Albentosa *et al.*, 1996). The role of ARA is less well established, although low amounts in the diet produced high mortalities and low growth rates of sea scallop, *P. magellanicus,* larvae (Pernet and Tremblay, 2004) and lower growth rates in larvae and juvenile bay scallop, *Argopecten irradians* (Milke *et al.*, 2006). Ratios of the PUFAs may also have nutritional significance as it is suggested that the ratio (DHA + EPA/ARA) is a key factor in determining larval performance of mussel (*Mytilus galloprovincialis*) (Pettersen *et al.*, 2010). Adequate concentrations of PUFA are also important in broodstock conditioning. Feeding the PUFA-deficient *D. tertiolecta* to *C. gigas* reduced their reproductive output compared to animals fed the same diet supplemented with a PUFA-rich oil (Caers *et al.*, 2002). Higher levels of DHA in conditioning diets fed to *P. margaritifera* were also associated with larger eggs and greater output of ensuing D-larvae (Ehteshami *et al.*, 2011).

Bivalves have a limited capacity to synthesize sterols *de novo* (Holden and Patterson, 1991) so dietary sterols from microalgae could be important in their nutrition. For post-larval *P. magellanicus*, the combination of *C. muelleri* with *Pavlova* spp. produced higher growth rates than other algal diets tested, which was attributed to the comparatively high proportion of cholesterol in sterols from *C. muelleri* (Milke *et al.*, 2008). Of the other

© Woodhead Publishing Limited, 2013

nutrients, microalgae supply a well-balanced composition of essential amino acid (Brown, 1991). Vitamins, which show higher variability between microalgae, potentially could contribute to nutritional value differences (Seguineau *et al.*, 1996), but this has been difficult to prove.

4.3.2 Abalone

In abalone hatcheries (e.g. *Haliotis* spp.), larvae are produced by spawning recently collected wild broodstock, or farmed or wild broodstock that have been maintained in conditioning tanks (Daume, 2006). After approximately one week, the non-feeding, free-swimming larvae actively seek out a settlement substrate. This is usually a collection of vertical plastic plates, colonized by a mix of microalgae (especially the benthic diatoms *Nitzschia* spp. and *Navicula* spp.), in association with natural bacteria (Daume, 2006). For settlement, diatom culture is first added into the larval culture unit with a series of plates (e.g. series of 10–20, 30 × 30 cm sheets arranged 3–5 cm apart) at a final concentration of 4×10^3 cells ml^{-1}, and incubated for 24 h to allow the diatoms to adhere to the plates, prior to the introduction of larvae (Anon., 1990). The final density of diatoms on plates is in the range of 10^4–10^5 cells cm^{-2}, depending on cell size (Kawamura *et al.*, 1998; Daume *et al.*, 2000). Coralline red algae, e.g. *Sporolithon durum* and *Phymatolithon repandum*, are also strong promoters of settlement (Daume *et al.*, 1999; Roberts, 2001), although these are not practical to culture at a commercial scale due to slow growth rates. The green alga *Ulvella lens* is also a strong enhancer of settlement, and is being used for this purpose in both Japanese and Australian farms (Daume, 2006).

After settlement, post-larvae commence grazing on the benthic diatom films and are maintained on diatom colonies up until a 3–13 mm juvenile stage (Suzuki *et al.*, 1987). Diatoms that have been successfully used as feed across a range of *Haliotis* spp. have included *Nitzschia* spp. (e.g. *N. laevis*, *N. incerta*, *N. grossestriata*, *N. closterium*), *Navicula* spp. (*N. jeffreyae*, *N. incerta*), *Amphiprora* spp., *Cylindrotheca* spp., *Cocconeis* spp. and *Achnanthes longpipes* (Kawamura *et al.*, 1998; Daume, 2006; Chen, 2007; Correa-Reyes *et al.*, 2009; de Vicose *et al.*, 2012). During the early post-larval stages, it is critical to maintain the diatom film at a density that meets feed demands. This is usually achieved through passive seeding, i.e. attachment and growth of new cells naturally present in inflowing sea water, and through manipulation of nutrient concentrations and light intensity (Daume, 2006). However, this provides poor control over the species composition and density on the biofilm. Alternatively, monocultures of diatoms could be mass cultured separately and introduced into the abalone tanks, but this approach is not widely used by the industry due to costs (Daume, 2006).

Digestion efficiency of diatoms is an important determinant of their nutritional value for abalone post-larvae; the former being a function of

© Woodhead Publishing Limited, 2013

attachment strength, structural strength and cell dimensions (Kawamura *et al.*, 1998). Early post-larvae may efficiently ingest diatoms up to 20 μm in width (Kawamura *et al.*, 1998). Differences in nutritional value may also be associated with biochemical composition; diatoms with higher percentages of lipid and soluble exo-polysaccharides produced the best survival rate of *H. diversicolor* post-larvae (Chen, 2007). Diatoms generally have high percentages of the PUFA EPA (Dunstan *et al.*, 1994), which is considered essential for normal development of abalone (Mai *et al.*, 1996) – although differences in content between species do not seem to correlate directly with nutritional value (Correa-Reyes *et al.*, 2009).

4.3.3 Shrimp larvae

Microalgae are usually added to penaeid shrimp larval culture tanks during the non-feeding nauplius stage so that they are available as a direct food source immediately upon larval moulting into the zoeal or protozoeal stages (Coutteau, 1996). Commonly used species include *T. chuii*, *C. gracilis*, *S. costatum* and *Isochrysis* sp. (T.ISO), but other representatives from these genera, and *Thalassiosira*, *Phaeodactylum* and *Chlorella* have also been used (Coutteau, 1996; Duerr *et al.*, 1998; Tredici *et al.*, 2009). Mixtures of diatoms and flagellates have produced good results in terms of growth and survival (Gaxiola *et al.*, 2010), although in some instances single species have performed satisfactorily, e.g. *T. suecica* for green tiger prawn (*Penaeus semisulcatus*) and giant tiger prawn (*P. monodon*), and *Isochrysis* sp. for *Litopenaeus vannamei* (D'Souza and Kelly, 2000; Richmond, 2004). Cell concentrations added to culture tanks during the herbivorous zoeal or protozoeal stages are typically 1–1.5 × 10^5 cells ml^{-1} (Duerr *et al.*, 1998; Anon., 2007). This level is maintained, or reduced slightly (e.g. to 0.8 × 10^5 cells ml^{-1}) during the transition to the carnivorous mysis stages and finally through the post-larval stages (Coutteau, 1996; Anon., 2007). *Artemia* are also introduced into the culture tanks during the latter two stages; hence microalgae may be consumed by the *Artemia* and their nutrients transferred indirectly to larvae feeding on the *Artemia*.

Information on the nutritional requirements of crustacean larvae has been based on a combination of the analysis of live prey (including microalgae) supporting growth, and from studies using formulated microdiets (which are accepted by some larval stages) either in addition to, or replacement for, live feeds (Jones *et al.*, 1997). Depending on species and larval stage, penaeid larvae require diets containing 30–56 % protein, 4–24 % lipid and 8–33 % carbohydrate (Jones *et al.*, 1997). Like other crustaceans, penaeids also require dietary cholesterol and phospholipid (Tredici *et al.*, 2009). Studies based on microalgae indicate requirements for the PUFAs, EPA and/or DHA, and ARA, adequate levels of which are provided by the commonly used mixed diets of diatoms and flagellates (Volkman and Brown, 2006).

© Woodhead Publishing Limited, 2013

4.3.4 Live prey

Microalgae play an important role in the culture and enrichment of live zooplankton prey (Table 4.1). They provide protein and energy to sustain the zooplankton, and also PUFAs, vitamins, sterols and pigments important for the next trophic level (Tredici *et al.*, 2009). Commonly used live prey range in size from 100 μm to 2.5 mm (Ritar *et al.*, 2003; Tredici *et al.*, 2009). Rotifers, especially *Brachionus plicatilis* (100–200 μm), are one of the most often reared prey. While rotifers may be reared on yeast and bacteria (Lubzens, 1987), these are generally deficient in PUFAs and vitamin C (Tredici *et al.*, 2009), so prior to on-feeding to larvae they are typically incubated in cultures containing $2–15 \times 10^6$ microalgal cells ml^{-1} (depending on cell size, and rotifer density) for 8–24 h (Lubzens, 1987; Tamaru *et al.*, 1993). *Nannochloropsis* spp. are frequently used for enriching rotifers with EPA, *Isochrysis* sp. (T.ISO) and *Rhodomonas* sp. are used for DHA enrichment, and *P. lutheri* can provide both EPA and DHA (Dhert *et al.*, 2001). *Chlorella* – a microalga naturally deficient in PUFA – has also been used after processing to incorporate PUFA into the cells by incubating with fish oil (Hagiwara *et al.*, 2001). Thraustochytrids, i.e. marine heterotrophs taxonomically related to microalgae and with high PUFA content (Lewis *et al.*, 1999), have also been utilised. For example, *Schizochytrium* sp. is commercially available as a dried-whole cell preparation (e.g. AlgaMac 3000 from Aquafauna Biomarine) and is frequently used to enrich rotifers with DHA (Harel *et al.*, 2002). Commercial oils (e.g. Selco products from INVE Aquaculture) enriched with PUFAs, vitamins and other nutrients may also be applied to enrich rotifers, either in conjunction with or as a replacement for microalgae (Dhert *et al.*, 2001).

Artemia spp. (0.8–2.5 mm) are fed to larvae either freshly hatched or, more commonly, following short-term enrichment (e.g. 3–24 h) with above-mentioned specialty PUFA-rich oils or products like AlgaMac (Harel *et al.*, 2002; Garcia *et al.*, 2008). Microalgae (e.g. *Isochrysis*, *Pavlova*, *Rhodomonas*, *Chaetoceros*, *Tetraselmis*) may also be used (Tredici *et al.*, 2009; Seixas *et al.*, 2009), although commercially this may be less attractive because of the associated higher labour cost. Where larger prey items are needed, *Artemia* may be ongrown to 2.5 mm (eight to nine days) using microalgae either as a complete diet, or for short-term enrichment prior to feeding to larvae (Ritar *et al.*, 2003; Seixas *et al.*, 2009).

Copepods are another live prey reared on microalgae, and they are frequently used when developing larval protocols for 'new' fish species being considered for aquaculture (Conceição *et al.*, 2010). The calanoid copepods, i.e. *Acartia* sp., *Eurytemora affinis*, *Centropages hamatus* and *Gladioferens imparipes*, are the most commonly used (Støttrup, 2003). Despite being more difficult to culture intensively, copepods are considered of superior nutritional value to rotifers and *Artemia*, probably because of their high natural levels of PUFAs, free amino acids and antioxidant pigments (Conceição *et al.*, 2010). Production usually requires feeding a combination

© Woodhead Publishing Limited, 2013

of two or more microalgae, and ensuring a high n-3 PUFA content (Knuckey *et al.*, 2005). Though dinoflagellates are not commonly used in aquaculture, some DHA-rich *Heterocapsa* and *Prorocentrum* spp. have been used within diets for copepods where their inclusion improves egg production (Knuckey *et al.*, 2005; Buttino *et al.*, 2009). Of other microalgae, *Isochrysis* sp. (T.ISO), *T. suecica*, *Rhodomonas* spp have been used successfully. Mixed results have been observed with diatoms, e.g. *Thalassiosira* spp., *Skeletonema* spp. (Dam and Lopes, 2003; Jones and Flynn, 2005) leading to a conclusion that they are toxic to copepods, and the presence of specific aldehydes inhibits egg production (Ianora *et al.*, 2003). An alternate view is that a nutritional deficiency in diatoms may be the cause of low egg production in copepods feeding on them (Jones and Flynn, 2005).

4.3.5 Microalgae for 'green-water' applications

The green water technique, i.e. whereby microalgae are added to intensive culture systems together with prey zooplankton, is a common practice in fish and crustacean larviculture that has been widely documented to improve larval growth, survival and feed ingestion, compared to similar systems without microalgal addition (clear water technique) (Nghia *et al.*, 2007; Rocha *et al.*, 2008). Concentrations added to culture systems range from 0.1 to 2×10^6 microalgal cells ml^{-1} (Carton, 2005; Nghia *et al.*, 2007). Microalgae used in this technique include *Nannochloropsis* spp., *Isochrysis* spp., *Chlorella* spp., *Tetraselmis* spp. and *Chaetoceros* spp. (Nghia *et al.*, 2007; Palmer *et al.*, 2007). This approach may also be applied to extensive outdoor pond production systems whereby the addition of fertiliser stimulates natural phytoplankton growth which, in turn, promotes zooplankton production as feed for the stocked larvae (Palmer *et al.*, 2007). Microalgal blooms developing in extensive systems will depend on the pre-existing natural phytoplankton assemblage, and environmental factors (particularly nutrient ratios), although diatoms and cyanobacteria often dominate (Cremen *et al.*, 2007).

There are a number of plausible reasons explaining the positive effects of microalgae in green water systems, and in any particular example it may be a combination of these acting either independently or synergistically. These include (from Palmer *et al.*, 2007):

- improved direct and indirect (through zooplankton enrichment) nutrition of larvae;
- improved water quality, through removal of nitrogenous substances and increased oxygenation;
- enhanced conditions for larval feeding, from increased turbidity, shading and visual contrast enhancement of prey items;
- chemical stimulants, aiding in feeding behaviour, digestion, immune function;
- antibacterial properties of algae.

© Woodhead Publishing Limited, 2013

4.4 Isolation of microalgae, and the establishment and maintenance of starter cultures

4.4.1 Isolation techniques

The isolation of microalgae from natural samples has a history of over 100 years, with success in the early years limited to fast growing 'weed' species (Jeffrey and Leroi, 1997). Isolation of species and strains suitable for use in aquaculture is not simple because of the requisite small cell size (e.g. 1–10 μm) and the association with other epiphytic species. Therefore it requires relevant knowledge and skills, patience and specific equipment. Several laboratory techniques have traditionally been used for isolating individual cells, including separation using micropipettes (this requires a good microscope, either a high magnification stereomicroscope or an inverted compound microscope), serial dilution cultures, successive plating on agar media and gravity/centrifugation techniques (Hoshaw and Rosowski, 1973; Andersen and Kawachi, 2005). These techniques may be supplemented with use of antibiotics, since it may be important for use in aquaculture to render cultures axenic. Antibiotics may also be applied to eliminate cyanobacteria, whereas inclusion of germanium dioxide inhibits development of diatoms (Andersen and Kawachi, 2005). More recently, automated technologies like flow cytometry with cell sorting have been applied for isolating small cells, which can then be sorted into multi-wall plates for establishing new microalgal cultures. The technique also allows for separation of co-occurring contaminants (e.g. bacteria) (Sieracki *et al.*, 2005).

4.4.2 Establishing and maintaining starter cultures

Once isolations are established, microalgae are grown and maintained in enriched culture media. The choice of medium is very species-dependent and will depend on whether the microalga comes from a fresh, brackish or a marine environment. Media contain macronutrients such as nitrogen and phosphorus as well as micronutrients such as iron, zinc, cobalt, cadmium and molybdenum and vitamins (e.g. B_1 and B_{12} and biotin). Some media, such as f medium and dilutions thereof, e.g. f/2 (Guillard and Ryther, 1962), have been used for over 40 years for diverse microalgae, and are available as concentrates from commercial suppliers. Walne and HESAW media are also commonly used (Laing, 1991; Helm and Bourne, 2004). There are other specialist media that have been developed for some oceanic species and long-term maintenance of fastidious microalgae, e.g. containing low levels of selenium; K medium (Keller *et al.*, 1987) and GSe medium (Blackburn *et al.*, 2001), with the latter also having the addition of an organic soil extract as does GP medium (Loeblich, 1975).

Stock cultures (or master cultures) are typically kept in cultures tubes (e.g. 10–20 ml) or small glass flasks (125–500 ml) fitted with a cotton wool

© Woodhead Publishing Limited, 2013

or packed cellulose plug at the neck (e.g. Steristoppers ®; Heinz Herenz, Germany). These are often maintained in specially allocated constant environment rooms or cabinets, where temperature and lighting can be strictly controlled to match physiological requirements. Stock cultures are used only to provide inocula for starter cultures when required, and are generally sub-cultured every two to six weeks by aseptically transferring a small volume to a sterilised flask with culture medium (e.g. 1–2 ml inoculum to 150 ml; Helm and Bourne, 2004).

To reduce risks, parent and sometimes grandparent stocks are retained as backup. On-going maintenance of stock cultures requires skilled staff and specialised equipment (e.g. autoclaves, UV-laminar flow cabinets, controlled-growth rooms). Many hatcheries choose to limit their in-house maintenance of stock cultures, and instead periodically purchase new stock cultures from dedicated culture collections that can guarantee their vitality and quality (Section 4.4.3).

4.4.3 Preservation of microalgal cultures

For longer term storage (e.g. from two months up to one year) of microalgal starter cultures, many of the traditional aquaculture strains may be maintained on agar slopes or plates enriched with nutrient media (Acreman, 1994; Youn and Hur, 2009). To initiate a liquid culture from these, individual colonies can be removed by scraping the slope or plate and transferred to a tube containing sterile medium. Cell suspensions may also be cryopreserved by the addition of 10–15 % dimethylsulphoxide as a cryoprotectant and then freezing with, and storage in, liquid or vapour-phase nitrogen (Day and Brand, 2005). To restore the culture, samples are thawed and transferred to fresh media. Not all microalgae are amenable to this process, although many of the standard aquaculture strains have been cryopreserved successfully, e.g. *C. muelleri*, *P. lutheri*, *Isochryis galbana*, *N. ovalis* and *Tetraselmis* spp. (Rhodes *et al.*, 2006), with cells of some strains retaining viability after one year or more in storage (Youn and Hur, 2009). Most of the large culture collections now routinely maintain a proportion of their strains in a cryopreserved form for one or more reasons: (i) to lower the maintenance costs and/or enable a larger number of strains to be held; (ii) to provide additional backup and reduce contamination risk; and (iii) to minimise any potential genetic drift (i.e. that may otherwise be associated with serial sub-culturing) (Rhodes *et al.*, 2006).

4.4.4 Role of culture collections

Living microalgae starter cultures are available from selected algal culture collections and other commercial suppliers (Table 4.2). Whilst there is still some exchange of microalgal strains between hatcheries, the industry is much more tuned to the importance of quality control in the microalgae

© Woodhead Publishing Limited, 2013

Table 4.2 Culture collections and suppliers of living microalgae cultures, for commercial or research purposes

Culture collection/supplier and website	Description
The Culture Collection of Algae – UTEX: web.biosci.utexas.edu/utex/	Approx. 3000 strains of marine and freshwater microalgae, including representatives from extreme environments
Culture Collection of Algae and Protozoa – CCAP: www.ccap.ac.uk	Approx. 2500 strains of marine and freshwater microalgae and protozoa
Microbial Culture Collection at the National Institute of Environmental Studies – NIES: mcc.nies.go.jp	Over 2300 strains (700 species) of marine and freshwater microalgae and protozoa
Provasoli-Guillard National Center for Culture of Marine Algae and Microbiota – NCMA: https://ncmg.bigelow.org	Over 2000 strains of marine microalgae; can also supply mass cultures (up to 200 L) as frozen concentrates
Australian National Algae Culture Collection – ANACC: www.csiro.au/ANACC	Over 1000 strains of microalgae (300 species); marine and freshwater; cold-water to tropical strains; majority Australian isolates
Fitoplancton Marino SL (Spain): www.easyalgae.com	Starter cultures of common aquaculture strains
Algae Depot (USA): www.algaedepot.com	Several aquaculture strains, i.e. *Isochrysis galbana*, *Tetraselmis chuii*, *Nannochloropsis oculata*, *Chlorella vulgaris*
Aquatic Eco-System Inc.: www.aquaticeco.com	Agar plates of common strains, i.e. *Chaetoceros gracilis*, *Thalassiosira weissflogii*, *Amphora* sp., *Dunaliella tertiolecta*, *Nannochloropsis salina*, *Pavlova* sp., *Isochrysis* spp. – shelf-life of up to 5 months, can be used to inoculate to liquid media

Note: Information on other algal culture collections is available from the World Federation of Culture Collections (http://www.wfcc.nig.ac.jp/wfcc.html).

used as feedstock. Culture collections provide strains that are usually well characterised for growth with physiological, biochemical and genetic information, highlighting their value as repositories of global microalgal biodiversity and providers of a reliable and on-going supply for industry (Blackburn *et al.*, 2005). From starter cultures (provided either as liquid culture or agar plates or slopes), the hatcheries often operate simple to quite sophisticated facilities to scale up microalgae production for use as live feeds (see Section 4.5). The CSIRO Microalgae Supply Service is one good example of an algal culture collection supporting the aquaculture industry for more than 25 years, a period of expansion and maturation of the Australian industry. Under the auspices of the Australian National

© Woodhead Publishing Limited, 2013

Algae Culture Collection (ANACC) the CSIRO Microalgae Supply Service started with research and development support from the Australian government. For many years now, operating in a cost recovery mode, the Supply Service has underpinned the Australian aquaculture live feed requirements and supplies over 60 countries globally. While many of the strains supplied are 'traditional' aquaculture strains developed in North America (see below), research associated with ANACC sought to develop Australian native strains as feeds for particular aquaculture animal species. Examples include strains of *S. pseudocostatum* used for shrimp larval culture, *N. closterium* for juvenile abalone and *T. pseudonana* for bivalves (Cathy Johnston, ANACC, pers. comm.).

More detailed information on methods for small-scale culture maintenance, preservation and on the different media can be found in Andersen (2005) as well as on the websites of some algal culture collections (Table 4.2). Additional information including an historical context is given by Stein (1973).

4.4.5 Biosecurity issues

Examples of the establishment of non-indigenous microalgae in waterways due to anthropogenic activity go back to the early 20th century (Hallegraeff and Gollasch, 2006). Although some introductions are benign, in other instances there may be profound impacts on the local ecology. One example is the introduction (purportedly through ships' ballast) of the toxin-containing dinoflagellate *Gymnodinium catenatum* to Australian waters. During bloom events the toxins of this microalga are bioaccumulated by shellfish which, upon human consumption, can lead to paralytic shellfish poisoning (Hallegraeff and Bolch, 1992). This example and others have contributed to a growing awareness of the potential risks associated with translocation of microalgae. Hence, nowadays there are much tighter controls by regulatory authorities on the importation of both living microalgae, and microalgae-derived products (e.g. concentrates) for use as hatchery feeds. Conditions differ between countries, but importation can require an import permit, which is issued conditional to the importer satisfying the authority of low quarantine risk through an assessment process. Therefore, hatchery operators need to be aware of the customs approval process before importation of non-endemic strains.

Apart from circumventing potential biosecurity issues, there may be other reasons for hatcheries to choose endemic microalgae strains. Firstly, local strains – especially when grown outdoors in extensive culture – may be better adapted to prevailing environmental conditions (e.g. temperature extremes, water chemistry) and therefore display better growth characteristics (Jeffrey *et al.*, 1992). Secondly, in some cases microalgae that are endemic to the same region of the bivalve species under cultivation have produced better growth and survival than conventional non-endemic

© Woodhead Publishing Limited, 2013

hatchery microalgae species (Marshall *et al.*, 2010 and references quoted therein).

Nonetheless, microalgal strains established as live feed species and proven in the North American hatcheries in the early days of aquaculture hatchery development and subsequently exported around the world (e.g. *Isochrysis* sp. (T.ISO), *C. calcitrans*, *P. lutheri*, *T. pseudonana*) remain the foundation of many operations globally today. In Australia, because the original importation of these strains predated the Convention of Biodiversity (CBD; ratified Rio de Janiero 1993) and they have had no history of adverse effects, they were given retrospective approval for use by the Australian Quarantine Inspection Service (AQIS). Hence, they remain primary live feed strains within Australia, with the CSIRO Microalgae Supply Service having AQIS approval to supply throughout Australia.

4.5 Mass scale production of microalgae

4.5.1 Scaling-up production and general principles

The intensive production of larval and juvenile animals within hatcheries requires a correspondingly large-scale production of microalgae as live feeds. The general process of scale-up involves the successive transfer of actively growing, dense cultures (e.g. mid-late logarithmic phase) to larger volume systems containing fertilised sea water, at inoculation ratios ranging from 1:5 to 1:100 (Coutteau, 1996). Exact protocols and culture end-points will vary between hatcheries, but a typical scale-up could be from 20–200 ml starter cultures, to 0.5–2 l flasks, to 10–20 l carboys, to 100–1000 l polythene bags or cylinders, to tubs, tanks, ponds or raceways of between 1000 and 100 000 l (Coutteau, 1996; Duerr *et al.*, 1998) (Plate II). Cultures may be operated in batch mode – where the entire volume is harvested once cells have achieved sufficient density. Alternatively, in continuous mode the culture is continuously harvested and media replenished at a rate to sustain a specific growth rate; in semi-continuous mode a proportion of the volume, e.g. 20–50 %, is removed every two to four days and replaced by fresh media (Laing, 1991).

In addition to having good nutritional qualities, a key criterion for a microalgal species to be successfully exploited in aquaculture is its amenability to mass culturing, including rapid growth rates, high cell densities and stability in culture. The key variables influencing growth and production rates of microalgae grown phototrophically are light intensity and photoperiod, CO_2 concentration, aeration, temperature, salinity, pH and nutrient concentrations and ratios (Jeffrey *et al.*, 1992; Renaud *et al.*, 2002; Richmond, 2004). As individual microalgae used in hatcheries have been originally isolated from varying geographic and environmental locations, not surprisingly their optimum conditions for growth can vary significantly. For practical reasons, hatcheries growing multiple species tend to use

© Woodhead Publishing Limited, 2013

standardised conditions, even though these may not be optimal for one or more species. One exception is for diatoms, where routinely the culture medium is supplemented with additional silicate (Helm and Bourne, 2004).

Culture unit design also has a profound influence on algal production (Table 4.3), including through interactions with above-mentioned culture variables. For example, systems such as PBRs (see below) with narrower light paths (and higher SA:V) can reach significantly higher cell densities before light may become limiting, compared to conventional bag or tank systems. However, for a given culture volume, the latter will be more thermally stable to ambient temperature fluctuations. Also, culture unit design will influence the level of turbulence associated with mixing, which in turn will affect in-culture light climate, nutrient transfer and level of hydrodynamic stress, all of which may impact on productivity (Grobbelaar, 2010). Theoretical aspects and biological principles of mass algal cultivation have been reviewed in detail by Richmond (2004) and Grobbelaar (2009).

4.5.2 Mass production systems for microalgae

Cultured microalgal monocultures have been widely used in hatcheries and other applications since midway through the 20th century. Systems were generally restricted to a scale of < 50 l (Bruce *et al.*, 1940; Tamiya *et al.*, 1953; Loosanoff and Davis, 1963), whereas mass culturing generally relied on natural blooming of mixed microalgae in fertilised sea water, e.g. in 100 000 l tanks (Loosanoff and Davis, 1963). The evolution and history of mass cultivation has been covered in detail in many publications (Soeder, 1986; Preisig and Andersen, 2005). Broadly speaking, systems can be separated into open systems or PBRs, the latter being defined as essentially closed systems that do not allow, or limit, direct exchange of gases and contaminants between the culture and the atmosphere, and where > 90 % of the light has to pass through the reactor's transparent wall to reach the culture (Tredici, 2004).

The most popular systems used within hatcheries include polythene bags, vertical cylinders (both forms of PBRs) and open tubs and tanks. Sleeves or bags, made from heavy gauge, transparent 'layflat' polyethylene, have been used for several decades (Laing, 1991). Smaller volume bags (e.g. < 100 l) may be hung on a suitable framework, intermediate volumes, e.g. 100–500 l, are typically arranged vertically supported by wire-mesh frames, 1000 l bags are placed horizontally supported by a frame (Tredici, 2004). These systems are usually operated inside buildings (with artificial illumination, with or without some supplementation by natural light), but larger volume units can also be placed outside (Zmora and Richmond, 2004). Polyethylene can be purchased from manufacturers as a roll that can be cut to size, or in pre-cut, sealed sleeves. One disadvantage is that the inner walls are prone to biofouling, but the bags are inexpensive and easy to replace (Tredici, 2004). The bags are operated predominantly sealed (apart from

© Woodhead Publishing Limited, 2013

Table 4.3 Examples of algal production systems used for the most popular microalgae genera in aquaculture, i.e. *Nannochloropsis* spp., *Isochrysis* spp. and *Chaetoceros* spp., and associated productivities

Genus/species	Culture system/ volume	Mode	Light/T conditions	Productivity (mg DW L^{-1} day^{-1})	Reference
Nannochloropsis spp.					
Nannochloropsis sp.	Annular PBR; 140 L	SC	Nat. + Art./25 °C	340	Zittelli *et al.*, 2003
Nannochloropsis sp.	Modular flat panels; 110 L	SC	Nat. + Art./25 °C	1450	Zittelli *et al.*, 2000
			Nat./ambient to ≤ 30 °C	360	Rodolfi *et al.*, 2009
Nannochloropsis sp.	Tubular PBR; 200 L	C	Art./23–34 °C	95	James and Al-Khars, 1990
Nannochloropsis sp.	Helical-tubular PBR; 200 L	C	Nat.+ Art./20–45 °C (avge: 36 °C)	1100–3030	Briassoulis *et al.*, 2010
Nannochloropsis sp.	Glass flat-plate PBR; 440 L	SC	Nat./14–27 °C	270	Zhang *et al.*, 2001
N. oculata	Polythene bags; 85 L	SC	Art./22 °C	17 ± 6	Brown *et al.*, 1993b
Isochrysis spp.					
Isochrysis sp. (T.ISO)	Horizontal tubular; 400 L	B	Nat./13–27 °C	20–169 (avge: 75)	Van Bergeijk *et al.*, 2010
Isochrysis sp. (T.ISO)	Acrylic columns; 50 L	B	Art./20 °C	76 ± 12	Van Bergeijk *et al.*, 2010
Isochrysis sp. (T.ISO)	Tanks; 9300 L	C	Art./25–29 °C	69	Rusch and Christensen, 2003
Isochrysis sp. (T.ISO)	Polythene bags; 85 L	SC	Nat./22 °C	13 ± 2	Brown *et al.*, 1993b
I. galbana	Outdoor ponds; 300 L	SC	Nat./18–34 °C	246 ± 20	Boussiba *et al.*, 1988
I. galbana	Column PBR; 25 L	SC	Nat./15–27 °C	1600	Qiang and Richmond, 1994
I. galbana	Tubular PBR; 10 L	SC	Nat./20 °C	28–32	Molina Grima *et al.*, 1994
Chaetoceros spp.					
C. muelleri	Glass flat-plate PBR; 300 L	SC	Nat./26 °C (daytime)	150 ± 20	Zhang and Richmond, 2003
C. muelleri	Open tubs; 200 L	B	Nat./19–35 °C	89	Becerra-Dórame *et al.*, 2010
C. calcitrans	Airlift column PBR; 17 L	C	Art./30 °C	850 (30 h duration)	Krichnavaruk *et al.*, 2007

PBR = photobioreactor; SC = semi-continuous; C = continuous; B = batch; Nat. = natural light; Art. = artificial light.

© Woodhead Publishing Limited, 2013

tubing lines for aeration and culture removal/addition); this also facilitates operation under very clean conditions. Systems of bags may be operated independently, or interconnected for continuous or semi-continuous addition of culture media and inoculum as well as harvesting, providing a high degree of automation. These systems are often constructed in-house by hatchery personnel, but there are commercial units available that have proven very popular, e.g. the SeaCAPS system of 40 × 500 l bags (Seasalter Shellfish (Whitstable) Ltd, UK). Transparent vertical cylinders, made from fibreglass, acrylic or Perspex, are also widely used both indoors and outdoors. These may range from 30–50 cm in diameter, and 1.5–2.5 m high (100–500 l) (Laing, 1991; Zmora and Richmond, 2004).

For larger production volumes of microalgae, shallow open tanks or ponds are used. These are usually constructed of fibreglass or concrete; ponds may also be constructed by simply lining excavated ground with a plastic membrane. Volumes range from 2000–10^9 l and culture depths from 0.8–1.5 m (Borowitzka, 1999; Zmora and Richmond, 2004), although for aquaculture hatchery applications volumes are more typically 20 000 l or less (Donaldson, 1991). The larger-scale, extensive open pond systems are low cost and relatively easy to build and operate. There are three major types: (i) inclined systems where mixing is achieved through pumping and gravity flow; (ii) circular ponds with agitation provided by a rotating arm; and (iii) raceway ponds constructed as an endless loop, in which the culture is circulated by paddle wheels (Tredici, 2004). One disadvantage is that cultures are readily contaminated (bacteria, other microalgae, protozoa) and maintenance of mono-cultures is more difficult (Duerr *et al.*, 1998; Helm and Bourne, 2004). The approach here by hatcheries is often to develop and maintain natural, mixed blooms (Helm and Bourne, 2004). A few examples of species that have been grown successfully as monocultures at this scale include *Chlorella* spp., *Scenedesmus* spp., *Arthrospira* spp., *D. salina*, *Pleurochrysis carterae* and *Nannochloropsis* spp. (Borowitzka, 1999; Tredici, 2004; Moheimani and Borowitzka, 2006). For most of these, the production of biomass has been applied to non-hatchery uses, such as high-value compounds like pigments (e.g. astaxanthin production from *Haematococcus* spp. by Cyanotech Corporation in Hawaii and Earthrise in California), human food products and biodiesel investigation. An exception is *Nannochloropsis* spp., where extensively grown pond cultures have been applied for rotifer enrichment as food for fish larvae (Zmora and Richmond, 2004).

The above-mentioned systems are generally considered low technology, but remain those of choice by most hatcheries because of their ease of use and maintenance, general reliability and low capital cost (or, in the case of bags, low replacement cost). However, these systems are typically characterised by low cell densities and productivities, i.e. for aquaculture strains often below 100 mg DW microalgae $l^{-1}d^{-1}$, compared to small volume cultures, or other PBR systems with significantly higher SA:V (Table 4.3 and Tredici *et al.*, 2009). Examples of PBRs include tubular (serpentine,

© Woodhead Publishing Limited, 2013

manifold, helical) and flat (alveolar panels, glass plate) systems (Tredici *et al.*, 2009). The common principle of their design is to reduce the light path (e.g. to 2–4 cm), thereby increasing the amount of light available per cell and enhancing productivity (Borowitzka, 1999). Other potential advantages are a greater ability to control culture conditions, less likelihood of contamination and increased amenability to continuous culture – leading to a more consistent product. However, such PBRs typically have higher construction costs compared to the traditional systems used in hatcheries and are more expensive and complex to operate, hence their application is usually more for high value bioproducts (Borowitzka, 1999). Also, not all species grow well in these systems – fragile species may be damaged by the hydrodynamics, and some species may be too 'sticky' and rapidly foul the inner culture wall (Borowitzka, 1999). Nonetheless, the growth of a number of the widely used aquaculture live feed species has been assessed in different types of PBRs. Summary details of the growth in different culture systems of three of the most popular genera used in aquaculture, i.e. *Nannochloropsis* spp., *Isochrysis* spp., and *Chaetoceros* spp., are given in Table 4.3. While these data are not intended to provide definitive data on the maximum productivity possible in each system (e.g. up to 3 g $l^{-1}d^{-1}$ for *Nannochloropsis*), they do allow for a general assessment and a source of further information.

Continuing research and development on algal mass cultivation systems is happening as part of the global interest in microalgae as a renewable source of biofuels. As production cost is also a major impediment to PBR adoption for biofuel applications, there are efforts to make lower cost systems (Rodolfi *et al.*, 2009). As some of the species of particular interest for biofuels are also those that are commonly used in aquaculture, e.g. *Nannochloropsis* spp. and *Isochrysis* spp. (the latter producing long-chain ketones; Volkman and Brown, 2006), there may be benefits to the aquaculture industry in the future availability of new, low cost technologies. Other benefits from the biofuel R&D activities are from the recognised economic imperative for co-products alongside the production of biofuels. Benemann (2010) considers that algal feeds, along with algal biofuels, require a great deal of R&D, and that biofuels from microalgae need to be part of a pathway that has initial focus on microalgae grown as animal feeds including aquaculture and speciality feeds for commodities production.

More comprehensive reviews on the different PBR and open culture systems are given in other publications (Chaumont, 1993; Borowitzka, 1999; Tredici, 2004; Zmora and Richmond, 2004).

4.5.3 Heterotrophic production

Heterotrophic growth in fermentor systems utilises organic carbon sources such as glucose and other sugars. Fermentor systems are based on well-established technologies that are used for growing micro-organisms such as

© Woodhead Publishing Limited, 2013

bacteria, yeast and fungi, mostly for industrial and medical applications (Stanbury *et al.*, 1995). The major advantages of this technology are greater process control and a cost reduction due to no lighting requirements, and associated higher cell densities (75 g l^{-1} DW) in large-scale systems (up to 500 000 l) (Running *et al.*, 1994). Production cost may be less than $US 5 kg^{-1} DW (Harel *et al.*, 2002; Richmond, 2004) compared to the phototrophic production cost of microalgae species commonly used in aquaculture, i.e. from $US 50–1000 kg^{-1} DW (Coutteau, 1996; Richmond, 2004; Tredici *et al.*, 2009). In general, the application of these organisms as hatchery foods is limited, and restricted to partial microalgae replacements with low to moderate nutritional value (Robert and Trintignac, 1997). However, thraustochytrids which are characterised by high PUFA yields (Lewis *et al.*, 1999) are an exception. The genus *Schizochytrium* is commercially produced (e.g. Aquafauna Biomarine Ltd) and distributed to hatcheries in a stable, spray-dried form that is widely used as a DHA enrichment for live feeds (Harel *et al.*, 2002).

Some microalgae can also be produced heterotrophically, but the number of species amenable to this form of growth remains relatively small, e.g. *Chlorella*, *Crypthecodinium*, *Navicula*, *Dunaliella*, *Tetraselmis* and *Haematococcus* species (Laing *et al.*, 1990; Lee and Lee, 2004; Harel and Place, 2004). Moreover, many of these strains have inferior food value as direct larval feeds; the possible exception is *Tetraselmis* but, when grown heterotrophically, its nutritional value is reduced by a reduction in its PUFA content (Day and Tsavalos, 1996). The utilisation of heterotrophically grown *Chlorella* – produced in centralised facilities (e.g. Pacific Trading Co Ltd, Japan) and shipped to hatcheries as concentrates – has increased significantly during the last decade, being used to rear live rotifers as larval feeds in fish hatcheries (Harel and Place, 2004).

In addition to applications as direct and indirect feeds mentioned above, the heterotrophic production of microalgae can potentially provide a cost-effective source of nutrient components, e.g. marine protein, PUFA or astaxanthin, that can be incorporated into formulated feeds for larval stages and beyond. Harel and Place (2004) provide more detail on the broad applications of heterotrophic production of marine algae for aquaculture.

4.6 Preserved microalgae as feed

The on-site production of microalgae for immediate use as live feeds remains a necessity for most hatcheries because of costs, logistics and the high nutritional value they can offer as feed during the critical larval-feeding stages (Tredici *et al.*, 2009). Nevertheless, there have been some significant developments in off-the-shelf alternatives during the last few decades that have seen their use increase as partial or, in some circumstances, total replacement for freshly cultured microalgae (Knauer and

© Woodhead Publishing Limited, 2013

Southgate, 1999; Nevejan *et al.*, 2007; Tredici *et al.*, 2009). Concentrated, preserved microalgae appear the most promising, with many studies demonstrating that they can satisfactorily substitute for fresh microalgae for greenwater applications (Rocha *et al.*, 2008) and in the culture of rotifers, shrimp larvae and larval and juvenile molluscs (Lubzens *et al.*, 1995; D'Souza *et al.*, 2000, Heasman *et al.*, 2000; Brown and Robert, 2002). Moreover, the growth and survival of certain larvae may even be improved when they are fed algal concentrates compared to equivalent algal cultures; this improvement may be associated with a reduction of the microbial and chemical load in the feed algae following the concentration process (Heasman *et al.*, 2000, and references cited therein).

Concentrates are developed by applying specific harvesting techniques to concentrate the microalgae culture, followed by a processing or storage protocol to preserve the preparation. Various harvesting techniques have been tested (Tredici *et al.*, 2009), but centrifugation and chemical flocculation are the most suitable for preparing concentrates for aquaculture feeds, based on cell density and harvesting efficiency. Centrifugation, when coupled with appropriate storage techniques, is effective for producing concentrates of *C. calcitrans*, *Tetraselmis* spp. and *N. oculata* with shelf-lives beyond one to two months (Heasman *et al.*, 2000; Tredici *et al.*, 2009), although the process does damage fragile flagellates such as *P. lutheri*, *Isochrysis* sp. (T.ISO) and larger diatoms, e.g. *C. muelleri*, reducing their shelf-life (Heasman *et al.*, 2000). Flocculation using pH adjustment and a non-ionic polymer is very efficient at harvesting diatoms, *P. lutheri* and *T. suecica*, with shelf-lives between one to three weeks (Ponis *et al.*, 2003a; Knuckey *et al.*, 2006). Various processes can extend the shelf-life of concentrates (e.g. addition of antioxidants, cryoprotectants, food acids, air-bubbling, storage at low positive temperature), although the results are very species-specific (Heasman *et al.*, 2000; Ponis *et al.*, 2008).

Commercially, concentrates are produced under two different scenarios. Some hatcheries with the infrastructure for microalgae mass culture prepare their own concentrates on-site, allowing them to limit their algal production to less busy periods of the year, better manage their microalgae requirements and also reduce costs associated with over-production (Donaldson, 1991). Alternatively, there are several companies with specialised facilities and economies of scale for mass algal culture that are producing concentrates for on-selling to hatcheries (Table 4.4). Of these, Reed Mariculture Inc. USA has been selling concentrates since 1998, and is currently the largest operation, distributing to over 70 countries. Their concentrates of *Isochrysis* sp., *Pavlova* sp., *T. weissfloggii*, *Tetraselmis* sp. and *Nannochloropsis* sp. have specified shelf-lives of at least four months when stored at – 1–4 °C.

Although dried preparation of microalgae can offer a prolonged shelf-life, the drying process can cause significant structural damage to the cell, especially to the strains commonly used in aquaculture. Dried *T. chuii* has

© Woodhead Publishing Limited, 2013

Table 4.4 Commercial providers of concentrated and dried preparations of microalgae

Provider and website	Product description	Target species
Reed Mariculture Inc. (USA): www.reed.mariculture.com	Microalgal concentrates including *Nannochloropsis oculata*, *Tetraselmis* sp., *Isochrysis* sp., *Pavlova* sp. and *Thalassiosira weissflogii*	Bivalve larvae and post-larvae; live feed enrichments for fish, shrimp larviculture
Innovative Aquaculture Products Ltd (Canada): www. innovativeaqua.com	Microalgal concentrates including *Nannochloropsis oculata*, *Isochrysis* sp. (T.ISO) and *I. galbana*, *Pavlova lutheri, Phaeodactylum tricornutum* and *Chaetoceros* sp.	Bivalve post-larvae; live feed enrichments for fish, shrimp larviculture
Necton SA (Portugal): www.phytobloom.com	Microalgae concentrates of *Nannochloropsis* sp.	Live feed enrichments for fish, shrimp larviculture; greenwater
Fitoplancton Marino SL (Spain): www. easyalgae.com	Microalgal concentrates of *Nannochloropsis* sp., *Tetraselmis* sp., *Chaetoceros* sp., *Isochrysis* sp. *Skeletonema* sp. *Phaeodactylum* sp. and *Rhodomonas* sp; freeze-dried *Nannochloropsis* sp. and *Tetraselmis* sp.	Bivalve larvae and post-larvae; live feed enrichments for fish, shrimp larviculture
Pacific Trading Co Ltd (Japan): www.pacific-trading.co.jp	Concentrates of freshwater *Chlorella* sp., with or without enrichment with EPA and DHA	Live feed enrichments for fish, shrimp larviculture; greenwater
Aquafauna Bio-Marine Inc.: www.aquafauna.com	Dried preparations of DHA-rich *Schizochytrium* sp. and astaxanthin-rich *Haematococcus* sp.	Specific nutrient enrichment of live feeds, for fish, shrimp larviculture

© Woodhead Publishing Limited, 2013

a moderate nutritional value for clam larvae (Laing *et al.*, 1990). Dried *Chlorella* powder has been incorporated into formulated feed and successfully fed to promote the growth of abalone *Haliotis diversicolor supertexta* post-larvae (Chao *et al.*, 2010). Heterotrophically grown organisms, i.e. the dinoflagellate *Crypthecodinium* and the thraustochytrid *Schizochytrium*, commercially available as spray-dried preparations (Aquafauna Biomarine Inc., USA), have a two-year shelf-life and a much higher content of DHA (over 40 % of fatty acids) than phototrophically grown microalgae (Harel *et al.*, 2002). They are now frequently used within experimental and commercial fish hatcheries as alternatives to microalgae and/or oil emulsions for enriching rotifers and *Artemia* (Faulk and Holt, 2005; Conceição *et al.*, 2010); however, they are only of moderate food value for juvenile molluscs (Brown and McCausland, 2000).

Other alternatives to living microalgae that have been investigated as hatchery feeds have included formulated microparticulate, microbound and encapsulated diets, and yeast and bacteria. These have proven useful in experimental investigations and in some instances as partial microalgal replacements for specific life stages, although their nutritional value is inferior to cultured microalgae (Robert and Trintignac, 1997; Knauer and Southgate, 1999).

4.7 Future trends

Aquaculture production is forecast to increase significantly in coming years (Bostock, 2011). Correspondingly, given the indispensible role that microalgae have as hatchery feedstock for many target species, this will place an increased demand on its global production. While construction of new hatchery infrastructure (e.g. based on traditional systems of tanks and bags) will undoubtedly be required, the adoption by industry of more efficient and intensive mass algal culture technology could also make a major impact. Significant progress has been made in the development of PBR systems in the last decade or so, but barriers to their widespread use are that not all aquaculture strains grow well in them (*Nannochloropsis* spp. are exceptions) together with their relatively high unit cost and operational cost. More research is required with respect to the former, to couple understanding of microalgal biology with system design and operation to improve production efficiency. An example of this approach comes from a recent study, whereby high density, continuous cultures of *Isochrysis* sp. (T.ISO) were assessed in a multi-factorial design that enabled a precise definition of irradiance, temperature, pH, nitrogen and phosphorus required for maximum productivity (Marchetti *et al.*, 2012). For PBRs, operational cost could also be reduced significantly by the adoption of innovative light sources (e.g. LED and optical fibre excited by solar energy) to replace conventional light systems (Chen *et al.*, 2011).

© Woodhead Publishing Limited, 2013

While at present open microalgal growth systems used by hatcheries are typically quite small open tubs and tanks, the interest in microalgae for biofuels and for high value products in other industry sectors, may result in a shift from the many hundreds of small systems worldwide to larger, more centralised options. Benemann (2010) considers open ponds the only economically practical growth systems for microalgae for all applications including aquaculture feeds. Technological developments in culture systems may flow from these associated efforts that have direct benefit for the future production of aquaculture strains. Centralised production, whether it be from extensive opens systems or PBRs, is likely to be a continuing trend. Several companies are already adopting this approach, through their production and distribution of microalgal concentrates for hatcheries.

Lower cost production of microalgae can also be achieved by adoption of fermentation technology. However, very few of the strains successfully used in aquaculture are amenable to such growth; examples of those that can be grown in fermenters include *Chlorella* spp., *Schizochytrium* spp. and *Tetraselmis* spp. Genetic modification of existing strains, e.g. to confer heterotrophic growth capability, or other improvements such as growth rate, stability in culture or chemical composition, are one potential avenue of investigation, although this approach would need to be met with public acceptance.

For the last several decades, most of the hatcheries' needs have been fulfilled by utilising a dozen or so key microalgae species. Nevertheless, there is likely to be an on-going investigation of new microalgae species or strains. These may prove useful to meet specific nutritional needs for new aquacultured species. Also, microalgal production or nutritional value may be improved from species or strain selection. Predicted climate change, both in terms of shifts in absolute mean temperatures, and temperature extremes, may impact on the productivity of outdoor culture systems in specific locations. In such circumstance, alternate species/strains may be needed, e.g. substitution of *P. lutheri* with *P. salina* – which has similar nutritional profile but a higher temperature tolerance (Jeffrey *et al.*, 1992). Finally, locally-isolated strains, better adapted to local ecology may prove to be better alternatives (Marshall *et al.*, 2010), particularly if the trend of increasing quarantine and biosecurity controls places some restriction on the future use and regional distribution of the standard currently used aquaculture species.

4.8 Sources of further information and advice

Books

- Andersen RA (2005) *Algal Culturing Techniques.* New York: Academic Press.

© Woodhead Publishing Limited, 2013

- Richmond A (2004) *Handbook of Microalgal Culture: Biotechnology and Applied Phycology*. Oxford: Blackwell Science.
- Støttrup J and McEvoy LA (2003) *Live Feeds in Marine Aquaculture*. Oxford: Blackwell Science.
- Subba Rao DV (2006). *Algal Culture, Analogues of Blooms and Applications.* Plymouth: Science Publishers.

Websites: algal culture and feeds and suppliers

- Refer to Tables 4.2 and 4.4.

Websites: culture systems and media

- Algamerica LLC (Pennsylvania, USA): www.algamerica.com
- AusAqua Pty Ltd: www.algaboost.com
- Fotosintetica & Microiologica Srl (Florence, Italy): www.femonline.it
- Seasalter SeaCAPS (Kent, UK): www.seasaltershellfish.co.uk

Websites: associations, taxonomic information, discussion groups

- AlgaeBase: www.algaebase.org/
- Algae-L: www.seaweed.ie/algae-l/
- International Phycological Association: www.intphycsoc.org/

4.9 References

ACREMAN J (1994) Algae and cyanobacteria – isolation, culture and long-term maintenance. *Journal of Industrial Microbiology*, 13, 193–194.

ALAGARSWAMI K, DHARMARAJ S, CHELLAM A and VELAYUDHAN T S (1989) Larval and juvenile rearing of black-lip pearl oyster, *Pinctada margaritifera* (Linnaeus). *Aquaculture*, 76, 43–56.

ALBENTOSA M, LABARTA U, FERNANDEZREIRIZ M J and PEREZCAMACHO A (1996) Fatty acid composition of *Ruditapes decussatus* spat fed on different microalgae diets. *Comparative Biochemistry and Physiology A–Physiology,* 113, 113–119.

ANDERSEN R A 2005. *Algal Culturing Techniques*, New York: Academic Press.

ANDERSEN R A and KAWACHI M (2005) Traditional microalgae isolation techniques, in Andersen R A (ed), *Algal Culturing Techniques.* New York: Academic Press, 83–100.

ANON (1990) *Training Manual on Artificial Breeding of Abalone* (Haliotis discus hannai) *in Korea DPR*, Project Reports, No. 7. Rome: FAO.

ANON (2007) *Improving* Penaeus monodon *hatchery practices. Manual based on experience in India*, FAO Fisheries Technical Paper. No. 446. Rome: FAO.

BECERRA-DÓRAME M, ANTONIO LÓPEZ-ELIAS J and MARTÍNEZ-CÓRDOVA L R (2010) An alternative outdoor production system for the microalgae *Chaetoceros muelleri* and *Dunaliella* sp. during winter and spring in Northwest Mexico. *Aquacultural Engineering*, 43, 24–28.

© Woodhead Publishing Limited, 2013

BENEMANN J (2010) Microalgae biofuels: A brief introduction, available at: www.adelaide.edu.au/biogas/renewable/biofuels_introduction.pdf (accessed September 2012).

BLACKBURN S I, BOLCH C J S, HASKARD K A and HALLEGRAEFF G M (2001) Reproductive compatibility among four global populations of the toxic dinoflagellate *Gymnodinium catenatum* (Dinophyceae). *Phycologia*, 40, 78–87.

BLACKBURN S I, FRAMPTON D M F, JAMESON I D, BROWN M R, MANSOUR M P, NEGRI A P, NICHOLS P D, PARKER N S, ROBERT S S, BOLCH C J and VOLKMAN J K (2005). The CSIRO collection of living microalgae: An Australian perspective on microalgal biodiversity and applications, in Kasai F, Kaya K and Watanabe M M (eds), *Algal Culture Collections and the Environment.* Hadano-shi Tokai University Press, 29–63.

BOROWITZKA M A (1999) Commercial production of microalgae: ponds, tanks, tubes and fermenters. *Journal of Biotechnology*, 70, 313–321.

BOSTOCK J (2011) The application of science and technology development in shaping current and future aquaculture production systems. *Journal of Agricultural Science*, 149, 133–141.

BOUSSIBA S, SANDBANK E, SHELEF G, COHEN Z, VONSHAK A, BENAMOTZ A, ARAD S and RICHMOND A (1988) Outdoor cultivation of the marine microalga *Isochrysis galbana* in open reactors. *Aquaculture*, 72, 247–253.

BRIASSOULIS D, PANAGAKIS P, CHIONIDIS M, TZENOS D, LALOS A, TSINOS C, BERBERIDIS K and JACOBSEN A (2010) An experimental helical-tubular photobioreactor for continuous production of *Nannochloropsis* sp. *Bioresource Technology*, 101, 6768–6777.

BROWN M and ROBERT R (2002) Preparation and assessment of microalgal concentrates as feeds for larval and juvenile Pacific oyster (*Crassostrea gigas*). *Aquaculture*, 207, 289–309.

BROWN M R (1991) The amino acid and sugar composition of 16 species of microalgae used in mariculture. *Journal of Experimental Marine Biology and Ecology*, 145, 79–99.

BROWN M R and FARMER C L (1994) Riboflavin content of 6 species of microalgae used in mariculture. *Journal of Applied Phycology*, 6, 61–65.

BROWN M R and JEFFREY S W (1995) The amino acid and gross composition of marine diatoms potentially useful for mariculture. *Journal of Applied Phycology*, 7, 521–527.

BROWN M R and MCCAUSLAND M A (2000) Increasing the growth of juvenile Pacific oysters *Crassostrea gigas* by supplementary feeding with microalgal and dried diets. *Aquaculture Research*, 31, 671–682.

BROWN M R and MILLER K A (1992) The ascorbic-acid content of 11 species of microalgae used in mariculture. *Journal of Applied Phycology*, 4, 205–215.

BROWN M R, DUNSTAN G A, JEFFREY S W, VOLKMAN J K, BARRETT S M and LEROI J M (1993a) The influence of irradiance on the biochemical composition of the prymnesiophyte *Isochrysis* sp. (clone T-ISO). *Journal of Phycology*, 29, 601–612.

BROWN M R, GARLAND C D, JEFFREY S W, JAMESON I D and LEROI J M (1993b) The gross and amino acid compositions of batch and semi-continuous cultures of *Isochrysis* sp. (clone T.ISO), *Pavlova lutheri* and *Nannochloropsis oculata. Journal of Applied Phycology*, 5, 285–296.

BROWN M R, DUNSTAN G A, NORWOOD S J and MILLER K A (1996) Effects of harvest stage and light on the biochemical composition of the diatom *Thalassiosira pseudonana. Journal of Phycology*, 32, 64–73.

BROWN M R, JEFFREY S W, VOLKMAN J K and DUNSTAN G A (1997) Nutritional properties of microalgae for mariculture. *Aquaculture*, 151, 315–331.

BROWN M R, MULAR M, MILLER I, FARMER C and TRENERRY C (1999) The vitamin content of microalgae used in aquaculture. *Journal of Applied Phycology*, 11, 247–255.

© Woodhead Publishing Limited, 2013

BRUCE J R, KNIGHT M and PARKE M W (1940) The rearing of oyster larvae on an algal diet. *Journal of the Marine Biological Association Plymouth*, 24, 337–374.

BUTTINO I, IANORA A, BUONO S, VITELLO V, SANSONE G and MIRALTO A (2009) Are mono-algal diets inferior to plurialgal diets to maximize cultivation of the calanoid copepod *Temora stylifera*? *Marine Biology*, 156, 1171–1182.

CAERS M, UTTING S D, COUTTEAU P, MILLICAN P F and SORGELOOS P (2002) Impact of the supplementation of a docosahexaenoic acid-rich emulsion on the reproductive output of oyster broodstock, *Crassostrea gigas*. *Marine Biology*, 140, 1157–1166.

CARTON A G (2005) The impact of light intensity and algal-induced turbidity on first-feeding *Seriola lalandi* larvae. *Aquaculture Research*, 36, 1588–1594.

CHAO W, HUANG C and SHEEN S (2010) Development of formulated diet for post-larval abalone, *Haliotis diversicolor supertexta*. *Aquaculture*, 307, 89–94.

CHAUMONT D (1993) Biotechnology of algal biomass production: A review of systems for outdoor mass-culture. *Journal of Applied Phycology*, 5, 593–604.

CHEN Y (2007) Immobilization of twelve benthic diatom species for long-term storage and as feed for post-larval abalone *Haliotis diversicolor*. *Aquaculture*, 263, 97–106.

CHEN G Q, JIANG Y and CHEN F (2008) Variation of lipid class composition in *Nitzschia laevis* as a response to growth temperature change. *Food Chemistry*, 109, 88–94.

CHEN C-Y, YEH K-L, AISYAH R, LEE D-J and CHANG J-S (2011) Cultivation, photobioreactor design and harvesting of microalgae for biodiesel production: A critical review. *Bioresource Technology*, 102, 71–81.

CHU F L E, DUPUY J L and WEBB K L (1982) Polysaccharide composition of 5 algal species used as food for larvae of the American oyster, *Crassostrea virginica*. *Aquaculture*, 29, 241–252.

CONCEIÇÃO L E C, YUFERA M, MAKRIDIS P, MORAIS S and DINIS M T (2010) Live feeds for early stages of fish rearing. *Aquaculture Research*, 41, 613–640.

CORREA-REYES J G, SANCHEZ-SAAVEDRA M D P, VIANA M T, FLORES-ACEVEDO N and VASQUEZ-PELAEZ C (2009) Effect of eight benthic diatoms as feed on the growth of red abalone (*Haliotis rufescens*) postlarvae. *Journal of Applied Phycology*, 21, 387–393.

COUTTEAU P (1996) Microalgae, in Lavens P and Sorgeloos P (eds), *Manual on the production and use of live food for aquaculture*, FAO Fisheries Technical Paper, No. 361. Rome: 7–48.

COUTTEAU P, CASTELL J D, ACKMAN R G and SORGELOOS P (1996) The use of lipid emulsions as carriers for essential fatty acids in bivalves: A test case with juvenile *Placopecten magellanicus*. *Journal of Shellfish Research*, 15, 259–264.

CREMEN M C M, MARTINEZ-GOSS M R, CORRE V L and AZANZA R V (2007) Phytoplankton bloom in commercial shrimp ponds using green-water technology. *Journal of Applied Phycology*, 19, 615–624.

D'SOUZA F M L and KELLY G J (2000) Effects of a diet of a nitrogen-limited alga (*Tetraselmis suecica*) on growth, survival and biochemical composition of tiger prawn (*Penaeus semisulcatus*) larvae. *Aquaculture*, 181, 311–329.

D'SOUZA F M L, LECOSSOIS D, HEASMAN M P, DIEMAR J A, JACKSON C J and PENDREY R C (2000) Evaluation of centrifuged microalgae concentrates as diets for *Penaeus monodon* Fabricius larvae. *Aquaculture Research*, 31, 661–670.

DAM H G and LOPES R M (2003) Omnivory in the calanoid copepod *Temora longicornis*: Feeding, egg production and egg, hatching rates. *Journal of Experimental Marine Biology and Ecology*, 292, 119–137.

DAUME S (2006) The roles of bacteria and micro and macro algae in abalone aquaculture: A review. *Journal of Shellfish Research*, 25, 151–157.

DAUME S, BRAND-GARDNER S and WOELKERLING W J (1999) Preferential settlement of abalone larvae: Diatom films vs. non-geniculate coralline red algae. *Aquaculture*, 174, 243–254.

© Woodhead Publishing Limited, 2013

DAUME S, KRSINICH A, FARRELL S and GERVIS M (2000) Settlement, early growth and survival of *Haliotis rubra* in response to different algal species. *Journal of Applied Phycology*, 12, 479–488.

DAUME S, LONG B M and CROUCH P (2003) Changes in amino acid content of an algal feed species (*Navicula* sp.) and their effect on growth and survival of juvenile abalone (*Haliotis rubra*). *Algal biotechnology: A sea of opportunities. 1st Congress of the International Society for Applied Phycology and 9th International Conference on Applied Algology*, 26–30 May, 2002, Almeria, 201–207.

DAY J G and BRAND J J (2005) Cryopreservation methods for maintaining microalgal cultures, in Andersen R A (ed), *Algal Culturing Techniques.* New York: Academic Press, 165–188.

DAY J G and TSAVALOS A J (1996) An investigation of the heterotrophic culture of the green alga *Tetraselmis*. *Journal of Applied Phycology*, 8, 73–77.

DE VICOSE G C, VIERA M P, HUCHETTE S and IZQUIERDO M S (2012) Improving nursery performances of *Haliotis tuberculata coccinea*: Nutritional value of four species of benthic diatoms and green macroalgae germlings. *Aquaculture*, 334, 124–131.

DHERT P, ROMBAUT G, SUANTIKA G and SORGELOOS P (2001) Advancement of rotifer culture and manipulation techniques in Europe. *Aquaculture*, 200, 129–146.

DONALDSON J (1991) Commercial production of microalgae at Coast Oyster Company, in Fulks W and Main K L (eds), *Rotifer and microalgae culture systems.* Honolulu, HI: The Oceanic Institute, 229–239.

DUERR E O, MOLNAR A and SATO V (1998) Cultured microalgae as aquaculture feeds. *Journal of Marine Biotechnology*, 6, 65–70.

DUNSTAN G A, VOLKMAN J K, JEFFREY S W and BARRETT S M (1992) Biochemical composition of microalgae from the green algal classes Chlorophyceae and Prasinophyceae. 2. Lipid classes and fatty-acids. *Journal of Experimental Marine Biology and Ecology*, 161, 115–134.

DUNSTAN G A, VOLKMAN J K, BARRETT S M and GARLAND C D (1993) Changes in the lipid composition and maximization of the polyunsaturated fatty acid content of three microalgae grown in mass culture. *Journal of Applied Phycology,* 5, 71–83.

DUNSTAN G A, VOLKMAN J K, BARRETT S M, LEROI J M and JEFFREY S W (1994) Essential polyunsaturated fatty-acids from 14 species of diatom (Bacillariophyceae). *Phytochemistry*, 35, 155–161.

DUNSTAN G A, BROWN M R and VOLKMAN J K (2005) Cryptophyceae and Rhodophyceae; chemotaxonomy, phylogeny, and application. *Phytochemistry*, 66, 2557–2570.

DURMAZ Y, DONATO M, MONTEIRO M, GOUVEIA L, NUNES M L, PEREIRA T G, GOKPINAR S and BANDARRA N M (2009) Effect of temperature on alpha-tocopherol, fatty acid profile, and pigments of *Diacronema vlkianum* (Haptophyceae). *Aquaculture International*, 17, 391–399.

EHTESHAMI F, CHRISTIANUS A, RAMESHI H, HARMIN S A and SAAD C R (2011) The effects of dietary supplements of polyunsaturated fatty acid on pearl oyster, *Pinctada margaritifera* L., gonad composition and reproductive output. *Aquaculture Research*, 42, 613–622.

ENRIGHT C T, NEWKIRK G F, CRAIGIE J S and CASTELL J D (1986a) Evaluation of phytoplankton as diets for juvenile *Ostrea edulis* L. *Journal of Experimental Marine Biology and Ecology*, 96, 1–13.

ENRIGHT C T, NEWKIRK G F, CRAIGIE J S and CASTELL J D (1986b) Growth of juvenile *Ostrea edulis* L. fed *Chaetoceros gracilis* Schütt of varied chemical composition. *Journal of Experimental Marine Biology and Ecology*, 96, 15–26.

FÁBREGAS J and HERRERO C (1986) Marine microalgae as a potential source of minerals in fish diets. *Aquaculture*, 51, 237–243.

© Woodhead Publishing Limited, 2013

FALK M, SMITH D G, MCLACHLA J and MCINNES A G (1966) Studies on chitan (β-(1- 4)-linked 2-acetamido-2-deoxy-D-glucan) fibers of diatom *Thalassiosira fluviatilis* Hustedt. *Canadian Journal of Chemistry*, 44, 2269–2281.

FAULK C K and HOLT G J (2005) Advances in rearing cobia *Rachycentron canadum* larvae in recirculating aquaculture systems: Live prey enrichment and greenwater culture. *Aquaculture*, 249, 231–243.

FISHER N S (1985) Accumulation of metals by marine picoplankton. *Marine Biology*, 87, 137–142.

GAGNÉ R, TREMBLAY R, PERNET F, MINER P, SAMAIN J F and OLIVIER F (2010) Lipid requirements of the scallop *Pecten maximus* (L.) during larval and post-larval development in relation to addition of *Rhodomonas sauna* in diet. *Aquaculture*, 309, 212–221.

GARCIA A S, PARRISH C C and BROWN J A (2008) Growth and lipid composition of Atlantic cod (*Gadus morhua*) larvae in response to differently enriched *Artemia franciscana*. *Fish Physiology and Biochemistry*, 34, 77–94.

GAXIOLA G, GALLARDO P, SIMOES N and CUZON G (2010) A red shrimp, *Farfantepenaeus brasiliensis* (Latreille, 1817), larvae feeding regime based on live food. *Journal of the World Aquaculture Society*, 41, 402–410.

GHOSH P, PATTERSON G W and WIKFORS G H (1998) Sterols of some marine Prymnesiophyceae. *Journal of Phycology*, 34, 511–514.

GROBBELAAR J U (2009) Factors governing algal growth in photobioreactors: The 'open' versus 'closed' debate. *Journal of Applied Phycology*, 21, 489–492.

GROBBELAAR J U (2010) Microalgal biomass production: challenges and realities. *Photosynthesis Research*, 106, 135–144.

GUEDES A C, MEIRELES L A, AMARO H M and MALCATA F X (2010) Changes in lipid class and fatty acid composition of cultures of *Pavlova lutheri*, in response to light intensity. *Journal of the American Oil Chemists Society,* 87, 791–801.

GUILLARD R R and RYTHER J H (1962) Studies of marine planktonic diatoms. 1. *Cyclotella nana* Hustedt, and *Detonula confervacea* (Cleve) Gran. *Canadian Journal of Microbiology*, 8, 229–239.

HAGIWARA A, GALLARDO W G, ASSAVAAREE M, KOTANI T and ARAUJO A B D (2001) Live food production in Japan: Recent progress and future aspects. *Aquaculture*, 200, 111–127.

HALLEGRAEFF G M and BOLCH C J (1992) Transport of diatom and dinoflagellate resting spores in ships ballast water – implications for plankton biogeography and aquaculture. *Journal of Plankton Research*, 14, 1067–1084.

HALLEGRAEFF G and GOLLASCH S (2006) Anthropogenic introductions of microalgae. *Ecology of Harmful Algae*, 189, 379–390.

HANDA N and YANAGI K (1969) Studies on water-extractable carbohydrates of particulate matter from northwest Pacific Ocean. *Marine Biology*, 4, 197–207.

HAREL M, KOVEN W, LEIN I, BAR Y, BEHRENS P, STUBBLEFIELD J, ZOHAR Y and PLACE A R (2002) Advanced DHA, EPA and ArA enrichment materials for marine aquaculture using single cell heterotrophs. *Aquaculture*, 213, 347–362.

HAREL M and PLACE A R (2004) Heterotrophic production of marine algae for aquaculture, in Richmond A (ed), *Handbook of Microalgal Culture: Biotechnology and Applied Phycology.* Oxford: Blackwell Science, 513–524.

HARRISON P J, THOMPSON P A and CALDERWOOD G S (1990) Effects of nutrient and light limitation on the biochemical composition of phytoplankton. *Journal of Applied Phycology*, 2, 45–56.

HEASMAN M, DIEMAR J, O'CONNOR W, SUSHAMES T and FOULKES L (2000) Development of extended shelf-life microalgae concentrate diets harvested by centrifugation for bivalve molluscs – a summary. *Aquaculture Research*, 31, 637–659.

HELM M M and BOURNE N (2004) Hatchery culture of bivalves. A practical manual, in Lovatelli A (ed), FAO Fisheries Technical Paper 471. Rome: FAO.

© Woodhead Publishing Limited, 2013

HELM M M and LAING I (1987) Preliminary observations on the nutritional value of Tahiti-*Isochrysis* to bivalve larvae. *Aquaculture*, 62, 281–288.

HOLDEN M J and PATTERSON G W (1991) Absence of sterol biosynthesis in oyster tissue-culture. *Lipids*, 26, 81–82.

HOSHAW R W and ROSOWSKI J R (1973) Methods for microscopic algae, in Stein JR (ed), *Handbook of Phycological Methods: Culture Methods and Growth Measurements.* London: Cambridge University Press, 53–68.

IANORA A, POULET S A and MIRALTO A (2003) The effects of diatoms on copepod reproduction: A review. *Phycologia*, 42, 351–363.

JAMES C M and AL-KHARS A M (1990) An intensive continuous culture system using tubular photobioreactors for producing microalgae. *Aquaculture*, 87, 381–393.

JAMES C M, AL-KHARS A M and CHORBANI P (1988) pH dependent growth of *Chlorella* in a continuous culture system. *Journal of the World Aquacuculture Society*, 19, 27–35.

JEFFREY S W and LEROI J M (1997) Simple procedures for growing SCOR reference microalgal cultures, in Jeffrey S W, Mantoura R F C and Wright S W (eds), *Phytoplankton Pigments in Oceanography.* Paris: UNESCO, 181–206.

JEFFREY S W, LEROI J M and BROWN M R (1992) Characteristics of microalgal species for Australian mariculture, in Allan G L and Dall W (eds), *Proceeding of the National Aquaculture Workshops*, Salamander Bay: NSW Fisheries Brackish Water Fish Culture Research Station, 164–173.

JONES R H and FLYNN K J (2005) Nutritional status and diet composition affect the value of diatoms as copepod prey. *Science*, 307, 1457–1459.

JONES D A, YULE A B and HOLLAND D L (1997) Larval nutrition, in D'Abramo L R, Conklin D E and Akiyama D M (eds), *Crustacean Nutrition.* Baton Rouge, LA: The World Aquaculture Society, 353–389.

KAWAMURA T, ROBERTS R D and NICHOLSON C M (1998) Factors affecting the food value of diatom strains for post-larval abalone *Haliotis iris*. *Aquaculture*, 160, 81–88.

KELLER M D, SELVIN R C, CLAUS W and GUILLARD R R L (1987) Media for the culture of oceanic ultraphytoplankton. *Journal of Phycology*, 23, 633–638.

KNAUER J and SOUTHGATE P C (1999) A review of the nutritional requirements of bivalves and the development of alternative and artificial diets for bivalve aquaculture. *Reviews in Fisheries Science*, 7, 241–280.

KNUCKEY R M, BROWN M R, BARRETT S M and HALLEGRAEFF G M (2002) Isolation of new nanoplanktonic diatom strains and their evaluation as diets for juvenile Pacific oysters (*Crassostrea gigas*). *Aquaculture*, 211, 253–274.

KNUCKEY R M, SEMMENS G L, MAYER R J and RIMMER M A (2005) Development of an optimal microalgal diet for the culture of the calanoid copepod *Acartia sinjiensis*: Effect of algal species and feed concentration on copepod development. *Aquaculture*, 249, 339–351.

KNUCKEY R M, BROWN M R, ROBERT R and FRAMPTON D M F (2006) Production of microalgal concentrates by flocculation and their assessment as aquaculture feeds. *Aquacultural Engineering*, 35, 300–313.

KRICHNAVARUK S, POWTONGSOOK S and PAVASANT P (2007) Enhanced productivity of *Chaetoceros calcitrans* in airlift photobioreactors. *Bioresource Technology*, 98, 2123–2130.

LAING I (1991) *Cultivation of marine unicellular algae*, Great Britain Ministry of Agriculture Fisheries and Food Laboratory Leaflet. Lowestoft: MAFF Directorate of Fisheries Research.

LAING I, CHILD A R and JANKE A (1990) Nutritional value of dried algae diets for larvae of Manila clam (*Tapes-philippinarum*). *Journal of the Marine Biological Association of the United Kingdom*, 70, 1–12.

© Woodhead Publishing Limited, 2013

LANGDON C J and WALDOCK M J (1981) The effect of algal and artificial diets on the growth and fatty acid composition of *Crassostrea gigas* spat. *Journal of the Marine Biological Association of the United Kingdom*, 61, 431–448.

LEBLOND J D and CHAPMAN P J (2000) Lipid class distribution of highly unsaturated long chain fatty acids in marine dinoflagellates. *Journal of Phycology*, 36, 1103–1108.

LEE Y and LEE Y K (2004) Algal nutrition: heterotrophic carbon nutrition, in Richmond A (ed) *Handbook of Microalgal Culture: Biotechnology and Applied Phycology*. Oxford: Blackwell Science, 116–124.

LEE B H and PICARD G A (1982) Chemical analysis of unicellular algal biomass from synthetic medium and sewage effluent. *Canadian Institute of Food Science and Technology Journal – Journal De L Institut Canadien De Science Et Technologie Alimentaires*, 15, 58–64.

LEWIS T E, NICHOLS P D and MCMEEKIN T A (1999) The biotechnological potential of thraustochytrids. *Marine Biotechnology*, 1, 580–587.

LIN Y H, CHANG F L, TSAO C Y and LEU J Y (2007) Influence of growth phase and nutrient source on fatty acid composition of *Isochrysis galbana* CCMP 1324 in a batch photoreactor. *Biochemical Engineering Journal*, 37, 166–176.

LOEBLICH A R (1975) Seawater medium for dinoflagellates and nutrition of *Cachonina niei*. *Journal of Phycology,* 11, 80–86.

LOOSANOFF V L and DAVIS H C (1963) Rearing of bivalve mollusks. *Advances in Marine Biology*, 1, 1–136.

LUBZENS E (1987) Raising rotifers for use in aquaculture. *Hydrobiologia*, 147, 245–255.

LUBZENS E, GIBSON O, ZMORA O and SUKENIK A (1995) Potential advantages of frozen algae (*Nannochloropsis* sp.) for rotifer (*Brachionus plicatilis*) culture. *Aquaculture*, 133, 295–309.

MAI K S, MERCER J P and DONLON J (1996) Comparative studies on the nutrition of two species of abalone, *Haliotis tuberculata* L and *Haliotis discus hannni* Ino. 5. The role of polyunsaturated fatty acids of macroalgae in abalone nutrition. *Aquaculture*, 139, 77–89.

MANSOUR M P, FRAMPTON D M F, NICHOLS P D, VOLKMAN J K and BLACKBURN S I (2005) Lipid and fatty acid yield of nine stationary-phase microalgae: Applications and unusual C-24-C-28 polyunsaturated fatty acids. *Journal of Applied Phycology*, 17, 287–300.

MARCHETTI J, BOUGARAN G, LE DEAN L, MÉGRIER C, LUKOMSKA E, KAAS R, OLIVO E, BARON R, ROBERT R and CADORET J P (2012) Optimizing conditions for the continuous culture of *Isochrysis affinis galbana* in commercial hatcheries. *Aquacuture*, 326–329, 106–115.

MARSHALL R, MCKINLEY S and PEARCE C M (2010) Effects of nutrition on larval growth and survival in bivalves. *Reviews in Aquaculture*, 2, 33–55.

MARTÍNEZ-FERNÁNDEZ E and SOUTHGATE P C (2007) Use of tropical microalgae as food for larvae of the black-lip pearl oyster *Pinctada margaritifera*. *Aquaculture*, 263, 220–226.

MARTÍNEZ-FERNÁNDEZ E, ACOSTA-SALMÒN H and RANGEL-DAVALOS C (2004) Ingestion and digestion of 10 species of microalgae by winged pearl oyster *Pteria sterna* (Gould, 1851) larvae. *Aquaculture*, 230, 417–423.

MARTÍNEZ-FERNÁNDEZ E, ACOSTA-SALMON H and SOUTHGATE P C (2006) The nutritional value of seven species of tropical microalgae for black-lip pearl oyster (*Pinctada margaritifera*, L.) larvae. *Aquaculture*, 257, 491–503.

MCCAUSLAND M A, BROWN M R, BARRETT S M, DIEMAR J A and HEASMAN M P (1999) Evaluation of live microalgae and microalgal pastes as supplementary food for juvenile Pacific oysters (*Crassostrea gigas*). *Aquaculture*, 174, 323–342.

© Woodhead Publishing Limited, 2013

MILKE L M, BRICELJ V M and PARRISH C C (2006) Comparison of early life history stages of the bay scallop, *Argopecten irradians*: Effects of microalgal diets on growth and biochemical composition. *Aquaculture*, 260, 272–289.

MILKE L M, BRICELJ V M and PARRISH C C (2008) Biochemical characterization and nutritional value of three *Pavlova* spp. in unialgal and mixed diets with *Chaetoceros muelleri* for postlarval sea scallops, *Placopecten magellanicus*. *Aquaculture*, 276, 130–142.

MOHEIMANI N R and BOROWITZKA M A (2006) The long-term culture of the coccolithophore *Pleurochrysis carterae* (Haptophyta) in outdoor raceway ponds. *Journal of Applied Phycology*, 18, 703–712.

MOLINA-GRIMA E, SANCHEZ-PEREZ J A, GARCIA-CAMACHO F, GARCIA-SANCHEZ J L, ACIEN-FERNANDEZ F G and LOPEZ-ALONSO D (1994) Outdoor culture of *Isochrysis galbana* ALII-4 in a closed tubular photobioreactor. *Journal of Biotechnology*, 37 (2), 159–166.

MOURENTE G, LUBIAN L M and ODRIOZOLA J M (1990) Total fatty acid composition as a taxonomic index of some marine microalgae used as food in marine aquaculture. *Hydrobiologia*, 203, 147–154.

NELL J A and O'CONNOR W A (1991) The evaluation of fresh algae and stored algal concentrates as a food source for Sydney rock oyster, *Saccostrea commercialis* (Iredale and Roughley), larvae. *Aquaculture*, 99, 277–284.

NEVEJAN N, DAVIS J, LITTLE K and KILIONA A (2007) Use of a formulated diet for mussel spat *Mytilus galloprovincialis* (Lamarck 1819) in a commercial hatchery. *Journal of Shellfish Research*, 26, 357–363.

NGHIA T T, WILLE M, BINH T C, THANH H P, VAN DANH N and SORGELOOS P (2007) Improved techniques for rearing mud crab *Scylla paramamosain* (Estampador 1949) larvae. *Aquaculture Research*, 38, 1539–1553.

PALMER P J, BURKE M J, PALMER C J and BURKE J B (2007) Developments in controlled green-water larval culture technologies for estuarine fishes in Queensland, Australia and elsewhere. *Aquaculture*, 272, 1–21.

PARSONS T R, STEPHENS K and STRICKLAND J D H (1961) On the chemical composition of 11 species of marine phytoplankters. *Journal of the Fisheries Research Board of Canada*, 18, 1001–1016.

PATIL V, KALLQVIST T, OLSEN E, VOGT G and GISLEROD H R (2007) Fatty acid composition of 12 microalgae for possible use in aquaculture feed. *Aquaculture International*, 15, 1–9.

PEIRSON W M (1983) Utilization of 8 algal species by the bay scallop, *Argopecten irradians concentricus* (Say). *Journal of Experimental Marine Biology and Ecology*, 68, 1–11.

PERNET F and TREMBLAY R (2004) Effect of varying levels of dietary essential fatty acid during early ontogeny of the sea scallop *Placopecten magellanicus*. *Journal of Experimental Marine Biology and Ecology*, 310, 73–86.

PETTERSEN A K, TURCHINI G M, JAHANGARD S, INGRAM B A and SHERMAN C D H (2010) Effects of different dietary microalgae on survival, growth, settlement and fatty acid composition of blue mussel (*Mytilus galloprovincialis*) larvae. *Aquaculture*, 309, 115–124.

PONIS E, ROBERT R and PARISI G (2003a) Nutritional value of fresh and concentrated algal diets for larval and juvenile Pacific oysters (*Crassostrea gigas*). *Aquaculture*, 221, 491–505.

PONIS E, ROBERT R, PARISI G and TREDICI M (2003b) Assessment of the performance of Pacific oyster (*Crassostrea gigas*) larvae fed with fresh and preserved *Pavlova lutheri* concentrates. *Aquaculture International*, 11, 69–79.

PONIS E, PROBERT I, VERON B, MATHIEU M and ROBERT R (2006) New microalgae for the Pacific oyster *Crassostrea gigas* larvae. *Aquaculture*, 253, 618–627.

© Woodhead Publishing Limited, 2013

PONIS E, PARISI G, ZITTELLI G C, LAVISTA F, ROBERT R and TREDICI M R (2008) *Pavlova lutheri*: Production, preservation and use as food for *Crassostrea gigas* larvae. *Aquaculture*, 282, 97–103.

PREISIG H R and ANDERSEN R A (2005) Historical review of algal culturing techniques, in Andersen R A (ed), *Algal Culturing Techniques.* New York: Academic Press, 1–9.

PRIETO A, CANAVATE J P and GARCIA-GONZALEZ M (2011) Assessment of carotenoid production by *Dunaliella salina* in different culture systems and operation regimes. *Journal of Biotechnology*, 151, 180–185.

QIANG H and RICHMOND A (1994) Optimizing the population-density in *Isochrysis galbana* grown outdoors in a glass column photobioreactor. *Journal of Applied Phycology*, 6, 391–396.

RAGG N L C, KING N, WATTS E and MORRISH J (2010) Optimising the delivery of the key dietary diatom *Chaetoceros calcitrans* to intensively cultured Greenshell (TM) mussel larvae, *Perna canaliculus. Aquaculture*, 306, 270–280.

RENAUD S M, THINH L V and PARRY D L (1999) The gross chemical composition and fatty acid composition of 18 species of tropical Australian microalgae for possible use in mariculture. *Aquaculture*, 170, 147–159.

RENAUD S M, LUONG-VAN T, LAMBRINIDIS G and PARRY D L (2002) Effect of temperature on growth, chemical composition and fatty acid composition of tropical Australian microalgae grown in batch cultures. *Aquaculture*, 211, 195–214.

RHODES L, SMITH J, TERVIT R, ROBERTS R, ADAMSON J, ADAMS S and DECKER M (2006) Cryopreservation of economically valuable marine micro-algae in the classes Bacillariophyceae, Chlorophyceae, Cyanophyceae, Dinophyceae, Haptophyceae, Prasinophyceae, and Rhodophyceae. *Cryobiology*, 52, 152–156.

RICHMOND A (2004) Biological principles of mass cultivation, in Richmond A (ed), *Handbook of Microalgal Culture: Biotechnology and Applied Phycology.* Oxford: Blackwell Science, 125–177.

RICO-VILLA B, COZ J R L, MINGANT C and ROBERT R (2006) Influence of phytoplankton diet mixtures on microalgae consumption, larval development and settlement of the Pacific oyster *Crassostrea gigas* (Thunberg). *Aquaculture*, 256, 377–388.

RICO-VILLA B, WOERTHER P, MINGANT C, LEPIVER D, POUVREAU S, HAMON M and ROBERT R (2008) A flow-through rearing system for ecophysiological studies of Pacific oyster *Crassostrea gigas* larvae. *Aquaculture*, 282, 54–60.

RICO-VILLA B, BERNARD I, ROBERT R and POUVREAU S (2010) A Dynamic Energy Budget (DEB) growth model for Pacific oyster larvae, *Crassostrea gigas*. *Aquaculture*, 305, 84–94.

RITAR A J, SMITH G G, DUNSTAN G A, BROWN M R and HART P R (2003) Artemia prey size and mode of presentation: Effects on the survival and growth of phyllosoma larvae of southern rock lobster (*Jasus edwardsii*). *Aquaculture International*, 11, 163–182.

ROBERT R and TRINTIGNAC P (1997) Substitutes for live microalgae in mariculture: a review. *Aquatic Living Resources*, 10, 315–327.

ROBERTS R (2001) A review of settlement cues for larval abalone (*Haliotis* spp.). *Journal of Shellfish Research*, 20, 571–586.

ROCHA R J, RIBEIRO L, COSTA R and DINIS M T (2008) Does the presence of microalgae influence fish larvae prey capture? *Aquaculture Research*, 39, 362–369.

RODOLFI L, ZITTELLI G C, BASSI N, PADOVANI G, BIONDI N, BONINI G and TREDICI M R (2009) Microalgae for oil: Strain selection, induction of lipid synthesis and outdoor mass cultivation in a low-cost photobioreactor. *Biotechnology and Bioengineering*, 102, 100–112.

RUNNING J A, HUSS R J and OLSON P T (1994) Heterotrophic production of ascorbic acid by microalgae. *Journal of Applied Phycology*, 6, 99–104.

© Woodhead Publishing Limited, 2013

RUSCH K A and CHRISTENSEN J M (2003) The hydraulically integrated serial turbidostat algal reactor (HISTAR) for microalgal production. *Aquacultural Engineering*, 27, 249–264.

SARGENT J, MCEVOY L, ESTEVEZ A, BELL G, BELL M, HENDERSON J and TOCHER D (1999) Lipid nutrition of marine fish during early development: current status and future directions. *Aquaculture*, 179, 217–229.

SEGUINEAU C, LASCHI-LOQUERIE A, MOAL J and SAMAIN J F (1996) Vitamin requirements in great scallop larvae. *Aquaculture International*, 4, 315–324.

SEIXAS P, COUTINHO P, FERREIRA M and OTERO A (2009) Nutritional value of the cryptophyte *Rhodomonas lens* for *Artemia* sp. *Journal of Experimental Marine Biology and Ecology*, 381, 1–9.

SIERACKI M, POULTON N and CROSBIE N (2005) Automated isolation techniques for microalgae, in Andersen R A (ed), *Algal Culturing Techniques*. New York: Academic Press, 101–116.

SOEDER C J (1986) *An Historical Outline of Applied Algology*. Boca Raton, FL: CRC Press.

STANBURY P F, HALL S and WHITAKER A (1995) *Principles of Fermentation Technology*. Oxford: Butterworth-Heinemann.

STEIN J R (1973) *Handbook of Phycological Methods. Culture Methods and Growth Measurements*. Cambridge: Cambridge University Press.

STØTTRUP J G (2003) Production and nutrition value of copepods, in Støttrup J G and Mcevoy L A (eds), *Live Feeds in Marine Aquaculture*. Oxford: Blackwell Publishing, 145–205.

SUZUKI H, IORIYA T, SEKI T and ARUGA Y (1987) Changes of algal community on the plastic plates used for rearing the abalone *Haliotis discus hannai*. *Nippon Suisan Gakkaishi*, 53, 2163–2167.

TAMARU C S, MURASHIGE R, LEE C S, AKO H and SATO V (1993) Rotifers fed various diets of baker's yeast and or *Nannochloropsis oculata* and their effect on the growth and survival of striped mullet (*Mugil cephalus*) and milkfish (*Chanos chanos*) larvae. *Aquaculture*, 110, 361–372.

TAMIYA H, HASE E, SHIBATA K, MITUYA A, IWAMURA T, NIHEI T and SASA T (1953) Kinetics of growth of *Chlorella*, with special reference to its dependence on quantity of available light and temperature, in Burlew J S (ed) *Algal Culture From Laboratory to Pilot Plant*. Washington, DC: Carnegie Institution of Washington Publication No. 600, Carnegie Institution, 204–232.

THOMPSON P A, GUO M X and HARRISON P J (1992) Effects of variation in temperature.1. On the biochemical composition of 8 species of marine phytoplankton. *Journal of Phycology*, 28, 481–488.

THOMPSON P A, GUO M and HARRISON P J (1993) The influence of irradiance on the biochemical composition of 3 phytoplankton species and their nutritional value for larvae of the Pacific oyster (*Crassostrea gigas*). *Marine Biology*, 117, 259–268.

THOMPSON P A, GUO M X and HARRISON P J (1996) Nutritional value of diets that vary in fatty acid composition for larval Pacific oysters (*Crassostrea gigas*). *Aquaculture*, 143, 379–391.

TREDICI M R (2004) Mass production of microalgae: Photobioreactors, in Richmond A (ed), *Handbook of Microalgal Culture: Biotechnology and Applied Phycology*. Oxford: Blackwell Science, 178–214.

TREDICI M R, BIONDI N, PONIS E, RODOLFI L and ZITTELLI G C (2009) Advances in microalgal culture for aquaculture feed and other uses, in Burnell G and Allan G (eds), *New Technologies in Aquaculture: Improving Production Efficiency, Quality and Environmental Management*. Boca Raton, FL: Woodhead Publishing, 610–676.

TREMBLAY R, CARTIER S, MINER P, PERNET F, QUERE C, MOAL J, MUZELLEC M L, MAZURET M and SAMAIN J F (2007) Effect of *Rhodomonas salina* addition to a standard

© Woodhead Publishing Limited, 2013

hatchery diet during the early ontogeny of the scallop *Pecten maximus*. *Aquaculture*, 262, 410–418.

TZOVENIS I, FOUNTOULAKI E, DOLAPSAKIS N, KOTZAMANIS I, NENGAS I, BITIS I, CLADAS Y and ECONOMOU-AMILLI A (2009) Screening for marine nanoplanktic microalgae from Greek coastal lagoons (Ionian Sea) for use in mariculture. *Journal of Applied Phycology*, 21, 457–469.

UTTING S D and MILLICAN P F (1997) Techniques for the hatchery conditioning of bivalve broodstocks and the subsequent effect on egg quality and larval viability. *Aquaculture*, 155, 45–54.

VAN BERGEIJK S A, SALAS-LEITON E and CANAVATE J P (2010) Low and variable productivity and low efficiency of mass cultures of the haptophyte *Isochrysis* aff. *galbana* (T-iso) in outdoor tubular photobioreactors. *Aquacultural Engineering*, 43, 14–23.

VOLKMAN J K (1986) A review of sterol markers for marine and terrigenous organic matter. *Organic Geochemistry*, 9, 83–99.

VOLKMAN J K and BROWN M R (2006) Nutritional value of microalgae and applications, in Subba Rao D V (ed), *Algal Cultures, Analogues of Blooms and Applications, Volume 1.* Science Publishers, Plymouth, 407–457.

VOLKMAN J K, JEFFREY S W, NICHOLS P D, ROGERS G I & GARLAND C D (1989) Fatty acid and lipid composition of 10 species of microalgae used in mariculture. *Journal of Experimental Marine Biology and Ecology*, 128, 219–240.

VOLKMAN J K, DUNSTAN G A, JEFFREY S W and KEARNEY P S (1991) Fatty acids from microalgae of the genus *Pavlova*. *Phytochemistry*, 30, 1855–1859.

VOLKMAN J K, BROWN M R, DUNSTAN G A and JEFFREY S W (1993) The biochemical composition of marine microalgae from the class Eustigmatophyceae. *Journal of Phycology*, 29, 69–78.

VOLKMAN J K, BARRETT S M, BLACKBURN S I, MANSOUR M P, SIKES E L and GELIN F (1998) Microalgal biomarkers: A review of recent research developments. *Organic Geochemistry*, 29, 1163–1179.

WEBB K L and CHU F E (1983) Phytoplankton as a food source for bivalve larvae, in Pruder G D, Langdon C J and Conklin D E (eds), *Proceedings of the second international conference on aquaculture nutrition: Biochemical and physiological approaches to shellfish nutrition.* Baton Rouge, LA: Louisiana State University, 272–291.

WHYTE J N C (1987) Biochemical composition and energy content of 6 species of phytoplankton used in mariculture of bivalves. *Aquaculture*, 60, 231–241.

WHYTE J N C, BOURNE N and HODGSON C A (1989) Influence of algal diets on biochemical composition and energy reserves in *Patinopecten yessoensis* (Jay) larvae. *Aquaculture*, 78, 333–347.

WIKFORS G H, PATTERSON G W, GHOSH P, LEWIN R A, SMITH B C and ALIX J H (1996) Growth of post-set oysters, *Crassostrea virginica*, on high-lipid strains of algal flagellates *Tetraselmis* spp. *Aquaculture*, 143, 411–419.

WRIGHT S W and JEFFREY S W (2006) Pigment markers for phytoplankton production. *Marine Organic Mattter: Biomarkers, Isotopes and DNA*, 2, 71–104.

YOUN J Y and HUR S B (2009) Cryopreserved marine microalgae grown using different freezing methods. *Algae*, 24, 257–265.

ZHANG C W and RICHMOND A (2003) Sustainable, high-yielding outdoor mass cultures of *Chaetoceros muelleri* var. *subsalsum* and *Isochrysis galbana* in vertical plate reactors. *Marine Biotechnology*, 5, 302–310.

ZHANG C W, ZMORA O, KOPEL R and RICHMOND A (2001) An industrial-size flat plate glass reactor for mass production of *Nannochloropsis* sp (Eustigmatophyceae). *Aquaculture*, 195, 35–49.

ZITTELLI G C, PASTORELLI R and TREDICI M R (2000) A Modular Flat Panel Photobioreactor (MFPP) for indoor mass cultivation of *Nannochloropsis* sp under artificial illumination. *Journal of Applied Phycology*, 12, 521–526.

© Woodhead Publishing Limited, 2013

ZITTELLI G C, RODOLFI L and TREDICI M R (2003) Mass cultivation of *Nannochloropsis* sp in annular reactors. *Journal of Applied Phycology*, 15, 107–114.
ZMORA O and RICHMOND A (2004) Microalgae for aquaculture: Microalgae production for aquaculture, in Richmond A (ed), *Handbook of Microalgal Culture: Biotechnology and Applied Phycology.* Oxford: Blackwell Science, 365–379.

© Woodhead Publishing Limited, 2013

5

Rotifers, *Artemia* and copepods as live feeds for fish larvae in aquaculture

J. Dhont and K. Dierckens, Ghent University, Belgium, J. Støttrup, Technical University of Denmark, Denmark and G. Van Stappen, M. Wille and P. Sorgeloos, Ghent University, Belgium

DOI: 10.1533/9780857097460.1.157

Abstract: This chapter provides an update on the most common zooplankton live feed species used in hatchery rearing of fish and shellfish larvae, namely rotifers, *Artemia* and copepods. Each section starts with a summary of the biology and ecology of these species. An overview of the most common techniques to culture, feed, harvest, disinfect and preserve these organisms is provided. Special attention is given to nutritional and microbiological aspects. Furthermore, new trends and developments are discussed.

Key words: live feed, rotifers, *Brachionus*, *Artemia*, copepods.

5.1 Introduction

Despite considerable progress in the development of formulated larval feeds and the adaptation of various techniques to reduce the quantities required (e.g. through improved zoo-techniques, feeding regimes and feeding practices), the worldwide use of live feed in the hatchery rearing of most fish and shellfish species is still essential and is expected to remain so in the (near) future. In contrast to the natural environment, where larvae have access to a large variety of plankton organisms, the artificial food web applied in hatcheries consists of only a few species. Apart from microalgae (which are dealt with in Chapter 4 of this book), rotifers (*Brachionus* spp.), the brine shrimp *Artemia* and (to a lesser extent) copepods are the main live feeds used. Rotifers are, because of their small size, suitable for the earliest stages of fish and shrimp larvae. However, they require a rather labour-intensive culture which, moreover, is not always reliable, and occasional culture crashes remain a reality. Nevertheless, in recent years

© Woodhead Publishing Limited, 2013

techniques have improved considerably with new feeds and culture systems (e.g. semi-continuous and high density systems) being developed. The popularity of *Artemia* as a live feed item is largely due to its convenience in use as it can be hatched within 24 hours from dormant cysts which can be easily distributed and stored for prolonged periods of time. Moreover, as soon as it was realized that its nutritional value was not optimal for all species of predator larvae, different techniques were developed to boost and manipulate its composition, originally mainly the vitamin content and fatty acid profile. Today, these enrichment techniques extend to a whole range of macro- and micronutrients and also encompass techniques to reduce or alter the microbial community or incorporate therapeutics (vaccines, drugs, immunostimulants, . . .). Copepods are considered to be superior in terms of nutritional composition (especially for cold water fish species) as they contain higher levels of highly unsaturated fatty acids and they have a better balance between triglyceride and phospholipid lipid classes. Although here also considerable progress has been made in recent years, techniques for reliable culture on a large scale remain the major bottleneck.

This chapter reviews currently used techniques to culture, harvest, treat and preserve these three groups of live feed. New trends are identified and possibilities for future research and development are set out.

5.2 Rotifers as live feed: culture and harvesting

Rotifers, mainly belonging to the genus *Brachionus*, have been used as live feed organism in aquaculture since the 1970s (Lubzens *et al.*, 2001). A lot of marine fish larvae are very small when they hatch. Consequently, they need small prey during the early life stages (Conceição *et al.*, 2010). Rotifers have become the most favoured live feed as they have a suitable size for most marine fish larvae, can be cultured at high densities, have a high reproductive rate (due to the possibility to reproduce parthenogenetically dependent on the environment), have a high temperature and salinity tolerance, swim slowly and are filter feeders (Lavens and Sorgeloos, 1996).

Thanks to this last characteristic, rotifers can be cultured on many feed sources (e.g. microalgae or yeast cells) and, even more important, their nutritional composition can be adjusted in a relatively short time (called enrichment) to better suit the nutritional requirements of the predator (e.g. fish or shrimp larvae). In the culture of fresh water rotifers, *B. calyciflorus*, pH is important due to the ammonia–ammonium equilibrium.

However, although rotifers are tolerant to many abiotic factors, sudden collapses of mass cultures do occur that may be related to the infection with rotifer birnavirus or a fungus (Comps and Menu, 1997). Hatcheries try to circumvent this problem by culturing more rotifers than are needed in separated tanks, and stock cultures of different volumes are kept as backup. In order to obtain sufficient numbers of rotifers to start up a large-scale

© Woodhead Publishing Limited, 2013

culture, rotifers are cultured in increasing volumes starting from as small as 50 mL (stock cultures) to 50 L. The rotifers harvested from one or several tanks of 50 L are used as inoculum for the large-scale cultures. In most cases, rotifers are cultured parthenogenetically all year round, which means effort/costs are also required outside the larvae (fish or crustacean) season.

Since 1995, the species *B. plicatilis* has been divided into two species, *B. plicatilis* and *B. rotundiformis* (Segers, 1995). Through the use of molecular techniques, the species *B. plicatilis* has been split into nine genetically different lineages (Gomez *et al.*, 2002). However, Suatoni *et al.* (2006) hypothesized that there are at least 14 species within the *B. plicatilis* complex based on COI and ITS1 patterns. From all genetic lineages, *B. plicatilis sensu strictu*, *B.* 'Nevada', *B.* 'Cayman' and *B.* 'Manjavacas' have been examined the most in aquaculture-related studies. Papakostas *et al.* (2006) and Dooms *et al.* (2007) have identified *B.* 'Nevada', *B.* 'Cayman' and *B.* 'Austria' as more common in European hatcheries. Although rotifers belonging to the *B. plicatilis* complex are still referred to as SS, S, SM or L-type according to their size, there are differences in feed conversion, growth rate and associated microbial communities amongst these lineages that can be important to aquaculture (Kostopoulou and Vadstein, 2007; Qi *et al.*, 2009a).

5.2.1 Culture

Both for freshwater and marine rotifers, dependent on the salinity of the predator environment, the size the predator can ingest is the most important characteristic in the selection of rotifer species/genotype that will be cultured in a hatchery. Commercial cultures of rotifers are nearly completely restricted to species of the genus *Brachionus* (Papakostas *et al.*, 2006). The parthenogenetic loop in the life-cycle (Fig. 5.1) with a high reproductive rate results in identical offspring with the same characteristics (for details, see Lubzens and Zmora, 2003). In order to determine the amount of feed that needs to be provided, several sub-samples (at least three) are taken daily.

Before counting the rotifers, it is best to observe their behaviour. Rotifers should be free swimming (not attached to the sides of the counting chamber), as low quality rotifers stop spending energy in swimming and filter particles while being attached. Also the swimming rate should be high, as slow moving rotifers can be an indication of starvation and/or bad water quality (Korstad *et al.*, 1995).

The most important characteristic of the physiological status of the rotifers is the egg ratio. This is the ratio of the number of eggs in a given sample over the total number of females in that sample. The egg ratio indicates in what stage the rotifer culture is (exponential or stationary) and, more important, it gives an estimation of the number of rotifers that is to be expected the next day. The egg ratio should be no lower than 10% as this may indicate a near future collapse (Snell *et al.*, 1987).

© Woodhead Publishing Limited, 2013

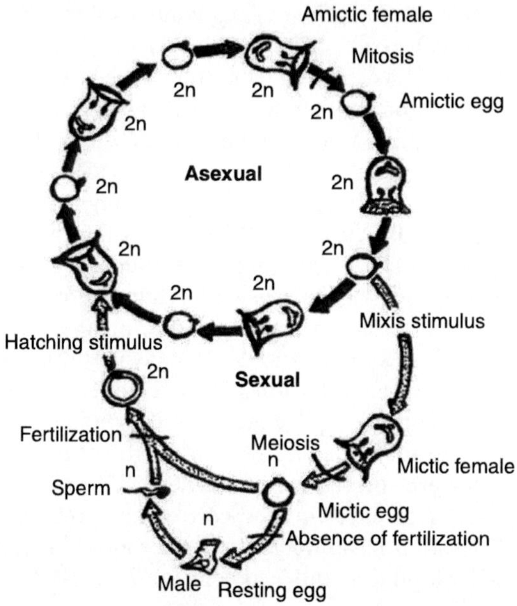

Fig. 5.1 Parthenogenetical and sexual reproduction of *Brachionus* sp. (modified from Hoff and Snell, 1987).

From a set up point of view, there are two main ways to culture rotifers in large scale that are important for hatcheries: batch culture and recirculation culture. From a culture management point of view, all rotifers from a culture can be harvested in one time or a portion of the cultured rotifers can be harvested frequently (e.g. daily).

Batch culture

Batch culture is the most widely used culture method (Dhert *et al.*, 2001). Within the batch culture method, one can distinguish two different ways: (i) Constant volume with increasing rotifer density; and (ii) Constant rotifer density with increasing volume.

When a constant volume is used, rotifers coming from a stock culture or from a previous batch culture are introduced in a clean tank at a density around 250 rotifers/mL (Suantika *et al.*, 2000). When rotifers are introduced in a tank at a low density (e.g. 50 rotifers/mL), the total filtration rate is low, which in its turn allows contaminating protozoans to proliferate.

The rotifers are fed microalgae, yeast cells or commercial products, based on one or both of the previous micro-organisms, according to the number of rotifers. The feed can be administered manually or through automated devices (e.g. peristaltic pumps). The rotifer growth rate can be around 1.5/day (Kostopoulou and Vadstein, 2007). As the culture period continues, the water quality decreases. This has a negative impact on the

© Woodhead Publishing Limited, 2013

rotifer performance leading to negative growth rates. In practice, rotifer batch cultures are run for two to four days. Then, the whole culture is harvested. The majority of the rotifers is used as live feed and the rest can be used as an inoculum for a following batch culture.

When rotifers are cultured at a constant rotifer density, a tank is filled only partially and rotifers are inoculated at a high density (750 rotifers/mL) (Kotani *et al.*, 2009). The next day, the increase in rotifer density is compensated by adding new clean water to the culture. The addition of new water keeps the water quality more constant, as metabolites, nitrite and ammonia are diluted daily. The culture can be maintained for 27 days (Kotani *et al.*, 2009).

The advantage of the latter form of batch culture is that the rotifer growth rate is more constant, resulting in rotifers that are physiologically more active (Kotani *et al.*, 2009). However, the culture period is limited due to accumulation of ammonia (De Araujo *et al.*, 2000, 2001). Bentley *et al.* (2008) have succeeded in mass culture of rotifers fed dead *Nannochloropsis oculata* by neutralizing ammonia using sodium hydroxymethanesulfonate in a continuous culture. Addition of gamma-aminobutyric acid (GABA) to the culture water led to better performing rotifer cultures even in the presence of 3.1 μg/mL free ammonia. GABA may increase the assimilation efficiency (Gallardo *et al.*, 1999).

Recirculating aquaculture systems (RAS)

Because the rotifer production decreases due to low water quality in batch cultures, the use of RAS to culture rotifers has been studied (Suantika *et al.*, 2000). These culture systems allow for a higher rotifer density and higher reproduction rate. The rotifer culture can be maintained for a longer period (e.g. several weeks). As a consequence, hatcheries can run smaller scale systems and produce rotifers at a cheaper cost compared to batch cultures (Suantika *et al.*, 2003). Most of the time, these culture systems are operated in a semi-continuous manner. As soon as the rotifer density surpasses a given density, part of the culture is harvested and part of the water is replaced (Suantika *et al.*, 2003).

This type of culture has enabled the running of super-intensive cultures where rotifer densities reach over 10 000 rotifers/mL (Yoshimura *et al.*, 1996; Suantika *et al.*, 2001). In order to maintain the dissolved oxygen levels above 4 ppm (Fulks and Main, 1991), pure oxygen (Yoshimura *et al.*, 1996) or ozone (Suantika *et al.*, 2001) has to be applied. The maintenance of super-intensive rotifer cultures is, however, very complicated and may involve computerised set ups (Yoshimura *et al.*, 2003).

5.2.2 Harvesting

A rotifer culture can be harvested all at once or partially (see above). As the rotifer culture water has high organic loads (rotifer feed, dead rotifers,

© Woodhead Publishing Limited, 2013

faecal matter), bacteria develop quickly to high densities (Skjermo and Vadstein, 1993). Therefore, it is necessary to separate the rotifers from the culture water and remove flocs containing feed particles, bacteria and uneaten feed, and the harvested rotifers must also be concentrated. This is achieved by filtering the rotifers and the culture water through two sieves, the first retaining bigger particles and the second retaining the rotifers while bacteria and rotifer feed pass. The mesh size depends on the size of the cultured rotifer species (Lavens and Sorgeloos, 1996). After harvesting, rotifers can be used immediately (depending on the nutritional composition) or can be stored at low temperature (4 °C) (Assavaaree *et al.*, 2001). Makridis and Olsen (1999) concluded that rotifers can even be stored at 4–5 °C up to five days without significant loss of their nutritional value.

5.3 Feed for rotifers: types, techniques and nutrition

5.3.1 Feed types

Rotifers can be fed different kinds of feeds as long the particle size ranges from 0.3 μm up to 21 μm in diameter (Vadstein *et al.*, 1993; Hansen *et al.*, 1997). Several species of microalgae can be used, of which *Nannochloropsis* spp., *Chlorella* spp., *Tetraselmis* spp. and *Isochrysis galbana* are the most common (Conceição *et al.*, 2010). *T. suecica* was found to result in optimal grazing (Lubzens and Zmora, 2003). Sayegh *et al.* (2007) have tested the effect of four different *I. galbana* strains and one *Nannochloropsis* sp. strain and found significant effects of these strains on the rotifer reproductive rate. The effects are not big and may not be important for commercial production. Still, they conclude that this strain effect may be considered to maximize productivity.

The microalgae can be cultured on site or condensed algae (algal paste) can be bought. Live microalgae have the advantage that they keep their nutritional composition longer, have a negative effect on some pathogenic bacteria (Natrah *et al.*, 2011) and have a positive effect on the water quality compared to algal paste. The use of algal paste, on the other hand, has the advantage that the supply and nutritional composition may be more stable and it allows the provision of enough feed for large-scale intensive or super-intensive rotifer cultures (Yoshimura *et al.*, 1996).

Baker's yeast is also a common feed. However, yeast contains less highly unsaturated fatty acids (HUFA), which results in rotifers with inferior nutritional quality for the larvae compared to microalgae fed rotifers (Conçeicão *et al.*, 2010). This is especially important for marine larvae as marine fish are not able to synthesize eicosapentaenoic acid (EPA) and docosahexaenoic acid DHA themselves from linolenic acid (for a review see Tocher, 2010). The nutritional composition of the rotifers can be adjusted through enrichment (i.e. inclusion of specific nutrients/chemicals essential to the larvae). There are many formulated feeds on the market that are

© Woodhead Publishing Limited, 2013

mostly based on baker's yeast and/or microalgae. As baker's yeast is deficient in HUFA, HUFA can be coated onto spray dried yeast during the manufacturing process, resulting in better adapted feed.

The different rotifer biotypes have different optimal growth requirements. At high food concentrations, *B.* 'Cayman' had a higher growth rate compared to *B. plicatilis* sensu strictu and *B.* 'Nevada', when fed on *N. oculata* (Kostopoulou and Vadstein, 2007). The same observations were made when these rotifer biotypes were fed a commercial yeast based diet (Qi *et al.*, 2009b).The higher growth rate was attributed to the faster egg development in *B.* 'Cayman' (Kostopoulou and Vadstein, 2007).

As rotifers are filter feeders, following a type-1 functional response (Holling, 1966), it is important to keep the feed suspension at a defined density. Too low a feed density will obviously result in lower ingestion and subsequently in lower growth rates. However, very high feed densities will lead to a reduced growth rate as well, since the rate of digestive enzyme production is too slow for the amount of feed entering the digestive tract (Rothhaupt, 1990).

5.3.2 Feeding techniques

In order to keep the feed density within the optimal range, the feed must be provided to the rotifers several times per day. This has led to systems with continuous feeding (Yoshimura *et al.*, 1996; Suantika *et al.*, 2000) using peristaltic pumps. Automated observation of the rotifer density, calculation of the required amount of feed and distribution of the feed to the rotifer tanks has been studied in Norway (Alver *et al.*, 2010).

In practice, many hatcheries, but also research institutes, provide a mixture of feeds to the rotifers: microalgae, yeast or yeast-based commercial diet and oil (Srivastava *et al.*, 2011). The mixture of these feeds is applied most of the time due to insufficient supply of live algae and in order, to reduce the feed costs and achieve rotifers which better nutritional quality.

The feed given also has an effect on the microbial communities associated with the cultured rotifers (see Section 5.4.1 on Microbial aspects).

5.3.3 Nutritional value and enrichment techniques

Nutritional value

HUFAs are very important during the development of marine larvae. DHA is a major component in the brain and nervous system development (Tocher, 2010). However, it is not only the amount of HUFA, but also the ratio between different HUFA (DHA, EPA and arachidonic acid (ARA)) which is important (Tocher, 2010) for marine larvae. Oie and Olsen (1997) observed that the protein and lipid content and the DHA/EPA ratio of the rotifers varied according to the growth rate and feed ration. This positive

© Woodhead Publishing Limited, 2013

correlation between the growth rate (physiological activity) and the fatty acid content was also observed by Kotani *et al.* (2009).

As well as the physiological status of the rotifers, the applied feed during the culture also has a major influence on their nutritional quality (see 'feed types' above). Rotifers cultured solely on baker's yeast are low in HUFA (Conçeicão *et al.*, 2010). The HUFA levels of algae species (e.g. *I. galbana* T-ISO strain) depend on the growth phase and culture protocol (Tzovenis *et al.*, 2003). The protein content of rotifers varies between 28 and 63 % of the dry weight (DW) (Lubzens and Zmora, 2003), of which about 50 % is soluble protein (Srivastava *et al.*, 2006).

The lipid content of rotifers is 9–28 % of the dry weight, of which 34–43 % is phospholipids and 20–55 % triacylglycerols (Lubzens and Zmora, 2003). Rotifers fed on microalgae have high concentrations of vitamin B_1, B_2, C and E (van der Meeren *et al.*, 2008). The vitamin content of the feed is reflected in that of the rotifers. Rotifers cultured on a mixture of *Chlorella* sp. and *I. galbana* can reach 2289 mg ascorbic acid/g DW in lab-scale cultures (Merchie *et al.*, 1995). Some bacteria present in rotifer cultures can produce vitamin B_{12} which is taken up by the rotifers (Yu *et al.*, 1989).

Enrichment

The composition of rotifers can be modified in a relatively short time by presenting particles within a suitable range (0.3–21 μm; Vadstein *et al.*, 1993; Hotos, 2002). These will be ingested due to their filter-feeding nature resulting in nutritional enrichment.

As HUFA are very important to marine larvae a lot of research has focused on the potential to augment the HUFA content of live feeds. Through an enrichment of 6 h, the caloric value of rotifers can nearly be doubled; from 1.34×10^{-3} cal/rotifer to 2.00×10^{-3} cal/rotifer (Fernandez-Reiriz *et al.*, 1993). There are several commercial HUFA enrichment products on the market (Haché and Plante, 2011). However, rotifers can also be enriched by applying microalgae (Kotani *et al.*, 2009; Ferreira *et al.*, 2009). Kotani *et al.* (2009) and Ferreira *et al.* (2009) used *N. oculata* and *N. gaditana*, respectively. Depending on the physiological status of the rotifers (see above), higher quantities of HUFA were obtained (Kotani *et al.*, 2009).

Although bacteria present in rotifer cultures can provide vitamin B_{12} (Yu *et al.*, 1989) and microalgae such as *Chlorella* sp. and *I. galbana* provide vitamin C (Merchie *et al.*, 1995), extra enrichment of thiamine (as thiamine HCl) and vitamin E (as DL-alpha-tocopherol) led to an increased concentration in the rotifers after four days. On the other hand, addition of vitamin A (as retinyl palmitate) in the diet did not lead to a significant increase in vitamin A content (Srivastava *et al.*, 2011). As copepods and their nauplii are the natural food for many marine larvae, rotifers are also enriched in iodine in order to obtain iodine levels similar to those of copepods (Srivastava *et al.*, 2006).

© Woodhead Publishing Limited, 2013

Selenium is important in the development of the thyroid and is only present in low amounts (around 0.08 mg/kg DW) in cultured rotifers compared to 2–5 mg/kg DW in wild copepods (Hamre *et al.*, 2008). Through enrichment during 3 h with selenized yeast, Ribeiro *et al.* (2011) were able to obtain Se levels of 35.9–104.0 mg/kg DW in rotifers. As this is much higher than the levels of wild copepods, lower amounts of selenized yeast could be used.

As rotifers can have 2.43×10^9 CFU/g rotifer wet weight and as the organic loads are very high during the enrichment process, the bacterial load of rotifers can increase significantly (Haché and Plante, 2011). Therefore, rotifers are rinsed thoroughly during harvesting and the enrichment process is kept as short as possible.

5.4 Rotifers as live feed: microbial aspects, hygiene and preservation techniques

5.4.1 Microbial aspects

Associated microbial community

As mentioned above, organic loads in rotifer cultures are very high. When the temperature is increased, high bacterial growth rates are obtained (Skjermo and Vadstein, 1993). This results in rotifers that carry a high bacterial load (e.g. 2.43×10^9 CFU/g wet weight rotifers for rotifers fed on a mixture of *I. galbana* and baker's yeast) (Nicolas *et al.*, 1989; Munro *et al.*, 1999; Haché and Plante, 2011).

Some of these bacteria can be pathogenic to rotifers (e.g. *Vibrio alginolyticus*; Yu *et al.*, 1990). There are also viruses, e.g. rotifer birnavirus (Comps *et al.*, 1991a, b; Comps and Menu, 1997), and fungi (Comps *et al.*, 1993) detected that are detrimental to rotifers and might be the cause of rotifer culture mass mortalities. Qi *et al.* (2009a) observed a correlation between the changes in microbial community composition and the change in rotifer growth rate. Lorenz curves and the Gini-coefficient indicated that good performing rotifer cultures have a more even microbial community structure in *B. plicatilis sensu strictu*, *B. plicatilis* 'Cayman' and *B. plicatilis* 'Nevada'. Hino (1993) had previously suggested that collapses in rotifer cultures were correlated to changes in the microbial community. The associated microbial community that Qi *et al.* (2009a) detected was different for rotifers belonging to the different genetic lineages.

However, the bigger issue lies in the fact that rotifers can be vectors of fish/shrimp pathogens (see microbial community management below) (Nicolas *et al.*, 1989; Muroga, 1995; Verdonck *et al.*, 1997; Yan *et al.*, 2007). On the other hand, bacteria can also have a positive effect towards rotifers as some produce vitamin B_{12} (Yu *et al.*, 1989) or enhance rotifer growth rate (Rombaut *et al.*, 1999) (see below).

© Woodhead Publishing Limited, 2013

Microbial community management

As rotifers are able to filter bacteria, the microbial community of the rotifers is defined by the microbial community present in the culture water. Selected bacteria can be added to the culture water in order to shift their microbial community towards one that results in a better performing culture (Gatesoupe *et al.*, 1989; Rombaut *et al.*, 1999) or to one that is beneficial for the predator (Gatesoupe, 1991).

Microalgae are known to shape the microbial community in their environment (Natrah *et al.*, 2011) by selectively enhancing and/or restraining certain bacteria. This means that one is able to shift the microbial community of the cultured rotifers through the applied feed. Qi *et al.* (2009b) found that the dominant bacterial species in rotifer cultures fed *N. oculata* were different from the ones fed a baker's yeast-based diet. When rotifers were cultured on a mixture of the above feeds, by simultaneous feeding or by feeding both feeds every other batch, the microbial fingerprints were similar.

Probiotic bacteria (probiotic was defined by Verschuere *et al.* (2000) as 'a live microbial adjunct which has a beneficial effect on the host by modifying the host-associated or ambient microbial community, by ensuring improved use of feed or enhancing its nutritional value, by enhancing the host response towards disease, or by improving the quality of its ambient environment) added at the start of the first batch culture were still retrieved after four consecutive batches of three days each. This was the case when rotifers were fed either solely *N. oculata*, a yeast-based diet, a mixture of both feeds fed simultaneously or fed every other batch (Qi *et al.*, 2009b).

The microbial community of the rotifers can also be shifted through a microbial enrichment (Makridis *et al.*, 2000). This enrichment takes place after the rotifers have been harvested and prior to feeding them to the predator larvae. Makridis *et al.* (2000) harvested and rinsed rotifers and exposed them to 5×10^7 CFU/mL of four probiotic species for 20–60 min. The probiotics remained associated with the rotifers for 4–24 h when placed in tanks with *Tetraselmis* sp. mimicking the conditions in turbot larvae tanks.

5.4.2 Hygiene: disinfection procedures

As rotifers are the first feed of fish/shrimp larvae, their bacterial load can have a major impact on their survival (Benavente and Gatesoupe, 1988). This has led to studies on the reduction of the number of bacteria after the rotifer harvest. Munro *et al.* (1999) found that UV light could reduce the bacterial load by more the 90 % after 2 min exposure. Suantika *et al.* (2001) used ozone in a recirculation system to reduce bacteria densities in the rotifer culture. However, in both studies there was no effect on the microbial composition. Tanasomwang and Muroga (1992) found that sodium nifurstyrenate could lower the bacterial load by 10–100 times, while the bacterial community shifted towards fewer *Vibrio* and more *Pseudomonas* and/or *Moraxella*. Takaoka *et al.* (2011) were able to reduce

© Woodhead Publishing Limited, 2013

the *V. anguillarum* load of rotifers cultured on baker's yeast and concentrated fresh water *Chlorella* during the enrichment procedure by using herbal extracts. Extract of *Crataegi fructus* and a mixture of *Massa medica*, *Crataegi fructus*, *Artemisia capillaries* and *Cndium officinale* reduced the bacterial load up to six times after 12 hours enrichment compared to a control.

In order to maintain healthy rotifer cultures, frequent cleaning/disinfecting of all used material is very important. After each harvest, the tanks, aeration tubing and harvesting devices must be disinfected using a sodium hypochlorite solution, a mixture of alkyldimethylbenzylammoniumchloride with glutaraldehyde or a mixture of hydrogenperoxide with peroxyacetic acid.

As new mass cultures are often started from a part of the harvest of the previous mass culture, contaminations (e.g. ciliates) accumulate. By frequently starting new mass cultures from the stock cultures, contaminations can be kept low. It is important to maintain the stock cultures in a separate room from the mass cultures to avoid cross-contamination. In the case where contamination persists or stock cultures are contaminated, a complete disinfection of rotifer stock cultures is necessary. A drastic way is to kill ovigerous females, separate the remaining amictic eggs and hatch them in axenic conditions. This can be achieved by exposure to 50–100 ppm glutaraldehyde for one to two hours at 28 °C for *B. plicatilis* sensu strictu (Tinh *et al.*, 2006), or by using trimetroprimsulfametoxasole at 10 mL/L (Martinez-Diaz *et al.*, 2003).

Rotifers can produce resting eggs through sexual reproduction. When these resting eggs were treated with 0.5 % NaOCl for 3 min, Douillet (1998) was able to obtain bacteria-free rotifers in 97 % of cases. Rombaut *et al.* (1999) successfully used glutaraldehyde to disinfect resting eggs. Suga *et al.* (2011) found that exposing resting eggs to 25 % NaOCl followed by growing the neonates in the presence of a mixture of four antibiotics led to an axenic rotifer culture.

5.4.3 Preservation techniques

As mentioned above, live rotifers can be stored at low temperature (4 °C) for several days without significant quality loss. Assavaaree *et al.* (2001) found that *B. rotundiformis* and smaller rotifer strains are more susceptible to cold conditions. Production of resting eggs can be a way to avoid continuous rotifer cultures outside the larvae production season. Hagiwara *et al.* (1997) have established a method with semi-continuous cultures to produce resting eggs of *B. rotundiformis* and *B. plicatilis*. They found that resting eggs could be stored for more than 20 years at 5 °C. However, the hatching rate decreased due to bacterial infection of the egg surface. Drying resting eggs through lyophilization and canning them at 61 kPa (Kilo Pascal) resulted in 35 % hatching after 12 months preservation. When the resting

© Woodhead Publishing Limited, 2013

eggs were treated with NaOCl or sodium nifurstyrenate, hatching increased to 68 % or 80 %, respectively (Balompapueng *et al.*, 1997).

5.5 *Artemia* as live feed: an overview

5.5.1 Morphology

Artemia is a primitive arthropod with a segmented body to which are attached broad leaf-like appendages named thoracopodes. Their length is ~8–10 mm for the adult male and ~10–12 mm for the female. The body is divided into head, thorax and abdomen. The head consists of one prostomial and five metameric segments which bear in order the median and compound eyes and labrum, first antennae, second antennae, mandibles, first maxillae or maxillulae, and second maxillae or maxillulae. The thorax is constructed of 11 segments, each provided with a pair of thoracopodes, while the abdomen is composed of eight segments. The anterior two abdominal segments are often referred to as the genital segments and of these the first bears the gonopods, either the brood pouch (uterus) of the female or the paired penes of the male. Abdominal segments 2–7 lack appendages. The final abdominal segment possesses the cercopods, also called the furca or telson. The entire body is covered with a thin, flexible exoskeleton of chitin to which muscles are attached internally. The exoskeleton is shed periodically and in females a moult precedes every ovulation, while in the male a correlation between moulting and reproduction has not been observed (see Criel and MacRae, 2002a, for more details on *Artemia* morphology and structure).

5.5.2 Taxonomy

The genus *Artemia* (Crustacea, Branchiopoda, Anostraca) comprises a number of sexually reproducing species ('bisexual species') and a number of parthenogenetically reproducing populations. There are very few macroscopically visible morphological differences between the various species of the genus. The identification of bisexual *Artemia* species has therefore been established by cross-breeding tests, morphological and morphometrical differentiation, cytogenetics and allozyme studies; presently, increasing importance is being given to nuclear and mitochondrial DNA analysis, including sequencing (Gajardo *et al.*, 2002). With the exception of cross-mating, all these techniques have also contributed to identifying the parthenogenetic types described as *A. parthenogenetica,* Barigozzi 1974. The phylogenetic relationship of populations and/or species within the genus is still a matter of discussion and the need for a multi-trait approach to identify species is generally recognized as essential (Baxevanis *et al.*, 2006). The name *A. salina* has caused considerable confusion worldwide as often authors have named (and still name) all brine shrimp *A. salina*, whereas this species name

© Woodhead Publishing Limited, 2013

should be restricted to one of the bisexual species, which is specifically found in the Mediterranean area (Triantaphyllidis *et al.*, 1997).

The differentiation of seven bisexual species, defined primarily by the criterion of reproductive isolation as found in laboratory tests, and of many parthenogenetic populations is currently acknowledged. Endemic to Europe, Africa and Asia (and also found in Australia) are the parthenogenetic populations (with different levels of ploidy). On these continents are also found the following bisexuals (see Gajardo *et al.*, 2002 for more details):

- *A. salina* (Leach, 1819): Mediterranean area
- *A. urmiana* (Günther, 1890): Lake Urmia, Iran and one Crimean site, Ukraine
- *A. sinica* (Cai, 1989): inland China and Mongolia
- *A. tibetiana* (Abatzopoulos *et al.*, 1998): Tibet
- *A. sp.* (Pilla and Beardmore, 1994): non-defined lake in Kazakhstan

Endemic to the Americas are:

- *A. persimilis* (Piccinelli and Prosdocini, 1968): southern South America
- *A. franciscana* (Kellogg, 1906): North, Central and South America, with *A. franciscana monica* being a special case of a population described for an ecologically unique habitat (Mono lake, California, USA).

5.5.3 Ecology

The brine shrimp *Artemia* is a zooplanktonic organism found globally in hypersaline habitats such as inland salt lakes, coastal salt pans and man-managed saltworks (see Van Stappen, 2002, for ecological aspects of *Artemia* distribution). Presently, more than 600 sites have been recorded, although such lists reflect systematic inventory work for specific areas rather than being an accurate reflection of true zoogeographical distribution, since many areas (e.g. sub-Saharan Africa) remain under-explored (Van Stappen, 2002). No *Artemia* is found in areas where year-round low temperatures exclude its development, but a lot of strains are found in the continental areas of North and South America and Asia with extremely cold winter temperatures, as long as sufficiently high summer temperatures allow cyst hatching and subsequent colonization of the environment.

Being extremely osmotolerant, brine shrimp survive in environments with salinities ranging between approximately 10 and 340 g/L with diverse ionic composition and temperature regimes; in general, the lower salinity threshold of its occurrence is determined by the salinity tolerance of its predators in the area, and abundant *Artemia* populations are consequentially only found at salinities elevated enough to eliminate (nearly) all predators or food competitors. *Artemia* is exceptionally adapted to such extreme environments, due to its unique osmoregulatory capacity and its capacity to synthesize highly efficient haemoglobins (see Clegg and Trotman,

© Woodhead Publishing Limited, 2013

2002, for more details on physiological and biochemical aspects of *Artemia* ecology). *Artemia* reproduces by two modes, involving either nauplius (ovoviviparous) or cyst (oviparous) production, depending on the prevailing ecological conditions (Criel and MacRae, 2002b). Ovoviviparity occurs under favourable ambient conditions: eggs (fertilized following mating in the case of bisexual species or non-fertilized in the case of parthenogenetic females) produce free-swimming larvae ('nauplii') released by the mother. On the other hand, oviparous reproduction occurs under unfavourable conditions usually characterized by factors such as high salinity, low oxygen levels, temperature stress, food depletion, etc. In this mode, the embryos only develop up to the gastrula stage and become surrounded by a thick shell (chorion) induced by hormonal secretions of the brown shell glands located in the uterus, thus forming what is referred to as a cyst. The embryo enters a state of metabolic arrest described as diapause and is spawned by the female (Lavens and Sorgeloos, 1987; Clegg *et al.*, 1996). Both oviparity and ovoviviparity are found in all *Artemia* strains, and female individuals can switch from one mode to the other between two reproduction cycles.

In nature, cysts can be produced in massive numbers, and the alveolar structure of the chorion ensures that large quantities float on the water surface, or may eventually be blown ashore by wind and waves. Upon dehydration, often in combination with other environmental cues such as hibernation, cyst diapause is deactivated, giving quiescent embryos with the ability to resume further embryonic development when hydrated in optimal hatching conditions. Once harvested and properly processed, the cysts can be stored for several years while the dried embryos stay in a state of arrested metabolism (Lavens and Sorgeloos, 1987; Clegg *et al.*, 1996). When quiescent cysts are immersed in lower salinity water, the biconcave cysts hydrate, becoming spherical and the shelled embryo resumes its interrupted metabolism. After a few more hours (depending on ambient conditions and strain) the cyst outer membrane breaks and the embryo appears, surrounded by a hatching membrane. At this point (umbrella stage), the embryo hangs underneath the empty shell, the development of the nauplius is completed and within a short period of time the hatching membrane ruptures (hatching) and the free-swimming instar I nauplius is born. This larva can be used as it is or, following a specific enrichment procedure to enhance its nutritional properties, as a convenient substitute for the natural plankton diet of fish and shrimp larvae (Dhont and Sorgeloos, 2002, and references herein; Støttrup and McEvoy, 2003). Under favourable ecological conditions, *Artemia* can live for several months, growing from nauplius to adult in only eight days and reproducing at up to 300 nauplii or cysts every four days.

5.5.4 Strain differences

The worldwide distribution of the brine shrimp *Artemia* in isolated habitats with specific ecological conditions has resulted in numerous geographical

© Woodhead Publishing Limited, 2013

strains, or genetically different populations within the same species (Van Stappen, 2002). Among these strains, a high degree of genetic variability as well as a unique diversity in various quantitative characteristics has been observed. Some of this variability is phenotypical, such as the nutritional composition of the cysts, and changes from batch to batch. Other characteristics such as cyst diameter, diapause characteristics, reproductive capacity and resistance to high temperature are considered strain-specific and remain relatively constant, i.e. they have become genotypical as a result of long-term adaptations of the strain to the local conditions (Vanhaecke and Sorgeloos, 1980, 1989). Knowledge of the characteristics (both genotypic and phenotypic) of a particular strain of cysts can greatly increase the effectiveness of its use in a fish or shrimp hatchery.

5.6 Diversification of *Artemia* resources

5.6.1 History of *Artemia* exploitation and rationale of diversification

The status of *Artemia* as an economic commodity dates back to the 1930s when some investigators adopted it as a convenient replacement for the natural plankton diet for fish larvae thus realizing the first breakthrough in the culture of commercially important fish species. In the 1950s, *Artemia* cysts were still predominantly marketed for the aquarium and pet trade; there were only two commercial sources: the coastal salt works in the San Francisco Bay (California, USA) and the Great Salt Lake (GSL) (Utah, USA). With fish and shrimp operations emerging from the early 1960s onwards, new marketing opportunities were created for *Artemia* cysts. However, by the mid-1970s increased demand, declining harvests from the GSL, high import taxes in certain developing countries and possibly artificial cyst shortages created by certain companies resulted in a severe price rise for *Artemia* cysts. The dramatic impact of the cyst shortage on the expanding aquaculture industry invigorated research on the rationalization of the use of *Artemia* and the exploration of new cyst resources. The cyst shortage simultaneously invigorated the search for alternatives for *Artemia* in an attempt to abandon its use as live feed in larval nutrition; a process that continues till today with slow but steady successes. Over the history of its exploitation, the GSL remained a natural ecosystem subject to climatic and other influences; this has been illustrated by unpredictable and fluctuating cyst harvests (see Dhont and Sorgeloos, 2002, and references herein, for more details). While at the end of the previous century, harvests from GSL were dramatically low, the situation has returned to normal since then with annual harvests making up more than 90 % of the world's cyst market. Despite this, the need for alternative resources and the increased demand from aquaculture has resulted in the occasional or regular exploitation of many other small and medium inland salt lakes, especially in southern Siberia, Kazakhstan and China and in coastal areas of the Bohai Bay, China,

© Woodhead Publishing Limited, 2013

along with further rationalization in the use of *Artemia* (Lavens and Sorgeloos, 2000; Dhont and Sorgeloos, 2002).

Along with the exploitation of natural resources, intensive cyst production in solar saltworks comprises an important market share in terms of high product quality and its importance in sustaining aquaculture development in many countries in the south. Often this production is carried out seasonally (e.g. in monsoon Southeast Asia) in areas where there is no natural occurrence of *Artemia*. This involves the deliberate transplantation of *Artemia*, not only for the production of *Artemia* cysts or biomass in itself but also because of the beneficial effect of *Artemia* presence on the salt production process. Depending on climatological conditions, inoculations can also be considered definitive when one or a few attempts of inoculation will lead to the permanent establishment of an *Artemia* population, as in Australia, China and East Africa. It is mainly in Vietnam that this activity has proven particularly successful. Since the first initiatives of the 1980s, interest in the seasonal culture of *Artemia* in the Mekong Delta has expanded and the know-how has gradually been transferred to artisanal salt farmers via local cooperatives. This region is currently an important supplier of high quality cysts for domestic use and for the international market (Nguyen Thi Ngoc Anh *et al.*, 2009, 2010).

5.6.2 Diversification of the resources

The bulk of the *Artemia* product reaching the world market is *A. franciscana* Kellogg 1906 from the GSL; product from central and eastern Asia consists of a variety of parthenogenetic strains and *A. sinica* Cai 1989. Harvests resulting from production in solar saltworks (such as in Vietnam) generally belong to the San Francisco Bay-type *A. franciscana*, as this strain has been used for the original inoculation material. This type of production may be the result of natural productivity (i.e. no or minimal human intervention to enhance productivity) or of intensive management procedures, as worked out in detail for seasonal solar saltworks in the Mekong delta, Vietnam (Baert *et al.*, 1997; Nguyen Thi Ngoc Anh *et al.*, 2009, 2010).

Depending on climatological conditions, an allochthonous strain may establish itself following deliberate or non-deliberate introduction by man. It should be noted that the recent gradual dispersion of *A. franciscana* into new environments competing with and eventually out-competing local populations is becoming an increasingly common pattern in various parts of the world (Mediterranean area, India, Sri Lanka, East Africa, Australia, coastal China; see for example Amat *et al.*, 2005). This adds to the complexity of the species status of cyst product originating from these areas (e.g. Bohai Bay, China) (Van Stappen *et al.*, 2007).

All *Artemia* product reaching the market is produced from feral strains (such as GSL) or from feral populations that have adapted following their introduction by man into a new environment (such as the Vietnamese

© Woodhead Publishing Limited, 2013

saltworks, Kappas *et al.*, 2004). So far no fully-fledged *Artemia* breeding or selection programmes have been launched, although research work on the heritability of commercially interesting characteristics is on its way (see for example Briski *et al.*, 2008).

Local harvesting procedures and regulations at GSL are strictly defined, enforced by local authorities and publically known. Harvesting regulations in other salt lakes (e.g. Russian Federation, Kazakhstan, China) are generally much less well defined (Van Stappen *et al.*, 2009). Although there may be a governmentally imposed quota system, the vastness and remoteness of the territory and the variety of lakes harvested make it difficult to eliminate all illegal harvesting. Generally the harvesting technology in these lakes is determined by the accessibility and topography of the site; the harvestable quantities and the predictability of harvests; the duration of the harvesting season; the characteristics of the local brine shrimp population; and the financial means for investment in harvesting logistics and infrastructure. Depending on local conditions, on the quality/quantity characteristics of harvests and on the business structure of harvesting companies, processing may be carried out partially on-site, and/or the product may be sent for domestic or overseas transport for final or complete processing elsewhere.

Production statistics are only available for GSL and for production in the Mekong Delta, Vietnam, where production is systematically monitored by local authorities or scientific organizations (see for example Wurtsbaugh and Gliwicz, 2001, for fluctuations in GSL production). In the absence of such public monitoring bodies in other areas, data on production lack completeness and/or reliability and/or are not openly accessible. At GSL, raw harvested quantities have been fluctuating over the past decade within the range of approximately 2000–12 000 tons. Raw cyst harvests include cysts, empty shells, *Artemia* biomass, algae and other material. The yield of dry, processed cysts from the raw product varies annually, but is typically 30–35 %.

Reliable estimates of future supplies remain difficult to obtain due to the lack of information on the ecology of new sites, the productivity of the local *Artemia* population, and on technical and economic studies related to accessibility, sustainable quantities, characteristics of the resource relevant to larviculture, etc. (see for example Marden *et al.*, 2012). Diversification of resources remains a most important issue, along with the further rationalization of the use of *Artemia*. As a net result of all the factors that play on the aquaculture market, the global demand for cysts, currently ~2500–3000 tons/year, is expected to increase further. China is the main consumer (and will continue to be so) with an annual consumption of ~1500 tons, of which about one half is imported from Russia and Kazakhstan and the other half produced domestically (mainly from inland lakes; Bohai Bay has had a relatively stable production over the past years of about 400 tons of raw product).

© Woodhead Publishing Limited, 2013

In the Mekong Delta there has been an initial expansion in the 1990s, thanks to the empirical development of the technique and the dissemination of knowhow to a wider group of salt farmers. Over the past few years, the culture area has more or less stabilized to 200–400 ha, producing 15–20 tons of raw product. Further consolidation of pond production technology into a fully scientifically sustained activity, combined with permanent and thorough dissemination of knowhow to the artisanal salt farmers, is a prerequisite for a really sustainable development of *Artemia* production in this region. Apart from cyst production, more emphasis will be given to *Artemia* biomass production (as a replacement for more traditional aquafeeds), thus providing more flexibility in integrated systems, and culture systems are being developed to make production of both products compatible (Baert *et al.*, 2002; Nguyen Thi Ngoc Anh *et al.*, 2009, 2010). Consolidation of the technology will also benefit from a governmental policy in Vietnam favourable for *Artemia* production (such as exists, for example, for aquaculture development). Only recently, a first econometric analysis of integrated *Artemia* production in the Mekong delta has been performed, formulating recommendations to increase the production efficiency from an economic point of view (Nguyen Phu Son, 2010). In parallel, it is expected that *Artemia* production will also be valorized as a means of extractive aquaculture (removal of excess nutrients through its grazing on phytoplankton, detritus and bacteria from effluents from – but not exclusively – aquaculture activities). Globally, it is expected that the model developed and the experience gained in Vietnam will be used for similar initiatives in other areas, albeit with the necessary adaptations, either as an entirely new activity (e.g. in sub-Saharan Africa) or to re-focus on the scientific management and valorization of *Artemia* production in areas where it has been introduced in the past (e.g. Bohai Bay area, China).

5.6.3 Diversification of the cyst end product

As early as the 1980s new insights in hatching characteristics and nutritional essentials gave rise to the segregation of different cyst qualities (Sorgeloos *et al.*, 1998, 2001; Sorgeloos, 1999). Nowadays cysts are offered on the market in a number of brands, corresponding with a variety of quality criteria, amongst which hatching quality, cyst size (and hence naupliar biomass) and nutritional composition (HUFAs, vitamins) are among the most important. These criteria, as well as practical aspects related to their daily manipulation in the hatchery (decapsulation behaviour, ease of nauplii harvesting, nauplius colour, enrichment kinetics, . . .) and microbiological aspects, all contribute to the sale prices of each brand. Consequently, favourable characteristics of the cyst product are used as sales argument, rather than its geographical origin. With the exploration and exploitation of resources other than GSL, further efforts are continuously made to expand and consolidate knowledge on how these strains can be applied in aquaculture with

© Woodhead Publishing Limited, 2013

maximal efficiency, i.e. optimal diapause deactivation strategy, state-of-the-art storage and processing technology, etc. The present trend of diversification of cyst products on the market, targeting at specific application aspects is expected to continue. Improvements related to general organizational issues such as reduction in manpower, space requirements for live feed production, smoothing out occasional variations in live feed quality, etc. are also a constant focus of cyst producers.

5.7 New developments in the use of *Artemia*

5.7.1 Hatching-related aspects

Protocols for hatching brine shrimp *Artemia* have been optimized over the years and are nowadays applied as a standard routine in commercial hatcheries and research institutes all over the world. In this chapter we will not reiterate established procedures. For a description of basic good practice of hatching, decapsulation and enrichment of *Artemia* we refer to Dhont and Van Stappen (2003). However, as the use of *Artemia* cysts in hatcheries requires a lot of labour and still makes up a considerable fraction of production costs, the manipulations needed to produce live feed out of the cysts are a focal R&D point for the cyst producing companies. The fact that a product or method allows automation or rationalization of daily procedures involved in decapsulation, hatching, nauplius harvesting, nauplius enrichment, nauplius storage, etc. is generally used as an important sales argument.

Moreover, in spite of the wide propagation within the sector of standard protocols for optimal *Artemia* use, practical working conditions in the hatcheries often force the cyst producing companies to permanent re-dissemination of knowhow and awareness-raising by means of training and demonstration sessions among their customers. The constant pressure to reduce production costs (e.g. by cutback in labour and/or in space required for live feed production) sometimes brings hatcheries to a reorganization of their live feed department leading to sub-optimal or mediocre output, for example by hiring cheap but unskilled labour, by unjustified simplifications of the standard procedures, etc.

The permanent need to rationalize live feed application procedures and to spread the related knowhow to the sector is further strengthened by the trend towards diversification. Firstly, the geographical origin of commercial cyst batches may be diverse (and new strains may occasionally – sometimes temporarily – make their appearance on the market). Secondly, application fields diversify with the increasing number of aquaculture species; this results in a permanent challenge to assess if the standard procedures (as worked out e.g. for Great Salt Lake *A. franciscana*, the most common strain on the market) also apply to cyst strains new in the market and to new ranges of application needed for larviculture of new species. Consequently,

© Woodhead Publishing Limited, 2013

favourable characteristics of the cyst product rather than its geographical origin are used as sales argument. As the availability of each strain might vary from harvest to harvest and as the characteristics of each batch are partially dependent on environmental conditions, the cyst producers perform regular biological, chemical, biometrical and nutritional analyses to determine the characteristics of the different cyst sources and batches, and to identify their proper fields of application.

When hatching *Artemia* cysts, considerable labour and sufficient skill is needed to separate the empty cyst shells and non-hatched full cysts from the instar I nauplii as much as possible and to concentrate and rinse the nauplii without damaging them. Thus rationalization of these procedures is a central point of attention. The technique of decapsulation, using a hypochlorite solution at elevated pH, has been worked out in the past to dissolve the chitinous cyst shell (chorion) prior to hatching, thus eliminating the need for cyst shell separation, and also as a prophylactic disinfection measure by reducing the bacterial and fungal load that is normally present on cyst shells (Van Stappen, 1996). As care should be taken to avoid prolonged exposure of the decapsulated *Artemia* embryo to the toxic hypochlorite solution, sufficient expertise is needed in the correct application of this technique, and its application modalities need amendment when using various *Artemia* strains which may differ in the characteristics of their chorion. Moreover, increasing restrictions imposed by authorities on discharge of effluents containing toxic compounds (such as hypochlorite) into the environment are creating a need to develop procedures minimizing the use of hypochlorite without, however, jeopardizing the efficiency of the treatment. The combination of R&D aimed at reduced labour and environmental friendliness is illustrated by the development of the Sep-*Art* SEPARATOR® (INVE, Baasrode, Belgium), a device which allows easy nauplii separation after hatching, independent of the origin or hatching quality, when using SEP-*Art*® *Artemia*, which is basically regular *Artemia* cysts provided with a magnetic coating on the cysts.

Microbiological aspects of live feed production and administration procedures are another focal point of new developments in the sector, in line with the increasing attention given to the microbiology of every conceivable aspect of larviculture in recent years (see Section 5.7.3). This includes zootechnical improvements in handling procedures (e.g. for optimal rinsing and concentration of instar I nauplii post hatching without causing naupliar damage and mortality which may enhance bacterial proliferation), but also development of new products claiming a positive effect by steering ('conditioning') of the microbial communities involved in live feed production (see Section 5.7.3). Some of these products claim additional beneficial zootechnical effects, which may be directly or indirectly related to the microbiological composition of the hatching medium: e.g. facilitation of separation of the nauplii post hatching, increasing naupliar 'vitality' and 'quality', reduction of foam formation during hatching and storage, etc. The active

© Woodhead Publishing Limited, 2013

agents may be commercialized either as a separate product (to be added to the hatching medium) or in combination with the cyst product itself.

The latter also applies to certain cyst batches which are offered on the market without diapause being properly broken during the processing line (e.g. for strains which require a very prolonged cold storage for optimal diapause breaking); in this case, prior to packing the cysts are mixed with a chemical agent (e.g. in crystalline form) which dissolves in the hatching medium and is claimed to break diapause during the hatching incubation process.

Quantitative criteria for the hatching output, such as 'hatching percentage' (the number of nauplii obtained from 100 full cysts) and especially 'hatching efficiency' (the number of nauplii obtained from one gram of cyst material) are universally utilized. Depending on the needs of specific fields of application, other quantitative information may be relevant, such as the amount of naupliar biomass obtained per weight unit of cysts; naupliar length at hatching; and naupliar growth during a standard hatching incubation period. Also hatching rate, expressed as the time lapse needed to obtain the hatching of a certain threshold, e.g. 85 or 90 % of hatchable cysts, or to obtain a certain amount of naupliar biomass, may be an important criterion of product performance.

5.7.2 Nutrition-related aspects

Soon after it became a popular live feed item in the early days of aquaculture, it was realized that *Artemia* was deficient in certain nutrients and therefore did not satisfy all requirements of the (especially marine) predator larvae it was fed to. Initially most attention was given to the HUFA content of *Artemia* as it was found that this was a crucial factor in the larval rearing success of fish and crustacean larvae. Sorgeloos *et al.* (2001) reviewed the history of research on dietary HUFAs. In the 1980s, focus was on the level of eicosapentaenoic acid (20:5n-3, EPA), while later on attention was shifted to docosahexaenoic acid (20:6n-3, DHA) and its ratio to EPA as this seemed to improve larval quality (e.g. stress resistance, pigmentation, etc.). In the late 1990s, it was then demonstrated that arachidonic acid (20:4n-6, ARA) and its ratio to EPA was also important. Information on the biochemical composition of *Artemia* can be found in Lavens and Sorgeloos (1996).

To overcome these deficiencies, different enrichment products and procedures were developed using selected microalgae, yeasts, (heterotrophically grown) bacteria, microencapsulated products, emulsified products and self-emulsifying concentrates or microparticulate products or combinations thereof. Enrichment techniques make use of the fact that meta-nauplii of *Artemia* are non-selective filter-feeders which take up practically all particles, as long as they are of an adequate dimension. Figure 5.2 shows an enriched *Artemia* meta-nauplius in which you can clearly see the small oil

© Woodhead Publishing Limited, 2013

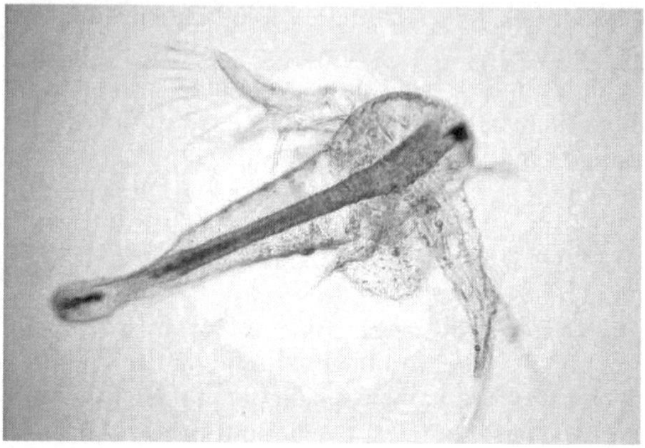

Fig. 5.2 Picture of an enriched *Artemia* metanauplius (courtesy Inve Aquaculture, Belgium).

droplets, collected from the enrichment product in the gut. Standard enrichment procedures and enrichment levels obtained therewith are described in detail in Lavens and Sorgeloos (1996).

While classical lipid emulsions and lipid concentrates remain very popular since they offer a simple and cost-effective way to boost nutrient levels in *Artemia* in a short time, they may also have some drawbacks as they provide HUFA mainly in the form of neutral lipids and therefore might cause an imbalance in protein/lipid ratio and an excess of trialcylglycerols. Enrichment products based on unicellular organisms offer the advantage that they deliver HUFA (partially) as polar lipids, which is regarded as beneficial for the predator larvae. A detailed discussion on this issue can be found in Conceição *et al.* (2010).

To date, modern commercial enrichment products not only boost essential fatty acid (EFA) levels, but aim to result in a more generally balanced live feed in terms of protein, phospholipid and energy content. Moreover, they generally include a wide range of essential or beneficial (micro-) nutrients such as vitamins, pigments, sterols, antioxidants, enzymes, etc. At the research level, the quest for the delivery of 'novel' nutrients (e.g. lysine, iodine, selenium, poly-β-hydroxybutyrate, . . .) through *Artemia* enrichment is on-going (e.g. Naz and Turkmen, 2009; Hawkyard *et al.*, 2011).

A trend in commercial enrichment products seems to be to include 'natural' ingredients such as microalgae or special yeasts in the formulation as these are thought to have, for example, additional nutritional or immune-stimulating action which cannot be provided through individual nutrients. Also, dry powdered enrichment products seem to become increasingly popular. Dry formulations might be less perishable, are more concentrated, can be more easily stored and are more convenient in use (e.g. dosage). New developments also include products that allow obtaining the same nutrient enrichment levels in a shorter time period or allow enrichment at

© Woodhead Publishing Limited, 2013

higher *Artemia* density, thus saving on time, space and labour requirements in hatcheries.

Lastly, enrichment procedures can also be engaged in the hygiene and disease control procedures in hatcheries through incorporating bacteriostatic products or serving as a vehicle for delivery of immunostimulants, drugs or possibly even vaccines. This will be discussed in more detail in the next section.

5.7.3 Microbiology-related aspects

Apart from the eggs or larvae themselves and the rearing water, the live feed is one of the main inputs during the larval rearing process, largely exceeding the larval biomass itself in quantity, and therefore potentially an important source of contamination. Several studies demonstrated that *Artemia* nauplii are a vector for introducing potentially harmful bacteria such as *Vibrio* spp. into the larval rearing system. Bacterial loads might be as high as 10^8 CFU/ml of *Artemia* homogenate.

Traditionally, the bacterial load of *Artemia* nauplii was controlled by either disinfecting the cysts with hypochlorite or removing the shell of the *Artemia* cysts altogether through the process of decapsulation (see Lavens and Sorgeloos, 1996). Another measure that is routinely taken to reduce the influx of bacteria into the larval rearing tanks is a thorough washing of the nauplii with clean (disinfected) water upon harvesting. Although these hygiene measures have been shown to reduce bacterial load to some extent, they are far from absolute. Bacteria numbers are known to increase dramatically during the *Artemia* hatching process. A subsequent enrichment step with lipid- and/or protein-rich products can further increase bacterial numbers. Moreover, bacteria are also taken up into the gut of the nauplii as was demonstrated by Høj *et al.* (2009). These authors showed that a washing step was effective to remove the bacteria that are loosely attached to the external surface of the nauplii; however, most bacteria were localized in the nauplii gut and these were not removed.

As mentioned earlier, larval feed companies, for these reasons, have developed specialized products which contain bacteriostatic agents to keep bacterial loads within acceptable levels. These include specially treated cysts, enrichment products with bacteriostatic properties, or separate formulations which can be added to the *Artemia* hatching or enrichment medium. Also at the research level, several attempts have been made to disinfect *Artemia* nauplii (e.g. Gatesoupe, 2002; Tolomei *et al.*, 2004; Gimenez *et al.*, 2006). These include treatment of either cysts or the hatched nauplii with biocides (e.g. formaldehyde), UV, ozone or peroxide-based products. Apart from the fact that these treatments might cause considerable mortality to the nauplii or reduce their vigour, they might also pose potential risk to the predator larvae they are fed to, because of residues or toxic by-products produced due to these treatments.

© Woodhead Publishing Limited, 2013

Therefore, a better technique seems to be to try to replace the existing, potentially pathogenic microflora by one that is harmless or even beneficial to the fish or shrimp larvae, by applying probiotic mixtures to the live feed cultures, eventually preceded by an initial disinfection treatment. Several studies (e.g. Skjermo and Vadstein, 1999; Gatesoupe, 2002) proved that this is feasible sometimes with enhanced performance of the predator larvae.

Finally, the non-selective filter-feeding behaviour of *Artemia* also offers potential to use them for delivery of immunostimants, drugs, antibiotics or even vaccines (although the latter might not be very effective through an oral route). For example, yeast cells or fragments thereof, many of which have been shown to have immunostimulatory properties, are readily taken up and accumulated by *Artemia* nauplii. Similarly, liposoluble drugs or antibiotics can be easily combined with standard enrichment products and bio-encapsulated into *Artemia*. Water-soluble compounds can be uploaded through the use of e.g. liposomes. In literature (Roiha *et al.*, 2010; Touraki *et al.*, 2010; Allender *et al.*, 2011; Kanjana *et al.*, 2011), many reports can be found on the use of medicated *Artemia* nauplii, e.g. with flumequine, florfenicol, metronidazole, seaweed-extracts, etc. Except maybe for immunostimulants, to our knowledge, this so far has only limited applications in practice however.

5.7.4 Live feed replacement and early weaning

The introduction of *Artemia* as live food in the early stages of cultured species was undeniably a crucial breakthrough in early aquaculture and it has played a decisive role in the impressive expansion of commercial aquaculture as witnessed since the early 1980s. Nonetheless, since the very first moment *Artemia* was introduced as live feed, relentless efforts have been made to get rid of it and replace it with formulated diets. Notwithstanding the unique characteristics of *Artemia* that have rightly procured it its ubiquitous use in larviculture, it remains a live organism that has and is frustrating farmers with its degree of unpredictability and variability, not least in supply and, subsequently, price. As the industry matured, the need for complete control over every stage and every input in the production process grew stronger. Dependence on live feed is increasingly a thorn in the side of farmers who nowadays have to comply with managerial rules if they wish to remain competitive. Hence the permanent efforts to develop formulated diets with comparable performance to the current live feed, rotifer and *Artemia*.

Ground-breaking new developments on formulated feed are not common. Progress tends to be stepwise while, in addition, the degree of successful replacement of live feed varies with the species. It is well known that for most commercially farmed freshwater fish, with their relatively larger larvae, complete substitution is possible; there, the whole culture can

© Woodhead Publishing Limited, 2013

Table 5.1 Overview of studies with live feed replacement diets

Cultured species	Reference
African catfish (*Clarias gariepinus*)	Chepirui-Boit *et al.*, 2011 Vandecan *et al.*, 2011
Barramundi (*Lates calcarifer*)	Curnow *et al.*, 2006
Atlantic cod (*Gadus morhua*)	Fletcher *et al.*, 2007
Atlantic halibut (*Hippoglossus hippoglossus*)	Murray *et al.*, 2010
Senegalese sole (*Solea senegalensis*)	Engrola *et al.*, 2010
Cobia (*Rachycentron canadum*)	Huy *et al.*, 2011
Mud crab (*Scylla serrata*)	Holme *et al.*, 2006
Ornate spiny lobster (*Panulirus ornatus*)	Johnston *et al.*, 2008
White shrimp (*Litopenaeus vannamei*)	Gamboa-Delgado and Le Vay, 2009
Brown shrimp (*Farfantepenaeus aztecus*)	Robinson *et al.*, 2005
American lobster (*Homarus americanus*)	Tlusty *et al.*, 2005
Various species	Conceição *et al.*, 2010
Various freshwater fish	Sales, 2011

be achieved without live feed albeit possibly with lower survival and/or growth than when feeding partially with live feed. In general, the replacement of live feed is typically achieved through a process of gradually increasing the ratio of formulated feed versus live feed and/or gradually shortening the live feed phase in favour of the earlier weaning to formulated feed. Table 5.1 presents an overview of studies with live feed replacement diets.

Evidently, the steady progress in formulated feeds will ultimately lead to the complete replacement of live feed but, although this has been predicted for decades, the utilization of *Artemia* in marine hatcheries remains a reality. In the end, commercial hatcheries adhere to a pragmatic approach: from a managerial point of view formulated feed is preferred for its consistency in quality, price and supply. Even a reasonable loss in hatchery output caused by a switch from live feed to formulated feed will be accepted for the sake of a stable production. Eventually, it is the price of *Artemia* cysts that will determine the choice between *Artemia* or formulated food according to each hatchery's specific cost–benefit structure. It is not uncommon to see hatcheries moving back and forth between live or formulated feed as prices of *Artemia* cysts fluctuate.

It is important to note that the scientific progress, as exemplified by the selection of articles above, does not necessarily reflect common procedures in commercial hatcheries worldwide. On the contrary, due to commercial value of the production, hatcheries tend to be reluctant to adopt new developments as long as consistency is not guaranteed or proven. It is interesting that the opposite is increasingly observed: current practice in hatcheries is frequently ahead of scientific developments. Modern hatcheries or feed companies are developing novel larval feeds and/or feeding strategies, but

© Woodhead Publishing Limited, 2013

this remains largely unnoticed by the academic world because it is not published, either because publishing is simply not part of the core business of private companies, or even because publishing can be against the commercial interest of the company.

Use of Artemia *umbrellae for earlier weaning*

The umbrella-stage is the stage of an *Artemia* embryo after it has broken out of the cyst shell but before it is released from the cyst membrane that envelopes the nauplius. It is therefore slightly smaller than the nauplius and immobile. For fish larvae with a critically small mouth opening, it offers the opportunity to offer a smaller live feed particle. In Vietnam, it seems to be common practice in some cobia hatcheries and at least one study demonstrates the possibility to skip the use of rotifers by using umbrella-stage *Artemia* in cobia (*Rachycentron canadum*) larviculture (Van Can Nhu *et al.*, 2010).

5.8 Copepods as live feed: an overview

In nature, copepods constitute the major part of the diet of early life stages of many marine fish species. Copepods have been demonstrated to be a nutritionally adequate live prey for rearing larvae of many marine fish species (McEvoy *et al.*, 1998; Shields *et al.*, 1999; Payne *et al.*, 2001; Støttrup, 2003; Busch *et al.*, 2011). However, larval rearing in intensive commercial hatcheries relies exclusively, or almost exclusively, on traditional live prey; enriched rotifers and enriched nauplii of *Artemia* sp. Despite significant progress in the development of copepod culture techniques (Støttrup *et al.*, 1986; Payne and Rippingale, 2001; Støttrup, 2003; Lee *et al.*, 2005; Drillet *et al.*, 2011a), the challenge still remains to develop an efficient, cost-effective large-scale production method that could match or compete with the established methods for culturing traditional live prey. The lack of copepods in aquaculture has for several species led to poor survival, low growth and/or inferior quality fry characterized by mal-pigmentation or other deformities (Næss *et al.*, 1995; Støttrup *et al.*, 1998; Hamre *et al.*, 2005; Busch *et al.*, 2011).

In what follows, progress on the culture techniques for copepods, recent revelations on biochemical content and developments for use of copepods in larval rearing are reviewed.

5.8.1 Biology, morphology and taxonomy

Copepod species in laboratory cultures usually belong to one of three of the ten orders of copepods, the Calanoida, Harpacticoida and Cyclopoida. Dussart and Defaye (2001) and Sazhina (2006) provide comprehensive

© Woodhead Publishing Limited, 2013

reviews on copepod biology and ecology, whereas Støttrup (2003) and Lee *et al.* (2005) provide reviews on copepod cultures and their use as live prey. Since the Calanoida are the most abundant in the marine environment, these have received most attention (Mauchline, 1998). The most widely studied calanoid species are *Acartia* spp. and *Calanus* spp., although much attention has also been given to *Temora*, *Paracalanus*, *Pseudocalanus* or *Centropages* spp. (Mauchline, 1998). In aquaculture, the most frequently used calanoid species for mono- or mixed cultures are *Acartia*, *Centropages* or *Eurytemora* spp. (Støttrup, 2003).

Copepods have a chitonous exoskeleton, with a fused head and thorax. Several appendages used for feeding or locomotion protrude from the ventral thorax section. Anteriorly, copepods have a typical naupliar eye and a conspicuous set of antennae. In adults, the size of the antennae is useful to identify whether the species belongs to one of the three orders of copepods mentioned above, the harpacticoids having the shortest antennae. The calanoids have long antennules, as long as the body itself or longer (Dussart and Defaye, 2001) and the adults are generally relatively easy to identify, at least to genus. The cyclopoids have antennae that rarely extend beyond the cephalothorax, thus shorter than calanoids and longer than those in harpacticoids. The calanoids are generally planktonic and require phytoplankton as food. Many harpacticoids are generally benthic grazers and are found on surfaces of grasses or other structures. They feed on a variety of substances including bacteria and bacterial films. Marine cyclopoids can be both benthic and planktonic, and are considered to be omnivorous.

Most copepods reproduce sexually. The genital opening is generally located in the first abdomen segment. Calanoids are generally broadcasters shedding eggs individually which sink slowly through the water column. Both harpacticoids and cyclopoids have one or two egg sacs borne by the female, in which the nauplii develop until they are hatched directly from the egg sac. The newly hatched nauplii are generally small in size. For example, the width of the first naupliar stage of *Acartia* spp was measured to 65 μm (Schipp *et al.*, 1999) and to ~38 μm for *Parvocalanus crassirostris* (McKinnon *et al.*, 2003) and are thus ideal prey for fish with small gapes. The nauplii molt through six stages, with a progressive increase in size and development of appendages. The final naupliar stage moults into a copepodite, resembling the adult form, but goes through six moults before reaching the final adult stage (Fig. 5.3).

5.8.2 Culture systems

Most cultures described are batch systems with regular water exchange, although there are promising developments with continuous or automated systems. A comprehensive list of mass culture systems is provided by

© Woodhead Publishing Limited, 2013

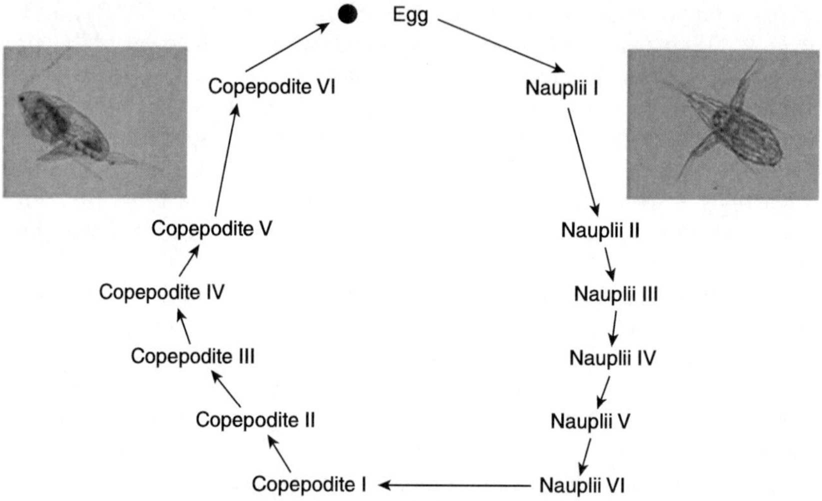

Fig. 5.3 Schematic life-cycle of a copepod.

Støttrup (2003), and Drillet *et al.* (2011a) reviewed the state of development of copepod cultures and made several proactive recommendations including a need to focus future work on automation and implementation of recirculation technology.

Extensive or semi-extensive cultivation

Copepod cultures in outdoor ponds or enclosed lagoons provide the basis for cost-effective, low technology extensive rearing of marine species such as turbot, flounder and cod, but have so far mostly been used commercially in Scandinavian countries (Støttrup *et al.*, 1994; Svåsand *et al.*, 1998; Engell-Sørensen *et al.*, 2004). These systems rely on bloom culture of phytoplankton and zooplankton to which the newly hatched larvae are added. They cannot operate during the winter due to the low prevailing air and sea temperatures. In the spring, the lagoons are filled with water filtered (20–40 μm) to allow phytoplankton to pass through but preventing potential predators from entering the system. Fertilizers may be added to enhance the phytoplankton bloom and careful administration of nutrients may be used to selectively favour growth of particular algal species, such as growth of diatoms (Naas *et al.*, 1991). Copepods are then filtered from the surrounding seas and added to the system. In some systems, the copepod production is started with diapause eggs that hatch from the sediment of the pond. These eggs were most likely produced by copepods towards the end of the previous season, as temperatures and day length started decreasing (Engell-Sørensen *et al.*, 2004). In these systems, a mixture of copepods is generally observed, providing a range of size and species representing all

© Woodhead Publishing Limited, 2013

three genera mentioned above (Svåsand *et al.*, 1998). Prey densities obtained may vary from 10–300 ind. L^{-1} in Norwegian poll systems (Svåsand *et al.*, 1998), to 374 nauplii L^{-1} of *Acartia* sp. and *Oithona* sp. (Colura *et al.*, 1987) to > 1000 nauplii L^{-1} of *Apocyclops panamensis* (Phelps *et al.*, 2005) in semi-extensive systems. Naturally, with such low prey densities, it is important not to load the systems with high fish larval densities. In contrast to intensive systems, these systems are normally stocked with fish larvae at densities well below 1 L^{-1} (e.g. ~30 larvae m^{-3} (Busch *et al.*, 2009) or 50—~80 larvae m^{-3} (Engell-Sørensen *et al.*, 2004). In periods with shortage of zooplankton production, zooplankton grown in separate ponds are added to the fish rearing ponds, or *Artemia* nauplii are supplied to meet the energy requirements of the growing larvae (Berg, 1997; Støttrup, 2003; Engel-Sørensen *et al.*, 2004).

Most outdoor systems are run as batch cultures, but Lemus *et al.* (2004) experimented with different harvest rates on production of wild harvested copepods (~80 % *Acartia tonsa*) grown in filtered brown-water (water from a nutrient-rich estuary). Highest naupliar harvest was observed in the group with the highest harvest rate (75 %) and where brown-water and rice bran were added at each harvest, which occurred every second day. These systems should be further refined for use in large-scale extensive culture systems. For experimental purposes, copepods are harvested from larger ponds through a filter system, separated into different size ranges and fed to intensive larval rearing systems (e.g. Busch *et al.*, 2011).

Intensive systems – batch cultures

Start-up procedures for cultures of calanoid copepods for intensive systems are described by Shields *et al.* (2005) and Støttrup (2006). Drillet *et al.* (2011a) describe an ideal automated large-scale copepod production system. The general idea is to exchange all labour-intensive processes with automated ones. An example of such an endeavour is provided by Alver *et al.* (2011), who developed a plankton counter able to estimate nauplii size, development stage and density for *A. tonsa*. This would allow for automated monitoring of culture development. Drillet *et al.* (2011a) further suggested applying re-circulation systems to maintain water quality, but filtration, treatment of incoming water (e.g. UV disinfection) and a good balance between food provision (phytoplankton) and water exchange rate may be sufficient to maintain water quality in copepod cultures.

Planktonic copepod species generally require a supply of suitably-sized phytoplankton at densities to obtain maximum filtration rates to yield maximum growth rates (Støttrup and Jensen, 1990; Payne and Rippingale, 2001; Støttrup, 2003, 2006). Maximum growth rates in adult copepods that do not build up large energy reserves are equivalent to egg production rates. A mixture of algae is often used to ensure the size range and nutritional value requirements for the different copepod stages (Knuckey *et al.*, 2005; Shields *et al.*, 2005; Gunvor *et al.*, 2011). Algal quality and quantity are

© Woodhead Publishing Limited, 2013

important for ensuring both high egg production, which is mostly dependent on energy, and high hatching success, which is tightly related to HUFA dietary content (Koski *et al.*, 2006). Dried or freeze-dried phytoplankton may be a potential substitute feed to fresh phytoplankton, but to utilize this food source effectively would require a system that would ensure it is maintained in the water column. The challenge for batch cultures for planktonic species is to administer the feed at sufficient levels to maintain egg production without compromising water quality and to ensure regular water exchange to counteract problems that may arise from sedimented feed and faecal pellets.

Densities for culture of calanoid species are usually within the range 100–2000 L^{-1} (Støttrup, 2003), although the paracalanid *Parvocalanus* sp. can be kept in culture at densities of up to 30 ml^{-1} (Shields *et al.*, 2005). In short-term experiments, adult stocking density for the calanoid *A. tonsa* ranging from 100–600 L^{-1} was demonstrated to be unrelated to adult mortality rate, egg production rate or egg hatching success (Jepsen *et al.*, 2007), although Peck and Holste (2006) reported a decrease in egg production from 40 to 16 to 10 eggs $female^{-1}$ day^{-1} when adult stocking density increased from 65 to 166 to 425 adults L^{-1}. This was despite the higher female to male ratio of 5:1 relative to 3:1 in the previous study. Female fecundity was also observed to decline in cultures of the paracalanid *Bestiolina similis*, but naupliar production was improved by better timing culture dilution (VanderLugt and Lenz, 2008). In most batch systems, decline in egg production with culture age is counteracted by periodic dilution either through the harvest procedure (Støttrup *et al.*, 1986; Payne and Rippingale, 2001), or regular reseeding with lower adult densities (Schipp *et al.*, 1999). There is thus further scope to examine density-dependent processes and improve production output in copepod cultures.

Production is through harvest of either eggs (mainly from broadcasting copepods) or nauplii (from egg bearing copepods). Eggs will sediment out and can be siphoned off the bottom. Nauplii can be separated from adults by sieving after harvesting a proportion of the culture volume. In some systems the production output is relatively high. The paracalanid *Parvocalanus* sp. can produce 3.7 nauplii ml^{-1} daily (Shields *et al.*, 2005) and the calanoid *Gladioferens imparipes* 440 000 nauplii day^{-1} (Payne and Rippingale, 2001). There is scope for development in automating the harvesting process.

Benthic copepod species can be fed a variety of inert feeds and are not dependent on a supply of fresh phytoplankton (Støttrup, 2003; Rhodes and Boyd, 2005). Many harpacticoids and some cyclopoids have generally a high tolerance to a wide range of environmental conditions. They are generally very productive, can attain high population densities in culture, are relatively tolerant to decreased water quality and to invasive species such as ciliates, nematodes and rotifers (Uhlig, 1984; Støttrup and Norsker, 1997; Pinto *et al.*, 2001; Cutts 2002; Rhodes, 2003; Støttrup, 2003). These traits,

© Woodhead Publishing Limited, 2013

together with a relatively short life-cycle such as 12 days at 18 °C in *Tisbe holothuriae* (Støttrup, 2000), 10–12 days at 20 °C in, *Nitokra lacustris* (Rhodes *et al.*, 2003) and four to five days in *Apocyclops royi* (Su *et al.*, 2005) make these species ideal for culture purposes. Benthic species require a surface area to graze on. Different substrates can be provided in the culture to increase the surface area per volume or per floor area. For example Støttrup and Norsker (1997) increased the surface area within a 150 L tank to an estimated 17 880 cm^{-2}. A disadvantage is their benthic behaviour which may make this prey unavailable to the fish larvae. This may however not apply to naupliar stages of all benthic species. *T. holothuriae* nauplii were shown to be available in the water column ten hours after addition, and thus available to the turbot larvae (Støttrup and Norsker, 1997).

Harpacticoid and cyclopoids can be grown at densities of 20–> 40 ind. ml^{-1} (Støttrup, 2003, 2005) and even as high as 115 ind. ml^{-1} (Kahan *et al.*, 1982). Thus they compensate for the lower fecundity relative to broadcast spawners (Kiørboe and Sabatini, 1995), with a higher reproductive capacity, higher culture densities and shorter life-cycles. Rhodes (2003) estimated that three 200 L batch cultures of copepod (harpacticoids) reaching a final density of 43 ind. ml^{-1} was sufficient to feed 4000 halibut in a 1500 L tank, at least through the sensitive phase. In a recent study, Ribeiro and Souza-Santos (2011) were able to harvest an average of 28 000 ind. L^{-1} day^{-1} of *T. biminiensis* over 130 days and in cultures of 4.5 L volume. Thus batch cultures could be a low-tech albeit in some cases labour-intensive short-term solution to providing copepods in marine fish rearing.

Continuous and automated cultures

Several advances have been made towards developing automated and continuous systems for the production and harvest of copepods for aquaculture, but there is still more research and development needed. An efficient system for the automated collection of eggs for broadcasting copepods was developed by Toledo *et al.* (2005), whereby adults were retained in the upper water layer by a mesh filter, and eggs and algae airlifted back into a retainer tank where eggs were concentrated within a sieve. Sun and Fleeger (1995) developed an automated system for the continuous culture of *Amphiascoides atopus* in a volume of 1440 L covering 4 m^2 floor space. In this system, about 500 000 individuals were harvested daily over five months. Støttrup and Norsker (1997) described an automated system for the culture and harvest of *T. holothuriae* nauplii with an average daily production of 500 000 nauplii and copepodites over 1.5 months. Also, Payne and Rippingale (2001) developed an automated double 500 L system for the culture of *G. imparipes* whose females carry an egg sac and harvest of up to ~880 nauplii L^{-1}. Despite the advancements in these endeavours, close collaboration with the industry is needed to take the final step towards commercial culture systems.

© Woodhead Publishing Limited, 2013

5.9 Copepods as live feed: nutritional value, microbiology and preservation techniques

5.9.1 Nutritional value

The availability of copepods seems to be especially crucial for rearing new species or species that require very small prey at first feeding, such as pink snapper (*Pagrus auratus*), West Australian dhufish (*Glauscosoma hebraicum*) (Payne *et al.*, 2001), grouper (*Epinephelus coioides*) (Toledo *et al.*, 2005), red snapper (*Lutjanus campechanus*) (Ogle *et al.*, 2005) and yellowtail clownfish (*Amphiprion clarkii*) (Olivotto *et al.*, 2008). The suitability of copepods as live prey for the more established species has also been demonstrated in a number of studies such as for halibut (*Hippoglossus hippoglossus*) (Shields *et al.*, 1999), cod (*Gadus morhua*) (Busch *et al.*, 2011) and seabass (*Lates calcarifer* Bloch) (Rajkumar and Vasagam, 2006). Lipids, both content and quality, seem to be the major factor determining nutritional value for marine fish larvae (e.g. Hamre *et al.*, 2005).

HUFAs are required by copepods, especially in planktonic species used to depending on a mixture of phytoplankton in nature. These may not be essential for egg production, but are apparently important to ensuring egg hatching success (Koski *et al.*, 2006). Copepods are unable to synthesize short-chain fatty acids to polyunsaturated fatty acids (PUFAs) at any significant rates and thus require these fatty acids through their diet (Bell *et al.*, 2007). There is some evidence that harpacticoids do have an ability to produce PUFA (Norsker and Støttrup, 1994; Nanton and Castell, 1999), possibly reflecting evolutionary pressures to develop the ability to synthesize EFA to compensate for the poor availability of these fatty acids in the sub-littoral benthic environment inhabited by these species (Bell *et al.*, 2007). While rotifers and *Artemia* can be enriched to imitate PUFA content in copepods, copepods contain most of their HUFA in the polar lipid fraction (Bell *et al.*, 2003), whereas in *Artemia* these are contained in storage lipids (Nanton and Castell, 1999). Furthermore, enriched rotifers and especially *Artemia* nauplii have higher neutral lipids than phospholipids (Fig. 5.4).

It has been suggested that dietary phospholipids are more biologically available to the developing fish larvae (McKinnon *et al.*, 2003), and more work is needed focusing on enhancing the phospholipid content of the traditional live prey by enrichment. DHA and EPA content is higher in both extensively and intensively reared copepods (adults and nauplii) and proportionally higher than in rotifers or *Artemia* nauplii (Fig. 5.4). These fatty acids have been demonstrated essential for the normal development, growth and survival of marine fish larvae (McEvoy *et al.*, 1998; Shields *et al.*, 1999; Hamre *et al.*, 2005; Busch *et al.*, 2011).

The developing fish larvae have low protein digestibility, but high assimilation capacity for free amino acids (FAA), which were twice as abundant in copepods as compared to enriched *Artemia* nauplii (van der Meeren

© Woodhead Publishing Limited, 2013

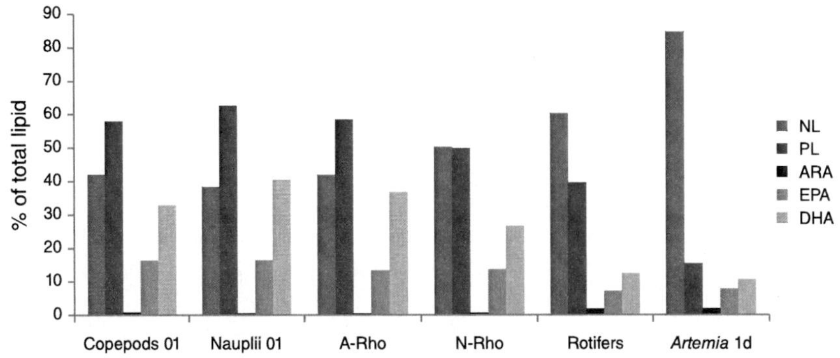

Fig. 5.4 Percentage content of neutral (NL) and polar (PL) lipids and of 20:4 n-6 (ARA), 20:5n-3 (EPA) and 22:6n-3 (DHA) in copepod adults (250–800 μm) and nauplii (80–150 μm) collected from an open Norwegian pond, rotifers enriched with *Isochrysis* and RotiMac™ and 1-day old *Artemia* nauplii enriched with DC-DHA Selco (data from van der Meeren *et al.*, 2008) and from *Acartia tonsa* adults (A-Rho) and nauplii (N-Rho) grown in culture on *Rhodomonas baltica* (data from Støttrup *et al.*, 1999).

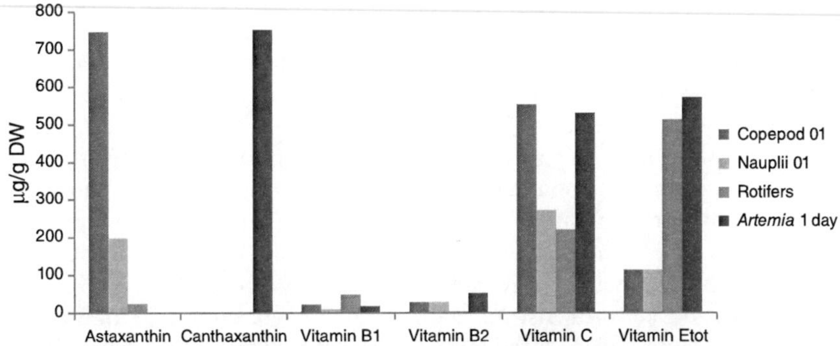

Fig. 5.5 Content of pigments and vitamins in copepod adults (250–800 μm) and nauplii (80–150 μm) collected from an open Norwegian pond, rotifers enriched with *Isochrysis* and RotiMac™ and 1-day old *Artemia* nauplii enriched with DC-DHA Selco (data from van der Meeren *et al.*, 2008).

et al., 2008). Digestion may be further aided by high autolytic activity of the ingested prey, and copepods were found to be more readily digested than *Artemia* by Atlantic halibut larvae (Luizi *et al.*, 1999).

Astaxanthin levels are generally high in copepods, very low in rotifers and not found in *Artemia* nauplii (Fig. 5.5). In *Artemia*, canthaxanthin is found in high amounts instead (van der Meeren *et al.*, 2008). Both these carotenoids may serve as precursors for vitamin A in fish. According to Bell *et al.* (2000), astaxanthin together with vitamin E work synergistically to suppress lipid peroxidation and are thus important in fish with high PUFA

© Woodhead Publishing Limited, 2013

requirements. Total levels of vitamin E were, however, higher in rotifers and *Artemia* nauplii than in copepods (Fig. 5.5; van der Meeren *et al.*, 2008). Vitamin C content was similar in adult copepods and in *Artemia* nauplii, but about half the amount in copepod nauplii and rotifers (Fig. 5.5).

Copepods may be provided during a sensitive period (Atlantic halibut: Næss and Lie, 1998) or as a supplement to the traditional feed – (e.g. turbot (*Scophthalmus maximus*): Støttrup and Norsker, 1997; Dover sole (*Solea solea*): Heath and Moore, 1997). In these systems, *Artemia* nauplii seem to meet energy requirements, while the copepods supplement the diet with essential nutrients for improved growth, survival or higher proportion of normally developed fry. Preserved copepods were also shown to be beneficial as a supplement to traditional live prey for an ornamental fish *Amphiprion clarkia* using frozen preserved copepods harvested from the wild (Olivotto *et al.*, 2010).

5.9.2 Microbiology

The use of freshly harvested copepods directly for fish larval rearing presents a risk of introducing parasites (Støttrup, 2003) and harmful bacteria to the larvae such as different *Vibrio* species (Gugliandolo *et al.*, 2008). Using only the first generation nauplii of freshly harvested copepods can minimize risk of parasitic infections in cases where the copepod act as intermediary host between compulsory hosts. In general, bacterial numbers associated with copepods seem to be lower in copepods than in *Artemia* nauplii (Verner-Jeffreys *et al.*, 2003). As copepods and nauplii seem to be well able to tolerate filtration, the copepod culture medium can be washed out thoroughly before concentrating and feeding the live prey to the fish, thus minimizing bacterial transfer. In some systems this is part of the harvest process, to minimize contaminants in the culture systems (Støttrup, 2003, 2006).

In copepod cultures, risk of ciliate infections seems to be one of the important factors for successful culture production (Støttrup, 2003). However, egg production and egg hatching success in *A. tonsa* cultures was shown to be improved with the addition of a commercially available probiotic (Sorbial A/S DANISCO) (Drillet *et al.*, 2011b), indicating that bacterial populations in the cultures may also play a significant role.

5.9.3 Preservation techniques

Olivotto *et al.* (2010) describes the use of frozen preserved copepods for co-feeding with traditional live prey as a feed supplement. The use of subitaneous and diapause eggs has been explored for use in aquaculture (Marcus and Murray, 2001; Drillet *et al.*, 2006, 2007; Ohs *et al.*, 2010) but so far only with some success in extensive systems (Engell-Sørensen *et al.*, 2004). Although diapause eggs can be induced (e.g. in *Centropages hamatus*: Marcus and Murray, 2001) and *A. tonsa* is known to produce diapause eggs

© Woodhead Publishing Limited, 2013

(Mauchline, 1998), it may prove too complicated for commercial purposes to produce diapause eggs in the laboratory. Subitaneous eggs of *A. tonsa* can be collected and stored in anoxic cool (~4 °C) conditions for a period up to six months (Peck and Holste, 2006) and 11 months (Drillet *et al.*, 2006). Egg hatching rates declined steadily with time during cold storage (Peck and Holste, 2006), but even after 11 months up to 70 % of those eggs considered viable hatched (Drillet *et al.*, 2006). The newly-hatched nauplii from cold stored eggs (three months) contained lower levels of EPA and especially DHA, whereas ARA levels had increased (Støttrup *et al.*, 1999). Attempts to prolong shelf-life for quiescent eggs (cold-stored) are in progress (e.g. Drillet *et al.*, 2007), but more work needs to be done towards a commercial product for cold stored eggs with reliable hatching success.

5.10 References

ALLENDER M C, KASTURA M, GEORGE R, BULMAN F, YARBROUGH J and COX S (2011) Bioencapsulation of metronidazole in adult brine shrimp (Artemia sp.). *Journal of Zoo and Wildlife Medicine* 42(2): 241–246.

ALVER M O, ALFREDSEN J A, OIE G, STOROY W and OLSEN Y (2010) Automatic control of growth and density in rotifer cultures. *Aquaculture Engineering* 43: 6–13.

ALVER M O, STORØY W, BARDAL T, OVERREIN I, ONSØYEN M K, TENNØY T and ØIE G (2011) Automatic measurement of *Acartia tonsa* nauplii density, and estimation of stage distribution. *Aquaculture* 313: 100–106.

AMAT F, HONTORIA F, RUIZ O, GREEN A J, SANCHEZ M I, FIGUEROLA J and HORTAS F (2005) The American brine shrimp as an exotic invasive species in the western Mediterranean. *Biological Invasions* 7: 37–47.

ASSAVAAREE M, HAGIWARA A, IDE K, MARUYAMA K and LUBZENS E (2001) Low-temperature preservation (at 4 °C) of marine rotifer *Brachionus. Aquaculture Research* 32: 29–39.

BAERT P, NGUYEN THI NGOC ANH, VU DO QUYNH, NGUYEN VAN HOA and SORGELOOS P (1997) Increasing cyst yields in *Artemia* culture ponds in *Vietnam*: the multi-cycle system. *Aquaculture Research* 28: 809–814.

BAERT P, NGUYEN THI NGOC ANH, BURCH A and SORGELOOS P (2002) The use of *Artemia* biomass sampling to predict cyst yields in culture ponds. *Hydrobiologia* 477: 149–153.

BALOMPAPUENG M D, HAGIWARA A, NOZAKI Y and HIRAYAMA K (1997) Preservation of resting eggs of the euryhaline rotifer *Brachionus plicatilis* O.F. Muller by canning. *Hydobiologia* 358: 163–166.

BAXEVANIS A D, KAPPAS I and ABATZOPOULOS T J (2006) Molecular phylogenetics and asexuality in the brine shrimp *Artemia. Molecular Phylogenetics and Evolution* 40: 724–738.

BELL J G, MCEVOY L A, TOCHER D R and SARGENT J (2000) Depletion of α-tocopherol and astaxanthin in Atlantic salmon (*Salmo salar*) affects autoxidative defense and fatty acid metabolism. *Journal of Nutrition* 130: 1800–1808.

BELL J G, MCEVOY L A, ESTEVEZ A, SHIELDS R J and SARGENT J (2003) Optimising lipid nutrition in first-feeding flatfish larvae. *Aquaculture* 227: 211–220.

BELL M V, DICK J R, ANDERSON T R and POND D W (2007) Application of liposome and stable isotope tracer techniques to study polyunsaturated fatty acid biosynthesis in marine zooplankton. *Journal of Plankton Research* 29: 417–422.

© Woodhead Publishing Limited, 2013

BENAVENTE G P and GATESOUPE F J (1988) Bacteria associated with cultured rotifers and Artemia are detrimental to larval turbot, *Scophthalmus maximus* L. *Aquaculture Engineering* 7: 289–293.

BENTLEY C D, CARROLL P M and WATANABE W O (2008) Intensive rotifer production in a pilot-scale continuous culture recirculating system using nonviable microalgae and an ammonia neutralizer. *Journal of the World Aquaculture Society* 39: 625–635.

BERG L (1997) Commercial feasibility of semi-intensive larviculture of Atlantic halibut (*Hippoglossus hippoglossus* L.). *Aquaculture* 155: 333–340.

BRISKI E, VAN STAPPEN G, BOSSIER P and SORGELOOS P (2008) Laboratory production of early hatching *Artemia* sp. cysts by selection. *Aquaculture* 282: 19–25.

BUSCH K E T, FOLKVORD A, OTTERÅ H, HUTCHINSON W F and SVÅSAND T (2009) Effects of female spawning experience and larval size on feeding and growth of cod larvae *Gadus morhua* L., reared in mesocosms. *Marine Biology Research* 5: 286–296.

BUSCH K E T, PERUZZI S, TONNING F and FALK-PETERSEN I-B (2011) Effect of prey type and size on the growth, survival and pigmentation of cod (*Gadus morhua*, L.) larvae. *Aquaculture Nutrition* 17 (2): 595–603.

CHEPKIRUI-BOIT V, NGUGI C C, BOWMAN J, OYOO-OKOTH E, RASOWO J, MUGO-BUNDI J and CHEROP L (2011) Growth performance, survival, feed utilization and nutrient utilization of African catfish (*Clarias gariepinus*) larvae co-fed Artemia and a micro-diet containing freshwater atyid shrimp (*Caridina nilotica*) during weaning. *Aquaculture Nutrition* 17(2): 82–89.

CLEGG J S and TROTMAN C N A (2002) Physiological and biochemical aspects of *Artemia* ecology, in Abatzopoulos Th J *et al.* (eds), *Artemia: Basic and Applied Biology*. Dordrecht: Kluwer Academic, 129–170.

CLEGG J S, DRINKWATER L and SORGELOOS P (1996) The metabolic status of diapause embryos of *Artemia franciscana* (SFB). *Physiological Zoology* 69: 49–66.

COLURA R L, MATLOCK G C and MACIOROWSKI A F (1987) Zooplankton abundance in unstocked mariculture ponds at three salinities. *Progressive Fish Culturist* 49: 253–259.

COMPS M and MENU B (1997) Infectious diseases affecting mass production of the marine rotifer *Brachionus plicatilis*. *Hydrobiologia* 358: 179–183.

COMPS M, MARI J, POISSON F and BONAMI J R (1991a) Biophysical and biochemical properties of an unusual birnavirus pathogenic to rotifers. *Journal of General Virology* 72: 1229–1236.

COMPS M, MENU B, BREUIL G and BONAMI J R (1991b) Viral infection associated with rotifer mortalities in mass culture. *Aquaculture* 93: 1–7.

COMPS M, MENU B and MOREAU V (1993) Massive infections with fungus of the rotifer *Brachionus plicatilis. Bulletin of the European Association of Fish Pathologists* 13: 28–29.

CONCEIÇÃO L E C, YÚFERA M, MAKRIDIS P, MORAIS S and DINIS M T (2010) Live feeds for early stages of fish rearing. *Aquaculture Research* 41: 613–640.

CRIEL G R J and MACRAE T H (2002a) *Artemia* morphology and structure, in Abatzopoulos Th J *et al.* (eds), Artemia: *Basic and Applied Biology*. Dordrecht: Kluwer Academic, 1–37.

CRIEL G R J and MACRAE T H (2002b) Reproductive biology of *Artemia*, in Abatzopoulos Th J *et al.* (eds), Artemia: *Basic and Applied Biology*. Dordrecht: Kluwer Academic, 39–138.

CURNOW J, KING J, BOSMANS J and KOLKOVSKI S (2006) The effect of reduced *Artemia* and rotifer use facilitated by a new microdiet in the rearing of barramundi *Lates calcarifer* (BLOCH) larvae. *Aquaculture*, 257 (1–4): 204–213.

CUTTS C J (2002) Culture of harpacticoid copepods: potential as live feed for rearing marine fish. *Advances in Marine Biology* 44: 295–316.

© Woodhead Publishing Limited, 2013

DE ARAUJO A B, SNELL T W and HAGIWARA A (2000) Effect of unionized ammonia, viscosity and protozoan contamination on the enzyme activity of the rotifer *Brachionus plicatilis*. *Aquaculture Research* 31: 359–365.

DE ARAUJO A B, HAGIWARA A and SNELL T W (2001) Effect of unionized ammonia, viscosity and protozoan contamination on the enzyme activity of the rotifer *Brachionus rotundiformis*. *Hydrobiologia* 446: 363–368.

DHERT P, ROMBAUT G, SUANTIKA G and SORGELOOS P (2001) Advancement of rotifer culture and manipulation techniques in Europe. *Aquaculture* 200: 129–146.

DHONT J and SORGELOOS P (2002) Applications of *Artemia*, in Abatzopoulos Th J *et al.* (eds), Artemia: *Basic and Applied Biology*. Dordrecht: Kluwer Academic, 251–277.

DHONT J and VAN STAPPEN G (2003) Biology, tank production and nutritional value of *Artemia*, in Støttrup J G and McEvoy L A (eds), *Live Feeds in Marine Aquaculture*. Oxford: Blackwell Science, 65–121.

DOOMS S, PAPAKOSTAS S, HOFFMAN S, DELBARE D, DIERCKENS K, TRIANTAFYLLIDIS A, DE WOLF T, VADSTEIN O, ABATZOPOULOS T J, SORGELOOS P and BOSSIER P (2007) Denaturing gradient gel electrophoresis (DGGE) as a tool for the characterisation of *Brachionus* sp. strains. *Aquaculture* 262: 29–40.

DOUILLET P (1998) Disinfection of rotifer cysts leading to bacteria-free populations. *Journal of Experimental Marine Biology and Ecology* 244: 183–192.

DRILLET G, IVERSEN M H, SØRENSEN T F, RAMLØV H, LUND T and HANSEN B W (2006) Effect of cold storage upon eggs of a calanoid copepod, *Acartia tonsa* (Dana) and their offspring. *Aquaculture* 254: 714–729.

DRILLET G, LINDLEY L C, MICHELS A, WILCOX J and MARCUS N H (2007) Improving cold storage of subitaneous eggs of the copepod *Acartia tonsa* Dana from the Gulf of Mexico (USA-Florida). *Aquaculture Research* 38: 457–466.

DRILLET G, FROUËL S, SICHLAU M H, JEPSEN P M, HØJGAARD J K, JOARDER A K and HANSEN B W (2011a) Status and recommendations on marine copepod cultivation for use as live feed. *Aquaculture* 315: 155–166.

DRILLET G, RABARIMANANTSOA T, FROUËL S, LAMSON J S, CHRISTENSEN A M, KIM-TIAM S and HANSEN B W (2011b) Do inactivated microbial preparations improve life history traits of the copepod *Acartia tonsa*? *Marine Biotechnology* 13: 831–836.

DUSSART B H and DEFAYE D (2001) *Introduction to the Copepoda*. Leiden: Backhuys.

ENGELL-SØRENSEN K, STØTTRUP J G and HOLMSTRUP M (2004) Rearing of flounder (*Platichthys flesus*) juveniles in semi-extensive systems. *Aquaculture* 230: 475–491.

ENGROLA S, DINIS M T and CONCEICAO L E C (2010) Senegalese sole larvae growth and protein utilization is depressed when co-fed high levels of inert diet and *Artemia* since first feeding. *Aquaculture Nutrition* 16 (5): 457–465.

FERNANDEZ-REIRIZ U, LABARTA U and FERREIRO M J (1993) Effects of commercial enrichment diets on the nutritional alue of the rotifer (*Brachionus plicatilis*). *Aquaculture* 112: 195–206.

FERREIRA M, COUTINHO P, SEIXAS P, FABREGAS J and OTERO A (2009) Enriching rotifers with 'Premium' microalgae *Nannochloropsis gaditana. Marine Biotechnology* 11: 585–595.

FLETCHER R C, ROY W, DAVIE A, TAYLOR J, ROBERTSON D and MIGAUD H (2007) Evaluation of new microparticulate diets for early weaning of Atlantic cod (*Gadus morhua*): Implications on larval performances and tank hygiene. *Aquaculture* 263(1–4): 35–51.

FULKS F and MAIN K L (1991) *Rotifer and microalgae culture systems. Proceedings US-Asia Workshop*. Honolulu, HI Oceanic Institute.

GAJARDO G, ABATZOPOULOS T J, KAPPAS I and BEARDMORE J A (2002) Evolution and speciation, in Abatzopoulos Th J *et al.* (eds), *Artemia: Basic and Applied Biology*. Dordrecht: Kluwer Academic, 225–250.

© Woodhead Publishing Limited, 2013

GALLARDO W G, HAGIWARA A, TOMITA Y and SNELL T W (1999) Effect of growth hormone and gamma-aminobutyric acid on *Brachionus plicatilis* (Rotifera) reproduction at low food or high ammonia levels. *Journal of Experimental Marine Biology and Ecology* 240: 179–191.

GAMBOA-DELGADO J and LE VAY L (2009) *Artemia* replacement in co-feeding regimes for mysis and postlarval stages of *Litopenaeus vannamei*: Nutritional contribution of inert diets to tissue growth as indicated by natural carbon stable isotopes. *Aquaculture* 297(1–4): 128–135.

GATESOUPE F J (1991) The effect of three strains of lactic bacteria on the production rate of rotifers, *Brachionus plicatilis*, and their dietary value for larval turbot, *Scophthalmus maximus*. *Aquaculture* 89: 139–148.

GATESOUPE F J (2002) Probiotic and formaldehyde treatments of *Artemia nauplii* as food for larval Pollack, *Pollachius pollachius*. *Aquaculture* 212(1–4): 347–360.

GATESOUPE F J, ARAKAWA T and WATANABE T (1989) The effect of bacterial additives on the production rate and dietary value of rotifers as food for Japanese flounder, *Paralichthys olivaceus*. *Aquaculture* 83: 39–44.

GIMENEZ G, PADROS F, ROQUE A, ESTEVEZ A and FURONES D (2006) Bacterial load reduction of live prey for fish larval feeding using Ox-Aquaculture ©. *Aquaculture Research* 37(11): 1130–1139.

GÓMEZ A, SERRA M, CARVALHO G R and LUNT D H (2002) Speciation in ancient cryptic species complexes: evidence from the molecular phylogeny of *Brachionus plicatilis* (Rotifera). *Evolution* 56: 1431–1444.

GULIANDOLO C, IRRERA G P, LENTINI V and MAUGERI T L (2008) Pathogenic *Vibrio*, *Aeromonas* and *Arcobacter* spp. associated with copepods in the Straits of Messina (Italy). *Marine Pollution Bulletin* 56: 600–606.

GUNVOR Ø, REITAN K I, EVJEMO J O, STØTTRUP J and OLSEN Y (2011) Live feeds, in Holt J G, (ed.), *Larval Fish Nutrition*. Chichester: Wiley-Blackwell, 307–334.

HACHÉ R and PLANTE S (2011) The relationship between enrichment, fatty acid profiles and bacterial load in culture rotifers (*Brachionus plicatilis* L-strain) and *Artemia* (*Artemia salina* strain *Franciscana*). *Aquaculture* 311: 201–208.

HAGIWARA A, BALOMPAPUENG M D, MUNUSWAMY N and HIRAYAMA K (1997) Mass production and preservation of the resting eggs of euryhaline rotifer *Brachionus plicatilis* and *B. rotundiformis*. *Aquaculture* 155: 223–230.

HAMRE K, MOREN M, SOLBAKKEN J, OPSTAD I and PITTMAN K (2005) The impact of nutrition on metamorphosis in Atlantic halibut (*Hippoglossus hippoglossus* L.), *Aquaculture* 250: 555–565.

HAMRE K, SRIVASTAVA A, RONNESTAD I, MANOR-JENSEN A and STOSS J (2008) Several micronutrients in the rotifer *Brachionus* sp. may not fulfil the nutritional requirements of marine fish larvae. *Aquaculture Nutrition* 14: 51–60.

HANSEN B, WERNBERG-MOLLER T and WITTRUP L (1997) Particle grazing efficiency and specific growth of the rotifer *Brachionus plicatilis* (Muller). *Journal of Experimental Marine Biology and Ecology* 215: 217–233.

HAWKYARD M, SAELE O, NORDGREN A, LANGDON C and HAMRE K (2011) Effect of iodine enrichment of *Artemia* sp. on their nutritional value for larval zebrafish (*Danio rerio*). *Aquaculture* 316: 37–43.

HEATH P L and MOORE C G (1997) Rearing Dover sole larvae on *Tisbe* and *Artemia* diets. *Aquaculture International* 5: 29–39.

HINO A (1993) Present culture systems of the rotifer (*Brachionus plicatilis*) and the function of micro-organisms, in Lee C S, Su M S and Liao I L (eds), *Proceedings of Finfish Hatchery in Asia 91*. Tungkang: Tungkang Marine Laboratory, 51–59.

HOFF F H and SNELL T W (1987) *Plankton Culture Manual*. Dade City, FL: Florida Aqua Farms Inc.

© Woodhead Publishing Limited, 2013

HØJ L, BOURNE D G and HALL M R (2009) Localization, abundance and community structure of bacteria associated with *Artemia*: Effects of nauplii enrichment and antimicrobial treatment. *Aquaculture* 293: 278–285.

HOLLING C S (1966) The functional response of invertebrate predators to prey density. *Memoirs of the Entomological Society of Canada* 48: 1–85.

HOLME M H, ZENG C S and SOUTHGATE P C (2006) Use of microbound diets for larval culture of the mud crab, *Scylla serrata*. *Aquaculture* 257(1–4): 482–490.

HOTOS G N (2002) Selectivity of the rotifer *Brachionus plicatilis* fed mixtures of algal species with various cell volumes and cell densities. *Aquaculture Research* 33: 949–957.

HUY Q N, REINERTSEN H, WOLD P, THIEN M T and KJORSVIK E (2011) Effects of early weaning strategies on growth, survival and digestive enzyme activities in cobia (*Rachycentron canadum* L.) larvae. *Aquaculture International* 19: 63–78.

JEPSEN P M, ANDERSEN N, HOLM T, JØRGENSEN A T, HØJGAARD J K and HANSEN B W (2007) Effects of adult stocking density on egg production and viability in cultures of the calanoid copepod *Acartia tonsa* (Dana). *Aquaculture Research* 38: 764–772.

JOHNSTON M D, JOHNSTON D J and JONES C M (2008) Evaluation of partial replacement of live and fresh feeds with a formulated diet and the influence of weaning *Panulirus ornatus phyllosomata* onto a formulated diet during early ontogeny. *Aquaculture International* 16 (1): 33–47.

KAHAN D, UHLIG G, SCHWENZER D and HOROWITZ L (1982) A simple method for cultivating harpacticoids copepods and offering them to fish larvae. *Aquaculture* 26: 303–310.

KANJANA K, RADTANATIP T, ASUVAPONGPATANA S, WITHYACHUMNARNKUL B and WONGPRASERT K (2011) Solvent extracts of the red seaweed *Gracilaria fisheri* prevent *Vibrio harveyi* infections in the black tiger shrimp *Penaeus monodon*. *Fish & Shellfish Immunology* 30(1): 389–396.

KAPPAS I, ABATZOPOULOS T J, VAN HOA N, SORGELOOS P and BEARDMORE J A (2004) Genetic and reproductive differentiation of *Artemia franciscana* in a new environment. *Marine Biology* 146: 103–117.

KIØRBOE T and SABATINI M (1995) Scaling of fecundity, growth and development in marine planktonic copepods. *Marine Ecology Progress Series* 120: 285–298.

KNUCKEY R M, SEMMENS F L, MAYER R J and RIMMER M A (2005) Development of an optimal microalgal diet for the culture of the calanoid copepod *Acartia sinjiensis*: effect of algal species and feed concentration on copepod development. *Aquaculture* 249: 339–351.

KORSTAD J, NEYTS A, DANIELSEN T, OVERREIN I and OLSEN Y (1995) Use of swimming speed and egg ratio as predictors of the status of rotifer cultures in aquaculture. *Hydrobiologia* 313/314: 395–398.

KOSKI M, KLEIN BRETELER W, SCHOGT N, GONZALEZ S and JAKOBSEN H H (2006) Life-stage-specific differences in exploitation of food mixtures: diet mixing enhances copepod egg production but not juvenile development. *Journal of Plankton Research* 28: 919–936.

KOSTOPOULOU V and VADSTEIN O (2007) Growth performance of the rotifers *Brachionus plicatilis*, *B.* 'Nevada' and *B.* 'Cayman' under different feed concentrations. *Aquaculture* 273: 449–458.

KOTANI T, GENKA T, FUSHIMI H, HAYASHI M, DIERCKENS K and SORGELOOS P (2009) Effect of cultivation methods on nutritional enrichment of euryhaline rotifer *Brachionus plicatilis*. *Fish Science* 75: 975–984.

LAVENS P and SORGELOOS P (1987) The cryptobiotic state of *Artemia* cysts, its diapause deactivation and hatching, in Sorgeloos P, Bengtson D A, Decleir W and Jaspers E (eds), *Artemia Research and its Applications. Vol. 3*. Wetteren: Universa Press, 27–63.

© Woodhead Publishing Limited, 2013

LAVENS P and SORGELOOS P (eds) (1996) *Manual on the production and use of live food for aquaculture.* FAO Fisheries Technical Paper No. 361. Rome: FAO.

LAVENS P and SORGELOOS P (2000) The history, present status and prospects of the availability of *Artemia* cysts for aquaculture. *Aquaculture* 181: 397–403.

LEE C-S, O'BRYEN P J and MARCUS N H (2005) *Copepods in Aquaculture.* Oxford: Blackwell Publishing.

LEMUS J T, OGLE J T and LOTZ J M (2004) Increasing production of copepod nauplii in a brown-water zooplankton culture with supplemental feeding and increased harvest levels. *North American Journal of Aquaculture* 66: 169–176.

LUBZENS E and ZMORA O (2003) Production and nutritional value of rotifers, in Støttrup J G and McEvoy L A (eds), *Live Feeds in Marine Aquaculture.* Oxford: Blackwell Science, 17-64.

LUBZENS E, ZMORA O and BARR Y (2001) Biotechnology and aquaculture of rotifers. *Hydrobiologia* 446/447: 337–353.

LUIZI F S, GARA B, SHIELDS R J and BROMAGE N R (1999) Further description of the development of the digestive organs in Atlantic halibut (*Hippoglossus hippoglossus*) larvae, with notes on differential absorption of copepod and *Artemia* prey. *Aquaculture* 176: 101–116.

MAKRIDIS P and OLSEN Y (1999) Protein depletion of the rotifer *Brachionus plicatilis* during starvation. *Aquaculture* 174: 343–353.

MAKRIDIS P, FJELLHEIM J A, SKJERMO J and VADSTEIN O (2000) Control of the bacterial flora of *Brachionus plicatilis* and *Artemia franciscana* by incubation in bacterial suspensions. *Aquaculture* 185: 207–218.

MARCUS N H and MURRAY M (2001) Copepod diapause eggs: a potential source of nauplii for aquaculture. *Aquaculture* 201: 107–115.

MARDEN B, VAN STAPPEN G, MUSAEV A, MIRABDULLAYEV I, JOLDASOVA I and SORGELOOS P (2012) Assessment of the production potential of an emerging *Artemia* population in the Aral Sea, Uzbekistan. *Journal of Marine Systems* 92: 42–52.

MARTINEZ-DIAZ S F, ALVAREZ-GONZALEZ C A, MORENO LEGORRETA M, VAZQUEZ-JUAREZ R and BARRIOS-GONZALEZ J (2003) Elimination of the associated microbial community and bioencapsulation of bacteria in the rotifer *Brachionus plicatilis*. *Aquaculture International* 11: 95–108.

MAUCHLINE J (1998) *Advances in Marine Biology. The biology of calanoid copepods.* San Diego, CA: Academic Press.

MCEVOY L, NÆSS T, BELL J G and LIE O (1998) Lipid and fatty acid composition of normal and malpigmented Atlantic halibut (*Hippoglossus hippoglossus*) fed enriched *Artemia*: a comparison with fry fed wild copepods. *Aquaculture* 163: 235–248.

MCKINNON A D, DUGGAN S, NICHOLS P D, RIMMER M A, SEMMENS G and ROBINO B (2003) The potential of tropical paracalanid copepods as live feeds in aquaculture. *Aquaculture* 223: 89–106.

MERCHE G, LAVENS P, DHERT P, DEHASQUE M, NELIS H, DELEENHEER A and SORGELOOS P (1995) Variation of ascorbic-acid content in different live food organisms. *Aquaculture* 134: 325–337.

MUNRO P D, HENDERSON R J, BARBOUR A and BIRBECK T H (1999) Partial decontamination of rotifers with ultraviolet radiation: the effect of changes in the bacterial load and flora of rotifers on mortalities in start-feeding larval turbot. *Aquaculture* 170: 229–244.

MUROGA K (1995) Viral and bacterial diseases in larval and juvenile marine fish and shellfish – a review. *Fish Pathology* 30: 71–85.

MURRAY H M, LALL S P, RAJASELVAM R, BOUTILIER L A, FLIGHT R M, BLANCHARD B, COLOMBO S, MOHINDRA V, YUFERA M and DOUGLAS S E (2010) Effect of early introduction of microencapsulated diet to larval Atlantic halibut, *Hippoglossus hippoglossus* L. assessed by microarray analysis. *Marine Biotechnology* 12(2): 214–229.

© Woodhead Publishing Limited, 2013

NAAS K E, VAN DER MEEREN T, AKSNES D L (1991) Plankton succession and responses to manipulations in a marine basin for larval rearing. *Marine Ecology Progress Series* 74: 161–173.

NÆSS T and LIE Ø (1998) A sensitive period during first feeding for the determination of pigmentation patterns in Atlantic halibut, *Hippoglossus hippoglossus* L., juveniles: the role of diet. *Aquaculture Research* 29: 925–934.

NÆSS T, GERMAIN-HENRY M and NAAS K E (1995) First feeding of Atlantic halibut (*Hippoglossus hippoglossus*) using different combinations of *Artemia* or wild zooplankton. *Aquaculture* 130: 235–250.

NANTON D A and CASTELL J D (1999) The effects of temperature and dietary fatty acids on the fatty acid composition of harpacticoid copepods, for use as a live food for marine fish larvae. *Aquaculture* 175: 167–181.

NATRAH F M I, DEFOIRDT T, SORGELOOS P and BOSSIER P (2011) Disruption of bacterial cell-to-cell communication by marine organisms and its relevance to aquaculture. *Marine Biotechnology* 13: 109–126.

NAZ M and TURKMEN M (2009) Changes in the digestive enzymes and hormones of gilthead seabream larvae (*Sparus aurata*, L. 1758) fed on Artemia nauplii enriched with free lysine. *Aquaculture International* 17(6): 523–535.

NGUYEN PHU SON (2010) *Socio-economic efficiency and development potential of Artemia production on salt-based area in the Mekong Delta, Vietnam.* PhD thesis, University of Antwerp, Belgium, 227 pp.

NGUYEN THI NGOC ANH, NGUYEN VAN HOA, VAN STAPPEN G and SORGELOOS P (2009) Effect of different supplemental feeds on proximate composition and *Artemia* biomass production in salt ponds. *Aquaculture* 286: 217–225.

NGUYEN THI NGOC ANH, NGUYEN VAN HOA, VAN STAPPEN G and SORGELOOS P (2010) Effect of partial harvesting strategies on *Artemia* biomass production in Vietnamese salt works. *Aquaculture Research* 41: 289–298.

NICOLAS J L, ROBIC E and ANSQUER D (1989) Bacterial flora associated with a trophic chain consisting of microalgae, rotifers and turbot larvae: Influence of bacteria on larval survival. *Aquaculture* 83: 237–248.

NORSKER N H and STØTTRUP J G (1994) The importance of dietary HUFA's for fecundity and HUFA content in the harpacticoid, *Tisbe holothuriae* Humes. *Aquaculture* 125: 155–166.

OGLE J T, NICHOLSON L C, BARNES D N and LOTZ J M (2005) Characterization of an extensive zooplankton culture system coupled with intensive larval rearing of red snapper *Lutjanus campechanus,* in Lee C-S, O'Bryen P J and Marcus N H (eds), *Copepods in Aquaculture.* Oxford: Blackwell Publishing, 225–244.

OHS C L, CHANG K L, GRABE S W, DIMAGGIO M A and STENN E (2010) Evaluation of dietary microalgae for culture of calanoid copepod *Pseudodiaptomus pelagicus. Aquaculture* 307: 225–232.

OIE G and OLSEN Y (1997) Protein and lipid content of the rotifer *Brachionus plicatilis* during variable growth and feeding condition. *Hydrobiologia* 358: 251–258.

OLIVOTTO I, CAPRIOTTI F, BUTTINO I, AVELLA A M, VITIELLO V, MARADONNA F and CARNEVALI O (2008) The use of harpacticoids copepods as live prey for *Amphiprion clarkia* larviculture: Effects on larval survival and growth. *Aquaculture* 274: 347–352.

OLIVOTTO I, TOKLE N E, NOZZI V, COSSIGNANI L and CARNEVALI O (2010) Preserved copepods as a new technology for the marine ornamental fish aquaculture: A feeding study. *Aquaculture* 308: 124–131.

PAPAKOSTAS S, DOOMS S, TRIANTAFYLLIDIS A, DELOOF D, KAPPAS I, DIERCKENS K, DE WOLF T, BOSSIER P, SORGELOOS P and ABATZOPOULOS T J (2006) Evaluation of DNA methodologies in identifying *Brachionus* species used in European hatcheries. *Aquaculture* 255: 557–564.

PAYNE M F and RIPPINGALE R J (2001) Intensive cultivation of the calanoid copepod *Gladioferens imparipes*. *Aquaculture* 201: 329–342.

© Woodhead Publishing Limited, 2013

PAYNE M F, RIPPINGALE R J and CLEARY J J (2001) Cultured copepods as food for West Australian dhufish (*Glauscosoma hebraicum*) and pink snapper (*Pagrus auratus*) larvae. *Aquaculture* 194: 137–150.

PECK M A and HOLSTE L (2006) Effects of salinity, photoperiod and adult stocking density on egg production and egg hatching success in *Acartia tonsa* (Calanoida: Copepoda): Optimizing intensive cultures. *Aquaculture* 255: 341–350.

PHELPS R P, SUMIARSA G S, LIPMAN E E, LAN H-P, MOSS K K and DAVIS A D (2005) Intensive and extensive production techniques to provide copepod nauplii for feeding larval red snapper *Lutjanus campechanus*, in Lee C-S, O'Bryen P J and Marcus N H (eds), *Copepods in Aquaculture*. Oxford: Blackwell Publishing, 151–168.

PINTO C S C, SOUZA-SANTOS L P and SANTOS P J P (2001) Development and population dynamics of *Tisbe biminiensis* (Copepoda: Harpacticoida) reared on different diets. *Aquaculture* 198: 253–267.

QI Z, DIERCKENS K, DEFOIRDT T, SORGELOOS P, BOON N, BAO Z and BOSSIER P (2009a) Analysis of the evolution of microbial communities associated with different cultures of rotifer strains belonging to different cryptic species of the *Brachionus plicatilis* species complex. *Aquaculture* 292: 3–29.

QI Z, DIERCKENS K, DEFOIRDT T, SORGELOOS P, BOON N, BAO Z and BOSSIER P (2009b) Effect of feeding regime and probionts on the diverting microbial communities in rotifer *Brachionus* culture. *Aquaculture International* 17: 303–315.

RAJKUMAR M and VASAGAM K P K (2006) Suitability of the copepod, *Acartia clausi* as a live feed for Seabass larvae (*Lates calcarifer* Bloch): Compared to traditional live-food organisms with special emphasis on the nutritional value. *Aquaculture* 261: 649–658.

RHODES A (2003) Methods for high density batch culture of *Nitokra lacustris*, a marine harpacticoid copepod, in Browman H I and Skiftesvik A B (eds), *The Big Fish Bang, Proceedings of the 26th annual larval fish conference*. Bergen: Institute of Marine Research, 449–465.

RHODES A and BOYD L (2005) Formulated feeds for harpacticoids copepods: implications for population growth and fatty acid composition, in Lee C-S, O'Bryen P J and Marcus N H (eds), *Copepods in Aquaculture*. Oxford: Blackwell Publishing, 61–73.

RIBEIRO A C B and SOUZA-SANTOS L P (2011) Mass culture and offspring production of marine harpacticoids copepod *Tisbe biminiensis*. *Aquaculture* 321: 280–288.

RIBEIRO A R A, RIBEIRO L, DINIS M T and MOREN M (2011) Protocol to enrich rotifers (*Brachionus plicatilis*) with iodine and selenium. *Aquaculture Research* 42: 1737–1740.

ROBINSON C B, SAMOCHA T M, FOX J M, GANDY R L and MCKEE D A (2005) The use of inert artificial commercial food sources as replacements of traditional live food items in the culture of larval shrimp, *Farfantepenaeus aztecus*. *Aquaculture*, 245, 135-147.

ROIHA I S, OTTERLEI E, LITLABO A and SAMUELSEN O B (2010) Uptake and elimination of florfenicol in Atlantic cod (*Gadus morhua*) larvae delivered orally through bioencapsulation in the brine shrimp *Artemia franciscana*. *Aquaculture* 310(1–2): 27–31.

ROMBAUT G, DHERT P, VANDENBERGHE J, VERSCHUERE L, SORGELOOS P and VERSTRAETE W (1999) Selection of bacteria enhancing the growth rate of axenically hatched rotifers (*Brachionus plicatilis*). *Aquaculture* 176: 195–207.

ROTHHAUPT K O (1990) Population growth rates of two closely related rotifer species: effects of food quantity, particle size, and nutrition quality. *Freshwater Biology* 23: 561–570.

SALES J (2011) First feeding of freshwater fish larvae with live feed versus compound diets: a meta-analysis. *Aquaculture International* 19(6): 1217–1228.

© Woodhead Publishing Limited, 2013

SAYEGH F A Q, RADI N and MONTAGNES D J S (2007) Do strain differences in microalgae alter their relative quality as a food for the rotifer *Brachionus plicatilis*? *Aquaculture* 273: 665–678.

SAZHINA L I (2006) *Breeding, growth rates, and production of marine copepods*. Hyderabad: Universities Press.

SCHIPP G R, BOSMANS J M P and MARSHALL A J (1999) A method for hatchery cultivation of tropical calanoid copepods, *Acartia* spp. *Aquaculture* 174: 81–88.

SEGERS H (1995). Nomenclatural consequences of some recent studies on *Brachionus plicatilis* (Rotifera, Brachionidae). *Hydrobiologia* 313/314: 121–122.

SHIELDS R J, BELL J G, LUIZI F S, GARA B, BROMAGE N R and SARGENT J R (1999) Natural copepods are superior to enriched Artemia nauplii as feed for halibut larvae (*Hippoglossus hippoglossus*) in terms of survival, pigmentation and retinal morphology: relation to dietary essential fatty acids. *Journal of Nutrition* 129: 1186–1194.

SHIELDS R J, KOTANI T, MOLNAR A, MARION K, KOBASHIGAWA J and TANG L (2005) Intensive cultivation of a subtropical paracalanid copepod, *Parvocalanus* sp., as prey for small marine fish larvae, in Lee C-S, O'Bryen P J and Marcus N H (eds), *Copepods in Aquaculture*. Oxford: Blackwell Publishing, 209–223.

SKJERMO J and VADSTEIN O (1993) Characterization of the bacterial-flora of mass cultivated *Brachionus plicatilis*. *Hydrobiologia* 255: 185–191.

SKJERMO J and VADSTEN O (1999) Techniques for microbial control in intensive rearing of marine larvae. *Aquaculture* 177(1–4): 333–343.

SNELL T W, CHILDRESS M J, BOYER E M and HOFF F H (1987) Assessing the status of rotifer mass cultures. *Journal of the World Aquaculture Society* 18: 270–277.

SORGELOOS P (1999) Challenges and opportunities for aquaculture research and development in the next century. *World Aquaculture* 30(3): 11–15.

SORGELOOS P, COUTTEAU P, DHERT P, MERCHIE G and LAVENS P (1998) Use of the brine shrimp, *Artemia* spp., in larval crustacean nutrition: a review. *Review in Fisheries Science* 6: 55–68.

SORGELOOS P, DHERT P and CANDREVA P (2001) Use of the brine shrimp *Artemia* spp., in marine fish larviculture. *Aquaculture* 200: 147–159.

SRIVASTAVA A, HAMRE K and STOSS J (2006) Protein content and amino acid composition of the live feed rotifer (*Brachionus plicatilis*): with emphasis on the water soluble fraction. *Aquaculture* 254: 534–543.

SRIVASTAVA A, STOSS J and HAMRE K (2011) A study on enrichment of the rotifer *Brachionus* 'Cayman' with iodine and selected vitamins. *Aquaculture* 319: 430–438.

STØTTRUP J G (2000) The elusive copepods: their production and suitability in marine aquaculture. *Aquaculture Research* 31: 703–711.

STØTTRUP J G (2003) Production and nutritional value of copepods, in Støttrup J G and McEvoy L A (eds), *Live Feeds in Marine Aquaculture,* Oxford: Blackwell Science, 145–205.

STØTTRUP J G (2006) A review on the status and progress in rearing copepods for marine larviculture. Advantages and disadvantages among calanoid, harpacticoid and cyclopoids copepods. *Avences en Nutrición Acuícola VIII* 333(5): 970–994.

STØTTRUP J G and JENSEN J (1990) Influence of algal diet on feeding and egg production of the calanoid copepod *Acartia tonsa* Dana. *Journal of Experimental Marine Biology and Ecology* 141: 87–105.

STØTTRUP J G and MCEVOY L A (eds) (2003) *Live Feeds in Marine Aquaculture*. Oxford: Blackwell Science.

STØTTRUP J G and NORSKER N H (1997) Production and use of copepods in marine fish larviculture. *Aquaculture* 155: 231–247.

STØTTRUP J G, RICHARDSEN K, KIRKEGAARD E and PIHL N J (1986) The cultivation of *Acartia tonsa* Dana for use as live food source for marine fish larvae. *Aquaculture* 52: 87–96.

© Woodhead Publishing Limited, 2013

STØTTRUP J G, NIELSEN R, KROG C and RASMUSSEN K (1994) Results on the extensive production of North Sea cod and their growth and distribution subsequent to release in the Limfjord, Denmark. *Aquaculture and Fishery Management* 25: 143–159.

STØTTRUP J G, SHIELDS R, GILLESPIE M, GARA M B, SARGENT J R, BELL J G, HENDERSON R J, TOCHER D R, SUTHERLAND R, NÆSS T, MANGOR-JENSEN A, NAAS K, VAN DER MEEREN T, HARBOE T, SANCHEZ F J, SORGELOOS P, DHERT P and FITZGERALD R (1998) The production and use of copepods in larval rearing of halibut, turbot and cod. *Bulletin of Aquaculture Association of Canada* 4: 41–45.

STØTTRUP J G, BELL J G and SARGENT J R (1999) The fate of lipids during development and cold-storage of eggs in the calanoid copepod *Acartia tonsa* Dana, and in response to different algal diets. *Aquaculture* 176: 257–269.

SU H-M, CHENG S-H, CHEN T-I and SU M-S (2005) Culture of copepods and application to marine finfish larval rearing in Taiwan, in Lee C-S, O'Bryen P J and Marcus N H (eds), *Copepods in Aquaculture*, Oxford: Blackwell Publishing, 183–194.

SUANTIKA G, DHERT P, NURHUDAH M and SORGELOOS P (2000) High-density production of the rotifer *Brachionus plicatilis* in a recirculation system: consideration of water quality, zootechnical and nutritional aspects. *Aquacultural Engineering* 21: 201–214.

SUANTIKA G, DHERT P, ROMBAUT G, VANDENBERGHE J, DE WOLF T and SORGELOOS P (2001) The use of ozone in a high density recirculation system for rotifers. *Aquaculture* 201: 35–49.

SUANTIKA G, DHERT P, SWEETMAN E, OBRIEN E and SORGELOOS P (2003) Technical and economical feasibility of a rotifer recirculation system. *Aquaculture* 227: 173–189.

SUATONI E, VICARIO S, RICE S, SNELL T and CACCONE A (2006) An analysis of species boundaries and biogeographic patterns in a cryptic species complex: The rotifer – *Brachionus plicatilis*. *Molecular Phylogenetics and Evolution* 41: 86–98.

SUGA K, TANAKA Y, SAKAKURA Y and HAGIWARA A (2011) Axenic culture of *Brachionus plicatilis* using antibiotics. *Hydrobiologia* 662: 113–119.

SUN B and FLEEGER J W (1995) Sustained mass culture of *Amphiascoides atopus*, a marine harpacticoid copepod in a recirculatoin system, *Aquaculture* 136: 313–321.

SVÅSAND T, KRISTIANSEN T S, PEDERSEN T, SALVANES A G V, ENGELSEN R and NØDTVEDT M (1998) *Havbeite med torsk – artsrapport*. Oslo: Norges forskningsråd.

TAKAOKA O, JI S-C, ISHIMARU K, LEE S-W, JEONG G-S, ITO J, BISWAS A, TAKIL K (2011) Effect of rotifer enrichment with herbal extracts on growth and resistance of red sea bream, *Pagrus major* (Temminck & Schlegel) larvae against *Vibrio anguillarum*. *Aquaculture Research* 42: 1824–1829.

TANASOMWANG V and MUROGA K (1992) Effect of sodium nifurstyrenate on the reduction of bacterial contamination of rotifers (*Brachionus plicatilis*). *Aquaculture* 103: 221–228.

TINH N T N, PHUOC N N, DIERCKENS K, SORGELOOS P and BOSSIER P (2006) Gnotobiotically grown rotifer *Brachionus plicatilis* sensu strictu as a tool for evaluation of microbial functions and nutritional value of different food types. *Aquaculture* 253: 421–432.

TLUSTY M F, FIORE D R, GOLDSTEIN J S (2005) Use of formulated diets as replacements for Artemia in the rearing of juvenile American lobsters (*Homarus americanus*). *Aquaculture* 250 (3–4): 781–795.

TOCHER D R (2010) Fatty acid requirements in ontogeny of marine and freshwater fish. *Aquaculture Research* 41: 717–732.

TOLEDO J D, GOLEZ M S and OHNO A (2005) Studies on the use of copepods in the semi-intensive seed production of grouper *Epinephelus coioides*, in Lee, C-S, O'Bryen P J and Marcus N H (eds), *Copepods in Aquaculture*. Oxford: Blackwell Publishing, 169–182.

© Woodhead Publishing Limited, 2013

TOLOMEI A, BURKE C, CREAR B and CARSON J (2004) Bacterial decontamination of on-grown *Artemia. Aquaculture* 232(1–4): 357–371.

TOURAKI M, NIOPAS I, LADOUKAKIS E and KARAGIANNIS V (2010) Efficacy of flumequine administered by bath or through medicated nauplii of *Artemia fransciscana* (L.) in the treatment of vibriosis in sea bass larvae. *Aquaculture* 306(1–4): 146–152.

TRIANTAPHYLLIDIS G V, CRIEL G R J, ABATZOPOULOS T J, THOMAS K M, PELEMAN J, BEARDMORE J A and SORGELOOS P (1997) International Study on *Artemia*. LVII. Morphological and molecular characters suggest conspecificity of all bisexual European and North African *Artemia* populations. *Marine Biology* 129: 477–487.

TZOVENIS I, DE PAUW N and SORGELOOS P (2003) Optimisation of T-ISO biomass production rich in esential fatty acids II. Effect of different light regime on the production of fatty acids. *Aquaculture* 216: 223–242.

UHLIG G (1984) Progress in mass cultivation of harpacticoids copepods for mariculture purposes, *Special Publication of the European Mariculture Society* 8: 261–273.

VADSTEIN O, OIE G and OLSEN Y (1993) Particle-size dependent feeding by the rotifer *Brachionus plicatilis*. *Hydrobiologia* 255: 261–267.

VAN CAN NHU, DIERCKENS K, NGUYEN HONG T, HOANG TUYET MINH, THANH LUU LE, MAI THIEN TRAN, NYS C and SORGELOOS P (2010) Effect of early co-feeding and different weaning diets on the performance of cobia (*Rachycentron canadum*) larvae and juveniles. *Aquaculture* 305(1–4): 52–58.

VAN DER MEEREN T, OLSEN R E, HAMRE K and FYHN H J (2008) Biochemical composition of copepods for evaluation of feed quality in production of juvenile marine fish. *Aquaculture* 274: 375–397.

VANDECAN M, DIALLO A and MELARD C (2011) Effect of feeding regimes on growth and survival of *Clarias gariepinus* larvae: replacement of *Artemia* by a commercial feed. *Aquaculture Research* 42(5): 733–736.

VANDERLUGT K and LENZ P H (2008) Management of nauplius production in the paracalanid, *Bestiolina similis* (Crustacea: Copepoda): Effects of stocking densities and culture dilution. *Aquaculture* 276: 69–77.

VANHAECKE P and SORGELOOS P (1980) International Study on *Artemia*. IV. The biometrics of *Artemia* strains from different geographical origin, in Persoone G, Sorgeloos P, Roels O and Jaspers E (eds), *The Brine Shrimp* Artemia. *Vol. 3. Ecology, culturing, use in aquaculture.* Wetteren: Universa Press, 393–405.

VANHAECKE P and SORGELOOS P (1989) International Study on *Artemia*. XLVII. The effect of temperature on cyst hatching, larval survival and biomass production for different geographical strains of brine shrimp *Artemia* spp. *Annals of the Royal Zoological Society of Belgium* 119(1): 7–23.

VAN STAPPEN G (1996) *Artemia*: Use of cysts, in Lavens P and Sorgeloos P (eds), *Manual on the production and use of live food for aquaculture*, FAO Fisheries Technical Paper No. 361. Rome: FAO.

VAN STAPPEN G (2002) Zoogeography, in Abatzopoulos Th J *et al.* (eds), *Artemia: Basic and Applied Biology.* Dordrecht: Kluwer Academic, 171–224.

VAN STAPPEN G, YU H, WANG X, HOFFMAN S, COOREMAN K, BOSSIER P and SORGELOOS P (2007) Occurrence of allochthonous *Artemia* species in the Bohai Bay area, PR China, as confirmed by RFLP analysis and laboratory culture tests. *Fundamental and Applied Limnology* 170(1): 21–28.

VAN STAPPEN G, LITVINENKO L I, LITVINENKO A I, BOYKO E G, MARDEN B and SORGELOOS P (2009) A survey of Artemia resources of Southwest Siberia (Russian Federation). *Reviews in Fisheries Science* 17(1): 117–148.

VERDONCK L GRISEZ L, SWEETMAN E, MINKOFF G, SORGELOOS P, OLLEVIER F and SWIS J (1997) Vibrios associated with routine productions of *Brachionus plicatilis*. *Aquaculture* 149: 203–214.

© Woodhead Publishing Limited, 2013

VERNER-JEFFREYS D W, SHIELDS R J and BIRKBECK T H (2003) Bacterial influences of Atlantic halibut *Hippoglossus hippoglossus* yolk-sac larval survival and start-feed response. *Diseases of Aquatic Organisms* 56: 105–113.

VERSCHUERE L, ROMBAUT G, SORGELOOS P and VERSTRAETE W (2000) Probiotic bacteria as biological control agents in aquaculture. *Microbiology and Molecular Biology Reviews* 64: 655–671.

WURTSBAUGH W A and GLIWICZ Z M (2001) Limnological control of brine shrimp population dynamics and cyst production in the Great Salt Lake, Utah. *Hydrobiologia* 466: 119–132.

YAN D C, FENG S Y, HUANG J and DONG S L (2007) Rotifer cellular membranes bind to white spot syndrome virus. *Aquaculture* 273: 423–426.

YOSHIMURA K, HAGIWARA A, YOSHIMATSU T and KITAJIMA C (1996) Culture technology of marine rotifers and the implications for culture of marine fish in Japan. *Marine Freshwater Research* 47: 217–222.

YOSHIMURA K, TANAKA K and YOSHIMATSU T (2003) A novel culture system for the ultra-high-density production of the rotifer *Brachionus rotundiformis* – a preliminary report. *Aquaculture* 227: 165–172.

YU J P, HINO A, USHIRO M and MAEDA M (1989) Function of bacteria as vitamin B12 producers during mass culture of the rotifer *Brachionus plicatilis*. *Nippon Suisan Gakkaishi* 55: 1799–1806.

YU J P, HINO A, NOGUCHI T and WAKABAYASHI H (1990) Toxicity of *Vibrio alginolyticus* on the survival of the rotifer *Brachionus plicatilis*. *Nippon Suisan Gakkaishi* 56: 1455–1460.

© Woodhead Publishing Limited, 2013

6

Microdiets as alternatives to live feeds for fish larvae in aquaculture: improving the efficiency of feed particle utilization

S. Kolkovski, Department of Fisheries, Australia

DOI: 10.1533/9780857097460.1.203

Abstract: Marine fish larvae fed microdiets have not, at this stage, matched the growth and survival performances demonstrated by larvae fed live feeds such as rotifers and *Artemia*. This chapter discusses the issues related to the use of microdiets as a sole or partial feed for marine fish larvae. The techniques and methods of manufacturing microdiet particles, chemical and physical properties of microdiet particles are described. The use of feeding systems and advantages and disadvantage of the current feeding systems is reviewed. The chapter also looks at advances in feeding regimes such as co-feeding and feeding systems.

Key words: marine fish larvae, feeding, weaning, nutrition, feeding systems, microdiets.

6.1 Introduction

Since the 1980s, enormous efforts have been made to develop microdiets[1] to replace live feed organisms, both rotifers and *Artemia*, as complete or partial replacements for marine fish larvae. However, although, there have been substantial achievements in reducing the reliance on live feeds (specifically the use of *Artemia*) and weaning the larvae earlier onto microdiets, microdiets still cannot completely replace live feeds for most species (Holt *et al.*, 2011; Langdon and Barrows, 2011).

Although weaning the larvae from *Artemia* onto a microdiet can be achieved at metamorphosis in some species such as European sea bass (*Dicentrarchus labrax*), the early introduction of prepared diets as the sole

[1]Microdiet is also referred in literature as formulated, inert, dry or weaning diet. It usually refers to first prepared (dry) diet fed to larvae. Usually particle size ranges between 150 μm and 800 μm.

© Woodhead Publishing Limited, 2013

replacement for live food has met with limited success. A clear example of the superiority of live food over commercial microdiets was demonstrated by Curnow *et al.* (2006a,b; Fig. 6.1). Barramundi (*Lates calcarifer*) larvae development was affected by rearing protocols, with co-feeding rotifers and commercial diet allowing complete replacement of *Artemia*. However, by including *Artemia* in the protocol with one of the commercial microdiets, survival was significantly improved. Furthermore, feeding protocols with earlier weaning from rotifers resulted in significantly reduced growth and survival.

The efficiency of utilization of feed particles (either live or formulated) by marine fish larvae is affected by many external and internal factors (Kolkovski *et al.*, 2009; Yúfera, 2011; Fig 6.2). Primarily, the searching, identification and ingestion processes are influenced by physical and chemical factors including colour, shape, size, movement and olfactory stimuli at a molecular level.

Substances secreted by live food organisms that act to stimulate a feeding response belong to a group of chemicals known as 'feed attractants' and some have been specifically identified for larvae (Kolkovski *et al.* 1997a,b, 2004; Kolkovski, 2001, 2006a,b; Table 6.1). Moreover, these physical and chemical factors affect the palette and influence the ingestion process, which is the precursor to the digestion process (Kolkovski *et al.*, 1995).

More and more hatcheries are using recirculating systems and more controlled and automatic systems including oxygen supply and live food and algae supply. Automation for feeding and weaning microdiets is still limited to a few systems and, in most cases, initiation of weaning to formulated microdiets is done by hand feeding following the use of belt feeders (Kolkovski *et al.*, 2009). In general, feeding inert diets presents several issues in the hatchery.

- Since the diet particles sink, feeding small amounts at frequent intervals is recommended.
- The microdiet particles tend to stick, creating large clumps that sink immediately to the bottom and are therefore unavailable to the larvae.
- Microdiet particles accumulating on the tank bottom and walls will encourage bacteria proliferation and reduced water quality.

During the past few decades, strong emphasis was given to the nutritional requirements of marine fish larvae focusing on the fatty acid requirements. In recent years, more information about the amino acid requirements has also been published. Burnell and Allen (2009) and more recently Holt (2011) published comprehensive reviews on many of the aspects related to marine larvae nutrition (the book edited by Holt being the first and only work dedicated to larvae nutrition). The relationship between microdiets and the digestive system of developing larvae was investigated with the understanding that, in most cases, early stage marine larvae lack the capability to digest and break down the microdiet hard particles due to incomplete

© Woodhead Publishing Limited, 2013

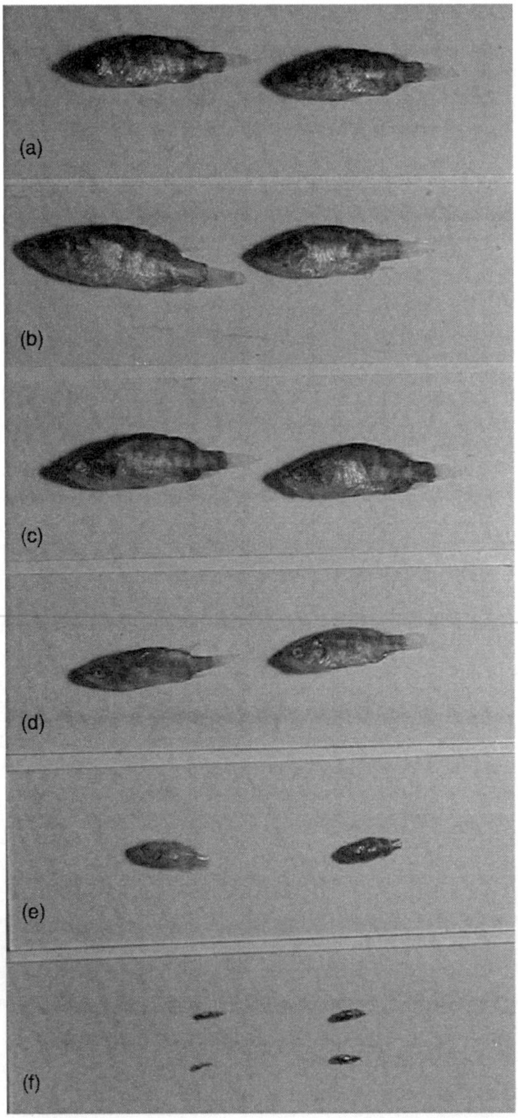

Fig. 6.1 Barramundi *Lates calcarifer* growth using different feeding protocols (Curnow *et al.*, 2006b). (a) 11 days rotifers, 9 days *Artemia* co-fed with Proton (INVE); (b) 11 days rotifers, 9 days *Artemia* co-fed with Micro-Gemma / Gemma (Skretting); (c) 11 days rotifers, Micro-Gemma / Gemma; (d) 7 days rotifers, Micro-Gemma / Gemma; (e) 3 days rotifers, Micro-Gemma / Gemma; (f) Micro Gemma / Gemma.

© Woodhead Publishing Limited, 2013

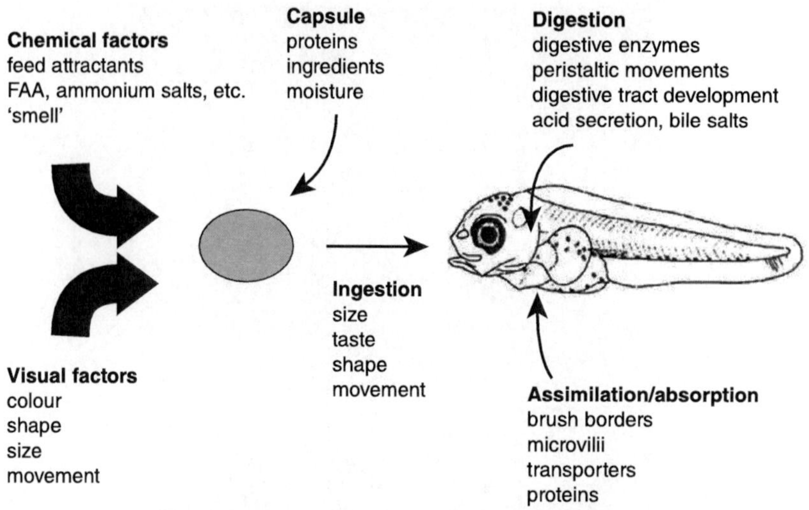

Fig. 6.2 Factors affecting food particle utilization.

digestive system (Kolkovski 2001, 2004, 2006b; Kolkovski *et al.*, 2009; Lazo *et al.*, 2011).

However, very little attention has been given to the process of feeding microdiets including feeding systems, specific design to deal with very small particles as microdiet particles, dispersion of particles in the water column, feeding strategies (continuous vs periodic), water hydrodynamics and the interaction between the diet particle and larvae in the rearing tank. In many commercial hatcheries, manual feeding is still very common. An optimal diet that supplies all the larva's nutritional requirements will not achieve optimal growth if the feeding frequencies, amounts and particle dispersal are not also optimised.

6.2 Diet manufacturing methods and microdiet characteristics

Several microdiet manufacturing methods are currently being used (Langdon and Barrows, 2011; Kolkovski *et al.*, 2009): microbound diets (MBD, Fig 6.3); microcoated diets (MCD); micro-encapsulated diets (MED, Fig. 6.4a,b); and micro-extrusion marumerization (MEM, Fig. 6.5).

All have been used extensively in nutritional studies with finfish larvae and are commercially manufactured. Aside from dedicated microdiet manufacturing methods as described above, many feed manufacturer are using extruders to produce large particles that are then crushed and sieved to the required particle sizes.

© Woodhead Publishing Limited, 2013

Table 6.1 Amino acids as feed attractants, reference list

Species	Amino acid found to have effect as feed attractant	Reference
Rainbow trout (*Salmo gairdneri*)	Mixture of L-amino acids	Adron and Mackie, 1978
Atlantic salmon (*Salmo salar*)	Glycine	Hughes, 1990
Sea bass (*Dicentrarchus labrax*)	Mixture of L-amino acids	Mackie and Mitchell, 1982
Pig fish (*Orthopristis chrysopterus*)	Glycine, betaine	Carr *et al.*, 1977, 1978
Red sea bream (*Chrysophrys major*)	Glycine, betaine	Goh and Tamura, 1980
	Glycine, alanine, lysine	Fuke *et al.*, 1981
	Valine, glutamic acid and arginine	Ina and Matsui, 1980
Gilthead sea bream (*Sparus aurata*)	Glycine, betaine, alanine, arginine	Kolkovski *et al.*, 1997
Turbot (*Scophthalmus maximus*)	Inosine and IMP	Mackie and Adron, 1978
Dover sole (*Solea solea*)	Glycine, betaine	Mackie *et al.*, 1980
	Glycine, inosine, betaine	Metaillet *et al.*, 1983
Puffer (*Fugu pardalis*)	Glycine, betaine	Ohsugi *et al.*, 1978
Japanese eel (*Anguilla japonica*)	Glycine, arginine, alanine, proline	Yoshii *et al.*, 1979
Cod (*Gadus morhua*)	Arginine	Doving *et al.*, 1994
Herring (*Clupea herangus*)	Glycine, proline	Damsey, 1984
Glass eel (*Anguilla anguilla*)	Glycine, arginine, alanine, proline	Mackie and Mitchell, 1983
	Alanine, glycine, histidine, proline	Kamstra and Heinsbroek, 1991
Lobster (*Homarus Americanus*)	Glutamate, betaine, taurine, ammonium chloride	Corotto *et al.*, 1992
Western Atlantic ghost crab (*Ocypode quadrata*)	Butanoic acid, carboxylic acid, trehalose, carbohydrates, homarine, asparagine	Trott and Robertson, 1984
Freshwater prawn (*Macrobrachium rosenbergii*)	Taurine, glycine, trimethylamine, betaine	Harpaz *et al.*, 1987
Abalone (*Haliotis discus*)	Mixture of L-amino acid and lecithin	Harada *et al.*, 1987

Source: Kolkovski *et al.*, 2009.

6.2.1 Microbound diets (MBD)

Currently, the manufacturing process for MBDs is the simplest and most commonly used method of preparation (aside from extruded particles described above). It consists of dietary components held within a gelled matrix or binder. They do not have a capsule and it is suggested that this

© Woodhead Publishing Limited, 2013

Fig. 6.3 Microbound diets.

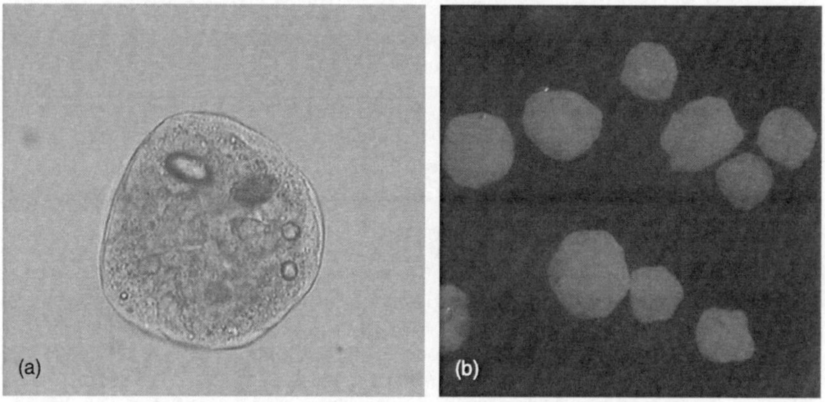

Fig. 6.4 (a) and (b) Microencapsulated diets (MED).

facilitates greater digestibility and increased attraction through greater nutrient leaching. All the ingredients are ground, mixed with a binder such as gelatine, alginate, zein, carrageenan or carboxymethyl-cellulose, activated by temperature or chemicals and then dried (drum drying or spray drying), ground and sieved to the required size. The final product is usually uneven and rough-edged particles (Fig. 6.6; Partridge and Southgate, 1999; Kolkovski, 2001; Kolkovski *et al.*, 2009; Langdon and Barrows, 2011).

6.2.2 Microcoated diets (MCD)

The MCD method is based on coating or binding small MBD particles to reduce leaching. The coating layer is usually lipids or lipoproteins. This method is not often used in commercial processes.

© Woodhead Publishing Limited, 2013

Fig. 6.5 Marrumerization diet (MEM).

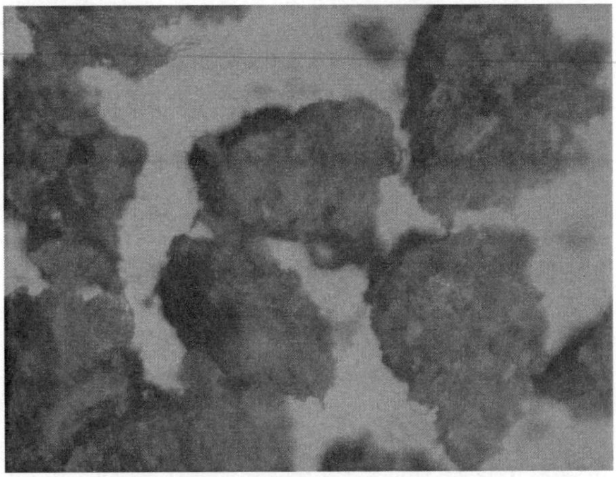

Fig. 6.6 Microbound diet close up.

6.2.3 Micro-encapsulated diets (MED)

MED particles are made using several different techniques. The particle usually has a membrane or capsule wall, which separates dietary materials from the surrounding medium. The capsule wall helps maintain the integrity of the food particle until it is consumed, preventing leaching and degradation of the nutritional ingredients in the water. However, this attribute may restrict leaching of water-soluble dietary components and therefore reduce the larvae's attraction to the food particles. The capsule wall is also thought to impair digestion of the food particle (Fig 6.7; Yúfera *et al.*, 1998, 2003;

© Woodhead Publishing Limited, 2013

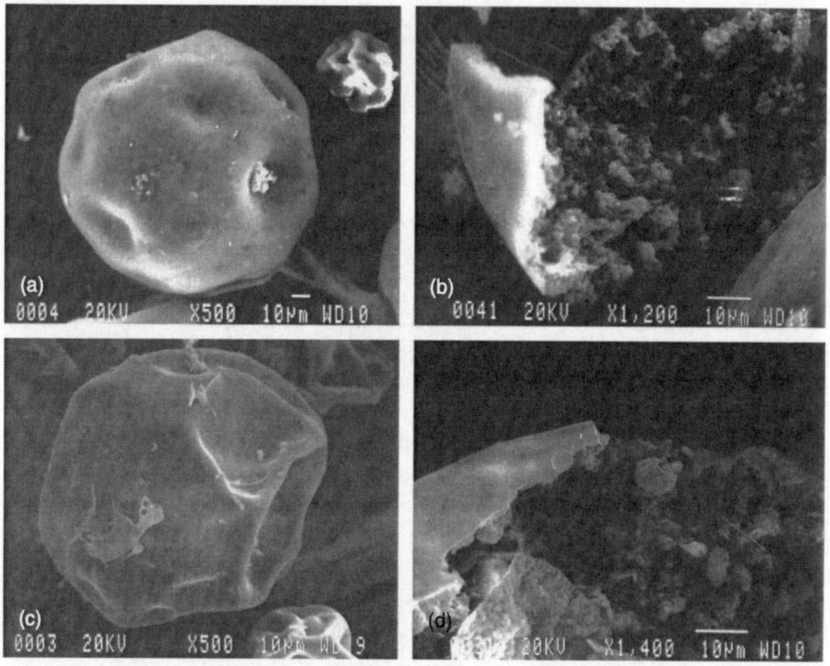

Fig. 6.7 Microdiets manufactured by MED: (a,b) thick particle wall; (c,d) thin particle wall (photo Manuel Yufera, CICS, Cediz, Spain).

Kolkovski, 2006b; Kolkovski *et al.*, 2009; Langdon and Barrows, 2011). There are several methods for micro-encapsulation (Benita, 2006). These include chemical processes and mechanical processes. In chemical processes, the capsules are made within a liquid, usually stirred or agitated. The capsules are formed by (i) spraying droplets of coating material on core ingredients; (ii) capsulating liquid droplets containing the nutritional ingredients by spraying into gas phase; (iii) creating gel capsules by spraying droplets, containing the nutritional ingredients and a binder, into liquid solution in order to activate the binder by polymerization reaction at a solid/gas or (iv) liquid interface. Protein cross-linking involves several stages of mixing and washing with organic solvents resulting in a very expensive diet process, which can be potentially toxic. These methods have never resulted in good growth rates due to the inability of larvae to digest and assimilate the particles and the high ratio of non-nutritional ingredients, mainly the capsule, to the essential nutrients.

Another method, complex co-acervation, involves mixing and activating, using electrical charges, two liquid phases differing in their viscosity resulting in very small capsules. These capsules then bind to create a larger capsule containing hundreds or thousands of microcapsules.

© Woodhead Publishing Limited, 2013

6.2.4 Micro-extrusion marumerization (MEM)

Mechanical encapsulation involves processes such as spray drying, fluidized bed drying, cold MEM and particle-assisted rotational agglomeration (Kolkovski *et al.*, 2009; Langdon and Barrows, 2011). The last two techniques have gained attention in the past few years with commercially available diets produced using these methods. Initially developed for pharmaceutical processes, these methods involve purpose-built machines. MEM is a two-step process of cold extrusion followed by marumerization (spheronization). The process has the capability of producing particles from 500–1000 μm and greater. Particle-assisted rotational agglomeration (PARA) is a single-step process capable of producing particles from 50–500 μm that are lower in density than particles produced by the MEM method due to the fact that the extrusion step is avoided. The method is based on a spinning disc (marumerizer). A wet mash of the ingredients is put into the marumerizer with or without inert beads. The rotation movement of the disc breaks down the mash into smaller spherical particles. The diameter of the particles depends on several factors including the disc rotation speed, the inert beads and the raw ingredients (Fig. 6.8a,b).

6.2.5 Microdiet characteristics

Very little attention has been given to the identification of the chemical and physical properties of diet particles (Kolkovski, 2006a,b; Holt *et al.*, 2011). Although these properties have a strong influence on both diet utilization and water quality, there are only a handful of scientific papers on leaching properties and even less on the buoyancy and the particle behaviour in the water column.

Leaching

One of the problems of MBD particles and most of the microdiet type particles is the high leaching rate of amino acids. Kvale *et al.* (2006) reported leaching of protein molecules (9–18 kD) after 5 min immersion in water (3 % NaCl, 12 °C) at a rate of 80–98 %, 43–54 % and 4–6 % for agglomerated, heat-coagulated and protein-encapsulated microdiet. Onal and Langdon (2004) compared the leaching of protein hydrolysate from three different microdiets (gelatine–alginate beads, spray-dried zein-bound particles and spray-water zein particles) and found that 20–30 % of the hydrolyaste leached out after 2 min immersion in water. Yufera *et al.* (2003) determined the rate of different amino acids leaching from both MBD and MED (microencapsulated diets). The authors found contrasting patterns between the two diet types. While hydrophilic amino acids leached the most from MBD, hydrophobic amino acids were found to leach from MED particles at a higher rate (Fig. 6.3). The leaching rates of the two diets were also significantly different. For instance, 70 % of free lysine leached from

© Woodhead Publishing Limited, 2013

Fig. 6.8 (a) Spin disc for MEM diet; (b) MEM diets after processing by spin disc (spheronizer).

MBD particles after less than 5 min, while less than 7 % leached from MED particles after 60 min. Heinen (1981) assessed water stability of formulated diets made from 11 different binders; MBD made from agar and alginate were amongst the most stable in terms of integrity, while carrageenan was amongst the poorest.

A diet particle needs to achieve a fine balance between leaching amino acids and other nutrients to act as feed attractant and digestibility of the particle to suit the undeveloped larvae digestive system. A particle that will be hard and leach resistant will also present a challenge to the larvae digestive system, whilst a particle that will digest easily in the gut will also disintegrate relatively quickly in the water.

© Woodhead Publishing Limited, 2013

Buoyancy

One of the most significant problems with microdiet particles is their negatively buoyant inert state. However, very few scientific studies have investigated this issue (Holt *et al.*, 2011). In addition, MBD particles don't move like live zooplankton. This specific movement acts as a visual stimulus for increased feeding activity (Kolkovski *et al.*, 1997a). Furthermore, the particles sink to the bottom of the tank where they are no longer available to the larvae and accumulate there, leading to bacterial proliferation and deterioration of water quality. Figure 6.9 shows the result of a poor feeding/cleaning regime in the larvae tank. The pink spots are bacteria accumulating on the tank floor. This emphasizes the need for effective feeding protocols and mechanisms coupled with hygienic control of the tank.

A change in the larvae behaviour during the weaning period is illustrated by the larvae's ability to recognize the inert particles as food and to more actively hunt for them during a relatively smaller window of opportunity, as the particles pass down through the water column. Figure 6.10 illustrates the sinking rates of several commercial microdiets (Jackson and Nimmo, 2005; Kolkovski *et al.*, 2009). Different attempts have been made to increase the time the microdiet particle spends in the water column, including increasing buoyancy by adjusting and modifying oil levels and manufacturing methods and also using rearing systems with upwelling currents (Kolkovski *et al.*, 2004). Knowledge of sinking and leaching rates of microdiets can and should be used to optimize feeding time in the larvae tank. The faster the diet particle sinks the shorter the feeding intervals should be coupled with smaller quantities of diet, in short, feeding less more often.

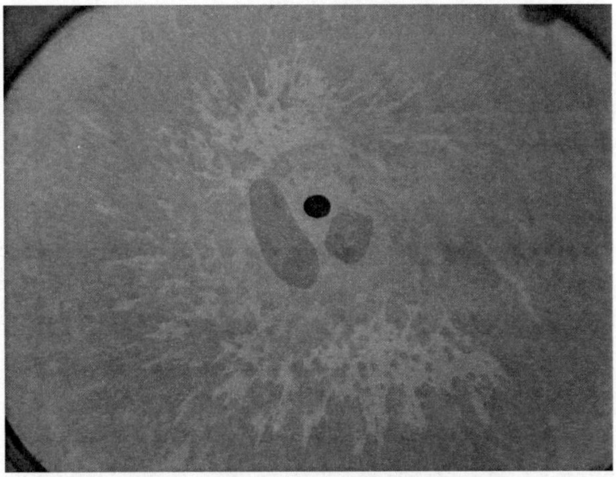

Fig. 6.9 Larvae tank bottom affected by bacteria.

© Woodhead Publishing Limited, 2013

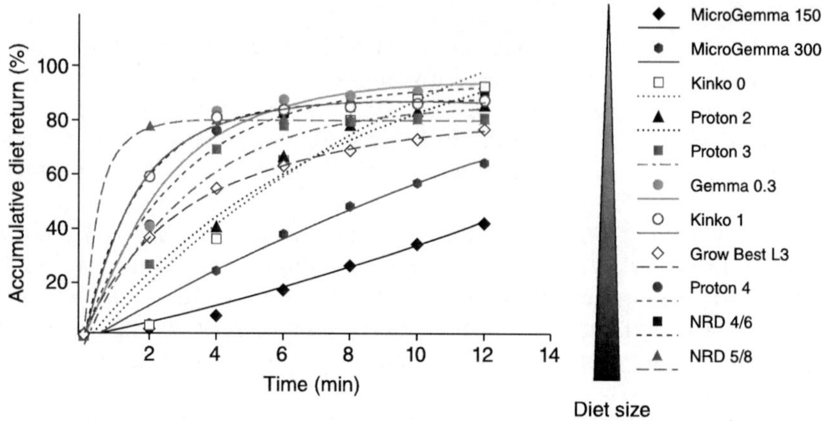

Fig. 6.10 Sinking patterns of commercial diets (Jackson and Nimmo, 2005).

Weaning and co-feeding methods

An important factor influencing the larvae's acceptance of a microdiet, which affects both their growth and survival, is the weaning process. In the past, early weaning has led to poor growth and inferior quality larvae with an increased risk of skeletal deformities (Cahu and Zambonino Infante, 2001; Kjorsvik *et al.*, 2011). Recent advances in microdiet formulation have considerably reduced the pre-weaning period allowing the introduction of specific larval diets to marine finfish culture as early as mouth opening. 'Co-feeding' weaning protocols, simultaneously using inert and live diets, allow an earlier and more efficient changeover period onto microdiet from live feeds. Co-feeding provides higher growth and survival than feeding solely live feeds or microdiets. Early co-feeding of an appropriate microdiet will improve larval nutrition and can condition the larvae to accept the microdiet more readily, thus preventing an adverse effect on subsequent growth following weaning. However, in spite of advances in diet manufacture technology and a better understanding of digestive ontogeny, complete live feed replacement in commercial hatcheries is still not routinely achieved for marine finfish species. In the past several years, many experiments have been conducted to address this problem and to attempt to wean larvae of different species at different ages to different microdiets. This research effort has met with limited success (Faulk and Holt, 2009; Tang *et al.*, 2010; Andréslow *et al.*, 2011; Vandecan *et al.*, 2010). Curnow *et al.* (2006a,b) demonstrated the effect of different weaning and co-feeding treatments on growth and survival of barramundi (*Lates calcarifer*) larvae (Fig. 6.1). The authors found that early weaning before the larvae were fully developed (in addition to the diet type and quality) not only influenced growth and survival, but also the occurrence of cannibalism. They concluded that co-feeding barramundi larvae on microdiet should be started no earlier

© Woodhead Publishing Limited, 2013

than three days prior to stomach differentiation and be continued post-metamorphosis. Co-feeding improves growth by 25–30 % over the previous standard method (shorter and earlier weaning), and mortality during weaning was reduced from 5 % to 1 %. Similar results were found with many freshwater and marine species. It is now clear that complete replacement of live food is still far from reality. Although replacement of *Artemia* has been achieved for some species such as red drum (*sciaenops ocellatus*) and European sea bass (Cahu and Zambonino Infante, 2001; Kolkovski, 2006b; Holt *et al.*, 2011), replacing *Artemia* is not yet considered to be realistic for most species. Replacement of rotifers is even more difficult.

Weaning protocols are almost standard in terms of the weaning steps, i.e. rotifers, *Artemia* (nauplii and then enriched *Artemia*), *Artemia* – microdiet co-feeding and complete weaning. However, these protocols vary according to species, temperature, microdiet type, rearing system, feeding system and intervals. Even in fish species where nutritional requirements could be considered similar, such as gilthead sea bream (*sparus aurata*) and European sea bass, weaning protocols are different. While sea bass is one of the only marine fish species for which complete replacement of *Artemia* has been achieved, sea bream protocols still rely on *Artemia*.

6.3 Feeding system

The digestibility and nutritional qualities of commercially available microdiets are improving due to continuous research and development. However, none of the currently available microdiets is used solely without *Artemia* and rotifers. Part of the reason is the microdiet distribution or the manner the particles are delivered to the larvae. The best formulated microdiet is only as good as the method with which it is dispensed into the larvae tank.

Compared to feeding systems and methods for ongrowing fish, larvae feeding systems have not been given much attention by either the scientific or the commercial sectors. Only a handful of automated microdiet feeding systems exist and almost no scientific papers on this topic have been published.

Hand feeding is the simplest and still the most widely used method. Hand feeding is usually undertaken using small devices (spoons, etc.) with relatively long periods between feeding events (30–60 min or longer). Feeding over long photoperiods (up to 24 h) is difficult due to labour requirements and logistics. Due to the high larval metabolic rates and fast evacuation rates, sporadic or feeding at long intervals results in sub-optimal results and reduces the benefits from a relatively expensive product (Kolkovski *et al.*, 2009).

With the modernization and move towards recirculating systems that require high mechanical expertise, there is a strong need for automation in all the production processes. Not only will this generate labour savings, but

© Woodhead Publishing Limited, 2013

it will also secure the production protocols and bring more repeatability to every step of the processes. There is a need for automatic dispensers of microparticles that will allow precise distribution and be easy to use in hatcheries.

6.3.1 Dosage system

The first requirement from a mechanical microdiet dispenser is that it must be able to deliver a stable quantity at each feeding event. Belt feeders (FIAP Aquaculture, Denmark) are currently the most popular feeders. These are driven by a motor or by a clock and are not capable of splitting a daily ration into equal aliquots of feed. They are handy and relatively cheap but, as they are not designed for microdiet particles, they suffer from microparticles sticking to the belt, especially in humid conditions.

Horizontal drums were developed by Arvotec (Finland). The small cavities on the drum external area can be loaded with relatively consistent quantities when the microparticles remain dry and not sticky. However, often, a thin layer of product accumulates between the rotating drum and its housing. This quickly becomes a contamination source as well as a factor generating inconsistency in the distribution.

Vertical hoppers with a rotating disc are also available commercially (Sterner, Norway). However, the self-compaction ('caking effect') of the microparticles in the hopper, coupled with great difficulty in cleaning the equipment, limits its value.

A different feeding mechanism was developed in Australia by the author ('AMD' [Automatic Microdiet Dispenser], Department of Fisheries, Western Australia, Figs 6.11 and 6.12). In this system, the dosage system is

Fig. 6.11 AMD feeder (Department of Fisheries, Western Australia).

© Woodhead Publishing Limited, 2013

Fig. 6.12 AMD feeder close up.

based on the rapid opening of a sluice-valve using a simple solenoid, allowing for a constant quantity of feed to be delivered at each feeding event. Cleaning the feeder is a simple and quick process. The feeder uses air from the hatchery blower to supply a built-in spreader preventing microdiet chunks dropping directly to the tank and sinking to the bottom.

6.3.2 Delivery to the rearing tank

Once a mechanism to deliver a reliable dose is established, its repetition over the tank surface or in the tank volume is necessary to increase the larvae particle interaction (i.e. before diet particles sink to the bottom of the tanks and become unavailable to the larvae).

The dose delivery can take place directly above the water surface. However, there is a risk that particles will aggregate and immediately sink to the tank bottom. This is especially true for the 100–300 μm particle size at the initiation of weaning and co-feeding.

To avoid this, Raunes, a cod hatchery in Norway, has developed an intermediate vessel where the microdiet is mixed with water and further distributed into the water column at different points of the tank. These are commonly referred to as 'spiders'. The vessel volume is only a few litres and the applied flow allows for 2–3 minutes of residence time in the vessel. In this vessel, the dose is delivered into the flow of water by any dispenser and the microparticles are separated and dispersed by the strong water movements into the vessel. The suspended particles are then pushed into the tanks through the 'spider legs', these being made from a number (6–12) of small rigid plastic (2–3 mm internal diameter) pipes. The lower end of each of them is set into the water column (2–3 cm below the tank surface). This means of dispersing the microparticles is very efficient, especially in large tanks (4–10 m^3 or more). It avoids trapping the microparticles in the surface

© Woodhead Publishing Limited, 2013

skimmers, which are often in use to capture the oil film at the surface of the tanks. However, due to the strong water movement it may also cause very strong leaching of the nutrients. The extent of this problem is not known.

Another way of spreading the microdiet over the surface is by air. An air-blade is formed under the dosage point (where the particles are delivered by the dispenser) and blows the light particles over an area that can reach 30–90 cm. Once separated from each other by the air current, the particles do not tend to clump and conglomerate over the surface. This is simpler than the 'spider' device described above, but requires either large surface tanks to avoid particles lost through the skimmers, or tanks without skimmers (only for <150 μm particles that tend to float rather than sink). Alternatively, shorter feeding intervals of small doses can compensate for the lost of particles through the outlet filters/skimmers.

6.3.3 Fractioning of the daily ration into multiple events

As mentioned above, fish larvae have a limited 'window of opportunity' to catch the microdiet particles before they sink to the bottom of the tank and become unavailable. It is also a current practice in modern hatcheries to establish long photoperiod during the larval production (from 16–24 h light). Therefore, it is extremely important that the distribution of the microdiet be split over time in very frequent feeding events. This will give the larvae a chance to catch fresh microparticles in the water column.

In that respect, it is an advantage for any control system for a microdiet feeding system to include the ability to have frequent and variable feeding events. This is vital in establishing the distribution regime and potential success of the microdiet feeding strategy. Knowing the sinking rates of specific diets and the minimum diet that can be dispersed at a feeding event will allow the design of an efficient and optimal feeding schedule to provide continuous availability of microdiet particles to the larvae. Programming the feeding regime through the day will also optimize the diet utilization. Usually, higher doses and more frequent feeding events are used until the larvae fill their digestive system. Thereafter, feeding frequencies are reduced to match the feeding habits of the larvae through the day.

6.3.4 Sparing the quantity of microdiet

Microdiets are expensive (up to $300 Kg) and are likely to remain so. Their production is difficult and the raw ingredients, in many cases, are very expensive. There is a relatively small market which limits the potential for cost reduction through large-scale production. Therefore, it is necessary to optimize the use and yield from such a product. In the ongrowing sector of fish farming, FCR (feed conversion ratio) is one of the most important factors to assess diet utilization. Using microdiet as a full or partial

© Woodhead Publishing Limited, 2013

substitute for the live feed should lead to a closer look at the FCR as a benchmark that can help hatchery managers improve their methods and efficiency. However, meaningful FCR figures for larvae feeds – live or formulated – do not exist. Estimating FCR with larvae is extremely difficult. Although it is possible to estimate commercial FCRs by measuring the amount of food (rotifers, *Artemia* and microdiets) fed to specific larvae tanks and calculating the final biomass of fish produced, measurements of these parameters during larvae production are always difficult to determine on a daily basis. During the past several years, the use of *Artemia* (per number of post-larvae produced) has dropped significantly due to better diets and weaning protocols. However, as mentioned previously, microdiet feeding is still far from optimal. Calculated microdiet FCRs (actual diet ingested / growth) could be as low as 0.6 or 0.8:1 due to high digestibility of the diet. However, this is rarely if ever achieved because of spoilage of excess feed being distributed due to feeding systems and strategies. Currently, one should consider that actual FCR values above 3:1 are due to poor feed distribution strategies.

Using the right feeding system to reach a precise and controlled distribution of feed will help spare microdiet. Some European hatcheries estimate 20–40 % less feed was used when an automated feeding system was used compared with hand distribution and that there were additional benefits of increased larval survival and growth.

6.3.5 Sparing cleaning time or cleaning efficiency

In most cases, the majority of modern hatcheries still have to siphon the larvae tanks daily. Automation for continuous cleaning of the tank bottom (for example, cod hatcheries, Norway (Alver *et al.*, 2009) and private and government hatcheries in Japan), has not yet gained full acceptance by the majority of European hatcheries. Therefore manual siphoning, to prevent accumulation of organic matter and proliferation of bacteria (Fig. 6.9) remains a time consuming task and an important factor affecting labor. Changing from live feed to microdiet may lead to a negative impact on water quality, particularly if large microdiet portions (e.g., via hand feeding) are given. In most cases, feed rations are higher than necessary, with the intention of maintaining particles in suspension between feeding events. If a good dispenser is used, an efficient feeding can be obtained with less feed and generating far less deposition on the bottom of the tank. Reduction in siphoning and feed input will help, in turn, to increase water quality in the tank and reduce bacteria proliferation resulting in a positive effect on larvae survival. This was demonstrated by Fletcher *et al.* (2007) when comparing different weaning diets to *Artemia* feeding. The authors didn't find significant differences between the treatments in either water turbidity or bacteriology tests. The experimental system consisted of 80 L tanks, far smaller than commercial hatchery tanks. However, as mentioned above,

© Woodhead Publishing Limited, 2013

with good feeding system, accurate feeding regimes adjusted to specific microdiet and hygiene protocols, using formulated diets can be as good if not better then using live food in terms of tank hygiene.

6.4 Future trends

It is clear that complete replacement of rotifers and *Artemia* with microdiets as first food items for marine finfish larvae has not yet been achieved commercially without reduced growth and survival. The reasons for this lack of success are due to several factors that need to be addressed. First and foremost, the particle need to be attractive to the larvae. Therefore, feed attractants need to be incorporated or coated onto the particles. This must involve diet-manufacturing techniques that can control leaching, particularly of amino acids.

The microdiet particles should be available to the larvae at all time, while limiting fouling of the tank. This will require research into new feeding systems or optimizing existing systems and integrating knowledge of sinking rates of specific diets. Optimizing tank hydrodynamics will also contribute to a better food particle distribution.

Following ingestion, easier to digest raw materials should be tested and adjusted to balance the amino acid requirements of specific species. This should be linked to diet manufacture methods that may increase or decrease the particle digestibility.

To date, larval nutritional requirements are only partially identified and much is still unknown. With the introduction of better microdiets with higher attractability and better digestibility, the nutritional requirements of marine fish larvae can be defined more easily.

6.5 References

ALVER M, OIE G, OLSEN Y and ALFREDSEN A (2009) 'Process monitoring and control in the next generation hatchery', in Hendry C, Van Stappen G, Willie M and Sorgeloos P (eds), *Larvi'09 – Fish & Shelfish Larviculture Symposium*, Special Publication No. 380. Oostende: European Aquaculture Society, 12–14.

ANDRÉSLOW M, ROTLLANT G, SASTRE M and ESTÉVEZ A (2011) 'Replacement of live prey by formulated diets in larval rearing of spider crab *Maja brachydactyla*', *Aquaculture*, 313(1–4), 50–56.

BENITA S (2006) *Microencapsulation: Methods and Industrial Applications*. Boca Raton, FL: CRC Press.

BURNELL G and ALLAN G (2009) *New Technologies in Aquaculture*. Boca Raton, FL: CRC Press.

CAHU C and ZAMBONINO INFANTE JL (2001) 'Substitution of live food by formulated in marine fish larvae', *Aquaculture*, 200, 161–180.

CURNOW J, KING J, PARTRIDGE G, BOSMANS J and KOLKOVSKI S (2006a) 'The effect of Artemia and rotifer exclusion during weaning on growth and survival of barramundi (*Lates calcarifer*) larvae', *Aquaculture Nutrition*, 12(4), 247–255.

© Woodhead Publishing Limited, 2013

CURNOW J, KING J, PARTRIDGE G and KOLKOVSKI S (2006b) 'The effect of various co-feeding and weaning regimes on growth and survival in barramundi (*Lates calcarifer*) larvae', *Aquaculture*, 257, 204–213.

FAULK C K and HOLT H (2009) 'Early weaning of southern flounder, *Paralichthys lethostigma*, larvae and ontogeny of selected digestive enzymes', *Aquaculture* 296, 213–218.

FLETCHER R C, ROY W, DAVIE A, TAYLOR J, ROBERTSON D and MIGAUD H (2007) 'Evaluation of new microparticulate diets for early weaning of Atlantic cod (*Gadus morhua*): Implications on larval performances and tank hygiene', *Aquaculture*, 263(1–4), 35–51.

HEINEN J M (1981) 'Evaluation of some binding agents for crustacean diets', *Progressive Fish Culturist*, 43(3), 142–145.

HOLT J (2011) *Larval Fish Nutrition*. Chichester: Wiley-Blackwell.

HOLT J, WEBB K and RUST M (2011) 'Microparticultae diets: testing and evaluating success', in Holt J (ed.), *Larval Fish Nutrition*. Chichester: Wiley-Blackwell, 353–372.

JACKSON A and NIMMO C (2005) Comparison of sinking and leaching rates of 11 commercially available microdiets, in Kolkovski S (Principal Scientist) *Development of marine fish larvae diets to replace Artemia*, FRDC final project report No. 2001/220. Perth: Fisheries Research and Development Corporation, 153–166.

KJORSVIK E, GALLOWAY T, ESTEVEZ A, SAELE O and MOREN M (2011) Effect of larval nutrition on development, in Holt J (ed.), *Larval Fish Nutrition*. Wiley-Blackwell, 219–248.

KOLKOVSKI S, ARIELI A and TANDLER A (1995) 'Visual and olfactory stimuli as determining factors in the stimulation of microdiet ingestion in gilthead seabream *Sparus aurata* larvae, in Lavens P, Jaspers E and Roelants I (eds), *Larvi '95 – Fish and Shellfish Symposium*, Special Publication vol. 24. Oostende: European Aquaculture Society, 289–292.

KOLKOVSKI S (2001) 'Digestive enzymes in fish larvae – Application and implication – a review', *Aquaculture*, 200, 181–201.

KOLKOVSKI S (2004) 'Marine fish larvae diets – current status and future directions', *11th International Symposium on Nutrition and Feeding in Fish*, 2–7 May, Phuket, Thailand.

KOLKOVSKI S (2006a) 'Amino acids as feed attractants for marine fish larvae', *World Aquaculture Symposium 2006*, 10–13 May, Florence.

KOLKOVSKI S (2006b) 'Marine fish larvae diets – current status and future directions', *International Workshop 'Producion de Larvas de Peces'*, 14–15 September, Temuco.

KOLKOVSKI S, KOVEN W M and TANDLER A (1997a) 'The mode of action of *Artemia* in enhancing utilization of microdiet by gilthead seabream *Sparus aurata* larvae', *Aquaculture*, 155, 193–205.

KOLKOVSKI S, ARIELI A and TANDLER A (1997b) Visual and chemical cues stimulate microdiet ingestion in seabream larvae, *Aquaculture International*, 5, 527–537.

KOLKOVSKI S, CURNOW J and KING J (2004) 'Intensive rearing system for fish larvae research – I. Marine fish larvae rearing system', *Aquaculture Engineering*, 31, 295–308.

KOLKOVSKI S, LAZO J P, LECLERCQ D and IZQUIERDO M (2009) 'Fish larvae nutrition and diet: new development', in Burnell G and Allan G (eds), *New Technologies in Aquaculture*. Boca Raton, FL: CRC Press, 315–370.

KVÅLE A, YÚFERA M, NYGÅRD E, AURSLand K, HARBOE T and HAMRE K (2006) 'Leaching properties of three different microparticulate diets and preference of the diets in cod (*Gadus morhua* L.) larvae', *Aquaculture*, 251, 402–415.

LANGDON C and BARROWS R (2011) 'Microparticulate diets: technology', in Holt J (ed.), *Larval Fish Nutrition*. Wiley-Blackwell, 335–351.

© Woodhead Publishing Limited, 2013

LAZO J P, DARIAS M and GISBERT E (2011) 'Ontogeny of the digestive tract', in Holt J (ed.), *Larval Fish Nutrition*. Wiley-Blackwell, 5–46.

ONAL U and LANGDON C (2004) 'Characterization of lipid spray beads for delivery of glycine and tyrosine to early marine fish larvae', *Aquaculture*, 233, 495–511.

PARTRIDGE G J and SOUTHGATE P C (1999) The effect of binder composition on ingestion and assimilation of microbound diets (MBD) by barramundi *Lates calcarifer* Bloch larvae. *Aquaculture Research*, 30, 879–886.

TANG B, CHEN G and HU Z (2010) 'Application of a microdiet in cobia *Rachycentron canadum* (Linnaeus, 1766) larvae rearing', *Aquaculture Research*, 41(2), 315–320.

VANDECAN M, DIALLO A and MELARD C (2010) 'Effect of feeding regimes on growth and survival of *Clarias gariepinus* larvae: replacement of *Artemia* by a commercial feed', *Aquaculture Research*, 42(5), 733–736.

YUFERA M (2011) 'Feeding behaviour in larval fish', in Holt J (ed.), *Larval Fish Nutrition*. Wiley-Blackwell, 285–306.

YÚFERA M, KOLKOVSKI S, FERNANDEZ-DIAZ C and THIES C (1998) Microencapsulated diets for fish larvae – current 'state of art'. *Bioencapsulation VII*, 20–23 November, Easton, MD.

YÚFERA M, KOLKOVSKI S, FERNANDEZ-DIAZ C and DABROWSKI K (2003) Free amino acid leaching from protein-walled microencapsulated diet for fish larvae. *Aquaculture*, 214, 273–287.

© Woodhead Publishing Limited, 2013

7

Management of finfish and shellfish larval health in aquaculture hatcheries

T. J. Bowden and I. R. Bricknell, University of Maine, USA

DOI: 10.1533/9780857097460.1.223

Abstract: This chapter looks at the issues surrounding health of finfish and shellfish larvae in the aquaculture environment. The chapter will examine issues such as biosecurity to see how it forms the cornerstone of effective hatchery management and how problems can arise when biosecurity fails. Then the chapter looks at how the health of the larvae can be manipulated and how the larvae can be protected from potential pathogens by good management.

Key words: larvae, vaccine, immmunostimulant, thymus, biosecurity, immunity, live feeds, hatchery, water quality, health.

7.1 Introduction

The increase in demand for seafood has major ecological and economic impacts. It contributes to the depletion of wild fish stocks causing fisheries collapses such as the collapse of wild cod stocks on the Grand Banks. The cod of the Grand Banks (Newfoundland) have failed to recover even after a 20 year moratorium on cod fishing in the region. In turn, this has led to the economic decline of many North Eastern fishing towns and villages, a depopulation of the working waterfront and a switch from vibrant economically robust communities to holiday homes and a seasonal economy. Often the traditional fishing families are forced out of their traditional communities due to escalating property prices.

Even though the wild cod of the Grand Banks have gone, the demand for seafood remains strong and the fish merchants have turned further afield exploiting populations of cod from Greenland, Iceland and Europe and these populations have also begun to collapse. Yet this insatiable demand for seafood, currently $18 billion for the USA alone, continues to be a significant contribution to the trade deficit (Sloane, 2010). It is clear that if seafood is going to be an affordable food item then aquaculture will

© Woodhead Publishing Limited, 2013

have to expand to fill the gap. Aquaculture already supplies around 50 % of all marine protein consumed, and this is forecast to rise by about 9 % per annum as wild fish species decline due to fisheries activity. According to the Food and Agriculture Organization of the United Nations (FAO) 'Aquaculture continues to be the fastest growing animal food-producing sector and to outpace population growth (Davies and Rangeley, 2010, FAO, 2012). This growth has delivered an increase from 0.7 kg per capita from aquaculture supply in 1970 to 7.8 kg in 2006, an annual growth rate of just under 7 %. Global production has risen from under 1 million tonnes in the 1950s to over 50 million tonnes by 2006 with a value of nearly 80 billion dollars. Clearly aquaculture is a growth business and it looks set to overtake capture fisheries as the main source of food fish. Given the increasing regulation of the capture fisheries in Europe and North America and an increasing global population, it would seem that aquaculture is well placed to become the dominant source of food fish.

The expansion of aquaculture to fill the wild catch deficit relies on providing suitable conditions for the development of the larvae in enclosed, recirculating systems and this provides one of the greatest challenges to the expansion of marine aquaculture. Not least as new species are brought into aquaculture, we have to meet the unique requirements of the larvae's environment and understand the diseases that they are susceptible to in high density larval systems. As many larvae have to undergo long periods of development from the egg to a robust free-living animal, the prevention of infectious disease is essential. For example, many fish species may have a period of many months before they metamorphose and develop an adaptive immune system of their own.

During this period, larval animals rely on immunological active material donated to the egg by their mother, limited innate defences produced by a finite number of immune cells or they just chance their arm and hope that the environment is of sufficient quality that they never encounter a pathogen. Aquaculture cannot rely on luck! To ensure that production systems are sufficient to meet demand, the active aquaculturist can take many steps to manipulate the environment, modulate the maternal and larval immune system and introduce immunotherapies to prevent disease outbreaks. This can take the form of pre- or probiotics, vaccination of broodstock to manipulate the Immunoglobulin (Ig) profile in eggs and yolk sac larvae, the use of immunostimulants to optimize the larvae's innate immune defences, and finally direct vaccination of the larvae when they have developed a fully functional immune system. Health issues for fish and shellfish larvae depend a great deal on the type of production facility (Olafson, 2001). Although aquaculture has been practised for millennia, the first evidence comes from Egypt during the Middle Kingdoms (2052–1786 BC) where they practised intensive fish culture using tilapia in open ponds following the River Nile's wet season. This was an earth pond system dug into the flood plane of the Nile. The animals were put in at the end of the wet season, fed on vegetable

© Woodhead Publishing Limited, 2013

matter (at least according to ancient stele depicting aquaculture), and the fish were harvested just before the next floods. The Romans also developed aquaculture practices, as they are known to have cultivated oysters. They may also have had a regional aquaculture industry in Byzantium where pools with amphora cemented into their sides have been found and these have been interpreted as being for a cave spawning fish or for the storage of octopus.

Although many authors credit China as the birthplace of aquaculture in 2000 BME, they were simply growing wild collected fish in ponds for commercial harvest, similar to Egyptian ancient aquaculture in a system whereby carp were collected as fry and transferred to special ongrowing ponds. In the 5th century BME, the culture of silver carp (*Hypophthalmichthys molitrix*) was detailed in a manuscript by Fan Li. This has been dated to 460 BME and is currently believed to be the first description of a species where the whole of the life-cycle was completed in captivity. This was the start of the domestication of fish and led to the first species being developed totally in aquaculture, the humble goldfish, which has never existed in the wild.

Larval marine fish production systems in the 1800s often relied on extensive systems where the larval fish were kept at low densities, in raw seawater ponds, pools or large tanks where the larvae could feed on natural prey items. In such a poorly controlled environment, disease outbreaks were common and the pathogens, bacterial, viral or parasitic, utilized the oral route to invade the larvae on infected food items. This is particularly true of pathogenic nematodes and pathogenic bacteria such as *Vibrio* spp. *Vibrio* spp are frequently associated with the invertebrate cuticle and colonize the surface of copepods (Kaneko and Colwell, 1975; Kirchner, 1995; Pruzzo *et al.*, 1996; Poulicek *et al.*, 1998), the preferred prey of most larval marine fish (Nunn *et al.*, 2012). Many nematode worms such as *Anisakis* spp. also use marine copepods as an intermediate host to invade fish (Klimpel *et al.*, 2008; Skov *et al.*, 2009). Although rare today, these primitive extensive systems are often used as the first stage in the domestication of a new species as they provide an environment where captive larvae can be studied and their life strategies optimized. This was particularly true of snook (*Centropomus undecimalis*) culture where the initial success with this species was in extensive pond system with the larvae requiring over six months in this system before they were robust enough to transfer to the ongrowing system (Tucker, 1987).

7.2 Diseases in hatcheries

The types of pathogen that may become problematic include bacteria, viruses, parasites and fungi. Depending on the type of pathogen, various routes are open to the pathogen to become problematic both within the

© Woodhead Publishing Limited, 2013

host and within the larval rearing system. Some of these routes include contamination of the egg, maternal transfer, infection through the micropyle, direct penetration of the skin, direct penetration of the egg, uptake through damaged skin, ingestion, and absorption through the gills (Smith *et al.*, 1999; Chu and Lu, 2008; Magi *et al.*, 2009; Menanteau-Ledouble *et al.*, 2011; Nagano *et al.*, 2011). Contamination of the egg can occur *in ovo* where the pathogen is incorporated into the egg during its development or ovagenesis. The pathogen will often lay dormant until the larva begins developing at which time the pathogen rapidly replicates overwhelming its host. This diseased egg can now act as a focal infection, contaminating nearby eggs. Additional routes of pathogen entry into the egg include direct penetration of the chorion in finfish or the egg membrane in aquatic invertebrates, or through the micropyle of fish eggs. An alternative situation can arise where the ovarian fluid is contaminated, while the eggs themselves are not, but the presence of the contaminated ovarian fluid results in pathogens entering the egg just prior to or following expulsion of the egg (Phelps and Goodwin, 2008; Bratland and Nylund, 2009; Fenichel *et al.*, 2009; Kai *et al.*, 2010; Kongtorp *et al.*, 2010; Kumagai and Nawata, 2010).

The epidermis is a key pathogen entry point. Some pathogens, especially parasites, can bore directly through the skin and infect the host. Other pathogens need the skin to be damaged, i.e. wounds, abrasions, etc. The pathogen colonizes the superficial lesion prior to a more substantial invasion of the host (Smith *et al.*, 1999; Chu and Lu, 2008; Menanteau-Ledouble *et al.*, 2011; Nagano *et al.*, 2011). In larval fish, ingestion is a major uptake route for pathogens (Munro *et al.*, 1993, 1995, 1999; Ringo *et al.*, 1996). This can occur either through ingestion of infected prey items or during drinking for osmoregulation (Verner-Jeffreys *et al.*, 2003b, 2004). Prey items such as live feeds are often heavily contaminated with bacteria, both pathogenic and non-pathogenic, as they are grown in non-sterile nutrient rich environments, suitable for the rapid replication of these bacteria. The bacteria aggregate on the live feed, especially crustacean live feeds such as *Artemia* and copepods. This occurs because bacteria often have receptors for the cuticle that assist in the colonization of the live feeds. Once contaminated, the live feeds can introduce an infectious dose of the pathogen when they are eaten. The elimination of pathogens from live feed production units is a critical control point in a competent management process that ensures the production of healthy larval animals (Davis and Arnold, 1997; Munro *et al.*, 1999).

The gills are incredibly delicate structures and are directly exposed to the environment. They are at risk of meeting pathogens directly from the water column. Some pathogens have receptors that specifically target the gill tissue. Once these pathogens are bound to the gill surface they gain access to the internal structures and tissues of the host and can become systemic in nature in a very short time. It is essential to ensure that the bacterial load of the water column is as low as reasonably possible.

© Woodhead Publishing Limited, 2013

Larval fish are very prone to infection with *Vibrio* spp. Larval Atlantic cod (*Gadus morhua*), for example, seem particularly susceptible to *V. anguillarum* O2β to the point that whole production runs have been lost to this pathogen (Verner-Jeffreys *et al.*, 2003b, 2004). There have been many attempts to control this disease, including the use of immunostimulants, probiotics, vaccines, abiotic culture, ozone and UV (Davis and Arnold, 1997; Theisen *et al.*, 1998; Bricknell and Dalmo, 2005; Hanif *et al.*, 2005; Battaglene and Cobcroft, 2007; Kai and Chi, 2008; Luzardo-Alvarez *et al.*, 2010; Chen *et al.*, 2011; Kumar *et al.*, 2012). However, most success has been seen with the initial use of immunostimulants followed by bath vaccination when the larvae reach about 500 mg in size (Bricknell and Dalmo, 2005; Hanif *et al.*, 2005, Kai *et al.*, 2010). In understanding the implications of these types of amelioration it is important to comprehend the particular physiology of the host. In the case of Atlantic cod, research had indicated that they did not respond to vaccination very well until they were immunologically mature around the time thymus develops (Samuelsen *et al.*, 2006). In addition, Atlantic cod are unusual in how their immune system functions. Instead of a increased level of highly specific antibodies being produced following vaccination, Atlantic cod seem to utilize a lower level of specificity and compensate with a broader range of antibody specificity, a situation that runs somewhat contrary to immunological dogma (Mikkelsen *et al.*, 2004; Sommerset *et al.*, 2005; Schroder *et al.*, 2006; Gudmundsdottir and Bjornsdottir, 2007).

Viral infections in hatcheries have the potential to be catastrophic as there are no effective treatments and very few effective viral vaccines. Often the virus is carried asymptomatically by the broodstock and is shed during periods of high stress, such as stripping and spawning (Saintjean *et al.*, 1991; Cutrin *et al.*, 2005; Munro and Ellis, 2008). The virus itself may infect the eggs both horizontally (i.e. egg to egg or fish to fish) or vertically (mother to embryo). Regardless, such infections are very difficult to control. Viral nervous necrosis (VNN), caused by nodaviruses, is a typical pathogen of finfish hatcheries. It affects both cold and warmwater species and has caused serious problems in Atlantic cod, Atlantic halibut (*Hippoglossus hippoglossus*), sea bass (*Dicentrarchus labrax*) and gilthead seabream (*Sparus aurata*) culture. Control measures usually include the screening of broodstock for the virus and the elimination of virus-positive animals from the breeding population (Grotmol and Totland, 2000; Breuil *et al.*, 2003; Athanassopoulou *et al.*, 2004; Manin and Ransangan, 2011). The broodstock are often vaccinated against the pathogen, and there has been some success in reducing the impact of this disease by incorporating immunostimulants into larval diets (Bricknell and Dalmo, 2005). Susceptible species have benefited from disease-resistant strain selection, especially for sea bass and gilthead seabream, and the recruitment of F1 and F2 generations for broodstock where their disease history is known and is VNN free (Ordas *et al.*, 2006; Odegard *et al.*, 2010).

© Woodhead Publishing Limited, 2013

An emerging disease that is potentially devastating is *Francisella* spp. It has a wide host range including: Atlantic salmon (*Salmo salar*), Atlantic cod and tilapia (*Oreochromis niloticus*), and causes a chronic granulomatous infection (Olsen *et al.*, 2006; Ostland *et al.*, 2006; Birkbeck *et al.*, 2007, 2011; Soto *et al.*, 2011; Zerihun *et al.*, 2011). This is often highly debilitating to the animal, leading to poor growth on an infected farm. Currently there are no effective vaccines against *Francisella* spp and spread of the disease is not well understood. It is certainly transmitted horizontally fish to fish (Birkbeck *et al.*, 2007, 2011), but it is unknown if it spreads vertically. Treatment is very difficult as *Francisella* spp is an intracellular pathogen and effectively evades antibiotic treatment (Soto *et al.*, 2011). Because of this, the eradication of carriers from broodstock has not been achieved and this pathogen remains a potentially serious risk to fin fish culture.

7.3 Development of immune systems in aquatic animals

Onset and ontogeny of the immune system in larval animals are important considerations in developing a robust management strategy that will mitigate health issues in the hatchery. Knowing when an animal becomes immunocompetent and capable of defending against pathogen attack can allow for timely changes in management that can reduce cost and increase productivity. Maintenance of high levels of biosecurity within a hatchery system can become expensive, and a good knowledge of the correct timing for reducing that biosecurity can make the difference between a commercial system running at a loss or making the production cycle cost-effective.

Development of invertebrate immune systems is poorly understood (Söderhäll, 2010). Nearly all our current understanding of invertebrate immunology focuses on animals that are juvenile or older, thus our comprehension of the immune function of invertebrate larvae is minimal. For instance, phagocytic cells are present in veliger larvae of bivalves and antimicrobial peptides are present in echinoderm larvae (Dyrynda *et al.*, 1995). Invertebrates possess a robust immune system that is similar to a vertebrate innate immune system. This includes antimicrobial proteins such as crustins, pattern recognition receptors such as glucan-binding proteins, and opsinins such as prophenoloxidase (Hauton, 2012). It has long been assumed that because invertebrates do not possess an adaptive response or immunoglobulin-like structure in their immune system that they would be unable to develop any form of immune memory. But occasional publications hint at the existence of some form of immune memory. Reports suggest that exposure to a parasite reduces the impact or severity of a subsequent exposure to the same parasite (Kurtz and Franz, 2003). This raises the possibility of developing a vaccine for use in invertebrates. Commercial vaccines are already available for this purpose and the efficacy of the concepts behind them seems to be well proven (Pope *et al.*, 2011). The response

© Woodhead Publishing Limited, 2013

seems to be linked to hemocyte cells which can produce a highly variable molecule that mimics vertebrate immunoglobulin diversity (Watson *et al.*, 2005). It remains to be seen if the mechanisms behind these observations are functional in larval animals.

A considerable body of research has been carried out to investigate the development of functional immune systems in fish larvae. A special edition of the journal *Fish & Shellfish Immunology* carried a set of review papers on the subject of fish larval immunity (Bowden *et al.*, 2005; Bricknell and Dalmo, 2005; Dalmo, 2005; Falk-Petersen, 2005; Magnadottir *et al.*, 2005; Rombout *et al.*, 2005). These looked at the development of various aspects of the immune system in the early life stages.

The thymus is considered an important organ in relation to the immune system of vertebrates as it is responsible for the production of self-restricted and self-tolerant T-cells (Manley, 2000). This is essential in developing an adaptive immune response. In following the development of the thymus in a range of cultured fish species, it is apparent that this development can be seen to occur before hatching in rainbow trout (*Oncorhyncus mykiss*) until about 30 days after hatching in Atlantic halibut (Grace and Manning, 1980; Bowden *et al.*, 2005).

Many egg laying fish include significant amounts of immunoglobulins in the yolk that are maternally derived. For many years, this was considered to be a nitrogen source for the growing embryo and they were not thought to be functional. More recent work has shown that these antibodies are functional and capable of binding to epitopes in the developing egg (Swain and Nayak, 2009). The composition of the antibody affinities reflects the antibodies circulating in the mother animal. Hence, the most recent immunological insult that has been seen by the mother will have the highest antibody titre in the egg. This can be of significant benefit in providing protection to larval fish as the mother can be vaccinated against pathogens present in the hatchery prior to spawning, which will result in the transfer of these specific antibodies to the developing egg. These antibodies can play a significant role in reducing the severity of a disease outbreak, providing the microbial flora in the hatchery is well understood (Bricknell *et al.*, 2000; Verner-Jeffreys *et al.*, 2003a).

It is clear that environmental influences can have significant impact on the immune function whether this is temperature, light, the presence of chemicals, oxygen, etc. For aquatic animals this is no different. Specific environmental factors can alter immune performance, improving it in some instances and reducing it in others (Bowden, 2008). Then again, it is also clear that natural cycling events such as the changing seasons can also impact on immune function (Bowden *et al.*, 2007). This provides another reason why competent hatchery managers must fully understand the requirements of their charges and provide the correct environment to ensure optimal health. While seasonal changes are unlikely to be relevant during the relative short timeframes of most larval development cycles,

© Woodhead Publishing Limited, 2013

other environmental parameters such as temperature must be considered to ensure optimal health (Langston *et al.*, 2002).

7.4 Management of larval health

7.4.1 Biosecurity

One of the major challenges in larval rearing systems is biosecurity. Biosecurity in its simplest terms is the protection from pathogen exposure. Larval animals appear to have a greater susceptibility to pathogens than adults (Bricknell and Dalmo, 2005). This may be a reflection of the immature status of their immune system (Bricknell and Dalmo, 2005). Protection of these larval animals at this delicate time is usually a function of good husbandry such that the management systems prevent contact with pathogens rather than reacting to the presence of pathogens within the system (Verner-Jeffreys *et al.*, 2003b, 2004). Good management would include a risk assessment of the larval rearing environment and the conditions that exist within it, to see what hazards are present and a definition of the associated risks. In discussing this aspect of biosecurity, it is important to understand some of the basic terms and their implications (Bergfjord, 2009; Ahsan and Roth, 2010). The principle concepts we need to consider in risk management are those of 'hazard' and 'risk'. The 'hazard' associated with an item is the dangers it can pose. Risk is the likelihood of the hazard occurring. Many things can be considered to be hazardous, but the risk assessment looks at the likelihood of the hazard. The risk can be mitigated or reduced by altering the circumstances. For biosecurity in a larval rearing facility we need to identify the hazards that exist and assess the risk associated with them. Good management would then seek either to remove the hazard or, if that is not possible, to reduce the risk to an acceptable level. One method of mapping out the risks and hazards is to produce a hazard matrix (Table 7.1).

Biosecurity is the control of pathogen entry and proliferation. It has become an essential aspect of any intensive animal production unit and is one of the most difficult challenges facing the emerging finfish and shellfish

Table 7.1 Hazard matrix for assisting in assessing risk

Approach	Where to apply it	Techniques to apply
Prevention	At source or upon arrival	Exclusion from the supply chain by vaccination/health screening, prophylaxis, etc. Isolation/quarantine
Long-term reduction of the impact and control of infection	On-site continual	As for prevention and detection of newly arrived unwanted pathogens

© Woodhead Publishing Limited, 2013

culture industry worldwide (Delabbio *et al.*, 2004; Pruder, 2004; Oidtmann *et al.*, 2011). Once introduced, pathogens can quickly proliferate within systems leading to short-term and long-term disease issues. These issues can sometimes result in a complete collapse of the production system. Pathogens can be extremely difficult to manage, especially where they become established in biofilms, within the filters, or in certain areas of the hatchery such as the live feed production unit. To eradicate a pathogen in these situations may require closure of the hatchery and a complete disinfection of the affected area (Skall *et al.*, 2005). This is easily achieved in a seasonal hatchery such as Atlantic salmon or cod, where the hatchery is normally fallowed when the fish are not spawning. However, in hatcheries for fish that spawn all year round, such as clown fish (*Amphiprion ocellaris*), sea bass or gilthead seabream, the hatchery has to be productive all year to be economically viable, and closure can cause economic problems.

Intermittent outbreaks or a more chronic or long-term, low level, type of outbreak can affect output reliability resulting in pulses of mortalities or a trickling mortality that can be difficult not only to define but to treat. Such problems can be exacerbated in warmer-water environments where there is often a greater opportunity for infections to arise. Often, these intermittent outbreaks of disease lead to hatchery managers describing a batch of fish with high mortality as 'poor doers', when in fact these animals are suffering from an intermittent infection.

Biosecurity requires an awareness of the entry points of potential pathogens. These are known as 'critical control points' and systems or procedures to either secure that point of entry or to minimize the risk of pathogen entry need to be developed (Fig. 7.1). One of the first entry points to consider is the water supply. A variety of water supply solutions exist for larval rearing systems. These include: inflow from an existing watercourse, lake or sea; pumping sea water; pumping fresh water from wells; partial recirculation systems; and, finally, complete recirculation systems. Inflow systems typically have the lowest cost but the greatest biosecurity threat (Delabbio *et al.*, 2004; Pruder, 2004). Recirculation systems have the highest cost but the fewest inherent biosecurity issues (Delabbio *et al.*, 2004). Filtration systems can reduce the pathogen load by removing pathogen particles. However, an infected filtration system can act as a pathogen incubator dosing the larval rearing environment repeatedly, leading to high levels of mortality. Alternative methods of reducing the pathogen load in inflow water include the use of ultraviolet light, which at specific wavelengths can lead to damage to DNA and the death of a pathogen (Battaglene and Cobcroft, 2007; Kumar *et al.*, 2012), or ozonation, which kills pathogens through the production of highly-reactive oxygen free radicals (Davis and Arnold, 1997; Grotmol and Totland, 2000).

As described previously, pathogens can infect eggs. Disinfection of eggs on release from the mother following fertilization is widely practised in certain species (Grotmol and Totland, 2000). Often this is done because

© Woodhead Publishing Limited, 2013

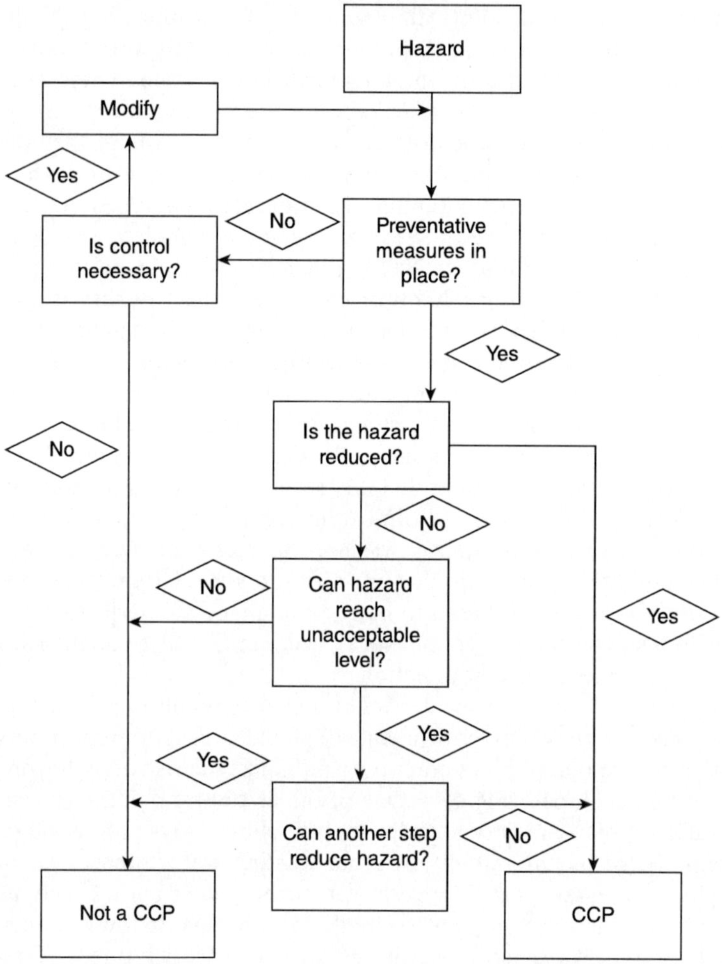

Fig. 7.1 A critical control point decision tree.

there is a requirement to water-harden the eggs following fertilization to complete that step of the development process. These techniques can include ozonation of the water following fertilization to eliminate bacterial and viral pathogens or immersion in a halogen-based disinfectant such as an iodophore, to reduce the pathogen burden (Yoshimizu *et al.*, 1993; Grotmol and Totland, 2000). The use of ozone as a treatment for intake and effluent water and for other processes in rearing systems is increasing as the technology offered for delivering ozone improves. Ozone is a powerful oxidant that has several advantages due to its rapid reaction rate and lower levels of harmful reaction by-products in fresh water, and because oxygen is produced as a reaction end product (Grotmol and Totland, 2000;

© Woodhead Publishing Limited, 2013

Summerfelt, 2003). In aquaculture it is used to inactivate pathogens, assist in the breakdown of organic materials and supplement the actions of other water treatment systems. More recently, ozone has been used to surface disinfect eggs in order to reduce vertical transfer of pathogen from parent broodstock to progeny (Buchan *et al.*, 2006; Ballagh *et al.*, 2011). It is often used in recirculating systems to reduce nitrite levels by the oxidation of nitrite to nitrate (Noble and Summerfelt, 1996). However, the use of ozone has the potential to reduce the bioavailability of iodine in salt water systems which can result in the larvae of Senegalese sole (*Solea senegalensis*) showing reduced growth rate and symptoms of goitre (Ribiero *et al.*, 2009). This is particularly important in animals like the sole which metamorphose during their early development where iodine is an important constituent of the thyroid hormones that drive the metamorphosis process (Einarsdóttir *et al.*, 2006). However, iodine monitoring and supplementation can easily overcome this issue.

Ultraviolet radiation or UVC is a widely used technology that disinfects by damaging the DNA of the target pathogen (Kim *et al.*, 2002; Mamane, 2008). Water from the larval system is diverted from the life support component after filtration to the UVC reactor chamber. Here the water is irradiated with high amounts of UVC from specialized light sources. The efficiency of UVC is dependent on the level of irradiation and the contact time. For example, a low wattage system with a long contact time is as effective as a high wattage system with a short contact time. The disadvantage is that the low wattage system cannot handle large water volumes (Gullian *et al.*, 2012). Very large hatchery systems often have banks of UVC light systems arranged in parallel to ensure the appropriate level of pathogen irradiation. Another drawback of this type of system is that the operating wavelength of these bulbs can drift over time, resulting in a change in the efficiency. To overcome this, most manufacturers recommend replacing the bulbs on a regular basis but this can make these systems expensive to maintain.

As already mentioned, biosecurity of live feeds in an important issue. The range of live feeds is expanding as our ability to culture the various types of live feeds improves. Live feeds such as *Artemia*, copepods and rotifers can now be routinely cultured on site without significant investment or cost. The ability to grow live feeds in a controlled environment allows for stricter control of the culture conditions and for the minimization of potential contamination. If required, live feeds such as *Artemia* and rotifers can be grown in what is essentially a pathogen-free environment (Lubzens *et al.*, 2001). Such growth conditions greatly reduce the biosecurity risk associated with live feeds but involve a considerable increase in their production costs.

Green water culture is the description of a variety of methods for the cultivation of larval fish and crustaceans where microalgae are included in the rearing environment. Naturally occurring phytoplankton have been positively encouraged in outdoor pond systems by the use of fertilization

© Woodhead Publishing Limited, 2013

and management strategies to encourage these species. More controlled systems have the desired microalgal species seeded or regularly pulsed into the system to provide continuous beneficial nutrition (Bosma and Verdegem, 2011). The main thrust of this technique has been the use of green microalgae. However, a range of phytoplankton is being used because of the different colour profiles. For example, some diatoms yield a brown shade when present in large numbers and have been used over many years for the larval culture of shrimp. The improvements in growth and survival are believed to be a result of better direct and indirect nutrition, lower stress, turbidity and contrast enhancement improving the environment, removal of nitrogenous waste and improved oxygenation, chemical and digestive stimulant, and antibacterial properties of microalgae (Battaglene and Cobcroft, 2007). There are clearly many benefits of green water culture systems for both finfish and shellfish in relation to growth rate, metamorphosis and live feed enrichment. All of these factors will help to produce healthy larvae. However, the main advantage from a specifically health point of view would be the reduction of pathogenic bacteria and viruses associated with green water culture systems and the probiotic effects that are concomitant with these types of systems.

Possibly the most important critical control point is the human interaction. Staff must be trained in good operating techniques. These include personal hygiene, equipment hygiene and restricting movement of both personnel and equipment between different units/buildings on a single site and between sites. Axenic rearing conditions are not widely practised at present. However, they were used to control microbial flora in certain types of larval rearing systems, notably, penaed shrimps and some highly delicate fish larvae that had long larval developmental cycles (He *et al.*, 2012). Usually antibiotics were added to the water column to reduce or eliminate the pathogens of concern and prevent infection. However, these conditions often selected for resistant strains of the pathogen, which then required a change in antibiotic allowing further production cycles before resistance reoccurred. Ultimately, this type of culture system developed resistant pathogens that were extremely difficult or impossible to treat with antibiotics. Often this led to the closure of the hatchery due not only to the pathogen load but also to the cost implications of the level of antibiotics required to maintain production.

Larval rearing systems have several methods of dealing with wastewater. In some systems, it can be channelled directly to waste. An alternative method involves reusing or recirculation of the water. This requires some form of processing to reduce contaminants, including pathogens from the current pass. Recirculation systems often use a variety of different filter types including; sand filters, biofilters and fluid beds. One issue that is common to all types of filter is the development of a biofilm that is essential to their operation whereby bacteria reduce ammonia to nitrite and then to nitrate (Ebeling, 2000; Piedrahita, 2003; Eding *et al.*, 2006; Martins *et al.*,

© Woodhead Publishing Limited, 2013

2010). Providing these biofilters only contain beneficial bacteria that are involved in this breakdown of ammonia then they pose little risk to the larvae. If the biosecurity breaks down due to say, operator error, and a pathogen is introduced it may become established in the biofilm with the biofilter and become a more permanent member of the bacterial community. In this situation, there is the potential for the filter to shed pathogens into the water column with subsequent infection of the production larvae. This situation is very difficult to eradicate. Control measures may include installation of UV or ozone after the biofilters to kill any shed bacteria before the water enters the production area or the breakdown and sterilization of the larval production system.

7.4.2 Therapeutics

There is a range of therapeutants available to the hatchery manager including vaccines, chemical treatments, antibiotics, immunostimulants and pre- and probiotics. Efficacy of these therapeutants varies depending on a range of factors, such as time of delivery, dose and pathogen.

Permitted chemical therapeutants are becoming increasingly restricted. Previously, chemicals such as malachite green hydrogen peroxide and formaldehyde were used to prevent parasite infections. However, issues such as ensuring the correct dose, residues and potentially harmful breakdown products have resulted in substantial restriction on the use of such compounds. Many of these compounds are used as treatments for surface parasites, such as *Ichthyophthirius multifillis*, for which they were extremely effective. Governmental regulation of the use of such compounds has substantially reduced the range of chemotherapeutants available. Malachite green is banned for use in aquaculture in the USA. When found in imported products, these are seized and prevented from being sold within the USA. Antibiotics are another group of compounds that have proved highly beneficial but which usually have a substantial financial cost. Concerns focus on several issues: the development of resistance in the target pathogen species; transfer of resistance genes to non-target pathogen species (especially human pathogens); and withdrawal periods that ensure residues and breakdown products are not present in the animal (Cabello, 2006). Again, statutory regulation of the use of such treatments has severely restricted the range of available compounds and when they can be used. In the USA, for example, only three compounds are currently licensed for use in aquaculture food products: oxytetracyclin, florfenicol and ormetoprim.

Vaccine technology is fairly well understood. A vaccine improves the body's ability to protect against a disease. A vaccine contains something that resembles the disease-causing agent and allows the host to recognize the disease agent as foreign, destroy it and remember what it looks like for future reference. The very first vaccine was made from cow pox which, when administered to people, provided protection from smallpox. Vaccines are

© Woodhead Publishing Limited, 2013

one area of interest for larval production as they offer the potential of a prophylactic treatment for many known disease issues. The major problem with their use is that most larval animals have an immature immune system, especially when considering the adaptive immune system to which vaccines will be targeted (Press and Lillehaug, 1995; Costello *et al.*, 2001; Lee, 2003; Simon and Leong, 2003; Rubio-Godoy, 2010). Diseases where vaccination in larger fish has proved successful include the bacterial diseases enteric redmouth disease, vibriosis, furunculosis and pasteurellosis. For viral diseases, there are vaccines against spring viremia of carp, infectious pancreatic necrosis, viral hemorrhagic septicemia, hematopoietic necrosis, infectious salmon anemia, iridoviral disease and channel catfish virus. The range of vaccines continues to expand as effective antigens are found. The efficacy of vaccines such as the furunculosis vaccine cannot be ignored. At the time that furunculosis started to seriously impact the Norwegian salmon industry in 1987, antibiotic use had peaked at 50 000 kg a year. Effective vaccines were introduced around 1992–3 and immediately antibiotic use fell and production climbed. This pattern was repeated in Scotland. The vaccine saved the industry by reducing cost and increasing production. As yet, there are no vaccines for parasites, although there has been considerable research in the field.

Inherent in this discussion is the concept that vaccination is a waste of time unless the larval fish has a functional thymus, which will allow the developing immune system to respond appropriately to the antigens contained in the vaccine (Fig. 7.2; Bowden *et al.*, 2005). Presented prior to that, there is a risk that the vaccine will not be recognized as an invading pathogen but instead will be recognized as 'self' and no immune response will ever be mounted. This is called immune-tolerance and forms a normal function in differentiating 'self' from 'non-self' in vertebrates. In addition, vertically transmitted pathogens, such as bacterial kidney disease (BKD), use this 'loop-hole' in the immune system to evade the normal immune response to its presence. It has also been seen that some parasites use systems that result in them becoming invisible to the host system to evade detection.

The use of vaccines within invertebrate culture is a recent phenomenon. However, commercial vaccines are becoming available for bacterial diseases in shrimp. The efficacy seems to be based on some form of 'immune-priming' (Pope *et al.*, 2011). These techniques are still relatively new and, while they do seem to be beneficial, the long-term effectiveness will need to be assessed. In addition, their application to the hatchery situation is still unclear, and this needs to be studied in order for them to become a routine part of the hatchery manager's arsenal.

An area of considerable interest in relation to larval rearing is the use of immunostimulants. An immunostimulant is a naturally occurring compound that modulates the immune system by increasing the host's resistance against diseases that in most circumstances are caused by pathogens (Bricknell and Dalmo, 2005). The classic immunostimulant consists of

© Woodhead Publishing Limited, 2013

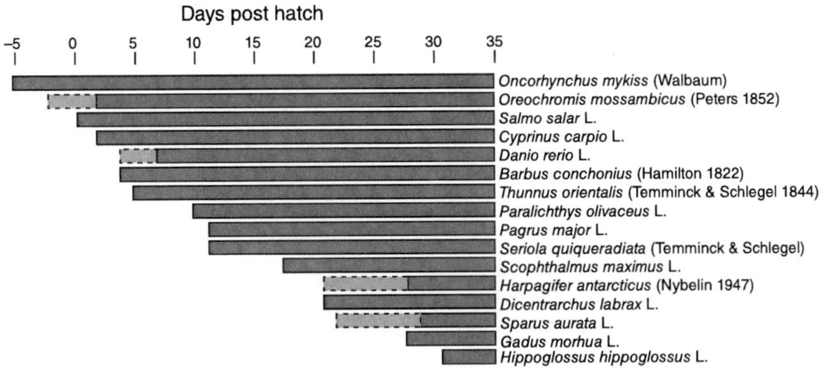

Fig. 7.2 Graph showing the variation in thymus development in teleosts (Bowden *et al.*, 2005). Reproduced with permission from Elsevier.

regular repeated units, usually a glucose-based polymer, such as chitosan or the β-glucans. These repeated units are recognized by a sub-group of the pattern recognition receptors (PRRs), the Toll-like receptors as potential pathogen. These molecules are often referred to as PAMPs (pathogen associated molecular patterns). Once bound to the receptor, the immunostimulant will trigger a series of non-specific immune responses that the host mounts against an infection. The concept is that immunostimulants ensure the host is always ready to repel a pathogen invasion; however, there is the risk that tolerance may develop. Tolerance occurs when immunostimulants are given over an extended timeframe, the infection never develops and the feedback loops within the host's immune system down-regulate the non-specific responses. To overcome this, the immunostimulants must be given strategically either as brief stimulation followed by a short withdrawal period, which can be repeated over the development cycle, or prior to a known or predictable stressor such as grading or vaccination.

The range of potential immunostimulants is wide. Most accepted immunostimulants are either β-glucans, bacterial cell-wall products, arthropod cuticle or plant products. Delivery of these products, especially to larval fish, can be challenging. Ideally, an oral delivery system, which places the required immunostimulant into the diet would provide the simplest method. Unfortunately, larval fish often only take live feeds and enriching live feeds with immunostimulants often results in them being degraded in the digestive tract of the live feed before they are eaten by the larvae. Bath delivery, where the animal is immersed in a solution of the immunostimulant, can be very effective, especially on mucosal or gill surfaces, which is a major uptake site for pathogens. The problem is this delivery method requires large amounts of immunostimulants which can become economically unrealistic (Bricknell and Dalmo, 2005; Kunttu *et al.*, 2009; Park, 2009; Suomalainen *et al.*, 2009; Aly and Mohamed, 2010; Ganguly *et al.*, 2010). The other

© Woodhead Publishing Limited, 2013

normal routes for delivering such material such as injection or intubation are technically challenging in larval fish given their very small size (Bricknell and Dalmo, 2005).

The immunostimulation of larval fish can be done at any stage following hatching. However, a major concern is the effect that the immunostimulant has on the developing immune system of the larvae. The immune system of a larval animal has a finite number of immune cells. Immunostimulation at a point early in the development process can result in the immune system becoming too focused on a very small set of antigens, which will compromise the development of a fully functional immune system in the mature animal that is capable of responding to a wide range of antigens. In extreme cases, such early exposure to an immunostimulant can result in 'burn-out' of the embryonic potential immune function, leaving the animal fatally compromised after this event.

The combination of immunostimulants and vaccination often increases the magnitude of protection, compared to the vaccine delivered on its own. This is due to the adjuvant effect of the immunostimulant. This is particularly useful in larval fish that are considered to be immunologically mature (i.e. they have a functional thymus) as this will increase the efficacy of bath vaccination, both in the magnitude of response and the duration of protection (Dalmo and Bogwald, 2008).

Finally, pre- and probiotics have been used in hatcheries to reduce the incidence of disease (Balcazar *et al.*, 2006). Prebiotics are nutrients that can be delivered to the host animal that promote the growth of beneficial bacteria. Prebiotics can provide the nutritional boost for a probiotic. These are live bacteria that can impart a health benefit to the recipient animal. Often administered in the feed, they can improve health through a variety of means, such as competitive exclusion, improving digestion, improving water quality and even antiviral actions. They can colonize the gastrointestinal tract resulting in exclusion of pathogenic bacteria. They can produce chemicals that can elevate the health of the host, sometimes through improving immune function but usually through a non-immune route. They can also reduce the impact of pathogenic bacteria by creating a environment that is less suitable. As a functional food component, pre- and probiotics are conceptually between food and drugs and thus receive a lower level of regulation that drugs.

7.4.3 Genetic improvement

Genetic management of the broodstock and progeny can improve the health of larvae. Selection of strains suitable for aquaculture often involves the inclusion of genes for rapid growth and slow maturation. It is only recently have attempts been made to include genes that enhance disease resistance. One good example is the resistance to infectious pancreatic necrosis (IPN) that one commercial supplier of Atlantic salmon applied to

© Woodhead Publishing Limited, 2013

their stock by developing strains with improved performance against the virus. However, genetic manipulation of broodstock is not limited to finfish. Oysters, especially the Eastern oyster (*Crassostrea virginica*), have numerous strains available that are resistant to diseases such as QX, MSX and Dermo, and these strains of oysters are routinely cultured where the diseases are endemic (Simonian *et al.*, 2009; Green *et al.*, 2011; Kan *et al.*, 2011).

7.5 Conclusion

Understanding of the larval immune system is still in its infancy. It is often the case that fish hatchery managers have to deal with neotonous animals (animals with extremely juvenile features). In higher vertebrates these developmental stages take place in the egg or uterus in a controlled and protected environment. Fish larvae are at the whim of the environment that they are reared in. Since we have control of these environments in the hatchery situation then we must assume control of all aspects of the hatchery, such as water chemistry, temperature, oxygen saturation and light and, most importantly, the microbial flora. A thorough understanding of the critical control points and the development of robust management protocols can provide the optimal environment for our aquatic charges, leading to happy animals, happy managers and a profitable business.

7.6 References

AHSAN D A and ROTH E (2010) Farmers' perceived risks and risk management strategies in an emerging mussel aquaculture industry in Denmark. *Marine Resource Economics,* 25, 309–323.

ALY S M and MOHAMED M F (2010) *Echinacea purpurea* and *Allium sativum* as immunostimulants in fish culture using Nile tilapia (*Oreochromis niloticus*). *Journal of Animal Physiology and Animal Nutrition,* 94, e31–e39.

ATHANASSOPOULOU F, BILLINIS C and PRAPAS T (2004) Important disease conditions of newly cultured species in intensive freshwater farms in Greece: first incidence of nodavirus infection in Acipenser sp. *Diseases of Aquatic Organisms,* 60, 247–252.

BALCAZAR J L, DE BLAS I, RUIZ-ZARZUELA I, CUNNINGHAM D, VENDRELL D and MUZQUIZ J L (2006) The role of probiotics in aquaculture. *Veterinary Microbiology,* 114, 173–186.

BALLAGH D A, PANKHURST P M and FIELDER D S (2011) Embryonic development of mulloway, *Argyrosomus japonicus*, and egg surface disinfection using ozone. *Aquaculture,* 318, 475–478.

BATTAGLENE S C and COBCROFT J M (2007) Advances in the culture of striped trumpeter larvae: A review. *Aquaculture,* 268, 195–208.

BERGFJORD O J (2009) Risk perception and risk management in Norwegian aquaculture. *Journal of Risk Research,* 12, 91–104.

BIRKBECK T H, BORDEVIK M, FROYSTAD M K and BAKLIEN A (2007) Identification of *Francisella* sp from Atlantic salmon, *Salmo salar* L., in Chile. *Journal of Fish Diseases,* 30, 505–507.

© Woodhead Publishing Limited, 2013

BIRKBECK T H, FEIST S W and VERNER-JEFFREYS D W (2011) *Francisella* infections in fish and shellfish. *Journal of Fish Diseases,* 34, 173–187.

BOSMA R H and VERDEGEM M C J (2011) Sustainable aquaculture in ponds: Principles, practices and limits. *Livestock Science,* 139, 58–68.

BOWDEN T J (2008) Modulation of the immune system of fish by their environment. *Fish & Shellfish Immunology,* 25, 373–383.

BOWDEN T J, COOK P and ROMBOUT J H W M (2005) Development of the thymus in telesots. *Fish & Shellfish Immunology,* 19, 413–427.

BOWDEN T J, THOMPSON K D, MORGAN A L, GRATACAP R M and NIKOSKELAINEN S (2007) Seasonal variation and the immune response: a fish perspective. *Fish & Shellfish Immunology,* 22, 695–706.

BRATLand A and NYLUND A (2009) Studies on the possibility of vertical transmission of Norwegian salmonid alphavirus in production of Atlantic Salmon in Norway. *Journal of Aquatic Animal Health,* 21, 173–178.

BREUIL G, THIERY R, PEPIN J F and BLANCHETON J P (2003) The control of nodavirus disease in sea bass *Dicentrarchus labrax*, from breeder to commercial size: Toward a new approach of organic aquaculture, in Lee C-S and O'Bryen P J (eds), *Biosecurity in Aquaculture Production Systems: Exclusion of Pathogens and Other Undesirables.* Baton Rouge, LA: The World Aquaculture Society, 65–79.

BRICKNELL I R and DALMO R A (2005) The use of immunostimulants in fish larval aquaculture. *Fish & Shellfish Immunology,* 19, 457–472.

BRICKNELL I R, BOWDEN T J, VERNER-JEFFREYS D W, BRUNO D W, SHIELDS R J and ELLIS A E (2000) Susceptibility of juvenile and sub-adult Atlantic halibut (*Hippoglossus hippoglossus* L.) to infection by *Vibrio anguillarum* and efficacy of protection induced by vaccination. *Fish & Shellfish Immunology,* 10, 319–327.

BUCHAN K A H, MARTIN-ROBICHAUD D J, BENFEY T J, MACKINNON A M and BOSTON L (2006) The efficacy of ozonated seawater for surface disinfection of haddock (*Melanogrammus aeglefinus*) eggs against piscine nodavirus. *Aquaculture Engineering,* 35, 102–107.

CABELLO F C (2006) Heavy use of prophylactic antibiotics in aquaculture: a growing problem for human and animal health and for the environment. *Environmental Microbiology,* 8, 1137–1144.

CHEN Y M, SHIH C H, LIU H C, WU C L, LIN C C, WANG H C, CHEN T Y, YANG H L and LIN J H Y (2011) An oral nervous necrosis virus vaccine using *Vibrio anguillarum* as an expression host provides early protection. *Aquaculture,* 321, 26–33.

CHU W H and LU C P (2008) *In vivo* fish models for visualizing *Aeromonas hydrophila* invasion pathway using GFP as a biomarker. *Aquaculture,* 277, 152–155.

COSTELLO M J, GRANT A, DAVIES I M, CECCHINI S, PAPOUTSOGLOU S, QUIGLEY D and SAROGLIA M (2001) The control of chemicals used in aquaculture in Europe. *Journal of Applied Ichthyology,* 17, 173–180.

CUTRIN J M, LOPEZ-VAQUEZ C, OLIVEIRA J G, CASTRO S, DOPAZO C P and BandIN I (2005) Isolation in cell culture and detection by PCR-based technology of IPNV-like virus from leucocytes of carrier turbot, *Scophthalmus maximus* (L.). *Journal of Fish Diseases,* 28, 713–722.

DALMO R A (2005) Ontogeny of the fish immune system. *Fish & Shellfish Immunology,* 19, 395–396.

DALMO R A and BOGWALD J (2008) Beta-glucans as conductors of immune symphonies. *Fish & Shellfish Immunology,* 25, 384–396.

DAVIES R W D and RANGELEY R (2010) Banking on cod: Exploring economic incentives for recovering Grand Banks and North Sea cod fisheries. *Marine Policy,* 34, 92–98.

DAVIS D A and ARNOLD C R (1997) Tolerance of the rotifer *Brachionus plicatilis* to ozone and total oxidative residuals. *Ozone-Science & Engineering,* 19, 457–469.

© Woodhead Publishing Limited, 2013

DELABBIO J, MURPHY B R, JOHNSON G R and MCMULLIN S L (2004) An assessment of biosecurity utilization in the recirculation sector of finfish aquaculture in the United States and Canada. *Aquaculture,* 242, 165–179.

DYRYNDA E A, PIPE R K and RATCLIFFE N A (1995) Host defence mechanisms in marine invertebrate larvae. *Fish & Shellfish Immunology,* 5, 569–580.

EBELING J M (2000) Engineering aspects of recirculating aquaculture systems. *Marine Technology Society Journal,* 34, 68–78.

EDING E H, KAMSTRA A, VERRETH J, HUISMAN E A and KLAPWIJK A (2006) Design and operation of nitrifying trickling filters in recirculating aquaculture: A review. *Aquacultural Engineering,* 34, 234–260.

EINARSDÓTTIR I, SILVA N, POWER D, SMÁRADÓTTIR H and BJÖRNSSON B (2006) Thyroid and pituitary gland development from hatching through metamorphosis of a teleost fish, the Atlantic halibut. *Anatomy and Embryology,* 211, 47–60.

FALK-PETERSEN I B (2005) Comparative organ differentiation during early life stages of marine fish. *Fish & Shellfish Immunology,* 19, 397–412.

FAO (2012) © 2005–2012. Aquaculture topics and activities. State of world aquaculture. Text by Rohana Subasinghe. In: *FAO Fisheries and Aquaculture Department* [online]. Rome. Updated 27 May 2005. [Cited 27 September 2012]. http://www.fao.org/fishery/topic/13540/en.

FENICHEL E P, TSAO J I and JONES M L (2009) Modeling fish health to inform research and management: *Renibacterium salmoninarum* dynamics in Lake Michigan. *Ecological Applications,* 19, 747–760.

GANGULY S, PAUL I and MUKHOPADHAYAY S K (2010) Application and effectiveness of immunostimulants, probiotics, and prebiotics in aquaculture: a review. *Israeli Journal of Aquaculture-Bamidgeh,* 62, 130–138.

GRACE M F and MANNING M J (1980) Histogenesis of the lymphoid organs in rainbow trout, *Salmo gairdneri* rich. 1836. *Developmental & Comparative Immunology,* 4, 255–264.

GREEN T J, RAFTOS D, O'CONNOR W, ADLARD R D and BARNES A C (2011) Disease Prevention strategies for QX disease (*Marteilia sydneyi*) of Sydney rock oysters (*Saccostrea glomerata*). *Journal of Shellfish Research,* 30, 47–53.

GROTMOL S and TOTLAND G K (2000) Surface disinfection of Atlantic halibut *Hippoglossus hippoglossus* eggs with ozonated sea-water inactivates nodavirus and increases survival of the larvae. *Diseases of Aquatic Organisms,* 39, 89–96.

GUDMUNDSDOTTIR B K and BJORNSDOTTIR B (2007) Vaccination against atypical furunculosis and winter ulcer disease of fish. *Vaccine,* 25, 5512–5523.

GULLIAN M, ESPINOSA-FALLER F J, NUNEZ A and LOPEZ-BARAHONA N (2012) Effect of turbidity on the ultraviolet disinfection performance in recirculating aquaculture systems with low water exchange. *Aquaculture Research,* 43, 595–606.

HANIF A, BAKOPOULOS V, LEONARDOS I and DIMITRIADIS G J (2005) The effect of sea bream (*Sparus aurata*) broodstock and larval vaccination on the susceptibility by *Photobacterium damsela* subsp *piscicida* and on the humoral immune parameters. *Fish & Shellfish Immunology,* 19, 345–361.

HAUTON C (2012) The scope of the crustacean immune system for disease control. *Journal of Invertebrate Pathology,* 110, 251–260.

HE X T, WANG Z H, NIE X P, YANG Y F, PAN D B, LEUNG A O W, CHENG Z, YANG Y T, LI K B and CHEN K C (2012) Residues of fluoroquinolones in marine aquaculture environment of the Pearl River Delta, South China. *Environmental Geochemistry and Health,* 34, 323–335.

KAI Y H and CHI S C (2008) Efficacies of inactivated vaccines against betanodavirus in grouper larvae (*Epinephelus coioides*) by bath immunization. *Vaccine,* 26, 1450–1457.

© Woodhead Publishing Limited, 2013

KAI Y H, SU H M, TAI K T and CHI S C (2010) Vaccination of grouper broodfish (*Epinephelus tukula*) reduces the risk of vertical transmission by nervous necrosis virus. *Vaccine,* 28, 996–1001.

KAN A, DOVE M C, O'CONNOR W A, NAIR S V and RAFTOS D A (2011) Mortality in single-pair mating families of QX disease-resistant and wild-type Sydney rock oysters (*Saccostrea glomerata*). *Aquaculture Research,* 42, 987–995.

KANEKO T and COLWELL R R (1975) Adsorption of *Vibrio parahaemolyticus* onto chitin and copepods. *Applied Microbiology,* 29, 269–274.

KIM B R, ANDERSON J E, MUELLER S A, GAINES W A and KENDALL A M (2002) Literature review – efficacy of various disinfectants against Legionella in water systems. *Water Research,* 36, 4433–4444.

KIRCHNER M (1995) Microbial colonization of copepod body surfaces and chitin degradation in the sea. *Helgolander Meeresuntersuchungen,* 49, 201–212.

KLIMPEL S, KELLERMANNS E and PALM H W (2008) The role of pelagic swarm fish (Myctophidae: Teleostei) in the oceanic life cycle of *Anisakis* sibling species at the Mid-Atlantic Ridge, Central Atlantic. *Parasitology Research,* 104, 43–53.

KONGTORP R T, STENE A, ANDREASSEN P A, ASPEHAUG V, GRAHAM D A, LYNGSTAD T M, OLSEN A B, OLSEN R S, ANDBERG M, SANTI N, WALLACE C and BRECK O (2010) Lack of evidence for vertical transmission of SAV 3 using gametes of Atlantic salmon, *Salmo salar* L., exposed by natural and experimental routes. *Journal of Fish Diseases,* 33, 879–888.

KUMAGAI A and NAWATA A (2010) Prevention of *Flavobacterium psychrophilum* vertical transmission by iodophor treatment of unfertilized eggs in salmonids. *Fish Pathology,* 45, 164–168.

KUMAR T T A, GOPI M, DHANEESH K V, VINOTH R, GHOSH S, BALASUBRAMANIAN T and SHUNMUGARAJ T (2012) Hatchery production of the clownfish *Amphiprion nigripes* at Agatti island, Lakshadweep, India. *Journal of Environmental Biology,* 33, 623–628.

KUNTTU H M T, VALTONEN E T, SUOMALAINEN L R, VIELMA J and JOKINEN I E (2009) The efficacy of two immunostimulants against *Flavobacterium columnare* infection in juvenile rainbow trout (*Oncorhynchus mykiss*). *Fish & Shellfish Immunology,* 26, 850–857.

KURTZ J and FRANZ K (2003) Innate defence: evidence for memory in invertebrate immunity. *Nature,* 425, 37–38.

LANGSTON A L, HOARE R, STEFANSSON M, FITZGERALD R, WERGELAND H and MULCAHY M (2002) The effect of temperature on non-specific defence parameters of three strains of juvenile Atlantic halibut (*Hippoglossus hippoglossus* L.). *Fish & Shellfish Immunology,* 12, 61–76.

LEE C S (2003) Biotechnological advances in finfish hatchery production: a review. *Aquaculture,* 227, 439–458.

LUBZENS E, ZMORA O and BARR Y (2001) Biotechnology and aquaculture of rotifers. *Hydrobiologia,* 446, 337–353.

LUZARDO-ALVAREZ A, OTERO-ESPINAR F J and BLANCO-MENDEZ J (2010) Microencapsulation of diets and vaccines for cultured fishes, crustaceans and bivalve mollusks. *Journal of Drug Delivery Science and Technology,* 20, 277–288.

MAGI G E, LOPEZ-ROMALDE S, MAGARINOS B, LAMAS J, TORANZO A E and ROMALDE J L (2009) Experimental *Pseudomonas anguilliseptica* infection in turbot *Psetta maxima* (L.): a histopathological and immunohistochemical study. *European Journal of Histochemistry,* 53, 73–79.

MAGNADOTTIR B, LANGE S, GUDMUNDSDOTTIR S, BØGWALD J and DALMO R A (2005) Ontogeny of humoral parameters in fish. *Fish & Shellfish Immunology,* 19, 429–439.

MAMANE H (2008) Impact of particles on UV disinfection of water and wastewater effluents: A review. *Reviews in Chemical Engineering,* 24, 67–157.

© Woodhead Publishing Limited, 2013

MANIN B O and RANSANGAN J (2011) Experimental evidence of horizontal transmission of Betanodavirus in hatchery-produced Asian seabass, *Lates calcarifer* and brown-marbled grouper, *Epinephelus fuscoguttatus* fingerling. *Aquaculture,* 321, 157–165.

MANLEY N (2000) Thymus organogenesis and molecular mechanisms of thymic epithelial cell differentiation. *Seminars in Immunology,* 12, 421–428.

MARTINS C I M, EDING E H, VERDEGEM M C J, HEINSBROEK L T N, SCHNEIDER O, BLANCHETON J P, D'ORBCASTEL E R and VERRETH J A J (2010) New developments in recirculating aquaculture systems in Europe: A perspective on environmental sustainability. *Aquacultural Engineering,* 43, 83–93.

MENANTEAU-LEDOUBLE S, KARSI A and LAWRENCE M L (2011) Importance of skin abrasion as a primary site of adhesion for *Edwardsiella ictaluri* and impact on invasion and systematic infection in channel catfish *Ictalurus punctatus*. *Veterinary Microbiology,* 148, 425–430.

MIKKELSEN H, SCHRODER M B and LUND V (2004) Vibriosis and atypical furunculosis vaccines; efficacy, specificity and side effects in Atlantic cod, *Gadus morhua* L. *Aquaculture,* 242, 81–91.

MUNRO E S and ELLIS A E (2008) A comparison between non-destructive and destructive testing of Atlantic salmon, *Salmo salar* L., broodfish for IPNV – destructive testing is still the best at time of maturation. *Journal of Fish Diseases,* 31, 187–195.

MUNRO P D, BIRKBECK T H and BARBOUR A (1993) Influence of rate of bacterial colonisation of the gut of turbot larvae on larval survival, in Reinertsen H, Dahle L A, Jõrgensen L and Tvinnereim K (eds), *Proceedings from International Conference on Fish Farming Technology*, August, Trondheim, 85–92.

MUNRO P D, BARBOUR A and BIRKBECK T H (1995) Comparison of the growth and survival of larval turbot in the absence of culturable bacteria with those in the presence of *Vibrio anguillarum*, *Vibrio alginolyticus*, or a marine *AEROMONAS* sp. *Applied and Environmental Microbiology,* 61, 4425–4428.

MUNRO P D, HENDERSON R J, BARBOUR A and BIRBECK T H (1999) Partial decontamination of rotifers with ultraviolet radiation: the effect of changes in the bacterial load and flora of rotifers on mortalities in start-feeding larval turbot. *Aquaculture,* 170, 229–244.

NAGANO I, OSHIMA S and KAWAI K (2011) In vivo analysis on the adherence and infection route of *Photobacterium damselae* subsp *piscicida* in Yellowtail. *Fish Pathology,* 46, 45–50.

NOBLE A C and SUMMERFELT S T (1996) Diseases encountered in rainbow trout cultured in recirculating systems. *Annual Review of Fish Diseases,* 6, 65–92.

NUNN A D, TEWSON L H and COWX I G (2012) The foraging ecology of larval and juvenile fishes. *Reviews in Fish Biology and Fisheries,* 22, 377–408.

ODEGARD J, SOMMER A I and PRAEBEL A K (2010) Heritability of resistance to viral nervous necrosis in Atlantic cod (*Gadus morhua* L). *Aquaculture,* 300, 59–64.

OIDTMANN B C, THRUSH M A, DENHAM K L and PEELER E J (2011) International and national biosecurity strategies in aquatic animal health. *Aquaculture,* 320, 22–33.

OLAFSON J A (2001) Interactions between fish larvae and bacteria in marine aquaculture. *Aquaculture,* 200, 223–247.

OLSEN A B, MIKALSEN J, RODE M, ALFJORDEN A, HOEL E, STRAUM-LIE K, HALDORSEN R and COLQUHOUN D J (2006) A novel systemic granulomatous inflammatory disease in farmed Atlantic cod, *Gadus morhua* L., associated with a bacterium belonging to the genus *Francisella*. *Journal of Fish Diseases,* 29, 307–311.

ORDAS M C, ABOLLO E, COSTA M M, FIGUERAS A and NOVOA B (2006) Molecular cloning and expression analysis of interferon regulatory factor-1 (IRF-1) of turbot and sea bream. *Molecular Immunology,* 43, 882–890.

© Woodhead Publishing Limited, 2013

OSTLAND V E, STANNARD J A, CREEK J J, HEDRICK R P, FERGUSON H W, CARLBERG J M and WESTERMAN M E (2006) Aquatic *Francisella*-like bacterium associated with mortality of intensively cultured hybrid striped bass *Morone chrysops* x *M. saxatilis. Diseases of Aquatic Organisms,* 72, 135–145.

PARK S I (2009) Disease control in Korean aquaculture. *Fish Pathology,* 44, 19–23.

PHELPS N B D and GOODWIN A E (2008) Vertical transmission of Ovipleistophora ovariae (Microspora) within the eggs of the golden shiner. *Journal of Aquatic Animal Health,* 20, 45–53.

PIEDRAHITA R H (2003) Reducing the potential environmental impact of tank aquaculture effluents through intensification and recirculation. *Aquaculture,* 226, 35–44.

POPE E C, POWELL A, ROBERTS E C, SHIELDS R J, WARDLE R and ROWLEY A F (2011) Enhanced cellular immunity in shrimp (*Litopenaeus vannamei*) after 'vaccination'. *PLoS ONE,* 6, e20960.

POULICEK M, GAILL F and GOFFINET G (1998) Chitin biodegradation in marine environments, in Stankiewicz B A and Vanbergen P F (eds), *Nitrogen-Containing Macromolecules in the Bio- and Geosphere.* Washington DC: Americom Chemical Society, 163–210.

PRESS C M and LILLEHAUG A (1995) Vaccination in European salmonid aquaculture – a review of practices and prospects. *British Veterinary Journal,* 151, 45–69.

PRUDER G D (2004) Biosecurity: application in aquaculture. *Aquacultural Engineering,* 32, 3–10.

PRUZZO C, CRIPPA A, BERTONE S, PANE L and CARLI A (1996) Attachment of *Vibrio alginolyticus* to chitin mediated by chitin-binding proteins. *Microbiology-UK,* 142, 2181–2186.

RIBIERO A R A, RIBIERO L, SÆLE Ø, HAMRE K, DINIS M T and MOREN M (2009) Iodine-enriched rotifers and *Artemia* prevent goitre in Senegalese sole (*Solea senegalensis*) larvae reared in a recirculation system. *Aquaculture Nutrition,* 17, 248–257.

RINGO E, BIRKBECK T H, MUNRO P D, VADSTEIN O and HJELMELAND K (1996) The effect of early exposure to *Vibrio pelagius* on the aerobic bacterial flora of turbot, *Scophthalmus maximus* (L) larvae. *Journal of Applied Bacteriology,* 81, 207–211.

ROMBOUT J H W M, HUTTENHUIS H B T, PICCHIETTI S and SCAPIGLIATI G (2005) Phylogeny and ontogeny of fish leucocytes. *Fish & Shellfish Immunology,* 19, 441–455.

RUBIO-GODOY M (2010) Teleost fish immunology. Review. *Revista Mexicana De Ciencias Pecuarias,* 1, 43–57.

SAINTJEAN S R, MINONDO M P V, PALACIOS M A and PRIETO S P (1991) Detection of infectious pancreatic necrosis in a carrier population of rainbow-trout, *Oncorhynchus mykiss* (RICHARDSON), by flow-cytometry. *Journal of Fish Diseases,* 14, 545–553.

SAMUELSEN O B, NERLAND A H, JORGENSEN T, SCHRODER M B, SWASAND T and BERGH O (2006) Viral and bacterial diseases of Atlantic cod *Gadus morhua*, their prophylaxis and treatment: a review. *Diseases of Aquatic Organisms,* 71, 239–254.

SCHRODER M B, MIKKELSEN H, BORDAL S, GRAVNINGEN K and LUND V (2006) Early vaccination and protection of Atlantic cod (*Gadus morhua* L.) juveniles against classical vibriosis. *Aquaculture,* 254, 46–53.

SIMON B E and LEONG J C (2003) A review of recombinant vaccines for aquaculture: Bacterial systems for vaccine production and delivery, in Bhojwani S S and Soh W-Y (eds), *Agrobiotechnology and Plant Tissue Culture.* Enfield, NH: Science Publishers, 169–189.

SIMONIAN M, NAIR S V, O'CONNOR W A and RAFTOS D A (2009) Protein markers of *Marteilia sydneyi* infection in Sydney rock oysters, *Saccostrea glomerata. Journal of Fish Diseases,* 32, 367–375.

SKALL H F, OLESEN N J and MELLERGAARD S (2005) Viral haemorrhagic septicaemia virus in marine fish and its implications for fish farming – a review. *Journal of Fish Diseases,* 28, 509–529.

© Woodhead Publishing Limited, 2013

SKOV J, KANIA P W, OLSEN M M, LAURIDSEN J H and BUCHMANN K (2009) Nematode infections of maricultured and wild fishes in Danish waters: A comparative study. *Aquaculture,* 298, 24–28.

SLOANE D (2010) *NASAC Newsletter*, April, available at: http://www.nasac.net/AprilNewsletter2010.pdf (accessed September 2012).

SMITH P A, PIZARRO P, OJEDA P, CONTRERAS J, OYANEDEL S and LARENAS J (1999) Routes of entry of *Piscirickettsia salmonis* in rainbow trout *Oncorhynchus mykiss*. *Diseases of Aquatic Organisms,* 37, 165–172.

SÖDERHÄLL K (2010) *Invertebrate immunity,* New York, Austin, TX: Springer Science+Business Media; Landes Bioscience.

SOMMERSET I, KROSSOY B, BIERING E and FROST P (2005) Vaccines for fish in aquaculture. *Expert Review of Vaccines,* 4, 89–101.

SOTO E, BAUMGARTNER W, WILES J and HAWKE J P (2011) *Francisella asiatica* as the causative agent of piscine francisellosis in cultured tilapia (*Oreochromis* sp.) in the United States. *Journal of Veterinary Diagnostic Investigation,* 23, 821–825.

SUMMERFELT S T (2003) Ozonation and UV irradiation – an introduction and examples of current applications. *Aquaculture Engineering,* 28, 21–36.

SUOMALAINEN L R, BANDILLA M and VALTONEN E T (2009) Immunostimulants in prevention of columnaris disease of rainbow trout, *Oncorhynchus mykiss* (Walbaum). *Journal of Fish Diseases,* 32, 723–726.

SWAIN P and NAYAK S K (2009) Role of maternally derived immunity in fish. *Fish & Shellfish Immunology,* 27, 89–99.

THEISEN D D, STANSELL D D and WOODS L C (1998) Disinfection of nauplii of *Artemia franciscana* by ozonation. *Progressive Fish-Culturist,* 60, 149–151.

TUCKER J W (1987) Snook and tarpon snook culture and preliminary evaluation for commercial farming. *Progressive Fish-Culturist,* 49, 49–57.

VERNER-JEFFREYS D W, SHIELDS R J and BIRKBECK T H (2003a) Bacterial influences on Atlantic halibut *Hippoglossus hippoglossus* yolk-sac larval survival and start-feed response. *Diseases of Aquatic Organisms,* 56, 105–113.

VERNER-JEFFREYS D W, SHIELDS R J, BRICKNELL I R and BIRKBECK T H (2003b) Changes in the gut-associated microflora during the development of Atlantic halibut (*Hippoglossus hippoglossus* L.) larvae in three British hatcheries. *Aquaculture,* 219, 21–42.

VERNER-JEFFREYS D W, SHIELDS R J, BRICKNELL I R and BIRKBECK T H (2004) Effects of different water treatment methods and antibiotic addition on larval survival and gut microflora development in Atlantic halibut (*Hippoglossus hippoglossus* L.) yolk-sac larvae. *Aquaculture,* 232, 129–143.

WATSON F L, PUTTMANN-HOGADO R, THOMAS F, LAMAR D L, HUGHES M, KONDO M, REBEL V I and SCHMUCKER D (2005) Extensive diversity of Ig-superfamily proteins in the immune system of insects. *Science,* 309, 1874–1878.

YOSHIMIZU M, NOMURA T, EZURA Y and KIMURA T (1993) Surveillance and control of infectious hematopoietic necrosis virus (IHNV) and *Oncorhynchus masou* virus (OMV) of wild salmonid fish returning to the northern part of Japan 1976–1991. *Fisheries Research,* 17, 163–173.

ZERIHUN M A, FEIST S W, BUCKE D, OLSEN A B, TANDSTAD N M and COLQUHOUN D J (2011) *Francisella noatunensis* subsp *noatunensis* is the aetiological agent of visceral granulomatosis in wild Atlantic cod *Gadus morhua*. *Diseases of Aquatic Organisms,* 95, 65–71.

© Woodhead Publishing Limited, 2013

8

Microbial management for bacterial pathogen control in invertebrate aquaculture hatcheries

E. F. Goulden, L. Høj and M. R. Hall, Australian Institute of Marine Science (AIMS), Australia

DOI: 10.1533/9780857097460.1.246

Abstract: In aquaculture production systems, the hatchery phase is the most challenging as larvae are prone to infections and disease from microbial agents causing mass mortalities. A holistic microbial management regime concept is developed with the hatchery managed as a set of compartments: (1) the water column (planktonic); (2) surfaces (biofilm); (3) the larvae themselves; and (4) feeds (live and formulated). Emphasis on the use of the microbial community as a whole to maintain larvae in high health with various prophylactic strategies is discussed. A range of treatment strategies are considered, including antibiotics, bacteriophage therapy, quorum sensing inhibition and microbial predation. Developing innovations and future trends in microbial management and biocontrol are outlined.

Key words: hatcheries, microbial management, pathogens, probiotics, quorum sensing.

8.1 Introduction

Aquaculture hatcheries are designed and operated for cost-effective mass rearing of specific marine larvae. Out of the entire production cycle, the hatchery phase is generally considered to be the most high risk component. Not only are the larvae reared in high density, increasing the risk of epizootics, but the larvae may only have a rudimentary immune system that is not yet fully competent, rendering them highly vulnerable to infectious disease (Jiravanichpaisal *et al.*, 2007) caused by viruses, bacteria, fungi, yeasts, protists and some metazoan parasites. Historically, hatcheries operated in a reactive mode, responding to observed disease symptoms using curative therapeutics to try to minimise larval mortalities and prevent an epizootic

© Woodhead Publishing Limited, 2013

event. However, by the time the first disease symptoms are observed, it may be too late to avoid contagion with significant biological and economic loss. Hence, there is strong incentive to be proactive and for the development of a holistic microbial management approach to obtain a stable hatchery environment that promotes high larval health and survival.

Although a hatchery is operated to produce larvae of a target aquaculture species, it is actually a complete ecosystem. Microbial organisms occupy every niche within the hatchery, including the water column, all surfaces (tanks and pipework), feed (live and/or artificial feed) and the larvae themselves. Indeed, a single larva may carry a higher population of microbes than there are conspecific larvae in the rearing tank. Microorganisms play important roles in nutrient cycling, and their associations with the aquaculture target organism and live feed organisms range from beneficial to pathogenic. In order to understand how this important part of the system functions and how it can be controlled or even manipulated, it is crucial to understand the composition, dynamics and variability of the microbial ecosystem.

The concept of viewing a hatchery as a complex multi-compartmental microbial ecosystem requires a change in mindset in order for managers to embrace and implement protocols that focus not so much on the larvae themselves, but upon the nurturing of conditions that favour a stable, benign and beneficial microbial community in order to minimise the risk of mass larval loss from microbial pathogens. An examination of all microbial life forms that may be found in the hatchery is beyond the scope of this chapter, which focuses on the new wave of enabling technologies to study and manage bacterial populations with the goal of minimising invertebrate larval mortalities.

8.2 Methods to study bacterial communities in hatchery systems

This section will briefly present approaches and methods that have been used to study the composition and dynamics of bacterial communities in aquaculture systems. A detailed description of the methodological approaches is outside the scope of this chapter and fine overviews are available elsewhere (Osborn and Smith, 2005; Munn, 2011). Traditionally, studies of bacteria in aquaculture systems used culture-based approaches in combination with histological preparations of host tissues affected by disease. An advantage of culture-based techniques is that isolates can be further analysed and characterised, and the traditional approach has generated a wealth of knowledge on the biology of many aquaculture pathogens. Some pathogens are, however, difficult to culture, in particular intracellular pathogens or those that can enter a so-called 'viable but non-culturable (VBNC) state under environmental stress.

© Woodhead Publishing Limited, 2013

Several recent studies have demonstrated the importance of understanding the general microbial communities of which aquaculture pathogens are members. For these studies, the culture-based approaches have severe limitations since only 0.01–10 % of bacteria in marine systems can be cultured by conventional techniques (Connon and Giovannini, 2002; Osborne and Smith, 2005). In addition, the appropriateness of the culture media used and culture conditions will affect the fraction of this 'culturable' community that is successfully recovered. For instance, many studies of fish gastrointestinal microbiota have not included anaerobic culture conditions and have therefore missed certain fastidious and obligate anaerobes completely (Nayak, 2010). In contrast, molecular studies based on direct analysis of deoxyribonucleic and ribonucleic acids (DNA or RNA) can provide a snapshot of the microbial community that is present (DNA) or active (RNA) in the sample independent of their. While molecular culturability methods have many advantages for studies of microbial communities it is, however, important to be aware of their limitations and biases (Osborn and Smith, 2005; Munn, 2011).

The polymerase chain reaction (PCR) has caused a revolution in the life sciences by providing a method to produce many copies of the genes of interest. These days, common routine techniques including conventional PCR or quantitative PCR enable the detection or quantification, respectively, of specfic genes with relatively high sensitivity. The techniques are used for a multitude of purposes including identification and quantification of pathogens (see Section 8.4). PCR is also an important step in many other molecular approaches including most fingerprinting and cloning techniques.

Molecular fingerprinting techniques for analysis of bacterial communities such as denaturing gradient gel electrophoresis (DGGE), temperature gradient gel electrophoresis (TGGE), terminal fragment length polymorphism (T-RFLP) and others can be used to analyse the general bacterial community, a subset of the bacterial community, or functional groups of interest depending on the primers used in the initial PCR step. DGGE in particular has been used extensively to study bacterial communities in aquaculture systems in the last decade resulting in several new insights (see Section 8.3). These fingerprinting methods are relatively reproducible and are suitable for comparisons between many samples. In the case of DGGE and TGGE, sequences can be recovered from individual bands to assist in the interpretation of community differences between samples or community changes with time (community dynamics).

Clone libraries of 16S ribosomal ribonucleic acid (rRNA) genes have been used to identify members of the bacterial community present in a sample with a much higher resolution than can be achieved by fingerprinting techniques. This approach is, however, more costly and generally only a limited number of samples is analysed. To reduce the cost of sequencing, it has been common to screen the clone libraries using restriction fragment

© Woodhead Publishing Limited, 2013

length polymorphism and then sequence representative clones from each restriction pattern (Payne *et al.*, 2007). An alternative approach is to use specific probes to estimate the proportion of a specific group, such as vibrios, within the general bacterial library and sequence only the clones of particular interest (Høj *et al.*, 2009). With the advancement of next generation sequencing technologies, clone libraries of 16S rRNA genes will be superseded by pyrotag sequencing. This will provide the potential to analyse a higher number of samples and/or detect less dominant community members. Fluorescence *in situ* hybridisation (FISH) combines the use of a probe targeting the 16S rRNA of a specific bacterial taxonomic group with microscopy, hence providing an opportunity to study the localisation of these bacteria in the sample. Probes with different specificity levels can be labelled with different fluorophores and used in combination, providing the potential to identify and quantify cells that bind both probes. While FISH is restricted by the specificity of the probes used, an important advantage of this technique is that it is not dependent on the initial extraction of nucleic acids or an initial PCR step. As it is an advantage to include methods subject to different biases, FISH is in many cases a good complementary method to verify trends observed by other approaches.

8.3 Hatchery microbial compartments

As health and survival of cultured aquatic larvae are influenced by microbial communities, there is merit in understanding the structure and dynamics of these communities in key ecological compartments of the hatchery in order to implement appropriate microbial management protocols (Golovlev, 2001; Grossart, 2010). The hatchery ecosystem can differ markedly from the natural environment of wild conspecifics and, as a consequence, can select for a diversity of microbiota non-indigenous to the host. An aquaculture hatchery consists of several interlinking microbial compartments (Fig. 8.1). These compartments include (i) the water column, (ii) the surfaces within the hatchery, (iii) the larvae themselves and (iv) the larval feed (Bourne *et al.*, 2004, 2006, 2007).

8.3.1 Water column

Sea water is a complex solution of more than 80 chemical elements, gases and organic substances and is an important niche for the microbial community. The organic component found in sea water may be separated into particulate organic matter (POM) and dissolved organic matter (DOM) fractions. The difference is purely empirical, with the general definition being that POM is > 0.45 μm in diameter and DOM is < 0.45 μm in diameter. Whereas most free-living bacteria (and all viruses) would be classified as DOM based on size, they are nevertheless discrete microscopic intact

© Woodhead Publishing Limited, 2013

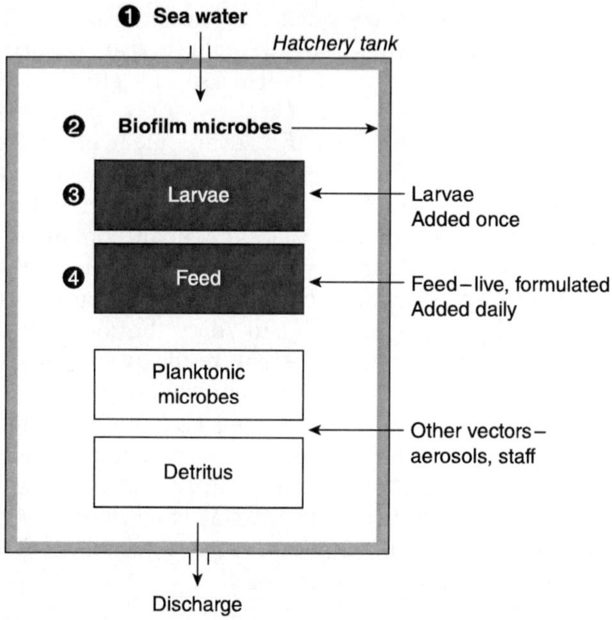

Fig. 8.1 Model of hatchery system showing the four interlinking microbial compartments of the hatchery tank: 1. sea water (water column); 2. surfaces (walls and pipework) on which microbial biofilms form; 3. larvae; and 4. larval feeds (live and/or formulated). Major inputs, each representing a potential biosecurity threat, include the sea water, the larvae – added on day of stocking, the feed – added on a frequent and daily basis, and other vectors such as aerosols, staff and equipment.

organisms, with the potential to be infective agents causing disease and mortality in larvae. Bacteria and viruses are also members of the plankton, more specifically the picoplankton (0.2–2 μm) and the femtoplankton (< 0.2 μm), respectively.

The pool of bacteria residing in the water column is not distributed homogeneously, and varies on both temporal and spatial scales. Many bacteria frequently alternate between free-living and particle- or host-associated life stages, but most concentrate around nutrient microgradients generated by organic particles, larvae, live feed and other microbes (Grossart, 2010). Increases in bacterial metabolic activity in these nutrient hotspots may lead to increases in bacterial biomass as well as their extracellular metabolic wastes, including nitrogenous compounds (e.g. nitrites and nitrates), phosphates and sulphates, as well as dissolved gases (e.g. carbon dioxide). These wastes, together with excretory products generated by the larvae themselves, become nutrients for further bacterial productivity. The aggregation of such enriched microbial communities and organic debris forms the nucleus of marine snow, which comprises a complex aggregate of viruses, bacteria, archaea, protists, yeasts and fungi, adhered together by

© Woodhead Publishing Limited, 2013

microbially-produced extracellular polymers. Marine snow typically harbours bacteria at concentrations of between 10^8 and 10^9 colony forming units per mL, which is up to 1000 times higher than that found in the water column itself (Munn, 2011). Around and within these larger particulates there may be steep gradients of oxygen and pH (Ploug *et al.* 1997). These complex aggregates of various micro-environments can become reservoirs from which pathogenic bacteria may flourish.

Recirculating aquaculture systems (RAS) are reliant on the proper functioning of biofilters for system success. Hence, the structure, dynamics and function of microbial communities in the biofilter is essential. Current knowledge on these aspects was recently reviewed by Schreier and co-workers (2010). In the context of pathogen control, one of the main functions of RAS is to control nutrient levels through processes such as nitrification, denitrification, dissimilatory nitrate reduction to ammonia, anaerobic ammonium oxidation (anammox), sulphide oxidation and methanogenesis (Schreier *et al.*, 2010), which provides a strong selection pressure determining the types of micro-organisms that are likely to proliferate in the system. On the other hand, the biofilters themselves may serve as pathogen refuges, with the microbes present in the deeper layers being protected against many disinfection methods. Studies have detected close relatives of pathogens including *Aeromonas*, *Coxiella*, *Erwinia*, and *Vibrio* spp. (Schreier *et al.*, 2010 and references therein). A later study (Matos *et al.*, 2011) showed temporal and spatial segregation of the biofilter bacterial community, and that common water quality descriptors could explain 70 % of the DGGE community profile variability.

So far, only a few studies have specifically targeted the bacterial community dynamics in the water column of invertebrate aquaculture hatcheries during a larval rearing run. The bacterial community in the water column of larval rearing tanks is diverse, and many of the retrieved sequences are closely related to species known to be involved in nutrient cycling in marine systems (Payne *et al.*, 2006; Johnson *et al.*, 2008). Payne *et al.* (2006) used a combination of flow cytometry and molecular methods to quantify and characterise the bacterial community in the water column during a larval rearing run of the ornate spiny lobster (*Panulirus ornatus*), the larvae of which are called phyllosomas. The bacterial community showed distinct changes during the first two days after stocking and around the time of moulting, likely related to changes in nutrient availability. *Vibrio* spp. are known to cause disease outbreaks in phyllosomas (Bourne *et al.*, 2007; Goulden *et al.*, 2012a) and *Vibrio* sequences were retrieved in the clone library generated from the water column following a mass mortality event. In contrast, *Vibrio*-related sequences were not retrieved from corresponding DGGE profiles suggesting that their relative abundance in the water column was below 1 %. Sandaa *et al.* (2003) used DGGE to analyse water samples from three systems used for rearing of scallop (*Pecten maximus*) larvae: a stagnant system, a flow-through system and a flow-through system

© Woodhead Publishing Limited, 2013

with ozonated sea water, and samples were collected from both inlet tanks and from the water pipes. Relatively similar bacterial communities were found in all water samples collected from larval rearing tanks and communities appeared to be relatively stable with time (days 0–10). In contrast, samples from the inlet tanks and from the water pipes revealed differences between water sources and also that changes occurred when water passed through the pipes, possibly due to sloughing of biofilm inside the pipes (see Section 8.3.2).

Several studies have demonstrated clear differences in the bacterial community of the rearing water as compared to that of the reared larvae (Sandaa *et al.*, 2003; Bourne *et al.*, 2004; Johnson *et al.*, 2008). However, at least for fish it has been shown that some bacteria present in the water may adhere to eggs and that newly hatched, unfed larvae become contaminated by bacterial communities present on egg debris and rearing water (Lauzon *et al.*, 2010 and references therein).

8.3.2 Surfaces

The surface to volume ratio in a hatchery is profoundly different from that in the ocean. There is effectively no physical containment effect in the natural habitat of planktonic marine larvae. In stark contrast, the tank walls and pipework in a hatchery means that there is a high degree of interaction between the water column and surfaces (Fig. 8.2). Bacteria have evolved complicated lifestyles which, in addition to the existence of planktonic forms in the water column, include surface attachment and incorporation into a biofilm. Biofilms are ubiquitous in marine environments and consist of a complex microbial consortium attached to a surface which is embedded in a self-produced extracellular polymeric matrix. The architecture of the biofilm community is shaped by numerous factors, namely surface texture, hydrodynamic conditions, shearing forces, substrate and nutrient availability, and interaction with other micro-organisms (Battin *et al.*, 2007). Importantly, in the hatchery biofilm formation it is not only restricted to inanimate surfaces like the tank and pipework, but also forms on ecto- and endobiotic surfaces of larvae. These surfaces support bacterial colonisation by providing a nutrient-rich micro-environment and protection from chemical and physical stressors, and by acting as a mechanism of dispersal and relocation (Tang *et al.*, 2010, 2011; Grossart, 2010).

Importantly, from a health risk perspective, bacteria residing in biofilms differ phenotypically from their planktonic conspecifics, most notably by showing increased resistance to antimicrobial agents and grazing by protozoans (Karunasager *et al.*, 1996; Matz *et al.*, 2008). Upon cell adhesion to the surface, bacteria secrete extracellular substances that can anchor them to the surface and enmesh neighbouring cells. As the biofilm develops, the thickness of the exopolymeric matrix increases such that it can trap or slow the diffusion of antimicrobial compounds leading to increased resistance

© Woodhead Publishing Limited, 2013

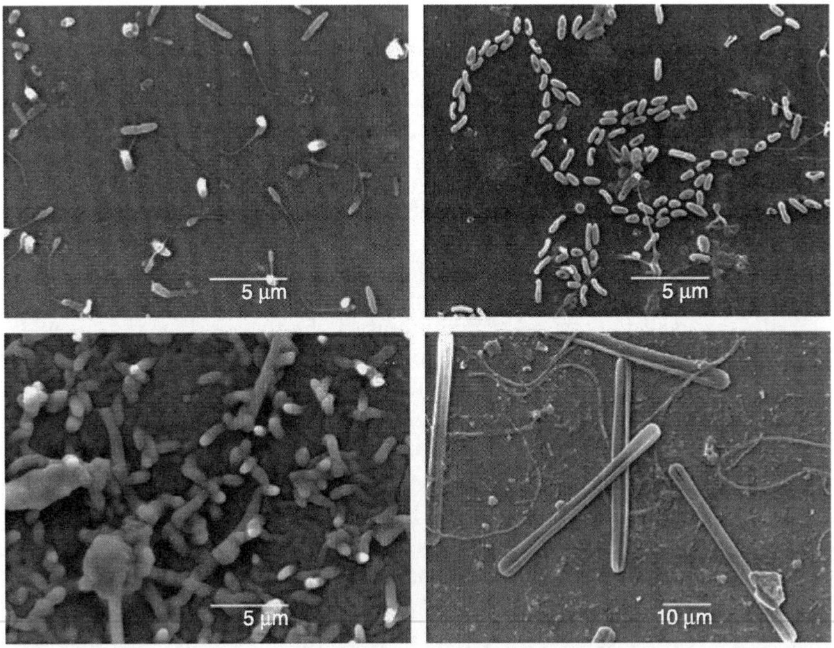

Fig. 8.2 Biofilm development on hatchery tank surface as revealed by scanning electron microscopy (SEM). Upper left: Founder populations of bacteria after 1 day of hatchery tank operation. Upper right: Bacterial proliferation after 3 days of hatchery tank operation. Bottom left: Fully developed biofilm with extensive extracellular matrix embedding and protecting incorporated bacteria cells. Lower right: Lower magnification (note scale) demonstrating complexity of microbial community, including single cells and filamentous bacteria as well as benthic diatoms. (Photos: M. Hall)

(Pasmore and Costerton, 2003; Burmølle *et al.*, 2006). Biofilms may also tolerate antimicrobial treatment through changes in genetic pathways, including up-regulation of genes encoding active transporters which facilitate the efflux of antimicrobials (Gillis *et al.*, 2005). In fact, it has been shown that the expression of over 800 bacterial genes can be altered once they become part of a biofilm (Sauer *et al.*, 2002). Moreover, the bacteria within a biofilm can act as a complex multicellular organism, using cell–cell chemical signalling called quorum sensing (see Section 8.6.3) to trigger changes in gene expression and metabolic activity (Ghannoum and O'Toole, 2004). Additionally, the micro-environmental conditions within the biofilm may facilitate horizontal transfer of virulence genes which can give rise to pathogenic bacteria capable of infections (Karunasager *et al.*, 1996; Hall-Stoodley *et al.*, 2004; Bourne *et al.*, 2006). In humans, it has been estimated that 60–80 % of microbial infections in the body are caused by bacteria in biofilms (Costerton *et al.*, 1999).

© Woodhead Publishing Limited, 2013

During biofilm development, regulated and passive detachment of cells means the water column is continuously seeded by planktonic bacteria (Kirov *et al.*, 2005). This mechanism enhances dispersal and seeding of a new biofilm, thus perpetuating the biofilm cycle (Grossart, 2010). An important epizootiological consequence of the free-living forms is that pathogenic micro-organisms may be transmitted through incidental ingestion, particularly in filter feeders like abalone (Travers *et al.*, 2008) and *Artemia* (Chair *et al.*, 1994; López-Torres and Lizárraga-Partida, 2001; Vaseeharan and Ramasamy, 2003) and zooplankton in general (Tang, 2005).

Biofilms forming in larval rearing tanks have a diverse bacterial community that is distinct from the communities of larvae and the rearing water (Bourne *et al.*, 2004; Jensen *et al.*, 2004). A dynamic succession of microbial species during biofilm development have been shown in tanks used for rearing *P. ornatus* phyllosomas (Bourne *et al.*, 2006) where prominent changes in the DGGE bacterial community profiles coincided with the decrease in bacterial numbers observed by scanning electron microscopy and agar plating due to sloughing of biofilm into the water column. The study also demonstrated that the biofilm is a potential reservoir for phyllosoma pathogens since DGGE bands affiliated with *Vibrio* became dominant just before a phyllosoma mortality event and a *V. harveyi* strain isolated from late biofilm caused increased phyllosoma mortalities in small-scale survival studies. This emphasises the importance of being able to control the formation of biofilms in larval rearing tanks. However, hatchery operators culturing sessile aquaculture species such as oysters, sea urchins or abalone need to be aware that biofilms can be crucial for efficient larval settlement and have to consider this aspect when choosing an approach for bacterial management (Hadfield, 2011).

8.3.3 Larvae

The bodies of larvae, in particular planktonic forms within the water column, are nutrient-rich micro-environments that support bacterial attachment and growth (Tang *et al.*, 2010), and thus represent major microbial compartments within the larval rearing ecosystem. Some ectobionts utilise a hitchhiking strategy on planktonic higher organisms to facilitate their dispersal (Grossart *et al.*, 2010), while others forge specialised and complex interactions with their eukaryotic partners (Goffredi, 2010). The gastrointestinal tract also exhibits many microniches harbouring microbial communities distinct from local ambient environments. Several studies have analysed the diversity of bacterial communities associated with eggs and larvae in fish hatcheries. Exogenous and endogenous factors that influence the establishment of fish gastrointestinal microbiota were recently reviewed by Nayak (2010). The composition of the fish larvae gut microbiota is influenced by the feed and feeding conditions, and rapid changes result from changes in diet (Brunvold *et al.*, 2007; McIntosh *et al.*, 2008; Bjornsdottir *et al.*, 2009).

© Woodhead Publishing Limited, 2013

Less detailed information is available for hatchery-reared invertebrates, though some information is available for Pacific white shrimp (*Litopenaeus vannamei*) (Johnson *et al.*, 2008; Thompson *et al.*, 2010). Johnson *et al.* (2008) used phospholipid fatty acid analysis and DGGE to confirm that the *L. vannamei* gut bacterial community differed from that of the water and the feed and to demonstrate clear differences between the foregut and the hindgut, with the latter generally showing a higher biomass, a lower diversity and *Vibrio* spp. as the dominant community members. Thompson and co-workers (2010) analysed the bacterial community in mid/hindgut and faecal samples showing a limited diversity and that significant changes could be induced after eight weeks' exposure to experimental diets amended with the probiotic candidate *Vibrio gazogenes*.

Importantly, the indigenous intestinal microbiota of larvae can act as a complementary first line of defence against ingested bacteria, which can include pathogens (Vine *et al.*, 2004). As such, it is important to define what is considered autochthonous and allochthonous in the larval rearing ecosystem. For example, Payne *et al.* (2008) showed the bacterial communities of cultured *P. ornatus* phyllosomas differed significantly from their wild conspecifics (Fig. 8.3). It was found that cultured phyllosomas harboured a more diverse community, probably reflecting the numerous environmental niches that exist within larval rearing systems. Species of bacteria isolated from wild larvae have been found to possess probiotic attributes and may protect such larvae in the wild from infection (Goulden *et al.* 2012b). The potential symbioses that exist between naturally occurring bacteria and wild larvae may provide insightful clues into developing microbial management approaches.

8.3.4 Larval feed

The composition and dynamics of the microbiota associated with aquatic larvae is heavily influenced by the microbial consortia associated with larval feeds and especially live prey. Planktonic bacteria are subject to bioaccumulation by live prey organisms such as *Artemia* (Chair *et al.*, 1994; López-Torres and Lizárraga-Partida, 2001; Vaseeharan and Ramasamy, 2003), which can harbour 10^2–10^6 colony forming units per *Artemia* nauplius (Gomez-Gil *et al.*, 1998; Makridis *et al.*, 2000; Soto-Rodriguez *et al.*, 2003; Høj *et al.*, 2009). It should be noted that enrichment of a live feed such as *Artemia* with a nutritional supplement can alter the associated bacterial load and community structure (Høj *et al.*, 2009; Hache and Plante, 2011). Vectored transmission of pathogens by live prey is a biosecurity risk in larviculture systems where animate prey is required to stimulate feeding (Johnston *et al.*, 2008). For example, vectored challenge is crucial to the infectivity of the specialist enteropathogen *V. owensii* DY05 in larvae of the ornate spiny lobster (*Panulirus ornatus*), causing more consistent and reproducible mortality rates than animals exposed by immersion (Goulden *et al.*

© Woodhead Publishing Limited, 2013

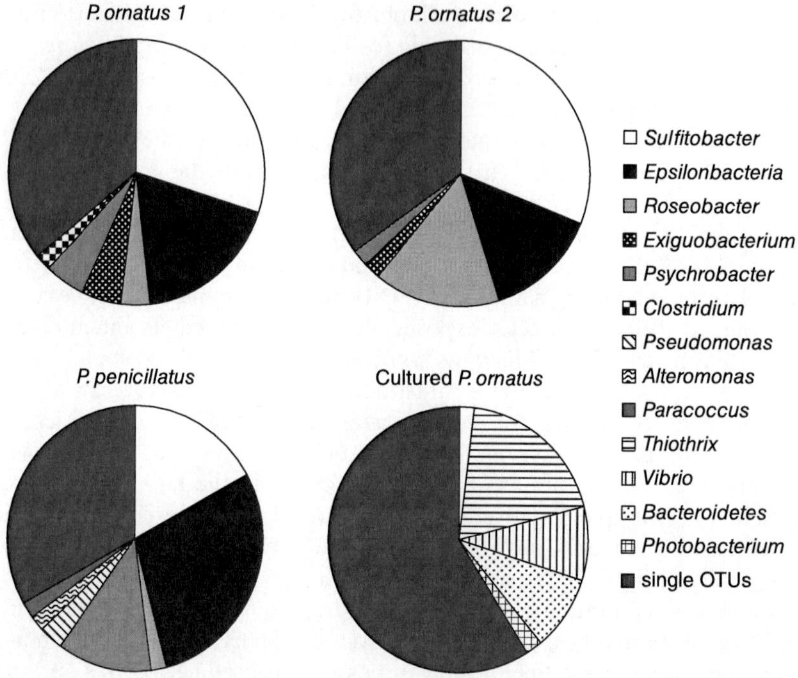

Fig. 8.3 Comparative bacterial community analysis (based on 16S rRNA gene clone libraries) of two species of Palinurid lobster larvae sampled directly from wild plankton of tropical ocean environment (*P. ornatus* 1, *P. ornatus* 2, *P. penicillatus*) or from a hatchery environment (cultured *P. ornatus*). Distinct community differences were detected between wild and aquaculture reared larvae. Also note the relatively high overall abundance of bacteria represented by single operational taxonomic units (OTUs).

2012a). *Artemia* were also recently identified as a potential vector for necrotising hepatopancreatitis bacteria in shrimp (*L. vannamei*) aquaculture (Avila-Villa *et al.*, 2011). Bacteria associated with *Artemia* are upon ingestion by the larval host partly protected against the potentially unfavourable gastrointestinal micro-environment, allowing passage through hostile locales (Grisez *et al.*, 1996). Furthermore, the *Artemia* gut can act as a safeguard from disinfectants and antimicrobials added to the water column, which complicates microbial control strategies (Høj *et al.*, 2009).

A few studies have analysed the bacterial community associated with live feeds. Molecular studies have shown that bacterial communities associated with *Artemia* can be highly variable with time (McIntosh *et al.*, 2008; Bjornsdottir *et al.*, 2009) and it is possible that this is the situation in most commercial hatcheries. Nevertheless, *V. alginolyticus* has consistently been reported as the dominant member of the culturable bacterial community of *Artemia* (Villamil *et al.*, 2003; Thomson *et al.*, 2005; Høj *et al.*, 2009). A

© Woodhead Publishing Limited, 2013

few studies have analysed the bacterial community associated with *Artemia* using culture-independent methods and all these studies detected *Vibrio* spp. as important components of the bacterial community (McIntosh *et al.*, 2008; Bjornsdottir *et al.*, 2009; Høj *et al.*, 2009). Interestingly, Høj *et al.* (2009) found that the relative dominance of *Vibrio* spp. increased after a disinfection step that decreased the overall bacterial abundance and also the abundance of culturable vibrios.

The bacterial community associated with rotifers has been studied extensively by culture-based approaches (Nicolas *et al.*, 1989; Skjermo and Vadstein, 1993; Blanch *et al.*, 1997). These studies indicate that bacteria isolated from rotifers depend on the protocols used for rotifer cultivation and enrichment as well as the non-selective uptake of bacteria present in the culture water (Skjermo and Vadstein, 1993). A recent molecular study suggested that the bacterial communities of rotifers are more stable than communities associated with *Artemia*, with a stable core microbiota being present in most rotifer samples (McIntosh *et al.*, 2008). Both culture-based and molecular studies have shown that *Vibrio* spp. are significant components of the bacterial community associated with rotifers (Blanch *et al.*, 1997; Nicolas *et al.*, 1989; Brunvold *et al.*, 2007).

8.4 Identification, detection and monitoring of pathogens

Detection and identification of pathogens using conventional culture-based techniques have several limitations. Selective media often allow growth of closely related species or strains and the identification of putative pathogen colonies by metabolic profiling is not always conclusive requiring additional confirmation assays, which can be time-consuming (Cano-Gomez *et al.*, 2009; Frans *et al.*, 2011). In addition, culture-based techniques are not able to detect pathogens in the VBNC state and may therefore result in false-negatives in periods of environmental stress. For these reasons, the use of culture-independent techniques for identification and detection of pathogens is growing rapidly. It is beyond the scope of this chapter to provide a detailed description of methods used for pathogen identification and typing, and these topics have recently been reviewed elsewhere (Cano-Gomez *et al.*, 2009; Adams and Thompson, 2011; Frans *et al.*, 2011). Commonly used approaches include immunological techniques, PCR-based techniques, isothermal amplification techniques, nucleic acid hybridisation techniques, fingerprinting techniques, sequence analysis of selected genes and DNA array technologies. Currently, most assays are specific for one pathogen, although some have been developed that can detect and even quantify multiple pathogens.

It is paramount to the focus and timing of appropriate biocontrol strategies that pathogens are identified, preferably using experimental infection models that provide information on infection routes and infection dynamics

© Woodhead Publishing Limited, 2013

(Saulnier *et al.*, 2000). Initial attempts to detect the most common pathogens associated with similar disease symptoms should be performed, and if isolates can be obtained they should be tested in infection models. If this approach is unsuccessful, a more comprehensive molecular study comparing healthy and diseased individuals may be necessary to identify disease agents. Based on such studies, targeted isolation strategies can in some cases be designed in order to obtain cultures of the pathogen(s) for detailed studies of their pathogenicity and virulence factors.

Fluorescently labelled bacteria, which permit *in vivo* tracking of their infection route and infection dynamics, have found increasing use in studies of aquaculture pathogen–host relationships. For instance, the infection route and infection dynamics of *V. anguillarum*, *Aeromonas hydrophilia* and *Edwardsiella tarda* in fish larvae (Ling *et al.*, 2001; O'Toole *et al.*, 2004; Chu and Lu, 2008), *V. harveyi* in abalone (Sawabe *et al.*, 2007; Travers *et al.*, 2008) and *V. owensii* in *P. ornatus* larvae (Goulden *et al.*, 2012a) have been determined using this approach.

Identification of pathogen ecological niches in the hatchery system and knowledge of typical infection sources and infection routes provide essential information for the development of successful biocontrol strategies. For instance, some pathogens typically enter through ingestion of contaminated ambient water or feed items, and others again through skin lesions or the gills if the water is contaminated. A live feed organism such as *Artemia* can bio-accumulate bacteria from the water column, and may provide a link between contaminated sea water and infections requiring direct transfer to the digestive system.

In hatcheries where the same pathogens or group of pathogens repeatedly cause mortalities it can be advisable to set up a strategy for pathogen monitoring of selected control points such as, for instance, the live feed or incoming sea water (Owens and Busico-Salcedo, 2006). The technique of choice would have to be rapid, accurate and reliable if economic losses are to be minimised (Frans *et al.*, 2011). In each case, it would be necessary to evaluate if there is a threshold level beyond which the pathogen is likely to become implicated in disease and the sensitivity of the assay of choice needs to be appropriate. It is also necessary to determine if it is most suitable to use an assay that targets a specific taxonomic group or whether specific genes involved in virulence should be targeted (Cano-Gomez *et al.*, 2009). For instance, a 4 h real-time PCR-based assay developed for detection of *V. penaeicida* in the Pacific blue shrimp (*L. stylostris*) and its aquaculture environment (Goarant and Merien, 2006) provides information on the infection level and has served as a decision tool for farmers as well as a research tool. Also, assays have been developed to detect the shrimp pathogen *V. nigripulchritudo* targeting three genetic levels: the species level, the pathogenic cluster level and the plasmid level (Walling *et al.*, 2010). The use of a pathogen monitoring strategy can aid the farmer in assessing the risk of using, for instance, a contaminated batch of live feed and in deciding

© Woodhead Publishing Limited, 2013

whether to discard a live feed batch, cull target larvae or use one of the treatment approaches described in Section 8.8.

8.5 Prophylactic strategies

8.5.1 General biosecurity and disinfection strategies

Biosecurity is a critical line of defence in a hatchery, involving the control of bacteria and increasing the probability of maintaining larvae in a status of high health and low morbidity. Bacterial control is applied through sterilisation, disinfection and antisepsis of equipment and other materials which enter and leave the hatchery. Physical treatment of hatchery equipment, such as desiccation, heat and ultra-violet (UV) exposure, are not applicable to the larvae themselves. Similarly, chemical treatment of hatchery equipment and of incoming sea water, by UV and/or ozone treatment, is widely practised, but cannot be directly applied to the larvae. On the other hand, disinfection methods can be generally applied in the hatchery as a whole at sub-lethal concentrations, but only in concentrations that do not harm the larvae, either acutely or chronically. As chemical disinfectants are often non-specific there is a fine balance between disinfection of the microbial community and death of larvae and, as discussed above, they can be ineffective in systems which possess biofilms. The treatment of bacteria on living tissue by antisepsis is rarely applicable in hatcheries due to the small size and sheer number of larvae.

The water column, tank and pipework surfaces require large-scale treatment methodologies, including filtration and UV sterilisation of incoming water. In contrast, the live feeds and the larvae themselves represent only a very small proportion of the hatchery and their gastrointestinal tracts especially lend themselves to more targeted treatments. This is of major economic importance when the treatment is by costly agents, such as antimicrobials and the new emerging generation of treatments that specifically target only pathogenic bacteria within a hatchery.

8.5.2 Green water: beneficial microalgae

The conditioning of the water column to improve larval survival includes the use of the green water technique (Shioda *et al.*, 1997). Green water culture application in hatcheries is an extension of the use of microalgae in the treatment of wastewater and bioremediation (Richmond, 2004). As the hatchery system maintains larvae at high density, there is a corresponding raised level of dissolved organic matter that can drive algal blooms. The microalgae compete against the bacterial planktonic community for nutrients and thereby limit bacterial growth. It is also believed that the complex microalgae and bacterial community results in a more stable environment that is less favourable to opportunistic bacteria, including pathogens, with high growth rates. One microalga used for green water

© Woodhead Publishing Limited, 2013

culture is *Nannochloropsis oculata*, as it grows at high concentrations for a long period and appears to extrude antimicrobial metabolites into the water column (Kittaka, 1994). A major problem with green water culture is preventing contamination from other microalgae. While managing a single microalgal species in a continuous log phase is challenging in its own right, it becomes nearly impossible to maintain a stable mixed species bloom due to different nutritional requirements of each species. Although green water culture has been promoted as one approach to minimise larval mortality risk due to pathogenic bacteria, it is being re-evaluated for invertebrate hatcheries due to variable results which are often not consistent with commercial production.

8.5.3 Probiotics: beneficial bacteria

The functionality and therefore definition of a probiotic is largely dependent on the animal production system in which it will be used. Fuller's widely cited definition of a probiotic as 'a live microbial feed supplement which beneficially affects the host animal by improving its intestinal balance' (Fuller, 1989) is too narrow for aquaculture production systems because aquatic animals are intimately associated with their ambient environment. Hence, microbial-induced improvements to the host ectobiota or host external environments may also be considered probiotic (Verschuere *et al.*, 2000). In their review, Irianto and Austin (2002) defined an aquaculture probiont as 'an entire or component(s) of a microorganism that is beneficial to the health of the host'. This all-encapsulating definition includes probiotic action on both internal and external environments, and introduces the prospect of microbial components, such as peptidoglycan and lipopolysaccharides, acting as immunostimulants (Smith *et al.*, 2003).

One of the main channels of a probiotic biocontrol approach is to pre-emptively colonise the hatchery environment with beneficial micro-organisms to reduce the incidence of primary colonisation by unfavourable micro-organisms, including pathogens (Verschuere *et al.*, 2000). Direct activity against pathogens may be facilitated through a number of possible mechanisms of action, which include, but are not limited to, antagonism (Verschuere *et al.*, 2000; Balcázar *et al.*, 2006; Kesarcodi-Watson *et al.*, 2008), competition for host adhesion sites and nutrients (Verschuere *et al.*, 2000; Vine *et al.*, 2004, 2006; Chabrillón *et al.*, 2005), quorum-sensing inhibition (Natrah *et al.*, 2011) and predation (Qi *et al.*, 2009). However, most of the perceived mechanisms of action are based on *in vitro* observations and there is still a need for *in vivo* approaches, such as gnotobiotic models, to elucidate probiont-host interactions and the effect of introduced probiotics on autochthonous microbiota (Tinh *et al.*, 2008).

Antagonism is a widespread trait among marine bacterial genera and is implicated in competitive interactions and ecological success of micro-organisms (Kalinovskaya *et al.*, 2003; Gram *et al.*, 2010; Wietz *et al.*, 2010;

© Woodhead Publishing Limited, 2013

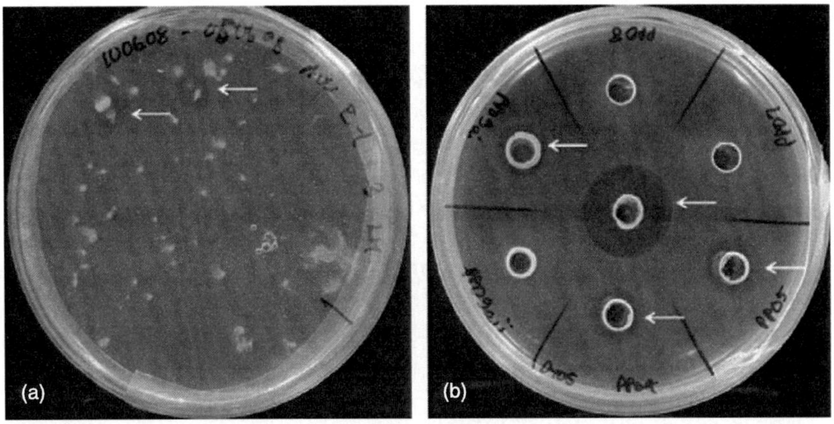

Fig. 8.4 Evaluation of the antagonistic potential of probiont candidates. (a) An initial screen for antagonistic bacteria can be performed by embedding a specific pathogen into agar. Plating of an environmental sample reveals bacterial strains that can interfere with pathogen growth, through the presence of clear zones (white arrows) in the otherwise cloudy agar. (b) Confirmation of antagonistic activity of isolates can be performed by adding broth culture to wells cut in agar with pathogen embedded. Clear zones are created where the candidate probiont interferes with pathogen growth. For the presented plate, three of six tested probiont candidates were antagonistic to the pathogen (white arrows). (Photos: E. Goulden)

Cao *et al.*, 2011). Antagonistic micro-organisms are commonly identified in probiotic screening processes by their production of bioactive compounds including antibiotics, bacteriocins, hydrogen peroxide and organic acids, which are all inhibitory towards other micro-organisms. Inhibitory activity is commonly identified using *in vitro* antagonism assays in which known pathogens are exposed to putative probiotics or their extracellular products (Fig. 8.4). Electron microscopy has revealed that the action of bioactive substances can induce morphological changes to pathogen cells, which disrupt physiology directly or make them susceptible to attack by other active compounds (Ringø, 2008; Ma *et al.*, 2009; Rattanachuay *et al.*, 2010).

There are many listings in the literature where antagonistic isolates or their products have conferred a protective benefit or enhanced survival of larval and post-larval crustaceans (Ravi *et al.*, 2007; Guo *et al.*, 2009; Swain *et al.*, 2009; Pai *et al.*, 2010) and molluscs (Ruiz-Ponte *et al.*, 1999; Riquelme *et al.*, 2001; Longeon *et al.*, 2004; Kesarcodi-Watson *et al.*, 2010). These studies indicate that a diversity of genera, including *Alteromonas, Arthrobacter, Bacillus, Enterococcus, Neptunomonas, Phaeobacter, Pseudomonas, Pseudoalteromonas, Roseobacter, Streptococcus, Thalassobacter* and *Vibrio,* harbour strains with probiotic potential.

The microbiota associated with larval hosts are good reservoirs of antagonistic bacteria (Hjelm *et al.*, 2004; Fjellheim *et al.*, 2007) and represent logical micro-environments in which to conduct initial searches for

© Woodhead Publishing Limited, 2013

probionts. It is expected such micro-organisms are intrinsically adapted to the dynamics of the culture system which thereby increases the chance of a desired probiotic effect (Kesarcodi-Watson *et al.*, 2008). However, *in vitro* antagonistic activity is not always concomitant with an *in vivo* effect. Ruiz-Ponte *et al.* (1999) found an antagonistic *Roseobacter* strain could not protect scallop (*Pecten maximus*) larvae from *V. anguillarum*. The authors postulated that one of the reasons for the lack of an effect was due to differential niche specialisation exhibited by the pathogen and probiont. This illustrates that selected probionts must share and exert their perceived mode of action in the ecological niche where the pathogen has its impact. The use of fluorescent-protein expressing bacteria is an alluring possibility in transparent larval forms to track probionts and pathogenic bacteria through the hatchery (Goulden *et al.*, 2012c).

Generally, a large pool of candidates will be generated from initial *in vitro* antagonism screening processes and it is highly impractical to validate the *in vivo* effects of all candidates due to the restricted accessibility and processing of large numbers of experimental animals (Kesarcodi-Watson *et al.*, 2008). To supplement *in vitro* antagonism screens, probiotic candidates can be assessed for adhesive and competitive potential on mucous or intestinal epithelial cells (Vine *et al.*, 2006), given the microbiota of the gastro-intestinal tract serve as an important barrier to invading pathogens in aquatic larvae (Olafsen *et al.*, 2001; Tinh *et al.*, 2007). Mechanisms by which micro-organisms adhere to intestinal epithelial cells include van der Waal's forces, electrostatic and hydrophobic interactions and specific receptor–ligand binding. However, the temporal attachment of microbes can be influenced by flushing rates and transforming physicochemical characteristics during larval development (Olafsen 2001; Denev *et al.*, 2009). As such, administered probionts may be evacuated from the gut before they can secrete antimicrobials which, for most aquaculture bacterial genera, occurs maximally in the stationary phase (Chythanya *et al.*, 2002; Campos *et al.*, 2006; Vijayan *et al.*, 2006; Ringø, 2008; Rattanachuay *et al.*, 2010).

Probiotic micro-organisms may also compete with pathogens for nutrients in the host or ambient environment. Siderophore-producing bacteria are prime examples of micro-organisms which deprive competitors of iron by chelation. In fact, recent research indicates siderophore-producing bacteria, including *Enterovibrio* spp., *Photobacterium* spp. and *Vibrio* spp., are widespread intestinal micro-organisms, at least in fish species, which could be used to suppress pathogens in culture environments (Sugita *et al.*, 2012).

8.5.4 Prebiotics

Prebiotics are nutrients in the environment that simulate the growth and/or activity of bacteria that are beneficial to the health of the host. It is now established, at least in humans, that specific areas of the body, such as skin, oral cavity, oesophagus and gastrointestinal tract, are micro-environments

© Woodhead Publishing Limited, 2013

that have specific bacterial communities (Costello *et al.*, 2009). For example, the gastrointestinal bacterial community is largely phylogenetically distinct from the external environmental microbiological community and most species found within it are rarely found in significant numbers outside of the gut (Ley *et al.*, 2008). It is likely that similar patterns in microbial communities occur in other organisms, including insects and marine larvae (Gusmao *et al.*, 2007; Reid *et al.*, 2009). The gut microbiota appears to be of significance to host health, and specific bacterial communities can alleviate disease and inhibit colonisation by pathogenic bacterial forms (Spor *et al.*, 2011). There is increasing evidence that modulation of the gut microbiota through the use of prebiotics can be beneficial in creating a micro-environment within the gut that decreases the risk of infection and may be given in the diet to condition the gastrointestinal tract for specific types of bacterial groups (Tuohy *et al.*, 2005). The most commonly used prebiotics in human gut conditioning include soluble fibre carbohydrates and, specifically, various oligosaccharides that are a nutritional source for bifidobacteria and lactic acid bacteria which produce short-chain fatty acids (SCFA). The exact mechanism by which SCFAs act is poorly understood, but they are known to have beneficial effects on calcium and mineral absorption and, in vertebrates, immune system effectiveness (Peppelenbosch and Ferreira 2009). Although the mechanism of action is even less well understood in marine organisms, it has been clearly demonstrated that SCFA, especially those produced from poly-β-hydroxybutyrate, protects zooplankton (*Artemia*) from such bacterial pathogens as *V. campbellii*, *V.harveyi* and *V. parahaemolyticus* (Defoirdt *et al.*, 2007, 2008; Immanuel *et al.*, 2010).

The establishment of probionts in the digestive tract is a possible means of controlling enteropathogen proliferation in invertebrates and fish (Sugita *et al.*, 1998; Balcázar and Rojas-Luna, 2007; Avella *et al.*, 2010). Another avenue in which to increase the colonisation potential of probionts is a synbiotic approach. Synbiotics is the synergistic action of prebiotics and probiotics on improving the survival and colonisation of live microbial supplements in the gastrointestinal tract which can promote the health and welfare of the host (Gibson and Roberfroid, 1995). Microbial community shifts resulting from synbiotic application can have a positive effect on health and pathogen inhibition and can stimulate innate immunity (Ringø *et al.*, 2010). Daniels *et al.* (2010) recently reported that the combined use of a mannan prebiotic and a commercial *Bacillus* spp. probiotic product had a stabilising effect on the bacterial community and increased the digestive capabilities of European lobster (*Homarus gammarus*) larvae. Further, Li *et al.* (2009) showed that shrimp (*L. vannamei*) administered a probiotic *B. megeterium* strain and an isomalto-oligosaccharide prebiotic harboured a bacterial community distinct from those fed a basal diet which was presumably involved in competitive exclusion of vibrios and enhanced immune response of the host.

© Woodhead Publishing Limited, 2013

8.5.5 Immunological protection

Until recently it was widely believed that the only line of immunological defence against pathogens for invertebrates was based on an innate, or non-specific, immune system. In contrast, vertebrates evolved humoral immune systems > 500 million years ago with a capability to produce clones of lymphocytes capable of recognising and neutralising, specific pathogens. It remains questionable whether invertebrates have functional immunological memory (Hauton *et al.*, 2007). While it has been generally considered that invertebrates only recognise and respond to microbial infections in a generic way, invertebrate immunological defence mechanisms are undoubtedly effective against pathogens as invertebrates are by far the most numerous and most diverse of all multi-cellular life forms, making up an estimated 1.6 million of the 1.8–1.9 million species described globally (Regier *et al.*, 2010).

It is now recognised that invertebrates use a range of cellular defences against pathogens (Rowley and Powell, 2007). Their most important cellular defence systems include circulating and sessile hemocyte-like cells that can phagocytose invasive microbes as well as nodule formation and encapsulation, while their most important humoral defence systems include the prophenoloxidase cascade and antimicrobial peptides (Ratcliffe *et al.*, 1985). The prophenoloxidase defence system, involving a proteolytic cascade of phenoloxidase with production of quinones and melanin, has been considered of significant importance in invertebrates to survive microbial infections. However, Leclerc and colleagues (2006), using *Drosophila* mutants lacking the prophenoloxidase-activating enzyme (PAE1), showed that mutants with a total absence of circulating phenoloxidase activity are just as resistant to microbial infection as their wild conspecifics (Leclerc *et al.*, 2006).

Further, in juvenile lobsters, transcript abundance of genes encoding prophenoloxidase (proPO), the β-1,3-glucan binding protein (β-GBP) and a 12.2 kDa antimicrobial peptide (amp), exhibit no significant change in expression upon exposure to opportunistic pathogens and a range of compounds claimed to induce immunological response (Hauton *et al.*, 2007).

Antimicrobial peptides (AMPs) are powerful antimicrobials and primarily act by forming pores or otherwise disrupting microbial membranes although targets may be intracellular, including heat shock proteins and DNA and protein synthesis (Brogden, 2005). More than 1518 AMPs have been identified and classified into three major groups including peptides with cysteine disulphide bonds, linear peptides capable of forming amphipatic α-helices and peptides with one or two amino acids over-represented, including proline, arginine, glycine and histidine (Wang *et al.*, 2009). In marine prawns and lobsters, AMPs include over 200 types of penaeidins, crustins and hemocyanin-derived peptides (Bachère *et al.*, 2004; Tincu and Taylor, 2004; Pisuttharachai *et al.*, 2009; Fusetani, 2010). From a managerial

© Woodhead Publishing Limited, 2013

viewpoint, decreasing the risk of larval mortality due to microbial infection may be implemented, to some degree, by priming the humoral defence of larvae through immunostimulants, such as lectins, as they act as both recognisers and effectors of immunity (Smith *et al.*, 2003; Rattanchai *et al.*, 2005; Chiu *et al.*, 2007; Powell and Rowley, 2007). However, little is known of the development of functional maturity of marine larvae immune systems. Competence of other larval processes, such as metamorphosis, exhibits high degrees of variability between individuals of the same age (Degnan and Degnan, 2010). In a study by Hauton *et al.* (2007), little immunity reactivity was detected in a population of lobster juveniles; however, some individuals were highly responsive to immunological stimulation which opens the intriguing possibility of whether selective breeding can be applied to develop highly disease-resistant populations. Since larvae are early developmental stages maturing towards adulthood, larvae may have limited immunological capacity upon exposure to bacterial pathogens (Jiravanichpaisal *et al.*, 2007). The development of immunological competence in marine larvae requires focused research in order to understand the capacity of larvae to defend themselves from microbial pathogens. Presently, the limited knowledge of larval immunological competence only reinforces the emphasis on biosecurity in the hatchery to obtain high survival of larvae. Due to the poorly understood humoral immunity (Shariff *et al.*, 2001) and, in particular, the controversy regarding immunological memory in invertebrates, vaccines against bacterial infections in marine invertebrate larvae are still at the developmental stage (Chiu and Liu, 2009). Rowley and Powell (2007) reported that there is one commercially available product, AquaVac™ Vibromax™ (Schering-Plough Animal Health), claimed to protect larval shrimp from pathogenic vibrios.

8.6 Treatment strategies

8.6.1 Antibiotics

Historically, antibiotics have been employed prophylactically and curatively to treat bacterial infections in aquaculture systems, in particular in the hatchery environment. Sulfamerazine has been used in US aquaculture since 1948 and antibiotics in general have undoubtedly been instrumental in the commercial success of many aquaculture hatcheries (Alderman and Michel, 1992). However, few antibiotics have been approved by authorities for use in seafood aquaculture and most are restricted to experimental use, or are used off-label (Lucchetti *et al.*, 2004). Indeed, in the USA only four antibiotics have been approved for use in aquaculture – oxytetracycline, florfenicol, ormethoprim/sulfadimethoxine and sulfamerazine – whereas in Japan over 30 antibiotics have been approved for use in aquaculture (Reimschuessel and Miller, 2006). Although antibiotics are effective control measures, there are significant environmental consequences relating to

© Woodhead Publishing Limited, 2013

indiscriminate usage of antibiotics which may exert selection pressures on bacterial communities within and outside the aquaculture area (Cabello, 2006; Heuer *et al.*, 2009). Increased circulation of antibiotic residues in the environment drives the selection of spontaneous mutants expressing antimicrobial resistance which may promote intra- and inter-genus dissemination of resistance determinants in bacterial populations through horizontal gene transfer (Cabello, 2006; Heuer *et al.*, 2009; Burridge *et al.*, 2010). One of the main risks to human health is the spread of antimicrobial resistance from aquatic bacteria to potential human pathogens including *Escherichia coli*, *Salmonella* spp. and *V. cholerae* (Weber *et al.*, 1994; Sørum and L'Abée-Lund, 2002; Molina-Aja *et al.*, 2002; Heuer *et al.*, 2009). Indeed, an epidemiological consequence of clinical antibiotic resistance is exemplified by methicillin-resistant *Staphylococcus aureus* (MRSA), which is a major public health problem implicated in the deaths of more than 18 000 people in the USA alone in 2005 (Klevens *et al.*, 2007). Moreover, human consumption of aquatic products with antibiotic residues could alter the composition of intestinal microbiota, increasing susceptibility to disease and generating allergic hypersensitivity and antibiotic toxicity (Cabello, 2006; Willing *et al.*, 2011).

Antibiotic usage is becoming increasingly obsolete in aquaculture as many economically important pathogens evolve resistance, including strains belonging to the genera *Aeromonas* and *Vibrio* (Akinbowale *et al.*, 2006; Sørum, 2006; Heuer *et al.*, 2009). Public health agencies are increasingly introducing legislation to restrict antibiotic use due to concerns about their impact on environmental bacteria and the development of resistance, which may impact on human health. It is likely that further restrictions will be introduced on the use of antibiotics in aquaculture, either within the country of production or by countries that import the product. Alternatives to bacterial management through antibiotics must be developed as seafood production through aquaculture becomes the most predominant supply to markets and the human food chain.

8.6.2 Bacteriophage therapy

Bacteriophage (phage) are viruses that infect. They have significant ecological relevance as mortality factors controlling bacterial populations in nature thereby shaping the composition of bacterial communities (Hambly and Suttle, 2005). Phage bind to bacterial cell surface receptors, inserting their DNA, and hijacking host cell machinery for phage DNA replication and protein synthesis (O'Flaherty *et al.*, 2009). Phage may undergo two types of life-cycles, lytic or lysogenic. Lytic phage propagate intracellularly by a self-assembly process and are released from the host by cell lysis; while lysogenic phage integrate and reside in host chromosomes (prophage) and are replicated along with the bacterial genome during cell replication (Fig. 8.5).

© Woodhead Publishing Limited, 2013

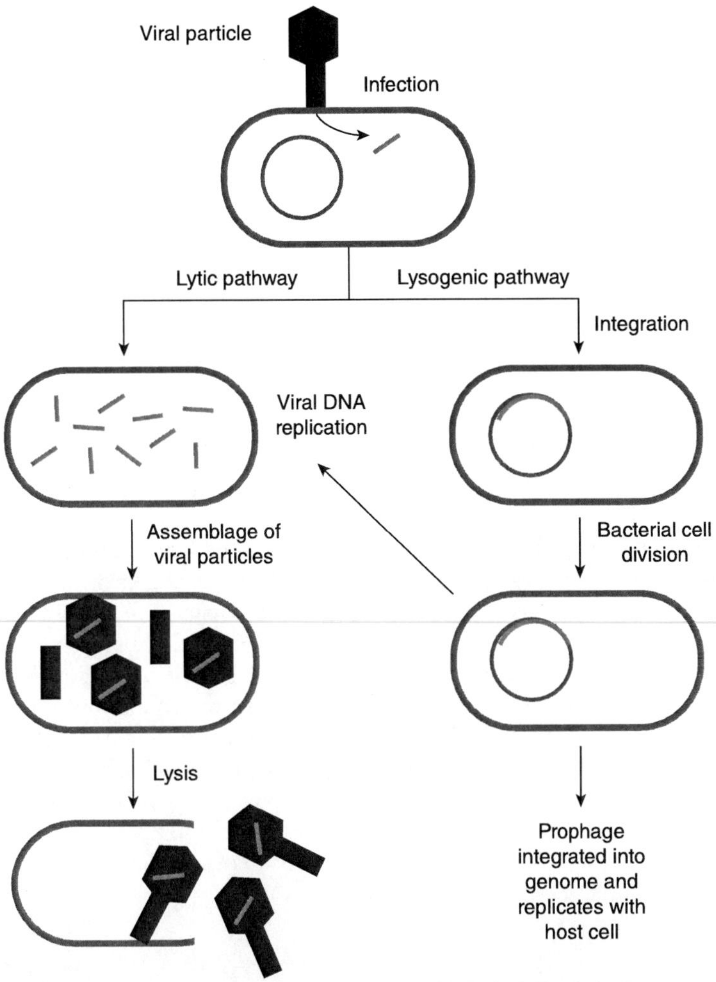

Fig. 8.5 Schematic diagram the two possible reproduction pathways of bacteriophages. Lysogeny occurs when the phage genome (short lines) is integrated into the host genome (circles) and replicates with bacterial cell division as a prophage. Alternatively, the lytic cycle is initiated with production and assemblage of new viral particles and death, by lysis, of the host cell resulting in release of bactreiophage into the environment.

The potential for phage therapy using lytic phage is supported by their ability (i) to self-replicate, thereby increasing autonomously in response to the number of bacterial hosts, and (ii) to infect a narrow host range, meaning non-targeted and potentially beneficial microbiota remain unaffected (Nakai and Park, 2002). Phage to be used for biocontrol in aquaculture systems should be capable of infecting a wide range of strains of the target pathogen species (Defoirdt *et al.*, 2011). Shivu *et al.* (2007) isolated a number

© Woodhead Publishing Limited, 2013

of phage from shrimp hatcheries and creek water which lysed 15–69 % of 183 *V. harveyi* isolates tested. Similarly, Karunusagar *et al.* (2007) found four phage with lytic activity against 55–70 % of 100 *V. harveyi* isolates originating from various environments. The authors further reported that two of the phage, belonging to the family Siphoviridae, enhanced survival of black tiger prawns (*Penaeus monodon*) larvae in hatchery trials. Vinod *et al.* (2006) also reported that a Siphoviridae phage with lytic activity against all 50 *V. harveyi* isolates tested could enhance survival of *P. monodon* larvae by 44–55 % and 17–86 % under experimental *V. harveyi* challenge and normal hatchery conditions, respectively. Survival of shrimp in the latter experiment also exceeded an analogous antibiotic treatment. In some cases, inter-species host-range is reported. For example, Crothers-Stomps *et al.* (2009) found a number of Siphoviridae phage isolated from prawn ponds that could lyse strains of *V. harveyi*, *V. campbellii*, *V. rotiferianus* and *V. parahaemolyticus*. All these studies demonstrate that invertebrate culture systems are good reservoirs for the isolation of phage as potential biocontrol agents.

There are a number of aspects of bacteriophage which could potentially render phage therapies unsuccessful. Firstly, phage are known to transfer virulence factors (Austin *et al.*, 2003; Pasharawipas *et al.*, 2005) and can integrate in the bacterial genome as prophage, and thus it is crucial to identify potential virulence factors in the phage genome. Furthermore, bacteria may adapt to local phage predation pressure. Production of extracellular polymeric substances (EPS) can mask phage attachment sites and a bacterial population may evade strain-specific phage by harbouring a diversity of EPS-gene encoding cassettes (Kunin *et al.*, 2010). The problem may be averted by continuous isolation of novel phage, and application of phage cocktails or phage lysins (Tanji *et al.* 2005; Crothers-Stomps *et al.*, 2009; Defoirdt *et al.*, 2011).

8.6.3 Quorum sensing inhibition

Quorum sensing (QS) is a type of bacterial cell–cell communication system used to coordinate the expression of genes in response to small signal molecules, known as autoinducers (Fig. 8.6). Several types of QS systems have been described, with different degrees of specificity. For instance, the LuxR/I-type systems using N-acylated homoserine lactones (AHLs) as the signal molecule have been found in a wide range of Gram-negative bacteria, while peptide signalling systems are common in Gram-positive bacteria. Interestingly, it has been suggested that a signalling system using the AI-2 autoinducer (a furanosyl borate diester) is universal and that the appearance of analogous QS systems among bacterial genera reflects possible roles in inter-species communication (Bassler, 1999; Waters and Bassler, 2005; Bassler and Miller, 2006; Boyer and Wisniewski-Dyé, 2009). However, there is some controversy regarding this (Turovskiy *et al.*, 2007), partly due to the

© Woodhead Publishing Limited, 2013

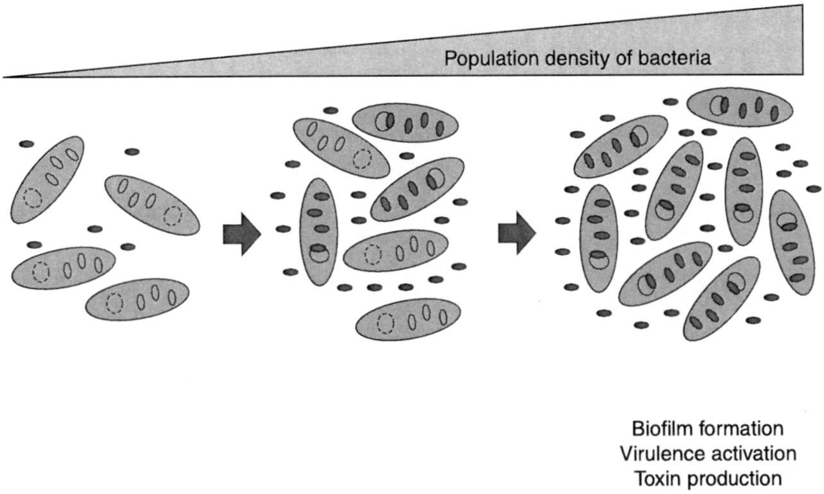

Fig. 8.6 Schematic diagram of the mode of action of cell density based quorum sensing (QS). Bacteria possess QS regulated genes (dotted circle inside cell), QS signal molecules (filled ovals) and QS signal receptors (open ovals). At low population densities (far left), QS signal molecules are in too low concentration to trigger expression of QS regulated genes. As the population density increases (middle), there is an increase in the concentration of QS signal molecules, which bind to intracellular QS receptors. At high density (far right), there is a high concentration of QS signal molecules with a concurrent increase in QS signal–receptor complexes resulting in up-regulation of QS regulated genes including genes associated with biofilm formation, virulence and toxin production.

discovery that not all bacteria use AI-2 as a signal molecule but rather release it as a waste product or use it as a metabolite (Winzer *et al.*, 2002; Turovskiy *et al.*, 2007). Multi-channel QS systems have been found in some aquaculture pathogens such as *V. harveyi* and *V. anguillarum* (Milton, 2006). For instance, in *V. harveyi* one pathway (LuxM/N system) uses an AHL autoinducer, a second pathway (LuxS/PQ system) uses AI-2 and a third pathway (CqsA/S system) uses the so-called cholera autoinducer, CAI-1 ((S)-3-hydroxytridecan-4-one). These three pathways converge at a downstream integrator protein, LuxU. A number of *Vibrio* virulence factors are regulated by QS, including a type III secretion system (Henke and Bassler, 2004a,b; Chen *et al.*, 2011), siderophores (Lilley and Bassler, 2000). Hence, interference with bacterial QS systems is an alluring possibility for controlling aquaculture pathogens.

Several studies have indicated AHL-mediated QS is implicated in disease of cultured aquatic invertebrates. Experimental exposure to mixed AHLs caused mortality in giant freshwater prawn (*Macrobrachium rosenbergii*) larvae, putatively due to the induction of virulence factors in associated microbial communities (Baruah *et al.*, 2009). Further, elevated levels of

© Woodhead Publishing Limited, 2013

AHL molecules were detected in conjunction with mass mortalities of *P. ornatus* phyllosomas (Bourne *et al.*, 2007), indicating that QS likely regulates virulence mechanisms of some phyllosoma pathogens. Recently, Defoirdt and co-workers have elucidated which QS pathways mediate virulence towards gnotobiotic *Artemia*. AI-2-deficient and CAI-1 deficient *V. harveyi* mutants were found to have attenuated virulence and no detectable *in vivo* QS activity, indicating that AI-2 and CAI-1 are the dominant QS pathways during *Artemia* infection (Defoirdt *et al.,* 2005; 2007; 2008; 2012).

Two different strategies for using QS interference as a microbial management strategy in aquaculture have been proposed. One utilises the ability of other marine organisms to interfere with the QS system of the pathogen (Natrah *et al.,* 2011), while the other uses addition of chemical compounds that can interfere with the pathogen's QS system (Manefield *et al.*, 2000; Rasch *et al.*, 2004; Defoirdt *et al.,* 2006; Brackman *et al.,* 2008). In both cases, it has to be considered whether the pathogen in question uses multiple QS pathways, in which case it may be necessary to use a cocktail of probiotic strains or compounds that can either interfere with each pathway or, preferably, can interfere with the QS system at a point after the pathways converge.

Several observations and studies support bacterially-mediated QS disruption as a probiotic mode of action. For instance, a number of aquatic bacteria are capable of QS signal antagonism and degradation, most notably the Gram-positive genera *Bacillus* and *Halobacillus* (Defoirdt *et al.*, 2011; Teasdale *et al.*, 2011). Screening for AHL-degrading bacteria is usually achieved by enrichment cultures selecting for bacteria that can utilise AHLs as the only carbon and/or nitrogen source. This approach has been used with samples from gastrointestinal tracts of shrimp (Tinh *et al.*, 2007) and fish (Dang *et al.*, 2009a, b). Strains derived from these efforts have neutralised the growth retarding effect of *V. harveyi* on rotifers (*Brachionus plicatilis*) (Tinh *et al.*, 2007) and improved survival in turbot (*Schopthalmus maximus*) larvae (Tinh *et al.*, 2008) and giant freshwater prawn (*M. rosenbergii*) larvae (Dang *et al.*, 2009b) when exposed to AHLs. Furthermore, AHL-acylases, encoded by the *aac* gene, were found in a fish-derived *Shewanella* strain which interrupted QS-regulated biofilm formation in icthyopathogenic *V. anguillarum* (Morohoshi *et al.*, 2008) and interfered with AHL-mediated production of exoproteases in an *Aeromonas* strain (Morohoshi *et al.*, 2005). Recently, Chen *et al.* (2010) reported high-yield recombinant production of a *Bacillus*-derived AHL-lactonase by expression in yeast (*Pichia pastoris*) and its use to reduce and delay mortalities of common carp (*Cyprinus carpio*) injected with *Aeromonas hydrophila*.

In vivo studies of QS disrupting compounds have included the use of halogenated furanones in black tiger prawns (*P. monodon*) (Manefield *et al.,* 2000), *Artemia* (Defoirdt *et al.,* 2006) and rotifers (Tinh *et al.*, 2007) and the use of cinnamaldehyde and cinnamaldehyde analogues in *Artemia*

© Woodhead Publishing Limited, 2013

(Brackman *et al.*, 2008) and the nematode *Caenorhabditis elegans* (Brackman *et al.*, 2011). Interestingly, in *V. harveyi* both compound types decrease the DNA binding activity of the master regulator $LuxR_{Vh}$ (not homologous to the *V. fischeri*-type LuxR present in many Gram-negative bacteria) (Defoirdt *et al.*, 2007; Brackman *et al.*, 2008, 2011), and therefore block all three QS channels in this organism. It should be noted that the treatment effect of the compounds differs between bacterial species and QS systems (Defoirdt *et al.*, 2006; Kastbjerg *et al.*, 2007). Also, issues with toxicity of some of the compounds towards the aquaculture target organism have been reported, especially for early work with halogenated furanones (Rasch *et al.*, 2004; Defoirdt *et al.*, 2006; Tinh *et al.*, 2007). This has been somewhat alleviated in recent years with rational design of active compounds with low toxicity (Persson *et al.*, 2005; McDougald *et al.*. 2007, Brackman *et al.*, 2008). Several QS inhibiting compounds have been developed that are currently covered by patents and not described in the scientific literature (Pan and Ren, 2009). An alternative approach, which has not yet been extensively explored for aquatic organisms, is the disruption of autoinducer synthesis (Alfaro *et al.*, 2004; Defoirdt *et al.*, 2007).

Unlike antibiotics, one of the main attractions of QS disruption is that pathogen growth is unaffected and selective pressures and risk of resistance are confined only to environmental situations where QS-regulated metabolism is essential (Defoirdt *et al.*, 2011). In this respect, knowledge of genetic and phenotypic factors regulated by QS in relevant pathogens is crucial prior to implementing such antivirulence strategies.

8.6.4 Predation

The use of the bacterivorous bacterial genus *Bdellovibrio* as biocontrol agents in aquaculture is another promising avenue (Qi *et al.*, 2009). These obligate predators of Gram-negative bacteria are highly motile, seeking prey cells by propulsion mediated by a polar flagellum. Once attached to the outer membrane of prey cells, *Bdellovibrio* cells penetrate the periplasm and grow using the cytoplasm as a substrate before replicating and releasing themselves from the host (Jurkevitch, 2007). Very recently, Cao *et al.* (2012) isolated a strain from sturgeon (*Acipenser baerii*) gut with broad spectrum lytic activity against pathogenic *Aeromonas* strains, supporting these bacteria as promising probiotic agents against pathogen infections.

8.7 Innovations and future trends

As our understanding of the microbial ecology and virulence mechanisms of important hatchery pathogens increases, a new world of specific biocontrol strategies will also emerge. Rapid developments in comparative genomics, transcriptomics and proteomics (Wu *et al.*, 2008) will enable

© Woodhead Publishing Limited, 2013

researchers to identify attributes that enable pathogens to become successful in the hatchery ecosystem and, further, identify physiological vulnerabilities in which to direct specific and novel biocontrol artillery. For instance, the protein (RctB) required for replication of *Vibrio* chromosome II can be blocked by the recently discovered compound vibrepin, inhibiting the growth of *V. cholerae*, *V. parahaemolyticus*, *V. vulnificus* and the shrimp pathogen *V. nigripulchritudo* (Yamaichi *et al.*, 2009). Since RctB is a defining feature of members of the Vibrionaceae (Egan and Waldor, 2003), this may prove a highly specific anti-*Vibrio* therapy. Furthermore, Hung *et al.* (2005) reported that the compound virstatin causes inhibition of ToxT, the transcriptional regulator of cholera toxin and toxin co-regulated pilus in *V. cholerae*, and could protect infant mice from intestinal colonisation by *V. cholerae*. This latter study exemplifies the need to genetically dissect pathogens and pursue antivirulence strategies (Natrah *et al.*, 2011). With the rapid development in sequencing technologies a rapid increase in the number of available full genome sequences of aquaculture pathogens is expected. This will provide a valuable resource for comparative genomics targeting the emergence of virulent strains and the identification of candidate genes that may be involved in host colonisation or virulence, and it may also provide important clues for the development of alternative management strategies.

A better understanding of probiotic modes of action and how gene expression of host and probiont mutually affect each other could be further enhanced through the development of gnotobiotic systems (Tinh *et al.*, 2008). Continuous monitoring of beneficial bacteria is required to detect chance occurrences of these agents developing virulence, which may occur in the intestinal microbiome if responses by the host are eventually outcompeted by changes in behaviour of the colonising micro-organism (Fjellheim *et al.*, 2010; Hooper and Gordon, 2001). For instance, some bacterial genomes, including *Pseudoalteromonas* and *Vibrio* species, possess dual function traits (Thomas *et al.*, 2008; Wietz *et al.*, 2010) which could see virulence emerge 'accidentally' through adaptation to new ecological niches (Casadevall and Pirofski, 2007). There is also a possibility that virulence traits may be transferred to the added probiotic strains by horizontal gene transfer (Kelly *et al.*, 2009).

It is unlikely that a single biocontrol strategy will efficiently reduce disease prevalence in larval hatcheries, and a combination of both prophylactic and curative treatment regimes are needed in the long term. As this chapter goes to publication, there are many new and exciting developments in biocontrol research, which seek to improve disease resistance of the host. For instance, Fu *et al.* (2011) recently demonstrated that recombinant *B. subtilis* spores can be used as carriers for expression and oral delivery of antigen proteins. They successfully engineered and delivered *B. subtilis* spores expressing a white spot syndrome virus (WSSV) envelope protein, which significantly induced phagocytic activity contributing to uptake of the

© Woodhead Publishing Limited, 2013

WSSV by haemocytes and resulting in substantially enhanced survival of *L. vannamei* challenged with WSSV. A recent review by Harikrishnan *et al.* (2011) attests to gaining momentum in the use of botanical extracts to stimulate immune responses in aquatic animals (including shellfish) and augmenting disease resistance. Although controversial, there are also some recent advances in transgenic methods to increase disease resistance in commercially important invertebrates including bivalves and shrimps (Rasmussen and Morrissey, 2007). Finally, and still very embryonic in its development, nanotechnology could revolutionise disease management in aquaculture through the use of lab-on-a-chip devices to rapidly detect pathogens and monitor and control the delivery of probiotics (Rather *et al.*, 2011).

Ultimately, prophylactic approaches should be expanded and developed within an integrated disease management strategy, including curative measures, for different larval stages and hatchery conditions. This could include concoctions of compatible and synergistic micro-organisms that exert multiple modes of attack on unfavourable micro-organisms. Indeed, the selection of multiple, mutually-compatible micro-organisms capable of exerting synergistic benefits forms the basis of the effective micro-organisms concept introduced by Japanese scientist Teruo Higa in the 1980s which has found wide use in agriculture and aquaculture systems in China (Zhou *et al.*, 2009). It can be expected that many selection pressures operating simultaneously would slow the rate of the numerous genetic rearrangements necessary to evolve resistance in a pathogen population. Optimising the microbial structure of the hatchery ecosystem is central to controlling pathogens; however, it is anticipated that biocontrol methods will need to continuously evolve during the infinite arms race with emerging pathogens.

8.8 References

ADAMS A and THOMPSON K D (2011) Development of diagnostics for aquaculture: challenges and opportunities, *Aquacult Res,* 42, 93–102.

ALDERMAN D J and MICHEL C (1992) *Chemotherapy in Aquaculture: From Theory to Reality*. Paris: O.I.E.

Alfaro J F, Zhang T, Wynn D P, Karschner E L and Zhou Z S (2004) Synthesis of LuxS inhibitors targeting bacterial cell-cell communications, *Org Lett*, 6, 3043–3046.

AKINBOWALE O L, PENG H and BARTON M D (2006) Antimicrobial resistance in bacteria isolated from aquaculture sources in Australia, *J Appl Microbiol,* 100, 1103–1113.

AUSTIN B, PRIDE A C and RHODIE G A (2003) Association of a bacteriophage with virulence in *Vibrio harveyi*, *J Fish Dis,* 26, 55–58.

AVELLA M A, GIOACCHINI G, DECAMP O, MAKRIDIS P, BRACCIATELLI C and CARNEVALI O (2010) Application of multi-species of *Bacillus* in sea bream larviculture, *Aquaculture,* 305, 12–19.

AVILA-VILLA L A, MARTINEZ-PORCHAS M, GOLLAS-GALVAN T, LOPEZ-ELIAS J A, MERCADO L, MURGUIA-LOPEZ A, MENDOZA-CANO Fand HERNandEZ-LOPEZ J (2011) Evaluation of

© Woodhead Publishing Limited, 2013

different microalgae species and *Artemia* (*Artemia franciscana*) as possible vectors of necrotizing hepatopancreatitis bacteria, *Aquaculture,* 318, 273–276.

BACHÈRE E, GUEGUEN Y, GONZALEZ M, DE LORGERIL J, GARNIER J and ROMESTand B (2004) Insights into the antimicrobial defense of marine invertebrates: the penaeid shrimps and the oyster *Crassostrea gigas*, *Immunol Rev,* 198, 149–168.

BALCÁZAR T and ROJAS-LUNA T (2007) Inhibitory activity of probiotic *Bacillus subtilis* UTM 126 against *Vibrio* species confers protection against vibriosis in juvenile shrimp (*Litopenaeus vannamei*), *Curr Microbiol,* 55, 409–412.

BALCÁZAR J L, DE BLAS I, RUIZ-ZARZUELA I, CUNNINGHAM D, VENDRELL D and MÚZQUIZ J L (2006) The role of probiotics in aquaculture, *Vet Microbiol,* 114, 173–186.

BARUAH K, CAM D T V, DIERCKENS K, WILLE M, DEFOIRDT T, SORGELOOS P and BOSSIER P (2009) *In vivo* effects of single or combined *N*-acyl homoserine lactone quorum sensing signals on the performance of *Macrobrachium rosenbergii* larvae, *Aquaculture,* 288, 233–238.

BASSLER B L (1999) How bacteria talk to each other: Regulation of gene expression by quorum sensing. *Curr Opin Microbiol,* 2, 582–587.

BASSLER BL and MILLER MB (2006) Quorum sensing, in Dworkin M, Falkow S, Rosenberg E, Schleifer KH, Stackebrandt E (eds), *The Prokaryotes.* New York: Springer, 336–353.

BATTIN T J, SLOAN W T, KJELLEBERG S, DAIMS H, HEAD I M, CURTIS T P and EBERL L (2007) Microbial landscapes: new paths to biofilm research, *Nat Rev Microbiol,* 5, 76–81.

BJORNSDOTTIR R, JOHANNSDOTTIR J, COE J, SMARADOTTIR H, AGUSTSSON T, SIGURGISLADOTTIR S and GUDMUNDSDOTTIR B K (2009) Survival and quality of halibut larvae (*Hippoglossus hippoglossus* L.) in intensive farming: Possible impact of the intestinal bacterial community, *Aquaculture,* 286, 53–63.

BLANCH A R, ALSINA M, SIMON M and JOFRE J (1997) Determination of bacteria associated with reared turbot (*Scopthalmus maximus*) larvae, *J Appl Microbiol,* 82, 729–734.

BOURNE D, YOUNG N, WEBSTER N, PAYNE M, SALMON M, DEMEL S and HALL MR (2004) Microbial community dynamics in a larval aquaculture system of the tropical rock lobster, *Panulirus ornatus*, *Aquaculture,* 242, 33–51.

BOURNE D G, HØJ L, WEBSTER N S, SWAN J and HALL M R (2006) Biofilm development within a larval rearing tank of the tropical rock lobster, *Panulirus ornatus*, *Aquaculture,* 260, 27–38.

BOURNE D, WEBSTER N, PAYNE M, HØJ L, SKINDERSO M, GIVSKOV M and HALL M R (2007) Aspects of the microbiology of phyllosoma rearing of the ornate rock lobster *Panulirus ornatus*, *Aquaculture,* 268, 274–287.

BOYER M and WISNIEWSKI-DYÉ F (2009) Cell–cell signalling in bacteria: not simply a matter of quorum, *FEMS Microbiol Ecol,* 70, 1–19.

BRACKMAN G, DEFOIRDT T, MIYAMOTO C, BOSSIER P, VAN CALENBERGH, S, NELIS H and COENYE T (2008) Cinnamaldehyde and cinnamaldehyde derivatives reduce virulence in *Vibrio* spp. by decreasing the DNA-binding activity of the quorum sensing response regulator LuxR. *BMC Microbiology*, 8, 149.

BRACKMAN G, CELEN S, HILLAERT U, VAN CALENBERGH S, COS P, MAES L, NELIS H J and COENYE T (2011) Structure-activity relationship of cinnamaldehyde analogs as inhibitors of AI-2 based quorum sensing and their effect on virulence of *Vibrio* spp. *PLoS ONE*, 6, e16084.

BROGDEN K A (2005) Antimicrobial peptides: pore formers or metabolic inhibitors in bacteria? *Nat Rev Microbiol,* 3, 238–250.

BRUNVOLD L, SandAA R-A, MIKKELSEN H, WELDE E, BLEIE H and BERGH Ø (2007) Characterisation of bacterial communities associated with early stages of intensively reared cod (*Gadus morhua*) using Denaturing Gradient Gel Electrophoresis (DGGE), *Aquaculture*, 272, 319–327.

© Woodhead Publishing Limited, 2013

BURMØLLE M, WEBB J S, RAO D, HANSEN L H, SØRENSEN S J and KJELLEBERG S (2006) Enhanced biofilm formation and increased resistance to antimicrobial agents and bacterial invasion are caused by synergistic interactions in multispecies biofilms, *Appl Environ Microbiol,* 72, 3916–3923.

BURRIDGE L, WEIS J S, CABELLO F, PIZARRO J and BOSTICK K (2010) Chemical use in salmon aquaculture: a review of current practices and possible environmental effects, *Aquaculture,* 306, 7–23.

CABELLO F C (2006) Heavy use of prophylactic antibiotics in aquaculture: a growing problem for human and animal health and for the environment, *Environ Microbiol,* 8, 1137–1144.

CAMPOS C A, RODRÍGUEZ O, CALO-MATA P, PRADO M and BARROS-VELÁZQUEZ J (2006) Preliminary characterization of bacteriocins from *Lactococcus lactis*, *Enterococcus faecium* and *Enterococcus mundtii* strains isolated from turbot (*Psetta maxima*), *Food Res Int,* 39, 356–364.

CANO-GOMEZ A, BOURNE D G, HALL M R, OWENS L and HØJ L (2009) Molecular identification, typing and tracking of *Vibrio harveyi* in aquaculture systems: Current methods and future prospects, *Aquaculture,* 287, 1–10.

CAO H, HE S, WEI R, DIONG M and LU L (2011) *Bacillus amyloliquefaciens* G1: a potential antagonistic bacterium against eel-pathogenic *Aeromonas hydrophila*, *J Evid Based Complementary Altern Med,* 2011, 1–7.

CAO H, HE S, WANGM H, HOU S, LU L and YANG X (2012) *Bdellovibrios*, potential biocontrol bacteria against pathogenic *Aeromonas hydrophila*, *Vet Microbiol,* 154, 413–418.

CASADEVALL A and PIROFSKI L-A (2007) Accidental virulence, cryptic pathogenesis, martians, lost hosts, and the pathogenicity of environmental microbes, *Eukaryot Cell,* 6, 2169–2174.

CHAIR M, DEHASQUE M, VAN POUCKE S, NELIS H, SORGELOOS P and DE LEENHEER A P (1994) An oral challenge for turbot larvae with *Vibrio anguillarum*, *Aquacult Int,* 2, 270–272.

CHEN R, ZHOU Z, CAO Y, BAI Y and YAO B (2010) High yield expression of an AHL-lactonase from *Bacillus* sp. B546 in Pichia pastoris and its application to reduce *Aeromonas hydrophila* mortality in aquaculture, *Microb Cell Fact,* 9, 39.

CHEN G, SWEM L R, SWEM D L, STAUFF D L, O'LOUGHLIN, C T, JEFFREY P D, BASSLER B L and HUGHSON F M (2011) A strategy for antagonizing quorum sensing, *Mol Cell,* 42, 199–209.

CHIU C S and LIU C-H (2009) Enhancement of immunity and disease resistance in the white shrimp, *Litopenaeus vannamei*, by the probiotic, *Bacillus subtilis* E20, *Fish Shellfish Immunol,* 26, 339–344.

CHIU C-H, GUU Y-K LIU C-H PAN T-Mand CHENG W (2007) Immune responses and gene expression in white shrimp, *Litopenaeus vannamei*, induced by *Lactobacillus plantarum, Fish Shellfish Immunol,* 23, 364–377.

CHU W H and LU C-P (2008) *In vivo* fish models for visualizing *Aeromonas hydrophila* invasion pathway using GFP as a biomarker, *Aquaculture,* 277, 152–155.

CHABRILLÓN M, RICO R M, ARIJO S, DÍAZ-ROSALES P, BALEBONZ M C, MORIÑIGO M A (2005) Interactions of microorganisms isolated from gilthead sea bream, *Sparus aurata* L., on *Vibrio harveyi*, a pathogen of farmed Senegalese sole, *Solea senegalensis* (Kaup). *J Fish Dis,* 28, 531–537.

CHYTHANYA R, KARUNASAGAR I and KARUNASAGAR I (2002) Inhibition of shrimp pathogenic vibrios by a marine *Pseudomonas* I-2 strain, *Aquaculture*, 208, 1–10.

CONNON S A and GIOVANNONI S J (2002) High-throughput methods for culturing microorganisms in very-low-nutrient media yield diverse new marine isolates, *Appl Environ Microbiol,* 68, 3878–3885.

© Woodhead Publishing Limited, 2013

COSTELLO E K, LAUBER C L, HAMADY M, FIERER N, GORDON J L and KNIGHT R (2009) Bacterial community variation in human body habitats across space and time, *Science,* 326, 1694–1697.

COSTERTON J W, STEWART P S and GREENBERG E P (1999) Bacterial biofilms: a common cause of persistent infections, *Science,* 284, 1318–1322.

CROTHERS-STOMPS C, HØJ L, BOURNE D, HALL M and OWENS L (2009) Isolation of lytic bacteriophage against *Vibrio harveyi, J Appl Microbiol,* 108, 1744–1750.

DANG T V C, NGUYEN V H,, DIERCKENS K, DEFOIRDT T, BOON N, SORGELOOS P and BOSSIER P (2009a) Novel approach of using homoserine lactone-degrading and poly-β-hydroxybutyrate-accumulating bacteria to protect *Artemia* from the pathogenic effects of *Vibrio harveyi, Aquaculture,* 291, 23–30.

DANG T V C, DIHN T N, CEUPPENS S, NGUYEN V H, DIERCKENS K, WILLE M, SORGELOOS P and BOSSIER P (2009b) Effect of *N*-acyl homoserine lactone degrading enrichment cultures on *Macrobrachium rosenbergii* larviculture, *Aquaculture,* 294, 5–13.

DANIELS C L, MERRIFIELD D L, BOOTHROYD D P, DAVIES S J, FACTOR J R and ARNOLD K E (2010) Effect of dietary *Bacillus* spp. and mannan oligosaccharides (MOS) on European lobster (*Homarus gammarus* L.) larvae growth performance, gut morphology and gut microbiota, *Aquaculture,* 304, 49–57.

DEFOIRDT T and SORGELOOS P (2012) Monitoring of *Vibrio harveyi* quorum sensing activity in real time during infection of brine shrimp larvae. *The ISME Journal,* doi:10.1038/ismej.2012.58.

DEFOIRDT T, BOSSIER P, SORGELOOS P and VERSTRAETE W (2005) The impact of mutations in the quorum sensing systems of *Aeromonas hydrophila, Vibrio anguillarum* and *Vibrio harveyi* on their virulence towards gnotobiotically cultured *Artemia franciscana, Environ Microbiol,* 7, 1239–1247.

DEFOIRDT T, CRAB R. WOOD T K, SORGELOOS P, VERSTRATE W and BOSSIER P (2006) Quorum sensing-disrupting brominated furanones protect the gnotobiotic brine shrimp *Artemia franciscana* from pathogenic *Vibrio harveyi , Vibrio campbelli* and *Vibrio parahaemolyticus* isolates. *Appl Environ Microbiol,* 72, 6419–6423.

DEFOIRDT T, BOON N, SORGELOOS P, VERSTRAETE W and BOSSIER P (2007) Alternatives to antibiotics to control bacterial infections: luminescent vibriosis in aquaculture as an example, *Trends Biotechnol,* 25, 472–479.

DEFOIRDT T, BOON N, SORGELOOS P, VERSTRAETE W and BOSSIER P (2008) Quorum sensing and quorum quenching in *Vibrio harveyi*: lessons learned from *in vivo* work. *The ISME Journal,* 2, 19–26. doi:10.1038/ismej.2007.92.

DEFOIRDT T, RUWandEEPIKA H A D, KARUNASAGAR I, BOON N and BOSSIER P (2010) Quorum sensing negatively regulates chitinase in *Vibrio harveyi, Environ Microbiol Rep,* 2, 44–49.

DEFOIRDT T, SORGELOOS P and BOSSIER P (2011) Alternatives to antibiotics for the control of bacterial disease in aquaculture, *Curr Opin Microbiol,* 14, 251–258.

DEGNAN S M and DEGNAN B M (2010) The initiation of metamorphosis as an ancient polyphonic trait and its role in metazoan life-cycle evolution, *Phil Trans R Soc B,* 365, 641–651.

DENEV S, STAYKOV Y, MOUTAFCHIEVA R and BEEV G (2009) Microbial ecology of the gastrointestinal tract of fish and the potential application of probiotics and prebiotics in finfish aquaculture, *Int Aquat Res,* 1, 1–29.

EGAN E S and WALDOR M K (2003) Distinct replication requirements for the two *Vibrio* cholerae chromosomes, *Cell,* 114, 521–530.

FJELLHEIM A J, PLAYFOOT K J, SKJERMO J, and VADSTEIN O (2007) Vibrionaceae dominates the microflora antagonistic towards *Listonella anguillarum* in the intestine of cultured Atlantic cod (*Gadus morhua* L.) larvae, *Aquaculture,* 269, 98–106.

FJELLHEIM A J, KLINKENBERG G, SKJERMO J, AASEN I M and VADSTEIN O (2010) Selection of candidate probionts by two different screening strategies from Atlantic cod (*Gadus morhua* L.) larvae, *Vet Microbiol,* 144, 153–159.

© Woodhead Publishing Limited, 2013

FRANS I, MICHIELS C W, BOSSIER P, WILLEMS K A, LIEVENS B and REDIERS H (2011) *Vibrio anguillarum* as a fish pathogen: virulence factors, diagnosis and prevention, *J Fish Dis,* 34, 643–661.

FU L-L, WANG Y, WU Z-C and LI W-F (2011) *In vivo* assessment for oral delivery of *Bacillus subtilis* harboring a viral protein (VP28) against white spot syndrome virus in *Litopenaeus vannamei*, *Aquaculture,* 322–323, 33–38.

FULLER R (1989) A review: probiotics in man and animals, *J Appl Bacteriol,* 66, 365–378.

FUSETANI N (2010) Antifungal peptides in marine invertebrates, *ISJ,* 7, 53–66.

GHANNOUM M and O'TOOLE G (2004) *Microbial Biofilms*. Washington, DC: ASM Press.

GIBSON G R and ROBERFROID M B (1995) Dietary modulation of the human colonic microbiota – introducing the concept of prebiotics, *J Nutr,* 125, 1401–1412.

GILLIS R J, WHITE K G, CHOI K.-H, WAGNER V E, SCHWEIZER H P and IGLEWSKI B H (2005) Molecular basis of Azithroycin-resistant *Pseudomonas aeruginosa* biofilms. *Antimicrob Agents Chemother,* 49, 3858–3867.

GOARANT C and MERIEN F (2006) Quantification of *Vibrio penaeicida*, the etiological agent of Syndrome 93 in New Caledonian shrimp, by real-time PCR using SYBR Green I chemistry, *J Microbiol Method,* 67, 27–35.

GOFFREDI S K (2010) Indigenous ectosymbiotic bacteria associated with diverse hydrothermal vent invertebrates, *Environ Microbiol Rep,* 2, 479–488.

GOLOVLEV E L (2001) Ecological strategy of bacteria: Specific nature of the problem, *Microbiology,* 70, 379–383.

GOMEZ-GIL B, HERRERA-VEGA M A, ABREU-GROBOIS F A, and ROQUE A (1998) Bioencapsulation of two different *Vibrio* species in nauplii of the brine shrimp (*Artemia franciscana*), *Appl Environ Microbiol,* 64, 2318–2322.

GOULDEN E F, HALL M R, BOURNE D G, PEREG L L and HØJ L (2012a) Pathogenicity and infection cycle of *Vibrio owensii* in larviculture of ornate spiny lobster (*Panulirus ornatus*), *Appl Environ Microbiol,* 78, 2841–2849.

GOULDEN E F, HALL M R, PEREG L L and HØJ L (2012b) Identification of an antagonistic probiotic combination protecting ornate spiny lobster *(Panulirus ornatus)* larvae against *Vibrio owensii* infection, *PLoS One,* 7, e39667.

GOULDEN E F, HALL M R, PEREG L L, BAILLIE B K and HØJ, L (2012c). Probiont niche specialization contributes to additive protection against *Vibrio owensii* in spiny lobster larvae. *Environ Microbiol Rep*. doi: 10.1111/1758-2229.12007.

GRAM L, MELCHIORSEN J and BRUHN J B (2010) Antibacterial activity of marine culturable bacteria collected from a global sampling of ocean surface waters and surface swabs of marine organisms, *J Mar Biotechnol,* 12, 439–451.

GRISEZ L, CHAIR M, SORGELOOS P and OLLEVIER F (1996) Mode of infection and spread of *Vibrio* anguillarum in turbot *Scophthalmus maximus* larvae after oral challenge through live feed, *Dis Aquat Org,* 26, 181–187.

GROSSART H-P (2010) Ecological consequences of bacterioplankton lifestyles: changes in concepts are needed, *Environ Microbiol Rep,* 2, 706–714.

GUO J-J, LIU K-F, CHENG S-H, CHANG C-I, LAY J-J, HSU Y-O, YANG J-Y and CHEN T-I (2009) Selection of probiotic bacteria for use in shrimp larviculture, *Aquacult Res,* 40, 609–618.

GUSMAO D S, SANTOS, A V, MARINI D C, RUSSO E S, PEIXOTO A M D, JUNIOR M C, BERBERT-MOLINA M A and LEMOS F J A (2007) First isolation of microorganisms from the gut diverticulum of *Aedes aegypti* (Diptera:Culicidae): new prespectives for an insect-bacteria association, *Mem Inst Oswaldo Cruz,* 102, 919–924.

HACHE R and PLANTE S (2011) The relationship between enrichment, fatty acid profiles and bacterial load in cultured rotifers (*Brachionus plicatilis* L-strain) and *Artemia* (*Artemia salina* strain Franciscana), *Aquaculture,* 311, 201–208.

HADFIELD M G (2011) Biofilms and marine invertebrate larvae: What bacteria produce that larvae use to choose settlement sites, *Ann Rev Mar Sci,* 3, 453–470.

© Woodhead Publishing Limited, 2013

HALL-STOODLEY L, COSTERTON J W and STOODLEY P (2004) Bacterial biofilms: from the natural environment to infectious diseases, *Nat Rev Microbiol,* 2, 95–108.

HAMBLY E and SUTTLE C A (2005) The viriosphere, diversity and genetic exchange within phage communities, *Curr Opin Microbiol,* 8, 444–450.

HARIKRISHNAN R, BALASUNDARAM C and HEO M-S (2011) Impact of plant products on innate and adaptive immune system of cultured finfish and shellfish, *Aquaculture,* 317, 1–15.

HAUTON C BROCKTON V and SMITH V J (2007) Changes in immune gene expression and resistance to bacterial infection in lobster (*Homarus gammarus*) post-larval VI following acute or chronic exposure to immune stimulating compounds, *Mol Immunol,* 44, 443–450.

HENKE J M and BASSLER B L (2004a) Quorum sensing regulates type III secretion in *Vibrio harveyi* and *Vibrio parahaemolyticus*, *J Bacteriol,* 186, 3794–3805.

HENKE J M and BASSLER B L (2004b) Three parallel quorum-sensing systems regulate gene expression in *Vibrio harveyi*, *J Bacteriol,* 186, 6902–6914.

HEUER O E, KRUSE H, GRAVE K, COLLIGNON P, KARUNASAGAR I and ANGULO F J (2009) Human health consequences of use of antimicrobial agents in aquaculture, *Clin Infect Dis,* 49, 1248–1253.

hjelm m, bergh ø, riaza a, nielsen j, melchiorsen j, jensen s, duncan h, ahrens p, birkbeck h, and gram l (2004) Selection and identification of autochthonous potential probiotic bacteria from turbot larvae (*Scophthalmus maximus*) rearing units, *Syst Appl Microbiol,* 27, 360–371.

HØJ L, BOURNE D G and HALL M R (2009) Localization, abundance and community structure of bacteria associated with *Artemia*: Effects of nauplii enrichment and antimicrobial treatment, *Aquaculture,* 293, 278–285.

HOOPER L V and GORDON J I (2001) Commensal host-bacterial relationships in the gut, *Science,* 292, 1115–1118.

HUNG D T, SHAKHNOVICH E A, PIERSON E and MEKALANOS J J (2005) Small-molecule inhibitor of *Vibrio cholerae* virulence and intestinal colonisation, *Science,* 310, 670–674.

IMMANUEL G, SIVAGNANAVELMURUGAN M and PALAVESAM A (2010) Antibacterial effect of medium-chain fatty acid: caprylic acid on gnotobiotic *Artemia* franciscana nauplii against shrimp pathogens *Vibrio harveyi* and *V. parahaemolyticus*, *Aquac Int,* 19, 91–101.

IRIANTO A and AUSTIN B (2002) Probiotics in aquaculture, *J Fish Dis,* 25, 633–642.

JENSEN S, ØVREÅS L, BERGH Ø and TORSVIK V (2004) Phylogenetic analysis of bacterial communities associated with larvae of the atlantic halibut propose succession from a uniform normal flora, *Syst Appl Microbiol,* 27, 728–736.

JIRAVANICHPAISAL P, PUANGLARP N, PÉTKON S, DONNUEA S, SÖDERHÄLL I and SÖDERHÄLL K (2007) Expression of immune-related genes in larval stages of the giant tiger shrimp, *Penaeus monodon*, *Fish Shellfish Immunol,* 23, 815–824.

JOHNSON C N, BARNES S, OGLE J, GRIMES D J, CHANG Y J, PEACOCK A D and KLINE L (2008) Microbial community analysis of water, foregut, and hindgut during growth of Pacific white shrimp, *Litopenaeus vannamei*, in closed-system aquaculture, *J World Aquac Soc,* 39, 251–258.

JOHNSTON M, JOHNSTON D and KNOTT B (2008) Ontogenetic changes in the structure and function of the mouthparts and foregut of early and late stage *Panulirus ornatus* (Fabricius, 1798) phyllosomata (decapoda: Palinuridae), *J Crustacean Biol,* 28, 46–56.

JURKEVITCH E (2007) Predatory behaviors in bacteria – diversity and transitions, *Microbe,* 2, 67–73.

Kalinovskaya NI, Ivanova E P, Alexeeva Y V, Gorshkova N M, Kuznetsova T A, Dmitrenok A S and NICOLAU D V (2003) Low-molecular-weight, biologically active

© Woodhead Publishing Limited, 2013

compounds from marine *Pseudoalteromonas* species, *Curr Microbiol,* 48, 441–446.

KARUNASAGAR I, OTTA S K and KARUNASAGAR I (1996) Biofilm formation by *Vibrio harveyi* on surfaces, *Aquaculture,* 140, 241–245.

KARUNASAGAR I, SHIVU M M, GIRISHA S K, KROHNE G and KARUNASAGAR I (2007) Biocontrol of pathogens in shrimp hatcheries using bacteriophages, *Aquaculture,* 268, 288–292.

KASTBJERG V G, NIELSEN K F, DALSGAARD I, RASCH M, CRUHN J B, GIVSKOV M and GRAM L (2007) Profiling acylated homoserine lactones in *Yersinia ruckeri* and influence of exogenous acyl homoserine lactones and known quorum sensing inhibitors on protease production, *J Appl Microbiol,* 102, 363–374.

KELLY B G, VESPERMANN A and BOLTON D J (2009) Horizontal gene transfer of virulence deterimants in selected bacterial foodborne pathogens, *Food Chem Toxicol,* 47, 969–977.

KESARCODI-WATSON A, KASPAR H, LATEGAN M J and GIBSON L (2008) Probiotics in aquaculture: The need, principles and mechanisms of action and screening processes, *Aquaculture,* 274, 1–14.

KESARCODI-WATSON A, KASPAR H, LATEGAN M J and GIBSON L (2010) *Alteromonas macleodii* 0444 and *Neptunomonas* sp. 0536, two novel probiotics for hatchery-reared Greenshell™ mussel larvae, *Perna canaliculus*, *Aquaculture,* 309, 49–55.

KITTAKA J (1994) 'Larval rearing'. in Phillips B F, Cobb J S and J. Kittaka J (eds) *Spiny Lobster Management.* Oxford, Fishing News Books, 402–423.

KIROV S M, WEBB J S, and KJELLEBERG S (2005) Clinical significance of seeding dispersal in biofilms, *Microbiology,* 151, 3452–3453.

KLEVENS R M, MORRISON M A, NADLE J, PETIT S, GERSHMAN K, RAY S, HARRISON L H, LYNFIELD R, DUMYATI G, TOWNES J M, CRAIG A S, ZELL E R, FOSHEIM G E, MCDOUGAL L K, CAREY R B and FRIDKIN S K (2007) Invasive methicillin-resistant *Staphylococcus aureus* infections in the United States, *J Am Med Assoc,* 298, 1763–1771.

KUNIN V, ENGELBREKTSON A, OCHMAN H and HUGENHOLTZ P (2010) Wrinkles in the rare biosphere: pyrosequencing errors can lead to artificial inflation of diversity estimates, *Environ Microbiol,* 12, 118–123.

JIRAVANICHPAISAL P, LEE SY, KIM YA, ANDREN T, SÖDERHÄLL I (2007) Antibacterial peptides in hemocytes and hematopoietic tissue from freshwater crayfish *Pacifastacus leniusculus*: characterization and expression pattern, *Dev Comp Immunol,* 31, 441–455.

JURKEVITCH E (2007) Predatory behaviors in bacteria—diversity and transitions, *Microbe,* 2, 67–73.

LAUZON H L, GUDMUNDSDOTTIR S, PETURSDOTTIR S K, REYNISSON E, STEINARSSON A, ODDGEIRSSON M, BJORNSDOTTIR R and GUDMUNDSDOTTIR B K (2010) Microbiota of Atlantic cod (*Gadus morhua* L.) rearing systems at pre- and posthatch stages and the effect of different treatments, *J Appl Microbiol,* 109, 1775–1789.

LECLERC V, PELTE N, CHAMY L E, MARTINELLI C, LIGOXYGAKIS P, HOFFRMANN J A and REICHHART J-M (2006) Prophenoloxidase activation is not required for survival to microbial infections in Drosophila, *EMBO Reports,* 7, 231–235.

LEY R E, LOZUPONE C A, HAMADY M, KNIGHT R and GORDON J I (2008) Worlds within worlds: evolution of the vertebrate gut microbiota, *Nat Rev Microbiol,* 6, 776–788.

LI J, TAN B and MAI K (2009) Dietary probiotics Bacillus OJ and isomaltooligosaccharides influence the intestine microbial populations, immune responses and resistance to white spot syndrome virus in shrimp (*Litopenaeus vannamei*), *Aquaculture,* 291, 35–40.

LILLEY B N and BASSLER B L (2000) Regulation of quorum sensing in *Vibrio harveyi* by LuxO and σ54, *Mol Microbiol,* 36, 940–954.

© Woodhead Publishing Limited, 2013

LING S H M, WANG X H, LIM T M and LEUNG K Y (2001) Green fluorescent protein-tagged *Edwardsiella tarda* reveals portal of entry in fish, *FEMS Microbiol Lett,* 194, 239–243.

LONGEON A, PEDUZZI J, BARTHÉLEMY M, CORRE S, NICOLAS J-L and GUYOT M (2004) Purification and partial identification of novel antimicrobial protein from marine bacterium *Pseudoalteromonas* species strain X153, *Mar Biotechnol,* 6, 633–641.

LÓPEZ-TORRES M A and LIZÁRRAGA-PARTIDA M L (2001) Bacteria isolated on TCBS media associated with hatched *Artemia* cysts of commercial brands, *Aquaculture,* 194, 11–20.

LUCCHETTI D, FABRIZI L, GUANDALINI E, PODESTA E, MARVASI L, ZAGHINI A and CONI E (2004) Long depletion time of enrofloxacin in rainbow trout (*Oncorhynchous mykiss*) *Antimicrob Agents Chemother,* 48, 3912–3917.

MA C-W, CHO Y-S and OH K-H (2009) Removal of pathogenic bacteria and nitrogens by *Lactobacillus* spp. JK-8 and JK-11, *Aquaculture,* 287, 266–270.

MAKRIDIS P, FJELLHEIM A J, SKJERMO J and VADSTEIN O (2000) Colonization of the gut in first feeding turbot by bacterial strains added to the water or bioencapsulated in rotifers, *Aquac Int,* 8, 367–380.

MANEFIELD M, HARRIS L, RICE S A, DE NYS R and KJELLEBERG S (2000) Inhibition of luminescence and virulence in the black tiger prawn (*Penaeus monodon*) pathogen *Vibrio harveyi* by intercellular signal antagonists, *Appl Environ Microbiol,* 66, 2079–2084.

MATOS A, BORGES M T, PEIXE C, HENRIQUES I, PEREIRA C M and CASTRO P M L (2011) A molecular and multivariate approach to the microbial community of a commercial shallow raceway marine recirculation system operating with a Moving Bed Biofilter, *Aquac Res,* 42, 1308–1322.

MATZ C, WEBB J S, SCHUPP P J, PHANG S Y and PENESYAN A (2008) Marine biofilm bacteria evade eukaryotic predation by targeted chemical defense, *PLoSONE* 3(7), e2744.

MCDOUGALD D, RICE S A and KJELLEBERG S (2007) Bacterial quorum sensing and interference by naturally occurring biomimics. *Anal Bioanal Chem,* 387, 445–453.

MCINTOSH D, JI B, FORWARD B S, PUVANENDRAN V, BOYCE D and RITCHIE R (2008) Culture-independent characterization of the bacterial populations associated with cod (*Gadus morhua* L.) and live feed at an experimental hatchery facility using denaturing gradient gel electrophoresis, *Aquaculture,* 275, 42–50.

MILTON D L (2006) Quorum sensing in vibrios: complexity for diversification, *Int J Med Microbiol,* 296, 61–71.

MOLINA-AJA A, GRACIA-GASCA A, ABRUE-GROBOIS A, BOLAN-MEJIA C, ROQUE A and GOMEZ GIL B (2002) Plasmid profiling and antibiotic resistance of *Vibrio* strains isolated from cultured penaeid shrimp, *FEMS Microbiol Lett,* 213, 7–12.

MOROHOSHI T, EBATA A, NAKAZAWA S, KATO N and IKEDA T (2005) *N*-Acyl homoserine lactone-producing or -degrading bacteria isolated from the intestinal microbial flora of Ayu fish (*Plecoglossus altivelis*), *Microbes Environ,* 20, 264–268.

MOROHOSHI T, NAKAZAWA S, EBATA A, KATO N and IKEDA T (2008) Identification and characterization of *N*-acylhomoserine lactone acylase from the fish intestinal *Shewanella* sp. strain MIB015, *Biosci Biotechnol Biochem,* 72, 1887–1893.

MUNN C (2011) *Marine Microbiology: Ecology and Applications.* New York: Garland Science.

NAKAI T and PARK S C (2002) Bacteriophage therapy of infectious diseases in aquaculture, *Res Microbiol,* 153, 13–18.

NATRAH F M, KENMEGNE M M, WIYOTO W, SOREGLOSS P, BOSSIER P and DEFOIRDT T (2011) Effects of micro-algae commonly used in aquaculture on acyl-homoserine lactone quorum sensing, *Aquaculture,* 317, 53–57.

© Woodhead Publishing Limited, 2013

NAYAK S K (2010) Role of gastrointestinal microbiota in fish, *Aquac Res*, 41, 1553–1573.

NICOLAS J L, ROBIC E and ANSQUER D (1989) Bacterial flora associated with a trophic chain consisting of microalgae, rotifers and turbot larvae: influence of bacteria on larval survival, *Aquaculture*, 83, 237–248.

O'FLAHERTY S, ROSS R P and COFFEY A (2009) Bacteriophage and their lysins for elimination of infectious bacteria, *FEMS Microbiol Rev*, 33, 801–819.

OLAFSEN J A (2001) Interactions between fish larvae and bacteria in marine aquaculture, *Aquaculture*, 200, 223–247.

OSBORN M and SMITH C (2005) *Molecular Microbial Ecology*. New York: Taylor & Francis.

O'TOOLE, R, VON HOFSTEN J, ROSQVIST R, OLSSON P.-E and WOLF-WATZ H (2004) Visualisation of Zebrafish infection by GFP-labelled *Vibrio anguillarum*, *Microb Pathog*, 37, 41–46.

OWENS L and BUSICO-SALCEDO N (2006) *Vibrio harveyi*: pretty problems in paradise, in Thompson F L, Austin B and Swings J (eds), *The Biology of Vibrios*. Washington DC: ASM Press, 266–280.

PAI S S, ANAS A, JAYAPRAKASH N S, PRIYAJA P, SREELAKSHMI B, PREETHA R, PHILIP R, MOHANDAS, A and SINGH I S B (2010) *Penaeus monodon* larvae can be protected from *Vibrio harveyi* infection by pre-emptive treatment of a rearing system with antagonistic or non-antagonistic bacterial probiotics, *Aquac Res*, 41, 847–860.

PAN J and REN D (2009) Quorum sensing inhibitors: a patent overview. *Expert Opin Therapeutic Patents*, 19, 1581–1601.

PASHARAWIPAS T, THAIKUA S, SRIURAIRATANA S, RUANGPAN L, DIREKBUSARAKUM S, MANOPVISETCHAREAN J and FLEGEL T W (2005) Partial characterization of a novel bacteriophage of *Vibrio harveyi* isolated from shrimp culture ponds in Thailand, *Virus Res*, 114, 63–69.

PASMORE M and COSTERTON J W (2003), Biofilms, bacterial signaling and their ties to marine biology, *J Ind Microbiol Biotechnol*, 30, 407–413.

PAYNE M S, HALL M R, BANNISTER R, SLY L and BOURNE D G (2006) Microbial diversity within the water column of a larval rearing system for the ornate rock lobster (*Panulirus ornatus*), *Aquaculture*, 258, 80–90.

PAYNE M S, HALL M R, SLY L and BOURNE D G (2007) Microbial diversity within early-stage cultured *Panulirus ornatus* phyllosomas, *Appl Environ Microbiol*, 73, 1940–1951.

PAYNE MS, HØJ L, WIETZ M, HALL M R, SLY L. and BOURNE D G (2008) Microbial diversity of mid-stage Palinurid phyllosoma from Great Barrier Reef waters, *J Appl Microbiol*, 105, 340–350.

PEPPELENBOSCH M P and FERREIRA C V (2009) Immunology of pre- and probiotic supplementation, *Br J Nutr*, 101, 2–4.

PERSSON T, HANSEN T H, RASMUSSEN T B, SKINDERSOE M E, GIVSKOV M, NIELSEN J (2005) Rational design and synthesis of new quorum sensing inhibitors derived from acylated homoserine lactones and natural products from garlic. *Org Biomol Chem*, 3, 253–262.

PISUTTHARACHAI D, YASUIKE M, AONO H, MURAKAMI K, KONDO H, AOKI T and HIRONO I (2009) Expressed sequence tag analysis of phyllosomas and hemocytes of Japanese spiny lobster *Panulirus japonicas*, *Fish Sci*, 75, 195–206.

PLOUG H, KUHL M, BUCHHOLZ B and JORGENSEN B B (1997) Anoxic aggregates – an ephemeral phenomenon in the pelagic environment. *Aquat Microbiol Ecol*, 13, 285–294.

POWELL A and ROWLEY A F (2007) The effect of dietary chitin supplementation on the survival and immune reactivity of the shore crab, *Carcinus maenas*, *Comp Biochem Physiol A Mol Integr Physiol*, 147, 122–128.

© Woodhead Publishing Limited, 2013

QI Z, ZHANG X-H, BOON N and BOSSIER P (2009) Probiotics in aquaculture of China – current state, problems and prospect, *Aquaculture,* 290, 15–21.

RASCH M, BUCH C, AUSTIN B, SLIERENDRECHT W J, EKMANN K S, LARSEN J L, JOHANSEN C, RIEDEL K, EBERL L, GIVSKOV M and GRAM L (2004) An inhibitor of bacterial quorum sensing reduces mortalities caused by vibriosis in rainbow trout (*Onchorhynchus mykiss, Walbaum*), *Syst Appl Microbiol*, 27, 350–359.

RASMUSSEN R S and MORRISSEY M T (2007) Biotechnology in aquaculture: transgenics and polyploidy, *Compr Rev Food Sci,* 6, 2–16.

RATHER M A, SHARMA R, AKLAKUR M, AHMAD S, KUMAR N, KHAN M and RAMYA VL (2011) Nanotechnology: A novel tool for aquaculture and fisheries development: A prospective mini-review, *Fish Aquac J* 2011, FAJ-16.

RATCLIFFE N A, ROWLEY A F, FITZGERALD S W and RHODES C P (1985) Invertebrate immunity: basic concepts and recent advances, *Int Rev Cytol,* 97, 183–350.

RATTANCHAI A, HIRONO I, OHIRA T, TAKAHASHI Y and AOKI T (2005) Peptidoglycan inducible expression of a serine proteinase homologue from kuruma shrimp (*Marsupenaeus japonicus*), *Fish Shellfish Immunol,* 18, 39–48.

RATTANACHUAY P, KANTACHOTE D, TANTIRUNGKIJ M, NITODA T and KANZAKI H (2010) Inhibition of shrimp pathogenic vibrios by extracellular compounds from a proteolytic bacterium *Pseudomonas* sp. W3 Electron, *J Biotechnol,* 13, 1–11.

RAVI A V, MUSTHAFA K S, JEGATHAMMBAL G, KATHIRESAN K and PandIAN S K (2007) Screening and evaluation of probiotics as a biocontrol agent against pathogenic *Vibrios* in marine aquaculture, *Lett Appl Microbiol,* 45, 219–223.

REGIER J C, SHULTZ J W, ZWICK A, HUSSEY A, BALL B, WETZER R, MARTIN J W and CUNNINGHAM C W (2010) Arthropod relationships revealed by phylogenomic analysis of nuclear protein-coding sequences, *Nature,* 463, 1079–1083.

REID H I, TREASURER J W, ADAM B and BIRKBECK T H (2009) Analysis of bacterial populations in the gut of developing cod larvae and identification of *Vibrio logie*, *Vibrio anguillarum* and *Vibrio splendidus* as pathogens of cod larvae, *Aquaculture,* 288, 36–43.

REIMSCHUESSEL R and MILLER R A (2006) Antimicrobial drug use in aquaculture, in Giguere S, Prescott J F, Baggot J D, Walker R D and Dowling P M (eds), *Antimicrobial Therapy in Veterinary Medicine*. Oxford: Blackwell Publishing, 593–606.

RICHMOND A (2004) *Handbook of Microalgal Culture: Biotechnology and Applied Phycology*. Oxford: Blackwell Publishing.

RINGØ E (2008) The ability of carnobacteria isolated from fish intestine to inhibit growth of fish pathogenic bacteria: a screening study, *Aquac Res,* 39, 171–180.

RINGØ E, OLSEN R E, GIFSTAD T Ø, DALMO R A, AMLUND H, HEMRE G-I and BAKKE A M (2010) Prebiotics in aquaculture: a review, *Aquac Nutr,* 16, 117–136.

RIQUELME C E, JORQUERA M A, ROJAS A I, AVENDAÑO R E and REYES N (2001) Addition of inhibitor-producing bacteria to mass cultures of *Argopecten purpuratus* larvae (Lamarck, 1819), *Aquaculture,* 192, 111–119.

ROWLEY A F and POWELL A (2007) Invertebrate immune systems – specific, quasi-specific or nonspecific? *The Journal of Immunology,* 179, 7209–7214.

RUIZ-PONTE C, SAMAIN J F, SÁNCHEZ J L and NICOLAS J L (1999) The benefit of a *Roseobacter* species on the survival of scallop larvae, *Mar Biotechnol*, 1, 52–59.

SandAA R A, MAGNESEN T, TORKILDSEN L and BERGH O (2003) Characterisation of the bacterial community associated with early stages of great scallop (*Pecten maximus*), using denaturing gradient gel electrophoresis (DGGE), *Syst Appl Microbiol,* 26, 302–311.

SAUER K, CAMPER A K, EHRLICH G D, COSTERTON J W and DAVIES D G (2002) *Pseudomonas aeruginosa* displays multiple phenotypes during development as a biofilm, *J Bacteriol*, 184, 1140–1154.

© Woodhead Publishing Limited, 2013

SAULNIER D, HAFFNER P, GOARANT C, LEVY P and ANSQUER D (2000) Experimental infection models for shrimp vibriosis studies: a review, *Aquaculture,* 191, 133–144.

SAWABE T, INOUE S, FUKU Y, YOSHIE K, NISHIHARA, Y and MIURA H (2007) Mass mortality of Japanese abalone *Haliotis discus hannai* caused by *Vibrio harveyi* infection, *Microbes Environ,* 22, 300–308.

SCHREIER H J, MIRZOYAN N and SAITO K (2010) Microbial diversity of biological filters in recirculating aquaculture systems, *Curr Opin Biotechnol,* 21, 318–325.

SHARIFF M, YUSOFF F M, DEVARAJA T N and SRINIVASA RAO P S (2001) The effectiveness of a commercial microbial product in poorly prepared tiger shrimp, *Penaeus monodon* (Fabricius), ponds, *Aquac Res,* 32, 181–187.

SHIODA K, IGARASHI M A and KITTAKA J (1997) Control of water quality in the culture of early-stage phyllosomas of *Panulirus japonicas, Bull Mar Sci,* 61, 177–189.

SHIVU M M, RAJEEVA B C, GIRISHA S K, KARUNASAGAR I, KROHNE G and KARUNASAGAR I (2007) Molecular characterization of *Vibrio harveyi* bacteriophage isolated from aquaculture environments along the coast of India, *Environ Microbiol,* 9, 322–331.

SKJERMO J and VADSTEIN O (1993) Characterization of the bacterial flora of mass cultivated *Brachionus plicatilis, Hydrobiologia,* 255/256, 185–191.

SMITH V J, BROWN J H and HAUTON C (2003) Immunostimulation in crustaceans: does it protect against infection? *Fish Shellfish Immunol,* 15, 71–90.

SØRUM H (2006) Antimicrobial drug resistance in fish pathogens, in Aarestrup F (ed.) *Antimicrobial Resistance in Bacteria of Animal Origin.* Washington DC: ASM Press, 213–238.

SØRUM H and L'ABÉE-LUND T M (2002) Antibiotic resistance in food-related bacteria – a result of interfering with the global web of bacterial genetics, *Int J Food Microbiol,* 78, 43–56.

SOTO-RODRIGUEZ S A, ROQUE A, LIZARRAGA-PARTIDA M L, GUERRA-FLORES A L and GOMEZ-GIL B (2003) Virulence of luminous vibrios to *Artemia franciscana* nauplii, *Dis Aquat Org,* 53, 231–240.

SPOR A, KOREN O and LEY R (2011) Unravelling the effects of the environment and host genotype on the gut microbiome, *Nature Reviews Microbiology,* 9, 279–290.

SUGITA H, HIROSE Y, MATSUO N and DEGUCHI Y (1998) Production of the antibacterial substance by *Bacillus sp.* strain NM 12, an intestinal bacterium of Japanese coastal fish, *Aquaculture,* 165, 269–280.

SUGITA H, MIZUKI H and ITOI S (2012) Diversity of siderophore-producing bacteria isolated from the intestinal tracts of fish along the Japanese coast, *Aquac Res,* 43, 481–488.

SWAIN S M, SINGH C and ARUL V (2009) Inhibitory activity of probiotics *Streptococcus phocae* PI80 and *Enterococcus faecium* MC13 against Vibriosis in shrimp *Penaeus monodon, World J Microbiol Biotechnol,* 25, 697–703.

TANG K W (2005) Copepods as microbial hotspots in the ocean: effects of host feeding activities on attached bacteria, *Aquat Microb Ecol,* 38, 31–40.

TANG K W, TURK V and GROSSART H-P (2010) Linkage between crustacean zooplankton and aquatic bacteria, *Aquat Microb Ecol,* 61, 261–277.

TANG K W, DZIALLAS C and GROSSART H-P (2011) Zooplankton and aggregates as refuge for aquatic bacteria: protection from UV, heat and ozone stresses used for water treatment, *Environ Microbiol,* 13, 378–390.

TANJI Y, SHIMADA T, FUKUDOMI H, MIYANAGA K, NAKAI Y and UNNO H (2005) Therapeutic use of phage cocktail for controlling *Escherichia coli* O157:H7 in gastrointestinal tract of mice, *J Biosci Bioeng,* 100, 280–287.

TEASDALE M E, DONOVAN K A, FORSCHNER-DANCAUSE S R and ROWLEY D C (2011) Gram-positive marine bacteria as a potential resource for the discovery of quorum sensing inhibitors, *Mar Biotechnol,* 13, 722–732.

© Woodhead Publishing Limited, 2013

THOMAS T, EVANS F F, SCHLEHECK D, MAI-PROCHNOW A, BURKE C, PENESYAN A, DALISAY D S, STELZER-BRAID S, SAUNDERS N, JOHNSON J, FERRIERA S, KJELLEBERG, S and EGAN S (2008) Analysis of the *Pseudoalteromonas tunicata* genome reveals properties of a surface-associated life style in the marine environment, *PLoS One,* 3, e3252.

THOMPSON J, GREGORY S, PLUMMER S, SHIELDS R J and ROWLEY A F (2010) An *in vitro* and *in vivo* assessment of the potential of *Vibrio* spp. as probiotics for the Pacific White shrimp *Litopenaeus vannamei*, *J Appl Microbiol,* 109, 1177–1187.

THOMSON R, MACPHERSON H L, RIAZA A and BIRKBECK T H (2005) *Vibrio splendidus* biotype 1 as a cause of mortalities in hatchery-reared larval turbot, *Scophthalmus maximus* (L.), *J Appl Microbiol,* 99, 243–250.

TINCU J A and TAYLOR S W (2004) Antimicrobial peptides from marine invertebrates, *Antimicrob Agents Chemother,* 48, 3645–3654.

TINH N T N, GUNASEKARA R A Y S A, BOON N, DIERCKENS K, SORGELOOS P and BOSSIER P (2007) *N*-acyl homoserine lactone-degrading microbial enrichment cultures isolated from *Penaeus vannamei* shrimp gut and their probiotic properties in *Brachionus plicatilis* cultures, *FEMS Microbiol Ecol,* 62, 45–53.

TINH N T N, DIERCKENS K, SORGELOOS P and BOSSIER P (2008) A review of the functionality of probiotics in the larviculture food chain, *Mar Biotechnol,* 10, 1–12.

TRAVERS M-A, BARBOU A, LE GOÏC N, HUCHETTE S, PAILLARD C and KOKEN M (2008) Construction of a stable GFP-tagged *Vibrio harveyi* strain for bacterial dynamics analysis of abalone infection, *FEMS Microbiol Lett,* 289, 34–40.

TUOHY K M, ROUZAUD G C M, BRUCK W M and GIBSON G R (2005) Modulation of the human gut microflora towards improved health using prebiotics: assessment of efficacy, *Curr Pharm Des,* 11, 75–90.

TUROVSKIY Y, KASHTANOV D, PASKHOVER B and CHIKINDAS M L (2007) Quorum sensing: Fact, fiction, and everything in between, *Adv Appl Microbiol*, 62, 191–234.

VASEEHARAN B and RAMASAMY P (2003) Abundance of potentially pathogenic microorganisms in *Penaeus monodon* larvae rearing systems in India, *Microbiol Res,* 158, 299–308.

VERSCHUERE L, ROMBAUT G, SORGELOOS P and VERSTRAETE W (2000) Probiotic bacteria as biological control agents in aquaculture, *Microbiol Mol Biol Rev,* 64, 655–671.

VIJAYAN K K, SINGH I S B, JAYAPRAKASH N S, ALAVandi S V, PAI SS, PREETHA R, RAJAN J J S and SANTIAGO T C (2006) A brackishwater isolate of *Pseudomonas* PS-102, a potential antagonistic bacterium against pathogenic vibrios in penaeid and non-penaeid rearing systems, *Aquaculture,* 251, 192–200.

VILLAMIL L, FIGUERAS A, PLANAS M and NOVOA B (2003) Control of *Vibrio alginolyticus* in *Artemia* culture by treatment with bacterial probiotics, *Aquaculture,* 219, 43–56.

VINE N G, LEUKES W D, KAISER H, DAYA S, BAXTER J and HECHT T (2004) Competition for attachment of aquaculture candidate probiotic and pathogenic bacteria on fish intestinal mucus, *J Fish Dis,* 27, 319–326.

VINE N G, LEUKES W D and KAISER H (2006) Probiotics in marine larviculture, *FEMS Microbiol Rev,* 30, 404–427.

VINOD M G, SHIVU M M, UMESHA K R, RAJEEVA B C, KROHNE G, KARUNASAGAR I and KARUNASAGAR I (2006) Isolation of *Vibrio harveyi* bacteriophage with a potential for biocontrol of luminous vibriosis in hatchery environments, *Aquaculture,* 255, 117–124.

WANG G, LI X and WANG Z (2009) APD2: the updated antimicrobial peptide database and its application in peptide design, *Nucleic Acids Res,* 37, D933-D937.

WALLING E, VOUREY E, ANSQUER D, BELIAEFF B and GOARANT C (2010) *Vibrio nigripulchritudo* monitoring and strain dynamics in shrimp pond sediments, *J Appl Microbiol*, 108, 2003–2011.

WATERS C M and BASSLER B L (2005) Quorum sensing: cell-to-cell communication in bacteria, *Annu Rev Cell Dev Biol*, 21, 319–346.

© Woodhead Publishing Limited, 2013

WEBER J T, MINTZ E D, CAÑIZARES R, SEMIGLIA A, GOMEZ I, SEMPÉRTEGUI R, DÁVILA A, GREENE K D, PUHR N D, CAMERON D N, TENOVER F C, BARRETT T J, BEAN N H, IVEY C, TAUXE R V and BLAKE P A (1994) Epidemic cholera in Ecuador: multidrug-resistance and transmission by water and seafood, *Epidemiol Infect,* 112, 1–11.

WIETZ M, MANSSON M, GOTFREDSEN C H, LARSEN T O and GRAM L (2010) Antibacterial compounds from marine Vibrionaceae isolated on a global expedition, *Mar Drugs,* 8, 2946–2960.

WILLING B P, RUSSELL S L and FINLAY B B (2011) Shifting the balance: antibiotic effects on host–microbiota mutualism, *Nat Rev Microbiol,* 9, 233–243.

WINZER K, HARDIE K R and WILLIAMS P (2002) Bacterial cell-to-cell communication: Sorry, can't talk now – gone to lunch!, *Curr Opin Microbiol,* 5, 216–222.

WU H-J, WANG A H J and JENNINGS M P (2008) Discovery of virulence factors of pathogenic bacteria, *Curr Opin Chem Biol,* 12, 93–101.

YAMAICHI Y, DUIGOU S, SHAKHNOVICH E A and WALDOR M K (2009) Targeting the replication initiator of the second Vibrio chromosome: towards generation of Vibrionaceae-specific antimicrobial agents, *PLoS Pathog,* 5, e1000663.

ZHOU Q, LI K, JUN X and BO L (2009) Role and functions of beneficial microorganisms in sustainable aquaculture, *Bioresour Technol,* 100, 3780–3786.

© Woodhead Publishing Limited, 2013

Part II

Closing the life-cycle and overcoming challenges in hatchery production for selected invertebrate species

9

Palinurid lobster larval rearing for closed-cycle hatchery production

M. R. Hall, M. Kenway, M. Salmon, D. Francis, E. F. Goulden and L. Høj, Australian Institute of Marine Science (AIMS), Australia

DOI: 10.1533/9780857097460.2.289

Abstract: Closed life-cycle breeding of aquaculture species is essential for sustainability. The primary bottleneck towards this goal is a robust commercial-scale hatchery technology. The larval phase of Palinurid lobsters is amongst the lengthiest of any marine invertebrates; hence a major leap forward in aquaculture hatchery technology is required for commercial-scale production. The main challenges for Palinurid hatchery technology development are outlined together including aspects of water quality and tank design. The larval biology of Palinurid lobsters is discussed as well as broodstock husbandry and spawning. A concise review of reported diseases is presented together with larval nutrition requirements and their relationship to final larval metamorphosis to juvenile.

Key words: palinurid lobsters, larval rearing, hatchery technology, health, nutrition.

9.1 Introduction

Lobsters of the sub-order Macrura Reptantia include the clawed (Astacidea) and clawless (Achelata) lobsters, and are amongst the most highly valued seafood groups in terms of demand and price in all parts of the world (Chan, 2010). The Astacidea (Neprhropoidea) and Achelata (Palinuroidea) represent approximately 57 % and 43 %, respectively, of global lobster production, with the vast majority of this from wild harvest. For the Achelata, the world landings from wild fisheries have largely plateaued with only a marginal increase in landings of approximately 14 % for the past 20 years (FAO, 2008). Wild populations are either at their maximum sustainable harvest levels, where there are fishery management regimes in place, or largely overfished, where they are lacking. Importantly, these stagnating or declining harvests of Palinurid lobsters are against a backdrop of consistently increasing demand, which is likely to grow very significantly over the coming

© Woodhead Publishing Limited, 2013

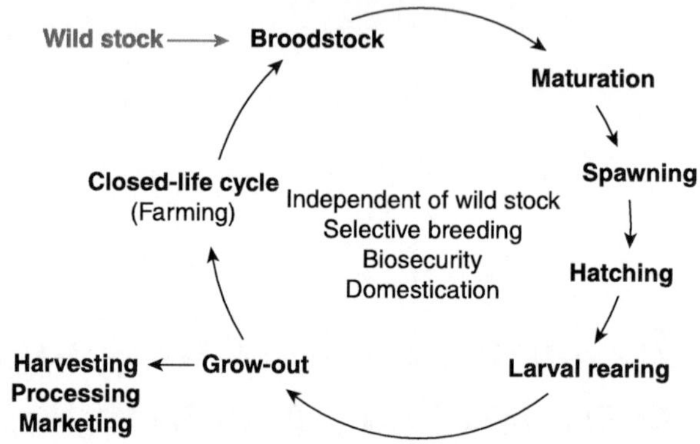

Fig. 9.1 Model of closed-life cycle aquaculture production system.

decades with China's rapid economic development and expanding middle classes (Kharas, 2010). Demand for Palinurid lobsters, particularly *Panulirus ornatus*, is predicted to explode (York and Gossard, 2004; Hart, 2009).

With the demand for Palinurid lobsters already high, and a favourite of seafood connoisseurs, wild populations are increasingly incapable of natural replenishment at a sufficient rate to withstand the persistent onslaught of harvesting effort. It is clear that an alternative production supply independent of wild harvest is required by means of closed-life-cycle aquaculture to meet increasing demand in a sustainable manner. Closed aquaculture production includes the husbandry of broodstock, the induction of spawning, egg hatching, larval rearing in specialised hatcheries, followed by grow-out to market size in nurseries and farms (Fig. 9.1).

The greatest bottleneck to closed-cycle aquaculture production of Palinurid lobsters is the lack of a robust commercially viable hatchery technology, which is primarily due to the unusual nature of their larval cycle (Section 9.1.3). The development of such technology is arguably one of the most challenging engineering and biological endeavours in aquaculture.

9.1.1 Palinurid lobsters as aquaculture candidates

Unlike the clawed lobsters, the clawless spiny and rock lobsters have many of the prerequisites and favourable attributes that are required for successful aquaculture production. Economically, they have very high value with good market size and demand. Most importantly, the most lucrative market is for live product, so the whole product is sold as is and the producer obtains the maximum price without any value-added investment. Biologically, they are omnivores with a reasonably low protein requirement (Smith *et al.*, 2009a). They appear to be fairly hardy with low mortality during grow-out if conditions are of good quality, and they readily breed in captivity with high fecundity (Thuy *et al.*, 2009). Most importantly, most species

© Woodhead Publishing Limited, 2013

are highly gregarious and can be grown at high density (Phillips and Melville-Smith, 2006). Many species of spiny and rock lobsters have favourable growth rates, which is particularly true of several species of the tropical *Panulirus* and *Palinurus* genera, but less so for the cold water species of the genus *Jasus* (Hooker *et al.*, 1997; Hart, 2009). Based on experiences from the *P. ornatus* aquaculture sector in Vietnam, this species is readily grown out in captivity to market size within a matter of months (Hart, 2009).

The present aquaculture sector for Palinurid lobsters is solely open-life-cycle production, based on capture of puerulus and juveniles from the wild, and primarily focused on the ornate rock lobster *P. ornatus* (Kenway *et al.*, 2009). 'Hold-and-fatten' facilities operated in Taiwan between 1985 and 2001 (Long Diann Marine Bio Technology Co. Ltd.) sourcing *P. ornatus* from the Philippines. Similarly, ventures in the Philippines, Singapore, Cuba and India have boomed and then collapsed (Phillips and Matsuda, 2011). In Vietnam, puerulus and early stage juveniles, with an optimal size of 4–6 cm, are caught in the wild by divers or using collectors and traps, and subsequently grown out in captivity for various periods before being shipped to market. As such, this open-life-cycle sector is still entirely dependent upon wild stocks and, although the industry is largely based on capture before the natural attrition that occurs between juvenile to adults, this sector is unsustainable in the long term and likely to be another classical 'boom and bust' endeavour. For example, the *P. ornatus* aquaculture industry in Vietnam started in 1992 as a high risk, high return industry. The industry peaked in 2005 with approximately 4000 mt of aquacultured lobster produced, with a value of approximately US$65 million. By 2007, it had declined to 1400 mt due to disease outbreaks and decreasing availability of wild juveniles. Other countries beginning to develop ranching and 'hold-and-fatten' lobster aquaculture ventures include Indonesia (primarily *P. homarus* and *P. penicillatus*), the Caribbean (*P. argus*) and Mexico (*P. interruptus*) with financial backing from China and Taiwan. As discussed above, as these ventures are based on the harvest of wild populations, the sustainable attributes of these sectors are questionable.

9.1.2 Overview of larval stages in decapoda

The larvae of decapod crustaceans are typically free-swimming planktonic forms that hatch from the egg, with larval development considered to be a continuum extending from hatching through a series of morphological stages and instars that vary according to species. The larvae periodically moult through a series of discrete developmental morphological stages until final transformational metamorphosis where the larvae enter a post-larval stage. There are four main larval forms which may be recognised, with variations in many, and these include (i) nauplius, (ii) prezoeal, protozea or nauplisoma, (iii) zoea or phyllosoma and (iv) megalopa, glaucothoe, mysis, nisto or puerulus.

© Woodhead Publishing Limited, 2013

The first and simplest form is the nauplius, which has been described as a precociously developing head with antennae, consisting of an oval cephalothorax, an elongated trunk and abdomen terminating in a caudal forked tail with setae (Foxon, 1936). Feeding, where it occurs, is through the rudimentary development of maxillule, maxilla and the first two pairs of maxillipeds. Locomotion is through the motion of the antennules and antennae. The free-swimming naupliar form is found in the widely aquacultured family of the Penaeidae (sub-order Dendrobranchiata). In Achelata (Palinurid) lobsters the naupliar stage is restricted to within the egg.

The second stage (prezoeal, protozea or nauplisoma) has been classified as an aberrant, premature hatched stage. Although it is considered a regular stage in larval development, it is not always observed. Where it does take place, it occurs just prior to the zoeal stage, as the name suggests (Gurney, 1942). The prezoeal stage has been observed in various crustaceans, but often only under certain environmental conditions. For example, in Atlantic blue crab (*Callinectes sapidus*) the prezoeal stage is observed when eggs are cooled for storage, when eggs are in high density, after exposure to decreased salinity, and after fungal and bacterial infections (Gore, 1985). The prezoeal stage is a transient and non-locomotory stage. The nauplisoma is a type of prezoeal larvae, but is an unusual prepelagic larval stage and only lasts for a few hours (Von Bonde, 1936; Crosnier, 1972). The prezoeal naupilosoma stage has been observed in Achelata lobsters, including *Jasus* and *Scyllarides* (Von Bonde, 1936; Crosnier, 1972).

The third stage (zoea or phyllosoma) represents the majority of the time in larval development of many crustaceans, and is characterised by successive moults and growth towards metamorphosis. This stage is characterised by a distinct cephalothorax and abdomen and eight pairs of appendages with buds of six more. Many penaeids hatch at the zoea stage. Locomotion is through thoracic appendages including long exopodites, which are often coated in fine setae on the three pairs of maxillipeds and pereiopods. Feeding anatomical morphology in the zoeal stage is variable, but involves mouthparts comprising a labrum, mandible, paragnath and maxillule. In the Achelata, the zoea phase is characterised by a greatly modified larval form termed a phyllosoma (from Greek: leaf-like).

The final larval stage in the Decapoda is represented by several morphological forms including the megalopa in crabs (Brachyura), the glaucothoe in hermit crabs (Paguroidea) and the mysis which is primarily found in shrimps and prawns (Penaeidae) but also in clawed lobsters (Nephropidae). Locomotion is through the abdominal pleopod movement and abdominal flexing. In the infraorder Achelata (Palinuridae) the post-larval moult is a metamorphic moult transforming into a puerulus, in the case of the spiny lobsters (Palinuridae), or into a nisto, as is the case of the slipper lobsters (Scyllaridae). Although the post-larval form resembles that of the adult form, the puerulus and nisto are believed to be a non-feeding planktonic form until benthic settlement.

© Woodhead Publishing Limited, 2013

9.1.3 The phyllosoma larvae

The larval development of Palinurid lobsters includes the main larval forms of decapods as described above; however, the naupliar stage and, when present, the prezoeal naupilosoma stage, both occur within the egg. Upon hatching, the zoeal stage, in the form of a phyllosoma, is characterised by a series of moults with associated growth that encompass the greatest proportion of the larval period. The final larval stage is represented by the puerulus stage. The uniqueness of larval development in the Palinurid lobsters lies in the total length of the entire larval period, which makes them different from all other crustaceans. The phyllosoma stage is in fact one of the longest larval stages of marine invertebrates (Thorson, 1950). For example, the larval development period is estimated to be between 12 and 22 months in the cold water species *Jasus edwardsii* (Booth, 2006), while the larval period is estimated to be 9–11 months for the sub-tropical *P. cygnus* (Phillips and Melville-Smith, 2006), and 4 to 7 months for the tropical *P. ornatus* (Dennis *et al.*, 2001). In stark contrast, penaeid prawns, which form the basis of a highly successful commercial crustacean aquaculture sector, have larval periods of between 10 and 30 days (Dall *et al.*, 1990). Other commercially viable crustacean sectors include the freshwater clawed lobsters (*Cherax*), which have a short maternal but no planktonic phase, and crabs, which have a larval phase of a few weeks (Fig. 9.2). Because of these lengthy time lines, the development of hatchery technology for Palinurid larvae focuses on minimising the attrition rate during larval rearing.

The leaf-like phyllosoma larval form is extremely thin, flat and transparent, and is restricted to the Palinuridae, Scyllaridae and Synaxidae. The phyllosoma larval primary morphological characteristics include flattened shape, transparency, long legs and eyes on long eyestalks. There are between 8 and 11 larval stages and the metamorphic transformation into the

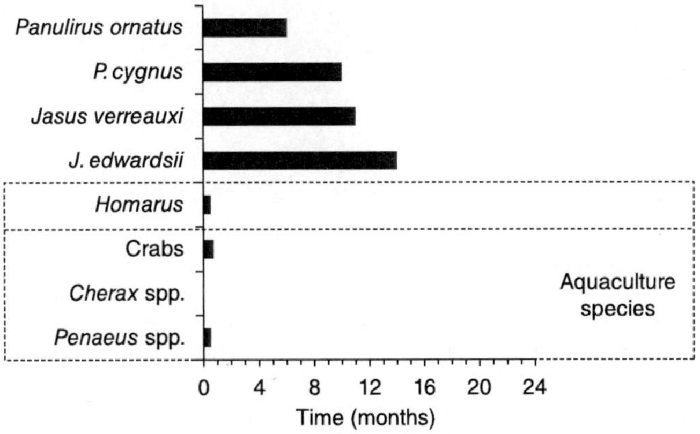

Fig. 9.2 Duration of the maternal (egg bearing) and planktonic larval phase in selected crustaceans.

© Woodhead Publishing Limited, 2013

puerulus stage represents perhaps the most profound transformation at a single moult in the Decapoda (Gurney, 1942) (Fig. 9.3). In addition to the 8 to 11 stages of morphological moults, there are often many additional supplemental moults (instars), the accumulation of which can readily double the number of moults (and increase the time) to complete the larval phase (Smith *et al.*, 2009b). Locomotion is primarily through movement of the cephalothoracic maxillipeds and associated natatorial setae on the well-developed and numerous exopods. Locomotion in the puerulus stage is through the abdominal pleopods and in particular by a rapid abdominal flex. The mouthparts of phyllosomas are characterised by a dorsoventrally flattened mandible located between a large labrum and paired paragnaths. More distally located from the mouth itself are the maxilla and maxillipeds. The puerulus stage is characterised by being a non-feeding stage. The larva

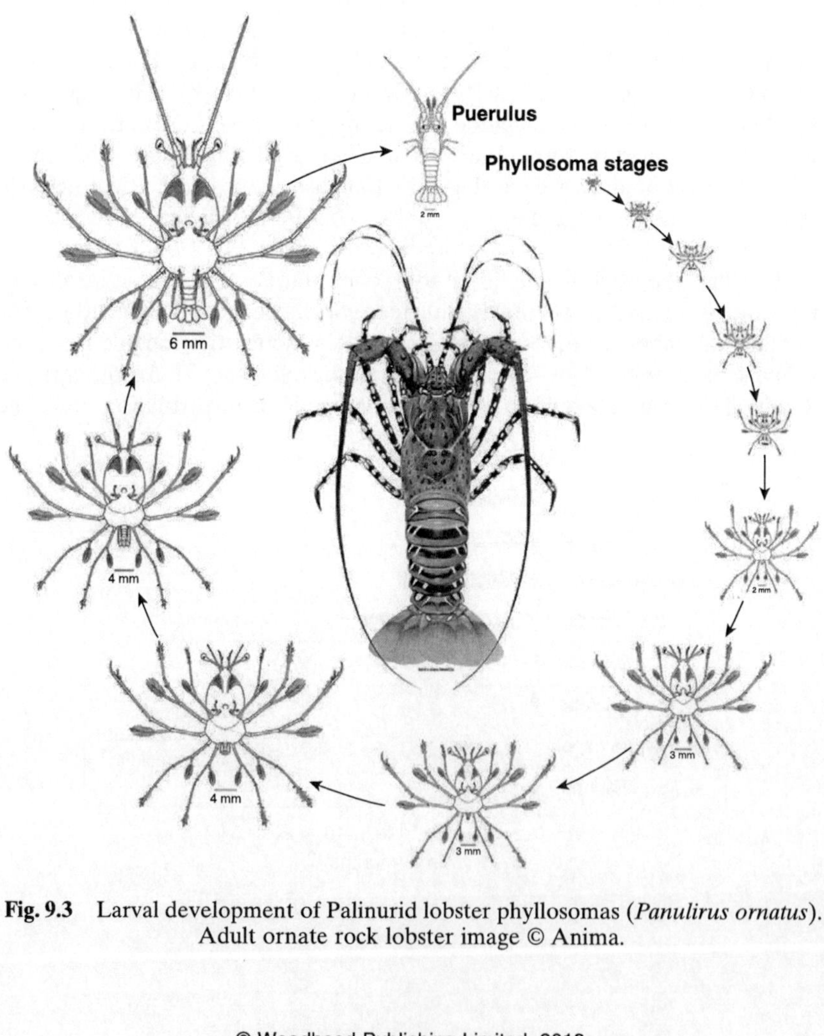

Fig. 9.3 Larval development of Palinurid lobster phyllosomas (*Panulirus ornatus*). Adult ornate rock lobster image © Anima.

© Woodhead Publishing Limited, 2013

will not eat again until benthic settlement and transformation to juvenile. Once settlement has occurred, the transparent chitin base carapace calcifies due to deposition of calcium carbonate (calcite) with the incorporation of pigmentation. Feeding resumes with the post-puerulus moult to juvenile. However, this scenario is largely based on the puerulus lacking a cardio-pyloric valve that divides the anterior cardio chamber from the posterior chamber of the proventriculus and the lack of a gastric mill (Johnston *et al.*, 2008). In contrast, the nisto does have a proventriculus that is at least partially developed. It is possible that puerulus feeding may take place if the food is primarily masticated by the mouthparts without further processing before ingestion. In any case, upon settlement, when the planktonic phase is terminated and replaced by a benthic existence after the moult into the first juvenile stage, feeding is resumed.

9.2 Development of hatchery technology for palinurid larvae and broodstock husbandry

Attempts to rear larvae of Palinurids have spanned three centuries. In 1898 larval rearing was attempted by Hattori and Oishi in Japan (Hattori and Oishi, 1899). Other attempts were made through the 1900s in Japan and other countries extending through to today, and these have been reported in the scientific literature (Table 9.1). There has been some private sector investment for the development of hatchery technologies for Palinurid lobsters, but it is impossible to verify the validity of any progress of these efforts and subsequent claims due to commercial confidentiality. Nevertheless, there have been reports in the general media of some success in the completion of the larval cycle in captivity. Whatever the case, there are as yet no established commercial-scale hatcheries for Palinurid spiny or rock lobsters.

Significant advances have been made in completing the larval cycle under hatchery conditions for Palinurid lobsters. The larval culture requirements for spiny lobsters, including *J. edwardsii, P. japonicus, P. ornatus* and others, are currently under investigation (for review see Jones, 2009; Phillips and Matsuda 2011). Complete larval development to metamorphosis has been reported in at least eight species: *P. japonicus, P. longipes, P. homarus, P. penicillatus, P. argus* (Kittaka and Booth, 2000; Goldstein *et al.*, 2008) *J. edwardsii, Sagmariasus* (*Jasus*) *verreauxi* (Kittaka *et al.*, 1997) and *P. ornatus* (Smith *et al.*, 2009b). The most extensive effort has occurred in Japan where several species have had their larval cycle completed in captivity but with a primary focus on *P. japonicus* (Matsuda and Takenouchi, 2007). From the first completed larval rearing attempt in 1988, Japanese researchers can now regularly produce several hundred juveniles per year (Matsuda and Takenouchi, 2007). Although this is far from commercial scale, it nevertheless demonstrates clear progress towards the final hurdles

© Woodhead Publishing Limited, 2013

Table 9.1 Species of Palinurid lobsters for which larval rearing research has been published

Species	Reference
Jasus	
*J. edwardsii**	Igarashi *et al.*, 1990; Illingworth *et al.*, 1997; Moss *et al.*, 1999; Kittaka and Booth, 2000; Ritar, 2001; Ritar and Smith, 2005
J. frontalis	Dupre and Guisado, 1996
*J. lalandii**	Kittaka and Ikegami, 1988
*J. verreauxi**	Kittaka *et al.*, 1997; Moss *et al.*, 2000; Ritar *et al.*, 2006
Panulirus	
*P. argus**	Moe, 1991; Goldstein *et al.*, 2008
P. echinatus	Carvalho and Ogawa, 2000
P. interruptus	Johnston, 1956; Dexter, 1972
*P. japonicus**	Oshima, 1936; Inoue, 1978; Inoue, 1965; Kittaka and Ikegami, 1988; Yamakawa *et al.*, 1989; Kittaka and Kimura, 1990; Souza *et al.*, 1996; Sekine *et al.*, 2000
P. homarus	Radhakrishnan and Vijayakumaran, 2000; Radhakrishnan, 1977
P. laevicauda	Carvalho and Ogawa, 2000
*P. longipes**	Matsuda and Yamakawa, 2000
*P. ornatus**	Smith *et al.*, 2009a
*P. penicillatus**	Matsuda *et al.*, 2006
P. polyphagus	Saisho, 1990; Sin, 1967
P. regius	Luis and Calado, 2009
P. stimpsoni	Wei and Lai, 2000; Chang-Sheng *et al.*, 2001
Palinurus	
*P. elephas**	Mercer *et al.*, 1997; Kittaka and Ikegami, 1988; Kittaka *et al.*, 2001

Note: Those marked with * have had their entire larval cycle completed in captivity.

for mass rearing. Research efforts elsewhere include New Zealand, where successful larval rearing of *J. edwardsii* and *S. verreauxi* was reported in the mid-1990s, but research has largely ceased (Booth, 1995). In Australia, research has focused on the cold water species of *Jasus*, including *J. edwardsii* and *S. Verreauxi*, and the tropical species *P. ornatus* (Phillips and Matsuda, 2011). It is believed that more recent initiatives on these and other species, primarily in the private sector, are being explored at various locations around the world.

The state-of-the-art in larval rearing of phyllosomas is commercially acceptable survival rates up to mid- to late-stage development (Phillips and Matsuda, 2011). At this stage, phyllosoma attrition rate is sufficiently high to result in low survival rate at the point of metamorphosis, hence final production is below commercially viable benchmarks. Research is focused on the two major pillars of health and nutrition. With incremental improvements in these key areas there is the expectation that puerulus and juvenile production will reach acceptable levels to make the closed-life-cycle production of Palinurid lobsters a commercial reality.

© Woodhead Publishing Limited, 2013

9.2.1 Broodstock husbandry and spawning

Palinurid lobster broodstock to supply larvae for hatcheries are primarily sourced from the wild from pre-breeding or breeding populations. However, second generation 'domesticated' broodstock have been produced for breeding programmes (Hall, unpubl.). Larval supply is typically secured by collecting females from the wild that are either already berried at the time of capture or are in a near-berried state. With only a short holding time in captivity, these females can be induced to spawn with minimal environmental manipulation (Sachlikidis *et al.*, 2005). However, long-term holding and induction of spawning relies on the use of environmental control and manipulation. Lobsters are seasonal breeders with photoperiod being the primary driver in bringing broodstock into and out of breeding condition. Temperature and social factors also have an influence (Sachlikidis *et al.*, 2005). In *P. japonicus* spawning has been induced in longer term captivity through a combination of photoperiod and temperature (Matsuda *et al.*, 2002).

In the wild, the breeding season of *P. ornatus* spans four months (Skewes *et al.*, 1994). In captivity, *P. ornatus* broodstock have been held for up to nine years and induced to breed through photoperiod and water temperature control to replicate the natural cycle (Hall, unpubl.). With photoperiod and temperature manipulation, it has been possible to have spawnings in every month of the year with three broodstock populations, with each breeding season lasting approximately four months as is the case in the wild (Skewes *et al.*, 1994; Hall *et al.*, unpubl.). Broodstock may be fed twice daily, in the morning and late afternoon, with a combination of squid, mussels and pilchards. Feeding rate is targeted to be approximately 5 % of the total biomass in the respective tank, but feeding intensity changes during the annual cycle. Feed rations may be adjusted daily depending on how much food is left in the tank on the morning after an afternoon feed. Most females in each broodstock population spawn three times within a single breeding season.

Fecundity varies between females for reasons that are not fully understood and may simply reflect natural variation in a population. Total phyllosomas produced per spawning of 1.5–2 kg females varies between 200 000 and 1 000 000 larvae. Phyllosomas produced per female is positively correlated with female body weight with 2–3.5 kg females producing up to 1 600 000 larvae (Fig. 9.4). The average weight per phyllosoma ranged between 253 and 378 mg (wet weight) with no significant decrease in average weight between the first and third spawning. Fertility is related to the size of the spermatophore (tar) patch, which itself is positively related to male size. The spermatophore reduces in size with each spawning, but a female can still maintain 100 % fertility on the third and fourth spawning if she has mated a second time during the breeding season. If the female has not mated a second time, fertility is typically < 50 % by the second spawning. Females brood eggs for approximately 27 days at 28 °C. Fertility

© Woodhead Publishing Limited, 2013

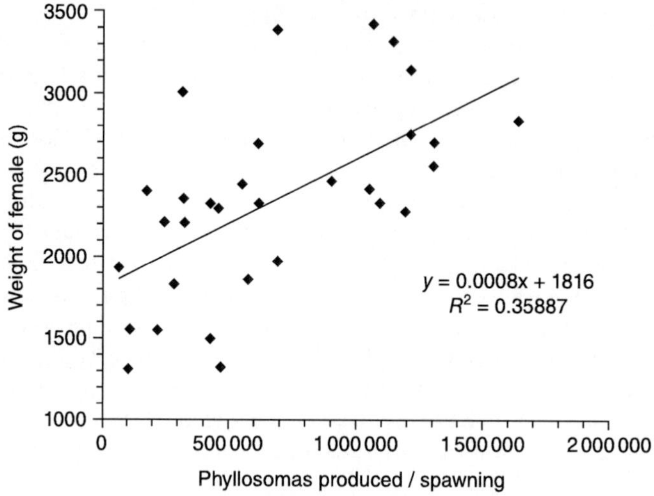

Fig. 9.4 Relationship between body mass of breeding female and number of phyllosomas per spawning.

can be determined within three to five days post ovi-positioning by whether cell division has occurred (Fig. 9.5). Embryo eye formation can be observed at approximately 15–16 days (Fig. 9.5).

The timing and frequency of moults can also be manipulated with photoperiod in *P. ornatus*. Typically, females moult three times per year, termed 'pre-reproductive' moult for the one that precedes the spawning period, the 'post-reproductive' moult for the one that terminates the spawning season, and the 'growth' moult (if it occurs) for the one that occurs after the post-reproductive moult but before the pre-reproductive one (Hall *et al.* unpubl.). The start of the reproductive season can be defined as being at the point of the pre-reproductive moult. It is at this moult that the female develops hairs on the pleopod setae, to which the ovipositioned eggs will adhere and be incubated by the female until hatching. Assuming that it is at this moult that the female begins to accumulate energy reserves for reproduction and the reproductive system develops to full functionality, it takes over three months for energy reserves to be built up before eggs are laid. It is likely that an adequate diet between the pre-reproductive moult and spawning itself is important to reproductive success.

The reproductive season is energetically demanding, in particular for the females. With each spawning event, a female exhibits an approximately 7 % decline of body mass. Although this loss is represented by larval production, the mass originally had to be assimilated by the female with energy partitioned to reproductive events. The drain on the female condition can be clearly observed when the weight gain between each of the three annual moults is compared (Fig. 9.6). The percentage body mass gain in females is

© Woodhead Publishing Limited, 2013

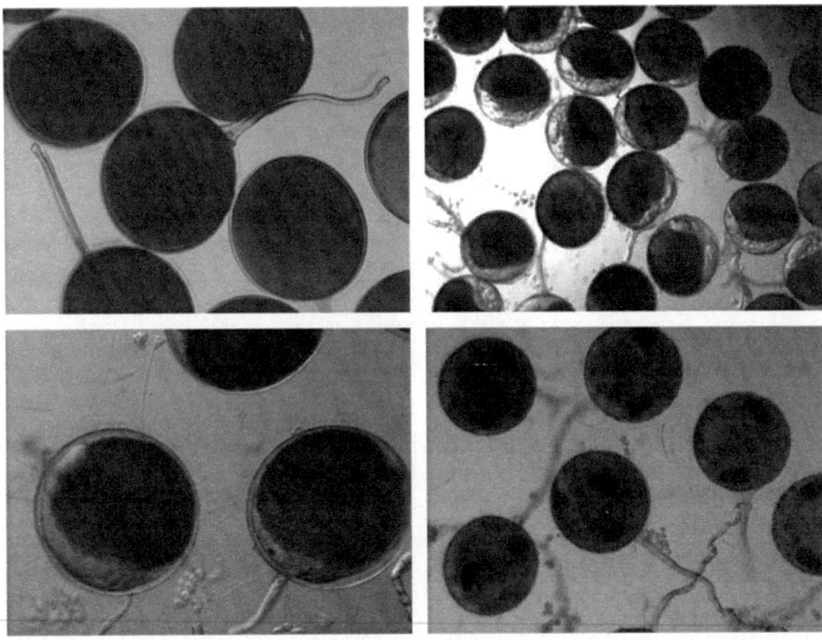

Fig. 9.5 Development of eggs in *P. ornatus*. Upper left: Newly ovipositioned eggs; Upper right: Eggs 4–6 days into development with embryo formation developing; Lower left: Formation of embryo eye at day 15–16; Lower right: Tightly curled fully developed embryo well-developed eyes immediately before hatching.

significantly reduced between the pre- and post-reproductive moults, expressed as percent weight gain per day, compared to the other moult intervals. In contrast, males do not exhibit such a pronounced decline in daily weight gain. However, compared to the other two moult intervals, it is between the pre- and post-reproductive moults that males gain the least amount of weight. Presumably this decline is associated with some aspect of reproductive activity in males, such as guarding or defence of females. Such information clearly indicates that if weight gain is the primary objective, then either pre-reproductive animal should be grown or reproduction in adults prevented.

9.3 Larval rearing, water quality and tank design

9.3.1 Larval rearing

The complete larval development of various Palinurid lobsters has been published for several species (Table 9.1). For *P. ornatus*, the complete description of larval morphology of hatchery reared phyllosomas benchmarked against wild conspecifics has been reported (Smith *et al.*, 2009b).

© Woodhead Publishing Limited, 2013

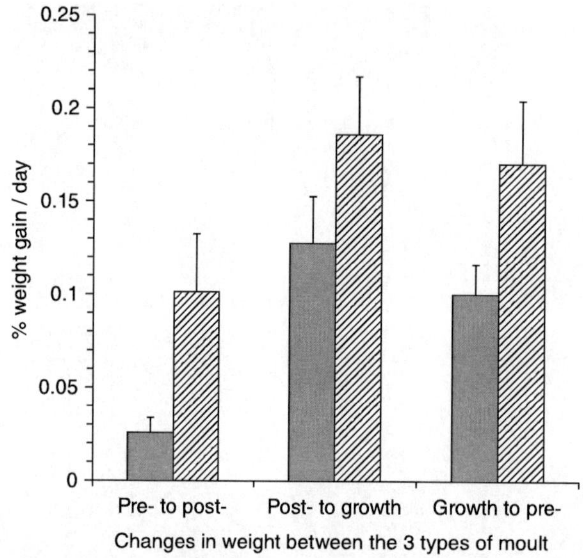

Fig. 9.6 Changes in average weight gain per day in female (solid) and make (diagonal hatched) broodstock between the three major moults per year: pre- (pre-reproductive), post- (post-reproductive) and growth moult.

Although there are 11 morphological stages, there are a total of 24 instars. Specimens exhibiting the same 24 increments have been found in plankton collections of *P. ornatus*, indicating that such plasticity in larval development occurs under both wild and hatchery conditions (Hall, unpubl.). This may represent the plasticity of larval development, as is demonstrated in decapod larvae by combinatorial staging and intercalated stages and in marked time moulting (Gore, 1985). Delayed or accelerated larval moults may be related to nutritional status (Smith *et al.*, 2009a). If insufficient reserves are present, then a 'mark time' moult may take place. In contrast, if sufficient reserves are present then the phyllosoma will moult directly to the next morphological stage (Knowlton, 1974). Knowlton suggested that nutritional intake is partitioned into moulting and growth, which was further developed into the concept of 'point of reserve saturation' in which there is a critical minimal feeding time before moult to the next stage (Anger and Dawirs, 1981). If a critical nutritional status is reached, an endocrine cascade involving ecdysone steroids is triggered accumulating in a moult (Wilson *et al.*, 2005a,b). A significant shortening of the larval cycle is frequently observed in hatchery reared phyllosomas compared to wild conspecifics (Kittaka, 1997; Matsuda and Takenouchi, 2007). Perhaps the most extreme has been observed in *P. elephas*, in which a larval period of 132 days has been reduced by nearly 50 % of that estimated in the wild to as short as 65 days in captivity (Kittaka *et al.*, 2001).

© Woodhead Publishing Limited, 2013

The general metabolism of phyllosomas is poorly known, but some aspects have been examined. For *P. ornatus*, oxygen consumption, ammonia excretion and chemical composition have been documented (Ikeda *et al.*, 2011). Dry mass (DM) specific oxygen consumption varied from 1.4–3.3 μl O_2 $(DM)^{-1}$ h^{-1} with a progressive decline during development. Ammonia secretion was variable ranging from 0.02–0.067 μg NH_4-N $(DM)^{-1}$ h^{-1}. Carbon content increased during development from 33 % to 41 % of DM whereas ash content decreased from 34 % to 17 % of DM. No consistent pattern with development was seen in either wild or captive phyllosomas from early to late stage development for water content (74–83 % of WM), N composition (6.9–9.5 % of DM) and C:N ratios (4.0–5.1). Overall, the study by Ikeda and colleagues indicated a preferential utilisation of dietary protein to somatic growth rather than metabolism. Aspects of nutrition are discussed later in this chapter.

In the hatchery, phyllosomas are fed *Artemia* up to four times per day at varying rates, with an average of eight *Artemia* nauplii per mL for the first 30 days. Weaning onto a formulated diet is possible after the first moult (day 6–8) and is complete by stage 3, instar 2 (day 35). Formulated larval diet is fed three to four times per day at a rate of 1 mg mL^{-1} per feed.

9.3.2 Water quality

Water quality is a critical parameter in the hatchery for phyllosoma larvae. In the wild, phyllosomas of Palinurid lobsters are primarily restricted to oligotrophic oceanic water where total suspended solids (TSS), as well as dissolved organic matter (DOM), are minimal. Locomotion in phyllosomas is driven by rapid movement in the thoracic appendages and long exopodites, which possess fine setae that are readily fouled by fine particulate matter in the water column. Any increase in DOM will result in an increase in the bacterioplankton community (Van Wambeke and Bianchi, 1985), some of which may be pathogens, increasing the risk that a disease outbreak will occur in the larval rearing tank. Because of these issues, particular attention has been given to producing the highest possible seawater quality.

One engineering approach to the production of high quality sea water is through the use of ozonation. Ozone is a powerful oxidising agent with acute disinfectant properties due to the rapid degradation from ozone (O_3) to oxygen (O_2), with the free oxygen atom causing oxidation to particulates in the water. Ozonation also produces a longer lasting disinfectant due to ozone reaction with halides present in sea water producing residual oxidation by-products (OBP). The OBPs, primarily bromine, as hypobromous acid and hypbromite, and bromate, are toxic to life (Crecelius, 1979). By careful application of the OBPs, ozonated sea water can provide an environment that is unfavourable to bacteria (Ritar *et al.*, 2006; Wietz *et al.*,

© Woodhead Publishing Limited, 2013

2009). However, as the OBPs are toxic to life in general, excessive application of OBPs can lead to morphological and histological deformities and death (Jensen *et al.*, 2011).

Alternative methods of maintaining high quality sea water and minimising bacterial proliferation include the use of ultraviolet (UV) treatment of sea water, which can be less costly and less complex than using ozone (Summerfelt, 2003). Although UV irradiation does not generate toxic residuals, the sea water must be particulate free to allow maximum UV transmittance and effectiveness. Removal of TSS can be obtained through application of various separation technologies, such as sedimentation, flocculation, electroflocculation and hydroclone (Gemende *et al.*, 2008). With advances in filtration technologies in the wastewater industry, various microfiltration membranes are also being implemented in aquaculture systems to obtain high quality sea water (Viadero and Noblet, 2002; Gemende *et al.*, 2008; Castaing *et al.*, 2011).

Aspects of temperature, salinity, photoperiod, light intensity and water flow were recently reviewed by Phillips and Matsuda (2011). As with aquaculture hatcheries in general, various dissolved organics in the water can be toxic to larvae. Ammonia is particularly toxic to phyllosoma larvae. In *J. edwardsii*, the median lethal concentration of ammonia after a 96-hour exposure, is between 32 and 52 mg L^{-1} (Bermudes and Ritar, 2008). This study concluded that ammonia levels must be held between or below 2.7–4.4 mg L^{-1} to avoid detrimental influences on the health of phyllosomas.

9.3.3 Tank design

The hydrodynamics of seawater flow through larval rearing tanks requires optimising to meet the specific requirements of the unique morphology of phyllosoma larvae. Although phyllosomas can be strong swimmers, they require some degree of suspension in the water column and this can be assisted by appropriate water flow characteristics and tank design. Consideration must also be made to their raptorial mode of feeding. At present, the standard diet for larval rearing of early-stage phyllosomas is based on the use of disinfected *Artemia* (Matsuda, 2006; Smith and Ritar, 2006). However, their use can be problematic due to the swarming behaviour of *Artemia* under certain conditions. The interaction between phyllosomas and *Artemia* and their ability to respond to differing water flow dynamics have important implications for the production of fast-growing, healthy larvae and may help set the focus for future designs of culture systems. To date, phyllosomas have been cultured in an array of vessels with distinctly different flows, including static beakers, tubs with circular horizontal flow, vessels with upwelling turbulence and plankton kreisel tanks (Illingworth *et al.*, 1997; Kittaka, 2000; Ritar, 2001; Smith *et al.*, 2003; Matsuda and Takenouchi, 2007).

© Woodhead Publishing Limited, 2013

Tank designs for larval rearing are diverse and range from experimental to commercial scales, and reported seawater flow regimes include static, periodic bulk water exchange, recirculation and flowthrough. In general, systems can be operated with either continuous flow or sporadic surge (by a sudden influx of water) or periodic pulsing (Carlson, 1996). Surging and pulsing of incoming water can partially mimic wave action and is often used in public aquaria to display marine organisms of the shoreline from the surf zone of high energy impact coasts. At least six basic tank designs have been used in larval rearing with many of them having been examined for suitability for phyllosoma larvae (Table 9.2).

Table 9.2 Larval rearing tank designs used or developed for phyllosomas

	Description	Characteristics	Reference
1	Horizontal gyre	Slow circular current generated by either wind surface sheer or multiple submersed microspray heads. Turbulent flow.	Dawson, 2000; Mikami, 1995; Reitan *et al.*, 1993; Sandier *et al.*, 1974
2	Vertical upweller – cone	Suspends inert and live objects in water column. Variable velocity throughout height of cone. Largely laminar flow.	Myhre *et al.*, 1993; Danielsberg *et al.*, 1993; Calado *et al.*, 2003
3	Vertical upweller – column	Suspends inert and live objects in water column. Constant velocity throughout height of column. Largely laminar flow.	Illingworth *et al.*, 1997; Soregloss and Persoone, 1972
4	Vertical kreisel	Creates 'dimensionless ocean'. Excellent for observational work and maintenance of delicate plankton. Flow velocity near zero towards centre of kreisel.	Greve, 1970; Greve, 1975; Hamner, 1990; Horita, 2007; Murakami *et al.*, 2007
5	Horizontal kreisel / raceway	Creates continuous circular, oval or elliptical current. Nutrient build-up can occur as water exchange only occurs gradually. Turbulent flow with some laminar flow along straight stretches.	Barrows *et al.*, 1993; Burrows and Chenoweth, 1970; Inoue, 1981; Mikami, 2003
6	Parabolic tank	Water flow primarily in upwelling mode. Largely turbulent flow with up- and down-welling zones.	Beard *et al.*, 1985; Beard and Wickins, 1992; Greve, 1968; Hughes *et al.*, 1974; Rice and Williamson, 1970

© Woodhead Publishing Limited, 2013

Horizontal gyre systems have been widely used for larval rearing (Sandifer *et al.*, 1974; Reitan *et al.*, 1993; Mikami, 1995; Dawson, 2000). Vertical cone upweller systems have been primarily used for finfish egg incubation and larvae (Danielsberg *et al.*, 1993; Myhre *et al.*, 1993; Calado *et al.*, 2003), but have also been used for Palinurid larval rearing as well as other marine invertebrates and fish larvae (Soregloss and Persoone, 1972; Illingworth *et al.*, 1997). Vertical kreisels have been widely used in zooplankton studies (Greve, 1970, 1975; Hamner, 1990; Horita, 2007; Murakami *et al.*, 2007). The kreisel design is widely used in public aquaria for display of delicate gelatinous zooplankton (Widmer, 2008). The flow dynamics in kreisels are such that a slow velocity current is created around the circular wall of the tank which creates a 'virtual' ocean in the central part of the tank where the water velocity is static. Phyllosomas either flow with the current around the periphery or enter and leave static water in the centre (Horita, 2007; Murakami *et al.*, 2007). In either case, the phyllosomas only momentarily make contact with tank surfaces. Raceway tank designs, or horizontal kreisels, are most widely used to simulate continuous water flow as found in rivers and streams (Burrows and Chenoweth, 1970; Inoue, 1981; Barrows *et al.*, 1993). Modifications have been made to the basic design, and one version, developed for the rearing of Scyllarid larvae, has been patented for phyllosoma rearing (Mikami, 2003). Parabolic tanks have been widely used in larval rearing of marine invertebrates and vertebrates, and these can be designed to produce a range of flow patterns including upwelling (Greve, 1968; Rice and Williamson, 1970; Hughes *et al.*, 1974; Beard *et al.*, 1985; Beard and Wickins, 1992). Parabolic tank designs of between 5000 and 15 000 L capacity are widely used in the commercial production of penaeid prawn larvae in commercial hatcheries (Wyban and Sweeney, 1991).

9.4 Health issues during larval rearing

High health is a prerequisite to optimal larval rearing in terms of final survival by the end of the larval rearing period and maximising growth, and in order to minimise the total time taken by the phyllosomas to complete all developmental stages by restricting the frequency of 'mark-time' moulting. Typically, if larval mortality builds gradually with an increasingly lengthy inter-moult period, it is likely that the problem is due to a nutritional deficiency. On the other hand, if mortality is transient and periodic, the problem is more likely to be a physical stressor such as low dissolved oxygen or, the most common mortality factor, opportunistic microbial disease agents. Acute and chronic stresses are major contributors to creating favourable conditions for opportunistic pathogens inducing mass mortality events. The hatchery environment must be managed to minimise the development of sub-optimal conditions leading to physiological compromise with

© Woodhead Publishing Limited, 2013

perturbations to homeostasis (Hall and de la Vega, 2004). Any conditions that induce stress require the larvae to spend energy to maintain homeostasis and place the organism at risk of opportunistic infection.

The primary agents of disease for phyllosomas are common throughout aquaculture and include viruses, bacteria, fungus, protozoans and metazoans (Shields, 2011). Many of these organisms are nearly ubiquitous in sea water and hence particular attention must be given to incoming water quality (see Section 9.3.2). Once in the hatchery proper, disease agents may go undetected whilst in a latent state causing little to no harm to the larvae; however, they are capable of causing significant mortality once certain triggers are activated, such as induction through stress. The initial signs of infection include reduced activity of larvae, colour change, interrupted feeding behaviour, body contortions and respiratory change. Once clinical signs are observable it may be too late for any action to be taken to avoid mass mortalities, although this depends on several factors including virulence, propensity for horizontal transmission and host resistance.

9.4.1 Viruses

Reports of viruses in Palinurid lobsters are rare but, as with all metazoans, they undoubtedly carry a variety of viruses including an unknown number of disease agents. To date, reports of mortalities due to viral infections have been limited to adult Palinurid lobsters (Quintana *et al.*, 2011). The only known viral disease agent specific to Palinurids is the *Panulirus argus* virus 1 (PaV1), described from the Caribbean lobster. The virus is believed to be a member of the Fuselloviridae. Infection takes place in phagocytes of the hepatopancreas followed by haemocytes (Behringer *et al.*, 2009). It has yet to be demonstrated in phyllosomas, but in juvenile lobsters the clinical signs include lethargy, a milky white hemolymph that does not clot, suppressed moulting and eventual death of the infected lobster (Quintana *et al.*, 2011). Experimental transmission via the water column, injection, ingestion or cohabitation has been demonstrated (Butler *et al.*, 2008). The ovaries of adult females can be heavily infected with PaV1, suggesting that vertical transmission from mother to larvae is a possibility if the oocytes are infected before ovopositioning (Quintana *et al.*, 2011). In adult lobsters, the disease takes weeks to months before mortality occurs so deaths may not be apparent in a hatchery. However, susceptibility decreases with increasing size so it remains a possibility that phyllosomas are susceptible to this virus (Butler *et al.*, 2008).

The only other virus known in lobsters is the white spot syndrome virus (WSSV), which has been the causative disease agent of major pandemics in penaeid prawn culture systems. WSSV has been demonstrated to be present in gills, stomach, hepatopancreas and epidermis of *P. versicolour*, *P. penicillatus*, *P. longipes* and *P. ornatus* in experimentally infected lobsters

© Woodhead Publishing Limited, 2013

(Chang *et al.*, 1998; Wang *et al.*, 1998). However, infected lobsters survived infection for at least 70 days without clinical signs of infection. Overall, whereas Palinurid lobster might be reservoirs of WSSV (and as such hatchery operators should be aware of this), they do not appear to be a natural host of this virus nor is there any evidence that this virus is pathogenic to lobsters so WSSV may not be an issue in a hatchery.

9.4.2 Bacteria

From the very early initial attempts at rearing phyllosomas, mass mortalities were a consistent characteristic with the likely disease agents being of bacterial origin (Kittaka and Abrunhosa, 1997). Bacteria form complex communities within the hatchery and occupy niches within the water column as bacterioplankton, on surfaces as sessile biofilms and externally and internally in larval feed and the larvae themselves. The bacterial community associated with aquaculture-reared phyllosomas differs significantly to that of their wild conspecifics (Payne *et al.*, 2008). This may be partially a reflection of the differences between the hatchery and the natural environments in which phyllosomas develop. Oceanic waters are an oligotrophic environment where the phyllosoma-associated bacterial community appears to be conserved. This is radically different to the hatchery environment, which is eutrophic with a high biomass density, offering ideal conditions for a diverse array of bacteria due to the availability of high nutrient concentrations, either consistently or periodically.

Using a combination of molecular methods, including clone libraries, denaturing gradient gel electrophoresis (DGGE) and fluorescence *in situ* hybridisation (FISH), it has been shown that a complex bacterial community exists in all compartments of the hatchery (Bourne *et al.*, 2004, 2007) including the water column (Payne *et al.*, 2006), the biofilm (Bourne *et al.*, 2006; Wietz *et al.*, 2009) the larval feed (Høj *et al.*, 2009) and the phyllosomas themselves (Payne *et al.*, 2007). Gammaproteobacteria, including Vibrionaceae-related species, and Alphaproteobacteria are dominant bacterial groups in all compartments. In addition, Actinobacteria, Bacteriodetes, Betaproteobacteria, Chlamydiales, Cytophagales, Epsilonproteobacteria, Firmicutes and Planctomycetales have been detected in one or more hatchery compartments. Studies have shown that the bacterial community associated with wild phyllosoma larvae is a conserved microbiota dominated by Alphaproteobacteria such as *Sulfitobacter* spp. which differs significantly from their hatchery conspecifics (Payne *et al.*, 2008).

Shell disease in lobsters has been linked primarily to Gram-negative, chitinoclastic, rod-shaped bacteria with *Vibrio*, *Aeromonas*, *Pseudomonas* and *Shewanella* being the most commonly isolated bacterial genera (Shields *et al.*, 2006). The ability to grow on chitin is limited to a number of bacterial groups within Proteobacteria, Bacteroidetes, Actinobacteria and Firmicutes (Hunt *et al.*, 2008), all of which have been detected in spiny

© Woodhead Publishing Limited, 2013

lobster hatcheries (see above). A molecular study of shell disease in *P. argus* demonstrated that the bacteria implicated in the disease were also present in healthy individuals and it appears likely that they act as opportunistic pathogens if abrasions or lesions occur in the carapace (Porter *et al.*, 2001). If phyllosomas suffer carapace damage or broken appendages, the hatchery environment is sufficiently rich in chitinoclastic groups of bacteria that they have the potential to infect carapace lesions with subsequent degradation in quality and possibly death of phyllosomas.

Vibriosis is one of the major diseases in aquaculture and is a particular problem during the rearing of Palinurid phyllosoma larvae (Webster *et al.*, 2006). Vibrios are ubiquitous in the marine environment and many pathogenic species have been isolated from adult Palinurid lobsters including *V. alginolyticus*, *V. anguillarum*, *V. harveyi*, *V. parahaemolyticus* amongst others (Brinkley *et al.*, 1976; Bowser and Rosemark, 1981; Jawajar *et al.*, 1996). *Vibrio* species isolated from early stage Palinurid phyllosomas include *V. alginolyticus*, *V. campbellii*, *V. harveyi*, *V. jasicida*, *V. natriegens*, *V. owensii*, *V. parahaemolyticus*, *V. proteolyticus*, *V. rotiferianus* and *V. tubiashii* (Diggles *et al.*, 2000; Bourne *et al.*, 2004; Payne *et al.*, 2007; Cano-Gomez *et al.*, 2010; Yoshizawa *et al.*, 2012). It has been reported that the relative abundance of Vibrionaceae in the hepatopancreas of phyllosomas increases rapidly four days before mass mortality events (Webster *et al.*, 2006). So far, only two detailed studies of Palinurid phyllosoma vibriosis have been presented (Diggles *et al.*, 2000; Goulden *et al.*, 2012a). In both cases rapid mortality rates were seen between day 2 and day 4 of experimental infection coinciding with rapid proliferation of vibrios in the hepatopancreas (Fig. 9.7). The disease agent associated with frequent mass mortalities in early stage phyllosoma larvae of *S. verreauxi* was originally identified as *V. harveyi* (Diggles *et al.*, 2000). However, subsequent molecular work has identified it as

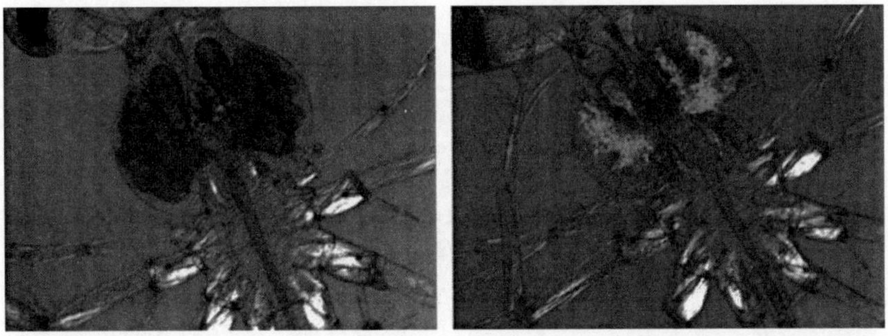

Fig. 9.7 Phyllosoma infected by *Vibrio owensii*, tagged with green-fluorescent protein (GFP) to visualise progress of infection. Healthy phyllosoma, early infection (left), with few GFP-labelled cells and late infection (right), with mass proliferation and infection of hepatopancreas tubules (light grey area within hepatopancreas). Photo: E. Goulden

© Woodhead Publishing Limited, 2013

representing a new species in the Harveyi clade, *V. jasicida* (Yoshizawa *et al.*, 2012). A highly virulent strain towards *P. ornatus* phyllosomas was identified as another new species in the Harveyi clade, *V. owensii*, which is highly virulent to phyllosomas and typically causes 80–90 % mortality within 72 hours (Cano-Gomez *et al.*, 2010; Goulden *et al.*, 2012a). There are no known antibiotic treatments which can fully prevent mortalities once the pathogen becomes established although a probiotic strategy appears promising. Due to the phenotypic similarities and genome plasticity within the Harveyi clade, traditional phenotypic identification methodologies do not offer sufficient resolution for accurate species identification and more specific molecular approaches have been developed (Cano-Gomez *et al.*, 2011). It has also been suggested that future approaches should include the detection of virulence genes rather than the bacterium *per se* (Cano-Gomez *et al.*, 2009).

Whereas Gram-positive bacteria constitute only approximately 5 % of marine bacteria they can be serious pathogens of lobsters. *Aerococcus viridans* is the most important bacterial pathogen in homarid lobsters (Stewart *et al.*, 2004). Upon entry into the lobster, bacteria colonise the hepatopancreas, haemocytes and heart. Clinical signs of infection in juveniles or adults include lethargy, appetite loss and a pinkish colouration with high mortality. *Aerococcus* has also been reported to be found in *P. argus* and experimentally infected *P. interruptus,* but may not be as virulent in spiny lobster as in homarid lobsters (Bobes *et al.*, 1988). In addition, the natural habitat of *Aerococcus* is benthic sediment (Wiik *et al.*, 1987) so it may not be an issue for planktonic larvae, although they may be agents of vertical transmission from parent to offspring if eggs are infected. Other Gram-positive bacteria found in phyllosoma hatcheries include *Bacillus* (Bourne *et al.*, 2004) and *Clostridium* (Payne *et al.*, 2007).

9.4.3 Fouling bacteria

Fouling of larvae by filamentous bacteria can be a major problem for phyllosomas (Kittaka, 1997). The inter-moult period of phyllosomas is typically many days to a week or more, and in combination with the multi-month length of the phyllosoma phase this gives ample time for heavy fouling by filamentous bacteria (Fig. 9.8). Although fouling in itself is not pathogenic, the ability of the phyllosomas to capture and properly masticate food becomes increasingly restricted. The interference with efficient feeding leads to a deteriorating nutritional status which, in turn, leads to a prolonged inter-moult period accentuating the fouling further (Fig. 9.9). The external fouling and poor nutrition of the phyllosomas leads to increased potential for opportunistic pathogen infection. The filamentous bacteria agent has frequently been attributed to *Leucothrix mucor*, but recent molecular studies have identified filamentous bacteria colonising *P. ornatus* phyllosomas as *Thiothrix* sp. (Payne *et al.*, 2007).

© Woodhead Publishing Limited, 2013

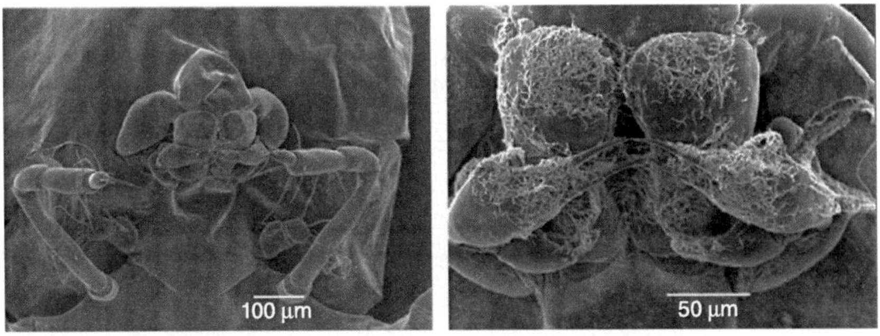

Fig. 9.8 Thiothrix fouling of phyllosoma larvae. Mouthparts of immediate post-moult unfouled individual (left) and pre-moult fouled individual (right). Photo: M. Hall

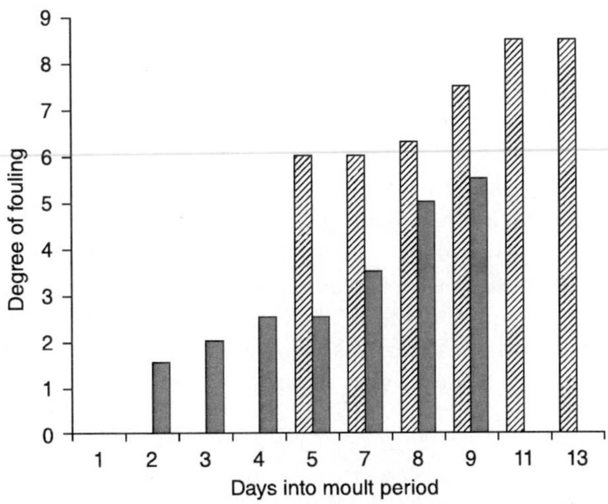

Fig. 9.9 Qualitative degree of fouling of phyllosomas across two intermoults starting from the day of moult. Phyllosoma developmental stage P1 to P2 (solid columns) and P3 to P4 (hatched columns).

9.4.4 Oomycota (lower fungi)

A range of pathogenic oomycetes have been isolated from crustaceans, including *Atkinsiella* spp., *Lagenidium callinectes* and *Haliphthoros milfordensis*. Species of *Atkinsiella* infect phyllosomas and can cause 100 % mortalities in crab larvae (Roza and Hatai, 1999). Species of *Lagenidium* infect and penetrate embryos of berried females as well as larvae (Nilson *et al.*, 1975). This genus has also been found to cause mortalities in larvae of *Homarus americanus* (Nilson *et al.*, 1976). The oomycete can kill larvae within one to two days of infection and can penetrate eggs by more than 3 mm, causing death. Its control is problematic, but bacteria of the genus

© Woodhead Publishing Limited, 2013

Alteromonas produce tyrosol, a natural antifungal compound, which may offer protection of eggs and larvae from infestation (Gil-Turnes *et al.*, 1989; Gil-Turnes and Fenical, 1992). Species of *Haliphthoros* can infect puerulus and juvenile spiny lobsters (*J. edwardsii*) when they show signs of lethargy and appetite loss (Diggles, 2001). Infection is primarily through invasion of gill tissue, with death typically associated with moult (Overton and Bland, 1981; Diggles, 2001). A range of chemotherapeutics were trialled as control agents in the study by Diggles, including malachite green, formalin, copper sulphate and trifuralin, but none, with the possible exception of the last, were particularly effective.

9.4.5 Fungi

Species of *Fusarium* have not been reported in larval lobsters, but it does occur in juveniles and adults (Shields *et al.*, 2006). The fungus causes lesions on the cuticle and in tissues, typically with melanisation. *Fusarium* has been a major problem in penaeid prawn hatcheries with no known effective control measure. Other fungal infections include those of microsporidians in *P. argus* and microsporidiosis lesions have also been reported in *P. cygnus* and *P. ornatus* (Dennis and Munday, 1994; Stephens *et al.*, 2003; Kiryu *et al.*, 2009). While the phylum Microsporidia is classified within the kingdom Fungi, their classification is open to debate (Hibbett *et al.*, 2007). Nevertheless, they are known pathogens in crustaceans (Findley *et al.*, 1981; Langon, 1991) and it is possible that the infective agents are species of the genus *Ameson* (Sprague and Couch, 1971). Clinical signs of microsporidian infection in juvenile and adult lobsters involve a change in muscle colouration from the normal translucent grey to cotton white.

9.4.6 Protozoa

Protozoans are a diverse group of single-cell eukaryotic organisms with a number of them known to be disease agents in lobsters. Protozoans associated with lobsters include species of ciliates, dinoflagellates, rhizopods and gregarines. The ciliate *Anophryoides haemophila* can invade the hepatopancreas hemolymph and gills of juvenile and adult lobsters but is not known from phyllosomas (Shields *et al.*, 2006). In contrast, peritrich ciliates of the genera *Vorticella* and *Zoothamnium* and the suctorian ciliates of the genera *Acineta* and *Ephelota* are frequently reported as fouling organisms on lobster including embryos and phyllosomas (Kittaka, 1997; Webster *et al.*, 2006). Whereas these ciliates are not pathogenic *per se*, they do interfere with normal locomotion and increase the energy costs for movement. In addition, the fouling can interfere with feeding and in combination this can lead to a deterioration of nutritional status opening the way for opportunistic pathogen infection. Chemotherapy trials have resulted in limited success in controlling these ciliates (Boghen, 1982). Haematodinium

© Woodhead Publishing Limited, 2013

dinoflagellates are known internal haemolymph parasites of lobsters and are highly pathogenic (Stentiford *et al.*, 2001). In juvenile and adult lobsters, clinical signs of infection include lethargy, a dull orange colouration, milky white haemolymph, watery muscles and damage to the hepatopancreas (Field *et al.*, 1992). Transmission is likely through microdinospores in the water column, and possibly cannibalism, and although not reported to date in phyllosomas it would represent a serious disease agent in a hatchery. Similarly, Rhizopoda amoebae have been reported in adult lobsters, in which the parasite invades the nerve cord, brain, eyes and neurosecretory organs, but so far these have not been reported in phyllosomas (Mullen *et al.*, 2005). Gregarines of the genus *Porospora* are reported to be the most common parasite of homarid lobsters, where they are found in the digestive system (Shields *et al.*, 2006). However, there are no reports of these parasites from Palinurids.

9.4.7 Helminths

Digenetic trematodes have been isolated from *P. cygnus* with high incidence rates and include the microphallid trematode *Thulakiotrema genitale*, which encysts in gonads (Deblock *et al.*, 1991). Other helminths include the brachycoeliid trematode *Cymatocarpus solearis* from *P. argus*, where it is found in muscle (Gomez del Prado-Rosas *et al.*, 2003). These helminths have not been reported in phyllosomas, perhaps due to their size, with cysts being 1.5 mm in diameter. However, helminth life-cycles are typically very complex involving several hosts. As some life-cycles include molluscs, it is plausible that helminths could be introduced into the hatchery through fresh feeds. Other helminths found in Palinurid lobsters include cestodes and nemerteans (Shields, 2011). Species of the genus *Carcinonemertes* have been isolated from the eggs of *P. interruptus* and *P. cygnus*, which is a potential avenue for their introduction into a hatchery. However, due to their size, 5–12 mm, adults should not be an issue for phyllosomas although they can cause mass egg mortalities of berried homarid females (Brattey *et al.*, 1985). To avoid introduction into the hatchery from berried females, short baths in fresh water have proven effective in homarid lobsters (Charmantier *et al.*, 1991). Other helminths of lobsters include polychaetes, nematodes and turbellarians but again, due to their size, they should not be an issue for larval rearing (Shields *et al.*, 2006; Shields, 2011).

9.5 Health: infections and nutrition

9.5.1 Primary and secondary infections

The maintenance of a stable hatchery environment to avoid stress in the larvae cannot be overemphasised. There is a high probability that the hatchery contains a reservoir of latent pathogens at any one time which, while under stable conditions, may not induce disease. However, a physiological

© Woodhead Publishing Limited, 2013

shock induced by stress can increase the larvae susceptibility and trigger an infection (Hall and de la Vega, 2004). This may cascade into secondary, or opportunistic, infections by pathogens which themselves may otherwise have limited ability as disease causing agents. The basic components of a disease agent include (i) occupation of the same environmental niche as the host, (ii) gaining access to the host, typically internalised, (iii) establishment in a particular niche within the host, (iv) proliferation within the host and finally (v) release from the host and dispersion into a new host to repeat the cycle. Portals of entry of potential microbial pathogens include ingestion, entry via gills or lesions, as well as active penetration. For example, a primary infection that causes the development of lesions, septicaemic response or ulceration removes one of the major barriers to host invasion and opens the way for bacterial colonisation of the whole organism.

The traditional management approach to controlling bacterial pathogens in a hatchery has been through antibiotics (FAO/OIE/WHO, 2006). Antibiotics can be very effective in controlling pathogens and infections, but their prophylactic use has raised health and environmental issues (Cabello, 2006). Due to reports of increasing incidence of antimicrobial resistance, and legislation limiting their use in food production systems, there is increasing interest in alternative control options. These include the use of probiotic bacteria (Verschuere *et al.*, 2000), the use of quorum sensing disruption (Defoirdt *et al.*, 2004) and the use of bacteriophages (Nakai and Park, 2002; Crothers-Stomps *et al.*, 2009). Due to the frequency of mass mortality events during larval rearing of phyllosomas, there is an urgent need to manage the bacterial population in the hatchery and in particular the *Vibrio* population. Some of these approaches are discussed in Chapter 8 on microbial management for bacterial pathogen control in invertebrate aquaculture hatcheries.

Moult death syndrome

A common cause of death in adult and larval crustaceans, including lobsters, is moult death syndrome (MDS) (Bowser and Rosemark, 1981). It is a very general term and used for any animal that dies during the moulting process and is likely to have several causative factors. It is typically expressed by a moulting individual being unable to completely free itself from the old exoskeleton, which is exacerbated by any carapace lesions incurred during the inter-moult period (Floreto *et al.*, 2000). There is some evidence that inadequate nutrition contributes to MDS, so an improved diet, in particular phosphatidylocholine, can minimise the occurrence of MDS (D'Abramo *et al.*, 1981). Equally, it has been suggested that MDS is indicative of a metabolic dysfunction in calcium metabolism (Dove *et al.*, 2004).

9.5.2 Nutrition

Common to the other aquaculture species, the development of a larval feed for phyllosomas requires at least a basic knowledge of what is the natural

© Woodhead Publishing Limited, 2013

diet. Observation of phyllosoma predator–prey interactions is virtually impossible as the larvae are in oceanic waters and are transparent making them extremely difficult to observe *in situ* in the wild. Some observations have been made of phyllosomas by photography and video recording by remote operated vehicles (ROV) (Lindsay, pers. com.) and fortuitous encounters by scuba divers. In the latter, there are several reports of phyllosomas being freely attached to medusae (Shojima, 1963; Thomas, 1963; Herrnkind *et al.*, 1976) although the significance of this association is unclear. An alternative approach is to observe predator–prey associations in aquaria with selected putative prey items that co-inhabit the same environment as the phyllosomas. Such an approach has been employed for more than a century, starting with Hattori and Oishi in 1899. These authors presented phyllosomas with a range of prey, and found that fish larvae were readily consumed (Hattori and Oishi, 1899). In other studies, phyllosomas could catch and consume fish and eel larvae, hydromedusae, ctenophores, chaetognaths, calanoid copepods, crab zoeas, trochophore veliger larvae, polychaetes and asicidian larvae (Lebour, 1925; Oshima, 1936; Johnston, 1956; Batham, 1967). Reviews on the dietary preferences of phyllosomas can be found in the literature (Provenzano, 1968; Tamura, 1970; Cox and Johnston, 2003).

Studies on phyllosoma diet preference have been restricted in that they are held in aquarium tanks and presented with a limited dietary choice and, as such, may not be a true representation of the natural diet. An alternative approach has been to examine the gut contents of wild phyllosomas. However, phyllosomas ingest prey through extensive processing; externally through thorough mastication and internally through a filter press before internalisation into the hepatopancreas, resulting in a highly homogenised amorphous mass and making microscope identification impossible (Smith DM *et al.*, 2009) (Figs 9.10 and 9.11). Nevertheless, numerous undigested nematocysts have been isolated from Scyllarid phyllosoma faecal analysis (Sims and Brown, 1968). Other approaches include molecular markers such as lipid signatures (Jeffs *et al.*, 2004). A more recent approach is the use of nucleotide sequence analysis of gut contents with the assumptions that DNA originating from prey is stable in the gut and that small quantities may be detectable by polymerase chain reaction (Chow *et al.*, 2011). This approach has met with limited success, partly because of the short transit time of food in the phyllosoma gut, being a matter of minutes to hour, such that most phyllosomas collected from plankton tows have evacuated guts as well as potential misleading prey consumption that may have taken place within the plankton net during trawling. Nevertheless, diet items identified by molecular approaches in Palinurid include fish larvae, gastropods, Scyllarid phyllosomas and chaetognaths (Suzuki *et al.*, 2006, 2008; Chow *et al.*, 2011). Overall, the evidence supports the assumption that phyllosomas are opportunistic carnivores capable of capturing and digesting a wide variety of primarily soft-bodied zooplankton and fish larvae.

© Woodhead Publishing Limited, 2013

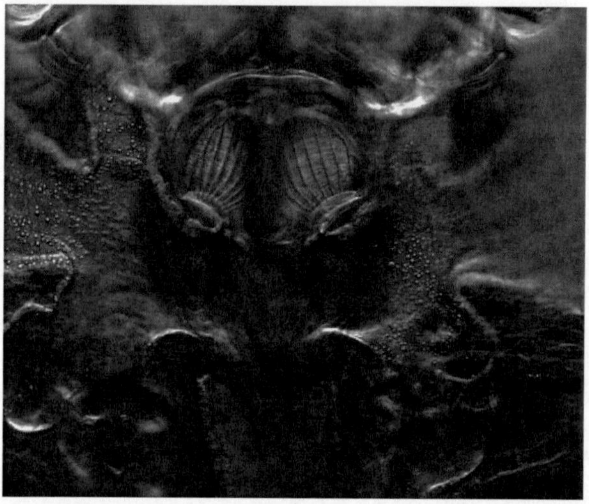

Fig. 9.10 Stereomicrograph of phyllosoma internal mouthparts, showing the filter press (centre) in which food which is finely masticated passes through to the hepatopancreas tubules (to the right and left of the filter press). Food particles that do not pass through the filter press are shunted through the sphincter valve (centre bottom) into the gut tubule and passed as faeces. Photo: M. Salmon

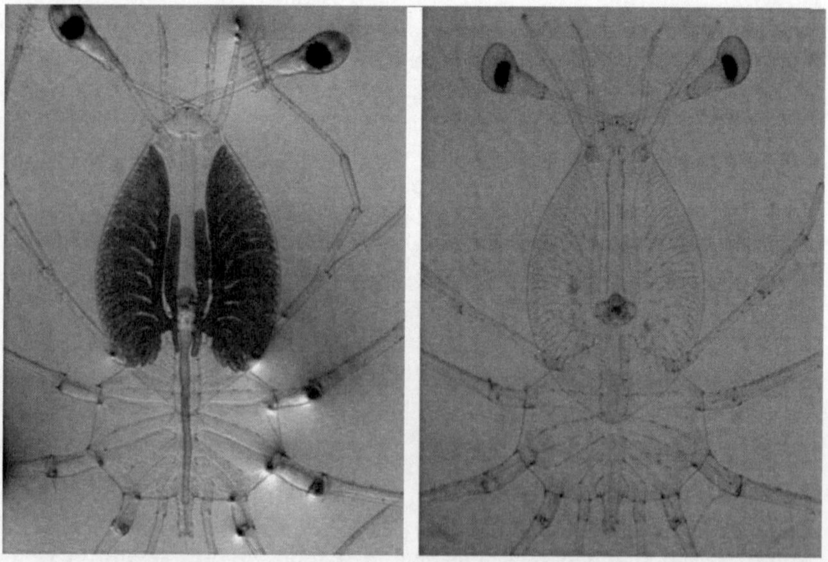

Fig. 9.11 Recently fed phyllosoma with full gut and hepatopancreas (left) compared to unfed empty gut and hepatopancreas individual (right). Photo: M. Salmon

© Woodhead Publishing Limited, 2013

Live and fresh feed

In the absence of a suitable commercially available artificial diet, live and fresh feed sources currently represent the most suitable vehicle for the administration of nutrition for the promotion of adequate growth and survival in phyllosomas. While phyllosomas will readily accept a variety of commercially available live and fresh feed sources, such as squid, mussel mantle and gonad, *Artemia* sp., oyster and clam, it is generally well accepted that *Artemia* sp. and mussel gonad, in combination, are the most common food sources for phyllosoma culture (Takeuchi and Murakami, 2007). However, fluctuations in the supply and nutritional quality of these feed sources results in an unreliable feed source, which potentially impedes the establishment of robust mass rearing protocols.

The use of *Artemia* sp. in commercial aquaculture hatcheries is an essential practice during early life stages, particularly for marine larvae. Stage 1 phyllosomas demonstrate a preference for, and readily take to, newly hatched *Artemia* nauplii and are seemingly able to draw enough nutrition to fuel the requirements of their basal metabolism and moulting through to subsequent developmental stages. However, the nutritional profiles of commercially available *Artemia* strains appear less than adequate for optimal growth and survival and, as such, represent a poor source of protein and essential long-chain fatty acids. Protein and marine oil loading is therefore a common application in marine hatcheries, whereby *Artemia* are subjected to a ~24-hour enrichment period post hatch as a means of providing fortified nutrition (Liddy *et al.*, 2003; Ritar *et al.*, 2003; Nelson *et al.*, 2004; Matsuda *et al.*, 2009). Bio-encapsulation allows *Artemia* to be nutritionally enhanced to provide a more beneficial, easily extractable nutrient source which can be produced on demand and ongrown to larger sizes as phyllosomas progress through their developmental stages. However, this practice can also lead to the introduction of harmful pathogens into the rearing environment as these are known to be more abundant in *Artemia* enriched with lipid emulsions (Høj *et al.*, 2009) and can further proliferate when the nutrient-dense enrichment is excreted by the *Artemia* into the water column.

Mussel gonad, primarily from *Mytilus* sp., is generally viewed as an acceptable nutritive fresh feed source available for phyllosomas. In comparison to *Artemia*, it is characterised by high protein levels (30–70 % on dry weight) and high concentrations of long-chain omega 3 fatty acids (1.5–3.9 % on dry weight) (Takeuchi and Murakami, 2007). Mussel gonad is generally fed during the later stages of larval development in synchrony with *Artemia*, presumably providing a more balanced supply of amino acids and fatty acids and other essential lipid classes. However, the preparation of mussel gonad is labour intensive, firstly requiring shell opening, gonad separation and removal, and subsequent fine chopping to a size suitable for phyllosoma capture and ingestion. While the possibility does exist to store this feed source in the freezer all year round, guaranteeing the nutritional

© Woodhead Publishing Limited, 2013

quality between different batches, it is extremely difficult to retain consistency due to large seasonal fluctuations in nutritional profile.

Under culture conditions, phyllosomas have successfully been grown from egg through the numerous larval stages to puerulus on diets consisting only of *Artemia* and mussel gonad (Kittaka, 1988). For example, Kittaka reported successful rearing from egg to puerulus in temperate *J. edwardsii* and *S. verreauxi* fed a combination of *Artemia* and mussel gonad from the blue mussel *Mytilus edulis* (Kittaka *et al.*, 1997, 2005). In this study, *Artemia* nauplii were fed exclusively over the first two stages followed by mussel gonad for all subsequent stages. Likewise, similar results have been obtained for *J. lalandii* and *P. elephas* by implementing similar feeding strategies (Kittaka, 1988; Kittaka and Ikegami, 1988). However, the extremely low numbers of phyllosomas successfully reaching metamorphosis in all reported studies clearly underlines the issues and shortfalls associated with the feeding of live and fresh feeds, where it is very difficult, to nearly impossible, to deliver complete nutrition for optimal larval development. This highlights the pertinence of artificial diet development where a broader array of nutritional factors can be controlled and fluctuations in the supply of essential nutrients can be eliminated.

Artificial diet development

The development of a suitable artificial diet requires knowledge of the nutritional requirements of the cultured species in order to deliver all nutritional needs. Requirements for individual nutrients are likely to vary between species and much insight into these requirements can be made via the investigation of natural feeding habits. For example, carnivorous species generally require a higher proportion of protein in their diet in comparison to omnivorous or herbivorous species. In the case of phyllosomas, investigation into natural feeding habits has revealed a preference for soft gelatinous zooplankton prey items. Such organisms are high in moisture (>90 %) and contain low levels of lipid (<1.5 %) and protein (<5 %). In consideration of their relative passive, free floating planktonic lifestyle, the interaction between phyllosomas and their prey items is likely dictated by chance, during which opportunistic encounters result in phyllosomas grasping prey in close proximity for consumption. However, phyllosomas have been observed in captivity to show some raptorial feeding, swooping down and capturing prey (Moe, 1991). Such information and observations provide valuable information not so much on the necessary contents/ ingredients of an artificial diet, but more so for the fundamental physical characteristics. For example, in order to maximise availability in the water column, these diets would need to be neutrally buoyant and sufficiently strong to withstand disintegration and minimise nutrient leaching. To date, there are no published studies relative to the quantitative nutritional requirements of phyllosomas. In this situation, it is common to base the nutritional composition of artificial diets on information published for similar species. In the

© Woodhead Publishing Limited, 2013

case of phyllosomas, penaeid prawns and crab species represent the closest proxy for diet development. However, given the large differences between these species, the use of any published data in dietary formulations for phyllosomas is highly speculative and ultimately individual nutrient requirements necessitate elucidation on a species-specific basis. Additionally, given the complexity of the phyllosoma larval cycle, it is also possible that nutritional requirements will change depending on the developmental stage. This is particularly emphasised towards the end of the larval cycle where energy requirements appear to elevate in preparation for the energy taxing process of metamorphosis and subsequent period of non-feeding throughout the puerulus stage. In general, based on published data for other crustacean species, penaeid prawns in particular, the protein requirement of phyllosomas could be expected to fall in the range 23–57 % dry weight and lipid from 5–10 %. Additionally, provision would also need to be made for other essential nutrients including cholesterol and phospholipids which crustaceans cannot produce *in vivo*.

9.6 Metamorphosis to puerulus and settlement to juvenile

The final phyllosoma metamorphic moult to puerulus marks the penultimate moult before the lobster becomes fully benthic. There has been much speculation on whether the metamorphosis moult is triggered by internal or external mechanisms (Dennis *et al.*, 2001). Metamorphosis, if it occurs, appears to occur spontaneously in hatcheries and there appears little need to seek critical triggers of the event. However, metamorphosis is often associated with high mortality, either through a type of MDS or other causes. With limited knowledge of the dietary requirements of phyllosomas, hatchery reared ones may be compromised and the cause of death may be a lack of some critical nutritional component. Nutritionally enriched diets have been shown to improve the number of final phyllosoma instars succeeding in metamorphosis as well as final survival (Kittaka, 2000). As the non-feeding puerulus stage relies exclusively on the energy reserves accumulated during the phyllosoma stages, nutrition is likely to be a critical factor in puerulus survival to settlement (Hayakawa and Nishida, 2005). Pueruli will readily settle onto available benthic structures and within a few days develop colouration as the carapace begins to calcify followed by a moult to juvenile and return to feeding.

9.7 Future trends

The larval cycle of several species of Palinurid lobsters has been successfully completed. In the spiny and rock lobsters of the family Palinuridae this has been restricted to research-scale production. In contrast, for slipper lobsters

© Woodhead Publishing Limited, 2013

(Theninae) of the family Scyllaridae, the technology has reached the scale of commercial larval rearing (Mikami, 2007). This is partly due to the fact that the larval period of *Thenus orientalis* is a fraction of that of Palinurid lobsters, being only 25–30 days (Mikami and Greenwood, 1997). The short larval phase combined with larval survival rates of 80 % have opened the way to commercial-scale production of Scyallrids. For Palinurid lobsters, the main bottleneck of successful commercial production is the poor production performance of late-stage phyllosomas and inferior quality of puerulus. Research is likely to be focused on the primary pillars of health and nutrition. These two factors interact, as poor health is often linked to a poor nutritional status, potentially due to an inadequate diet leading to limited growth. Likewise, poor nutrition is likely to lead to increased mortality by opportunistic pathogens which can breach the limited defences of phyllosomas. On the other hand, phyllosomas exposed to a high load of potential pathogens or biofouling organisms may have a limited ability to catch and process feed items leading to a reduced nutritional status. Health and nutrition must be optimised to produce high larval survival and metamorphosis from phyllosoma to puerulus to juvenile to be reliable and consistent to meet the requirements of a commercial-scale production.

9.8 Acknowledgements

Much of the knowledge that has been accumulated for the development of hatchery protocols for phyllosomas has been due to the professionalism and dedication of the Tropical Aquaculture staff at the Australian Institute of Marine Science (AIMS) and includes Justin Hochen, Katie Holroyd, Grant Milton and Greg Smith amongst others. Research into microbiological and nutritional aspects of hatchery rearing have been made possible by the efforts of Rochelle Soo, Emmanulle Botte and Rose Cobb and a suite of students and volunteers. Several granting agencies, including some private companies, have been instrumental in support of research in this area and include the Fisheries Research and Development Corporation (FRDC), the McCloy Group Ltd., Lobster Harvest Ltd. and others.

9.9 References

ANGER K and DAWIRS R R (1981) 'Influence of starvation on the larval development of *Hyas araneus* (Decapoda, Majidae)', *Helgol and wiss Meer*, 34, 287–311.

BARROWS F T, ZITZOW R E and KINDSCHI G A (1993) 'Effects of water spray, diet and phase feeding on swim bladder infation, survival and cost of production of intensively reared larval walleyes', *Prog Fish-Cult*, 55, 224–228.

BATHAM E J (1967) 'The first three larval stages and feeding behaviour of phyllosomas of the New Zealand Palinurid crayfish', *Trans Royal Soc NZ*, 9, 53–64.

© Woodhead Publishing Limited, 2013

BEARD T W and WICKINS J W (1992) *Techniques for the production of juvenile lobsters (Homarus gammarus)*, Fisheries Research Technical Report 92. Lowestoft: MAFF Directorate of Fisheries Research.

BEARD T W, RICHARDS P R and WICKINS J F (1985) *The techniques and practicability of year-round production of lobster, Homarus gammarus, in laboratory recirculation systems*, Fisheries Research Technical Report 79. Lowestoft: MAFF Directorate of Fisheries Research.

BEHRINGER D C, BUTLER M J and SHIELDS J D (2009) 'A review of the lethal spiny lobster virus PaV1 – Ten years after its discovery', *Proceedings of the 62nd Gulf and Caribbean Fisheries Institute*, 62, 370–375.

BERMUDES M and RITAR A J (2008) 'Response of early stage spiny lobster *Jasus edwardsii* phyllosoma larvae to changes in temperature and photoperiod', *Aquaculture*, 281, 63–69.

BOBES R, DIAZ J and DIAZ E (1988) 'Aislamiento e identificacion de *Aerococcus virdans ver. homari.* en la langosta *Panulirus argus* con sintomas de septicemia', *Revista De Invertigaciones Marinas*, 9, 97–103.

BOGHEN A D (1982) 'Effects of Wescodyne and malachite green on parasitic ciliates of juvenile American lobsters', *Prog Fish-Cult*, 44, 97–99.

BOOTH J D (1995) 'Phyllosoma reared to settlement', *The Lobster Newsletter*, 8, 1–12.

BOOTH J D (2006) *Jasus* species, in Phillips B F (ed.), *Lobsters: biology, management, aquaculture and fisheries*. Oxford: Blackwell Publishing, 340–358.

BOURNE D, YOUNG N, WEBSTER N, PAYNE M, SALMON M, DEMEL S and HALL M R (2004) 'Microbial community dynamics in a larval aquaculture system of the tropical rock lobster, *Panulirus ornatus*', *Aquaculture*, 242, 33–51.

BOURNE D, HØJ L, WEBSTER N, SWAN J and HALL M R (2006) 'Biofilm development within a larval rearing tank of the tropical rock lobster, *Panulirus ornatus*', *Aquaculture*, 260, 27–38.

BOURNE D, WEBSTER N, PAYNE M, L. H, SKINDERSO M, GIVSKOV M and HALL M R (2007) 'Aspects of the microbiology of phyllosoma rearing in the ornate rock lobster *Panulirus ornatus*', *Aquaculture*, 268, 274–287.

BOWSER P R and ROSEMARK R (1981) 'Mortalities of cultured lobsters, *Homarus*, associated with a molt death syndrome', *Aquaculture*, 23, 11–18.

BRATTEY J, CAMPBELL A, BAGNALL A E and UHAZY L S (1985) 'Geographic distribution and seasonal occurrence of the nemertean *Pseudocarcinonemertes homari* on the American lobster, *Homarus americanus*', *Can J Fish Aqua Sci*, 42, 360–367.

BRINKLEY A W, ROMMEL F A and HUBER T W (1976) 'The isolation of *Vibrio parahaemloyticus* and related vibrios from moribund aquarium lobsters', *Can J Fish Aqua Sci*, 22, 315–317.

BURROWS R and CHENOWETH H H (1970) 'The rectangular circulating rearing pond', *Prog Fish Cult*, 32, 67–80.

BUTLER M J, BEHRINGER D C and SHIELDS J D (2008) 'Transmission of Panulirus argus virus 1 (PaV1) and its effect on the survival of juvenile Caribbean spiny lobster', *Dis Aquat Org*, 79, 173–182.

CABELLO F C (2006) 'Heavy use of prophylactic antibiotics in aquaculture: a growing problem for human and health and for the environment', *Environ Microbiol*, 8, 1137–1144.

CALADO R, NARCISO L, MORAIS S, RHYNE A L and LIN J (2003) 'A rearing system for the culture of ornamental decapod crustacean larvae', *Aquaculture*, 218, 329–339.

CANO-GOMEZ A, BOURNE D, HALL M R, OWENS L and HØJ L (2009) 'Molecular diagnosis of Vibrio harveyi in aquaculture: current methods and future prospects', *Aquaculture*, 287, 1–10.

CANO-GOMEZ A, GOULDEN E F, OWENS L and HØJ L (2010) 'Vibrio owensii sp. nov., isolated from cultured crustaceans in Australia', *FEMS Microbiol Lett*, 302, 175–181.

© Woodhead Publishing Limited, 2013

CANO-GOMEZ A, HØJ L, OWENS L and ANDREAKIS N (2011) 'Multilocus Sequence Analysis (MLSA) provides basis for fast and reliable indentification of *Vibrio-harveyi*-related species and confirms previous misidentification of important marine pathogens', *Syst Appl Microbiol*, 34, 561–565.

CARLSON B A (1996) 'How to build a powerful surge device', *Seascope*, 13, 1–2.

CARVALHO R C A and OGAWA M (2000) Economic assessment of marine grow-out cage for spiny lobster in Ponta Grossa Beach, Ceara, Brazil. *Sixth International Conference and Workshop on Lobster Biology and Management*, 10–15 September, Key West, FL, 34.

CASTAING J B, MASSE A, SECHET V, SABIRI N-E, PONTIE M, HAURE J and JAOUEN P (2011) 'Immersed hollow fibres microfiltration (MF) for removing undersirable micro-algae and protecting semi-closed aquaculture basins', *Desalination,* 276, 386–396.

CHAN T-Y (2010) 'Annotated checklist of the world's marine lobsters (Crustacea: Decapoda: Astacidea, Glypheidea, Achelata, Polychelida)', *The Raffles Bull Zool*, 23, 153–181.

CHANG-SHENG C, CHEN Z, HU J, YE Z, JI D and LAN L (2001) 'Study on halotolerance of phyllosoma larva of Chinese spiny lobster (*Panulirus stimpson* Holthuis)', *Acta Oceanological SinicaHaiyang Xuebao*, 20, 409–415.

CHANG P S, CHEN H C and WANG Y C (1998) 'Detection of white spot syndrome associated with baculovirus in experimentally infected wild shrimp, crab and lobsters by *in situ* hybridisation', *Aquaculture*, 164, 233–242.

CHARMANTIER G, CHARMANTIER-DAURES M, WADDY S L and AIKEN D E (1991) 'Salinity tolerance and osmoregulation in the nermertean *Pseudocarcinonemertes homari,* an egg predator of American lobster, *Homarus americanus*', *Can J Fish Aqua Sci*, 48, 209–214.

CHOW S, SUZUKI S, MATSUNAGA T, LAVERY S, JEFFS A and TAKEYAMA H (2011) 'Investigation on natural diets of larval marine animals using peptide nucleic acid (PNA)-directed PCR clamping', *Mar Biotechnol*, 13, 305–313.

COX S L and JOHNSTON D J (2003) 'Developmental changes in the structure and function of mouthparts of phyllosoma larvae of the packhorse lobster, Jasus verreauxi (Decapoda : Palinuridae)', *J Exp Mar Bio Eco*, 296, 35–47.

CRECELIUS E A (1979) 'Measurements of oxidants in ozonized seawater and some biological reactions', *J Fish Res Board Can*, 36, 1006–1008.

CROSNIER A (1972) 'Naupliosoma, phyllosomes et pseudibacus de *Scyllarides herklotsi* (Herklots) (Crustacea, Deacapoda, Scyllaridae) recoltes par l'Ombango dans le sud de Golfe de Guinee', *Cah ORSTOM, Ser Oceanogr*, 10, 139–149.

CROTHERS-STOMPS C, HØJ L, BOURNE D, HALL M and OWENS L (2009) 'Isolation of lytic bacteriophage against *Vibrio harveyi*', *J Appl Microbiol*, 18, 1744–1750.

D'ABRAMO L R, BORDNER C E, CONKLIN D E and BAUM N A (1981) 'Essentiality of dietary phosphatidylcholine for the survival of juvenile lobsters', *J Nut,* 111, 425–431.

DALL W, HILL B J, ROTHLISBERG P C and STAPLES D J (1990) *The Biology of the Penaeidae.* London: Academic Press.

DANIELSBERG A, BERG A and LUNDE T (1993) Design of inlets for incubation of yolk sac larvae of Atlantic halibut (*Hippoglossus hippoglossus*), in Reinertsen H, Dahle L A, Jorensen L and Tvinnereim K (eds), *Fish Farming Technology.* Rotterdam: Balkema, 425–428.

DAWSON M N (2000) 'Variegated mesocosms as alternatives to shore-based planktokreisels: notes on the husbandry of jellyfish from marine lakes', *J Plankton Res*, 22, 1673–1682.

DEBLOCK S, WILLIAMS A and EVANS L H (1991) 'Contribution a l'etude des Microphallidae Travassos 1920 (Trematoda). Description de *Thulakiotrema genitae* n. gen. n. sp., metacercaire parasite de langoustes australiennes', *Bull du Museum Nationale D'Historie Naturelle, Paris*, 12, 563–276.

© Woodhead Publishing Limited, 2013

DEFOIRDT T, BOON N, BOSSIER P and VERSTRAETE W (2004) 'Disruption of bacterial quorum sensing: an unexplained strategy to fight infections in aquaculture', *Aquaculture*, 240, 69–88.

DENNIS D M and MUNDAY B L (1994) 'Microsporidiosis of palinurid lobsters from Australian waters', *Bull Eur Assoc Fish Pathol*, 14, 16–18.

DENNIS D M, PITCHER C R and SKEWES T D (2001) 'Distribution and transport pathways of *Panulirus ornatus* and *Panulirus* spp. larvae in the Coral Sea', *Aus Mar Freshwater Res*, 52, 1175–1185.

DEXTER D M (1972) 'Molting and growth in laboratory reared phyllosomas of the California spiny lobster *Panulirus interruptus*', *Calif Fish Game*, 58, 107–115.

DIGGLES B K (2001) 'A mycosis of juvenile spiny rock lobster, *Jasus edwardsii* (Hutton, 1875) caused by *Haliphthoros* sp., and possible methods of chemical control', *J Fish Dis*, 24, 99–110.

DIGGLES B K, MOSS G A, CARSON J and ANDERSON C D (2000) 'Luminous vibriosis in rock lobster *Jasus verreauxi* phyllosoma larvae associated with infection by *Vibrio harveyi*', *Dis Aquat Org*, 43, 127–137.

DOVE A D M, LOBUE C, BOWSER P and POWELL M (2004) 'Excretory calcinosis: a new fatal disease of wild American lobsters *Homarus americanus*', *Dis Aquat Org*, 58, 215–221.

DUPRE E and GUISADO C (1996) 'Early stages of phyllosoma of the spiny lobster of Juan Fernandez *Jasus frontalis* maintained in laboratory conditions', *Invest Mar*, 29, 39–50.

FAO (2008) *FAO Yearbook. Fishery and Aquaculture Statistics. 2008.* Rome: FAO.

FAO/OIE/WHO (2006) *Antimicrobial Use in Aquaculture and Antimicrobial Resistance*, Geneva.

FIELD R H, CHAPMAN C J, TAYLOR A C, NEIL D M and VICKERMAN K (1992) 'Infection of the Norway lobster *Nephrops norvegicus* by a *Hematodinium*-like species of dinoflagellate on the west coast of Scotland', *Dis Aquat Org*, 13, 1–15.

FINDLEY A M, BLAKENEY E W and WEIDNER E H (1981) 'Ameson michaelis (microsporidia) in the blue crab, *Callinectes sapidus*: parasite-induced alterations in the biochemical composition of host tissues', *Biol Bull*, 161, 115–125.

FLORETO E A T, PRINCE D L, BROWN P B and BAYER R C (2000) 'The biochemical profiles of shell-diseased American lobsters, *Homarus americanus*', *Aquaculture*, 188, 247–262.

FOXON G E H (1936) 'A note on recapitulation in the larvae of the Decapoda Crustacea', *Ann Mag Nat Hist*, 18, 117–123.

GEMENDE B, GERBETH A, PAUSCH N and VON BRESINSKY A (2008) 'Test for the application of membrane technology in a new method for intensive aquaculture', *Desalination*, 224, 57–63.

GIL-TURNES M S and FENICAL W (1992) 'Embryos of *Homarus americanus* are protected by epibiotic bacteria', *Biolog Bull*, 182, 105–108.

GIL-TURNES M S, HAY M E and FENICAL W (1989) 'Symbiotic marine bacteria chemically defend crustacean embryos from a pathogenic fungus', *Science*, 246, 116–118.

GOLDSTEIN J S, MATSUDA H, TAKENOUCHI T and BUTLER M J (2008) 'The complete development of larval Caribbean spiny lobster *Panulirus argus* (Latreille, 1804) in culture', *J Crustacean Biol*, 28, 306–327.

GOMEZ DEL PRADO-ROSAS M C, ALVAREZ-CADENA J N, LAMOTHE-ARGUMEDO R and GRANO-MALDONAO M I (2003) '*Cymatocarpus solearis* a brachycoeliid metacercaria parasitizing *Panulirus argus* (Crustacea: Decapoda) from the Mexican Caribbean Sea', *An Inst Biol Univ Na Auton Mex Ser Zool*, 74, 1–10.

GORE R H (1985) Molting and growth in decapod larvae, in Wenner A M (ed.), *Crustacean Issues*. Rotterdam: Balkema, 1–66.

© Woodhead Publishing Limited, 2013

GOULDEN E, HALL M R, BOURNE D G, PEREG L L and HØJ L (2012a) 'Pathogenicity and infection cycle of *Vibrio owensii* in larval culture of ornate spiny rock lobster (*Panulirus ornatus*)', *Appl Environ Microbiol*, 78, 2841–2849.

GOULDEN E F, HALL M R, PEREG L L and HØJ L (2012b) Identification of an antagonistic probiotic combination protecting ornate spiny lobster *(Panulirus ornatus)* larvae against *Vibrio owensii* infection, *PLoS One*, 7, e39667.

GREVE W (1968) 'The "planktonkreisel". a new device for culturing zooplankton', *Mar Biol*, 1, 201–203.

GREVE W (1970) 'Cultivation experiments on North Sea ctenphores', *Helgol Wiss Meer*, 20, 304–317.

GREVE W (1975) 'The "Meteor Planktoncuvette": A device for the maintenance of macrozooplankton aboard ships', *Aquaculture*, 6, 77–82.

GURNEY R (1942) *Larvae of Decapod Crustacea.* Dorking: Adlard & Son Ltd.

HALL M R and DE LA VEGA E (2004) Physiological response to stress and health implications in crustacea, in Goarant C, Harache Y, Herblan C and Mugnier C (eds), *Styli 2003: Thirty Years of Shrimp Farming in New Caledonia*. Poluzane: INRA, 39–56.

HAMNER W M (1990) 'Design developments in the planktonkreisel, a plankton aquarium for ships at sea', *J Plankton Res*, 12, 397–402.

HART G (2009) *Assessing the South-East Asian tropical lobster supply and major market demands*, SADI-ACIAR Research Report, FR2009–06, 55. Canberra: ACIAR.

HAYAKAWA Y and NISHIDA S (2002) 'Diel behavior of the puerulus and the first instar of the red rock lobster, *Jasus edwardsii*', *Fish Sci*, 68, 393–396.

HATTORI T and OISHI Y (1899) 'Hatching experiment on Ise lobster', *Rep Imp Fish Inst*, 1, 76–132.

HERRNKIND W, HALUSKY J and KANCIRUK P (1976) 'A further note on phyllosoma larvae associated with medusae', *Bull Mar Sci*, 26, 110–112.

HIBBETT D S, BINDER M, BISCHOFF J F and BLACKWELL M (2007) 'A higher-level phylogenetic classification on the fungi', *Mycol Res*, 111, 509–547.

HØJ L, BOURNE D G and HALL M R (2009) 'Localisation, abundance and community structure of bacteria associated with Artemia: effectso nauplii enrichment and antimicrobial treatment', *Aquaculture*, 293, 278–285.

HOOKER S H, JEFFS A G, CREESE R G and SIVAGURU K (1997) 'Growth of captive *Jasus edwardsii* (Hutton) (Crustacea:Palinuridae) in north-east New Zealand', *Mar Freshwater Res*, 48, 903–909.

HORITA T (2007) Challenge of culturing phyllosomata, in Nishi G and Saruwatari T (eds), *Work at Aquariums*. Hatano: Tokai University Press, 84–98.

HUGHES J T, SHLESER R A and TCHOBANOGLOUS G (1974) 'A rearing tank for lobster larvae and other aquatic species', *Prog Fish-Cult*, 36, 129–133.

HUNT D E, GEVERS D, VAHORA N M and POLZ M F (2008) 'Conservation of the chitin utilization pathway in the *Vibrionaceae*', *Appl Environ Microbiol*, 74, 44–51.

IGARASHI M A, KITTAKA J and KAWAHARA E (1990) 'Phyllosoma culture with inoculation of marine bacteria', *Bull Jap Soc Sci Fish*, 56, 1781–1786.

IKEDA T, SMITH G G, MCKINNON A D and HALL M R (2011) 'Metabolism and chemical composition of phyllosoma larvae, with special reference to the tropical rock lobster *Panulirus ornatus* (Decapod; Palinuridae)', *J Exp Mar Biol Ecol*, 405, 80–86.

ILLINGWORTH J, TONG L J, MOSS G A and PICKERING T D (1997) 'Upwelling tank for culturing rock lobster (Jasus edwardsii) phyllosomas', *Mar Freshwater Res*, 48, 911–914.

INOUE M (1965) 'On the relation of amount of food taken to the density and size of food and water temperature in rearing the phyllosoma of the Japanese spiny lobster, *Palinurus japonicus*', *Bull Jap Soc Sci Fish*, 31, 902–906.

© Woodhead Publishing Limited, 2013

INOUE M (1978) 'Studies on the cultured phyllosoma larvae of the Japanese spiny lobster *Panulirus japonicus*', *Bull Jap Soc Sci Fish*, 44, 457–475.

INOUE M (1981) 'Studies on the cultured phyllosoma larvae of the Japanese spiny lobster, *Panulirus japonicus* (V. Siebold)', *Special Rep Kanagawa Pref Fish Exp Stn*, 1, 1–19 (in Japanese).

JAWAJAR A, KALEEMUR R and LEEMA J (1996) 'Bacterial disease in cultured spiny lobster, *Panulirus homarus* (Linnaeus)', *J Aquac Trop*, 11, 187–192.

JEFFS A G, NICHOLES P D, MOONEY B D, PHILLIPS K L and PHLEGER C F (2004) 'Identifying potential prey of the pelagic larvae of the spiny lobster *Jasus edwardsii* using lipid signatures', *Comp Biochem Physiol Part B*, 137, 487–507.

JENSEN M A, RITAR A J, BURKE C and WARD L R (2011) 'Seawater ozonation and formalin disinfection for the larval culture of eastern rock lobster *Jasus* (Sagmariasus) *verreauxi*, phyllosoma', *Aquaculture*, 318, 213–222.

JOHNSTON M W (1956) 'The larval development of the California spiny lobster, *Panulirus interruptus* with notes on *Panulirus gracilis*', *Proc Calif Acad Sci*, 29, 775–793.

JOHNSTON M D, JOHNSTON D J and KNOTT B (2008) 'Ontogenetic changes in the structure and function of the mouthparts and foregut of early and late stage *Panulirus ornatus* phyllosomata (Decapoda: Palinuridae)', *J Crustcean Biol* 28, 46–56.

JONES C M (2009) Advances in the culture of lobsters, in Burnell G and Allen G L (eds), *New Technologies in Aquaculture: Improving production efficiency, quality and environmental management.* Cambridge: Woodhead Publishing and CRC Press, 822–844.

KENWAY M, SALMON M, SMITH G G and HALL M (2009) Potential of sea cage aquaculture of *Panulirus ornatus* in Australia, in Williams K C (ed.), *Spiny Lobster Ecololgy and Exploitation in the South China Sea Region*, Canberra: ACIAR, 18–25.

KHARAS H (2010) *The emerging middle class in developing countries*, OECD Development Centre, Working Paper No. 285. Paris: OECD.

KIRYU M Y, BEHRINGER D C, LANDSBERG J H and PETTY B D (2009) 'Microsporidiosis in the Caribbean spiny lobster *Panulirus argus* from southeast Florida, USA', *Dis Aquat Org*, 84, 237–242.

KITTAKA J (1988) 'Culture of the Palinurid *Jasus lalandii* from egg stage to puerulus', *Nippon Suisan Gakk*, 54, 87–93.

KITTAKA J (1997) 'Culture of larval spiny lobsters: a review of work done in northern Japan', *Mar Freshwater Res*, 48, 923–930.

KITTAKA J (2000) Culture of larval spiny lobsters, in Phillips B F and Kittaka J (eds), *Spiny Lobsters: Fisheries and Culture.* Oxford: Fishing News Books, 508–532.

KITTAKA J and ABRUNHOSA F A (1997) 'Characteristics of palinurids (Decapoda; Crustacea) in larval culture', *Hydrobiologia*, 358, 305–311.

KITTAKA J and BOOTH J D (2000) Prospectus for aquaculture, in Phillips B F and Kittaka J (eds), *Spiny Lobsters: Fisheries and Culture.* Oxford: Fishing News Books, 465–473.

KITTAKA J and IKEGAMI E (1988) 'Culture of the palinurid *Palinurus elephas* from egg stage to puerulus', *Nippon Suisan Gakk*, 54, 413–417.

KITTAKA J and KIMURA K (1990) 'Culture of the Japanese spiny lobster *Panulirus japonicus* from egg to juvenile stage', *Nippon Suisan Gakkaishi*, 55, 963–970.

KITTAKA J, ONO K and BOOTH J D (1997) 'Complete development of the green rock lobster, *Jasus verreauxi* from egg to juvenile', *Bull Mar Sci*, 61, 57–71.

KITTAKA J, KUDO R, ONODA S, KANEMARU K and MERCER J P (2001) 'Larval culture of the European spiny lobster Palinurus elephas', *Mar Freshwater Res*, 52, 1439–1444.

KITTAKA J, ONO K, BOOTH J D and WEBBER W R (2005) 'Development of the red rock lobster, *Jasus edwardsii*, from egg to juvenile', *NZ J Mar Freshwater Res*, 39, 263–277.

© Woodhead Publishing Limited, 2013

KNOWLTON R E (1974) 'Larval development processes and controlling factors in decapod Crustacea, with emphasis on Caridea', *Thalassia Jugosl*, 10, 138–158.

LANGON J S (1991) 'Microsporidiosis due to a pleistophorid in marron, *Cherax tenuimanus* (Smith)', *J Fish Dis*, 14, 33–44.

LEBOUR M V (1925) 'Young anglers in captivity and some of their enemies. A study in a plunger jar', *J Mar Bio Assoc UK*, 13, 721–734.

LIDDY G C, PHILLIPS B F and MAGUIRE G B (2003) 'Survival and growth of instar 1 phyllosoma of the western rock lobster, *Panulirus cygnus*, starved before or after periods of feeding', *Aquac Int*, 11, 53–67.

LUIS O J and CALADO R (2009) 'Captive breeding of the Eastern Atlantic spiny lobster *Panulirus regus* in a recirculation system: a candidate for aquaculture in temperature zones?', *World Aquac*, 40, 22–24.

MATSUDA H (2006) 'Studies on the larval culture and development of Panulirus lobsters', *Nippon Suisan Gakk*, 72, 827–830.

MATSUDA H and TAKENOUCHI T (2007) 'Development of technology for larval *Panulirus japonicus* culture in Japan', *Bull Fish Res Agen*, 20, 77–84.

MATSUDA H and YAMAKAWA T (2000) 'The complete development and morphological changes of larval Panulirus longipes under laboratory conditions', *Fish Sci*, 66, 278–293.

MATSUDA H, TAKENOUCHI T and YAMAKAWA T (2002) 'Effects of photoperiod and temperature on ovarian development and spawning of the Japanese spiny lobster *Panulirus japonicus*', *Aquaculture*, 205, 385–398.

MATSUDA H, TAKENOUCHI T and GOLDSTEIN J S (2006) 'The complete larval development of the pronghorn spiny lobster *Panulirus pencillatus* (Decapod: Palinuridae) in culture', *J Crustaean Biol*, 26, 579–600.

MATSUDA H, TAKENOUCHI T, TANAKA S and WATANABE S (2009) 'Relative contribution of *Artemia* and mussel as food for cultured middle-stage *Panulirus japonicus* phyllosomata as determined by stable nitrogen isotope analysis', *NZ J Mar Freshwater Res*, 43, 217–224.

MERCER J P, MADDOCK T, BROWNE R and O'CEIDIGH (1997) 'Solving the crawfish (*Palinurus elephas*) enigma', *Aquac Ireland*, 79, 13–15.

MIKAMI S (1995) *Larviculture of Thenus (Decapoda. Scyllaridae), the Moreton Bay Bugs,* PhD thesis, University of Queensland.

MIKAMI S (2003) *Crustacean larva raising method and apparatus.* US Patent 6,561,134.

MIKAMI S (2007) 'Prospects of aquaculture on bay lobsters (*Thenus* spp.)', *Bull Fish Res Agency*, 20, 45–50.

MIKAMI S and GREENWOOD J G (1997) 'Influence of light regimes on phyllosomal growth and timing of moulting in *Thenus orientalis* (Lund) (Decapod: Scyallridae)', *Mar Freshwater Res*, 48, 777–782.

MOE M A (1991) *Lobsters – Florida, Bahamas, the Caribbean.* Plantation, FL: Green Turtle Publication.

MOSS G A, TONG L J and ILLINGWORTH J (1999) 'Effects of light intensity and food density on the growth and survival of early-stage phyllosoma larvae of the rock lobster Jasus edwardsii', *Mar Freshwater Res*, 50, 129–134.

MOSS G A, JAMES P J and TONG L J (2000) '*Jasus verreaux*i cultured', *The Lobster Newsletter*, 13, 9–10.

MULLEN T E, RUSSELL R, TUCKER M T, MARATEA J L, KOERTING C, HINCKLEY L, DE GUISE S, FRASCA S J and FRENCH R A (2005) 'Paramobiasis associated with mass mortality of American lobster *Homarus americanus* in Long Island Sound, USA', *J Aqua Anim Health*, 16, 29–38.

MURAKAMI K, JIMBO T and HAMASAKI K (2007) 'Aspects of the technology of phyllosoma rearing and metamorphosis form phyllosoma to puerulus in the Japanese spiny lobster *Panulirus japonicus* reared in the laboratory', *Bull Fish Res Agen*, 20, 59–68.

© Woodhead Publishing Limited, 2013

MYHRE P, DANIELSBERG A and BERG L (1993) Experiments on upwelling tank systems for halibut larvae (*Hippoglossus hippoglossus*), in Reinertsen H, Dahle L A, Jorensen L and Tvinnereim K (eds), *Fish Farming Technology*. Rotterdam: Balkema, 421–424.

NAKAI T and PARK S C (2002) 'Bacteriophage therapy of infectious diseases in aquaculture', *Res Microbiol*, 153, 13–18.

NELSON M M, CREAR B J, NICHOLS P D and RITZ D A (2004) 'Growth and lipid composition of phyllosoma of the southern rock lobster, *Jasus edwardsii*, fed enriched *Artemia*', *Aquac Nut*, 10, 237–246.

NILSON E H, FISHER W S and SHLESER R A (1975) 'Filamentous infestations observed on eggs and larvae of cultured crustaceans', *Proc World Maricul Soc,* 6, 367–375.

NILSON E H, FISHER W S and SHLESER R A (1976) 'A new mycosis of larval lobster (*Homarus americanus*)', *J Inverteb Patho*, 27, 177–183.

OSHIMA Y (1936) 'Feeding habit of Ise lobster', *Suisan Gakkai Ho*, 7, 16–21.

OVERTON S V and BLAND C E (1981) 'Infection of *Artemia* saline by *Haliphthoros milfordensis*: A scanning and transmission electron microscope study', *J Inverteb Pathol*, 37, 249–257.

PAYNE M, HALL M R, BANNISTER R, SLY L and BOURNE D (2006) 'Microbial diversity within the water column of a larval rearing system for the ornate rock lobster (*Panulirus ornatus*)', *Aquaculture*, 258, 80–90.

PAYNE M S, HALL M R, SLY L and BOURNE D G (2007) 'Microbial diversity within early stage cultured *Panulirus ornatus* phyllosomas', *Appl Envir Microbiol*, 73, 1940–1951.

PAYNE M S, HOJ L, WIETZ M, HALL M R, SLY L and BOURNE D G (2008) 'Microbial diversity of mid-stage palinurid phyllosoma from great barrier reef waters', *J Appl Microbiol*, 105, 340–350.

PHILLIPS B F and MATSUDA H (2011) A global review of spiny lobster aquaculture, in Fotedar R K and Phillips B F (eds), *Recent Advances and New Species in Aquaculture*, Oxford: Wiley-Blackwell, 22–84.

PHILLIPS B F and MELVILLE-SMITH R (2006) Panulirus species, in Phillips B F (ed.), *Lobsters: Biology, Management, Aquaculture and Fisheries*. Oxford: Blackwell Publishing, 359–384.

PORTER L, BUTLER M and REEVES R (2001) 'Normal bacterial flora of the spiny lobster Panulirus argus and its possible role in shell disease', *Mar Freshwater Res*, 52, 1401–1405.

PROVENZANO A J (1968) 'Recent experiments on the laboratory rearing of tropical rock lobster larvae', *Gulf Carib Fish Inst*, 21, 152–157.

QUINTANA Y C, CANUL R R and MARTINEZ V M V (2011) 'First evidence of Panulirus argus Virus 1 (PaV1) in spiny lobster from Cuba and clinical estimation of its prevalence', *Dis Aquat Org*, 92, 141–147.

RADHAKRISHNAN E V (1977) 'Breeding of laboratory reared spiny lobster *Panulirus homarus* (Linnaeus) under controlled conditions', *Indian J Fish*, 24, 269–270.

RADHAKRISHNAN E V and VIJAYAKUMARAN M (2000) Problems and prospects for lobster farming in India, in Pillai V N and Menon N G (eds) *Marine Fisheries Research and Management*. Kerala: Central Marine Fisheries Research Institute, 753–764.

REITAN K I, EVJEMO J O, OLSEN Y, SALVESEN I, SKJERMO J, VADSTEIN O and OIE G (1993) Comparison of incubator concepts for yolk sac larvae of Atlantic halibut (*Hippoglossus hippoglossus*), in Reinertsen H, Dahle L A, Jorensen L and Tvinnereim K (eds) *Fish Farming Technology*. Rotterdam: Balkema, 51–56.

RICE A L and WILLIAMSON D I (1970) 'Methods for rearing larval decapod Crustacea', *Helgol Wiss Meer*, 20, 417–434.

RITAR A J (2001) 'The experimental culture of phyllosoma larvae of southern rock lobster (*Jasus edwardsii*) in a flow-through system', *Aquacul Eng*, 24, 149–156.

© Woodhead Publishing Limited, 2013

RITAR A and SMITH G G (2005) 'Hatchery production of southern rock lobster in Tasmania', *Austasia Aquaculture*, February/March, 42–43.

RITAR A J, DUNSTAN G A, CREAR B J and BROWN M R (2003) 'Biochemical composition during growth and starvation of early larval stages of cultured spiny lobster (*Jasus edwardsii*) phyllosoma', *Comp Biochem Physiol A Mol Integr Physiol*, 136, 353–370.

RITAR A, SMITH G G and THOMAS C W (2006) 'Ozonation of seawater improves the survival of larval southern rock lobster, *Jasus edwardsii*, in culture from egg to juvenile.', *Aquaculture*, 261, 1014–1025.

ROZA D and HATAI K (1999) 'Atkinsiella dubia infection in the larvae of Japanese mitten crab, *Eriocheir japonicus*', *Myoscience*, 40, 235–240.

SACHLIKIDIS N G, JONES C M and SEYMOUR J E (2005) 'Reproductive cues in *Panulirus ornatus*', *NZ J Mar Freshwater Res*, 39, 305–310.

SAISHO T (1990) 'A note of the phyllosoma stages of spiny lobster', *Bull Planktol Soc Japan*, 13, 69–71.

SANDIER P A, ZIELINSKI P B and CASTRO W E (1974) 'A simple airlift-operated tank for closed-system culture of decapod crustacean larvae and other small aquatic animals', *Helgol Wiss Meer*, 26, 82–87.

SANDIFER P A, SMITH T I J and CALDER D R (1974) 'Hydrozoans as pests in closed-system culture of larval decapod crustaceans', *Aquaculture*, 4, 55–59.

SEKINE S, SHIMA Y, FUSHIMI H and NONAKA M (2000) 'Larval period and molting in the Japanese spiny lobster Panulirus japonicus under laboratory conditions', *Fish Sci*, 66, 19–24.

SHIELDS J D (2011) 'Diseases of spiny lobsters: A review', *J Invertr Pathol*, 106, 79–91.

SHIELDS J D, STEPHENS F J and JONES B (2006) Pathogens, parasites and other symbionts, in Phillips B F (ed.), *Lobsters: Biology, Management, Aquaculture and Fisheries*. Oxford: Blackwell Publishing, 146–204.

SHOJIMA Y (1963) 'Scyllarid phyllosomas' habitat of accompanying the jelly-fish', *Nippon Suisan Gakk*, 29, 349–353.

SIMS H W and BROWN C L (1968) 'A giant Scyllarid phyllosoma larva taken north of Bermuda (Palinuridae)', *Crustaceana Suppl*, 2, 80–82.

SIN O K (1967) 'A preliminary study of the early larval development of the spiny lobster *Panulirus polyphagus*', *Malaysian Agricult J*, 46, 183–190.

SKEWES T D, PITCHER C R and TRANDALL J T (1994) 'Changes in size structure, sex ratio and moulting activity of a population of ornate rock lobsters Panulirus ornatus caused by an annual maturation and migration', *Bull Mar Sci*, 54, 38–48.

SMITH D M, IRWIN S J and MANN D (2009) Optimising the physical form and dimensions of feed pellets for tropical spiny lobsters, in Willims K C (ed.), *Spiny Lobster Aquaculture in the Asia-Pacific Region*. Canberra: ACIAR, 157–162.

SMITH G G and RITAR A (2006) 'The influence of animal density and water turbulence on growth and survival of cultured spiny lobster (*Jasus edwardsii*) larvae', *Aquaculture*, 258, 404–411.

SMITH G G, RITAR A J, CARTER C G, DUNSTAN G A and BROWN M R (2003) 'Morphological and biochemical characteristics of phyllosoma after photothermal manipulation of reproduction in broodstock of the spiny lobster, *Jasus edwardsii*', *Aquaculture*, 220, 299–311.

SMITH G G, SALMON M and HALL M R (2009a) 'Artificial diets for *Panulirus ornatus* spiny lobster larvae', *NZ J Mar Freshwater Res*, 43, 205–215.

SMITH G G, SALMON M, KENWAY M and HALL M R (2009b) 'Description of the larval morphology of captive reared Panulirus ornatus spiny lobsters, benchmarked against wild-caught specimens', *Aquaculture*, 295, 76–88.

SOREGLOSS P and PERSOONE G (1972) 'Three simple culture devices for aquatic invertebrates and fish larvae with continuous recirculation of the medium', *Mar Biol*, 15, 251–254.

© Woodhead Publishing Limited, 2013

SOUZA R, SEKINE S, SUZUKI S, SHIMA Y, STRUESSMAN C A and TAKASHIMA F (1996) 'Usefulness of histological criteria for assessing the adequcy of diets for *Panulirus japonicus* phyllosoma larvae', *Aquac Nutr*, 2, 133–140.

SPRAGUE V and COUCH J (1971) 'An annotated list of protozoan parasites, hyperparasites and commensals of decapod crustacean', *J Protozool*, 18, 526–537.

STENTIFORD G D, NEIL D M and ATKINSON R J A (2001) 'The relationship of Hematodinium infection prevalence in a Scottish *Nephrops norvegicus* population to seasonality, moulting and sex', *ICES J Mar Sci*, 58, 814–823.

STEPHENS F, EVANS L H and JONES B (2003) Disease of mature spiny and clawed lobster, in Evans L H (ed.), *A Review of Lobster Diseases, Their Investigation and Pre-disposing Factors*. Curtin University of Technology, Perth: Fisheries Research and Development Corporation, 42–67.

STEWART J E, CORNICK J W, ZWICKER B M and ARIE B (2004) 'Studies on the virulence of *Aerococcus viridans* (var.) *homari*, the causative agent of gaffkemia, a fatal disease of homarid lobsters', *Dis Aquat Org*, 60, 149–155.

SUMMERFELT S T (2003) 'Ozonation and UV irradiation – an application and examples of current applications', *Aquacult Eng*, 28, 21–36.

SUZUKI N, MURAKAMI K, TAKEYAMA H and CHOW S (2006) 'Molecular attempt to identify prey organisms of lobster phyllosoma larvae', *Fish Sci*, 72, 342–349.

SUZUKI N, HOSHINO K, MURAKAMI K, TAKEYAMA H and CHOW S (2008) 'Molecular diet analysis of phyllosoma larvae of the Japanese spiny lobster *Panulirus japonicus* (Decapoda : Crustacea)', *Mar Biotechnol*, 10, 49–55.

TAKEUCHI T and MURAKAMI K (2007) 'Crustacean nutrition and larval feed, with emphasis on Japanese spiny lobster, *Panulirus japonicus*', *Bull Fish Res Agen*, 20, 15–23.

TAMURA T (1970) *Marine Aquaculture,* Washington, DC: National Science Foundation.

THOMAS L R (1963) 'Phyllosoma larvae associated with medusa', *Nature*, 198, 200.

THORSON G (1950) 'Reproductive and larval ecology of marine bottom invertebrates', *Biol Rev*, 25, 1–45.

THUY N T B, HA N N and DANH D V (2009) Effect of environmental conditions during holding and transport on survival of *Panulirus ornatus* juveniles, in Williams K C (ed.), *Spiny Lobster Aquaculture in the Asia-Pacific Region*. Canberra: ACIAR, 79–84.

VAN WAMBEKE F and BIANCHI M A (1985) 'Bacterial biomass production and ammonium regeneration in Mediterranean sea water supplemented with amino acids. 2. Nitrogen flux through heterotrophic microplankton food chain', *Mar Ecol Prog Ser*, 23, 117–128.

VERSCHUERE L, ROMBAUT G, SORGELOOS P and VERSTRATE W (2000) 'Probiotic bacteria as biological control agents in aquaculture', *Micro Mol Bio Rev*, 64, 655–671.

VIADERO R C and NOBLET J A (2002) 'Membrane filtration for removal of fine solids from aquaculture process water', *Aquacult Eng*, 26, 151–169.

VON BONDE C (1936) 'The reproduction, embryology and metamorphosis of the Cape crawfish (*Jasus lalandii*)', *Investl Rep Fish Mar Biol Surv Div Un S Aftr*, 6, 1–25.

WANG Y-C, LO C-F, CAHNG P-S and KOU G-H (1998) 'Experimental infection of white spot baculovirus in some cultured and wild decapods in Taiwan', *Aquaculture*, 164, 221–231.

WEBSTER N, BOURNE D and HALL M R (2006) 'Vibrionaceae infection in phyllosomas of the tropical rock lobster *Panulirus ornatus* as detected by fluorescence *in situ* hybridisation', *Aquaculture*, 255, 173–178.

WEI S and LAI B (2000) 'Preliminary experiment on the nutrition of *Panulirus stimpsoni* phyllosoma', *Mar Sci Bull*, 19, 36–41.

WIDMER C L (2008) *How to Keep Jellyfish in Aquariums: An Introductory Guide for Maintaining Healthy Jellies*, Tuscon, AZ: Wheatmark.

© Woodhead Publishing Limited, 2013

WIETZ M, HALL M R and L. H (2009) 'Effects of seawater ozonation on biofilm development in aquaculture tanks', *Syst Appl Microbiol*, 32, 266–277.

WIIK R, EGIDIUS E and GOKSOYR J (1987) 'Screening of Norwegian lobsters *Homarus gammarus* for the lobster pathogen *Aerococcus viridans*', *Dis Aquat Org*, 3, 97–100.

WILSON K, HALL M R, DAVEY M, KENWAY M and COREN D (2005a) Biotechnology to improve reproductive performance and larval rearing in prawns and rock lobsters, in Pandian C A and Marin M P (eds), *Fish Genetics and Aquaculture Biotechnology*. Enfield, NH: Science Publishers, Inc., 103–117

WILSON K, SWAN S and HALL M R (2005b) *Reducing Rock Lobster Larval Rearing Time Through Hormonal Manipulation,* Townsville: FRDC.

WYBAN J A and SWEENEY J N (1991) *Intensive Shrimp Production Technology (The Oceanic Institute Shrimp Manual)*. Makapu'u Point, Waimanalo, HI: The Oceanic Institute.

YAMAKAWA T, NISHIMURA M, MATSUDA H, TSUJIGADO A and KAMIYA N (1989) 'Complete larval rearing of the Japanese spiny lobster *Panulirus japonicus*', *Nippon Suisan Gakk*, 55, 745.

YORK R and GOSSARD M H (2004) 'Cross-national meat and fish consumption: exploring the effects of modernization and ecological context', *Ecol Econ*, 48, 293–302.

YOSHIZAWA S, TSURUYA Y, FUKUI Y, SAWABE T, YOKOTA A, KOGURE K, HIGGINS M, CARSON J, THOMPSON L (2012) *Vibrio jasicida* sp. nov., a member of the Harveyi clade, from marine animals (packhorse lobster, abalone and Atlantic salmon), *Int J Syst Evol Microbiol*, 62, 1864–1870.

© Woodhead Publishing Limited, 2013

10

Biosecurity measures in specific pathogen free (SPF) shrimp hatcheries

J. Wyban, High Health Aquaculture Inc., USA

DOI: 10.1533/9780857097460.2.32a

Abstract: Global shrimp farming more than tripled production over the last decade. The major driver of that expansion was the introduction, adoption and expansion of farming *P. vannamei* as the shrimp species of choice. A key element determining farmers' preference for *P. vannamei* was widespread availability of High Health post-larvae (PL) produced from specific pathogen free (SPF) broodstock that outperformed other shrimp species in reliability and profitability. Production of High Health PL requires use of SPF broodstock in hatcheries using biosecurity. This chapter reviews key management practices of securing SPF broodstock and biosecurity in *P. vannamei* shrimp hatchery systems.

Key words: shrimp, hatchery, biosecurity, SPF.

10.1 Introduction

A consistent supply of quality seed is essential to all agriculture systems. In shrimp farming, seed are called post-larvae or PL. Production technology for marine shrimp PL is referred to as hatchery technology. Shrimp hatchery technology was first developed in Japan in the 1940s and was subsequently transferred to and refined by the shrimp farming industries in the eastern and western hemispheres during the 1970s/1980s.

SPF (specific pathogen free) shrimp are shrimp that have passed through a rigorous pathogen screening and multi-generational quarantine process. They must be certified SPF or free of known disease-causing pathogens by a qualified third party. This technology was developed in the USA in the early 1990s (Wyban *et al.*, 1992). Subsequent commercial production of SPF white shrimp broodstock in Hawaii led to worldwide availability of SPF broodstock which led to the introduction, successful production and widespread adoption of SPF *P. vannamei* in Asia from 1999. Widespread use of

© Woodhead Publishing Limited, 2013

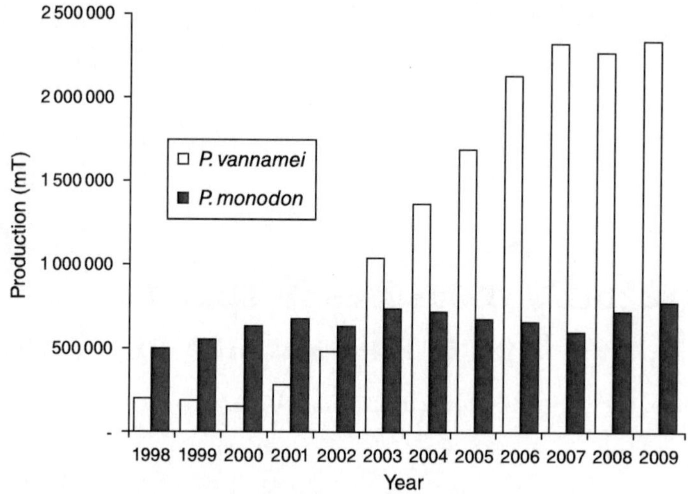

Fig. 10.1 World shrimp farming production by species using FAO 2011 statistics.

SPF *Penaeus vannamei* in Asia resulted in dramatic increases (a tripling) in total shrimp farming production between 2000 and 2009 (Fig. 10.1).

This rapid expansion of white shrimp production in Asia resulted from widespread availability of High Health PL derived from SPF broodstock. To produce High Health PL requires use of SPF broodstock combined with good biosecurity at the hatchery. If shrimp larvae produced from SPF broodstock are infected with disease during their hatchery phase, the benefits of using SPF shrimp are lost. Thus best use of SPF broodstock in a shrimp hatchery requires special attention to biosecurity to produce true SPF or High Health PL.

10.1.1 Historical development of shrimp hatchery technology

Shrimp farming developed in both the eastern and western hemispheres using wild PL from the ocean (Wyban, 2009). Coastal ponds flooded with estuarine waters containing local shrimp PL produced early crops of shrimp that led to profitable harvests. These encouraged early developers to catch shrimp PL in coastal waters to stock into constructed ponds. Stocking wild-caught PL led to further success and shrimp farming was established as an enterprise. With further success, interest in captive reproduction of shrimp emerged. Hudinaga (1942) in Japan first closed and described the life-cycle of marine shrimp. His foundational work gave rise to the global industry known today as shrimp hatchery technology.

Using hatchery-produced seed was expected to increase the reliability of shrimp farming, and shrimp hatcheries were developed in both hemispheres. PL were produced in land-based hatcheries from wild-caught broodstock. These PL were genetically wild animals because their parents

© Woodhead Publishing Limited, 2013

Table 10.1 Shrimp farming eras based on FAO data of global production

Era name	Years	Region – Species	Annual production (1000 t) Start	Finish	Gain	Production growth rate (%/yr)
Start-up	1982–1988	Americas – *P. vannamei* Asia – *P. monodon*	84	604	520	103
Hatchery	1988–1996	Americas – *P. vannamei* Asia – *P. monodon*	604	693	89	2
SPF *P. vannamei*	1996–2011	All – *P. vannamei*	693	3100	2407	23

Source: Wyban, 2009.

were wild-caught broodstock gathered from the sea. During this era, shrimp farming in each hemisphere used native species. The West used the white shrimp, *P. vannamei*, while in Asia, shrimp farming was based on the black tiger shrimp, *P. monodon*.

The historical development of shrimp farming can be divided into three distinct eras (Wyban, 2009) (Table 10.1). During the hatchery development era, global shrimp production only increased from 604 000 to 693 000 MT resulting in an average annual gain of just 2 %/year (Wyban, 2009). Thus there was very little industry growth during this era compared to the rapid growth in the start-up era. The main obstacle to industry growth in this era was widespread shrimp disease. The distribution of hatchery-produced PL was the primary vector in shrimp disease transmission during this era because the hatcheries paid little or no attention to broodstock health status or biosecurity (Lightner, 1996). Diseases carried by wild-sourced broodstock were passed to the PL offspring in the hatcheries and then transferred to the farms with the PL.

The other obstacle to industry growth in this era was the continued use of wild animals. Shrimp farming production during the hatchery era reached a 'carrying capacity' for use of wild, non-domesticated, non-SPF animals. While farmers tried increasing stocking densities to increase yields and profits, their use of diseased, wild animals precluded these attempts and prevented industry growth.

10.2 SPF shrimp and the development of hatchery technology

In the late 1980s, *P. vannamei* was the preferred species farmed by US shrimp farmers, but they were suffering a variety of serious disease

© Woodhead Publishing Limited, 2013

problems. To reduce these problems, the US Shrimp Farming Program developed a population of specific pathogen free (SPF) shrimp in Hawaii (Wyban *et al.*, 1992).

Commercial production trials comparing SPF and non-SPF stocks were conducted by the US industry. Biosecurity protocols were developed and implemented to prevent disease contamination of the SPF broodstock in order to produce High Health PL (Jaenike *et al.*, 1992). More than 50 million High Health PL were produced in 1991 and stocked into commercial US ponds for field trials of the new stock. High Health PL-stocked ponds were found to be more than twice as profitable as the non-High Health crop. More uniform harvest-size distribution was seen in the High Health crop (Carpenter and Brock, 1992). More uniform-sized shrimp at harvest translates directly into improved profitability.

Based on the excellent results of pond trials in 1991, all US shrimp ponds were stocked with High Health PL in 1992. Total production of the US industry doubled as a direct result of this innovation. Use of High Health shrimp in commercial farms increased production and survival, improved feed conversion ratio (FCR) and narrowed harvest size distribution all of which contributed to increased profitability. In addition to increased production, use of High Health shrimp reduced incidence of shrimp disease. It was predicted that the shrimp farming problems that were solved by use of SPF shrimp in the US industry could be duplicated elsewhere (Wyban *et al.*, 1992).

Development of SPF *P. vannamei* in the USA required development of supporting technologies, including SPF broodstock culture technology, long-distance broodstock shipping technology and biosecure SPF shrimp hatchery technology. The key innovation of SPF hatchery technology is biosecurity. Biosecurity is technology to reduce the risk of introduction and transmission of infectious diseases and pests. In order for PL produced from SPF broodstock to enjoy the benefit of the SPF broodstock, the PL must be kept free of disease. This is the job of biosecurity.

10.3 Biosecurity in SPF shrimp hatcheries

Biosecurity includes the suite of technology designed to exclude infectious diseases from SPF animals. Biosecurity includes design and operating parameters to exclude disease causing agents (pathogens) from the system. Effective biosecurity requires knowledge of diseases with reliable diagnostic tools and the use of SPF shrimp stocks. Environmental controls and effective culture practices prevent the introduction and spread of pathogens. Biosecurity also includes procedures for dealing with a disease outbreak if it occurs.

Prevention of disease introduction requires comprehensive analysis and management of all aspects of the facility design and operations. A critical

© Woodhead Publishing Limited, 2013

control point approach modeled on hazard analysis and critical control points (HAACP) can be applied. In a hatchery, broodstock, incoming water, people, feeds, vehicles, vermin or airborne aerosols are critical control points which can introduce a pathogen.

10.3.1 Location

It is often said that the three most important factors in real estate valuation are location, location and location. This is also true for an SPF hatchery. The first step in an SPF hatchery biosecurity plan is site selection. A shrimp hatchery requires abundant supplies of high quality sea water. High quality sea water is not polluted with industrial or urban pollutants and is far from other shrimp farming activities. In some countries where a successful hatchery develops others soon follow, and this is a negative factor in SPF hatchery site selection. An extreme example of this can be found in Wenchang, China, a small coastal town on Hainan Island. More than 600 shrimp hatcheries operate in a coastal area less than 50 km in length. It is very difficult for a hatchery operator to maintain good biosecurity in such a location.

10.3.2 Broodstock

Authentic certified SPF broodstock is the key element in producing High Health PL. A neutral third party must certify broodstock. In Hawaii, the Shrimp Surveillance and Certification Program (SSCP) administered by the State's Department of Agriculture provides the third party screening and certification. It is conducted by the State's veterinarian and is based on guidelines of the World Organisation for Animal Health (OIE). Key elements of the program are listed below (after Riggs, 2009):

- SPF shrimp are subject to twice per year surveillance using polymerase chain reaction (PCR) testing.
- Sites must test negative for listed pathogens for a minimum of 24 months continuous to achieve SPF status.
- PCR testing is performed by USDA- and OIE-certified laboratories for crustacean diseases.
- SPF certification status is site-specific.
- Listed pathogens are based on availability of PCR tests (see Table 10.2).
- The entire shrimp production facility or site must continuously (semi-annually) test negative to maintain SPF Certification status.
- Samples from all shrimp life stages on site must be collected by State veterinarian to achieve 'chain of custody' and 'third party' documentation of sampling.

10.3.3 Sea water

Incoming sea water is the most likely vector for introducing a pathogen to an SPF hatchery. If the hatchery is located in a shrimp-farming region (most

© Woodhead Publishing Limited, 2013

Table 10.2 Current listed pathogens in Hawaii SPF shrimp program (2012)

Pathogen abbreviation	Full name
WSSV	White spot syndrome virus
YHV	Yellow head virus
TSV	Taura syndrome virus
IHHNV	Infectious hematopoietic virus
MBV	Monodon baculovirus
BP	Baculovirus penaei type-A
IMNV	Infectious myonecrosis virus
HPV	Hepatopancreatic parvo-like virus
NHP	Necrotizing hepatopancreatitis

are), shrimp farms nearby may harbor shrimp pathogens. These are potential sources of shrimp pathogens which must be protected against. An SPF hatchery must take special care to treat the incoming water to prevent disease. To begin, the best source of sea water for an SPF hatchery is from a well. By pumping water from a well rather than from surface water, some natural filtration provided by the land above the well may occur. In such a case, it is unlikely that live or dead shrimp from local sources could infest the well. In most cases, high quality sea water can be pumped from a sub-sand well at the beach side.

In large hatcheries where massive quantities (up to 3000 m^3/day) of sea water are required, the sea water from the well is usually first held in a land-based reservoir system. In some cares, the large reservoirs maybe lined ponds up to 2000 m^2 in area. Such a reservoir must be lined with plastic or other pond liner material to prevent introduction of contaminants from soil. They should be well aerated so water and disinfectants can be thoroughly mixed. This is often done with a paddlewheel or aspirator aerator like those used in grow-out ponds. The reservoir ponds should be surrounded by low fencing to keep walking crabs out of the reservoir. Crabs are well-known vectors of white spot syndrome virus (WSSV).

At the reservoir stage, sea water has been filtered by the natural filtration system of the well so few or no particles should be present. If significant particles or sand are entrained in the sea water, a filtering system must be added to the system to keep the particles out of the reservoir. A large bag-type filter of 250 μm or less can be used.

SPF hatcheries should apply chlorination to the reservoir water. An applied dose of 5 ppm chlorine overnight is advised. The water should be well mixed during chlorination so that all the water in the reservoir is treated. Usually, if sea water in the reservoir is well mixed, the chlorine will dissipate by morning. Chlorine concentration can be measured using the chemical O-toluidine (same as used to test swimming pools). If detectable

© Woodhead Publishing Limited, 2013

chlorine remains, the sea water should be dechlorinated using sodium thiosulfate.

After the sea water has been chlorine treated in a reservoir overnight, it is pumped through a filtration system and into a secondary, indoor reservoir or directly to the production tanks in the hatchery. The indoor secondary reservoir usually has heaters to adjust water temperature. Different components of a SPF shrimp hatchery operate at different temperatures and thus a reservoir for each section is advised. For example, maturation systems are operated at 27–28 °C while larval rearing is operated at 30–32 °C. Some industrial-scale hatcheries use passive solar water heating systems (like solar swimming pool heaters) between the outdoor reservoir and indoor reservoirs to preheat the water and reduce energy costs of heating water.

10.3.4 People

People are a great risk for introducing disease to a SPF hatchery. They come and go on a daily basis and it is difficult to control their movements. Therefore, their entry into the hatchery must be controlled. An SPF hatchery should have a staff entry center where each staff member removes their street clothes and shoes and puts on their hatchery clothes and rubber boots. In addition, they wash their hands with soap and hot water.

No outside foods can be brought into the hatchery by staff. Thus the hatchery must have a canteen that prepares meals for the staff. In the canteen, no crustaceans are ever used for food. Raw crustaceans are a very high risk disease vector and must always be kept out of the facility. Food must be kept in the canteen and not carried into culture areas of the hatchery. No outside equipment such as fishing equipment can be brought into the hatchery. When establishing a new SPF hatchery, only new equipment should be used to avoid inadvertent contamination.

10.3.5 Feeds

Feeds are a significant biosecurity risk. There is no single pelletized feed that produces high quantities and quality of shrimp nauplii. Successful reproduction in marine shrimp requires the use of fresh feeds such as marine polychaetes, squid and shellfish. This daily addition of fresh feeds to the SPF broodstock is a large risk to biosecurity. Sourcing of the fresh feeds is important in trying to prevent any disease introduction.

Fresh feeds should be fresh. They must be food grade; if the staff wouldn't eat the feed, they shouldn't feed it to the broodstock. Some companies use an iodine rinse to the fresh feeds to try to disinfect them. While a good idea in principle, the daily addition of even small amounts of iodine to the broodstock diet can be very stressful. Other groups wash the fresh feeds in ozone-treated seawater. This is a better option than daily iodine use.

© Woodhead Publishing Limited, 2013

10.3.6 Vehicles

Vehicles coming from a shrimp farm to the hatchery (either owned by the hatchery or by the shrimp farm) to pick up PL are a significant risk of pathogen introduction. Trucks may carry viral particles in mud stuck on the wheels or body of the truck. When driving on shrimp farms, trucks often crush shrimp that have been dropped on the farm roads and may carry the shrimp carcass.

A biosecure hatchery should use a tire bath at the entrance of the hatchery. A tire bath is a small concrete pool all vehicles must drive through as they enter the hatchery compound. The truck tire bath usually contains water with permanganate or iodine solution to disinfect the tires. Outside trucks should not enter into the production area of the hatchery but only access the packing and shipping building where PL packages can be loaded into the truck.

10.3.7 Vermin

Vermin are unwanted animals that may come into the hatchery. The highest risk vermin are walking crabs which are known WSSV vectors. A biosecure hatchery must have a perimeter fence that excludes any type of vermin such as crabs, rats, mice and feral cats. All of these animals can be pathogen vectors.

The SPF hatchery should use bird netting overhead over all open areas, including reservoirs, to exclude birds from entering the hatchery. Birds are known vectors of shrimp disease because they can carry diseased shrimp over long distances. Some shrimp disease may remain a viable pathogen after passage through the bird's digestive system.

10.3.8 Airborne aerosols

Aerosols may carry virus particles and this is why an SPF hatchery must be isolated from neighboring non-SPF hatcheries.

10.4 Industry impact

Introduction of SPF *P. vannamei* to Asia in the late 1990s, produced dramatic increases in shrimp production (Fig. 10.1) and rapid spread through Southeast Asia that continues today (Wyban, 2003). Rapid and sustained increases in Asian shrimp production resulted from *P. vannamei*'s widespread adoption. This was the primary factor responsible for a three-fold increase in global shrimp production in the last decade (Wyban, 2009). By 2004, *P. vannamei* emerged as the leading shrimp species in worldwide production contributing more than 50 % of total world farmed-shrimp production. In 2009, *P. vannamei* production accounted for more than 80 % of

© Woodhead Publishing Limited, 2013

Table 10.3 Estimated *P. vannamei* broodstock and PL requirements for world shrimp farming in 2011

Total *P. vannamei* production (kg)	2700000000
Mean harvest size (pcs/kg)	75
Mean grow-out survival	50 %
Total PL required = (production x size)/survival	405000000000
Broodstock pairs required:	
Spawns per pair per month	3
Nauplii per spawn	200000
Nauplii per pair per month	600000
Months production per pair	3.5
Total nauplii per pair	2100000
PL production: nauplii to PL hatchery survival	40 %
PL per broodstock pair	840000
Broodstock pairs required = total PL/(PL/pair)	482143

total world production and was the dominant species farmed in Thailand, China and Indonesia – the world's three leading production countries.

Estimates of *P. vannamei* broodstock and PL requirements for world shrimp farming are listed in Table 10.3. Approximately 500000 pairs of SPF *P. vannamei* broodstock are required to produce the 2.5 m MT of shrimp grown annually.

Over the last decade, shrimp farming annual crop value has tripled as a result of the widespread use of domesticated *P. vannamei*. In 1997, global farmed-shrimp annual production of 700000 MT had a total crop value of about $3.5 billion based on an average price of $5/kg (FAO Fisheries and Aquaculture Department, 2010). Widespread adoption of *P. vannamei* over the last 10 years resulted in sustained production increases of 23 %/year (Wyban *et al.*, 2005). While shrimp prices declined as a result of increasing supply, total shrimp crop value today is worth more than $11 billion based on production of 3.1 m MT and a worldwide price of $3.50/kg. This tripling of industry crop value over 15 years directly resulted from domestication, breeding and widespread adoption of *P. vannamei*.

10.5 References

CARPENTER N and BROCK J (1992) Growth and survival of virus-infected and SPF *P. vannamei* on a shrimp farm in Hawaii, in: Fulks W and Main K (eds), *Disease of Cultured Penaeid Shrimp in Asia and the United States*. Honolulu, HI: The Oceanic Institute, 285–293.

FAO FISHERIES and AQUACULTURE DEPARTMENT (2010) *The State of World Fisheries and Aquaculture 2010*. Rome: FAO, available at: http://www.fao.org/docrep/013/i1820e/i1820e.pdf (accessed September 2012).

HUDINAGA M (1942) Reproduction, development and rearing of *Penaeus japonicus* Bate. *Jap. J. Zool.* 10:305–393.

JAENIKE F, GREGG K and HAMPER L (1992) Shrimp production in Texas using Specific Pathogen Free stocks, in: Fulks W and Main K (eds), *Disease of Cultured Penaeid*

© Woodhead Publishing Limited, 2013

Shrimp in Asia and the United States. Honolulu, HI: The Oceanic Institute, 295–302.

LIGHTNER D V (1996) *A Handbook of Shrimp Pathology and Diagnostic Procedures for Diseases of Cultured Penaeid Shrimp*. Baton Rouge, LA: World Aquaculture Society.

RIGGS A (2009) *Shrimp Surveillance & Certification Program (SSCP)*, online Powerpoint presentation. College of Tropical Agriculture and Human Resources, University of Hawaii, available from: www.ctahr.hawaii.edu (accessed September 2012).

WYBAN J A (2003) Recent developments in *P. vannamei* seedstock production Asia. *Global Aquaculture Advocate* 6(6):78–79.

WYBAN J (2009) World shrimp farming revolution: Industry impact of domestication, breeding and widespread use of SPF *P. vannamei*, in Browdy C L and Jory D E (eds), *The Rising Tide, Proceedings of a Special Session on Sustainable Shrimp Farming*. Baton Rouge, LA: World Aquaculture Society, 12–21.

WYBAN J, SWINGLE J, SWEENEY J and PRUDER G (1992) Development and commercial performance of High Health shrimp using specific pathogen free (SPF) broodstock *Penaeus vannamei*, in: Wyban J (ed), *Proceedings of the Special Session on Shrimp Farming*, Baton Rouge, LA: World Aquaculture Society, 254–260.

WYBAN J A, WHITE B and LIGHTNER D (2005) TSV Challenges advance selective breeding in Pacific white shrimp. *Global Aquaculture Advocate* 7(6):40–41.

© Woodhead Publishing Limited, 2013

11

Blue mussel hatchery technology in Europe

P. Kamermans, IMARES, The Netherlands, T. Galley, Bangor University, Wales, P. Boudry, IFREMER, France, J. Fuentes, CIMA, Spain, H. McCombie and F. M. Batista, Bangor University, Wales, A. Blanco, IMARES, The Netherlands, L. Dominguez, CIMA, Spain, F. Cornette, IFREMER, France, L. Pincot, Grainocean hatchery, France and A. Beaumont, Bangor University, Wales

DOI: 10.1533/9780857097460.2.339

Abstract: To date, European mussel culture has relied entirely on wild seed from suspended collectors or mussel beds. One problem faced by blue mussel producers is the unpredictability of seed supply, the amounts of wild seed available being extremely variable from year to year. A second problem is that recently spawned mussels cannot be sold due to insufficient meat. Hatcheries can complement wild seed supply. Hatcheries also allow triploid induction that produces non-maturing mussels. In this chapter, the different steps in hatchery production of mussel seed are described. A final section addresses future trends.

Key words: *Mytilus*, broodstock, larvae, spat, hatchery, triploid.

11.1 Introduction

In the last decade, Europe has produced in the region of one million tons of bivalves from natural fisheries and aquaculture annually (http://www.fao.org/fishery/). In Europe, the blue mussels *Mytilus edulis* (common mussel) and *M. galloprovincialis* (Mediterranean mussel) are the bivalves with the highest production output. The other species in this genus are of lower or no aquaculture interest. FAO statistics show that in 2009 natural fisheries and aquaculture provided 213 891 tons of *M. edulis* and 120 607 tons of *M. galloprovincialis* (www.fao.org). No data were submitted to the FAO for *Mytilus* by Spain or China. Production in Spain was 225 091 tonnes in 2009 (www.pescadegalicia.com). China is the largest producer with more than 663 000 tonnes in 2002 (www.fao.org). The vast majority of European production is derived from aquaculture activities, although natural fisheries

© Woodhead Publishing Limited, 2013

Fig. 11.1 The common blue mussel *Mytilus edulis* (left) and the Mediterranean mussel *Mytilus galloprovincialis* (right).

provide a small but significant contribution of *M. edulis*, which mainly comes from Denmark and is supplemented by the UK and Ireland (97 956 tonnes in 2009; www.fao.org).

The two species are very closely related and have similar morphology (Fig. 11.1). In the Northern hemisphere, *M. edulis* is mainly present in the North Atlantic while *M. galloprovincialis* ranges from the Mediterranean Sea to the coasts of Spain, France, the UK and Ireland. *M. galloprovincialis* is presumed to have expanded northward, outcompeting *M. edulis*. Wherever the distributions of the two species overlap, hybridisation occurs and this results in a patchy mosaic of pure species and hybrids along thousands of kilometres of Atlantic coast line from France in the south to the Shetlands in the north (Bierne *et al.*, 2003). In Norway, Sweden, Denmark, Germany, the Netherlands, Atlantic France and the UK, the farmed species is commonly identified as *M. edulis*. Populations in the Irish Sea are pure *M. edulis* whereas most other farmed mussels, e.g. Bantry Bay, consist of both *M. edulis* and *M. galloprovincialis* and hybrids of these two. In Spain, Mediterranean France, Italy and Greece, it is *M. galloprovincialis* that is cultured, although the FAO figures for production still regard Spanish Atlantic mussels as *M. edulis*.

To date, in a manner similar to most mussel farming worldwide, European mussel culture has relied entirely on wild seed. In the Netherlands, 65 000 tons of seed are needed to produce 100 000 tons of mussels (Kamermans and Smaal, 2002). The seed is fished from natural beds in the Wadden Sea. For this purpose, wild beds are dredged (scraping mussels from intertidal hard surfaces). Fishing wild stocks of shellfish in this way is criticised because of the possible damage to bottom habitats and resulting food shortages for shellfish-eating birds. In addition to these environmental concerns, the supply of seed shows large fluctuations. It is extremely difficult to collect sufficient amounts of mussel seed each year. As a result, Dutch mussel

© Woodhead Publishing Limited, 2013

farmers recently started using seed mussel collectors to obtain mussel seed (Kamermans *et al.*, 2002). Spain is an important producer of mussels in the world and the top European producer of *M. galloprovincialis* (www.pescadegalicia.com), of which 225 091 tonnes is produced in Galicia in 2009 (NW Spain) and where mussels are an important economic and social resource. Cultures in Spain also depend on a yearly seed supply from natural settlement on both surrounding rocks and collecting ropes.

Compared to the collection of natural spatfall, which is highly seasonal, hatcheries may produce seed on a year-round basis, on the condition that ripening of the gonads can be achieved under controlled conditions. Hatcheries can, therefore, complement wild seed supply. Hatcheries also allow the development of genetic improvement through selective breeding and/or polyploidisation. Methods for the hatchery culture of oysters, scallops and clams are well established and there are currently around 40 commercial shellfish hatcheries present along the European Atlantic coast. None of these hatcheries produces mussel seed commercially, but two Dutch companies have started hatchery production of *M. edulis* at pilot scale. In addition, one hatchery in Tasmania and another in North America are known to undertake commercial mussel seed production. In Tasmania, Spring Bay Seafoods, to our knowledge, has developed the only modern hatchery facilities principally dedicated to mussel seed production, and Shellfish Culture Ltd produces both Pacific oysters and blue mussels. In the USA, at the time of writing, the company Taylor Shellfish similarly produces several bivalve species including *M. galloprovincialis* at a price of US$ 15 per foot of seed-covered rope for a local rope culture industry where there is insufficient natural spatfall. An evaluation with a bio-economic model compared production costs of 5 mm seed produced in a hatchery, collected with ropes and fished, showing these to be €0.31 per kg mussel seed for fished seed, €1.35 per kg mussel seed for collector seed and €430 per kg mussel seed for hatchery seed. (BLUE SEED Final Report, www.blueseedproject.com). Some areas were identified where economies could be made to bring hatchery production costs more into line with the potential sale value of mussel seed: (i) use low-tech algal culture; (ii) restrict activities to the natural season; (iii) scale up culture volumes during this restricted period of activity. This result also indicates that the added value of the hatchery-produced mussel seed must be large in order to make it an economically viable product. Hatchery seed could have higher added value for a number of reasons: (i) there is no reliable or no sufficient wild seed source (Tasmania, North America); (ii) wild seed is composed of different species of different economic value (e.g. *M. trossolus* in Scotland); (iii) hatchery seed is available earlier in the season which provides an opportunity to reduce the growing period; or (iv) hatchery seed possesses superior qualities through genetic improvement or triploid induction.

From 2005–2007 an EU-funded project called BLUE SEED was carried out. The objectives were to secure a reliable supply of blue mussel seed and

© Woodhead Publishing Limited, 2013

to develop techniques allowing farmers to market blue mussels year round. A problem blue mussel producers face is the unpredictability of seed supply. Seed is mostly harvested from suspended collectors or mussel beds, and the amounts of wild seed available are extremely variable from year to year. A reliable supply of seed from hatchery sources will allow mussel farmers to overcome this. A second problem is that recently spawned mussels cannot be sold due to insufficient meat content. Producers will benefit greatly from a hatchery-based technique, such as triploid induction, that produces non-maturing mussels that can be marketed year round. In this project, mussel farmers and sellers, a network for training and technology transfer, universities and research institutes collaborated. Attention was given to broodstock conditioning and larval rearing, production of triploid larvae and tetraploid broodstock, spat settlement and on rearing of diploid and triploid spat to seed size.

In the following sections, outcomes of the BLUE SEED project in the different steps in hatchery production of mussel seed are described. A final section addresses future trends.

11.2 Broodstock: holding, conditioning and management

Unlike oysters, mussels are dioecious. This means they are either female or male and do not change sex during their life. They can become sexually mature in their first year, although the size at sexual maturity will depend on the growth rate at the production site. The reproductive tissue of the mussel is located within the mantle, which extends along the inside of the shell valves, enveloping the rest of the body. Activation of the gonad generally starts before winter and gametogenesis takes place over winter. Up to 40 million eggs can be released by an individual female, but in hatchery conditions from one to eight million eggs per female is more normal.

Ripe mussels for spawning in hatcheries can only be collected from the wild over a short period of the year, just prior to their natural spawning period. It is desirable to extend the availability of ripe mussels on either side of this natural spawning period to allow a greater flexibility in production timing or to increase the number of larval batches produced and thus optimise the use of hatchery installations. There are two main methods for such extension: the 'cold and hold' method, which extends the spawning period into the summer, and the 'heat and treat' method, which allows spawning before the natural period.

11.2.1 The 'cold and hold' method

Within the hatchery environment, it is possible to extend the natural spawning period of mussels by maintaining naturally ripened mussels at low temperature on a minimal or irregular live algal ration. This is a simple method for providing a reliable broodstock that can be spawned when

© Woodhead Publishing Limited, 2013

required over several months, thereby extending the period over which spat can be produced.

The ideal temperature for holding mussels is 6 °C for *M. edulis* (Beaumont *et al.*, 2004; Galley *et al.*, 2010) and 9 °C for *M. galloprovincialis*, as these cool temperatures prevent the mussels from spawning. *M. edulis* have been successfully maintained in spawning condition using this procedure from late March until September.

11.2.2 The 'heat and treat' method

Evidence suggests that, like other species of bivalves, enhanced ripening of gonad material in broodstock mussels (*M. edulis* and *M. galloprovincialis*) is possible within a hatchery environment by holding them under appropriate conditions. Mussels can be brought into condition during the winter and early spring before adults in the wild reach their peak of maturity. However, the effect of broodstock conditioning can be specific to different populations.

The provision of a live algae diet in conjunction with a gradual rise in temperature or compressed 'winter-to-spring' temperature profile is successful for enhancing spawning condition and gamete quality of *M. edulis* (Pronker *et al.*, 2008).

The duration of conditioning is also critical, a 6–10 week period having been found appropriate. The precise value of light regime on conditioning remains speculative as conditioning can be achieved under both constant illumination and variable photoperiods. Although constant illumination permitted maturation, it can reduce spawning and should, therefore, be avoided (Dominguez *et al.*, 2010). Figure 11.2 illustrates the effect of warm

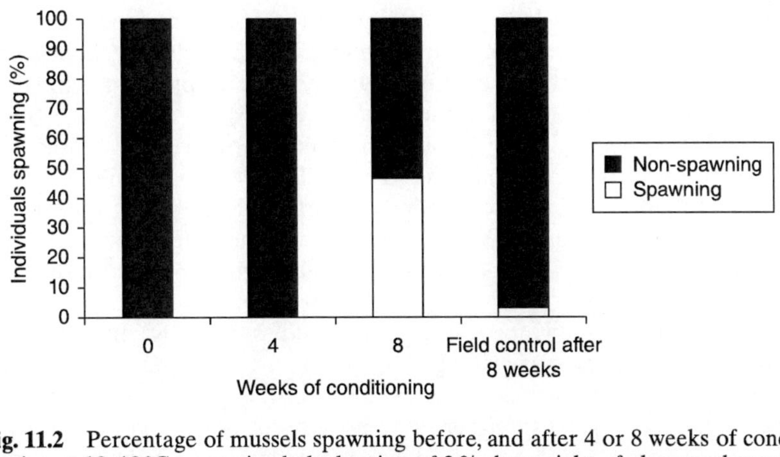

Fig. 11.2 Percentage of mussels spawning before, and after 4 or 8 weeks of conditioning at 18–19 °C on a mixed algal ration of 3 % dry weight of algae to dry meat weight of broodstock per day and a field control after 8 weeks (BLUE SEED Final Report).

© Woodhead Publishing Limited, 2013

water broodstock conditioning on a population of mussels conditioned over the winter (November–January).

'Cold and hold' is thought to be less expensive than 'heat and treat' because it does not require a lot of algal rearing. Also, on the condition that the mussels are well cleaned beforehand, this technique has fewer problems with growth of epibiota and extraneous algae, a lesser requirement for tank cleaning and less potential for disease as all biological processes are slowed down. As the animals are kept on a minimal or irregular live algal ration, egg quality is generally retained up to September. After that quality declines.

11.3 Spawning, fertilisation, embryo development, early D-larvae and triploid and tetraploid induction

11.3.1 Spawning

Spawning is the release of mature eggs and sperm. In mussels, this release is made directly from their genital ducts into the open water, where fertilisation takes place in the natural environment. Mussels can be artificially encouraged to spawn using a number of procedures, including thermal cycling, intermittent exposure to air and introduction of 0.5 M potassium chloride (KCl) into the mantle cavity. Generally, mussels are spawned in individual vessels in order to prevent uncontrolled fertilisation. This is especially important when producing triploid mussels, to increase the temporal uniformity of embryo development stages with regard to induction treatment. Individualisation is also useful for optimising gamete quality (enabling gametes with abnormal morphology to be selectively discarded). 'Strip' spawning (as used in Pacific oysters) furnishes poor-quality mussel gametes (especially in females) as these may not be mature, and these give poor-quality embryos. Male spawn disperses after release producing a dispersed milky white appearance in spawning containers, whilst female spawn is very pale orange/brown, sometimes almost white in appearance and tends to fall gently to the bottom of the vessel (Fig. 11.3).

The various spawning protocols are not necessarily definitive though, i.e. not guaranteed to work in all cases. They will only be successful if mussels are in a suitable condition to respond to any stimuli offered. It appears that the condition of the animals is paramount and, when they are ripe, a range of different stimuli can trigger spawning, although some are more efficacious than others.

An alternative spawning technique is to induce mussels collectively in a group. Known as mass spawning, this method involves mussels being held together in the same vessel and exposed to thermal cycling (Fig. 11.4). Using this method, mussels can either be removed once they begin to spawn or spawned gametes can be collected via an overflow into a submerged mesh sieve by running sea water through the spawning vessel/tank. This second approach invariably results in the collection of developing embryos rather

© Woodhead Publishing Limited, 2013

Fig. 11.3 The release of eggs by female mussels (left) and spermatozoa by male mussels (right).

Fig. 11.4 Spawning of the mussel *Mytilus edulis* under group conditions.

than separate unfertilised gametes. This method of spawning is typically applied in hatcheries for the mass production of diploid larvae. However, because of the associated risk of uncontrolled fertilisation, the use of mass spawning is not recommended if the aim is to use the gametes for the production of triploid mussels or for selective breeding experiments.

11.3.2 Fertilisation and development of D-larvae

The mixing of eggs and sperm results in fertilisation. As in the case of most bivalves, mussel eggs are released when they are at the developmental stage

© Woodhead Publishing Limited, 2013

metaphase I of meiosis I and require activation by the addition of spermatozoa in order for meiosis to proceed (Fig. 11.5a). The first sign of successful fertilisation is the extrusion from the egg of a small, transparent, dome-like structure, which is the first polar body during meiosis I (Fig. 11.5b). This is followed by the extrusion of a second polar body during meiosis II, after

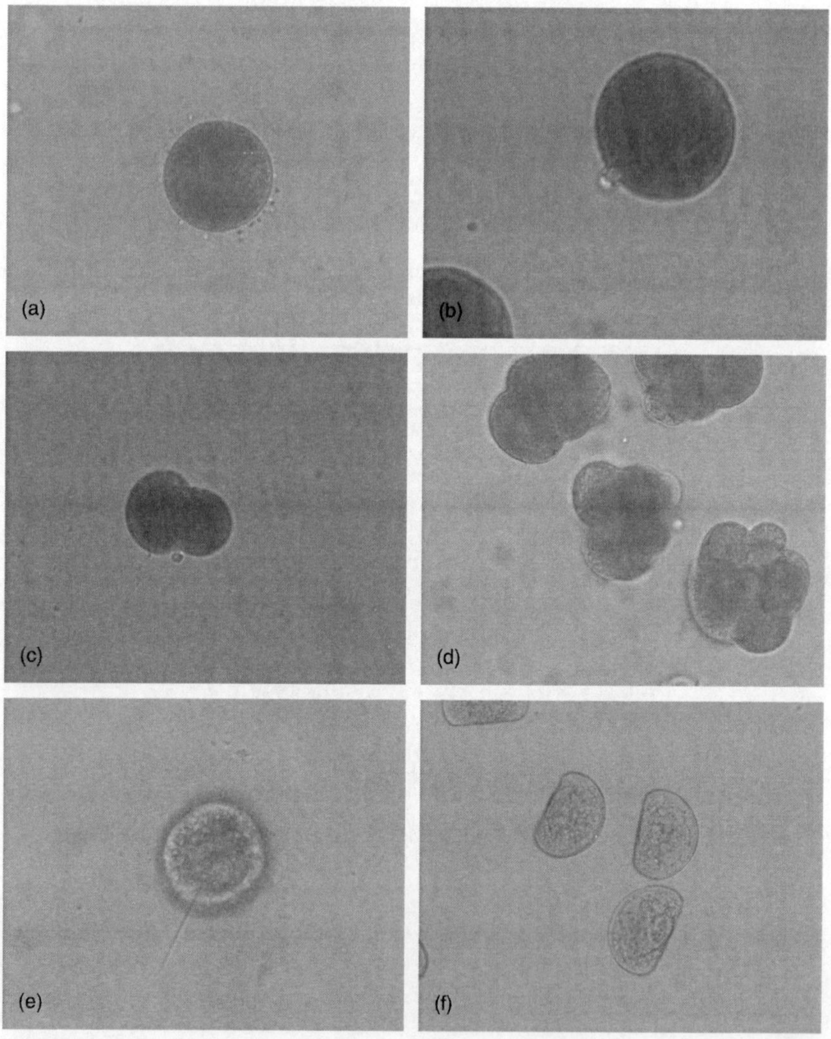

Fig. 11.5 Fertilisation and development of mussel embryos: (a) unfertilised egg and sperm; (b) polar bodies developing, 24 min post-fertilisation; (c) two cells with polar lobe ('trefoil stage'), 50 min post-fertilisation; (d) 4–8 cell embryos, 1 h 40 min post-fertilisation; (e) trochophore stage, 23 h post-fertilisation; (f) D-veliger, 48 h post-fertilisation.

© Woodhead Publishing Limited, 2013

which the fertilised eggs begin to divide, approximately 1 h post fertilisation. The first division is almost equal into two cells (1st cleavage), followed by a 'trefoil' stage (Fig. 11.5c) and an unequal division into four cells where one large cell is observed capped by three much smaller ones. Subsequent divisions lead to the development of an embryo (Fig. 11.5d) that begins to swim when cilia appear after 4–5 h. The embryo then develops into a non-shelled trochophore by 24–48 h after fertilisation (Fig. 11.5e), before forming a shell and developing into the characteristic D-shaped larvae with two shells (Fig. 11.5f).

The earliest embryonic stages last for approximately 48–72 h. In the hatchery environment, fertilisation can be successfully initiated by combining eggs and sperm at a suitable concentration (approximately 200 sperm per egg) and at a suitable temperature for the species in a plastic or glass vessel, or even in the larval rearing tank. There is some evidence that the use of glass vessels improves the yield of D-larvae compared with new or abrasion-cleaned plastic vessels due to potential contamination from plasticisers (pers. comm. A. Beaumont). Eggs and sperm should be used as soon after spawning as possible. Eggs should not be older than 1 h, sperm not older than 30 min, although holding gametes at low temperature, for example in a fridge, can extend their effective life. Assessment of whether fertilised eggs are developing normally can be made using a relatively low power microscope (×20–40 magnification). Embryos should be placed into development vessels at a density of approximately 200 embryos per cm^2 to be left undisturbed to develop to 'D'-larva stage. Although eggs can be stocked at greater densities, which can be more practical for hatcheries, the proportion of normal D-larvae maybe reduced (Galley *et al.*, 2010). By this strategy of increased density, however, even though the proportion of normality is reduced, more D-larvae may be obtained overall. The effect of egg density on the development of embryos into D-larvae is also thought to be influenced by egg quality. Alternatively, developing embryos can be transferred directly to larval rearing tanks (see below) once correct development has been confirmed on the day of fertilisation. After 48–72 h the larvae can be assessed for the percentage development of normal D-larvae and yield of D-larvae before being cultured through the larval stage. The development period of mussel embryos is temperature dependent: at 17.5 °C *M. edulis* reach the D-larvae stage within 48 h while at 12 °C this can take up to 72 h.

All sea water used in mussel reproduction and larval rearing must first be filtered to a high level (e.g. 1 µm filtered and preferably also UV irradiated) to ensure it is clean, i.e. pathogen- and particulate-free. Additionally, the cleanliness of glassware and other materials which come into contact with the gametes, embryos and larvae is also important. All glassware and equipment used should be cleaned using dilute 'Chloros' (a commercial sodium hypochlorite solution) or alternative disinfectant. As noted earlier, strong abrasion-cleaning of plastic equipment can release potentially toxic

© Woodhead Publishing Limited, 2013

plasticisers and subsequent leaching in sea water before use is recommended. Furthermore, prior to use it is recommended to pass spawned egg and sperm suspensions through a suitable mesh-size sieve (180 μm aperture or greater for eggs and 20 μm for sperm) to remove contaminating faecal pellets from the adults and reduce risk of the subsequent proliferation of bacteria and other micro-organisms during subsequent stages of the culture process. Gametes must, however, be maintained submerged at all times. Typically more than 50 % of eggs will develop into larvae, of which almost 100 % can develop into the normal D-larvae stage at 48–72 h post fertilisation, although this is highly dependent upon gamete quality.

11.3.3 Triploid induction

Triploids are organisms with three sets of homologous chromosomes in their somatic cells instead of the usual two in diploids. In mussel, diploids have 14 pairs of chromosomes. Polyploidy can occur naturally in some species, mainly plants showing vegetative reproduction, although in animals it is more commonly associated with simpler life forms such as flat worms, earth worms, leeches and brine shrimp (e.g. White, 1940; Zhang and King, 1992; Vsevolodova-Perel and Bulatova, 2008).

The induction of a triploid state in bivalves has received a significant amount of attention because triploids generally exhibit increased growth as a result of retarded gonadal development or virtual sterility (Beaumont and Fairbrother, 1991; Normand *et al.*, 2008, 2009). This retarded gonadal development can also prolong the annual period over which they are in a marketable condition. In the case of Pacific oysters (*Crassostrea gigas*), triploids are marketable all year round.

Triploidy can be artificially induced by suppressing meiosis I (inhibiting the extrusion of first polar body, PB1), or meiosis II (inhibiting the extrusion of second polar body, PB2), essentially forcing the oocyte to retain the extra genetic material. This can be achieved by applying an interference in the form of a physical or chemical shock at key stages during early embryonic development (Fig. 11.6). In practice, suppression of PB2 is preferred because of higher triploidy yields and larval survival.

Initially, the chemical cytochalisin B (CB), a fungal derived cell-permeable mycotoxin that inhibits cytoplasmic division, was used to induce triploidy in oysters. However, CB is both highly cytotoxic and a carcinogen. The puromycin analogue 6-dimethylaminopurine (6-DMAP) has been recommended as a safer alternative and protocols using this chemical have been explored for the mussels *M. edulis* and *M. galloprovincialis*. Other methods used to induce triploidy include physical shocks such as temperature changes and application of pressure, although these work better in fish species than in molluscs.

The BLUE SEED project adopted the use of 6-DMAP and worked to optimise its use in producing triploid mussels by targeting the second

© Woodhead Publishing Limited, 2013

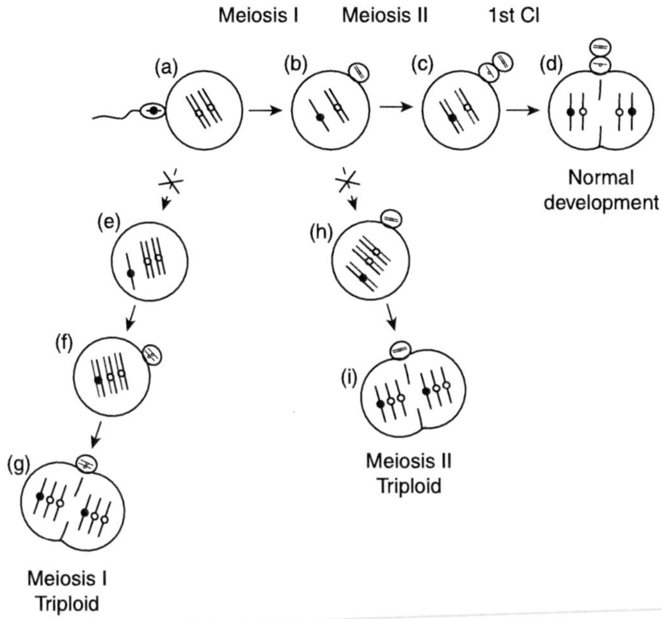

Fig. 11.6 Manipulation of ploidy in developing embryos by targeting meiosis I or meiosis II with a physical or chemical stimulus. For simplicity only one homologous pair of chromosomes is shown. (a–d) Normal development: (a) primary oocyte at release activated by sperm; (b) meiosis I, 1st polar body extruded; (c) meiosis II, 2nd polar body extruded and syngamy occurs; (d) 1st cleavage. (e–g) Following a shock at meiosis I female chromosomes are retained in the egg and a meiosis I triploid is produced. (h,i) Shock at meiosis II produces meiosis II triploids. (Modified from Beaumont and Fairbrother, 1991)

meiotic division. This method involves the immersion of eggs on a sieve into a 300 μM 6-DMAP L^{-1} solution for 20 min beginning just prior to second polar body extrusion (Brake *et al.*, 2002). This technique can produce yields of up to 100 % triploids, although the proportion of embryos developing into normal D-larvae and the percentage yields can be substantially reduced compared with diploid counterparts, because of the toxic secondary effects of the chemical treatment. The surface area of the sieve will limit the number of eggs that can be treated overall as there must be space for efficient draining. We therefore recommend that an absolute maximum of 40000 eggs cm^{-2} of 20 μm sieve should not be exceeded.

The timing of the application of the shock is critical. In mussels, the first polar body is usually extruded 15–30 min after egg activation, with the second polar body appearing 30–45 min after egg activation. The rate of this process is temperature sensitive, with higher temperatures increasing the speed of polar body extrusion.

© Woodhead Publishing Limited, 2013

11.3.4 Tetraploid induction

Tetraploid broodstock offers the possibility to produce triploid (3n) progenies directly through tetraploid (4n) × diploid (2n) crosses. As such progenies will consist of 100 % or close to 100 % triploid individuals without the need for chemical treatment, they offer advantages over the chemical induction method in terms of efficiency, ease of crossing procedure and safety.

Tetraploidy has been induced in several bivalve species, notably *C. gigas*, where induction was achieved by a cross of rare fertile triploid oysters with diploid males followed by chemical polar body retention (Guo and Allen, 1994). This process was patented by the company 4Cs Breeding Technology (Strathmere, New Jersey, USA) and is in commercial use, but is presently inappropriate for mussels due to the step requiring fertile triploid females. A previous study had shown all tested triploid mussels to be male (Brake *et al.*, 2004), so we chose to adapt a direct induction technique that had already produced tetraploid *M. galloprovincialis* in an experimental trial (Scarpa *et al.*,1993).

The BLUE SEED project induced tetraploidy by using CB on crosses between diploid mussels, 5 min 0.5 $\mu L\ L^{-1}$ CB shock at 3 min post fertilisation (McCombie *et al.*, 2009): a very short treatment intended to reduce toxic effects. A treatment that begins and ends before the extrusion of the first polar body may nevertheless alter embryogenesis and the ploidy of resulting offspring. The shock may still have effects on embryogenesis even after the treatment has ended and 'applied this early' may notably affect interactions between the spermatozoid and oocyte due to effects on the permeability of the oocyte wall. Embyogenesis examinations both in the BLUE SEED project and in a previous study on Pacific oyster (Ledu and McCombie, 2003) implied that such early treatments may prevent spermatozoids from penetrating the oocytes but still enable oocyte activation.

Ploidy levels of the resulting progenies were mixed, as in any chemical induction, because embryonic development is not uniform between oocytes and thus different ploidy levels were achieved. Pair matings also showed different results in terms of representation of ploidy classes, indicating parental influences. A difference in size between ploidy levels, as observed in oyster by McCombie *et al.* (2005a), was used to concentrate tetraploid offspring into 'pseudocohorts' as they developed.

As success is variable, the production of tetraploid mussels requires ploidy testing, which can be made by flow cytometry. Early-stage batches have to be tested destructively on small samples of larvae. In BLUE SEED project trials, the technique produced tetraploidy in D-larvae in five separate experiments, covering both *M. edulis* and *M. galloprovincialis*, with tetraploid percentages between 18 % and >60 %, based on samples of 50–200 ground larvae. Later, non-destructive tests could be made on juveniles and adults by biopsy under $MgCl_2$ anaesthesia. It was found that

© Woodhead Publishing Limited, 2013

tetraploidy decreased over time, as a level of 7.5 % was found in a successful pseudocohort at five months. It is not clear whether this reduction in tetraploidy was due to poor survival or reversion to lower ploidy levels (e.g. triploid or aneuploid, McCombie *et al.*, 2005b). Maintenance of tetraploid mussels requires a secure facility with effluent treatment to prevent any possible escape of gametes or young of these polyploid bivalves. Such precautions must be imposed from the earliest age, through the motile stages and into adulthood. Tetraploid spat produced in BLUE SEED was held in the secure facility in Ifremer La Tremblade with ozone treatment of effluents.

The studies performed in the BLUE SEED project have incited continued work on polyploidisation of mussel species at Ifremer. The ability to produce all-female populations using females that produce exclusively daughters can be an advantage of hatchery-produced seed since females grow faster than males. The advantages offered by tetraploid broodstock must be weighed up against the costs or their maintenance, including security aspects, and the work required to produce such a population, notably the repeated testing and selection according to ploidy level.

11.3.5 Sex determination in mussels

Monosex populations may have desirable aquacultural characteristics, such as higher growth rates in females compared to males – as observed in *M. edulis* (Mills and Côté, 2003). Sex can be determined by genetic factors, non-genetic factors or both. In bivalve molluscs, little is known about sex determination and to date no sex chromosome have been identified. This is the case in mussels of the genus *Mytilus* which have 14 putative pairs of autosomal chromosomes ($2n = 2$) (Ahmed and Sparks, 1970). However, different studies have provided insights about sex determination in these bivalve species and how it can be manipulated. In *M. edulis* and *M. galloprovincialis* it was observed that the gonads of all laboratory-produced triploid mussels were of the male type (Kiyomoto *et al.*, 1996). In these two species, the sex ratio among progeny from pair matings can differ greatly from 1:1, a phenomenon known as sex-ratio bias (SRB) (Zouros *et al.*, 1994a). Some females almost exclusively produce daughters, whereas other females produce mainly sons (Kenchington *et al.*, 2009; Batista *et al.*, unpublished data). A third type of females produces approximately the same number of sons and daughters.

Another unusual phenomenon in mussels of the genus *Mytilus* is the occurrence of two types of mitochondrial DNA, one being transmitted from the female parent to both daughters and sons through the egg (F type), and the other being transmitted from the male parent to sons only through the sperm (M type). This phenomenon is known as doubly uniparental inheritance (DUI) (Zouros *et al.*, 1994b). Under DUI, the M type cannot be

© Woodhead Publishing Limited, 2013

passed to females and, based on this finding, it was hypothesised that the presence of the M type is causally linked to maleness (Saavedra *et al.*, 1997). However, recent findings from hybrid crosses, from triploid mussels and from observations of the 'behaviour' of sperm mitochondria in embryos showed that maleness and the presence of the M type are not causally linked (Kenchington *et al.*, 2009). Moreover, these authors proposed a model for sex determination in which the sex is determined by a locus with two alleles that have different 'dosage' effects ($S_1 = 1$ and $S_2 = 2$). Under this model, a dosage of two or higher (in case of the triploids) is required for maleness and only the maternal allele is expressed during the determination of the sex; therefore, the sex is determined by the maternal genotype.

During the final meeting of the BLUE SEED project, an evaluation was conducted to compare the sensory properties (colour/appearance, odour, taste/flavour and texture) of diploid and triploid *M. edulis*. As a whole, both types of mussels had a good acceptance by the panellists. It was agreed by most of the panelists that diploid mussels had a slightly stronger marine flavour than triploid mussels grown under the same conditions. However, some of the panel considered that the less intense and more neutral flavour of the triploid mussel could favour their acceptance by consumers not accustomed to eating mussels or marine products.

11.4 Larval rearing

Veliger mussel larvae growth, as in most bivalve larvae, is dependent on temperature, salinity and phytoplankton availability and species composition. Other considerations include larval density, concentration of food and size, bacterial control and shape of the rearing vessels. Loosanoff and Davis (1963) described the basic methodologies for handling and rearing the larvae of bivalves. This technology is well known and has been in commercial use for several decades.

Once mussel larvae develop to the D-larvae stage (Fig. 11.7) they can be reared to metamorphosis in either flat-bottom vessels or conically-based vessels with bottom drains. If conically-based containers are employed then aeration should be provided because it has been shown to improve larval survival in the early stages. In flat-bottom vessels aeration is not necessary, but is not detrimental. The larvae are grown in UV irradiated sea water filtered to 0.2–5 μm. Although some evidence suggests that UV treatment is not essential, this depends upon the quality and background bacterial content of the sea water at any particular site. There is a recent and increasing concern about reducing seawater treatment (filtering and UV irradiation) to permit the presence of beneficial/probiotic bacterial strains. In this regard, Kesarcodi-Watson *et al.* (2010) showed beneficial effects of two bacterial strains for rearing larvae of the greenshell mussel (*Perna*

© Woodhead Publishing Limited, 2013

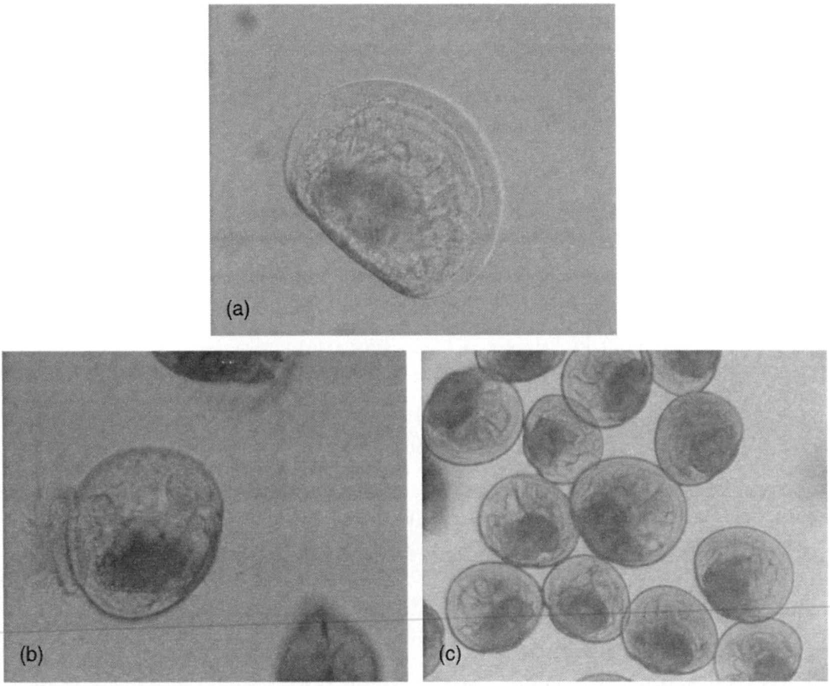

Fig. 11.7 Larvae of the mussel *Mytilus edulis*: (a) 7 days; (b) 13 days (swimming); (c) 23 days.

canaliculus). The source of sea water can have a significant effect on growth and survival of larvae, an issue which can be influenced by the location of the hatchery. Only natural sea water should be used, artificial sea water made up from synthetic salt mixes is inadequate and causes complete larvae mortality.

Initial stocking densities of approximately 10 larvae ml^{-1} appear to be optimal, allowing good survival rates, although larvae can be cultured at higher concentrations. The growth rate of larvae is temperature sensitive, with higher temperatures resulting in higher growth rates. Suitable culture temperatures are 15–17 °C for *M. edulis* larvae, and 16–18 °C for *M. galloprovincialis* larvae. Below these optima, growth appears slow and above them there is increased risk of infection by opportunistic bacteria, particularly of the genus *Vibrio*. For example, at 12 °C *M. edulis* larvae take approximately 36 days to reach metamorphosis, whereas at 17 °C they take about 21 days.

The mussel larvae are fed with unicellular algae. A mixed diet is recommended (Marshall *et al.*, 2010) to be added to rearing vessels at a concentration of 25 cells μL^{-1} day^{-1}. Food can be added daily or every two or three days at a ration equivalent to 25 cells μL^{-1} day^{-1} without affecting growth

© Woodhead Publishing Limited, 2013

or survival. Alternatively, a dynamic staged feeding regime can be employed that changes as the larvae develop, incorporating additional algal species as described below for *M. galloprovincialis,* although this does not appear necessary.

Although larval stages require only relatively small quantities of food, it must be of a high microbiological and nutritional quality. The larvae of both *M. edulis* and *M. galloprovincialis* grow successfully through their D-larval phase on a number of mixed algal diets consisting of two or three high nutritional value species. Recommended diet compositions for the larvae of the two mussel species include the following:

- For *M. edulis*:
  - *Pavlova lutheri* + *Rhinomonas reticulata* (80:20 ratio based on cells per mL)
  - *Isochrysis galbana* + *Chaetoceros calcitrans* (50:50)
  - *Pavlova lutheri* + *Chaetoceros calcitrans* (50:50)

The diet composed of two flagellate species *Pavlova lutheri* and *Rhinomonas reticulata* has been used to regularly rear veliger larvae of *M. edulis* to metamorphosis (Beaumont *et al.*, 2004; Galley *et al.*, 2010). However, this diet has not been compared directly with alternative diets to ascertain its relative suitability.

- For *M. galloprovincialis*:
  - *P. lutheri* + *Chaetoceros gracilis* (50:50)
  - *P. lutheri* + *C. gracilis* + *Tetraselmis suecica* (50:25:25)
  - *I. galbana* + *C. gracilis* + *Tetraselmis suecica* (50:25:25)
  - *I. galbana* + *P.utheri* (week 1) + *C. gracilis* (week 2) + *T. suecica* (week 3)
  - *I. galbana* + *I. galbana (T-ISO)* + *P. lutherii* + *C. gracilis* (week 1) + *T. suecica* (week 2 and 3)

Evidence gathered during the BLUE SEED project suggests the food value of different algae varies for the two species of mussels. If a *Chaetoceros* spp. is included in the diet, then *C. calcitrans* should be included for veliger larvae of *M. edulis*, but for *M. galloprovincialis* larvae, *C. gracilis* should replace *C. calcitrans*.

Maintaining clean water conditions is critical to maintaining healthy growing larvae. To this end, water in the rearing vessels is changed three times per week (Monday, Wednesday and Friday), at which point all rearing vessels are also cleaned (with Chloros and hot water). Larvae are removed by siphon/tap and retained on nylon mesh sieves of appropriate sizes during their development. Recently, flowthough systems have been developed for scallops (*Pecten maximus*) (Andersen *et al.*, 2000) and *C. gigas* larvae (Rico Villa *et al.*, 2008) and recirculation systems are being now developed for mussel larvae in the framework of the EU project REPROSEED (www.REPROSEED.com).

© Woodhead Publishing Limited, 2013

11.5 Metamorphosis and spat settlement

Towards the end of the planktonic larval phase, mussel larvae develop a characteristic pigmented 'eye spot' (Fig. 11.8a), usually at a shell length of 200–260 μm. They then also develop a foot, which rapidly becomes functional in crawling (Fig. 11.8b). The velum starts to degenerate and is replaced by buds that will later form the gills (gill buds) (Fig. 11.8c). Larvae at this stage are known as pediveligers. The eye, foot and, later, gill filaments are visible indications that the larvae are ready to terminate their planktonic life, settle and metamorphose into spat. During this time, due to the degeneration of the velum, feeding activity stops and larvae descend towards the bottom of the culture tanks.

Settlement is a reversible, exploratory and behavioural response of mature larvae as they descend and undertake active site selection, and this differentiates 'settlement' from the process of metamorphosis (Pawlik, 1990; Widdows, 1991; Lutz and Kennish, 1992; Gosling, 2003). Mussel

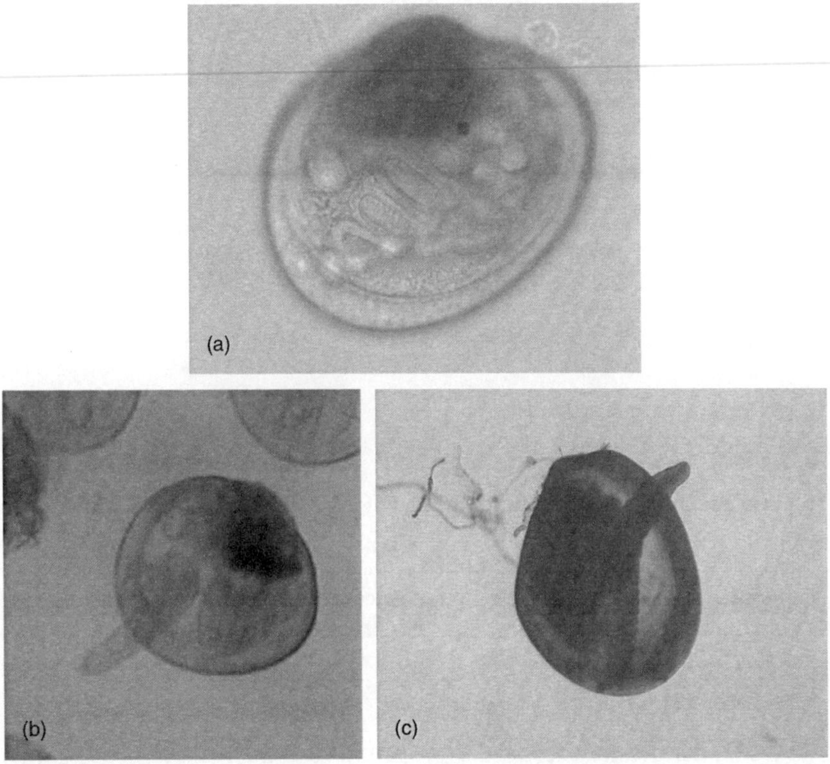

Fig. 11.8 (a) *Mytilus* larvae with distinctive black 'eye-spot' and gill filaments. (b) Pediveliger larvae with developed foot. (c) Metamorphosed mussel spat with extended foot and many gill filaments.

© Woodhead Publishing Limited, 2013

pediveliger larvae explore potential places to settle by crawling around on the substrate using their foot. If the pediveliger does not come into contact with a suitable substrate, it withdraws its foot and may swim away while the velum is still intact. Larvae are able to delay metamorphosis for several weeks enabling them to seek more suitable locations. Settlement culminates in attachment to the substratum once a suitable substrate for settlement, with the appropriate physical and chemical cues, triggers crawling behaviour, which is followed by metamorphosis.

Metamorphosis is a process encompassing the morphological changes which occur within the larvae, including the reorientation of internal structures, increasing complexity of the organ systems and secretion of the adult shell – known as the dissoconch – as they develop into juveniles (Bayne, 1965; Widdows, 1991; Gosling, 2003). A significant and distinctive change during the onset of metamorphosis is the degeneration of the velum and the development of ciliated gill filaments, which become the feeding organ of the animal (Gosling, 2003). This marks the end of the pelagic phase and is the transition from the larval to the juvenile/adult stage. Metamorphosis is a critical stage in development, during which high mortalities can occur.

The success of metamorphosis is heavily dependent upon the production of good quality, healthy larvae with large energy reserves accumulated during the larval phase. Because of the transition from the velum to the much more efficient gill system, a marked increase in filtering capacity and food requirements follows metamorphosis. At this point, spat need to be transferred to more appropriate structures or out into the natural environment. In *M. edulis*, the settlement of mature larvae is known to be influenced by water agitation, the presence of filamentous substrate (Bayne, 1964; Eyster and Pechenik, 1987) and, in some areas, by adult mussel beds (McGrath *et al.*, 1988).

11.5.1 Settlement systems

Generally mussel larvae in the hatchery are transferred from larval rearing vessels to settlement systems prior to the onset of settlement and metamorphosis. This transition is usually made when 50 % of larvae have developed eye spots, at which point they are deemed 'ready to settle'. This stage of larval development occurs at approximately the end of the third week of larval rearing and, besides the appearance of the eye spot, is characterised by a reduction in larval swimming, an increasing in crawling activity by foot extension and by a tendency of the larvae to clump together forming mucous masses in the bottom and walls of the larval rearing tanks. This tendency causes some problems when cylindrical tanks with conical bottoms are used, due to the excessive aggregation of the larval clumps, together with faeces and microalgae, in the bottom of the larval rearing tanks. In this case, it is recommended to start transfer to the settlement systems earlier, when 10–20 % of the larvae are eyed and 'ready to settle'. In some cultures,

© Woodhead Publishing Limited, 2013

it may be necessary to separate larvae which are not ready to settle from those which are. This can be done by using a series of graded mesh sieves (180–220 μm, depending on the batch), allowing the continued growth of larvae yet to develop an eye spot or foot under larval rearing conditions.

The type of settlement system is somewhat dependent upon the eventual spat cultivation system to be deployed, although suitable types include both down-welling systems and vessels incorporating settlement materials. Many mussel spat will also spontaneously settle upon the sides of most rearing vessels. Tanks of different types and sizes can be used for initial settlement. These can be rectangular or circular tanks with either a flat or a conical bottom. Rectangular tanks are used when spat are to be settled onto sieves in down-welling systems while circular tanks are preferable when settlement on collector ropes is the chosen method. Because larval rearing tanks are usually inspected every other day, several batches of competent larvae can be obtained from each larval rearing tank. It is recommended to transfer these batches separately to different settlement units.

In down-welling systems (Fig. 11.9) pediveligers are placed within the confines of a nylon mesh bottomed cylinder (120–150 μm mesh) suspended within a holding tank. A reservoir can be connected if the holding tank is not of sufficient volume. Sea water containing a source of food is circulated by a water pump. It flows into the enclosure from above the water surface at a constant rate, moving downwards over the spat and out through the mesh base of the cylinder to return to the reservoir via an overflow that maintains the water level constant in the holding tank. Such a system can easily be set up in a closed recirculating system with regular water changes and addition of unicellular algae. The flow of the water encourages the larvae to settle upon the surface of the sieve. An example of this type of system is shown in Fig. 11.10.

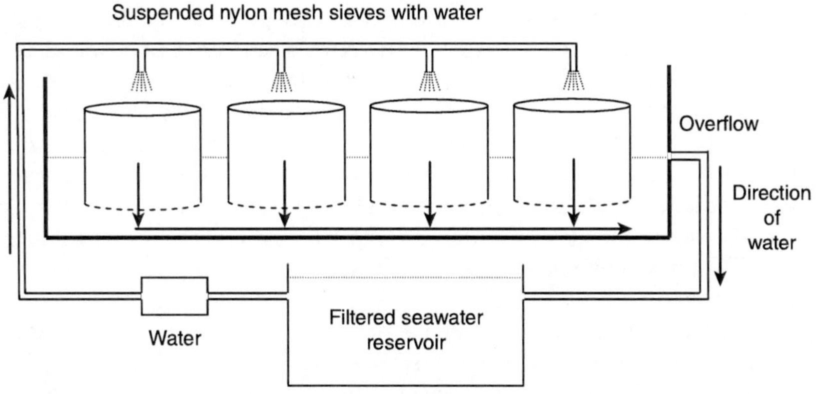

Fig. 11.9 Schematic of basic down-welling system for mussel settlement.

© Woodhead Publishing Limited, 2013

Fig. 11.10 Experimental scale downwelling system for mussel spat.

Alternatively, substrates onto which the mussels will settle can be introduced directly into a rearing environment similar or identical to that used during the larval phase (Fig. 11.11a). This approach is useful for hatcheries producing mussels for rope culture as this material can easily be suspended in the rearing vessel for larvae to settle upon. In this case, all the larvae retained on 150 μm sieves are transferred to settlement tanks, preferably of circular shape (Fig. 11.11a). Size and volume of the tanks will depend on the length and number of collector ropes used. Different types of ropes can be used depending on the availability and practices in each culture area: straight or looped trim polypropylene Xmas tree ropes, as used in long-lines systems in New Zealand and the Netherlands; natural coconut rope, as used in the traditional 'bouchot' system in France; several types of Galician ropes, manufactured from used fishing nets and composed of different proportions of nylon and polyethylene strands; and also new types of ropes such as the hairy rope made 'ad hoc' for this purpose by the company Itsaskorda (Bizkaia, Spain) (Fig. 11.11b). Ropes can either be suspended in the tanks, straight or coiled, or laid down at the bottom. In this last case, flat-bottomed tanks are recommended. Other types of collecting substrates can also be used such as corn material net, a type of net which is very suitable for on-bottom cultivation because it disintegrates into the sea water after six months, and other synthetic nets and screen meshes. Larval density must be adjusted to approximately 1 larva mL^{-1} or lower and spat collector ropes should be maintained within settlement tanks for a period of approximately one to four weeks.

Settlement tanks should be filled with sea water heated to a temperature no higher than 18 °C. Sea water has always to be treated, but the type of treatment will depend on cost considerations and quality of the incoming supply. Thus, sea water could be filtered, using a series of several cartridge

© Woodhead Publishing Limited, 2013

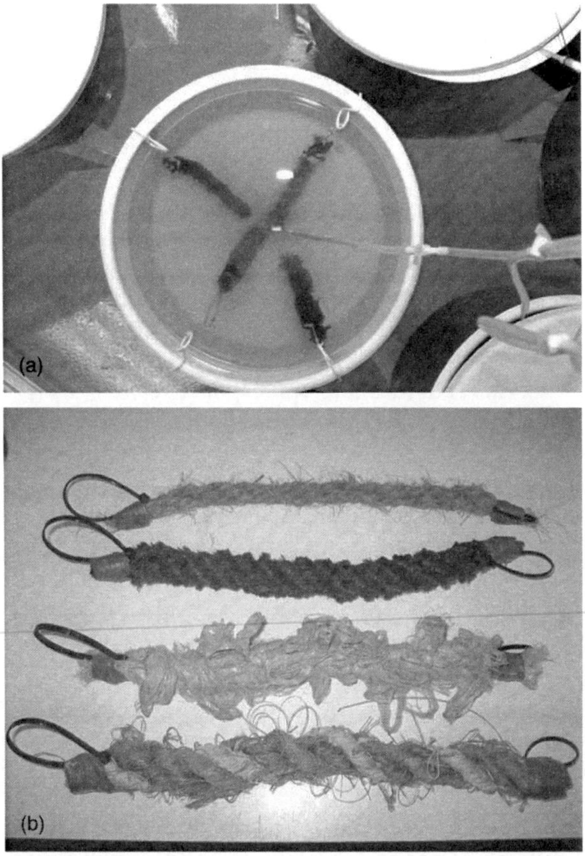

Fig. 11.11 (a) Settlement tanks containing rope substrate and (b) different rope substrates – from top to bottom: natural coconut rope, hairy rope by Itsaskorda, polypropylene Xmas tree rope, and a traditional Galician rope.

filters, down to sizes of 25 µm, 10 µm, 1 µm or even 0.20 µm. After filtration, it can also be disinfected with either UV-light or ozone treatment. During settlement, sea water in the tanks must be renewed, but different schedules can be followed, also depending on the characteristics of the nursery areas and facilities. Thus, sea water can be renewed every day, every other day or even only once a week. In all cases, tanks should always be cleaned before adding new sea water. Cleaning can be done using filtered sea water alone or fresh water previously treated with peroxide.

Settling larvae need to be fed a daily ration of 50000 cell mL^{-1} of a mix of microalgae, such as *I. galbana, C. gracilis, Skeletonema costatum* and *T. suecica*. Sea water in conical-bottomed tanks must be agitated throughout the settlement period, both to avoid concentration of larvae and microalgae in the conical bottom and to enhance attachment of spat to collectors.

© Woodhead Publishing Limited, 2013

Agitation can be achieved by bubbling air from the bottom of the tanks. After a settlement period, the material can optionally be transferred to larger ongrowing vessels supplied with filtered sea water and food, prior to ongrowing at sea.

Both the down-welling and the substrate approach have been applied successfully to mussels (Dominguez, unpublished results). Typical survival from eyed larvae to four weeks post settlement is 50 %.

11.5.2 Chemical induction of settlement and metamorphosis

The control of settlement and metamorphosis in many marine invertebrate species in general is thought to be controlled by a combination of neuronal and neuroendocrine activities, and is a response to complex signals originating from a wide range of sources in the natural environment, as reviewed by Pawlik (1990, 1992) and Rodríguez *et al.* (1993). Such stimuli, or cues, have been attributed to physical and biological elements. However, the primary stimulus is believed to be chemical in nature, originating from biological sources associated with the substratum (Morse and Morse, 1984).

For many years the use of exogenous chemicals to influence settlement and metamorphosis in a range of marine invertebrate species has been studied. In an aquaculture context, the ability to influence this transitional period represents a powerful tool (Baloun and Morse, 1984; Mesías-Gansbillar *et al.*, 2008). The identification of simple chemical inducers could provide routine, inexpensive and effective culture techniques for the settlement of larvae onto a substrate in a controlled system (Cooper, 1982; Davis *et al.*, 1990).

A wide range of pharmacological agents has been tested on a wide range of mollusc and other marine invertebrate species with varying effects on settlement and metamorphosis, ranging from the induction of normal settlement and metamorphosis, to abnormal or partial development, death or no effect at all (Morse *et al.*, 1979; Hadfield, 1984; Pawlik, 1990; Pires and Hadfield, 1991; Boettcher and Targett, 1998; Carpizo-Ituarte and Hadfield, 1998; Nicolas *et al.*, 1998; Dobretsov and Qian, 2003; Zhao *et al.*, 2003; García-Lavandeira *et al.*, 2005; Yang *et al.*, 2008). These agents have been suggested to act either as functional analogues of natural inducers, as precursors or as active components within a signalling pathway (Yool *et al.*, 1986; Pawlik, 1990; García-Lavandeira *et al.*, 2005).

It has been determined that the settlement and metamorphosis of both *M. edulis* and *M. galloprovincialis* can be influenced by the use of a number of exogenously applied chemical agents, thereby increasing the rate of development over that of untreated larvae. In *M. edulis*, it has been found that L-DOPA, a precursor to the neurotransmitters dopamine, norepinephrine and epinephrine, induces metamorphosis (Cooper, 1981), whilst isobutyl-1-methylxanthine (IBMX) and acetylcholine chloride have been found to induce settlement (Eyster and Pechenik, 1987; Dobretsov and

© Woodhead Publishing Limited, 2013

Qian, 2003). The xanthine derivative IBMX has so far proven particularly effective at inducing settlement, with up to 83 % of larvae settling after a 48 h exposure at concentrations 10^{-4}M (Dobretsov and Qian, 2003).

Effective agents identified for *M. galloprovincialis* include the catecholamine neurotransmitter epinephrine which has proven highly effective at inducing metamorphosis, with up to 90 % success within an exposure time of 48 h at 10^{-4} and 10^{-5}M (García-Lavandeira *et al.*, 2005; Satuito *et al.*, 2005; Yang *et al.*, 2008). A second effective chemical is γ-aminobutyric acid, commonly known as GABA, which has demonstrated inductive properties for both settlement and metamorphosis over an exposure of 48 h at a concentration of 10^{-4}M (García-Lavandeira *et al.*, 2005). Another effective chemical is ammonium chloride (NH_4Cl) which is effective at inducing metamorphosis at 10^{-2}M after 24 h exposure (Yang *et al.*, 2008). However, care must be taken as many of these chemical agents are toxic to larvae at high concentrations or over prolonged periods.

11.6 Nursery rearing of mussel spat up to seed

Once post-larvae (spat) are attached to the settlement material (nets, screens or ropes), they should be transferred as soon as possible to the final outdoor ongrowing areas. This provides them with a more natural environment with higher quality food and, more importantly, avoids the high costs of rearing them indoors. Indoor rearing requires the pumping and heating of sea water and the highly expensive production of several species of microalgae.

Nevertheless, it is best to permit the spat a gradual adaptation to the environmental conditions of the final ongrowing areas to reduce the risk of premature mortality due to dislodgement from the settlement substrates, predation or competition by other sessile organisms, particularly tunicates (fouling). After settlement, post-larvae are usually reared in indoor facilities until they reach an adequate size to be transferred to the definitive ongrowing areas. Although this size depends on the particular environmental conditions of each growing area, mussel seed is usually considered ready for transfer to ongrowing sites at a size of 0.5–1 cm shell length. There is a tradeoff between early transfer to the field, associated with greater losses of mussels, and the greater cost of a longer nursery period. As mussels grow older, byssus formation occurs which is an issue for cleaning and maintenance of equipment. Byssus can encourage other organisms to grow and thus decrease the overall sanitary status. This problem favours the early transfer of settled spat to sea. Different approaches have been used for rearing spat up to seed size. Different types of down-welling units placed inside flowthrough rectangular tanks can be used (Fig. 11.12).

In Fig. 11.12, each down-welling unit consists of a plastic or PVC cylinder with an initial 150 μm nylon screen mesh fitted to it by covering one of their

© Woodhead Publishing Limited, 2013

Fig. 11.12 (a, b) Down-welling units for spat rearing, in rectangular flowthrough tanks.

ends and an air-lift device to circulate the sea water. Cylinders with different mesh size screens (from 150 to 200, 500, 1000 and 2000 μm) can be gradually changed depending on the growth rate of the spat. Spat can be grown without using any type of collector just directly attached to the screen on the bottom of each cylinder (Fig. 11.12). When cylinders are changed, the spat can be easily removed from the screen by scraping, without any major damage.

Another method involves rearing the spat, previously settled on different types of collector ropes, in circular or rectangular flowthrough tanks into which the collector ropes are either directly hung or introduced in down-welling units (Fig. 11.13).

During rearing of spat on collector ropes, a considerable proportion of individuals can become detached from the ropes and then settle on the walls and bottom of the tanks. This recurrent behaviour of mussel spat creates management problems, particularly when ropes containing spat are hung directly into flowthrough tanks. A third approach for spat rearing is to spread the mussel spat, previously detached from the settlement substrates, on 250 μm mesh size square-frame screens which are horizontally introduced into rectangular tanks in a indoor micronursery (Fig. 11.14).

In the three types of approach, spat culture tanks were supplied with a flowthrough circulation of filtered sea water. Size of filter used can vary, depending on the specific environmental conditions of each culture situation: the amount of seston or the presence of wild fertilised mussel eggs, embryos or even larvae in the incoming sea water, for example. Thus, while

© Woodhead Publishing Limited, 2013

Fig. 11.13 Collector ropes with mussel spat: mussel spat on a white-coloured rope (a), hanging directly into a circular tank (b) and within downwelling units in a rectangular tank (c).

in some cases only a single 50 μm filter is sufficient, in other cases a series of smaller filter sizes from 25 to 5 μm are needed. Spat in the rearing tanks can be fed an initial daily ration from 40 000–100 000 cell mL^{-1} of mixtures of microalgae such as *I. galbana, C. gracilis, S. costatum* and *T. suecica*. This ration can be progressively increased depending on the requirements of the seed. Spat culture tanks can be cleaned only with filtered sea water without using any type of chemical treatment, but in some cases fresh water treated with peroxide is needed. Cleaning operations can be carried out every day, every other day or once a week, depending on the facilities.

The rearing density (number of individuals per length or surface unit) and the average size of the spat on ropes or screens are two important variables that should be estimated regularly during the rearing process, to adjust culture densities, but especially at the end of the process, just before the transference of the seed to the final ongrowing sites. These two variables are very difficult to assess at the beginning of the rearing process due to the small size, fragility and aggregation of individuals but, as the spat grows, they can be more easily estimated. Calculation of the average size of spat is a straightforward method, accomplished by measuring the individuals contained in one or more random samples taken from the ropes or screens under a binocular microscope equipped with an eyepiece micrometer.

© Woodhead Publishing Limited, 2013

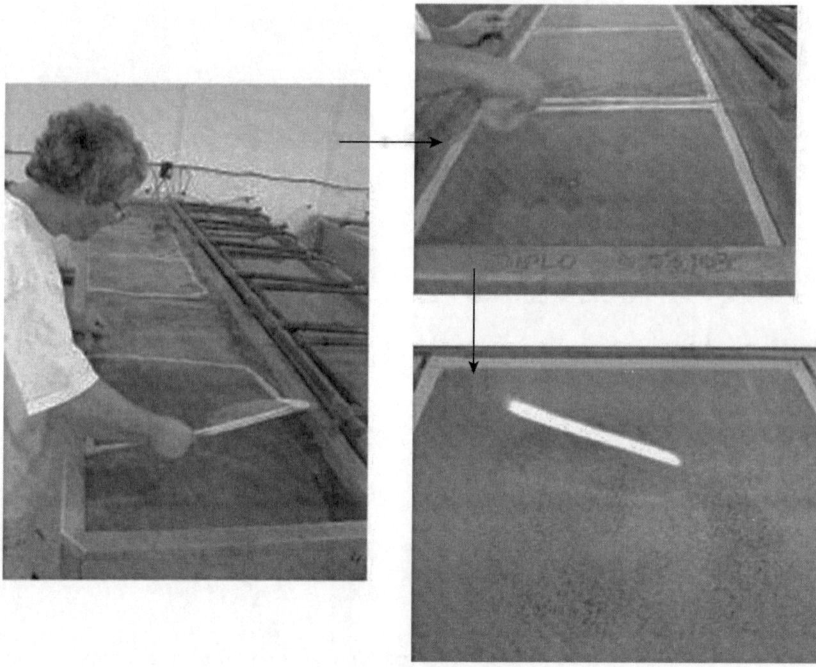

Fig. 11.14 Rearing of mussel spat on screens in a micronursery.

Estimation of the rearing density is a more complex process and depends upon the type of substrate used for rearing the spat: ropes or screens.

11.7 Grow-out of mussel seed in land- and sea-based facilities

Once mussel spat have reached a suitable seed size (5–10 mm), they are usually ready to be transferred to definitive grow-out sites where the pre-ongrowing, thinning out and final mussel ongrowing phase are undertaken. However, in some circumstances, such as very adverse environmental conditions or a high risk of predation on small mussels, it could be preferable or necessary to carry out the pre-ongrowing of mussel seed in land-based or sea-based nurseries as a previous step to their transfer to the final grow-out sites. In European mussel cultivation, the final grow-out sites are always natural marine environments where different types of sea-based facilities are established. In France, for example, some inter-tidal grounds of the Atlantic Coast are used for the ongrowing of mussels tied to poles or 'bouchots' placed in rows perpendicular to the coastal line. In other countries, like the Netherlands, Wales, England and part of Ireland, mussels are

© Woodhead Publishing Limited, 2013

traditionally seeded and ongrown on bottom plots situated in inter-tidal and sub-tidal areas of shallow coastal areas. However, the most efficient environment for mussel grow-out is the water column of seston-rich estuaries and sheltered coastal areas where mussels are ongrown tied to ropes or within plastic socks suspended from rafts or long-line systems. Raft cultivation is the system traditionally used in Galicia (NW, Spain). Long-line culture is a more recent method for mussel cultivation, successfully developed some years ago in New Zealand for the Greenshell mussel, now introduced in some European countries, including the Netherlands, Ireland, Scotland and Denmark.

11.7.1 Seed grow-out in lab-scale or land-based nurseries

Most of lab-scale culture systems to test the grow-out of mussel seed in land-based facilities are variations on the same basic structure: downwellers placed within holding tanks (Fig. 11.15).

These seed culture systems are similar to those described above for spat rearing. Down-wellers are PVC or plastic cylinders of different lengths equipped with bottom nylon nets to hold the mussel seed and with air-lift tubes to move the sea water through (Fig. 11.16). As mussel seed grows, the bottom nylon net can be replaced by a larger mesh size to permit a better seawater circulation through the system.

Because food requirements in this step of the mussel cultivation are very high, it is recommended to place all seed-containing downwellers into large flowthrough tanks (Fig. 11.15) through which sea water with high seston content (only filtered by 50 μm) is continuously pumped together with a mix of several species of bag-cultured microalgae (*T. suecica* and

Fig. 11.15 Different downwellers in a rectangular holding tank type raceway.

© Woodhead Publishing Limited, 2013

Fig. 11.16 Downweller with mussel seed on the bottom.

S. costatum) contained in a reservoir tank. Down-wellers and holding tanks need to be cleaned frequently (every day or every other day) with fresh water to remove all the retained seston material as well as the faeces and pseudo-faeces. Research on recirculation systems for mussel spat rearing is in progress to reduce the amount of algae needed (www.REPROSEED.com).

Clearly, a different, more industry-focused approach is needed to undertake the pre-ongrowing of the mussel seed in large outdoor tanks or land-based facilities (Fig. 11.17). In this approach, the mussel seed, previously attached to polyester net frames in an indoor micronursery, is transferred to large outdoor raceway tanks in a land-based nursery. Naturally-produced microalgae are pumped to the raceway tanks to feed the seed. To promote a more homogeneous distribution of the mussel seed on the frames, it is recommended to grow them in darkness by covering the raceways with a light-reflecting material (Fig. 11.18). After a month under these rearing conditions mussel seed between 4 and 10 mm long can be separated from the frames and transferred to a long-line systems situated in the final ongrowing offshore environment.

11.7.2 Seed grow-out on ropes suspended from rafts

In the BLUESEED Project, seed was transferred from a lab-scale micronursery after two to three months of growth. Mussel seed from several batches (around 8 mm average size) was wrapped to different types of culture ropes with degradable cotton net (Fig. 11.19), and transferred to a mussel raft placed in an inner culture polygon in the Ría de Arousa (NW Spain).

After three months of raft pre-ongrowing, the mussels, now approximately 40 mm long (Fig. 11.20a), were thinned out and tied again to

© Woodhead Publishing Limited, 2013

Fig. 11.17 (a) Polyester net frames in an outdoor raceway tank in a land-based nursery. (b) and (c) Net frames showing attached mussel seed.

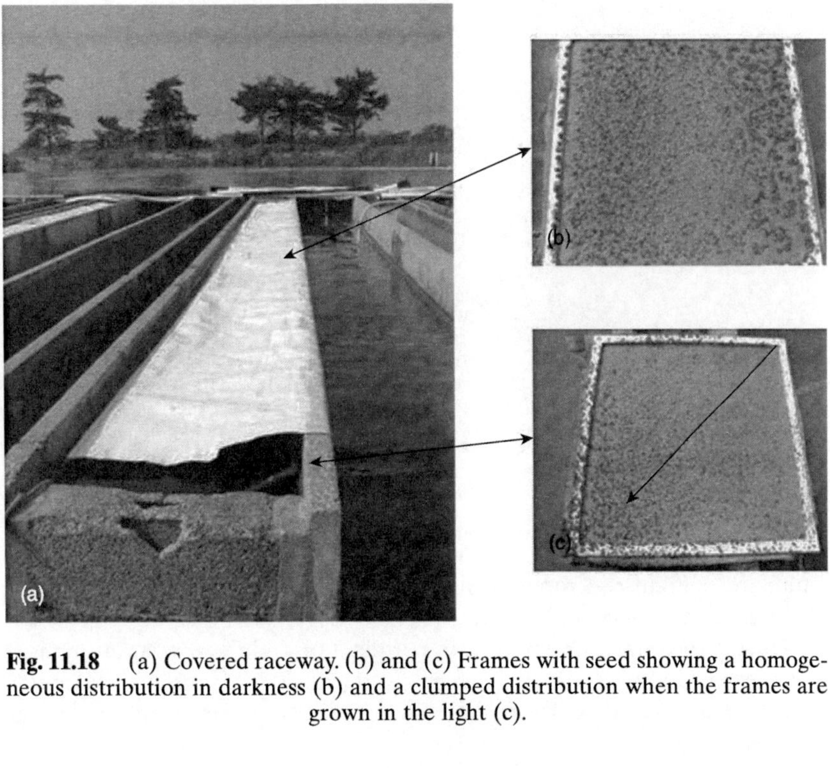

Fig. 11.18 (a) Covered raceway. (b) and (c) Frames with seed showing a homogeneous distribution in darkness (b) and a clumped distribution when the frames are grown in the light (c).

© Woodhead Publishing Limited, 2013

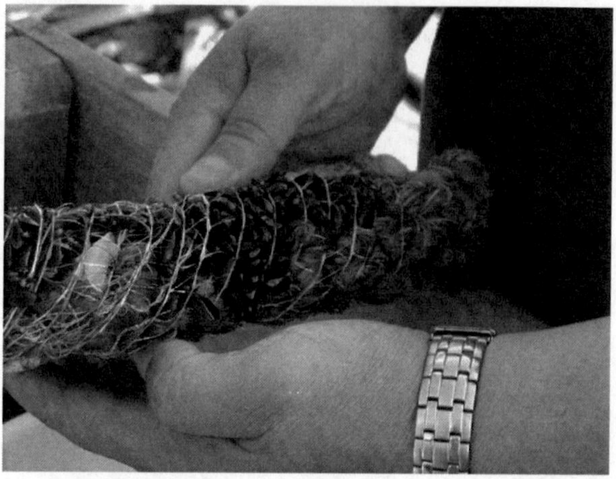

Fig. 11.19 Wrapping seed up on ropes.

Fig. 11.20 (a) Mussels after pre-ongrowing and (b) mussels tied to standard rope.

standard commercial ropes with plastic pegs (Fig. 11.20b), then ongrown for a complete year until market size.

During the ongrowing period, the production traits (growth rate, density changes and biomass) of the hatchery-produced diploid mussels and similar-size mussels obtained from the wild were compared. After a year of

© Woodhead Publishing Limited, 2013

ongrowing on the raft, hatchery-produced diploid mussels were similar (not significantly different) to their wild counterparts (BLUE SEED Final Report).

11.8 Future trends

One of the reasons why hatchery production of mussel seed is less developed than for other bivalves, such as oysters or scallops, is most probably the fact that demand for the industry has been limited until now and that very large-scale production is required to make hatchery seed competitive with wild seed. As a result, future developments of mussel seed supply will depend on the evolution of wild seed availability and demand from the mussel industry.

Significant investments are needed to develop hatchery production at a scale that would be economically viable. The potential of flowthrough systems for larval rearing is currently being tested as part of the EU-supported project 'ReproSeed' (http://www.reproseed.com). Development of efficient settlement and nursery systems remains the main challenge. One aspect of mussel hatchery technology that is of great interest is the study of the mechanisms and factors involved in larval settlement and metamorphosis. A good knowledge of the two stages of the mussel life-cycle will permit more efficient technology to be developed. In this regard, future studies should be focused on the role of biochemical mechanisms linked to neuroactive compounds (potassium ions, GABA, dopamine, microbial films, etc.) as well as the contribution of genetic characteristics (inter- and intra-species differences) on both settlement and metamorphosis. Better control of reproductive conditioning is also needed, to provide methodology similar to what is currently available in Pacific oysters.

Future trends will be focused on the production of higher added-value products. Otherwise, the production of 'normal' seed by hatchery techniques will be not be profitable in most cases compared with the cost of obtaining the wild counterparts. The advantages of triploidy, in terms of added-value for the producers and the consumers, remain to be demonstrated and a recent review reveals many of the constraints that have limited the potential value of ploidy manipulation in fish and shellfish (Piferrer *et al.*, 2009).

Further information concerning heritability of traits of interest (e.g., mussels selected to be less affected or unaffected by toxic blooms, mussels with a specific commercial characteristic) is needed to assess the potential of breeding programmes. It is really only possible to undertake artificial selection for useful traits in mussels when there are large-scale hatchery facilities routinely available to carry out controlled crosses and to maintain mussel broodstock with clearly identified pedigrees based on DNA analysis. Such facilities are already available for oysters, and increased commercial

© Woodhead Publishing Limited, 2013

pressure for mussel culture could lead to the development of selected mussel lines with increased aquaculture value.

11.9 Sources of further information

COUTTEAU P and SORGELOOS P (1992) The use of algal substitutes and the requirement for live algae in the hatchery and nursery rearing of bivalve molluscs: an international survey. *J Shell Res* 11, 467–476.

HELM M M, BOURNE N and LOVATELLI A (2004) *Hatchery Culture of Bivalves*. FAO Fisheries Technical Paper 471. Rome: FAO.

LAING I, LEES D N, PAGE D J and HENSHILWOOD K (2004) *Research on Shellfish Cultivation*, Technical Report No. 112. Lowestoft: CEFAS.

LAVENS P and SORGELOOS P (1996) *Manual on the Production and Use of Live Feeds for Aquaculture*, FAO Fisheries Technical Paper 361. Rome: FAO.

LOOSANOFF V L and DAVIS H C (1963) Rearing of bivalve molluscs, in Russell F S (eds), *Advances in Marine Biology. Vol. 1*, Academic Press, London. 1–136.

MILLICAN P F (1997) *The Hatchery Rearing of the King Scallop* (*Pectin maximus*). Lab leaflet. Lowestoft: CEFAS.

MULLER-FEUGA A, MOAL J and KASS R (2003) The microalgae of aquaculture, in: Støttrup J G and McEvoy L A (eds), *Live Feeds in Marine Aquaculture*. Oxford; Malden, MA: Blackwell Publishing, 206–252.

MULLER-FEUGA A, ROBERT R, CAHU C, ROBIN J and DIVANACH P (2003) Uses of microalgae in aquaculture, in: Støttrup J G and McEvoy L A (eds), *Live Feeds in Marine Aquaculture*. Oxford; Malden, MA: Blackwell Publishing, 253–299.

PONIS E, PROBERT I, VÉRON B, MATHIEU M and ROBERT R (2006) New microalgae for the Pacific oyster *Crassostrea gigas* larvae. *Aquaculture* 253, 618–627.

ROBERT R and GÉRARD A (1999) Bivalve hatchery technology: The current situation for the Pacific oyster *Crassostrea gigas* and the scallop *Pecten maximus* in France. *Aquat Living Resour* 12(2), 121–130.

UTTING S D and SPENCER B E (1991) *The hatchery culture of bivalve mollusc larvae and juveniles*, Lab. Leaflet 68. Lowestoft: MAFF Directorate of Fisheries Research.

11.10 References

AHMED M and SPARKS A K (1970) Chromosome number, structure and autosomal polymorphism in the marine mussels *Mytilus edulis* and *Mytilus californianus*. *Biol Bull* 138, 1–13.

ANDERSEN S *et al.* (2000) Flow-through systems for culturing great scallop larvae. *Aquac Int* 8, 249–257.

BALOUN A J AND MORSE D E (1984) Ionic control of settlement and metamorphosis in larval *Haliotis rufescens* (Gastropoda). *Biol Bull* 167, 124–138.

BAYNE B L (1964) Primary and secondary settlement in *Mytilus edulis* L. (Mollusca). *J Anim Ecol* 33, 513–523.

BAYNE B L (1965) Growth and the delay of metamorphosis of the larvae of *Mytilus edulis* (L.). *Ophelia* 2, 1–47.

BEAUMONT A R and FAIRBROTHER J E (1991) Ploidy manipulation in molluscan shellfish: a review. *J. Shellfish Res.*, 10, 1–18.

BEAUMONT A R, TURNER G, WOOD A R and SKIBINSKI D O F (2004) Hybridisations between *Mytilus edulis* and *Mytilus galloprovincialis* and performance of pure

© Woodhead Publishing Limited, 2013

species and hybrid veliger larvae at different temperatures. *J Exp Mar Biol Ecol* 302, 177–188.

BIERNE N, BORSA P, DAGUIN C, JOLIVET D, VIACA F, BONHOMME F and DAVID P (2003) Introgression patterns in the mosaic hybrid zone between *Mytilus edulis* and *M. galloprovincialis. Mol Ecol* 12, 447–461.

BOETTCHER A A and TARGETT N M (1998) Role of chemical inducers in larval metamorphosis of queen conch, *Strombus gigas* Linnaeus: relationship to other marine invertebrate systems. *Biol Bull* 194, 132–142.

BRAKE J, DAVIDSON J and DAVIS J (2002) Induction of triploidy in *Mytilus edulis* using 6-DMAP. *J Shellfish Res* 21, 503–508.

BRAKE J, DAVIDSON J and DAVIS J (2004) Field observations on growth, gametogenesis, and sex ratio of triploid and diploid *Mytilus edulis*. *Aquaculture* 236, 179–191.

CARPIZO-ITUARTE E and HADFIELD M G (1998) Stimulation of metamorphosis in the polychaete *Hydroides elegans* Haswell (Serpulidae). *Biol Bull* 194, 14–24.

COOPER K (1982) A model to explain the induction of settlement and metamorphosis of planktonic eyed-pediveligers of the blue mussel *Mytilus edulis* L. by chemical and tactile cues. *J Shellfish Res* 2, 117.

DAVIS M, HEYMAN W D, HARVEY W and WITHSTANDLEY C A (1990) A comparison of two inducers, KCl and *Laurencia* extracts, and techniques for the commercial scale induction of metamorphosis in queen conch *Strombus gigas* Linnaeus, 1758 larvae. *J Shellfish Res* 9, 67–73.

DOBRETSOV S V and QIAN P-Y (2003) Pharmacological induction of larval settlement and metamorphosis in the blue mussel *Mytilus edulis* L. *Biofouling* 19, 57–63.

DOMÍNGUEZ L, VILLALBA A and FUENTES J (2010) Effects of photoperiod and the duration of conditioning on gametogenesis and spawning of the mussel *Mytilus galloprovincialis* (Lamarck). *Aquac Res* 41, e807–e818.

EYSTER L S and PECHENIK J A (1987) Attachment of *Mytilus edulis* L. larvae on algal and byssal filaments is enhanced by water agitation. *J Exp Mar Biol Ecol* 114, 99–110.

GALLEY T H, BATISTA F M, BRAITHWAITE R, KING J and BEAUMONT A R (2010) Optimisation of larval culture of the mussel *Mytilus edulis* (L.) *Aquac Int* 18, 315–325.

GARCÍA-LAVANDEIRA M, SILVA A, ABAD M, PAZOS A J, SANCHEZ J L and PEREZ-PARALLE M L (2005) Effects of GABA and epinephrine on the settlement and metamorphosis of the larvae of four species of bivalve molluscs. *J Exp Mar Biol Ecol* 316, 149–156.

GOSLING E M (2003) *Bivalve Molluscs: Biology, Ecology and Culture*. Oxford: Blackwell Science.

GUO X and ALLEN S K JR (1994) Viable tetraploids in the Pacific oyster (*Crassostrea gigas* Thunberg) produced by inhibiting polar body 1 in eggs from triploids. *Mol Mar Biol Biotechnol* 3, 42–50.

HADFIELD M G (1984) Settlement requirements of molluscan larvae: new data on chemical and genetic roles. *Aquaculture* 39, 283–298.

KAMERMANS P and SMAAL A C (2002) Mussel culture and cockle fisheries in the Netherlands: finding a balance between economy and ecology. *J Shellfish Res* 21, 509–517.

KAMERMANS P E, BRUMMELHUIS E and SMAAL A (2002) Use of spat collectors to enhance supply of seed for bottom culture of blue mussels (*Mytilus edulis*) in the Netherlands. *World Aquac* 33, 12–15.

KENCHINGTON E L, HAMILTON L, COGSWELL A and ZOUROS E (2009) Paternal mtDNA and maleness are co-inherited but not causally linked in Mytilid mussels. *Plos One* 4, e6976.

KESARCODI-WATSON A, KASPAR H, LATEGAN M J and GIBSON L (2010) *Alteromonas macleodii* 0444 and *Neptunomonas* sp. 0536, two novel probiotics for hatchery-reared Greenshell™ mussel larvae, *Perna canaliculus*. *Aquaculture* 309, 49–55.

© Woodhead Publishing Limited, 2013

KIYOMOTO M, KOMARU A, SCARPA J, WADA K T, DANTON E *et al.* (1996) Abnormal gametogenesis, male dominant sex ratio, and Sertoli cell morphology in induced triploid mussels, *Mytilus galloprovincialis*. *Zool Sci* 13, 393–402.

LEDU C and MCCOMBIE H (2003) Effects of cytochalasin B on fertilization and ploidy in the Pacific oyster *Crassostrea gigas*. *Inverteb Reprod Dev* 42, 131–137.

LOOSANOFF V L and DAVIS H C (1963) Rearing of bivalve mollusc, in Russell, F. S. (ed.), *Advances in Marine Biology*, Vol. 1. London: Academic Press, 1.136.

LUTZ R A and KENNISH M J (1992) Ecology and morphology of larval and early postlarval mussels, in Gosling E (ed.) *The Mussel Mytilus: ecology, physiology, genetics and culture*. Amsterdam: Elsevier Science, 53–85.

MARSHALL R, MCKINLEY S and PEARCE C M (2010) Effects of nutrition on larval growth and survival in bivalves. *Rev Aquac* 2, 33–55.

MCCOMBIE H, LEDU C, PHELIPOT P, LAPÈGUE S, BOUDRY P and GERARD A (2005a) A complementary method for production of tetraploid *Crassostrea gigas* using crosses between diploids and tetraploids with cytochalasin B treatments. *Mar Biotechnol* 7, 318–330.

MCCOMBIE H, LAPÈGUE S, CORNETTE F, LEDU C and BOUDRY P (2005b) Chromosome loss in bi-parental progenies of tetraploid Pacific oyster *Crassostrea gigas*. *Aquaculture* 247, 97–105.

MCCOMBIE H, CORNETTE F and BEAUMONT A R (2009) Short sharp shock produces viable tetraploids in crosses of diploid blue mussels *Mytilus edulis*. *Aquac Res* 40, 1680–1682.

MCGRATH D, KING P A and GOSLING E M (1988) Evidence for the direct settlement of *Mytilus edulis* larvae on adult mussel beds. *Mar Ecol Prog Ser* 47, 103–106.

MESÍAS-GANSBILLAR C, EL AMINE BENDIMERAD M, ROMÁN G, PAZOS A J, SÁNCHEZ J L and PÉREZ-PARALLE M L (2008) Settlement behaviour of black scallop larvae (*Chlamys varia*, L.) in response to GABA, Epinephrine and IBMX. *J Shellfish Res* 27, 261–264.

MILLS S C AND CÔTÉ I M (2003) Sex-related differences in growth and morphology of blue mussels. *J Mar Biol Assoc UK* 83, 1053–1057.

MORSE A N C and MORSE D E (1984) Recruitment and metamorphosis of *Haliotis* larvae induced by molecules uniquely available at the surfaces of crustose red algae. *J Exp Mar Biol Ecol* 75, 191–215.

MORSE D E, HOOKER N, DUNCAN H and JENSEN L (1979) γ-aminobutyric acid, a neurotransmitter, induces planktonic abalone larvae to settle and begin metamorphosis. *Science* 204, 407–410.

NICOLAS L, ROBERT R and CHEVOLOT L (1998) Comparative effects of inducers on metamorphosis of the Japanese oyster *Crassostrea gigas* and the great scallop *Pecten maximus*. *Biofouling* 12, 189–203.

NORMAND J, LE PENNEC M AND BOUDRY P (2008) Comparative histological study in diploid and triploid Pacific oysters (*Crassostrea gigas*) reared in an estuarine farming site in France during the 2003 heatwave. *Aquaculture* 282, 124–129.

NORMAND J, ERNANDE B, HAURE J, MCCOMBIE H AND BOUDRY P (2009) Reproductive effort in *Crassostrea gigas*: comparison of 5-month-old diploid and triploid oysters issued from natural crosses or chemical induction. *Aqua Biol* 7, 229–241.

PAWLIK J R (1990) Natural and artificial induction of metamorphosis of *Phragmatopoma lapidosa californica* (Polychaeta: Sabellariidae), with a critical look at the effects of bioactive compounds on marine invertebrate larvae. *Bull Mar Sci* 46, 512–536.

PAWLIK J R (1992) Chemical ecology of the settlement of benthic marine invertebrates. *Oceanogr Mar Biol Ann Rev* 30, 273–335.

PIFERRER F, BEAUMONT A R, FALGUIÈRE J-C, FLAJŠHANS M, HAFFRAY P and COLOMBO, L (2009) Polyploid fish and shellfish: Production, biology and applications to

© Woodhead Publishing Limited, 2013

aquaculture for performance improvement and genetic containment *Aquaculture* 293, 125–156.

PIRES A and HADFIELD M G (1991) Oxidative breakdown products of catecholamines and hydrogen peroxide induce partial metamorphosis in the nudibranch *Phestilla sibogae* Bergh (Gastropoda: Opisthobrachia). *Biolo Bull* 180, 310–317.

PRONKER A E, NEVEJAN N M, PEENE F, GEIJSEN P and SORGELOOS P (2008) Hatchery broodstock conditioning of the blue mussel *Mytilus edulis* (Linnaeus 1758). Part I. Impact of different micro-algae mixtures on broodstock performance. *Aquacult Int* 16, 297–307.

RICO-VILLA B, WOERTHER P, MINGANT C, LEPIVER D, POUVREAU S, HAMON M AND ROBERT R (2008) A flow-through rearing system for ecophysiological studies of Pacific oyster *Crassostrea gigas* larvae. *Aquaculture* 282, 54–60.

RODRÍGUEZ S R, OJEDA F P and INESTROSA N C (1993) Settlement of benthic marine invertebrates. *Mar Ecol Prog Ser* 97, 193–207.

SAAVEDRA C, REYERO M I and ZOUROS E (1997) Male-dependent doubly uniparental inheritance of mitochondrial DNA and female-dependent sex-ratio in the mussel *Mytilus galloprovincialis*. *Genetics* 145, 1073–1082.

SATUITO C G, BAO W, YANG J and LITAMURA H (2005) Survival, growth, settlement and metamorphosis of refrigerated larvae of the mussel *Mytilus galloprovincialis* Lamarck and their use in settlement and antifouling bioassays. *Biofouling* 21, 217–225.

SCARPA J, WADA K T and KOMURU A (1993) Induction of tetraploidy in mussels by suppression of polar body formation. *Nippon Suisan Gakk* 59, 2017–2023.

VSEVOLODOVA-PEREL T S and BULATOVA N SH (2008) Polyploid races of earthworms (Lumbricidae, Oligochaeta) in the East European plain and Siberia. *Biol Bull* 35, 385–388.

WHITE M J D (1940) Evidence for polyploidy in the hermaphrodite groups of animals. *Nature* 146, 132–133.

WIDDOWS J (1991) Physiological ecology of mussel larvae. *Aquaculture* 94, 147–163.

YOOL A J, GRAU S M, HADFIELD M G, JENSEN R A, MARKELL D A and MORSE D E (1986) Excess potassium induces larval metamorphosis in four marine invertebrate species. *Biol Bull* 170, 255–266.

YANG J-L, SATUITO C G, BAO W-Y and KITAMURA H (2008) Induction of metamorphosis of pediveliger larvae of the mussel *Mytilus galloprovincialis* Lamarck, 1819 using neuroactive compounds, KCl, NH_4Cl and organic solvents. *Biofouling* 24, 461–470.

ZHANG L and KING C E (1992) Genetic variation in sympatric populations of diploid and polyploid brine shrimp. *Genetica* 85, 211–221.

ZHAO B, ZHANG S and QIAN P-Y (2003) Larval settlement of the silver- or goldlip pearl oyster *Pinctada maxima* (Jameson) in response to natural biofilms and chemical cues. *Aquaculture* 220, 883–901.

ZOUROS E, BALL A O, SAAVEDRA C and FREEMAN K R (1994a) An unusual type of mitochondrial DNA inheritance in the blue mussel *Mytilus*. *Proc Natl Sci USA* 91, 7463–7467.

ZOUROS E, BALL A O, SAAVEDRA C and FREEMAN K R (1994b) Mitochondrial DNA inheritance. *Nature* 368, 818.

© Woodhead Publishing Limited, 2013

12

Research on the production of hatchery-reared juveniles of cephalopods with special reference to the common octopus (*Octopus vulgaris*)

J. Iglesias and L. Fuentes, Spanish Institute of Oceanography, Vigo, Spain

DOI: 10.1533/9780857097460.2.374

Abstract: This chapter deals with the different methods used in the world to produce hatchery-reared juveniles of *Octopus vulgaris*, including capture and transport methods, broodstock conditioning and paralarval rearing technology. First, a review is made of the different systems used for hatchery production of the various Cephalopod species in which there is the greatest commercial interest. Then, specifically for *Octopus vulgaris*, conclusions and standardized methods related to the spawning process, the control of embryonic development and, finally, a protocol for paralarval rearing are also given. The main bottlenecks and future research trends are pointed out, and important sources of further information and advice are referenced. A brief review of the state-of-the-art on larvae and juvenile rearing of other cephalopods species is also included in this chapter.

Key words: cephalopoda, *Octopus vulgaris*, culture, reproduction, paralarvae rearing.

12.1 Introduction

12.1.1 Research on cephalopod culture in the world

The class Cephalopoda is the most complex taxa in the phylum Mollusca, consisting of a large group of marine organisms of more than 700 species. They appeared in the Cambrian period (over 500 million years ago), but the majority are now extinct.

Living cephalopods are grouped into two 'sub-classes': the Coleoidea, which include squids, cuttlefishes, octopods and vampires; and the Nautiloidea, containing two genera, *Nautilus* and *Allonautilus*, both still with an external shell. The systematics and classification of the living Cephalopoda

© Woodhead Publishing Limited, 2013

is still under consideration; however, for practical purposes the FAO has put together a new catalogue (Jereb and Roper, 2010).

Global cephalopod capture was 3 458 410 t in 2009 (FAO, 2010), which represents 3.9 % of the total world fishery capture, and it was distributed among countries such as China (1 056 165 t), Japan (392 440 t), Italy (34 727 t), Republic of Korea (432 752 t) and Spain (43 093 t). The total capture of the different octopus species was 363 428 t in 2009, 40 609 t of which corresponded to *Octopus vulgaris*, distributed as follows: 4.6 % Africa, 41.1 % America, 1.6 % Asia and 52.6 % Europe.

In relation to cephalopod consumption, China was the biggest consumer with 1 383 545 t in 2007 followed by Japan (616 146 t), Italy (228 919 t), Republic of Korea (221 337 t) and Spain (200 995 t). Japan, Italy, Spain and China are great importer countries; for example, 59 100 t of squid and 56 200 t of octopus were imported by Japan in 2009.

Cephalopods are marine organisms with a very short life-cycle and, in consequence, they present very high growth and conversion rates (Mangold and Boletzky, 1973; Wells, 1978; Mangold, 1983). Other advantages to culture cephalopods are their high protein content (O'Dor and Wells, 1987) and, in general, their high fecundity (Wells, 1978; Mangold, 1983). The greatest inconvenience in their culture is that, in the majority of cases, the paralarvae are carnivores and they require live prey for the first two months of life. They also have very poor tolerance of low salinity and oxygen levels, although they are highly tolerant of ammonium, nitrite and nitrate levels (Hanlon and Messenger, 1996).

In order to differentiate the activities carried out with cephalopods in captivity, Boletzky and Hanlon (1983) distinguish three different categories: the first one is 'maintenance', which means the conditioning of juveniles of the same age of one species in captivity; the second was defined as 'rearing', or the growth of specimens over a period of time but without reaching the second generation; and the third was 'culture', when, from paralarvae obtained in captivity, the growth of the juveniles continues until the second generation, thus closing the culture cycle of the species.

Although about 10 % of cephalopod species (approximately 70 of the 700 known species), have been kept under captive conditions, only around 15 of these have been cultured but, in general, the results obtained did not attain an industrial scale.

One of the most widely studied species is the cuttlefish *Sepia officinalis*; cuttlefish has been maintained, reared and cultured in the laboratory for many years (Richard, 1971; Pascual, 1978; Forsythe *et al.*, 1994; Domingues and Marques, 2010, etc.). It is highly adaptable to life in captivity, has large eggs, a high hatchling survival and sedentary behaviour, tolerates high culture densities and handling, presents little cannibalism, accepts dead prey (Forsythe *et al.*, 1994) and, most importantly, readily reproduces in captivity (Domingues *et al.*, 2002). Sykes *et al.* (2006) describe a complete view of the state-of-the-art and future trends of this species.

© Woodhead Publishing Limited, 2013

However, *S. officinalis* culture experiences several bottlenecks that prevent fully commercial culture. Those are: lower fertility and fecundity under culture conditions; semelparous life history, therefore requiring a new group of breeders for each cycle; hatchlings requiring live food; and juveniles and adult stages refusing dry pellets. The routine production of a suitable live food is not yet developed so the cost of food supply is high. The fact that the species is cannibalistic and has a basic immunological system (Forsythe *et al.*, 1987, 1990) can generate problems in intensive culture.

The existence of appropriate and inexpensive artificial diets is a vital requirement for the viability of its commercial culture. Although *S. officinalis* has been maintained in captivity fed on artificial diets (Castro, 1991; Hanlon *et al.*, 1991; Castro *et al.*, 1993; Castro and Lee, 1994; Domingues, 1999) these gave very low growth rates. The best growth rates were obtained when feeding with wild caught frozen shrimp (Domingues *et al.*, 2003, 2004, 2006; Sykes *et al.*, 2006).

With relation to other cuttlefish species, Nabhitabhata (1997) and Nabhitabhata and Nilaphat (1999) have published culture conditions for *Sepiella inermis* and *S. pharaonis,* respectively, in Thailand.

Considerable effort has been made to improve laboratory maintenance, rearing and culturing of loliginid squid with the main propose of providing a reliable supply for neuroscientists (Boletzky and Hanlon, 1983; Hanlon, 1990; Lee *et al.*, 2000). Nonetheless, most major attempts to culture squid species of the genus *Loligo* have been unsuccessful. The reasons for failure have been attributed to extremely high mortality rates (80–99 %) during the first weeks after hatching (Turk *et al.*, 1986; Yang *et al.*, 1986; Hanlon *et al.,* 1987, 1989), caused by starvation and inadequate tank design that led to skin damage and infection (Hanlon *et al.*, 1983, 1991). The first completion of a squid cycle was accomplished by Yang *et al.* (1983, 1986) working with *L. opalescens*. However, mortality was still quite high and the best survival rate obtained during the first two months after hatching was 25 %. Vidal *et al.* (2002a, b) designed a study to optimise survival of squid (*L. opalescens*) hatchlings, especially through tank design, temperature and feeding.

Advances have been made in squid culture using the loliginid species *Sepioteuthis lessoniana*. This species has been cultured through multiple generations in the laboratory (Lee *et al.* 1994, 1998). Improvements in the culture system resulted in higher survival, mating and fecundity in captivity. The major advantage in culturing *S. lessoniana* is the extremely large size of the hatchlings (approx. 1 cm), making them easier to feed. The life-cycle of the bobtail squid *Euprymna hyllebergi* has also been described by Nabhitabhata *et al.* (2005) in Thailand. Nabhitabhata and Suwanamala (2008) study the reproductive behaviour, mating and spawning in captivity of the squid species *Idiosepius biserialis* and *I. thailandicus* (Cephalopoda: Idiosepiidae).

© Woodhead Publishing Limited, 2013

In relation to different species of octopuses, *O. maya* has large eggs and direct embryonic development, in contrast to the majority of the octopus species that have planctonic larvae (Boletzky, 1974), making them easier to culture in the laboratory; *O. maya* is highly adaptable for laboratory conditions (Boletzky and Hanlon, 1983), and it accepts dead or artificial feeds early in the life-cycle, compared with other cephalopods. *O. maya* has been cultured in the laboratory (Solis, 1967; van Heukelem, 1976, 1977; Hanlon and Forsythe, 1985; DeRusha *et al.*, 1989) for up to four (Hanlon and Forsythe, 1985) or five (van Heukelem, 1983) generations. It has been maintained on artificial diets (Lee *et al.*, 1991), but growth obtained was moderate compared with natural diets. Domingues *et al.* (2007) also determined the effects of an artificial diet on growth and survival. Many investigations have been carried out on the species *O. maya* in Mexico (Rosas *et al.*, 2006, 2011; Domingues *et al.*, 2007, 2012; Quintana *et al.*, 2011). Despite this theoretical easy rearing process, the species also shows low fecundity, and very high mortality during the first few months of culture due fundamentally to the lack of a suitable pelleted feed for ongrowing.

The marbled octopus, *Amphioctopus aegina*, is the second octopus species with planktonic hatchlings, following *O. vulgaris*, for which the life-cycle has been completed through rearing in the laboratory (Promboon *et al.*, 2011). Growth is allometric, consisting of three phases in terms of body weight. The period from hatching to spawning takes 74 % of the entire lifespan and the reproductive phase took 34 %, similar to those of other benthic cephalopods with planktonic hatchlings.

Concerning *O. mimus,* data have been published on the reproduction and embryonic development (Zúñiga *et al.*, 1995; Warnke, 1999; Castro-Fuentes *et al.*, 2002), and growth (Cortez *et al.*, 1999) of this species in Peru and Chile. However, very few publications exist on *O. mimus* culture (Olivares *et al.*, 1996; Zúñiga *et al.*, 1996; Baltazar *et al.*, 2000). Its growth rate (1–1.5 kg in four to six months) is much less than that of the common octopus (*O. vulgaris*), and the main problem for its culture also lies in the high mortality of paralarvae during the first month of culture. There is still not a pellet to feed the sub-adults, which represents an obstacle to its industrial application. A project on juveniles production based on the use of microdiets is currently underway in the Second Region of Chile.

Uriarte *et al.* (2009, 2010, 2011a) describe the characteristics of the embryonic and paralarval stages and explore the feasibility of obtaining *Robsonella fontaniana* juveniles under controlled conditions and show that *R. fontaniana* can be reared from hatching through the final paralarval stage on a diet of *Lithodes santolla* (king crab) zoeae; after settlement, the juveniles can be reared on a diet of crab such as *Petrolisthes* sp. This species, a small-sized octopus traded in the international markets as 'baby octopus', is abundant in the south of Chile and it is distributed over nearly the whole southern coast of South America, both on the Pacific and Atlantic coasts.

© Woodhead Publishing Limited, 2013

The Patagonian red octopus, *Enteroctopus megalocyathus*, has been studied in Chile at the Austral and Los Lagos Universities; they have focused on the effect of broodstock diet on the fecundity and biochemical composition of the eggs (Farias *et al.*, 2011) and on the study of female maturation and the embryonic development (Ortiz *et al.*, 2006). They found that females can produce 3000 eggs by spawning and that the embryonic development was very long, which gave a high risk of microbial contamination, but even with that situation they attained hatching. Some studies were carried out focusing on growth of juveniles in captivity using moist diets like fresh mytilids (Pérez *et al.*, 2006) or fresh fish and crab paste (Farias *et al.*, 2010), indicating that the range of growth expected for the Patagonian red octopus could be close to 2 % per day. In any case more research is required in this field.

Other species of octopus, such as *O. bimaculoides* in the USA and Mexico (Solorzano *et al.*, 2009), *O. ocellatus* in Japan (Segawa and Nomoto, 2002), *O. minor* in China (Zheng, personal communication), *O. tehuelchus* in Argentina (Klaich *et al.*, 2008) and *O. rex* in Thailand (Nabhitabhata *et al.*, 2003), are also being studied. Uriarte *et al.* (2011b) analyse the current status of Octopod aquaculture in Latin America.

12.2.1 Historical review of the octopus *O. vulgaris* culture

O. vulgaris is one of the most studied species of cephalopod in the world. Diverse aspects of its biology, physiology, behaviour and even its culture have been the subject of broad reviews (Wells, 1978; Mangold, 1983; O'Dor and Wells, 1987; Hanlon and Messenger, 1996; Vaz-Pires *et al.*, 2004; Iglesias *et al.*, 2007a; Villanueva and Norman, 2008).

Attempts to rear octopus paralarvae started in Japan in the 1960s, and Itami *et al.* (1963) were the first in managing to obtain benthic juveniles with a 5 % survival rate after two months, using shrimp (*Palaemon serrifer*) zoeae as prey. Later on, Imamura (1990) and Hamasaki *et al.* (1991) reported survival rates of 25–28 % at 25 days, using *Artemia* enriched with microalgae as the only live prey. They report for the first time the possibility to produce octopus paralarvae on a mass production scale. Current research in Japan focuses on the production of juveniles for enhancement programs (Okumura *et al.*, 2005), using *Artemia* and frozen slices of Pacific sandeel, *Ammodytes personatus*, as paralarvae feed.

In the 1990s, researchers from the Spanish Scientific Research Council (CSIC) took the first steps in researching the rearing of larvae, obtaining benthic juveniles for the first time in Europe, with a survival rate of 8.9 and 0.8 % at 52 and 60 days, respectively (Villanueva, 1994, 1995). They used crustacean zoeae as prey and a temperature of 21 °C. Later on, Moxica *et al.* (2002) used larger preys: adult *Artemia* of up to 2 mm and spider crab zoeae and megalops, increasing survival and dry weight of the paralarvae.

© Woodhead Publishing Limited, 2013

The most important driving force behind the industrial development of the culture of this species started in Spain and came from the very high growth rates obtained in tanks at the Spanish Institute of Oceanography (IEO) in Vigo (Iglesias *et al.*, 1997), which were later also confirmed in floating cages (Rama-Villar *et al.*, 1997). In the mid-1990s, the first companies engaged in ongrowing wild octopus sub-adults were set up in the Galician Rias (NW Spain). Small cooperatives of fishermen were pioneers in this field. Their approach was the fattening of sub-adults of 750 g (the minimum legal size then) to 2.5 kg in weight over a period of four months, using crabs and fish of low commercial value (Rama-Villar *et al.*, 1997; Rey-Méndez *et al.*, 2003; Chapela *et al.*, 2006; Iglesias *et al.*, 2007b).

From 2001 to 2004, a Spanish programme for the culture of octopus was granted by the Spanish Ministry for Agriculture, Fisheries and Food, focusing basically on two sub-projects: the paralarval rearing and the ongrowing of sub-adults. Concerning larval rearing, the major advance was brought about by the use of live prey, both cultivated (*Artemia*) and from the natural environment (crustacean zoeae), thus improving growth and survival rates. In 2001, a survival rate of 31.5 % was obtained at 40 days of age using these preys and, for the first time anywhere in the world, it became possible to complete the culture cycle on an experimental scale (Iglesias *et al.*, 2004). Later, in 2002–2003 Carrasco *et al.* (2005) obtained similar results using the same prey, but with very different cultivation systems. As a result of this programme, an international workshop for world experts in the rearing of cephalopod larvae was held in Vigo (Spain) in November 2005, with the aim of discussing the different culture systems used, analysing the causes of larval mortality and establishing future research priorities. This chapter presents many of the conclusions drawn from this international workshop (Iglesias *et al.*, 2007a), and also includes many personal recommendations emanating from more than 15 years of experience working on *O. vulgaris* culture.

Currently, two national research programmes are being carried out in Spain; one of them is titled 'Nutrition and feeding of *Octopus vulgaris* paralarvae and subadult', and the other 'Nutritional physiology and stress of the first life stages of common octopus (*Octopus vulgaris*)'; both of them are an attempt to solve the problem of high paralarvae mortality. Parallel to the research on octopus culture made in Spain, in the period 2002–2007, an Italian group has also carried out a very comprehensive investigation on this field in Italy (Lenzi *et al.*, 2006; De Wolf *et al.*, 2011). They focus their research mainly on paralarvae rearing, using as preys rotifers and *Artemia* with different enrichment products; they obtained the first juveniles in 2007, attaining an age of 160 days.

A large group of Spanish, Portuguese and Latin American researchers have recently applied for a thematic network to CYTED (Ibero-American Programme for Science, Technology and Development) to keep on working together on the culture of different species of octopus (see Section 12.5).

© Woodhead Publishing Limited, 2013

12.2 Broodstock conditioning and reproduction process

A wide range of conditions have been used to maintain the octopus broodstock in captivity (Table 12.1). The method of capture, transport conditions, food supply and light intensity are similar among different research groups, but the most obvious differences are the male:female ratios and the broad range of temperatures used (14–25 °C).

In general, there are no major problems regarding the acclimatisation of spawners and obtaining viable egg-laying under culture conditions. This process can be carried out either by creating a broodstock of spawners, mixing males and females (Iglesias *et al.*, 2000; Roo, personal communication), or individually, separating already fertilised wild females captured at sea (Villanueva, 1995; Okumura *et al.*, 2005). In general, 100 % of the females mature and lay eggs in captivity, with egg hatching percentages of over 80 % being obtained. Based on a comparative analysis of the different systems used in *O. vulgaris* culture throughout the world, it is possible to establish a set of general recommendations or protocols for the reproduction phase.

12.2.1 Broodstock capture

It is recommended to use selective methods like creels or individual traps for the capture of spawners, avoiding trawl nets or other fishing methods that may harm the specimens caught and increase the environmental impact.

12.2.2 Adult transport

Mortality during transport is reduced if individual mesh bags or PVC tubes are used (Fuentes *et al.*, 2005). Additionally simple aeration or oxygen supplies are recommended in the case of low and high densities, respectively.

12.2.3 Broodstock feeding

The broodstock diet influences considerably the biochemical composition of the paralarvae (Quintana *et al.,* 2009) and, in general, crustaceans and fish of low commercial value are used as food with optimal results (Quintana, 2009; Estefanell, 2012). In order to enhance the biochemical profile of the paralarvae, it is recommended that at least 30 % of the diet should be composed of crustaceans (Iglesias *et al.*, 2007a).

12.2.4 Sex ratio of spawners

During the spawning period in nature, it would be enough to maintain only wild mature females (> 1 kg) in captivity, because they would already be fertilised. These females can preserve the spermatophores viable for long

© Woodhead Publishing Limited, 2013

Table 12.1 Summary of *Octopus vulgaris* broodstock capture methodology, transport and maintenance used by the different research groups

	Barcelona ICM-CSIC	Vigo IEO	Canary Island ICCM	Asturias CEP	Andalusia IFAPA	Brazil FURG	Japan YS	Seixas *et al.* (2010) USC	De Wolf *et al.* (2011) MRS	Estévez *et al.* (2009) IRTA
Capture method	Traps	Traps	Traps	Traps	Traps and hooks	Pots and diving	Pots, hooks and lines	Mature female donated by IEO (Vigo)	Professional fishermen during fall	Mature female donated by IEO (Vigo)
Transport	Oxygen supply Low density	Oxygen supply Low density	Oxygen supply	<15 min no water >15 min aeration	Aeration	Oxygen supply, aeration	Oxygen supply	35 L container (1 h)		
Sex ratio (male : female)	1:5–8	1:2	1:3	1:2–3	1:1	1:2–3	No males			
Food	Crab, fish Reduce to induce spawning	Crab, fish, mussel Reduce to induce spawning	Crab, fish	Crab, fish	Crab, fish	Crab, fish	Crab, mantis shrimp, fish		Frozen crabs, mussel, shrimp and trash fish	
Temperature (°C)	14–23	15–20		15–20	15–22	19–23	15–25		Natural (heating if < 15 °C)	
Light	Darkness	Dim		Dim	Shaded	Dim	Shaded (90 % reduction)		Natural photo periods (shaded)	

Note: ICM-CSIC (Instituto de Ciencias del Mar-Consejo Superior de Investigaciones Científicas), IEO (Instituto Español de Oceanografía), ICCM (Instituto Canario de Ciencias Marinas), CEP (Centro de Experimentación Pesquera), IFAPA (Instituto de Investigación y Formación Agraria y Pesquera), FURG (Universidade Federal do Rio Grande), YS (Yashima Station), USC (Universidad Santiago de Compostela, Spain), MRS (Maricoltura di Rosignano Solvay s.r.l., Italy), IRTA (Recerca I Tecnologia Agroalimentaries, Spain)
Source: Adapted from Iglesias *et al.*, 2007a. Reproduced with permission from Elsevier.

© Woodhead Publishing Limited, 2013

periods. The rest of the year, males and females should be kept in a ratio of one male to three females.

12.2.5 Spawning females and egg strings care

When the spawning has been completed, and all the strings of eggs have been laid, the *O. vulgaris* females, together with their egg masses, must be transferred to another tank, thus avoiding females being disturbed by other spawners. However, special incubators for egg masses without the female have been developed for other species like *O. maya* (Rosas *et al.*, 2006, 2010) and they are currently being tested in *O. vulgaris* in Chile.

12.2.6 Control of ageing eggs

Each egg batch must be checked daily to establish in accordance with the temperature, the time in days necessary to hatch. Using this method the required concentrations of phytoplankton and amount of *Artemia* for larval rearing can be estimated in advance. Figure 12.1 shows a summary of the embryonic development of *O. vulgaris* at 18 °C; the different stages have been assigned according to Naef (1928).

12.2.7 Egg handling

The transport and handling of the egg mass between different localities must be done together with the female and in the same spawning shelter. This process must be performed before the second embryonic inversion occurs (Fig. 12.1e), thus avoiding premature hatching. In addition, drastic changes (e.g. in temperature, salinity, pH, light intensity) in the aquatic environment containing the female and its eggs mass must be avoided.

12.2.8 Paralarvae collection and transfer

The transfer of the newly hatched paralarvae to another tank must be carried out with extreme care in order to avoid causing them stress or irreversible damage. It is recommended to use a 30 cm PVC collector with a 0.5 mm bottom plankton mesh to concentrate the paralarvae into and then remove them gently with a 1 L jar. Sampling following hatching must be also carried out carefully, either by individual counting or by volumetric estimation.

12.3 Paralarvae rearing

At hatching, *O. vulgaris* paralarvae (defined by Young and Harman, 1988) show a total length of 3 mm and an individual dry weight of 0.20–0.35 mg (Iglesias, personal communication; Fig. 12.2).

© Woodhead Publishing Limited, 2013

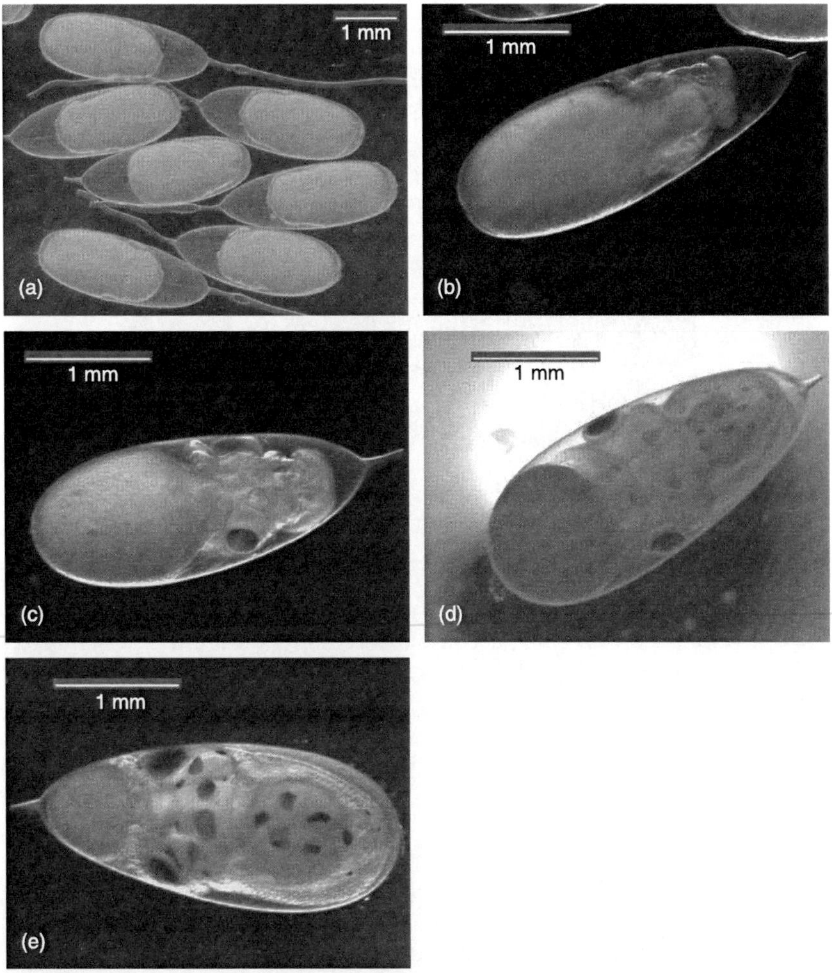

Fig. 12.1 Embryonic development of octopus *Octopus vulgaris* at 18 °C. (a) Eggs after 6 days of embryonic development; first inversion is beginning (Naef VIII stage). (b) 14-day embryo; mantle, eyes and arms formation is beginning (Naef X stage). (c) 23-day embryo; mantle, eyes and arms are completely formed; yolk sac is a half the size of the egg (Naef XIV stage). (d) 30-day embryo; paralarvae are fully formed; yolk sac is third of the size of the egg (Naef XVII-XVIII stage). (e) 34-day embryo; second inversion has happened and paralarvae will hatch in four days (Naef XIX-XX stage). (Photographs by M. Nande)

The main problem impeding commercial culture lies in the high mortality rate observed in the paralarval rearing during the first two months of life. Despite the limited knowledge on the fundamental biological and physical requirements of cephalopod paralarvae and their behaviour, many attempts have been made to explore the feasibility of their culture.

© Woodhead Publishing Limited, 2013

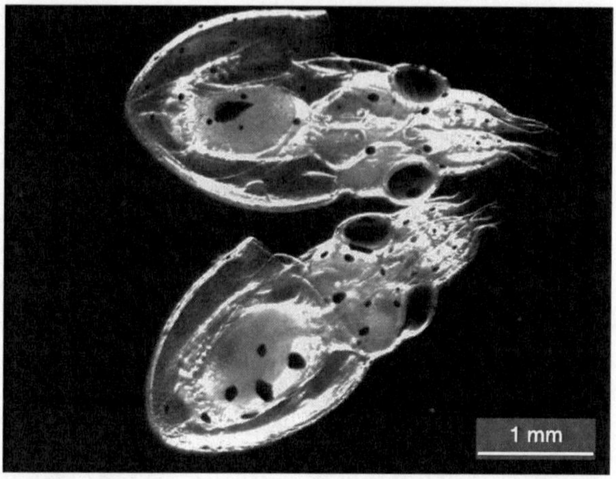

Fig. 12.2 *Octopus vulgaris* paralarvas at hatching (total length: approx. 3 mm; 3 suckers per arm). (Photograph by M. Nande).

Consequently, a great variety of types, volumes and colours of tanks, lighting conditions, water turbidity and supply systems have been used in this phase of cultivation (Iglesias *et al.*, 2007a). Furthermore, different types and sizes of prey have been tested in various feeding regimes (concentrations, number of intakes, food processing) (Table 12.2). As a result, there are currently no standard methods or protocols for the rearing of paralarvae.

For many years, there have been discussions about what causes the high larval mortality observed during the first two months of life. Two possible factors have been considered as responsible of this mortality (Iglesias *et al.*, 2007a): the lack of a standardised culture system and the absence of a suitable live food source to cover the paralarvae requirements. The following recommendations about paralarvae rearing parameters are made based upon 15 years of research and practical experience.

12.3.1 Prey and paralarvae distribution

There is general agreement that there must be a homogeneous distribution of both live prey and paralarvae in the water column of the culture tank. In order to achieve this, some authors (Iglesias *et al.*, 2004; Okumura *et al.*, 2005) use gentle aeration at the bottom of the tank, produced by diffuser stones, avoiding the production of small bubbles that could get into the paralarvae mantle. Other authors employ special systems for the distribution and diffusion of incoming water; for example, Villanueva (1995) used multiple superficial intakes and Carrasco *et al.* (2005, 2006) installed a water inlet at the bottom of the tank and a superficial outlet, which produced upward vertical circulation as is commonly used for lobster culture. However, it is important to mention that with high water flow cephalopod paralarvae

© Woodhead Publishing Limited, 2013

can suffer arm erosions and skin damage that significantly increase mortality (Vidal *et al.*, 2002a).

12.3.2 Tank colour and volume

Black or dark tanks are recommended by the majority of authors; however Carrasco *et al.* (2005) using white tanks also attained very good results on growth and survival. The use of tanks between 500 and 1000 L is strongly recommended because experience shows that lower results in growth and survival are attained with tanks smaller than 100 L. In fact, Sánchez *et al.* (2011) conclude that tank volume significantly influences the growth of octopus paralarvae reared in an intensive system; the greater the volume, the better the growth obtained (Fig. 12.3).

The same trend has been observed by other researchers; Moxica *et al.* (2006) and Viciano *et al.* (2011), working in 1000 L tanks and using the same diet (enriched *Artemia*), obtained 1.76 mg and 1.88 mg of dry weight, respectively, for one month-old paralarvae, while Seixas (2009) and Villanueva *et al.* (2002), using smaller rearing tanks (25–50 L), reached values of 0.83 and 0.90 mg, respectively, for larvae of the same age. De Wolf *et al.* (2011) also attained better results using larger tanks; they suggest that fluctuations in physical conditions are less dramatic in larger volumes than in small ones.

12.3.3 Prey size

With respect to food supply, it is generally accepted that *Artemia* constitutes an attractive prey that is well accepted by octopus larvae. However, there are discrepancies concerning the optimal size of the *Artemia* to be supplied to the culture tanks. Some authors (Navarro and Villanueva, 2003) use *Artemia* nauplii of 450–750 μm, in the first few weeks of culture, while others (Moxica *et al.*, 2002; Iglesias *et al.*, 2004; Carrasco *et al.*, 2005; Estévez *et al.*, 2009) use *Artemia* bigger than 2 mm in length which are better accepted than the nauplii at first feeding (Iglesias *et al.*, 2006).

In order to investigate this concept, Fuentes *et al.* (2009) compared the use of small *Artemia* enriched for 24 h (A1) with *Isochrysis galbana* with larger *Artemia* grown for four days with the same microalgae (AN) during a one-month larval rearing experiment. Statistical analysis of the results showed no significant differences in growth over the first 15 days, but the growth was significantly faster when larger AN was used as diet from 15 to 30 days of culture (Fig. 12.4). Despite the fact that octopus paralarvae, at first feeding, preferred adult *Artemia* to nauplii (Iglesias *et al.*, 2006) for experimental purposes and ease of management processes in larval rearing, it is recommended to use one day-*Artemia* of 0.5–0.7 mm length for the first 15 days and *Artemia* of 2 mm (previously ongrown for four days with phytoplankton) from then onwards.

© Woodhead Publishing Limited, 2013

Table 12.2 Summary of paralarvae rearing conditions of *Octopus vulgaris* carried out by different research groups

Parameters	Barcelona ICM-CSIC	Vigo IEO	Canary Islands ICCM	Asturias CEP	Andalusia IFAPA
Tank volume (L)	25–50	1000	100	30	400
Tank colour	Black	Black	Grey	White	Black, grey
Tank shape	Cylindrical and parabolic	Cylindrical	Cylindrical	Parabolic	Cylindrical and rectangular
Water system	Open	1st week stagnant, 2nd semi-open (3–4 h = 100 %)	Open 25 % day^{-1}	Open (recirculation)	Open
Aeration	No	Yes, intermediate	Yes, gentle	Yes, gentle (cleaning)	Yes, gentle
Light	24 h bulb 60 w 900 lx at 1 cm under the water surface	24 h fluorescent (2) 36 W 2000 lx	Natural photoperiod	12 h light-12 h darkness; fluorescent (1)40 W	Natural photoperiod
Temperature (°C)	19–23	20–22	21.5–22.5	20–22	19–22
Clear/green water	Clear	Green *Isochrysis* + *Nannochloropsis* sp.	Clear	Clear	Green *Tetraselmis* + *Isochrysis*
Paralarvae density (ind. L^{-1})	13–48	5	25	25	20
Type and prey density (ind. mL^{-1})	Zoeae (*Liocarcinus* and *Pagurus*) nauplii *Artemia* (2–6) and *Artemia* biomass	Zoeae *Maja* (0.01–0.1) (when available) + *Artemia* (0.05–0.1)	Zoeae *Grapsus*_ (15 ind. L^{-1}) + *Artemia* (72 h) (2)	Zoeae *Maja* (0.7–1) + *Artemia* (3 times/week) (0.5–0.7)	Zoeae (*Carcinus*, *Palaemon* and *Maja*) (<0.1) + *Artemia* + *Moina* (4–5 day-old) (1.0)
Size of prey	Zoea (1.3–3.1 mm TL) *Artemia* nauplii to 1–3 mm *Artemia* biomass	Zoeae: 1 mm TL *Artemia*: 2–3 mm TL	Zoeae: 1.5 mm TL *Artemia*: 0.85 mm TL	Zoeae: 1 mm TL *Artemia* retained in 300 μm sieve	Zoeae: 0.8–1.0 mm Moina: 1.0–1.2 mm *Artemia*: 1–3 mm
Artemia enrichment	DCSuperSelco, Metionine	Reared in commercial cereal flour, enriched with *Nannochloropsis* (5.10^6 cell mL^{-1})	*Artemia* enrichment (A_1 Selco Inve)	Reared and enriched with *Tetraselmis*	Reared and enriched with *Tetraselmis* + *Isochrysis* SuperSelco, Prolon
Sampling	Every 7–10 days	Every 7 days	Every 7 days	Every 10 days	Every 7–10 days
Survival	0.8 % at day 60 with zoeae, and 54 % at day 20 with *Artemia* nauplii (with poor growth)	31.5 % at day 40	11–27 % at day 30	89.6–93.5 % at day 20 and 3.4 % at day 60	5–15 % at day 35
Cleaning	Daily tank bottom siphoning		No bottom cleaning	Every 20 days changing tank by pipetting and checking the survival	Daily tank bottom siphoning

Source: Adapted from Iglesias *et al.*, 2007a. Reproduced with permission from Elsevier.

© Woodhead Publishing Limited, 2013

Brazil FURG	Japan YS	P. Seixas *et al.* (2010) USC	T. De Wolf *et al.* (2011) MRS	A. Estévez *et al.* (2009) IRTA
100	500	50	100–6000	500
Black	Orange	White	Light-grey (100 L) and black walls and white bottoms (larger volumes)	Black
Cylindrical	Cylindrical	Conical	Circular or slightly cylindroconical	Cylindroconical
Closed (recirculation)	5 first days stagnant then open system	Semi-closed circuit (600 L; 10 % renewal rate)	Semi-closend system (20 % of new sea water daily in system; 100–200 % in tanks). Spray inlet (shower-type)	Recirculation unit ('green water' during first week and then water was renewed daily (0.7–1 L min^{-1})
No	Yes, gentle		Yes, gentle	
10 h light-14 h darkness; natural + cold light	Fluorescent (1) 36 W	14 h light-10 h darkness; fluorescent day light lamp	14 h light-10 h darkness, artificial light (60–250 lx)	16 h light-8 h darkness (500 lx)
19–24	25	19–20 (stable climate room)	Regulated by means of a heat exchanger (18.5–25)	18
Clear	Fresh water *Chlorella* sp.	Clear	700 000 cells mL^{-1} (*Nannochloropsis* sp., *Isochrysis galbana* and a mix)	Green during first week
5–30	3	10	1–15	20
Crustacean zoeae, copepods, mysids, nauplii and adult *Artemia* 0.15–0.3 (4–5 takes)	*Artemia* nauplii + fish flakes from 5th day	Enriched *Artemia* juveniles (0.05 ind. day^{-1})	Rotifers (5 ind mL^{-1}) and *Artemia* nauplii (1–2 ind mL^{-1}) and adult (0.05–0.1 ind mL^{-1})	*Artemia* metanauplii (1–3 ind mL^{-1}) alone or mixed with zooplankton (0.05–1 *Palaemon* sp. zoeae and copepods)
0.4–8 mm	650 µm (*Artemia*) 10–20 mm diameter, 0.5–1 mm thickness (FF)	1.5–2.8 mm	750–850 µm (*Artemia* nauplii); 12–20 mm (adult *Artemia*)	5 days *Artemia* and zooplankton fraction <0.4 mm
Super SELCO and DHA SELCO Inve	Fish egg powder (Plus Aquaran, BASF Japan)	Reared with *Rhodomonas lens* and *Isochrysis galbana* and then enriched with different procedures	A1DHASelco® by Inve; *Isochrysis* and Prolon®	*Artemia* metanauplii reared for 5 days with *Tetraselmis suecica* and *Isochrysis galbana*
Daily up to day 7 and every 5 days thereafter	Every 5 days	Days 15, 25 and 35	Every 10 days	Every 10 days
1–20 % at day 40 with *Artemia* and from 20–39 % at day 40 with *Artemia* and copepods	10–30 % at day 30	35–53 % at day 15 7–20 % at day 25	8 % at 45 dph (best survival)	
Bottom siphoning daily or every other day	Daily tank bottom siphoning after 5th day	Siphoning		

© Woodhead Publishing Limited, 2013

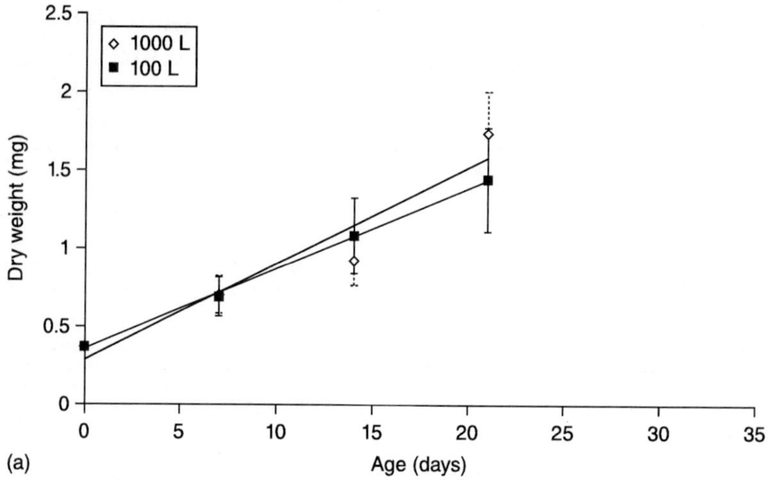

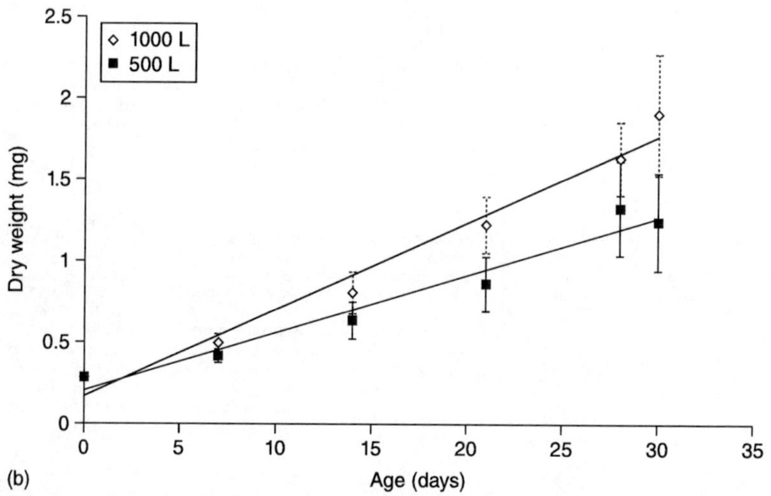

Fig. 12.3 (a) Growth of *Octopus vulgaris* paralarvae reared in 1000 L (◇) and and 100 L (○) tanks. (b) Growth of octopus paralarvae reared in 1000 L (◇) and 500 L (■) tanks.

12.3.4 Paralarval nutritional requirements

Some authors have studied the biochemical composition of paralarvae and their live preys. Navarro and Villanueva (2000, 2003) reported a deficiency in polyunsaturated fatty acids (PUFA) of the diets utilised (usually enriched *Artemia*) as a cause of the low growth and survival in cultivation. They indicated the importance of the DHA in the composition of the juveniles of octopus, sepia and squid. Okumura *et al.* (2005) increased the proportion of PUFA in the diet using *Artemia* supplemented with flakes of *A.*

© Woodhead Publishing Limited, 2013

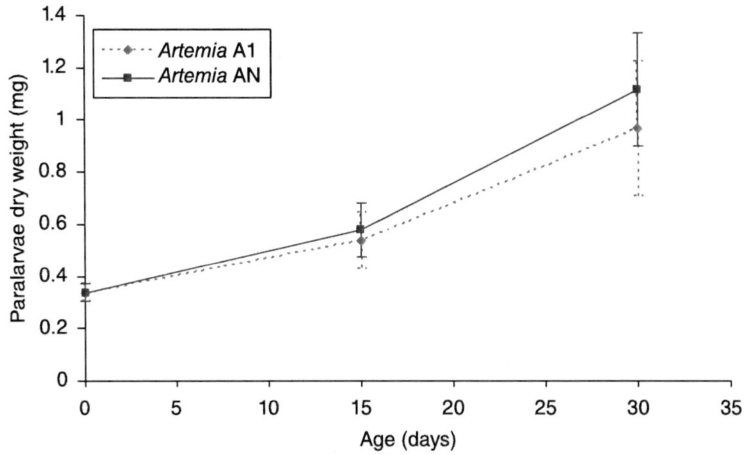

Fig. 12.4 *Octopus vulgaris* paralarvae growth using *Artemia* of different size as diet: A1 (small *Artemia*) and AN (larger *Artemia*).

personatus. Other authors use live copepods and other zooplankton groups (Bersano, 2003; Estévez *et al.*, 2009) to increase the PUFA contents of the reared paralarvae.

Another important nutritional item to be considered is the amino acid (AA) composition of the preys. Villanueva *et al.* (2004) found that lysine, arginine and leucine represented about half of the total essential amino acids (EAA) in paralarvae. More recently, Villanueva and Bustamante (2006) also studied the possible importance of essential elements such as copper in the diet.

12.3.5 *Artemia* enrichment

Concerning the *Artemia* enrichment process, Seixas *et al.* (2010) indicate *I. galbana* and *Rhodomonas lens* as the best microalgae combination, but the best results in paralarval growth and survival after one month of culture were attained when *Artemia* is cultured three to four days with the microalgae *I. galbana* and further enriched with *Nannochloropsis* sp. (Hamasaki *et al.*, 1991; Moxica *et al.*, 2006; Viciano *et al.*, 2011); that was the only method able to achieve the octopus adult stage using only microalgae to enrich the *Artemia* (Moxica *et al.*, 2006; Fuentes *et al.*, 2011). The reasons for the best growth results being observed with this diet (*I. galbana* plus *Nannochloropsis* sp.) could be that the combination of the DHA provided by *I. galbana* and the high EPA content present in *Nannochloropsis* sp. cover the basic PUFA requirements of octopus paralarvae and also that the higher protein:lipid ratio present in the *Artemia* can positively affect the paralarvae growth (Fuentes *et al.*, 2011). Seixas *et al.* (2010) find similar conclusions and suggest that, in order to sustain a good paralarvae growth, a minimum

© Woodhead Publishing Limited, 2013

dietary protein: lipid ratio should be attained. Additionally, the high concentration of *Nannochoropsis* sp. used in the paralarvae rearing tank could also produce an inhibitor effect on the growth of harmful microflora. Several authors have also indicated that improvement in growth and survival associated with green water is likely a result of the indirect effects that microalgae might have on the nutritional quality of the food, or on the larval physiology (Skiftesvik *et al.*, 2003).

12.3.6 Optimal preys

Although *Artemia* is well accepted and, when enriched, can produce good survival rates up to the first month of life, it does not have a suitable biochemical profile to meet the nutritional requirements of octopus larvae. This is very evident in relation to the level of PUFAS, particularly EPA and DHA (Navarro *et al.*, 1992, 1993), which are found to be almost non-existent in newly hatched nauplii. Crustacean zoeae from the natural environment have a higher content of PUFA, particularly DHA (Navarro and Villanueva, 2000; Moxica *et al.*, 2002). Other authors use frozen flakes of *A. personatus* (Okumura *et al.*, 2005), cultured copepods Bersano (2003) or wild zooplankton (Estévez *et al.*, 2009; Fuentes *et al.*, 2009) to provide an adequate nutritional profile.

Best results for growth and survival were recorded in those cultures where *Artemia* complemented by live spider crab (*Maja brachydactyla*) zoeae were used as prey. Using this method, it was possible to complete the whole cultivation cycle of the octopus in captivity, obtaining juveniles that reach the adult stage and lay fertilised eggs. These world pioneering results were obtained for the first time in 2001 at the Spanish Institute of Oceanography (IEO) in Vigo (Iglesias *et al.*, 2004) achieving a paralarvae dry weight of 9.5 ± 1.9 mg and a survival rate of 31.5 % at 40–45 days (Fig. 12.5).

Consequently a mixed live diet, composed of enriched *Artemia* and crustacean zoeae, is currently the most balanced diet for producing the best growth and survival results in the larval phase (Fig. 12.6), increasing six-fold the dry weight obtained with *Artemia* enriched with microalgae.

Later on, the octopuses reached a weight of 0.5–0.6 kg at six months of age and a final weight of 1.4–1.8 kg at eight months, the time at which they reached maturity and began to spawn. However, this method, used again later by other authors (Carrasco *et al.,* 2005), is not transferable to a commercial level as there is limited availability of live zoeae and it is difficult to scale up.

12.3.7 Summary in paralarval rearing

The success in the rearing experiments carried out with *Artemia* and zoeae could be explained by the higher level of phospholipids observed in the

© Woodhead Publishing Limited, 2013

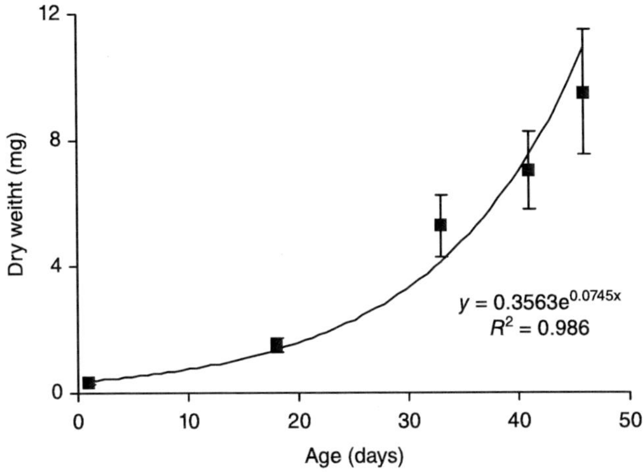

Fig. 12.5 Increase in dry weight (mg) of *Octopus vulgaris* paralarvae from hatching to day 45 using large *Artemia* and crab zoeae, (from Iglesias *et al.*, 2004). Reproduced with permission from Springer

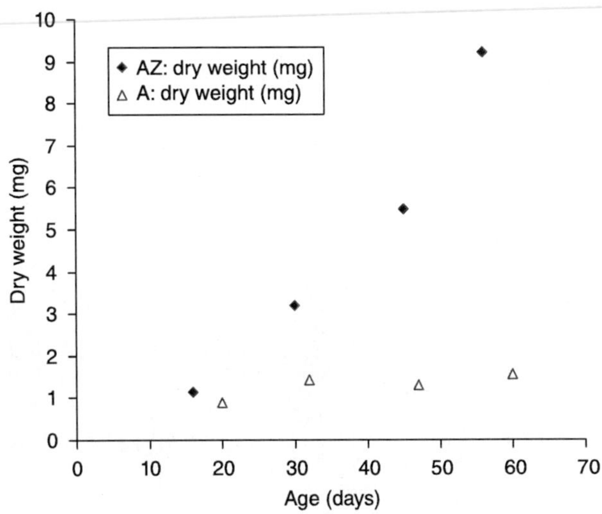

Fig. 12.6 Growth of *Octopus vulgaris* paralarvae fed on enriched *Artemia* and zoeae of spider crab (AZ) compared with paralarvae fed on enriched artemia (A).

paralarvae (51.73 % of the total lipids in Moxica *et al.* (2002)). This value reflects the composition of the food since, in the zoeae, phospholipids represent 69.72 % of the total lipids, a value that is clearly higher than the normal diets used in paralarvae rearing experiments. It was also noteworthy that the highest phospholipids percentage in paralarvae delivered the best

© Woodhead Publishing Limited, 2013

survival, which could indicate the importance of this lipid class for *O. vulgaris*. Bearing in mind that, as well as PUFAS, the zoeae also contain other basic nutritional components (e.g. amino acids and cholesterol), further research is required in this field and also in the formulation of a mixed diet (*Artemia* and microdiets) that could be used on an industrial scale. On the other hand, considering that paralarvae easily accept an inert diet from day 15 on, it is recommended that the weaning process be started earlier (around day 20 of life) than in works previously carried out (day 40–60). In this way, the amount of live prey needed can be reduced and, consequently, the production costs lowered.

12.3.8 Proposal for a standardised paralarval rearing method

The following protocol to produce good quality of one-month old octopus paralarvae is recommended:

In accordance with previous studies, a 500 or 1000 L volume tank has been selected as the most appropriate for the semi-industrial scale assays (Fig. 12.7). The circuit will be closed (stagnant water) during the first week and from then on opened for 4 h day^{-1}. Temperature will be kept at 21–23 °C, and salinity at 32–35 ppt. Drainage for water exchange, consisting of a

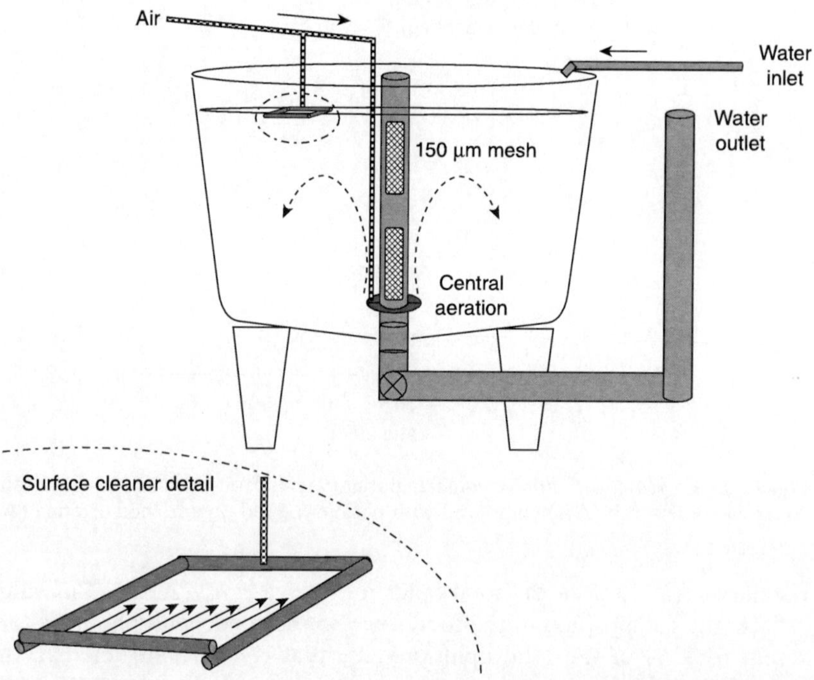

Fig. 12.7 Diagram of *Octopus vulgaris* paralarvae rearing tank.

© Woodhead Publishing Limited, 2013

cylindrical PVC pipe with 150 μm mesh size, will be located at the centre of the tank. Walls and bottom tanks will be black. Surface cleaners and moderate central aeration based on air pressure will be used. The surface light intensity in the tank will be of 500–700 lx during 24 h. *Nannochloropsis* sp. will be used in the culture medium at a concentration of 1.0 10^6 cel.mL^{-1} and paralarvae concentration will be of 10 indiv.L^{-1}. Aside from experimental diets based on specific *Artemia* enrichments, paralarval food will consist of 24 h *Artemia* enriched with *I. galbana* at a concentration of 0.75 10^6 cel.mL^{-1}(A1) for the first 15 days followed by bigger *Artemia* (1.5–2 mm TL) cultured for four to five days with *I. galbana* and further enriched over the last 24 h with *Nannochloropsis* sp., to a concentration of 10^6 cel.mL^{-1}, keeping a prey concentration of 0.3 art.mL^{-1}. When zoeae of spider crab are used in co-feeding with the *Artemia*, they will be added at a concentration of 0.05–0.1 ind.mL^{-1}, three to four days per week. Under these conditions relatively high survival rates and paralarvae dry weights between 1.3 and 1.8 mg can be attained after one month of rearing process when enriched *Artemia* is the only diet, and 2–4 mg when zoeae are used in co-feeding with *Artemia*.

12.4 Conclusions and future trends

Analysing the available hatchery technology of *O. vulgaris* culture, it is possible to conclude that reproduction process has reached an adequate level of development, with the greatest number of problems being encountered in the larvae rearing phase. The current high larval mortality observed in the first few weeks of life prevents the completion of the culture cycle for this species on an industrial scale.

The whole life-cycle was only successful at an experimental scale using *Artemia* and live crabs zoeae (Iglesias *et al.*, 2004), but in order to develop this technique to a commercial level is necessary to better define the nutritional requirement of the species and to use an adequate prey (enrichment for *Artemia* or inert diet) for first feeding.

A standardised method for *O. vulgaris* paralarvae rearing, which allows obtaining good quality individuals (dry weight and survival) to start the weaning process is proposed for the first time. However, further research is needed to cover the nutritional requirements of *O. vulgaris* paralarvae.

Considering the positive biological characteristics and the worldwide industrial interest in the market (high demand and market price), cephalopods are considered strong candidates for diversification in the near future in marine aquaculture (Iglesias, 2010). The following future research topics are proposed in relation to *O. vulgaris*:

- *Zootechnical improvements and feeding requirements*: Three lines are suggested with regard to culture methods: (i) devise an enrichment

© Woodhead Publishing Limited, 2013

for *Artemia* that gives them similar properties to zoeae; (ii) produce a microdiet to use in co-feeding with *Artemia* from day 20 on; (iii) start weaning around day 20–30 after hatching, in order to reduce live prey requirements.

- *Nutrition*: It is necessary to continue research into the nutritional requirements of paralarvae (lipids, proteins, aminoacids, essential elements and vitamins) both in wild and culture conditions to solve the problem of larval rearing.
- *Microbiology and water quality*: Analyse in more depth the suitable water quality factors (e.g. microbial population, use of pro- and prebiotics, recirculation, gas supersaturation, tolerance limits to different metabolites, etc.).
- *Histology and paralarval development*: Research on aspects related to paralarval morphology, physiology, growth dynamics and internal rhythms is required to understand proper paralarvae fitness.
- *Pathology*: Study of specific pathogens and bacterial growth in the rearing systems should be taken into consideration.

Close coordination between different research and industrial groups is considered essential to achieve a holistic view, optimise progress and reduce overlap in the study of *O. vulgaris* paralarvae rearing.

12.5 Sources of further information and advice

Books

- Boyle P R (ed.) (1987), *Cephalopod life cycles. Volume 1 and 2. Comparative reviews*, London: Academic Press.
- Hanlon R T and Messenger J B (1998), *Cephalopod behaviour*, New York: Cambridge University Press.
- Jereb P and Roper CFE (2010), *Cephalopods of the world. An annotated and illustrated catalogue of cephalopod species known to date*, FAO Species Catalogue for Fishery Purposes. N° 4. Rome: FAO.
- Villanueva R and Norman M D (2008), Biology of the planktonic stages of benthic octopuses, in Gibson R N, Atkinson R J A and Gordon J D M (eds), *Oceanography and Marine Biology: An Annual Review*, vol. 46. Taylor and Francis, 105–202.

Research groups and websites

- Cephalopod International Advisory Council: http://www.abdn.ac.uk/CIAC/
- http://invertebrates.si.edu/cephs/
- http://www.thecephalopodpage.org/
- ICES (International Council for the Exploration of the Sea) Working Group on Cephalopod Fisheries and Life History (WGCEPH): www.ices.dk

© Woodhead Publishing Limited, 2013

Table 12.3 Participants in the Working Group on paralarvae rearing of *Octopus vulgaris*. Vigo (Spain), 7–11 November 2005

Name	Institution
Erica Vidal ericavidal2000@yahoo.com.br	Universidad Federal do Paraná (Brazil)
F. Javier Sánchez javier.sanchez@vi.ieo.es	Instituto Español de Oceanografía (Spain)
Fátima Linares flinares@cimacoron.org	Centro de Investigacións Mariñas. Xunta de Galicia (Spain)
Francisco Carrasco josefrancisco.carrascofidalgo@asturias.org	Centro de Experimentación Pesquera (Spain)
Javier Roo jroo@iccm.rcanaria.es	Instituto Canario de Ciencias Marinas (Spain)
Jean Dhont jean.dhont@UGent.be	Ghent University (Belgium)
José Iglesias jose.iglesias@vi.ieo.es	Instituto Español de Oceanografía (Spain)
José Luis Muñoz jluis.munoz@juntadeandalucia.es	Instituto Andaluz de Investigación y Formación Agraria, Pesquera, Alimentaria y de la Producción Ecológica (Spain)
Roger Villanueva roger@icm.csic.es	Instituto de Ciencias del Mar (CSIC) (Spain)
Shigenobu Okumura (Deceased)	Yashima Station (Japan)
Terje v d Meeren terje.van.der.meeren@imr.no	Institute of Marine Research (Norway)
Francesco Lenzi f.lenzi@inveaquaculture.com	Maricoltura di Rosignano Solvay SRL (Italy)

- Latin American Groups working on cephalopods culture (see Table 12.4).
- Working Group on paralarvae rearing of *Octopus vulgaris*, Vigo (Spain), 7–11 November 2005 (see Table 12.3).

12.6 Acknowledgements

This chapter provides a review of the last 15 years of research on the culture of the common octopus made at the Spanish Institute of Oceanography in Vigo. Besides the two authors, other members of the team, F.J. Sánchez, J.J. Otero, M.J. Lago and C. Moxica, have actively participated as co-authors in most of previous publications, so we would like to mention them for their invaluable contribution.

© Woodhead Publishing Limited, 2013

Table 12.4 Latin American Groups working on the culture of cephalopods

Head of the Group E-mail	Country	Institution
Carlos Rosas crv@ciencias.unam.mx	México	Universidad Nacional Autónoma de México
César Lodeiros cesarlodeirosseijo@yahoo.es	Venezuela	Instituto Oceanográfico de Venezuela, Universidad de Oriente
Érica A.G. Vidal ericavidal2000@yahoo.com.br	Brazil	Universidade Federal do Paraná
Fernando José Valenzuela fvalenzuela@uantof.cl	Chile	Universidad de Antofagasta
Íker Uriarte iuriarte@spm.uach.cl	Chile	Universidad Austral de Chile
John Gabriel Ramírez johnramirez@fundacionecosfera.com	Colombia	Fundación Ecósfera
Jose Iglesias jose.iglesias@vi.ieo.es	España	Instituto Español de Oceanografía
Luísa Valente lvalente@icbas.up.pt	Portugal	Centro Interdisciplinar da Investigação Marinha e Ambiental
Luz Adriana Velasco molmarcol@gmail.com	Colombia	Universidad del Magdalena
María De Lourdes Jiménez ljimenez@uv.mx	México	Universidad Veracruzana. Instituto de Ciencias Marinas y Pesquerías
María Luisa González malugon@ulagos.cl	Chile	Universidad de los Lagos-Osorno
Óscar Zúñiga ozuniga@uantof.cl	Chile	Universidad de Antofagasta
Paul Martín Baltasar paulbaltazar2005@yahoo.es	Perú	Universidad Científica del Sur

12.7 References

BALTAZAR P, RODRÍGUEZ P, RIVERA W and VALDIVIESO V (2000) 'Cultivo experimental del pulpo (*Octopus mimus*) en el Perú', *Rev Peru Biol*, 7(2), 151–160.

BERSANO J G F (2003) 'Intensive cultivation of the calanoid copepod *Acartia tonsa*. A potential source of live food for aquaculture', *Book of Abstracts. World Aquaculture*, 1, 95.

BOLETZKY S (1974) 'Elevage de Céphalopodes en aquarium', *Vie Milieu*, 24, 309–340.

BOLETZKY S and HANLON R (1983), 'A review of the laboratory maintenance, rearing and culture of cephalopod mollusks', *Mem Nat Mus Vic*, 44, 147–187.

CARRASCO J F, RODRÍGUEZ C and RODRÍGUEZ M (2005), 'Cultivo intensivo de paralarvas de pulpo (*Octopus vulgaris*, Cuvier) utilizando como base de la alimentación zoeas vivas de crustáceos', *Libro de Actas IX Congreso Nacional de Acuicultura*,

© Woodhead Publishing Limited, 2013

12–16 May 2003, Cadiz. Sevilla: Consejería de Agricultura y Pesca, Junta de Andalucuía, 219–222.

CARRASCO J F, ARRONTE J C and RODRÍGUEZ C (2006), 'Paralarval rearing of the common octopus, *Octopus vulgaris* (Cuvier)', *Aquac Res*, 37, 1601–1605.

CASTRO B (1991), 'Can *Sepia officinalis* L. be reared on artificial food?' *Mar Behav Physiol*, 19, 35–38.

CASTRO B G and LEE P G (1994) 'The effects of semi-purified diets on growth and condition of *Sepia officinalis* L. (Mollusca: Cephalopoda)', *Comp Biochem Physiol*, 109A, 1007–1016.

CASTRO B G, DIMARCO F P, DE RUSHA R H and LEE P G (1993), 'The effects of surimi and pelleted diets on the laboratory survival, growth and feeding rate of the cuttlefish *Sepia officinalis* L.', *J Exp Mar Biol Ecol*, 170, 241–252.

CASTRO-FUENTES H, OLIVARES-PAZ A, QUINTANA-FELLAY A and ZÚÑIGA-ROMERO O (2002) 'Descripción del desarrollo embrionario y paralarva de *Octopus mimus* (Gould, 1852) (Mollusca: Cephalopoda) en cautiverio', *Estudios Oceanológicos*, 21, 13–25.

CHAPELA A, GONZÁLEZ A F, DAWE E G, ROCHA F J and GUERRA A (2006) 'Growth of common octopus (*Octopus vulgaris*) suspended from rafts in cages', *Sci Mar*, 70 (1), 121–129.

CORTEZ T, GONZÁLEZ A and GUERRA A (1999), 'Growth of cultured *Octopus mimus* (Cephalopoda, Octopodidae)', *Fish Res*, 40, 81–89.

DERUSHA R H, FORSYTHE J W, DIMARCO F P and HANLON R T (1989) 'Alternative diets for maintaining and rearing cephalopods in captivity', *J Lab Anim Sci*, 39, 306–312.

DE WOLF T, LENZI S and LENZI F (2011) 'Paralarval rearing of *Octopus vulgaris* (Cuvier) in Tuscany, Italy', *Aquac Res*, 42, 1406–1414.

DOMINGUES P M (1999) *Development of alternative diets for the mass culture of European cuttlefish* Sepia officinalis, Ph D Thesis, University of the Algrave, Faro.

DOMINGUES P and MARQUES L (2010) 'Effects of culture density and bottom area on growth and survival of the cuttlefish *Sepia officinalis* (Linnaeus, 1758)', *Aquac Int*, 18, 361–369.

DOMINGUES P, SYKES A and ANDRADE J P (2002) 'The effects of temperature in the life cycle of two consecutive generations of the cuttlefish *Sepia officinalis* (Linnaeus, 1758), cultured in the Algarve (South Portugal)', *Aquac Int*, 10, 207–220.

DOMINGUES P M, POIRIER R, DICKEL L, ALMANSA E, SYKES A and ANDRADE J P (2003) 'Effects of culture density and live prey on growth and survival of juvenile cuttlefish, *Sepia officinalis*', *Aquac Int*, 11, 225–242.

DOMINGUES P, SYKES A, SOMMERFIELD A, ALMANSA E, LORENZO A and ANDRADE J P (2004) 'Growth and survival of cuttlefish (*Sepia officinalis*) of different ages fed crustaceans and fish. Effects of frozen and live prey', *Aquaculture*, 229, 239–254.

DOMINGUES P M, BETTENCOURT V and GUERRA A (2006) 'Growth of *Sepia officinalis* in captivity an in nature', *Vie Milieu*, 56(2), 109–120.

DOMINGUES P M, LÓPEZ N, MUÑOZ J A, MALDONADO T, GAXIOLA G and ROSAS C (2007) 'Effects of a dry pelleted diet on growth and survival of the Yucatan octopus *Octopus maya*', *Aquac Nutr*, 13, 273–280.

DOMINGUES P, LÓPEZ N and ROSAS C (2012) 'Preliminary trials on the use of large outdoor tanks for the ongrowing of *Octopus maya* juveniles', *Aquac Res*, 43, 26–31.

ESTEFANELL J A (2012) *Optimización de las condiciones de engorde y avances en el conocimiento de los requerimientos nutricionales del pulpo común* Octopus vulgaris *(Cuvier, 1797)*, Doctoral Thesis, Universidad de Las Palmas de Gran Canaria, Gran Canaria.

ESTÉVEZ A, GAIRIN I and BERGER E (2009) 'Wild zooplancton for *Octopus vulgaris* larval rearing', in Hendry G, Van Stappen G, Wille M and Sorgeloos P (eds),

© Woodhead Publishing Limited, 2013

LARVI 09, Fish & Shellfish Larviculture Symposium, Special Publication No. 38. Oostende: European Aquaculture Society, 88–91.

FAO (2010) *Fishery and Aquaculture Statistics. Yearbook 2008*. Romea: FAO.

FARIAS A, PEREDA S V, URIARTE I, DOERNER J, CUZON G and ROSAS C (2010), 'Evaluating the effects of formulated moist diets on juveniles of patagonian octopus *Enteroctopus megalocyathus* (Gould 1852)', *J Shellfish Res*, 29(4), 793–798.

FARIAS A, NAVARRO J C, CERNA V, PINO S and URIARTE I (2011) 'Effect of broodstock diet on the fecundity and biochemical composition of eggs of the Patagonian red octopus (*Enteroctopus megalocyathus* Gould 1852)', *Cienc Mar*, 37(1), 11–21.

FORSYTHE J W, HANLON R T and LEE P G (1987) 'A synopsis of cephalopod pathology in captivity', *Proc Int Assoc Aquat Anim Med*, 1(4), 130–135.

FORSYTHE J W, HANLON R T and LEE P G (1990), 'A formulary for treating cephalopod mollusc diseases', in Perkins FO and Cheng TC (eds), *Pathology in Marine Science*. San Diego, CA: Academic Press, 51–63.

FORSYTHE J W, DE RUSHA R H and HANLON R T (1994), 'Growth, reproduction and life span of *Sepia officinalis* (Cephalopoda: Mollusca) cultured through seven consecutive generations', *J Zool (Lond)*, 233, 175–192.

FUENTES L, IGLESIAS J, SÁNCHEZ F J, OTERO J J, MOXICA C and LAGO M J (2005), 'Técnicas de transporte de paralarvas y adultos de pulpo (*Octopus vulgaris*)', *Bol Inst Esp Oceanogr*, 21 (1–4), 155–162.

FUENTES L, SÁNCHEZ F J, OTERO J J, LAGO M J and IGLESIAS J (2009) 'Utilización de zooplancton natural y artemia en el cultivo de paralarvas de pulpo *Octopus vulgaris*', in Beaz Paleo, D, Villarroel Robinson, M and Cardenas Rojas S (eds), *Libro de Resúmenes, XII Congreso Nacional de Acuicultura*, 24–26 Noviembre, Madrid. Madrid: CICEGRAF Artes Graficas, 122–123 (in Spanish, abstract in English).

FUENTES L, SÁNCHEZ F J, LAGO M J, IGLESIAS J, PAZOS G and LINARES F (2011) 'Growth and survival of *Octopus vulgaris* (Cuvier 1797) paralarvae fed on 3 *Artemia*-based diets complemented with frozen fish flakes, crushed zooplankton and marine microalgae', *Scientia Marina*, 75(4), 771–777.

HAMASAKI H, FUKUNAGA K, YOSHIDAY and MARUYAMA K (1991) 'Effects of marine microalgae *Nannochloropsis* sp. on survival and growth on rearing pelagic paralarvae of *Octopus vulgaris*, and results of mass culture in the tank of 20 metric tons', *Saibai-giken*, 19, 75–84.

HANLON R T (1990) 'Maintenance, rearing and culture of teuthoid and sepioid squids', in Gilbert D L, Adelman Jr W J and Arnold J M (eds), *Squid as experimental animals*. New York: Plenum, 35–62.

HANLON R T and FORSYTHE J W (1985) 'Advances in the laboratory culture of octopuses for biomedical research, *Lab Anim Sci*, 35, 33–40.

HANLON R T and MESSENGER J B (1996) *Cephalopod Behaviour*. New York: Cambridge University Press.

HANLON R T, HIXON R F and HULET W H (1983) 'Survival, growth and behaviour of the loliginid squids *Loligo plei*, *Loligo pealei*, *Lolliguncula brevis* (Mollusca: Cephalopods) in closed sea water systems', *Biol Bull*, 165(3), 637–685.

HANLON R T, TURK P E and YANG W T (1987) 'Laboratory rearing of the squid *Loligo pealei* to the juvenile stage: growth comparisons with fishery data', *Fish Bull (Wash DC)*, 85, 163–167.

HANLON R T, YANG W T, TURK P E, LEE P G and HIXON R F (1989) 'Laboratory culture and estimated life span of the eastern Atlantic squid *Loligo forbesi* (Mollusca: Cephalopoda)', *Aquac Fish Manag*, 20, 15–33.

HANLON R T, TURK P E and LEE P G (1991) 'Squid and cuttlefish mariculture: An updated perspective', *J Cephalopod Biol*, 2, 31–40.

IGLESIAS J (2010) 'Species diversification – Other species', in Parallel Session Summaries, *Sea Farming Tomorrow, Summary Document of Aquaculture Europe 2010*, 5–8 October, Porto (Portugal), 18–20.

© Woodhead Publishing Limited, 2013

IGLESIAS J, SÁNCHEZ F J and OTERO J J (1997) 'Primeras experiencias sobre el cultivo integral del pulpo (*Octopus vulgaris*) en el Instituto Español de Oceanografía', in Costa J, Abellan E, García B, Ortega A and Zamara S (eds), *Actas del VII Congreso Nacional de Acuicultura*. Madrid: Ministerio de Agricultura, Pesca y Alimentación, 221–226.

IGLESIAS J, SÁNCHEZ F J, OTERO J J and MOXICA C (2000) 'Culture of octopus (*Octopus vulgaris*, Cuvier). Present knowledge, problems and perspectives', *Cah Options Méditer*, 47, 313–321.

IGLESIAS J, OTERO J J, MOXICA C, FUENTES L and SÁNCHEZ F J (2004) 'The completed life cycle of the octopus (*Octopus vulgaris*, Cuvier) under culture conditions: paralarval rearing using Artemia and zoeae, and first data on juvenile growth up to 8 months of age', *Aquac Int*, 12, 481–487.

IGLESIAS J, FUENTES L, SÁNCHEZ F J, OTERO J J, MOXICA C and LAGO M J (2006) 'First feeding of *Octopus vulgaris* Cuvier, 1797 paralarvae using Artemia: effect of prey size, prey density and feeding frequency', *Aquaculture*, 261(2), 817–822.

IGLESIAS J, SÁNCHEZ F J, BERSANO J G F, CARRASCO J F, DHONT J, FUENTES L, LINARES F, MUÑOZ J L, OKUMURA S, ROO J, VAN DER MEEREN T, VIDAL E A G and VILLANUEVA R (2007a) 'Rearing of *Octopus vulgaris* paralarvae: Present status, bottlenecks and trends', *Aquaculture*, 266, 1–15.

IGLESIAS J, SÁNCHEZ F J, OTERO J J, LAGO M J, MOXICA C, FUENTES L and MARTÍNEZ F J (2007b) 'Engorde industrial de pulpo *Octopus vulgaris* en batea en la Ría de Vigo', *XI Congreso Nacional de Acuicultura, Libro de resúmenes*, 24–28 Septiembre, Vigo. Pontevedra: Gráficas Salnés SL, 683–686 (in Spanish, abstract in English).

IMAMURA S (1990) 'Larval rearing of Octopus (*Octopus vulgaris* Cuvier). The progress of technological development and some problems remained', *Collect Breed* 52, 339–343.

ITAMI K, IZAWA Y, MAEDA S and NAKAI K (1963) 'Notes on the laboratory culture of the Octopus larvae', *Bull Jap Soc Sci Fish*, 29(6), 514–520.

JEREB P and ROPER C F E (2010) *Cephalopods of the world. An annotated and illustrated catalogue of cephalopod species known to date. Vol. 1, Chambered Nautiluses and Sepioidsolume*, FAO Species Catalogue for Fishery Purposes. N° 4,. Rome.

KLAICH M J, RE M E and PEDRAZA S N (2008) 'Gross growth efficiency as a function of food intake level in the "Pulpito" *Octopus tehuelchus*: a multimodel inference appliaction', *Aquaculture*, 284 (1–4), 272–276.

LEE P G, FORSYTHE J W, DIMARCO F P, DERUSHA R and HANLON R T (1991) 'Initial palatability and growth trials on pelleted diets for cephalopods', *Bull Mar Sci*, 49, 362–372.

LEE P G, TURK P E, YANG W T and HANLON R T (1994) 'Biological characteristics and biomedical applications of the squid *Sepioteuthis lessoniana* cultured through multiple generations', *Biol Bull (Woods Hole)*, 186, 328–341.

LEE P G, TURK P E, FORSYTHE J W and DIMARCO F P (1998) 'Cephalopod culture: physiological, behavioral and environmental requirements', *Suisanzoshoku*, 46, 417–422.

LEE P G, WALSH L, TURK P E and DIMARCO F P (2000) 'Large-scale culture of the loliginid squid, *Sepioteuthis lessoniana*', in Tiensongrassmee B, Tedengren M and Jarayabhand P (eds), *Proceedings of the special session on mollusc research in Asia, 5th Asian Fisheries Forum*. Bangkok: Thailand Research Fund, 217–225.

LENZI F, CAPIFERRI U and DE WOLF T (2006) 'Paralarval rearing of the common octopus *Octopus vulgaris*: state of the art in Italy', Aqua 2006 Abstract CD, 9–13 May, Firenze, 523.

MANGOLD K (1983) '*Octopus vulgaris*', in Boyle PR (ed.), *Cephalopod Life Cycles, Vol. I*. London: Academic Press, 335–364.

© Woodhead Publishing Limited, 2013

MANGOLD K M and BOLETZKY S V (1973) 'New data on reproductive biology and growth of *Octopus vulgaris*', *Mar Biol*, 19, 7–12.

MOXICA C, LINARES F, OTERO J J, IGLESIAS J and SÁNCHEZ F J (2002) 'Cultivo intensivo de pararlarvas de pulpo, *Octopus vulgaris* Cuvier, 1797, en tanques de 9 m^3, *Bol Inst Esp Oceanogr*, 18 (1–4), 31–36.

MOXICA C, FUENTES L, HERNÁNDEZ J, IGLESIAS J, LAGO M J, OTERO J J and SÁNCHEZ F J (2006) 'Efecto de *Nannochloropsis* sp. en la supervivencia y crecimiento de paralarvas de pulpo *Octopus vulgaris*', *IX Foro dos Recursos Mariños e da Acuicultura das Rías Galegas*, 10–11 Octubro, O Grove.

NABHITABHATA J (1997) 'Life cycle of three cultured generations of spineless cuttlefish, *Sepiella inermis* (Ferrusac & d'Orbigny, 1848)', *Phuket Mar Biol Cen Spec Pub*, 17, 289–298.

NABHITABHATA J and NILAPHAT P (1999) 'Life cycle of cultured pharaoh cuttlefish, *Sepia pharaonis* Ehrenberg, 1831', *Phuket Mar Biol Cent Spec Pub*, 19, 25–40.

NABHITABHATA J and SUWANAMALA J (2008), 'Reproductive behaviour and cross-mating of two closely related pygmy squids *Idiosepius biserialis* and *Idiosepius thailandicus* (Cephalopoda: Idiosepiidae)', *J Mar Biol Assoc UK*, 88 (5), 987–993.

NABHITABHATA J, NILAPHAT P, JAROONGPATTANANON C and PROMBOON P (2003) 'Culture, growth and behaviour of king octopus, *Octopus rex* Nateewathana and Norman, 1999', *Rayong Coast Fish Rese Develop Cent Contrib*, 26, 17.

NABHITABHATA J, NILAPHAT P, PROMBOON P and JAROONGPATTANANON C (2005) 'Life cycle of cultured bobtail squid, *Euprymna hyllebergi* Nateewathana, 1997', *Phuket Mar Biol Cent Res Bull,* 66, 351–365.

NAEF A (1928) 'Cephalopoda embryology. Part I, Volume II (final part of monograph n°. 35)'. in *Fauna and Flora of the Bay of Naples*, translated by the Smithsonian Institution Libraries, Washington, 35, 1–461.

NAVARRO J C and VILLANUEVA R (2000) 'Lipid and fatty acid composition of early stages of cephalopods: an approach to their lipid requirements', *Aquaculture*, 183, 161–177.

NAVARRO J C and VILLANUEVA R (2003) 'The fatty acid composition of *Octopus vulgaris* paralarvae reared with live and inert food: deviation from their natural fatty acid profile', *Aquaculture*, 219, 613–631.

NAVARRO J C, AMAT F and SARGENT J R (1992) 'Fatty acid composition of coastal and inland *Artemia* sp. populations from Spain', *Aquaculture*, 102, 219–230.

NAVARRO J C, AMAT F and SARGENT J R (1993) 'The lipids of the cysts of freshwater and marine-type Artemia', *Aquaculture*, 109, 327–336.

O'DOR R K and WELLS M J (1987) 'Energy and nutrient flow', in Boyle PR (ed.), *Cephalopod Life Cycles, Vol. 2.* London: Academic Press, 109–133.

OKUMURA S, KURIHARA A, IWAMOTO A and TAKEUCHI T (2005) 'Improved survival and growth in *Octopus vulgaris* paralarvae by feeding large type Artemia and Pacific sandeel, *Ammodytes personatus*', *Aquaculture*, 244, 144–157.

OLIVARES A., ZÚÑIGA O, CASTRO G, SEGURA C and SÁNCHEZ J (1996) 'Bases biológicas para el manejo de *Octopus mimus*: reproducción y crecimiento', *Estud Oceanol*, 15, 61–74.

ORTIZ N, RÉ M E and MÁRQUEZ F (2006) 'First description of eggs, hatchlings and hatchling behaviour of *Enteroctopus megalocyathus* (Cephalopoda: Octopodidae)', *J Plankton Res*, 28 (10), 881–890.

PASCUAL E (1978) 'Crecimiento y alimentación de tres generaciones de *Sepia officinalis* en cultivo', *Inv Pesq*, 42, 421–442.

PÉREZ M C, LÓPEZ D A, ÁGUILA K and GONZÁLEZ M L (2006) 'Feeding and growth in captivity of the octopus *Enteroctopus megalocyathus* Gould, 1852', *Aquac Res*, 37(6), 550–555.

© Woodhead Publishing Limited, 2013

PROMBOON P, NABHITABHATA J and DUENGDEE T (2011) 'The life cycle of the marbled octopus, *Amphioctopus aegina* (Gray) (Cephalopoda: Octopodidae) reared in laboratory', *Scientia Marina*, 75 (4), 811–821.

QUINTANA D (2009) *Valoración de los requerimientos nutricionales de reproductores de pulpo común Octopus vulgaris*. Doctoral Thesis, Universidad de La Laguna, Tenerife.

QUINTANA D, MARQUEZ L, SUAREZ H, RODRÍGUEZ E, JEREZ S and ALMANSA E (2009) 'Efecto de la dieta de los reproductores de pulpo común (*Octopus vulgaris*) sobre la composición de aminoácidos de huevos y paralarvas: Relación con la calidad de puesta', Poster, *XII Congreso Nacional de Acuicultura*, 24–26 Noviembre, Madrid.

QUINTANA D, ROSAS C and MORENO-VILLEGAS E (2011) 'Relation between nutrition and rearing parameters of *Octopus maya* juveniles fed with different rations of crab paste', *Aquac Nutr*, 17, 379–388.

RAMA-VILLAR A, FAYA V, MOXICA C and REY-MENDEZ M (1997) 'Engorde de pulpo (*Octopus vulgaris*) en batea', in Costa J, Abellán E, García B, Ortega A and Zamara S (eds), *Actas del VII Congreso Nacional de Acuicultura*. Madrid: Ministerio de Agricultura, Pesca y Alimentación, 245–250.

REY-MÉNDEZ M, TUÑÓN E and LUACES-CANOSA M (2003) 'Estudio de los efectos del peso inicial y el sexo sobre el comportamiento, la mortalidad y el crecimiento del pulpo (*Octopus vulgaris*, Cuvier 1797) en cultivo industrial', *IX Congreso Nacional de Acuicultura,* 12–16 May 2003, Cadiz. Sevilla: Consejeria de Agricultura y Pesca, Junta de Andalacuía, 276–277 (in Spanish, abstract in English).

RICHARD A (1971) *Contribution à l'étude expérimentale de la croissance et de la maturation sexuelle de* Sepia officinalis L. *(Mollusque, Céphalopode)*, Thèse État, nr 248. University of Lille.

ROSAS C, CAAMAL C, CÁZARES R, RODRÍGUEZ D, ROMERO M and DARWIN C (2006) *Manual Preliminar para el cultivo del pulpo Octopus maya*, Unidad Multidisciplinaria de Docencia e Investigación. Facultad de Ciencias, UNAM, Sisal, Yucatán.

ROSAS C, CAAMAL C and CAZARES R J (Universidad Autónoma de México) (2011) *Incubation process for octopuses and incubator*. Organización Mundial de la Propiedad Intelectual, WO 2010/030155 A1.

ROSAS C, SÁNCHEZ A, PASCUAL C, AGUILA J, MALDONADO T and DOMINGUES P (2011) 'Effects of two dietary protein levels on energy balance and digestive capacity of *Octopus maya*', *Aquac Int*, 19, 165–180.

SÁNCHEZ F J, FUENTES L, OTERO J J, LAGO M J, LINARES F, PAZOS G and IGLESIAS J (2011) 'Effect of tank volume on the growth and survival of reared *Octopus vulgaris* paralarvae', *Aquac Res*, 1–4. doi: 10.1111/j.1365-2109.2011.03049.x.

SEGAWA S and NOMOTO A (2002) 'Laboratory growth, feeding, oxygen consumption and ammonia excretion of *Octopus ocellatus*', *B Mar Sci*, 71, 801–813.

SEIXAS P (2009) *Composición bioquímica y crecimiento de paralarvas de pulpo (*Octopus vulgaris *Cuvier, 1797), alimentadas con juveniles de Artemia enriquecidos con microalgas y otros suplementos nutricionales*, PhD Thesis. University of Santiago de Compostela.

SEIXAS P, REY-MÉNDEZ M, VALENTE L M P and OTERO A (2010) 'High DHA content in *Artemia* is ineffective to improve *Octopus vulgaris* paralarvae rearing', *Aquaculture*, 300, 156–162.

SKIFTESVIK A B, BROWMAN H I and ST-PIERRE J F (2003) 'Life in green water: the effect of microalgae on the behaviour of Atlantic cod (*Gadus morhua*) larvae', in Browman H I and Skiftesvik A B (eds), *The Big Fish Bang. Proceedings of the 26th Annual Larval Fish Conference*. Bergen: Institute of Marine Research, 97–103.

SOLIS M J (1967) *Aspectos biológicos del pulpo Octopus maya Voss y Solis*, Publicación núm 18, Inst Nacional Investig Biol Pesqueras (México).

© Woodhead Publishing Limited, 2013

SOLORZANO Y, VIANA M T, LÓPEZ L M, CORREA J G, TRUE C C and ROSAS C (2009) 'Response of newly hatched *Octopus bimaculoides* fed enriched *Artemia salina*: growth performance, ontogeny of the digestive enzyme and tissue amino acid content', *Aquaculture*, 289, 84–90.

SYKES A V, DOMINGUES P M, CORREIA M and ANDRADE J P (2006) 'Cuttlefish culture. State of the art and future trends', *Vie Milieu*, 56(2), 129–137.

TURK P E, HANLON R T, BRADFORD L A and YANG W T (1986) 'Aspects of feeding, growth and survival of the European squid *Loligo vulgaris* Lamark, 1799, reared through the early growth stages', *Vie Milieu*, 36, 9–13.

URIARTE I, ZÚÑIGA O, OLIVARES A, ESPINOZA V, CERNA V, FARIAS A, and ROSAS C (2009) 'Morphometric changes and growth rate during embryonic development of *Robsonella fontaniana*', *Vie Milieu*, 59, 315–323.

URIARTE I, HERNÁNDEZ J, DORNER J, PASCHKE K, FARIAS A, CROVETTO E and ROSAS C (2010) 'Rearing and growth of the octopus *Robsonella fontaniana* (Cephalopoda: Octopodidae) from planktonic hatchlings to benthic juveniles', *Biol Bull*, 218 (2), 200–210.

URIARTE I, FARIAS A, PASCHKE K, NAVARRO J C and ROSAS C (2011a) 'Observations on feeding and biochemical characteristics to improve larviculture of *Robsonella fontaniana* (Cephalopoda: Octopodidae)', *Aquaculture*, 315 (1–2), 121–124.

URIARTE I, IGLESIAS J, DOMINGUES P, ROSAS C, VIANA M T, NAVARRO J C, SEIXAS P, VIDAL E, AUSBURGER A, PEREDA S, GODOY F, PASCHKE K, FARÍAS A, OLIVARES A, and ZÚÑIGA O (2011b) 'Current status and bottleneck of octopod aquaculture: the case of American species', *J World Aquac Soc*, 42(6), 735–752.

VAN HEUKELEM W F (1976) *Growth, bioenergetics and life span of* Octopus cyanea *and* Octopus maya, PhD Thesis, University of Hawaii.

VAN HEUKELEM W F (1977) 'Laboratory maintenance, breeding, rearing and biomedical research potential of the Yucatan octopus (*Octopus maya*)', *Lab Anim Sci*, 27, 852–859.

VAN HEUKELEM W F (1983) '*Octopus maya*', in Boyle PR (ed.), *Cephalopod Life Cycles, Vol. I.* London: Academic Press, 311–323.

VAZ-PIRES P, SEIXAS P and BARBOSA A (2004) 'Aquaculture potential of the common octopus (*Octopus vulgaris Cuvier*, 1797): a review', *Aquaculture*, 238(1–4), 221–238.

VICIANO E, IGLESIAS J, LAGO M J, SÁNCHEZ F J, OTERO J J and NAVARRO J C (2011) 'Fatty acid composition of polar and neutral lipid fractions of *Octopus vulgaris* Cuvier, 1797 paralarvae reared with enriched on-grown Artemia', *Aquac Res*, 42, 704–709.

VIDAL E A G, DIMARCO F P, WORMUTH J H and LEE P G (2002a) 'Optimizing rearing conditions of hatchling loliginid squid', *Mar Biol*, 140, 117–127.

VIDAL E A G, DIMARCO F P, WORMUTH J H and LEE P G (2002b) 'Influence of temperature and food availability on survival, growth and yolk utilization in hatchling squid', *B Mar Sci*, 71 (2), 915–931.

VILLANUEVA R (1994) 'Decapod crab zoeae as food for rearing cephalopod paralarvae', *Aquaculture*, 128, 143–152.

VILLANUEVA R (1995) 'Experimental rearing and growth of planktonic *Octopus vulgaris* from hatching to settlement', *Can J Fish Aquat Sci*, 52, 2639–2650.

VILLANUEVA R and BUSTAMANTE P (2006) 'Composition in essential and non-essential elements of early stages of cephalopods and dietary effects on the elemental profiles of *Octopus vulgaris*', *Aquaculture*, 261, 225–240.

VILLANUEVA R and NORMAN M D (2008) 'Biology of the planktonic stages of benthic octopuses', in Gibson RN, Atkinson RJA and Gordon JDM (eds), *Oceanography and Marine Biology: An Annual Review*, 46, Boca Raton, FL: CRC Press – Taylor and Francis, 105–202.

© Woodhead Publishing Limited, 2013

VILLANUEVA R, KOUETA N, RIBA J and BOUCAUD-CAMOU E (2002) 'Growth and proteolytic activity of *Octopus vulgaris* paralarvae with different food rations during first-feeding, using *Artemia* nauplii and compound diets', *Aquaculture*, 205, 269–286.
VILLANUEVA R, RIBA J, RUÍZ-CAPILLAS C, GONZÁLEZ A V and BAETA M (2004) 'Amino acid composition of early stages of cephalopods and effect of amino acid dietary treatments on *Octopus vulgaris* paralarvae', *Aquaculture*, 242, 455–478.
WARNKE K (1999) 'Observations on the embryonic development of *Octopus mimus* (Mollusca: Cephalopoda) from Northern Chile', *Veliger*, 42(3), 211–217.
WELLS M J (1978) *Octopus. Physiology and Behaviour of an Advanced Invertebrate.* London: Chapman and Hall.
YANG W T, HIXON R F, TURK P E, KREJCI M E, HANLON R T and HULET W H (1983) 'Culture of California market squid from hatching-completion of the rearing cycle to second generation hatchlings', *Aquabiology*, 5, 328–339.
YANG W T, HIXON RF , TURK P E, KREJCI M E, HULET W H and HANLON R T (1986) 'Growth, behavior and sexual maturation of the market squid, *Loligo opalescens*, cultured through the life cycle', *Fish Bull (Wash DC)*, 84, 771–798.
YOUNG R E and HARMAN R F (1988) 'Larva, paralarvae and subadult in cephalopod terminology', *Malacologia*, 29, 201–207.
ZÚÑIGA O, OLIVARES A and OSSANDON L (1995) 'Influencia de la luz en la maduración sexual de hembras *Octopus mimus*', *Est Oceanol*, 14, 75–76.
ZÚÑIGA O, OLIVARES A and MUÑOZ L (1996) 'Resultados preliminares de la producción de larvas de pulpos *Octopus mimus* (II región, Chile)', *VII Congreso Latinoamericano sobre Ciencias del Mar, Libro de resúmenes*, Santos, 567–569.

© Woodhead Publishing Limited, 2013

13

Jellyfish as products and problems of aquaculture

J. E. Purcell, Western Washington University, USA, E. J. Baxter, Vet-Aqua International, Ireland and V. L. Fuentes, Instituto de Ciencias del Mar (CSIC), Spain

DOI: 10.1533/9780857097460.2.404

Abstract: This chapter begins by reviewing the fisheries and culture of jellyfish for human food, multi-million-dollar industries with markets currently centered in Asia. Second, we present guidelines for culture conditions and tank construction for display or study of 27 jellyfish species. Most types of jellyfish (scyphomedusae, hydromedusae, siphonophores and ctenophores) also damage the aquaculture industry by causing fish gill disorders and by fouling net pens. We review the lifecycles of these groups and the damage they cause. Finally, we offer recommendations on how to minimize this damage. Ironically, aquaculture may be inadvertently exacerbating the problems with jellyfish blooms.

Key words: jellyfish blooms, fish farming, fouling, public aquaria, jellyfish stings.

13.1 Introduction

Jellyfish present both opportunities and challenges for the aquaculture industry. They can produce economic gains or losses. In the first section, we review the fisheries and culture of jellyfish for human food, multi-million-dollar industries with markets currently centered in Asia. Interest in jellyfish has increased around the world in recent decades, both for public aquarium exhibition, where they offer economic and educational opportunities, and for scientific study, because the problems jellyfish cause have increased in coastal waters. Thus, in the second section, we present guidelines for culture conditions and tank construction for display or study of 27 jellyfish species. Although some jellyfish species benefit human health and economics, most are considered to be harmful, resulting in many millions of dollars in costs annually. Human problems with jellyfish include stinging and the costs associated with prevention, treatment, and loss of tourism,

© Woodhead Publishing Limited, 2013

clogging seawater intakes of power and desalination plants that increases maintenance costs and can cause closures with resulting societal costs, interference with commercial fisheries and resulting catch reduction and gear costs, and mortality in fish farms. In the third section, we review the damage caused by jellyfish to the aquaculture industry, and recommend how to minimize such damage. Ironically, aquaculture may be inadvertently exacerbating the problems with jellyfish blooms.

13.2 Jellyfish as human food, their fisheries and aquaculture

13.2.1 Jellyfish as human food

The Chinese have harvested jellyfish for 1700 years (Omori and Nakano, 2001). Jellyfish are popular foods in Chinese and Japanese cooking, and also are known for their medicinal benefits for high blood pressure, bronchitis, and arthritis (Hsieh *et al.*, 2001). Usually only the swimming bell is processed, although the oral arms are also used for some species. Jellyfish are processed with a mixture of salt and alum, and the semi-dried products are marketed, almost exclusively to China, Korea, and Japan.

Only certain types of jellyfish are considered suitable for human food. All of the species belong to the sub-order Daktyliophorae in the order Rhizostomeae, class Scyphozoa (Table 13.1). Most of the species are distributed in sub-tropical and tropical waters. These jellyfish are characterized by a firm body texture with high proportions of protein (collagen) (Hsieh *et al.*, 2001). The most valuable qualities are a crunchy-elastic texture and white color for this commodity. The preferred species is *Rhopilema esculentum*, for which the life-cycle, reproductive biology, and response to environmental factors have been studied in China for 30 years (Dong *et al.*, 2009); however, nothing is known about the biology of the other species, to our knowledge. Like most scyphozoans, it has a two-stage life-cycle (Fig. 13.1). The polyps live attached to surfaces and asexually produce more polyps by budding or creating small cysts (podocysts). They also 'strobilate' to asexually produce one to many new medusae (ephyrae 1–2 mm in diameter), which mature into sexually-reproductive medusae. In temperate waters, strobilation usually occurs during the seasonal transition to spring. Fertilization can occur in the water, or within the female medusae, which then brood the larvae (planulae).

13.2.2 Jellyfish fisheries

Jellyfish are caught by set-net, drift-nets, push-nets, beach-seines, weirs and hooks, with no differences among species. The fishery catch of *R. esculentum* was characterized by large interannual fluctuations, caused by natural variations in the jellyfish populations, and a short fishing season (2–4 months) and consequently remained at a small scale before 1970 (Fig. 13.2).

© Woodhead Publishing Limited, 2013

Table 13.1 Commercially-exploited edible jellyfish species, their distributions, and the fishing grounds

Taxon	Trade name	Distribution	Countries fished
Asia			
Cephea cephea	?	Red Sea to Touamotu Archipelago	?
Lobonemoides gracilis, L. robustus	White	Tropical waters in Indo-West Pacific	Philippines, Vietnam, Thailand, Malaysia, Indonesia, Myanmar
Lomonema smithi	?White	Same	Same
Acromitus hardenbergi	River	Malaysia, Indonesia, Thailand	Malaysia, Indonesia, Thailand
Crambione mastigophora	Prigi	Malay Archipelago, Java and Truk Island	Indonesia
Crambionella annandalei	Ball, Sunflower	Madras, India, Andaman Sea, Bay of Bengal off Myanmar	Myanmar
Crambionella sp.	Cilacap, Sunflower	Central Java, Indian Ocean	Indonesia
Cassiopea ndrosia			Philippines
Rhopilema esculentum	?Red, ?China	Chinese Yellow, East, and South China Seas, Vietnam, Ariake Sea, Japan	Vietnam, Malaysia?, China, Japan
Rhopilema hispidum	Sand	S Japan, S China, Philippines, Malaysia, Indonesia, Indian Ocean, Red Sea	Vietnam, Indonesia, China
Neopilema nomurai		East China Sea, Japan Sea	China
Other locations			
Stomolophus meleagris		SE USA, Mexico, Panama	United States
Rhizostoma pulmo		Mediterranean, Bay of Biscay, North and Black Seas, Sea of Marmara	Turkey
Catostylus mosaicus		Philippines, New Guinea, W Australia	Australia

Sources: Omori and Nakano, 2001; Kitamura and Omori, 2010.

© Woodhead Publishing Limited, 2013

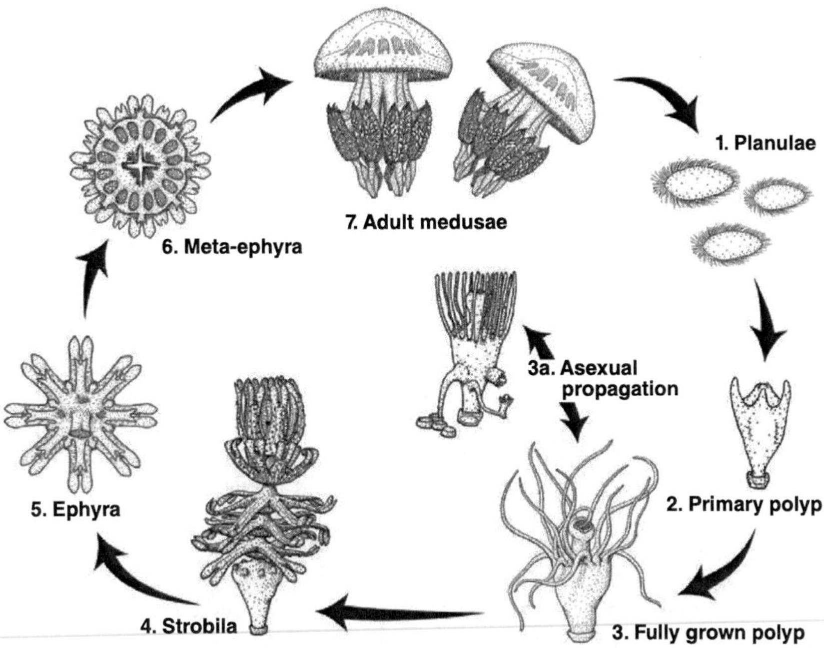

Fig.13.1 The life cycle of the scyphozoan *Rhizostoma pulmo* is typical for most scyphozoans, and similar to those of many hydrozomedusae (from Fuentes *et al.*, 2011).

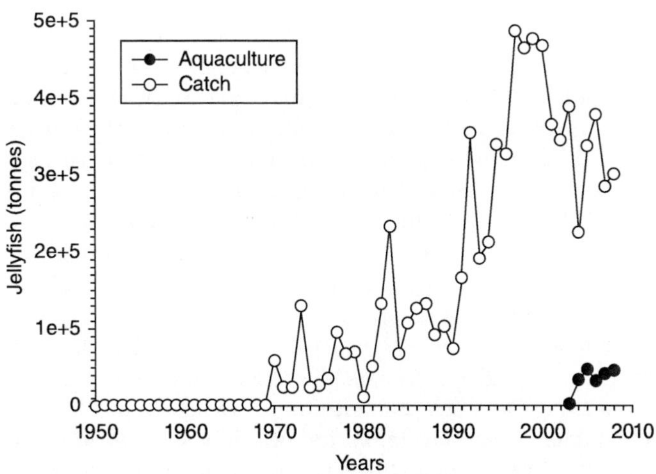

Fig.13.2 Global jellyfish aquaculture (•) and total (○) production from 1950 to 2008 (FAO 2011). Aquaculture production was zero before 2003 (from Purcell *et al.*, 2007).

© Woodhead Publishing Limited, 2013

Thereafter, the Japanese demand for jellyfish product and its price increased, and the fishery grew and expanded from mostly China to southeast Asia and included more species (Table 13.1).

13.2.3 Husbandry of *R. esculentum*

R. esculentum was selected as the species to be studied at the Liaoning Ocean and Fisheries Science Research Institute of China, cultured, and released for commercial harvest (Dong *et al.*, 2009). The technology of artificial breeding, pond culture, and stock enhancement in nature were developed along the coast of northern China.

Dong *et al.* (2009) summarize many studies of *R. esculentum* previously published in Chinese. The medusae are dioecious and fertilization and embryogenesis occur in the sea water. In culture, swimming planulae appear 7 h after fertilization and then the planulae metamorphose into polyps (scyphistomae) with four tentacles in three to four days. The scyphistomae continuously form podocysts that each will become a polyp. The polyps grow to have 16 tentacles in 15–20 days and strobilation occurs at 18–20 °C two months later. Each strobila produces 6–10 ephyrae, which grow to 50 mm diameter within 30 days. In Liaodong Bay, ephyrae grow into mature jellyfish 250–450 mm diameter in two to three months.

Light, temperature, salinity, and food are important factors for asexual reproduction by the benthic stages scyphozoans (e.g., Purcell *et al.*, 2009), as they are for *R. esculentum* (reviewed by Dong *et al.*, 2009). Weak light stimulates planula metamorphosis, while dark conditions promote podocyst excystment but prolong or prevent strobilation. Temperature also is important; the podocysts do not excyst below 10 °C and the excystment rate increases from 15 to 30 °C. Increasing temperature from 2–10 °C to 22 °C in winter induces strobilation in two weeks. The ephyrae grow at temperatures between 16 and 28 °C and 24 °C is optimal. Low salinities limit the benthic stages. No podocysts are produced at salinities < 6 and the optimum range for podocyst production is 20–22‰ salinity. The lower limits for survival are 12‰ for planulae, 10‰ for scyphistomae (optimal 14–20‰), 8‰ for ephyrae (optimal 14–20‰), and 8‰ for young medusae. Planulae of *R. esculentum* are the most favourable food for its early scyphistomae, supplemented with early larvae of oysters and sea urchins. Older stages can be fed with *Artemia* spp. nauplii and zooplankton.

Thus, abundant juvenile jellyfish can be obtained in a short time in the correct culture conditions, and the techniques were applied to large-scale production for stock enhancement (Dong *et al.*, 2009). In 2005, the total volume of rearing tanks for juvenile *R. esculentum* was about 30 000–40 000 m^3 in Liaoning Province. In recent years, over 500 million juvenile jellyfish were cultivated in breeding centres there annually.

The objectives of enhancement were to stabilize and increase the catch of *R. esculentum* jellyfish (Dong *et al.*, 2009). Enhancement trials were

© Woodhead Publishing Limited, 2013

conducted 11 times during 1984–2004. Small-scale releases (2–5 × 10^5) of ephyrae 5–15 mm diameter were made in June–July. The recapture rates in in Heishijiao Bay were estimated to vary from 1.2 to 2.5 % in two years. Recapture rates from large-scale releases (5–17 × 10^6) in coastal waters near the Dayang River estuary were estimated to be 0.1 to 1.0 %. Releases of one to five million juvenile jellyfish 20 mm in diameter were released in Jinpu Bay with recapture rates of 1.2 %; releases of juveniles in the coastal location yielded recapture rates of 0.2 %.

Large-scale enhancements were made in Liaodong Bay, China during 2005 and 2006 (Dong *et al.*, 2009). Juvenile jellyfish were raised in more than 20 breeding centres and 414 million (> 10 mm diameter) were released in June into shallow (3–5 m) waters where they mixed with the existing population of similar sizes.

The jellyfish were monitored five times during the summer to determine mortality and growth rates (Dong *et al.*, 2009). Initial handling mortality was estimated at 6 %. An additional 79 % died two to three days after transfer. An estimated 55 % of the surviving jellyfish died during July before the official harvest in August, which included undetermined losses from poaching before the season opened. The recapture rates of released *R. esculentum* jellyfish were ~3.0 % in 2005 and 2006.

The ratio of the cost of culturing juvenile jellyfish (input) to the value of the sales (output) was about 1:18, making the enhancement very successful economically (Dong *et al.*, 2009).

13.2.4 Future trends

- Large populations of potentially edible jellyfish exist in several regions, such as the Irish Sea and the Mediterranean Sea. These populations may be harvested, but the local human consumer bases are small.
- The high return on *R. esculentum* enhancement in Liaodong Bay makes this a very appealing enterprise that is spreading to other regions. It is very likely that jellyfish farming will become increasingly common in countries of southeast Asia.
- With jellyfish becoming greater nuisances to humans in many locations, utilization of less-preferred food species and increased efforts at new product development are likely.

13.3 Culture of jellyfish for aquaria and research

In recent decades, interest in rearing jellyfish has increased, in part due to success of the incredible exhibit at the Monterey Bay Aquarium that opened in 1992. Jellyfish exhibits have become a relatively common sight in public aquaria around the world and hobbyists are interested in keeping them as pets. More recently, concern in the public, industrial, and government

© Woodhead Publishing Limited, 2013

sectors has grown because human problems caused by jellyfish seem to have increased (Purcell *et al.*, 2007; Purcell, 2012). Consequently, scientists around the world now need to culture jellyfish for research purposes.

In spite of this interest, keeping these delicate animals in captivity requires special methods. Most effort has been put into the most easily cultured species *Aurelia aurita* (the common moon jellyfish), which originally was the only jellyfish in aquaria. Even today with many new techniques available, breeding most other jellyfish species is a challenge for researchers and aquarists. Research on the basic ecology is necessary for each species to create an adequate habitat, including knowledge of water temperatures, salinity, light conditions, and diet. Although we summarize the information available today on more than 20 species (Table 13.2), there are still hundreds of important species about which little is known. Such studies will greatly increase the success of captive breeding jellyfish in the future.

13.3.1 Establishing a culture

When jellyfish are needed for display or research, the institution may require animals all year, which makes establishing a continuous culture necessary. The variability of environmental conditions, the seasonality of most species, and lack of knowledge about their life-cycles result in uncertainty when searching for them in nature.

A culture can be started with healthy adult medusae collected from their natural environment. From them, gametes or planulae can be obtained and the polyp of scyphomedusae or hydroid of hydromedusae stage can be raised and easily maintained. Raskoff *et al.* (2003) proposed several methods to gather fertilized eggs, which will hatch into planulae (the larvae of cnidarian jellyfish) (Fig.13.1). The most common method is to promote natural spawning by grouping mature individuals of both sexes together in a confined space. Hydrozoan planulae can be obtained placing some adult medusae in a beaker (1 or 2 L is sufficient for small hydromedusan species) and leaving them overnight at an appropriate temperature. Hundreds of fertilized eggs can result. Many species of hydrozoans, scyphozoans, cubozoans, and ctenophores spawn after several hours in the dark followed by light. The larger scyphozoan jellyfish require some different methods to obtain planulae. Some species (*A. aurita*, *Cyanea capillata*, and *Cotylorhiza tuberculata*) brood planulae on the oral arms, which can be retrieved by leaving females in a large container. For non-brooding species, several adults can be placed together in a sufficiently large container. If mature, the females will soon release their eggs, which then will be fertilized by sperm from the males and the fertilized eggs can be collected. Gonads from males and females also can be extracted, homogenized, and placed together in a dish of sea water. If the gonads are ripe, fertilized eggs will result. Adult jellyfish should be housed together in adequate tanks and fertilized eggs

© Woodhead Publishing Limited, 2013

Table 13.2 Conditions for jellyfish and ctenophore species that have been successfully cultured

Species	T (°C)	Salinity	Tanks					Foods				
			K	P-K	CT	RMT	RST	R	AN	WP	FC	GZ
Scyphomedusae												
Aurelia spp.	9–29	28–38	M	M	–	M	J	E, J	P, E, M	A	A	–
Cassiopea spp.*	18–29	33 best, tolerant	–	–	–	M[e]	–	E, J	P, E, M	A	–	–
Cyanea spp.	6–15	34 best	A	A	–	A	J	E, J	P, E, M	A	A	E, M
Chrysaora fuscescens	8–15	33 best	A[a]	M	–	A	A	E, J	P, E, M	A	A	E, M
Chrysaora achylos	15–25	33 best	A[a]	M	–	–	J	E, J	P, E, M	A	A	E, M
Chrysaora melanaster	10–14	33 best	A[a]	M	–	A	J	E, J	P, E, M	A	A	E, M
Chrysaora colorata	15–21	33 best	A[a]	M	–	A	J	E, J	P, E, M	A	A	E, M
Pelagia noctiluca	9–24	33–38	J	J	–	M	–	E, J	E, M	M	–	M
*Cotylorhiza tuberculata**	22–28	38 best, tolerant	–	–	A	J[f]	J	P, E	M	–	–	–
Rhizostoma pulmo	18–28	34–38	J	J	–	A[b]	–	P, E	P, E, M	M	–	–
*Phyllorhiza punctata**	24–26	33–35	–	M	A	–	J	P, E	M	–	–	–
*Mastigias papua**	20–29	33 best	–	J	A	A[f]	J	P, E	M	–	–	–
Hydromedusae												
Eutonina indicans	7–15	33 best	A	M	–	–	–	P, E, J	M	A	–	low light
Polyorchis penicillatus	10–15		M	M	–	–	–	–	P, E, M	–	–	–
Craspedacusta sowerbii	~27	0	M	–	–	M[f]	–	–	–	P, E, M	P, E, M	–
Tima formosa	10–15		M	M	–	–	–	P, E, M	P, E, M	–	–	–
Mitrocoma cellularia	7–15	33 best	A	M	–	–	–	P, E, J	M	A	–	low light
Olindias phosphorica	15–24	37–38	–	J	A[g]	–	–	P, E, M	M	–	–	–
Aequorea victoria	6–15	33 best	A	M	–	A	–	P, E, J	M	–	–	A
Aequorea forskalea	12–22	37–38	–	M	–	A	J	P, E, J	M	–	–	A

(Continued)

© Woodhead Publishing Limited, 2013

Table 13.2 *Continued*

Species	T (°C)	Salinity	Tanks K	P-K	CT	RMT	RST	Foods R	AN	WP	FC	GZ
Cubomedusae												
Carybdea marsupialis	18–28	33–38 Low best	–	–	–	M^{c}	–	–	E	E, A	A	–
Ctenophores												
Pleurobrachia sp.	7–15	34	A	M	–	–	–	–	all	all	–	–
												J, large jars
Mnemiopsis leidyi	15–31	Very tolerant	M	M	–	A	–	–	all	all	–	–
Bolinopsis infundibulum	10–15		M	M	–	–	–	–	all	all	–	–
Leucothea sp.	15–21		M	–	–	M	–		all	all		
Beroe gracilis	15		M^{d}	–	–	–	–	–	–	–	–	all
Beroe cucumis	15		M^{d}	–	–	–	–	–	–	–	–	all

Notes: T = temperature. **Tanks**: Kreisel (K), Pseudo-Kreisel (P-K), Rectangular modified tank (RMT), rectangular screened-in flow-through tank (RST), stretch kreisel (S-K), cylindrical tanks (CT). [a] = also stretch kreisel; [b] = large volume; [c] = special design; [d] = modified; [e] = rectangular tank with gentle water flow; [f] = gentle aeration; [g] = algae on bottom. **Foods**: Rotifers (R); *Artemia salina* nauplii (AN), Wild plankton (WP), Frozen crustaceans (FC), blended fish, diced gelatinous zooplankton (GZ). **Stages**: P = polyps or hydroids, E = ephyrae or new medusae, J = juveniles; A = adults; M = juveniles and adults. *Species with zooxanthellae all stages also require full-spectrum light. Liquid coral food provides proper nutrients for zooxanthellae.
Sources: Sommer, 1992, 1993; Raskoff *et al.*, 2003; Widmer, 2007; Gili *et al.*, 2010; Acevedo *et al.*, in press.

© Woodhead Publishing Limited, 2013

may be seen in the water after gamete release or polyps may be found on the tank walls. The eggs or polyps be carefully can then removed for continued development or re-settlement in a dish.

Because hydroids and polyps reproduce asexually to expand the colony, a large culture can be established within a few months from a small sample. If the polyps or hydroids are kept with consistent feeding (Table 13.2) and in a debris- and competitor-free environment, healthy polyps will generally grow and produce juvenile medusae.

Although several treatments can be tried to induce production of juvenile medusae at any time, it is best to recognize when polyps are ready and let them produce medusae naturally. If the polyps are not physiologically ready, deformed, unhealthy young medusae and high mortality result. Knowledge about the life history of the species is necessary to understand this process. Scyphozoan polyps typically produce juvenile medusae by strobilation which can be induced by briefly increasing the water temperature, reducing the temperature for weeks and subsequent return to a normal temperature over a few days, and withholding food for one to two weeks followed by feeding. Other inducers that had some success are changes in illumination and pH, increases in salinity, and treatment with iodine (Raskoff *et al.*, 2003). For jellyfish with symbiotic algae (e.g., *Mastigias papua*, *C. tuberculata*, or *Phyllorhiza punctata*) light is an important factor affecting strobilation.

13.3.2 Maintaining the jellyfish in culture

Maintaining jellyfish in aquaria presents a host of challenges. The following is a list of basic aspects that must be considered for successful breeding:

- Knowledge about the natural history and basic ecology of the jellyfish species.
- Adequate tank design for each species and developmental stage (polyp, small or large medusae). Main considerations are that the volume of the tank, the water flow, and currents be adjusted for each species or developmental stage.
- Proper water characteristics for each species (filtration, temperature, salinity).
- Appropriate diets.
- Careful collection and transport.

Tank design

Tank design, water flow patterns, and good regulation of suction velocity of the outflow are critical determinants for success in the captive husbandry of jellyfish. Several tank system designs have been developed that enable many species to thrive in captivity (Fig. 13.3, Table 13.2). Tank sizes have not been given because that will differ depending on the size and number

© Woodhead Publishing Limited, 2013

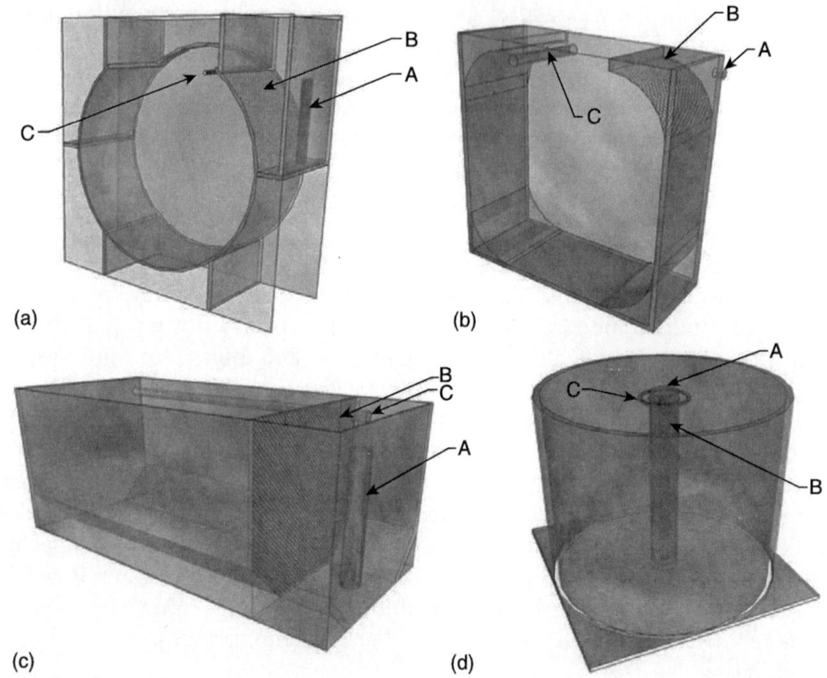

Fig.13.3 Types of culture aquaria for jellyfish. (a) Kreisel tank; (b) Pseudo-kreisel; (c) rectangular tank; and (d) cylindrical tank. A represents the water outlet, B represents the mesh preventing jellyfish suction and C represents the water inlet

of animals. The tank should allow free movement of the jellyfish and those with long tentacles might need larger tanks.

The most well-known tank for keeping jellyfish is the kreisel tank (Fig. 13.3a; Greve, 1968). A kreisel is circular, which helps the water to flow constantly in a circle. With its unique inlet and drain, the filtration system uses a special screen to prevent the jellyfish from getting sucked out of the tank. Separate inlet and outlet chambers help the jellyfish to stay away from the sides of the tank. The combination of the filter and tank design creates equal water pressure throughout the tank, enabling the jellyfish to swim freely. The kreisel was first modified by the Monterey Bay Aquarium to create the pseudo-kreisel (Fig. 13.3b), which is easy to construct and has proven to be versatile and satisfactory for several jellyfish species. This tank follows the same principle as the kreisel and is basically a square tank with curved corners. The pattern of water currents in this tank is generated through a spray bar (with holes) and the current is mainly regulated by the position of the bar.

Another variation on the kreisel is the stretch kreisel, in which two inlet/outlet chambers located on each side of a rectangular tank (twice as wide as tall) send flow upward (Raskoff *et al.*, 2003; Widmer, 2008). The current

© Woodhead Publishing Limited, 2013

produces two gyres, one clockwise and the other counterclockwise, that join in the middle. In each type of kreisel, the proportions are important in maintaining a laminar flow. If the tank is too wide, areas of turbulence areas develop.

Regular rectangular tanks can be modified in various ways to make them suitable for many species of jellyfish. One modification is when the water enters the tank by a flow-bar with holes, which can be positioned at the top or bottom of the tank (Fig. 13.3c). Different configurations may work best for different species. The water goes through an outflow screen made of some rigid material with regularly-spaced holes (Wrobel, 2007). Another type of tank is the screened-in flowthrough tank (Widmer, 2008), in which a parabolic screen of a semi-rigid material located along one side of a normal box tank allows the water leaving the tank to pass through many small holes. In these designs, suction at the outflow screen is diffused over a large surface area and water enters the tank from a tube on the opposite side of the tank.

Cylindrical tanks are good for displaying some jellyfish species, which can be observed from all possible angles (Fig. 13.3d). A variety of models have been developed, but all maintain an updraft that keeps the animals away from the suction area at the bottom of the tank. There are some jellyfish species that cannot be kept successfully in cylindrical tanks (V. L. Fuentes pers. obs.).

The ephyrae of most species of scyphomedusae can be kept in the above aquaria. A variety of containers have been designed to maintain ephyrae until they develop into small medusae and need to be changed to another kind of tank. A small rectangular 3–6 L aquarium with central aeration is adequate for many species (Table 13.2). Widmer (2008) found glass culture dishes adequate for most scyphozoan ephyrae and some small hydromedusae. *Pelagia noctiluca* ephyrae cannot be kept in this kind of aquarium.

Because *P. noctiluca*, which lacks a polyp stage, is an important blooming species throughout the Mediterranean Sea and in the North Atlantic Ocean, several research institutions want to maintain this species in captivity to help understand its ecology. Ephyrae of *P. noctiluca* can be obtained by keeping at least six healthy mature adults together in a closed system and letting them spawn. Kreisel tanks are not good for long-term maintenance of this species (Table 13.2); however, they are adequate to keep adults a couple of days for spawning. In the centre of a kreisel is a current-free area that accumulates the gametes of both sexes and accelerates fertilization of the eggs. The fertilized eggs can be gently transferred to a 3–5 L jar with a gentle aeration until the planulae appear. The planulae will develop into ephyrae in a few days and more rapidly in warmer water temperatures. The ephyrae can be maintained in a kreisel tank with very slow current. When the ephyrae develop marginal tentacles, they must be transferred to a specially-designed tank (Fig. 13.3).

© Woodhead Publishing Limited, 2013

Probably any jellyfish species can be kept in aquaria if enough time is devoted to observing its behavior. This might imply the construction of a totally new tank or the modification of a known one, to be adequate for the species.

Feeding jellyfish in captivity

To feed captive jellyfish requires basic knowledge of the feeding behaviour of the species in their natural environment (Table 13.2). Most jellyfish and ctenophores are predators and most require living prey. The best is to feed with fresh zooplankton, but few caretakers have daily access. Feeding jellyfish with a variety of items is very important to ensure the growth. Newly-hatched *Artemia salina* nauplii are used to feed most jellies. This serves well as the basis during the entire life-cycle; however, nutritional supplements are necessary. Without additions to their diets, captive jellyfish become deformed and may die. Dietary additives (encapsulated diets, e.g. produced by INVE) for *A. salina* nauplii are commercially available.

A variety of supplemental foods that are available online are preferable to the commercial supplements. These foods may enhance the growth and vitality of certain species. Frozen crustaceans, such as eupahusiids, are an excellent and convenient food for some jellyfish species. Live copepods obtained from eggs that can be grown at the laboratory are very helpful for enhancing the survival of several species. Rotifers (*Brachionus* sp.) are smaller than *A. salina* nauplii, can be grown in culture, and are adequate for feeding ephyrae, small medusae, polyps, and hydroids. Some jellyfish species feed on other jellies in nature (e.g., Purcell, 1997). This is important because medusivorous jellyfish will not grow properly without other jellies in the diet. Moon jellyfish, *Aurelia* spp., are often used to supplement captive jellyfish diets. Some hydromedusae (species) are good for feeding ephyrae of some scyphomedusae and salps are important to the diet of *Aequorea forskalea* and *P. noctiluca.*

13.3.3 Future trends

- Because jellyfish have provided beautiful and popular aquarium exhibits, and because their populations and the problems they cause for humans appear to be increasing around the world, interest in rearing and exhibiting jellyfish is likely to increase.
- As more research is conducted and the needs of each species are elucidated, more species will be kept in culture.
- As techniques for successful culture of important species becomes available, it would be helpful for a network of institutions working on jellyfish husbandry to be created in the near future.

© Woodhead Publishing Limited, 2013

13.4 Problems with aquaculture caused by jellyfish

The negative interactions between jellyfish (primarily cnidarians – those with stinging cells) and fish in aquaculture appear to be an increasing problem through the intensification of aquaculture operations in many coastal areas worldwide. Over the last three decades there have been numerous fish kill events or health problems in marine-farmed fish that have been associated with jellyfish (Purcell *et al.*, 2007). The majority of problems have occurred with marine-farmed salmonids in northwest Europe (Rodger *et al.*, 2011a). Nevertheless, aquaculture operations in other regions such as Asia, North America, and Australia have also been affected (Yasuda, 1988; Willcox *et al.*, 2008; Rodger *et al.*, 2011a). Although large-scale jellyfish blooms can sometimes be the obvious cause (in the case of conspicuous scyphomedusae), often problems are associated with small, transparent species. A lack of sampling and investigation around these types of event has previously inhibited the identification of the causative agents (Cronin *et al.*, 2004). Furthermore, whilst mass mortality events are commonly reported in the literature and the media, lower level health problems are very poorly reported despite the scale of the problems in some regions (Rodger, 2007).

13.4.1 Gill disorders caused by jellyfish

Almost all of the initial gill damage caused by cnidarians is due to their stings. Cnidarian jellyfish are characterized by having millions of microscopic stinging cells, primarily in their tentacles. Inside the cell is a specialized stinging capsule called a nematocyst (most 10–20 μm long) that contains a coiled, harpoon-like hollow tubule often armed with spines. Nematocysts can be triggered by mechanical or chemical stimulation. Toxins can be injected from the nematocyst into the prey to immobilize them. Each jellyfish can have a variety of nematocysts that differ in size, shape, and toxicity among species. When a nematocyst fires mechanical damage is caused when the tubule penetrates the tissue, as well as toxic damage from the activities of the enzymes, neurotoxins, myotoxins, and hemolytic compounds contained within (Mariottini and Pane, 2010). As such, nematocysts are especially damaging to delicate fish gills.

Gill disorders (damage to the gill arch and filaments caused by pathogens and parasites) are emerging as one of the most serious causes of mortality in marine-farmed salmonids (Rodger, 2007) (Fig. 13.4), with an average of 12 % mortality of farmed fish recorded annually from Irish fish farms (Rodger and Mitchell, 2005). In temperate regions where salmonids are cultivated, gill disorders typically occur in late summer (giving rise to the term 'summer syndrome'). Information on the interactions with other fish species is even scarcer. Gill disorders are thought to be multi-factorial, often caused by a primary attack from agents like jellyfish, leaving the gills open

© Woodhead Publishing Limited, 2013

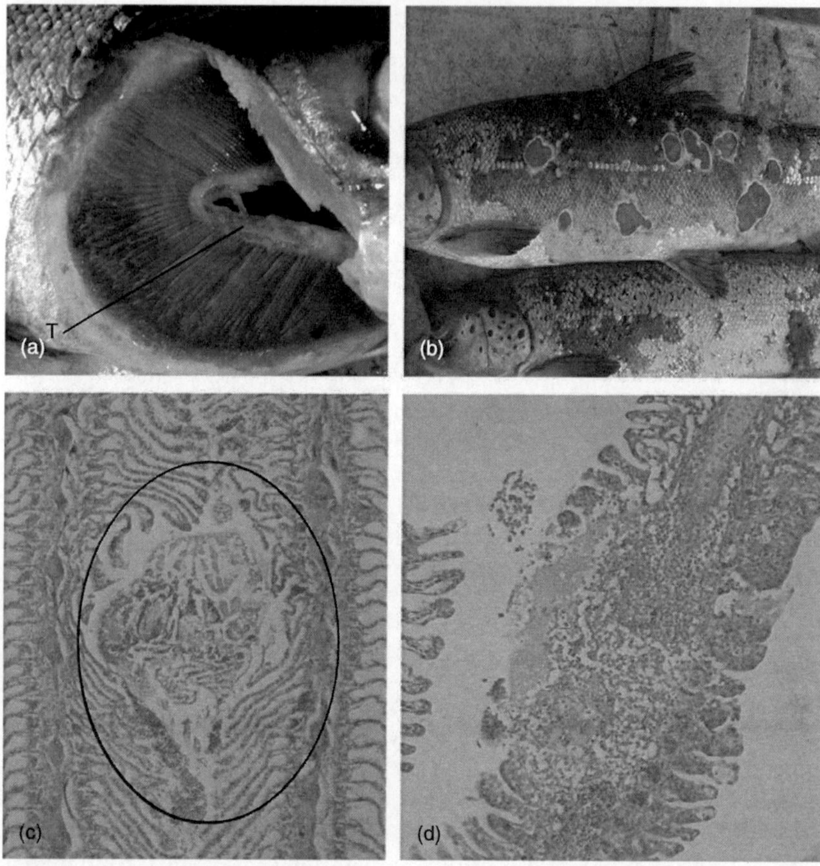

Fig.13.4 Skin and gill damage to fish caused by jellyfish. (a) Sloughed gill raker epithelium, necrosis of the filaments and *Tenacibaculum* sp. infection (T) in Atlantic salmon associated with *Muggiaea atlantica* and *Solmaris corona*. (b) Skin ulceration in Atlantic salmon as a result of exposure to *Pelagia noctiluca*. (c) Unidentified gelatinous agent found in the gills under histopathology. (d) Significant fusion of the gill lamellae, necrosis and inflammation 38 hours after exposure to *Aurelia aurita*. (a–c) courtesy of Hamish Rodger; (d) from Baxter *et al.* 2011b. Gill damage to Atlantic salmon (*Salmo salar*) caused by the common jellyfish (*Aurelia aurita*) under experimental challenge. *Plos One* **6**, (4): e18529. Copyright © 2011 Baxter *et al.* Reprinted by permission.

to secondary bacterial infections which exacerbate the problem (Rodger *et al.*, 2011a) (Fig. 13.4). Although gill disorders can be caused by a variety of agents (e.g. jellyfish, phytoplankton, amoeba, parasites, bacteria, and viruses), the type of damage exhibited by the gills is observed with a limited number of responses (i.e. hemorrhage, necrosis, edema). Whilst some causative agents can be easily diagnosed via microscopy or histopathology, others cannot. With respect to jellyfish-induced damage, recent

© Woodhead Publishing Limited, 2013

monitoring studies and experimental trials have significantly increased the knowledge on the pathogenesis of the disorder (Baxter *et al.*, 2011a,b).

13.4.2 Types of jellyfish causing aquaculture problems

The term jellyfish is most often used to describe the free-swimming (medusoid) stage of the phylum Cnidaria (classes Hydrozoa, Scyphozoa, and Cubozoa), although it has been used to include ctenophores (comb jellyfish without stinging cells). In addition to the fact that they sting, problems from jellyfish also derive from their rapid population growth and their tendency to become concentrated in shallow waters (Graham *et al.*, 2001). Most species have seasonal peaks in abundance, usually in spring through summer, that coincide with local patterns of plankton production. Several types of jellyfish have been previously linked to fish kill events and gill damage in marine-farmed fish including hydromedusae, siphonophores, scyphozoans, and ctenophores (Table 13.3 and Fig. 13.5).

Hydromedusae

Hydromedusae are a diverse group (> 800 species worldwide), most of which are < 1 cm in size and transparent, thereby going mostly unnoticed by humans. Because of their small sizes, they can pass through the nets and some directly into the opercula of the fish. Many of the species produce medusae asexually from the hydroid stage, which lives attached to hard surfaces, including aquaculture structures. Other species lack an attached, benthic stage (holoplanktonic) (e.g., *Solmaris corona*); still other species (e.g., *Ectopleura larynx* syn. *Tubularia larynx*) produce swimming larvae, but not medusae (Fig. 13.6).

A bloom of holoplanktonic *S. corona* (~0.5–20 mm) medusae was implicated as the causative agent in the mortality of ~900 000 salmon at Scottish aquaculture sites in August/September 1997 (Båmstedt *et al.*, 1998) and again in 2002 when there were around 650 000 mortalities in two days (Rodger *et al.*, 2011a). This species has also been recently linked to more chronic mortalities and severe gill damage at a salmon farm in Ireland (Baxter *et al.*, 2011a).

The neritic medusae of *Phialella quadrata* have been implicated in fish kill events and as a vector of bacterial gill disease. In the 1980s 1500 salmon died at farms on the Shetland Isles, Scotland. Histopathological examination revealed severe epithelial stripping and necrosis of the lamellae amongst other damage and some of the fish were reported to have up to 40 *P. quadrata* in their stomach contents (Bruno and Ellis, 1985). Recent research has identified *P. quadrata* as a potential vector for the bacterial pathogen *Tenacibaculum maritimum*. These filamentous bacteria were found on both the manubrium of *P. quadrata* and on the gills of salmon from a Scottish salmon farm (Ferguson *et al.*, 2010). This bacterium is common to farmed fish and has also been found on parasitic sea lice (Barker

© Woodhead Publishing Limited, 2013

Table 13.3 Published reports of jellyfish interfering with aquaculture operations around the world.

Jellyfish species	Dates	Locations	Type of damages	Reference
Asia				
Aurelia aurita s.l.	Jul–Sep 1950	Lake Hachirogata, Akita Prefecture	Mass mortality of fish and bivalves	Yasuda, 1988
*Porpita porpita**	Aug–Oct 2000	Kyoto, Fukui Prefectures	Mortality of penned fish	Yasuda, 2003
*Pelagia noctiluca**	Apr 2004	Ehime Prefecture	Mortality of penned fish	Uye (pers. info. from local fisherman)
Australia/Indo-Pacific				
Aurelia sp.	Summer 1998–2001	Tasmania	Atlantic salmon	Tasmanian Aquaculture and Fisheries Institute, 2003
Unidentified	<1995	India	Giant tiger prawns	Rajagopal *et al.* 1989
Rhizostome scyphozoan	2006	Goa, India	Shrimp	RA Sreepada (pers. com.)
Europe				
Aurelia aurita	June 2010	Northwest Ireland	Acute gill lesions and mortality in Atlantic salmon	Mitchell *et al.*, 2011
*Pelagia noctiluca**	1994	Brittany, France	Salmon and trout	Merceron *et al.*, 1995
*Pelagia noctiluca**	Nov 2007	Western Ireland	250000 salmon killed	Doyle *et al.*, 2008
Cyanea capillata	1996	Loch Fyne, Scottland	Thousands of salmon killed, £250000 loss	Anon, 1996

© Woodhead Publishing Limited, 2013

*Solmaris corona** *Muggiaea atlantica**	Aug–Nov 2009	Western Ireland	Severe gill damage and potential mortalities in Atlantic salmon	Baxter *et al.*, 2011a
*Solmaris corona**	Summer 1997	Shetland Isles	Salmon killed	Anon, 1997a
*Solmaris corona**, *Phialidium* sp., *Leuckartiara octona*, *Catablema vesicarium*	Aug 2001–2002	Isle of Lewis in the Outer Hebrides, Scotland	2 747 680 salmon killed in 11 incidents, £5 mil loss	Johnson, 2002
*Apolemia uvaria**, *Aurelia aurita*, *Cyanea capillata*, *Bolinopsis infundibulum**	Nov 1997–Feb 1998 and 2003; 1994 and 1995; May–June 1986	West coast of Norway and Sweden	Mass mortalities of Atlantic salmon	Båmstedt *et al.*, 1998; Heckmann, 2004
Phialella quadrata	Aug 1984	Shetland Isles, Scotland	Killed 1500 salmon and severe gill damage	Bruno and Ellis, 1985
*Muggiaea atlantica**	Aug 2002	Norway	Killed > 100 000 salmon	Fosså *et al.*, 2003
North America				
Moerisia lyonsi	1970s; May–Oct 1994–1997	Mesocosms, Louisiana and Maryland USA	Killed decapods; ≤ 13.6 med. L^{-1}	Sandlifer *et al.*, 1974; Purcell *et al.*, 1999

Source: Updated from Purcell *et al.*, 2007.
*Holoplanktonic, all others have a benthic stage.

© Woodhead Publishing Limited, 2013

Fig.13.5 Examples of harmful jellyfish species that have caused gill problems and/or mortalities in marine-farmed salmon. (a) A bloom of *Aurelia aurita* (and a single *Chrysaora hysoscella* in the centre) on an Atlantic salmon in Ireland. (b) *Pelagia noctiluca* taken from the surface waters during the 2007 bloom in Northern Ireland. (c) a single polyp head of the biofouling hydroid species, *Ectopleura larynx*, found on aquaculture nets in Ireland. (d) *Muggiaea atlantica* and (e) *Solmaris corona*. (c–e) only a few mm in length (a) and (c–e) photos courtesy of Emily Baxter; (b) photo courtesy of John Russell.

et al., 2009); however, importantly it may have the potential to exacerbate jellyfish-induced gill damage (Rodger, 2007) as it has been shown to be a secondary opportunistic pathogen (Handlinger *et al.*, 1997).

Siphonophores

Siphonophores are colonial, holoplanktonic cnidarians that range in size from centimetres to several metres in length. Their colonies can break up

© Woodhead Publishing Limited, 2013

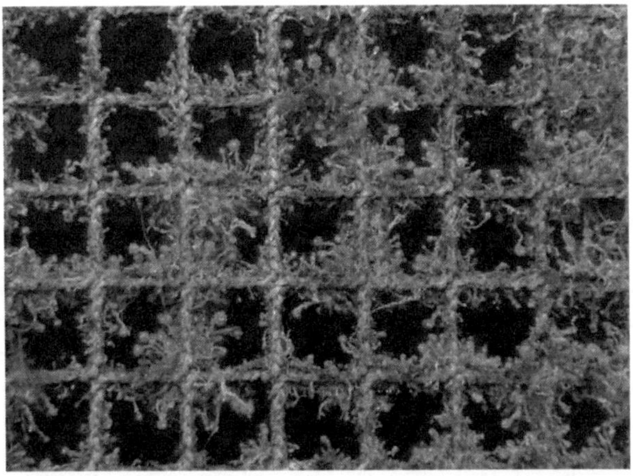

Fig.13.6 *Ectopleura larynx* hydroids fouling aquaculture netting (courtesy of Jana Guenther).

into small pieces that can pass through net pens. In addition, some species (e.g., *Muggiaea atlantica*) produce many free sexual stages (eudoxids) that are generally < 1 cm in size) (Fig. 13.5d).

The mass occurrence of the siphonophore *Apolemia uvaria*, which can be several meters long, on the west coast of Norway in the winter of 1997/1998 caused severe lesions and the mortality of farmed fish. Fish were stung on the skin, eyes, and gills; fish that were blinded had the highest mortality (Båmstedt *et al.*, 1998). *Muggiaea atlantica*, a small, coastal species, has also been associated with mortalities and health problems in marine-farmed fish in Norway and Ireland. In August 2002, a bloom of *M. atlantica* with maximum densities of 13 000 individuals m^{-3} caused the death of >100 000 salmon in Norway, again with stings to the skin and gills (Fosså *et al.*, 2003). Furthermore, the mortality of over 1 000 000 farmed salmon and trout on the west coast of Ireland in September 2003 was thought to be attributable to *M. atlantica*, although a lack of investigation at the time of the event (before and after) prevented confirmation of this (Cronin *et al.*, 2004). This species is common throughout Europe and had recently been linked to severe gill damage in farmed salmon at densities less than above (Baxter *et al.*, 2011a).

Scyphomedusae

Scyphomedusae generally are large jellyfish that have an asexual stage attached to hard surfaces that produce tiny jellyfish (ephyrae) (e.g., *Aurelia* spp.) (Fig. 13.1). The attached stage of *Aurelia* spp. is known to occur abundantly in harbors and aquaculture facilities (Lo *et al.*, 2008; Hoover and Purcell, 2009). *P. noctiluca* is a holoplanktonic species that lacks an asexual

© Woodhead Publishing Limited, 2013

benthic stage in the water column. The new ephyrae of all species are 1–3 mm in diameter and can easily pass through nets.

The lion's mane jellyfish, *Cyanea capillata*, has been implicated in causing thousands of mortalities at fish farms in Scotland on more than one occasion (Bruno and Ellis, 1985; Purcell *et al.*, 2007), as well as sporadic problems in Norway (Båmstedt *et al.*, 1998).

Ephyrae and small medusae of the common jellyfish, *A. aurita* are thought to have caused huge losses of farmed fish in Norway and Scotland in the mid-1990s through suffocation (Båmstedt *et al.*, 1998). Incidents of *A. aurita* causing mortalities of farmed fish in Tasmania and Asia also exist (Yasuda, 1988; Willcox *et al.*, 2008). Nevertheless, until recently the potential damage caused by the common jellyfish has been under-estimated. Large aggregations of *A. aurita* are a common occurrence around sites of aquaculture in northwest Europe although problems associated with them often go overlooked due to their mild sting to human skin. This may not be the case for more delicate fish gills. Mitchell *et al.*, (2011) documented the pathological damage to Atlantic salmon caused by such an aggregation at salmon farm in Ireland. The authors were the first to show intravascular hemolysis combined with multifocal hemorrhage and lamellar thickening of the gills over the duration of the event (~3000 mortalities in the first week after the occurrence of *A. aurita*) and damage persisted at two months after the event. Controlled challenge experiments where Atlantic salmon were exposed to *A. aurita* also described the pathogenesis of the gill damage caused by this species (Baxter *et al.*, 2011b). The damage presented as widespread multifocal hemorrhage, necrosis, and fusion of the lamellae. There was also a lag in the peak of the damage after the jellyfish had been removed and a delayed inflammatory response. As with the field observations, the damage was still present three weeks after the challenge. Moreover, up to an 85% reduction in gill cell viability was observed from *in vitro* experiments where an assay of rainbow trout gill cells was exposed to toxins extracted from *A. aurita* and *C. capillata* stinging capsules (Helmholz *et al.*, 2010).

Another detrimental scyphozoan species is the mauve stinger, *P. noctiluca*. This species has been noted to irritate the gills of salmon and trout in France (Merceron *et al.*, 1995) and killed all 250000 Atlantic salmon at a farm in Northern Ireland, as well as similar incidents in Scotland and Asia (Purcell *et al.*, 2007; Doyle *et al.*, 2008; Hay and Murray, 2008).

Ctenophores

There has been only one reported incident of a ctenophore causing the death of farmed fish; this occurred when a bloom of the ctenophore *Bolinopsis infundibulum* was thought to have caused the death of farmed fish through direct suffocation (clogging of the gills) in Norway in 1986 (Båmstedt *et al.*, 1998).

© Woodhead Publishing Limited, 2013

13.4.3 Potential problems associated with net fouling hydroids

A further threat to the finfish aquaculture industry is posed by the hydroid stage of hydrozoans. These sessile animals are known to rapidly and extensively foul aquaculture structures (pontoons and netting) (Carl *et al.*, 2010) (Fig. 13.6). Before now it was unknown whether hydroids (which also possess stinging cells) affect farmed fish health and play a role in gill disorders (Rodger *et al.*, 2011a). The risk of damage caused by biofouling hydroids was briefly mentioned in an examination of clubbing and gill necrosis syndrome in Atlantic salmon in Tasmania (Clark *et al.*, 1997). Some hydroid species, such as *Ectopleura larynx*, have become the dominant fouling organisms of the fish farming industry in Norway where they cause problems by reducing water flow and quality (Guenther *et al.*, 2009; Carl *et al.*, 2010). This fast-growing species is common and abundant in many European coastal waters (Browne, 1897; Boero and Fresi, 1986; Östman *et al.*, 1995; Galea, 2007) and is now becoming prevalent on Irish finfish farms.

Organic farms, which are not permitted to use copper-based anti-foulants that other farms use (IOFGA, 2006), have to clean the nets every three to four weeks during the peak period of hydroid growth. *In situ* net washing is an increasingly common means to deal with such biofouling. This practice uses rotating discs that jet-wash the nets at high pressure (as described by Guenther *et al.*, 2010). This blasts pieces of hydroids into the water that then can damage the fish. After the first net wash of the production cycle, hydroids can grow at an even faster rate than before because remnants remain on the net (Guenther *et al.*, 2010). However, efforts to investigate this potential problem and the effects of net washing have been neglected until recently. Baxter (2012) quantified the rapid growth of the biofouling hydroid *E. larynx* on aquaculture nets and used experimental challenge trials to identify possible damage they cause to fish in aquaculture. Fish were exposed to hydroid-fouled nets and to loose hydroids, to simulate the process of net washing. Indeed, the hydroids caused gill damage to Atlantic salmon smolts in both treatment groups over the duration of the experiment. This was visible under histopathology as focal areas of epithelial sloughing and necrosis. Such damage has been linked to lethargy and decreased appetite in farmed salmon (Rodger *et al.*, 2011b). Some medusa-producing hydroids are known to live on aquaculture structures and nets (Chaplygina, 1993), although little documentation exists. *Obelia* spp. are ubiquitous in harbours globally and produce great numbers of 1 mm medusae that might damage fish gills, although no direct evidence yet exists.

13.4.4 Enhancement of jellyfish blooms by aquaculture

Speculation continues about whether jellyfish populations are increasing worldwide, at least in part due to anthropogenic degradation of the environment (e.g., Purcell *et al.*, 2007; Condon *et al.*, 2012; Purcell, 2012). There is

© Woodhead Publishing Limited, 2013

mounting evidence that populations fluctuate with climatic cycles and that temperate species may increase their abundances, spatial, and seasonal distributions in a warming climate (Purcell, 2005). 'Fishing down the food web' reduces piscivorous fish (Pauly and Palomares, 2001), which favours zooplanktivorous fish that are heavily fished to produce oil and meal for aquaculture feeds (Metian and Tacon, 2009); the catches of forage fish increased 8.8-fold between 1950 and 2006 (http://www.seaaroundus.org). This further 'fishing down' removes competitors of jellyfish, possibly increasing their success (e.g., Purcell *et al.*, 2007, Purcell, 2012).

Aquaculture operations also can have several potentially beneficial effects on jellyfish populations at local scales (Purcell, 2005). Lo *et al.* (2008) discuss how increased nutrients around the farms, due to excess fish food and waste products, could create eutrophic conditions that may favour jellyfish over fish, such as smaller zooplankton that are less beneficial foods for fish and decreased dissolved oxygen and water clarity (Arai, 2001; Purcell *et al.*, 2007). Furthermore, the structures of aquaculture facilities (e.g. cages, pontoons, and feed barges) may restrict water flow which could act to retain medusae, as well as provide substrates for the benthic stages (Båmstedt *et al.*, 1998; Lo *et al.*, 2008). Problems associated with jellyfish are likely to arise in all areas of finfish production where jellyfish are abundant (Rodger, 2007). Many highly productive aquaculture operations in regions such as Asia, north western Europe, Australia, and South America have already been affected (Båmstedt *et al.*, 1998; Doyle *et al.*, 2008; Lo *et al.*, 2008; Willcox *et al.*, 2008). Therefore, the expansion of the aquaculture industry globally is of increasing concern due to its potential to enhance jellyfish populations. In the future this could mean more frequent and significant losses of finfish as a result of interactions with jellyfish.

13.4.5 Recommendations for problem minimization

It has become commonplace and necessary for extensive phytoplankton monitoring to be conducted at many shellfish and finfish aquaculture sites to rapidly identify harmful species which may pose a threat to the health of humans and the stock (Marine Institute, 2009); as yet no such system exists for monitoring jellyfish populations. Recent studies (Baxter *et al.*, 2011a,b; Mitchell *et al.*, 2011) have shown that widespread, routine monitoring for jellyfish around marine fish farms is necessary and an understanding of the links between jellyfish blooms and detrimental effects on the fish is needed. Information at a site-specific level is essential to identify the seasonal and interannual abundance and occurrence of detrimental species, highlighting risk periods for each location. As yet, no reliable mitigation methods exist to prevent blooms of jellyfish passing through finfish cages (Rodger, 2007; Hay and Murray, 2008). Rodger *et al.* (2011a) list and suggest several mitigation methods (e.g., early warning systems and predictive models, oxygenation/aeration of cages, rapidly installed protective

© Woodhead Publishing Limited, 2013

enclosures/bubble curtains, and submersion or towing cages out of the bloom area) that could be developed and tested in the future although cost, effectiveness, and site-specific suitability need to be thoroughly investigated. Investigation into non-toxic surfaces that deter settlement of cnidarian (both scyphozoan and hydrozoan) planulae offers promise for minimizing their use of aquaculture structures as habitat. To combat the problem of net-fouling hydroids, non-toxic anti-foulant suitable for use on organic farms could be developed. For example, silicone-coated netting significantly reduced fouling at a salmon farm in Tasmania (Hodson *et al.*, 2000). Furthermore, the development of such products should be combined with detailed investigations on the effects of net-washing practices on fish health before further reliance of the industry on this potentially detrimental method.

13.5 References

ANON (1996) 'Jellyfish stings kill Scottish salmon', *Fish Farm Int* 23.

ANON (1997) 'Shetland jellyfish plague', *Fish Farm Int* 24,5.

ACEVEDO M J, FUENTES V L, OLARIAGA A, CANEPA A, BELMAR M B, BORDEHORE C and CALBET A (submitted) 'Maintenance, feeding and growth of *Carybdea marsupialis* (Cnidaria: Cubozoa) in the laboratory', *J Exp Mar Biol Ecol*, in press.

ARAI M N (2001) 'Pelagic coelenterates and eutrophication: a review', *Hydrobiologia*, 451, 69–87.

BÅMSTEDT U, FOSSÅ J H, MARTINUSSEN M B and FOSSAHAGEN A (1998) 'Mass occurence of the physonect siphonophore *Apolemia uvaria* (Lesueur) in Norwegian waters', *Sarsia*, 83, 79–85.

BARKER D E, BRADEN L M, COOMBS M P and BOYCE B (2009) 'Preliminary studies on the isolation of bacteria from sea lice, *Lepeophtheirus salmonis*, infecting farmed salmon in British Columbia, Canada', *Parasitol Res*, 105, 1173–1177.

BAXTER E J, RODGER H D, MCALLEN R and DOYLE T K (2011a) 'Gill disorders in marine-farmed salmon: investigating the role of hydrozoan jellyfish', *Aquacult Env Interact*, 1, 245–257.

BAXTER E J, STURT M M, RUANE N M, DOYLE T K, MCALLEN R, HARMAN L and RODGER H D (2011b) 'Gill damage to Atlantic salmon (*Salmo salar*) caused by the common jellyfish (*Aurelia aurita*) under experimental challenge', *PLoS One*, 6, (4): e18529.

BAXTER E J, STURT M M, RUANE N M, DOYLE T K, MCALLEN R and RODGER H D (2012) 'Biofouling of the hydroid *Ectopleura larynx* on aquaculture nets in Ireland: Implications for finfish health', *Fish Vet J*, in press. doi: 10.1371/journal.pone.0018529.

BOERO F and FRESI E (1986) 'Zonation and evolution of a rocky bottom hydroid community', *Mar Ecol*, 7, 123–150.

BROWNE E T (1897) 'The hydroids of Valencia Harbour, Ireland', *Irish Naturalists' J*, 6, 241–246.

BRUNO D W and ELLIS A E (1985) 'Mortalities in farmed atlantic salmon associated with the jellyfish *Phialella quadrata*', *Bull Eur Assoc Fish Pathol*, 5, 64–65.

CARL C, GUENTHER J and SUNDE L M (2010) 'Larval release and attachment modes of the hydroid *Ectopleura larynx* on aquaculture nets in Norway', *Aquac Res*, 42, 1056–1060.

CHAPLYGINA S F (1993) 'Hydroids in the fouling of mariculture installations in Peter the Great Bay, Sea of Japan', *Russ J Mar Biol*, 19, 29–36.

© Woodhead Publishing Limited, 2013

CLARK A, NOWAK B, HANDLINGER J, MUNDAY B L and PERCIVAL S (1997) 'Clubbing and necrosis gill (CNG) syndrome in sea-caged Atlantic salmon, *Salmo salar* L., in Tasmania: an initial report', *J Fish Diseases*, 20, 59–68.

CONDON R H, GRAHAM W M, DUARTE C M, PITT K A, LUCAS C H, HADDOCK S H D, SUTHERLAND K R, ROBINSON K L, DAWSON M N, DECKER M B, MILLS C E, PURCELL J E, MALEJ A, MIANZAN H, UYE S-I and GELCICH S (2012) 'Questioning the rise of gelatinous zooplankton in the world's oceans', *BioScience*, 62, 160–169.

CRONIN M, CUSACK C, GEOGHEGAN F, JACKSON D, MCGOVERN E, MCMAHON T, O'BEIRN F, O'CINNEIDE M AND SILKE J (2004) *Salmon mortalities at Inver Bay and Mc Swynes Bay finfish farms, County Donegal, Ireland during 2003.* Marine Environment and Health Serves No. 15. Marine Institute.

DONG J, JIANG L-X, TAN K-F, LIU H-Y, PURCELL J E, LI P-J and YE C-C (2009) 'Stock enhancement of the edible jellyfish (*Rhopilema esculentum* Kishinouye) in Liaodong Bay, China: a review', *Hydrobiologia*, 616, 113–118.

DOYLE T K, DE HAAS H, COTTON D, DORSCHEL B, CUMMINS V, HOUGHTON J D R, DAVENPORT J and HAYS G C (2008) 'Widespread occurrence of the jellyfish *Pelagia noctiluca* in Irish coastal and shelf waters', *J Plankton Res*, 30, 963–968.

FERGUSON H W, DELANNOY C M J, HAY S, NICOLSON J, SUTHERLAND D and CRUMLISH M (2010) 'Jellyfish as vectors of bacterial disease for farmed salmon (*Salmo salar*)', *J Vet Diagn Investig*, 22, 376–382.

FOSSÅ J H, FLOOD P R, OLSEN A B and JENSEN F (2003) Små og usynlige, men plagsomme maneter av arten *Muggiaea atlantica*', *Fisken og Havet*, 2, 99–103 (in Norwegian).

FUENTES V, STRAEHLER-POHL I, ATIENZA D, FRANCO I, TILVES U, GENTILE M, ACEVEDO M, OLARIAGA A and GILI J-M (2011) 'Life cycle of the jellyfish *Rhizostoma pulmo* (Scyphozoa: Rhizostomeae) and its distribution, seasonality and inter-annual variability along the Catalan coast and the Mar Menor (Spain, NW Mediterranean)', *Mar Biol*, 158, 2247–2266.

GALEA H R (2007) 'Hydrozoa, La Ciotat and nearby areas, Mediterranean coast of France', *Check List*, 3, 193–199.

GILI JM, FUENTES V, ATIENZA D and LEWINSKY I (2010) *Report of the Medusa Project*, Tech Rep No. 8. Barcelona: Agencia Catalana de l'Aïgua, Generalitat de Catalunya.

GRAHAM W M, PAGES F and HAMNER W M (2001) 'A physical context for gelatinous zooplankton aggregations: a review', *Hydrobiologia*, 451, 199–212.

GREVE W (1968) 'The planktonkreisel, a new device for culturing zooplankton', *Mar Biol*, 1, 201–203.

GUENTHER J, CARL C and SUNDE L M (2009) 'The effects of colour and copper on the settlement of the hydroid *Ectopleura larynx* on aquaculture nets in Norway', *Aquaculture*, 292, 252–255.

GUENTHER J, MISIMI E and SUNDE L M (2010) 'The development of biofouling, particularly the hydroid *Ectopleura larynx*, on commercial salmon cage nets in mid-Norway', *Aquaculture*, 300, 120–127.

HAMNER W M (1990) 'Design developments in the planktonkreisel, a plankton aquarium for ships at sea', *J Plankton Res*, 12, 397–402.

HANDLINGER J, SOLTANI M and PERCIVAL S (1997) 'The pathology of *Flexibacter maritimus* in aquaculture species in Tasmania, Australia', *J Fish Dis*, 20, 159–168.

HAY S and MURRAY A (2008) 'Jellyfish problems faced by the aquaculture industry', *Fish Farmer*, September/October, 40–41.

HECKMANN R (2004) 'What else can happen? Other problems for fish production', *Aquaculture Mag*, Mar/Apr, 1–8.

HELMHOLZ H, JOHNSTON B, RUHNAU C and PRANGE A (2010) 'Gill cell toxicity of northern boreal scyphomedusae *Cyanea capillata* and *Aurelia aurita* measured by an in vitro cell assay', *Hydrobiologia*, 645, 223–234.

© Woodhead Publishing Limited, 2013

HODSON S L, BURKE C M and BISSETT A P (2000) 'Biofouling of fish-cage netting: the efficacy of a silicone coating and the effect of netting colour', *Aquaculture*, 184, 277–290.

HOOVER R A and PURCELL J E (2009) 'Substrate preferences of scyphozoan *Aurelia labiata* polyps among common dock-building materials', *Hydrobiologia*, 616, 259–267.

HSIEH Y-H P, LEONG F-M and RUDLOE J (2001) 'Jellyfish as food', *Hydrobiologia*, 451, 11–17.

IOFGA (2006) *Irish Organic Farmers and Growers Association Standards for Organic Aquaculture*, Certified marine products (salmonid species), 15 pp.

JOHNSON P (2002) Jellyfish, algae take toll on Scottish salmon. *Fish Information Service*, Europe.

KITAMURA M and OMORI M (2010) 'Synopsis of edible jellyfishes collected from Southeast Asia, with notes on jellyfish fisheries', *Plankton Benthos Res*, 5, 106–118.

LO W-T, PURCELL J E, HUNG J-J, SU H-M and HSU P-K (2008) 'Enhancement of jellyfish (*Aurelia aurita*) populations by extensive aquaculture rafts in a coastal lagoon in Taiwan', *ICES J Mar Sci*, 65, 453–461.

Marine Institute (2009) *Phytoplankton monitoring*, available at: www.marine.ie/home/services/operational/phytoplankton (accessed September 2012).

MARIOTTINI G L and PANE L (2010) 'Mediterranean jellyfish venoms: a review on scyphomedusae', *Mar Drugs*, 8, 1122–1152.

MERCERON M, LE FEVRE-LEHOERFF G, BIZOUARN Y and KEMPF M (1995) 'Fish and jellyfish in Brittany (France)', *Equinoxe*, 56, 6–8.

METIAN M and TACON A G J (2009) 'Fishing for feed or fishing for food: increasing global competition for small pelagic forage fish', *Ambio*, 38, 294–302.

MITCHELL S O, BAXTER E J and RODGER H D (2011) 'Gill pathology in farmed salmon associated with the jellyfish *Aurelia aurita*', *Vet Rec*, 169, 609, doi: 10.1136/vetrec-2011-100045.

OMORI M and NAKANO E (2001) 'Jellyfish fisheries in southeast Asia', *Hydrobiologia*, 451, 19–26.

ÖSTMAN C, MYRDAL M, NYVALL P, LINDSTROM J, BJORKLUND M and AGUIRRE A (1995) 'Nematocysts in *Tubularia larynx* (Cnidaria, Hydrozoa) from Scandinavia and the northern coast of Spain', *Scientia Mar*, 59, 165–179.

PAULY D and PALOMARES M L D (2001) 'Fishing down marine food webs: an update', in Bendell-Young L and Gallaugher P (eds), *Waters in Peril*. Norwell, MA: Kluwer Academic, 47–56.

PURCELL J E (1997) 'Pelagic cnidarians and ctenophores as predators: selective predation, feeding rates and effects on prey populations', *Ann Inst Oceanogr Paris*, 73, 125–137.

PURCELL J E (2005) 'Climate effects on formation of jellyfish and ctenophore blooms: a review', *J Mar Biol Assoc UK*, 85, 461–476.

PURCELL J E (2012) 'Jellyfish and ctenophore blooms coincide with human proliferations and environmental perturbations', *Ann Rev Mar Sci*, 4, 209–235.

PURCELL J E, BÅMSTEDT U and BÅMSTEDT A (1999) 'Prey, feeding rates, and asexual reproduction rates of the introduced oligohaline hydrozoan *Moerisia lyonsi*', *Mar Biol*, 134, 317–325.

PURCELL J E, UYE S-I and LO W-T (2007) 'Anthropogenic causes of jellyfish blooms and their direct consequences for humans: a review', *Mar Ecol Prog Ser*, 350, 153–174.

PURCELL J E, HOOVER R A and SCHWARCK N T (2009) 'Interannual variation of strobilation of the scyphozoan *Aurelia labiata* in relation to polyp density, temperature, salinity, and light conditions', *Mar Ecol Prog Ser*, 375, 139–49.

RAJAGOPAL S, NAIR K V K and AZARIAH J (1989) 'Some observations on the problem of jelly fish ingress in a power station cooling system at Kalpakkam, east coast of

© Woodhead Publishing Limited, 2013

India', *Mahasagar Quart J Oceanogr, Natl Inst Oceanogr Goa, India*, 22, 151–158.

RASKOFF K A, SOMMER F A, HAMMER W M and CROSS K M (2003) 'Collection and culture techniques for gelatinous zooplankton', *Biol Bull*, 204, 68–80.

RODGER H D (2007) 'Gill disorders: an emerging problem for farmed Atlantic salmon (*Salmo salar*) in the marine environment?', *Fish Vet J*, 9, 38–48.

RODGER H D and MITCHELL S O (2005) Pancreas disease in Ireland: epidemiological survey results for 2003 and 2004, in Ruane *et al.* (eds), *Research on pancreas disease in Irish farmed salmon 2004/2005 – Current and future initiatives*, Marine Environment and Health Series, 22. Marine Institute, 5–26.

RODGER H D, HENRY L and MITCHELL S O (2011a) 'Non-infectious gill disorders of marine salmonid fish', *Rev Fish Biol Fisheries*, 21, 423–440.

RODGER H D, MURPHY K, MITCHELL S O and HENRY L (2011b) 'Gill disease in marine farmed Atlantic salmon at four farms in Ireland', *Vet Rec*, 168, 668.

SANDLIFER P A, SMITH T L J and CALDER D R (1974) 'Hydrozoans as pests in closed-system culture of larval decapod crustaceans, *Aquaculture*, 4, 55–59.

SOMMER F A (1992) 'Husbandry aspects of a jellyfish exhibit at the Monterey Bay Aquarium', *American Associations of Zoological Parks and Aquariums Annual Conference*. Rockville, MD: AAZ PA, 362–369.

SOMMER F A (1993) 'Jellyfish and beyond: husbandry of gelatinous zooplankton at the Monterey Bay Aquarium', *Proceedings of the Third International Aquarium Congress*, 25–29 April, Boston, MA, 249-261.

Tasmanian Aquaculture and Fisheries Institute (2003) *Research report for 2001 and 2002*. Hobart: IAFI/University of Tasmania, 74 pp.

WIDMER C L (2008) *How to Keep Jellyfish in Aquariums: An Introductory Guide for Maintaining Healthy Jellies*, Tucson, AZ: Wheatmark.

WILLCOX S, MOLTSCHANIWSKYJ N A and CRAWFORD C M (2008) 'Population dynamics of natural colonies of *Aurelia* sp. scyphistomae in Tasmania, Australia', *Mar Biol*, 154, 661–670.

WROBEL D *Captive Jellies: Keeping Jellies in an Aquarium*, The Jellies Zone, available at: http://jellieszone.com/captivejellies.htm (accessed Semtember 2012).

YASUDA T (1988) *Studies on the common jelly-fish,* Aurelia aurita *(Linné)*. Otokyo: Japanese Fisheries Resource Conservation Association.

© Woodhead Publishing Limited, 2013

14

Sea cucumber aquaculture: hatchery production, juvenile growth and industry challenges

A. Mercier, Memorial University, Canada and J.-F. Hamel, Society for the Exploration and Valuing of the Environment (SEVE), Canada

DOI: 10.1533/9780857097460.2.431

Abstract: Sea cucumber fisheries have existed for centuries, driven primarily by the Chinese markets. In recent decades, overfishing has severely depleted commercial sea cucumber populations worldwide. Consequently, a growing number of countries are becoming interested in developing aquaculture programs. Every year billions of larvae and millions of juveniles are successfully grown in aquaculture facilities and new initiatives are booming. Here we introduce briefly the main species that have so far been successfully cultivated: *Apostichopus japonicus*, *Holothuria scabra* and *Isostichopus fuscus*. We describe common techniques for the culture of larvae and juveniles and outline some of the main challenges of this industry.

Key words: sea cucumber, beche-de-mer, aquaculture, nursery, sea ranching, *Apostichopus japonicus*, *Holothuria scabra*, *Isostichopus fuscus*.

14.1 Introduction: historical background

Sea cucumbers belong to the phylum Echinodermata, class Holothuroidea, and are thus also called holothuroids. Of the approximately 1400 species that have been described so far, some 77 are harvested commercially at various scales around the world (Purcell *et al.*, in press). While the football-shaped slimy creatures are not necessarily attractive at first glance, they are nonetheless a prized commodity.

Sea cucumber fisheries have existed for centuries, driven primarily by the Chinese markets (Chen, 2003; Lovatelli *et al.*, 2004; Toral-Granda, 2008). They constitute a long-standing tradition in many countries of the tropical Indo-Pacific, where sea cucumbers are typically collected by hand, either

© Woodhead Publishing Limited, 2013

using simple tools (net, spear, hook) or snorkeling in shallow waters (Anderson *et al.*, 2010; Purcell *et al.*, in press). Diving or the use of hookah compressors is common in deeper waters to about 40 m. In some temperate regions where the densest populations occur in even deeper and colder water, sea cucumbers are trawled (Hamel and Mercier, 2008). Low technology is required to processes sea cucumbers into the main product found on the market: beche-de-mer or trepang (the eviscerated, boiled and dried body wall; Fig. 14.1a, b). Quick-frozen and fresh products (body wall, gonads and muscle bands) are available in local stores (Fig. 14.1c), whereas salted intestines and dried aquapharyngeal bulbs can be found on certain markets. Moreover, scientists and processors have developed ways to valorize the by-products of sea cucumbers through nutraceuticals (Hamel and Mercier, 2008). Several extracts concentrated in pills are demonstrated or suspected to have beneficial effects against arthritis and other pains, viral infections and some cancers. They can also be used as nutritional supplement for humans (Fig. 14.1d) and pets and in cosmetic products (Hamel and Mercier, 2008). Extracts from sea cucumbers have also been found to impair the development of the malaria parasite (Yoshida *et al.*, 2007).

Fig. 14.1 (a) Sea cucumbers processed as beche-de-mer (eviscerated, boiled and dried) in the Philippines. (b) Dry sea cucumber products (beche-de-mer) for sale in Qingdao, China. (c) Fresh sea cucumbers (*Apostichopus japonicus*) sold in Dalian, China. (d) Sea cucumber-based nutritional supplement sold in specialized boutiques in Yantai, China. Photos courtesy of J.-F. Hamel and A. Mercier.

© Woodhead Publishing Limited, 2013

Overfishing has now severely decreased the biomass of several sea cucumber populations (e.g. Conand, 2004; Toral-Granda, 2008). Catch trends of most fisheries have followed boom-and-bust patterns since the 1950s, declining nearly as quickly as they expanded (Anderson *et al.*, 2010). Since then, and especially recently with the economic boom in China, the number of consumers of sea cucumber (an expensive delicacy) has increased dramatically, applying direct pressure on this resource in a growing number of localities. Even with the entry of countries like Russia, Iceland, the USA and Canada in this market over recent decades, available sea cucumber resources are apparently not meeting the growing demand. According to Anderson *et al.* (2010), regional assessments show that sea cucumber declines from overfishing occurred in 81 % of fisheries, with collected size declining by 35 %, and that fishers moved from near- to offshore in 51 % of the fisheries and from high to low value species in 76 % of them.

In addition to the ecological importance of sea cucumbers (e.g. as bioturbators), they are of great social and economic importance to many coastal communities and the loss of this resource, especially in the island nations of the Indo-Pacific, could have catastrophic outcomes. Globally, the status of most fished populations of sea cucumber is 'not healthy and very vulnerable' (Purcell *et al.*, in press). Their late age at maturity, slow growth and low rates of recruitment [at least for some species (Hamel and Mercier, 1996)] make for sluggish replenishment of natural populations (Uthicke *et al.*, 2004; Bruckner, 2005). At low population densities, these broadcast-spawning species may face Allee effects (Stephens *et al.*, 1999; Uthicke *et al.*, 2009), resulting in population collapse and inhibiting recovery (Uthicke and Benzie, 2000; Bruckner, 2005).

For this reason, an ever growing number of countries are becoming interested in the development of aquaculture programs. Hatchery production, sea farming, sea ranching, restocking and stock enhancement are being contemplated for a larger number of species and in more locations. In the present chapter, we introduce briefly the main species for which aquaculture techniques are already well developed. We describe common techniques for the collection of broodstock, the culture of larvae and grow-out of juveniles as well as the main challenges faced by the industry, including diseases and cost-effectiveness. We also explore co-culture scenarios.

14.1.1 *Apsotichopus japonicus, Holothuria scabra* and *Isostichopus fuscus*

These three species are from three distinct parts of the world oceans and are considered the most valuable sea cucumber species on the markets today; they are also the species on which aquaculture programs have focused. *Apostichopus japonicus* is a temperate species found on rocky bottoms (sometimes on sandy or muddy substrates near the shore) along

© Woodhead Publishing Limited, 2013

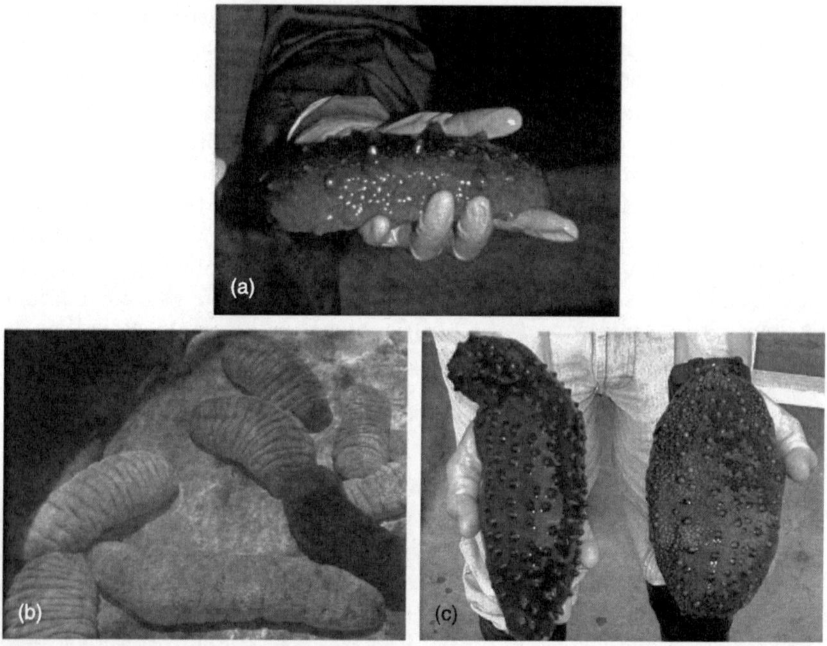

Fig. 14.2 (a) *Apostichopus japonicus* (~15 cm long). (b) *Holothuria scabra* (~25 cm long). (c) *Isostichopus fuscus* (~22 cm long). Photos courtesy of J.-F. Hamel and A. Mercier.

the coasts of southern far-east Russia, Korea, northern China and Japan down to 80–100 m depth (Fig. 14.2a). *Holothuria scabra* is a shallow-water tropical species (<20 m depth) found throughout the Indo-Pacific in sandy and muddy habitats as well as seagrass beds (Fig. 14.2b). *Isostichopus fuscus* is found on rocky substrates and coral habitats of the eastern tropical Pacific between Mexico and Ecuador, including the Galapagos Islands, down to 40 m depth (Fig. 14.2c). For a global review of the biology, harvesting and trade of these and other commercial sea cucumber species, refer to Toral-Granda *et al.* (2008).

14.2 Hatchery production

14.2.1 Broodstock collection, handling and maintenance

For *H. scabra* and *I. fuscus,* broodstock generally comes from the field and individuals are spawned within a few hours or days of their collection. To our knowledge, F1 broodstock is rare. The use of hatchery-reared individuals as broodstock is only mentioned for *A. japonicus* in China (Xiyin *et al.*, 2004) and *H. scabra* in Vietnam (Pitt and Duy, 2004). Moreover, despite attempts to develop proper diets and expose adults to fluctuating

© Woodhead Publishing Limited, 2013

environmental conditions to stimulate normal gametogenesis, inconsistent results were obtained with long-term maintenance and spawning of wild-caught individuals. Nevertheless, a growing number of farms can rely on monthly spawning of captive individuals kept in large tanks or ponds for several reproductive cycles (Agudo, 2007). Typically, fully operational hatcheries might require the spawn of hundreds of individuals every year. Hence, becoming independent of wild populations should be among the objectives of sustainable aquaculture programs in years to come.

The stocking density for broodstock is generally considered a minor concern; in several cases, they are returned to the sea after spawning (or kept in large tanks without any special care). As already mentioned, some laboratories are trying to condition broodstock for long-term use. These conditions are obviously species-specific and vary across locations, but all include parameters that promote the maintenance of high quality food and water. Adults should be maintained in large tanks or ponds with a minimum water depth of 1 m under flowthrough conditions that allow a minimum of two complete water changes daily at densities that have yet to be optimized, but should probably not be over ~1 ind. m^{-2}. Agudo (2007) refers to 15–30 individuals per 1000 L tank for *H. scabra*, with 10–15 cm of sand on the bottom. Rocks or stacks placed in the tanks have the potential to increase the surface area available to the sea cucumbers (on which they will find food) and are thus a good option, especially for *A. japonicus* and *I. fuscus*.

Numerous diets have been tested. The best results in *A. japonicus* were obtained when adults were fed 200–400 g kg^{-1} protein with 20 g kg^{-1} lipid (Seo and Lee, 2011). Various elements like soya beans, ground macrophytes, fish meals, mud, lees, yeast and bran are used in combination with mineral and vitamin supplements [developed for *A. japonicus* in China (Huiling *et al.*, 2004)]. Some facilities also use a supplement of Algamac® (Bio-Marine, Inc., Aquafauna, Hawthorne, CA, USA), or residues from moribund phytoplankton cultures. Finally, the broodstock tanks should receive high quality water, preferably unfiltered, and be located to receive some sunlight and moonlight. When reared in covered tanks, *I. fuscus* was demonstrated to interrupt its natural lunar spawning cycle (Mercier *et al.*, 2007).

14.2.2 Spawning

The natural spawning periodicity of *A. japonicus* remains poorly understood. However, it seems to spawn between March and August, with variations among locations (Ito and Kitamura, 1998; Xiyin *et al.*, 2004). Fully mature broodstock will spawn naturally in the tanks between 1900 h and 2000 h. Moreover, some laboratories in China are keeping individuals out-of-phase temperature-wise to stimulate spawning outside the normal season (Xilin, 2004). *I. fuscus* spawns mainly in the fall in Mexico and every month closer to the equator (Mercier *et al.*, 2004). Gametes are released between sunset and 2100 h, a few days after the new moon (Mercier *et al.*, 2007).

© Woodhead Publishing Limited, 2013

H. scabra was also demonstrated to be affected latitudinally, with individuals spawning almost every month in regions close to the equator and inside a narrower season closer to the tropics (Hamel *et al.*, 2001). Natural spawning occurs at sunset generally around the full moon in the Solomon Islands (Mercier *et al.*, 2000a; Hamel *et al.*, 2001) and a day after the first quarter moon around mid-day in the Philippines (Olavides *et al.*, 2011). Unfortunately, sexual dimorphism is not marked in any of the three species, preventing the identification of male and female individuals, except when gamete release is imminent and the shape of the bulging gonopore can be used to distinguish them.

Male gametes can be extracted surgically from the gonad, but female gametes (oocytes) must be obtained from natural spawning in order to be fully competent (fertilizable). For the three species, natural spawning can be observed in tanks without any stimulation following a more or less precise periodicity (see above). The most predictable breeding periodicity is exhibited by *I. fuscus* (Mercier *et al.*, 2007) and natural spawning is thus considered the best method to obtain competent oocytes in this species (Mercier *et al.*, 2004). For *A. japonicus* and *H. scabra*, inducing spawning in ripe adults is the norm. Similar techniques are used with both species, including desiccation, water-jetting and applying thermal shocks, with or without exposure to *Spirulina* algae. Results are very variable and success rate is unpredictable. Agudo (2007) indicated that *H. scabra* maintained in captivity seem to be more prompt to spawn than newly caught ones.

In order to improve the availability of fully competent oocytes, various compounds have been developed to induce their final maturation. Kishimoto and Kanatani (1980) successfully induced final oocyte maturation in *Parastichopus (= Stichopus) californicus* with disulfide-reducing agents such as dithiothreitol (DTT) and 2, 3-dimercapto-1-propanol (BAL). DTT, BAL and L-cysteine also induced ovulation in surgically collected oocytes of *H. scabra* (Rasolofonirina *et al.*, 2009), although subsequent rates of fertilization were generally low and embryonic development abnormal. A natural complex called 'nirina' has recently been found to be highly effective in inducing maturation of extracted oocytes in several aspidochirotid species from the Indian Ocean and the Mediterranean Sea including *H. scabra* (Rasolofonirina *et al.*, 2009); fertilization rates >90 % were usually obtained. However, 'nirina' was ineffective on the gonadal tubules or via injection into the coelomic cavity (Rasolofonirina *et al.*, 2009). Kato *et al.* (2009) purified a gonadotropic neural peptide (NGIWYamide) from the buccal ring nerve of *A. japonicus* that induced gamete release 60–80 min after injection into mature males and females. A synthetic derivative was ten times more potent than the natural NGIWYamide. This neuropeptide is called cubifrin. The gametes obtained through this technique were successfully fertilized, developed normally and metamorphosed into young sea cucumbers.

© Woodhead Publishing Limited, 2013

14.2.3 Fertilization

This step is often overlooked, especially in *H. scabra*. However, the ratio of gametes at fertilization can have major repercussions on this and later stages of development. In *A. japonicus* and *I. fuscus*, adults exhibiting signs of imminent spawning are commonly removed from the tank and isolated. Males and females will continue to spawn in isolated tanks. Mixing of gametes will then be carefully controlled, using 3–5 spermatozoa oocyte^{-1} in *A. japonicus* in China (Xiyin *et al.*, 2004), $5–10 \times 10^4$ spermatozoa mL^{-1} in Japan and no more than 500 spermatozoa mL^{-1} in *I. fuscus* (Mercier *et al.*, 2004). The spermatozoa used should be a mix from at least three males. When fertilization is confirmed under a microscope (through elevation of the fertilization membrane or first cleavage), generally after ~15 min, the oocytes are gently sieved to remove excess spermatozoa before transferring them into culture vessels. One of the main problems in several laboratories that cultivate *H. scabra* is that male and female adults are left to spawn together under static conditions. The water soon clouds up, indicating concentrations of millions of spermatozoa per mL, i.e. well over the natural capability of oocytes to block polyspermy (fertilization by more than one spermatozoon), which can lead to abnormal development and high mortality rates.

14.2.4 Larval culture

Larval culture is generally straightforward, although sudden crashes are commonly reported in all three species. Two main schools have developed: most laboratories use static conditions and water changes at regular intervals, whereas a smaller number of laboratories use flowthrough conditions. In both cases, the quality of the water supply is the primary criterion for success.

The developmental biology of the three species is very similar. All develop into a planktotrophic larva called the auricularia (Fig. 14.3a) that will metamorphose through a doliolaria stage (Fig. 14.3b) into a pentactula (Fig. 14.3c). The various steps of the development have been outlined for *I. fuscus* (Mercier *et al.*, 2004), *H. scabra* (Hamel *et al.*, 2001) and *A. japonicus* (Ito and Kitamura, 1998; Renbo and Yuan, 2004). Rearing the pelagic stages in the three species requires considerable attention and constant monitoring of the culture. In most hatcheries, the use of opaque conical tanks that drain through the center is the standard (Fig. 14.4a). For *H. scabra*, Agudo (2007) mentioned that fertilized oocytes or embryos need to be stocked at densities not exceeding 0.3–1 ind. mL^{-1}. This density is adjusted around 5 ind. mL^{-1} for *A. japonicus* and up to 3–5 ind. mL^{-1} for *I. fuscus.* Density-dependent mortalities have been described for *A. japonicus* by Sui (1990) who mentioned that a higher density can dramatically increase malformations and mortality. Densities should take into account the size increment of larvae and their survival rate over time.

© Woodhead Publishing Limited, 2013

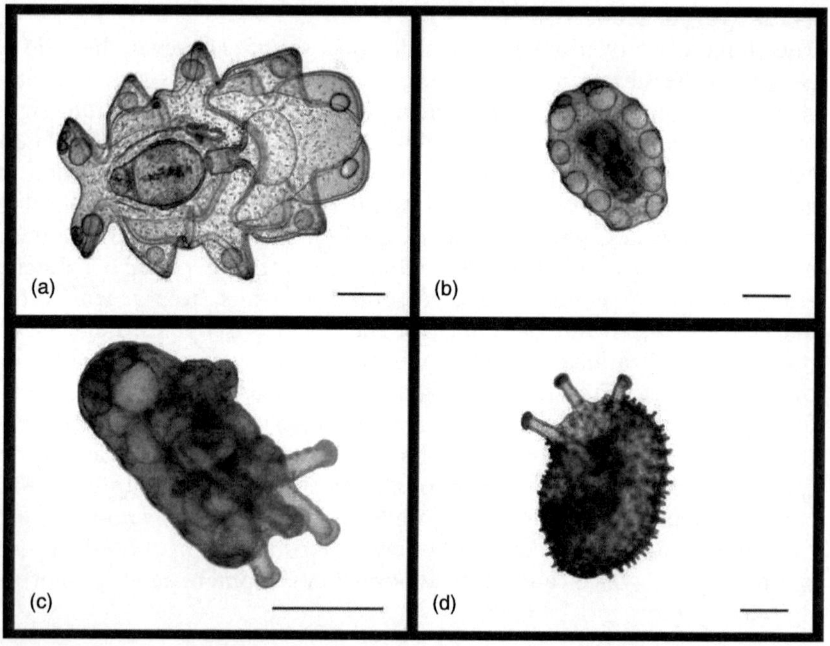

Fig. 14.3 Development of *Isostichopus fuscus*. (a) Fully developed auricularia larva. (b) Doliolaria larva. (c) Pentactula undergoing settlement. (d) Newly settled juvenile. Scale bars represent 200 μm. Photos courtesy of J.-F. Hamel and A. Mercier.

Under static conditions, 30–100 % of the water needs to be changed on a regular basis (24–48 h). The sea water used is generally filtered and UV-sterilized. Great care must be taken to avoid stressing the larvae, mechanically or chemically, during the process. Slowly siphoning out the water containing the larvae at a rate <5 L min^{-1} (Agudo, 2007) and transferring the larvae back to the culture tanks is a common technique, but it may result in significant mortalities. Good results were obtained under flowthrough with 200–300 % water renewal every day. Under both static and flowthrough conditions, siphoning the bottom to remove dead larvae, detritus and traces of bacterial proliferation (sometimes pinkish in color on the side and bottom of the tank) is imperative. Light aeration with large bubbles is recommended. Fine bubbles can clog the larvae's digestive tract and should be avoided. Under static conditions, ethylenediamenetetraacetic acid (EDTA) and antibiotics such as erythromycin are sometimes added as a preventive measure during water changes. Physicochemical parameters (pH, temperature, salinity, dissolved oxygen) should be adjusted to meet the specifics needs of each species. The presence of copepods in the larval culture appears to be a recurrent problem. They are presumed to compete for food and cause damage to the sea cucumber larvae. In China, EDTA and water changes are used to contain them (Renbo and Yuan, 2004). Ito and Katimura

© Woodhead Publishing Limited, 2013

Fig. 14.4 (a) Tanks used to rear the larvae of *Isostichopus fuscus* (Ecuador). (b) Conditioning of settlement sheets for *I. fuscus* in large tanks exposed to the sun (Mexico). (c) In some cases, post-settled juveniles are allowed to feed on the conditioned substrates (polypropylene cloth) for the first days or weeks of their growth. Juveniles of *Apostichopus japonicus* are visible as dark spots on the exposed cloth (China). (d) Undulated plastic sheets are also used to favor settlement of sea cucumber larvae (China, Ecuador, many countries of the Indo-Pacific). Photos courtesy of J.-F. Hamel and A. Mercier.

(1998) mentioned the use of trichlorfon (0.5–1.0 ppm) to eliminate the copepods.

A critical factor related to food ration and quality is the accumulation of reserves throughout the larval phase. In all species, auricularia larvae convey unicellular algae and suspended fragments of organic matter into the alimentary canal through ciliary action. They need to be fed microalgae until they reach the doliolaria (non-feeding) stage (Fig. 14.3b). The algal cultures should be of high quality, free of contamination from bacteria and copepods. In most laboratories, technicians use what is locally available rather than a diet optimized for the larvae, a practice that may lead to weaker cultures. Xilin (2004) mentioned that species like *Isochrysis* should

© Woodhead Publishing Limited, 2013

never be provided alone but rather should be used as a supplement. Moreover, Xilin (2004) indicated that the use of different species of microalgae is crucial for the development of *A. japonicus*.

The most common species of microalgae used are *Dunaliella euchlaia*, *Chaetoceros gracilis*, *C. muelleri*, *C. calcitrans*, *Phaeodactylum tricornutum*, *Rhodomonas salina* and *Tetraselmis chuii*. In *H. scabra* larvae fed with different algae, growth declined as follows: *C. cerastosporum* > *T. chuii* > *Nannochloropsis oculata* (Morgan, 1999, 2001). In similar studies by Battaglene (1999) and Battaglene *et al.* (2002), growth declined as follows: *R. salina* > *C. muelleri* > *C. calcitrans*. Agudo (2007) mentioned that a concentration between 20 000 and 40 000 cells mL^{-1} should be maintained in the culture for *H. scabra* larvae fed twice daily. Similar procedures are used in China for *A. japonicus*, with lower algal concentrations at the beginning and a progressive increase as the larvae grow, i.e. from 10 000 to 40 000 cells mL^{-1} (Xilin, 2004). However, no justification for the cell counts proposed in the literature could be found. It is likely that the larvae will not be able to consume such a large quantity of algae; many dead algal cells will therefore foul the culture vessels, enhancing bacterial proliferation, especially under static conditions. To avoid this, we believe the concentration of algal cells provided should be carefully adjusted to better fit the needs of the larvae without exceeding them. This can be done by monitoring the larvae: as long as algal cells are found in their digestive tract, adding more algae is not required. Based on algal concentrations in nature and the fact that field-captured larvae of sea cucumbers do not typically exhibit full digestive tracts, cultivated larvae can probably go without food for a few hours or days without negative impacts. Under flowthrough conditions, two systems can be used: a continuous drop-by-drop supply with a constant flow of algae or several daily meals. Again, regular monitoring of algal densities in the digestive tract is the best option to maintain water conditions and enhance survival rates.

A growing number of laboratories report success using alternative sources of food when live phytoplankton is not available. Preliminary results obtained by Hair *et al.* (2011) indicate that marine microalgae grown under laboratory conditions and concentrated for easy storage and long life, such as Shellfish Diet® and Instant Algae® (Reed Mariculture Inc., Campbell, CA, USA), could be used without compromising growth and survival in *H. scabra*. Use of these prepared concentrates as larval food would be a cheaper, easier and more efficient option for hatchery culture in several countries.

14.2.5 Larval settlement

The goal of the nursery stage is to obtain larvae competent for settlement. Most studies indicate that settlement occurs at the pentactula stage, although some teams whose work focused on aquaculture techniques reported

© Woodhead Publishing Limited, 2013

settlement at the doliolaria stage (Fig. 14.3b). This is unlikely, as the larva needs the first functional primary tentacles to settle (sometimes with one or two primary podia), at which time the larva is called a pentactula (Fig. 14.3c). Settlement occurs after 9–17 days in *A. japonicus* depending on the location (Renbo and Yuan, 2004). This period is 12–16 days in *H. scabra* (Hamel *et al.*, 2001) and 21–27 days in *I. fuscus* (Mercier *et al.*, 2004). Generally, settlement plates are added at the end of the pelagic phase when 10–20 % of the larvae reach the doliolaria stage. While some of the larvae in the three species will settle on the bottom of the tanks without specific stimulations, it was demonstrated that using conditioned surfaces increases the number of larvae that enter the final phase of metamorphosis and undergo settlement.

According to Xilin (2004) the Chinese use fine flexible polyethylene or polypropylene cloth or sheets for the settlement of *A. japonicus* (Fig. 14.4b, c). This technique is also used for *I. fuscus* in Mexico. However, some hatcheries of *A. japonicus* also use corrugated plates affixed together to form stacks (Ito and Katimura, 1998) (Fig. 14.4d). For *H. scabra*, Agudo (2007) reported the use of plates made out of various plastic and fiberglass materials. For *I. fuscus* in Ecuador technicians also use undulated Plexiglas plates.

When artificial substrates are used to promote metamorphosis and settlement, surfaces are first conditioned for several days to develop a biofilm that will attract the larvae. The protocols vary among the different laboratories and according to the species under study. Xiyin *et al.* (2004) mentioned that benthic diatoms are inoculated on the polyethylene sheets 7–10 days prior to settlement in China. Agudo (2007) reported for *H. scabra* that the plates may be immersed in diatom cultures (*Nitzschia* sp., *Navicula* sp., *Platymonas* sp.), or covered either with *Spirulina* or with filtered extracts from crushed *Sargassum* or seagrass (*Thalassia hemprichii*, *Enhalus acoroides*). For *I. fuscus* in Ecuador and Mexico, exposing the sheets or plates to sunlight in running sea water for two to three weeks is reportedly working well. Once covered with a natural biofilm the plates are rinsed thoroughly to remove any excess of deposited material and potential predators, and transferred in the tanks with the doliolaria larvae.

In experimental studies of *H. scabra*, leaves of the seagrass *T. hemprichii*, with or without their natural biofilm, were demonstrated to yield the highest settlement rates (4.8–10.5 %) (Mercier *et al.*, 2000b). The seagrass *T. hemprichii* was preferred as a settlement substratum over sand, crushed coral, several other plant species and artificial seagrass leaves, with or without a biofilm. Only settlement on another seagrass, *E. acoroides*, was similar to that recorded for *T. hemprichii* (Mercier *et al.*, 2000b). In the absence of a suitable substrate, the larvae delayed settlement for nearly 96 h and survival was <0.5 %.

After settlement, the larva turns into a miniature of the adult (Fig. 14.3d) with a functional digestive system. The biofilm available on the plates, which promoted settlement, will also provide food to the juveniles for the first

© Woodhead Publishing Limited, 2013

days of their benthic life (Fig. 14.4c, d). Adding food usually becomes necessary ~3–5 days post settlement (sometimes later). Note that the three species discussed here are mostly nocturnal, being more active at sunset and during the night and remaining mostly static during the day when exposed to natural light cycles. To promote their growth, some aquaculture facilities in China keep *A. japonica* in darkness condition 24 h a day.

14.3 Juvenile growth

14.3.1 Early growth

Post-larval development and initial juvenile growth have been well described. While methods vary according to the different species, laboratories and countries, the central goal of this step is to bring the juveniles to a size at which they can be transferred to larger enclosures for further growth (i.e. grow-out).

Time needed for juveniles to reach a size of ~10 mm is 30–40 days post fertilization in *H. scabra* (Agudo, 2007), about 47–61 days post fertilization in *I. fuscus* (Mercier *et al.*, 2004, 2012) and roughly three months post settlement in *A. japonicus* (Ito and Kitamura, 1998). This phase of the growth can be achieved in the tanks used for larval culture (Fig. 14.3a), in dedicated indoor tanks (sometimes divided into smaller units), or in outdoor tanks. Apart from conditioned settlement plates, the enclosures are generally bare. As small juveniles are very fragile, it is often easier to keep them in tanks 1–1.5 m deep that allow regular monitoring of their health and of the physical and chemical parameters of the sea water. Agudo (2007) mentions that 6–10 m^3 tanks are used for *H. scabra*. At that stage, running sea water should be provided equivalent to 200–400 % daily water change and tanks should be exposed to natural light. Stocking density can influence survival, feeding, growth and general health of the juveniles. Renbo and Yuan (2004) indicated optimal juvenile densities of 30–100 ind. m^{-2} for *A. japonicus*, similar to those used for *I. fuscus* in Ecuador and Mexico. In *H. scabra*, Pitt and Duy (2004) and Agudo (2007) reported initial densities around 500 ind. m^{-2}, that should be decreased to 100–300 ind. m^{-2} after a couple of months. In Mexico, small juveniles of *I. fuscus* (~5 mm long) are sometimes transferred directly to 'hapas' enclosures in the ocean. This method increases survival rates by decreasing the incidence of diseases.

For *A. japonicus*, Xiyin *et al.* (2004) and Ito and Katimura (1998) mentioned a salinity between 27 and 32 and a temperature of 18–26 °C. Movement and feeding activity of young sea cucumbers sharply decreases when sea water temperature exceeds 23 °C (aestivation period) (Renbo and Yuan, 2004). Aestivation in *A. japonicus* coincides with gut degeneration and activity shutdown, including cessation of feeding, a phenomenon that was demonstrated to be cued by natural seasonal increases in temperature (Yang *et al.*, 2005). Generally, significant weight loss is expected in the

© Woodhead Publishing Limited, 2013

aestivation phase, slow growth during the winter phase, and rapid growth in spring and autumn. In *H. scabra*, a temperature of 26–28 °C is recommended and a salinity >28 (Agudo, 2007); however, the species can tolerate a salinity as low as 20 for short periods of time (Hamel *et al.*, 2001). *I. fuscus* should be grown at a temperature of 24–27 °C and a salinity around 30. Temperatures below 22–23 °C will decrease growth rates significantly. Moreover, *I. fuscus* does not cope well with salinity decreases and low levels of dissolved oxygen.

Food availability and diets considerably influence growth in sea cucumbers. During the first few days, juveniles of *A. japonicus* and *I. fuscus* will generally feed on the plates provided for settlement. Afterwards, they will need to be fed fresh sea bottom mud and/or a mixed microalgae diet, while ground and filtered *Sargassum* become necessary as the juveniles develop. Some laboratories may use yeast, fishmeal, kelp powder as well as *Spirulina platensis* powder. Crushed *Sargassum* can be used at a concentration of 20–50 g m^{-3} of water daily, increasing to 50–100 g m^{-3} day^{-1} when the juveniles reach 2–5 mm in length. When tanks are dedicated to the juveniles of *H. scabra* (as opposed to settlement tanks being used), they are inoculated with benthic diatoms, equivalent to 6–7 % of the total volume. In addition, 5 g m^{-3} of metasilicate sodium and 7 g m^{-3} of fertilizer are added, under lighted conditions, to promote the multiplication of algae (Agudo, 2007). In *I. fuscus*, crushed cultures of phytoplankton as well as finely ground goat manure and Algamac® are also used. Using settlement plates regularly replaced by freshly conditioned ones, either alone or in combination with the previously described diet, reportedly gives good results.

During the growth phase, juveniles will generally need to be moved from one tank to another, or removed from and returned to their original tank, for the purpose of replenishing the food supply, cleaning the enclosure, grading the juveniles or removing injured and unhealthy individuals. At the very least, juveniles will eventually need to be moved to larger more appropriate environments: ponds, ocean enclosures or field sites. Therefore, removal techniques have been devised to minimize damage to the juveniles. Sea cucumbers possess adhesive podia and tentacles that serve in locomotion and attachment. As water level decreases in the holding tanks, most juveniles will detach passively. A gentle jet of water may also be used. Note that the body wall of *I. fuscus* is particularly sensitive (it is by far the most fragile of the three species); juveniles of this species should be manipulated with great care to avoid injuries and infections (that may lead to mortalities). They should under no circumstance be directly exposed to air (a precaution best followed for all species). Adding potassium chloride (KCl) at a concentration of 1 % before lowering the water level will favor easy detachment of the juveniles (Battaglene and Seymour, 1998).

Of the three species, only *H. scabra* will need a soft substrate (sand or mix of sand and mud) when reaching ~10–15 mm in length, as the typical feeding and burrowing behavior develops. In contrast, *A. japonicus* and

© Woodhead Publishing Limited, 2013

I. fuscus are found mainly on hard substrates and do not express any kind of burrowing behavior. Sand and mud should be added to the *H. scabra* tanks when individuals are >20 mm (Agudo, 2007). Mercier *et al.* (1999) showed that the burrowing cycle of small juveniles of *H. scabra* (>10–40 mm) was linked to the light regime. These juveniles begin to burrow around sunrise and emerge close to sunset, and their burrowing behaviour is inhibited in continuous darkness. Individuals measuring 40–140 mm responded to changes in temperature, burrowing earlier at around 0300h as temperature declined, and emerging at mid-day. Constant high temperature prevented their burrowing (Mercier *et al.*, 1999). Hence, the burrowing pattern could differ in the various locations of the distribution range (Purcell, 2010).

14.3.2 Juvenile grow-out

After reaching a size between 10 and 20 mm in length (sometimes larger), juveniles are typically transferred to larger enclosures with more or less contact with the ocean, until they reach a suitable size for harvest (Fig. 14.5). Each laboratory has its own protocol and the information provided here is for general guidance. Different methods are used that may involve cages or hapas (Fig. 14.5a), direct release into the field, ponds (Fig. 14.5c, d) or seabed enclosures (Fig. 14.5e), often a combination of two or more of these. This is a stage where high mortality rates can be observed due to environmental conditions (e.g. heavy rain, storms) and predation. A number of sea cucumber growers have tried to recycle ponds from the shrimp industry; many are now realizing that not all shrimp ponds are suitable for sea cucumbers on the basis of location (coastal vs inland), substrate, salinity fluctuations or frequency of water changes. While using existing installations might work (and be more cost-effective) under the right circumstances and in the right location, it is generally better to aim for a dedicated sea cucumber farm. Again, water quality and presence of natural predators are prime considerations in choosing suitable ponds, or the location for hapas, seabed enclosures and/or release sites. Conditions that closely match those found in a species' natural habitat should be the goal. Xilin (2004) highlighted that several hatcheries failed to ensure good water quality due to poor design and inappropriate selection of farming site. This can be a costly mistake.

In *A. japonicus*, harvestable sizes are reached in about 10–24 months in China (although it was reported to take five to six years in some locations in Japan) (Renbo and Yuan, 2004). In *H. scabra*, ~12 months are required, but survival rates remain very low to this day (Agudo, 2007). Around 18–24 months are required for *I. fuscus*, for which reported survival rates are higher (Mercier *et al.*, 2012). In most cases, the grow-out phase does not entail addition of food, especially if low densities are maintained and a proper environment/habitat is supplied.

In *A. japonicus*, ponds (2–6 ha, 1.5–2.5 m depth) usually located in the inter-tidal zone are used (Yaqing *et al.*, 2004). Stones or artificial hard

© Woodhead Publishing Limited, 2013

Fig. 14.5 (a) Juvenile sea cucumbers may be transferred to hapas (cages at sea) for growout, here shown from a site in the Philippines. (b) Offloading of large boulders along the coast of northern China to increase the suitability of the habitat for the growout of *Apostichopus japonicus.* (c) An old shrimp pond is prepared to accommodate juveniles of *Isostichopus fuscus* with the addition of concrete blocks (Mexico). (d) The same pond filled with sea water and covered with mesh to provide shade. (e) Sea bed enclosures used to grow *Holothuria scabra* (Madagascar). Photos (a, b) courtesy of J.-F. Hamel and A. Mercier; photos (c, d) courtesy R. Espinoza, Acuacultura dos mil, Mexico; photo (e) courtesy of I. Eeckhaut, Mons-Hainaut University, Belgium.

© Woodhead Publishing Limited, 2013

materials covering 50–70 % of the bottom are placed in the ponds to provide adequate substrate or shelter to the sea cucumbers. The number of juveniles released in a pond is size-dependent: 15–40 ind. m^{-2} at 2–5 cm, 15–25 ind. m^{-2} at 5–10 cm and 5–8 ind. m^{-2} at 10–15 cm (Yaqing *et al.*, 2004). However, Xilin (2004) mentioned that the best results are obtained at a density <10 ind. m^{-2}. About 10–60 % of the sea water is renewed by opening and closing the sluice gates with the change of tide. China and Japan are also growing juveniles to the harvestable size directly in the ocean, preferably in rocky sites with abundant macrophytes (sometimes rocks are added to make the habitat suitable for the sea cucumbers). These areas should be protected against strong winds and tidal action (Renbo and Yuan, 2004). A simple enhancement technique was tested in the Shandong and Liaoning Provinces (China), by adding stones into a selected field site (Fig. 14.5b): production increased up to 16 times following such practice.

In *H. scabra*, juveniles are either released in ponds (free or contained in hapas), in seabed enclosures (Fig. 14.5e), or directly in the ocean. Pascal and Robinson (2011) summarize useful criteria for choosing a site in Madagascar. For the setup of sea pens or seabed enclosures, areas prone to influxes of fresh water, strong currents or waves should be avoided. The substrate (soft, sandy-muddy sediment, which is often found close to mangroves) should be at least 40 cm deep to allow the construction of pens and should never dry out completely and have a minimum water depth of 10–50 cm during the lowest tide. Agudo (2007) talks about 150 ind. m^{-2} at the beginning of the grow-out phase with later densities of about 4 ind. m^{-2} (at 1–20 g) in New Caledonia. When the sea cucumbers become bigger, Pitt and Duy (2004) suggest 1 ind. m^{-2} or 200–240 g m^{-2} (from 30 g they will reach 300 g in about three months). In New Caledonia, individuals of 1 g stocked at 1.4 ind. m^{-2} had an average growth rate of 0.8 g day^{-1}. The enclosures used varied from 500 to 2000 m^{-2}. According to Pascal and Robinson (2011), farmers in Madagascan villages buy batches of juveniles from a hatchery at an average size of 15 g and stock them at 0.5 ind. m^{-2}. The grow-out period for 15 g juveniles to reach market size (400 g) is eight months, at a site with a carrying capacity of 600 g m^{-2} (Pascal and Robinson, 2011). Several types of seabed enclosure with a mesh screen <1 mm were used in water 1.5–2.5 m deep on substrates composed of mud and sand or coral rubbles (and sometimes seagrass beds). Irrespective of the type of enclosure, unwanted fouling organisms (algae, sponges, etc.) and harmful organisms (e.g. crabs) should be removed on a regular basis. Lavitra *et al.* (2009) mentioned that crabs *Thalamita crenata* are the fiercest predators of juvenile sea cucumbers in Madagascar, sometimes decimating the entire stock in a month. However, adults >250 g are not affected. Also, isopods *Cymodoce* sp. were seen to infest *H. scabra* in outdoor ponds during the warm season, giving rise to high mortality rates (Lavitra *et al.*, 2009). Common problems highlighted in pond systems include the presence of crabs that may injure the sea cucumbers (Pitt and Duy, 2004), and decreasing water salinity and/

© Woodhead Publishing Limited, 2013

or stratification of the water column that lowers dissolved oxygen levels (Agudo, 2007).

In Mexico, the use of an experimental shrimp pond was tested for the grow-out of hatchery-reared juveniles of *I. fuscus* (starting with juveniles of 3–5 mm, 4–7 mg, two to three months old) (Fig. 14.5c, d). During the monitoring phase, the animals grew from an average of 1.07 g to 42.07 g in three months, but survival was low due to outbreaks of a skin disease. Using mesh cages at sea (1.8 m × 1.8 m × 1.8 m) stocked with 3000 juveniles resulted in a more conservative growth rate (from 2.66 g to 26.92 g in three months), but greater survival (40–90 %). The presence of sponges and other fouling organisms on the mesh (500–1000 μm) and clogging caused by sedimentation might have hampered the flow of fresh deposits serving as food to juveniles. Presence of crabs in the cages was also noted, which might have caused stress (slowing growth) and possibly mortalities. Juvenile *I. fuscus* were also successfully reared in shrimp ponds in Ecuador. A preliminary experiment was conducted to determine if small sea cucumbers (~100–150 g) collected from the wild would grow in ponds from different locations. Enclosures of 1 m^2 were used to facilitate recapture. These sea cucumbers grew an average of 17 g $week^{-1}$ with 98 % survival, suggesting that shrimp ponds along the coast can provide a good environment to grow *I. fuscus* juveniles to adult size in a reasonable time period. However, juveniles grown in shrimp ponds in Ecuador may develop skin diseases leading to mass mortalities, especially during months with warmer temperatures and heavy rains (Mercier *et al.*, 2012).

14.4 Co-culture

Co-culture of sea cucumber with various other species has been investigated in a number of settings to improve the cost-effectiveness of sea cucumber farming, to take advantage of the waste generated by another culture (e.g. feces of molluscs recycled by sea cucumber as food), or to combine cost-effectiveness and environmental relief.

Co-culture of *A. japonicus* with shrimps, molluscs, finfishes and algae (e.g. *Gracilaria lemaneiformis*) has been successful in China. The most common polyculture is with shrimps (e.g. *Penaeus japonicus*) (Xilin, 2004; Yaqing *et al.*, 2004) and other less common combinations include the abalone, *Haliotis discus hannai* (Kang *et al.*, 2003), and bivalves (Zhou *et al.*, 2006), the latter study showing that the sea cucumbers can feed on bivalve biodeposits.

A variety of experiments have also examined the feasibility of rearing *H. scabra* within polycultures. Most projects were developed in Indonesia, where wild-caught and transplanted juvenile *H. scabra* positively grew with the seaweed, *Eucheurna cottonii* (Madeali *et al.*, 1993), *Gracilaria* sp. or *Eucheuma* sp. (Rachmansyah *et al.*, 1992; Tangko *et al.*, 1993). The co-culture

© Woodhead Publishing Limited, 2013

of juvenile *H. scabra* with juveniles of the blue shrimp, *Litopenaeus stylirostris*, was tested and showed interesting but mitigated results, highlighting the need for further research to validate the feasibility of this co-culture (Purcell *et al.*, 2006). Pitt and Duy (2004) mentioned the co-culture of Babylon snails, *Babylonia areolata*, with *H. scabra* in sea pens and cages in Vietnam.

To our knowledge, co-culture involving *I. fuscus* has never been attempted.

14.5 Diseases

Diseases and other problems related to embryonic, larval and juvenile development are common features of sea cucumber aquaculture, which tend to increase with the scale of the cultures. The list of potential diseases is too long to be reviewed here, hence only the major threats will be highlighted. In some cases these diseases are so severe and virulent that larvae or juveniles can be largely or totally decimated within a few days. The main goal is therefore to avoid the initial appearance of the problem. Yin-Geng *et al.* (2004) mentioned that intensification of sea cucumber farming in China has led to the occurrence of various diseases, causing serious economic losses and becoming one of the limiting factors in the sustainable development of this industry.

A. japonicus is by far the most extensively cultivated species with billions of larvae produced annually. It is also the one for which the most diseases and problems have been detected. As the accumulation of excess food and feces increases, harmful micro-organisms tend to multiply, causing serious diseases outbreak among larvae and juveniles. At the larval stage, symptoms of 'rotting edges', stomach ulceration (auricularia stage) and gas bubble diseases are the most common. Diseases of juveniles include skin ulceration (at the aestivation stage), body wall ulceration, fungal disease and body edema. They may be triggered by various pathogens including bacteria, fungi and parasites (Yin-Geng *et al.*, 2004). The best way to avoid or mitigate these problems is prevention, careful monitoring to remove the first affected individuals and/or use of a treatment when available, which is generally not the case (Yin-Geng *et al.*, 2004). Eeckhaut *et al.* (2004) mentioned that the most pathogenic agents reported in cultures of *H. scabra* are bacteria causing skin ulceration. Morgan (2001) observed that *H. scabra* broodstock became diseased after a few weeks in captivity. The first visual signs were pigmentation loss and presence of copious amounts of mucus, coinciding with a period of weight loss. Although the most common bacterium present on diseased individuals was *Vibrio harveyi*, there were also low numbers of motile Gram-negative rod bacteria. Surface damage to the epidermis was not a prerequisite for infection, but infection progressed more rapidly from an existing wound. Most infected animals died within three to seven days

© Woodhead Publishing Limited, 2013

(Morgan *et al.*, 2001). More recently, Becker *et al.* (2004) showed that the skin ulceration disease (or closely related forms) occurs worldwide: it has been recorded in *A. japonicus* in China, *I. fuscus* in Ecuador and *H. scabra* in Australia and New Caledonia. The highly contagious skin ulceration begins with the appearance of a white lesion that extends over the whole body surface. It destroys the cuticle and the epidermis, then the connective tissue and ossicles (Becker *et al.*, 2004). The disease is due to a bacterial infection, but the agent that triggers the skin ulceration has not been identified yet: biomolecular techniques identified three species of bacteria in the lesions: *Vibrio* sp., *Bacteroides* sp. and a-Proteobacterium (Becker *et al.*, 2004).

The most common problem observed during the culture of *I. fuscus* is the development of a disease in the digestive system of early larvae (Becker *et al.*, 2009). The first stage is the appearance of opaque cells around the digestive tract, followed by the contraction of the intestine and stomach. In the worst cases, the digestive tract completely shrivels up and disappears. When it becomes visible, the condition is usually fatal to the larvae. Upon close examination of the affected larvae under the microscope, the disease was determined to be caused by protozoan parasites. During the first stage of the disease, the parasites can be seen entering through the body wall and the digestive tract, probably inducing the observed contraction. Later on, the parasites become larger and are present everywhere around the intestine, both inside and outside. The parasites appear to feed on the intestinal contents or tissues, slowly making it shrink and disappear, typically causing the death of the larvae (Becker *et al.*, 2009). Another common affliction occurs in shrimp ponds in Ecuador and Mexico, where juveniles can exhibit a disease of the body wall (not yet characterised). This condition may degenerate and cause the animals to eviscerate and eventually die.

14.6 Conclusions and future trends

The present review outlined major breakthroughs in the aquaculture of sea cucumbers but also highlighted common problems. Particularly evident is the poor understanding of the diseases and parasites that affect the various life stages. For the species that have been successfully reared to the juvenile stage, cost-effectiveness is not always achieved, especially in *H. scabra* where grow-out techniques and global survival rates need to be improved. Problems related to the quality of food and water are common due to variable and often poor environmental conditions along the coast. They are a concern in many aquaculture facilities, affecting both the sea cucumbers and the microalgae cultures serving as larval food. While recycling certain shrimp infrastructures has potential and co-culture of sea cucumbers with other commercial species is being investigated, the former option is not always suited and the latter option needs to be researched more thoroughly

© Woodhead Publishing Limited, 2013

in *I. fuscus* and *H. scabra*. The putative environmental effects of commercial-sized aquaculture facilities have not received much attention.

In countries where production is well mastered, restocking and stock enhancement are now being contemplated with the aims of sustaining active fisheries, restoring depleted populations or increasing yields in certain locations (Bell and Nash 2004). A form of stock enhancement (bordering on sea ranching) is on-going along the coast of China where appropriate habitats are sometimes modified to fit the needs of increased numbers of *A. japonicus* being grown in the ocean (Bell *et al.*, 2008). There is interest in restocking *H. scabra* populations in the Indo-Pacific, but it remains at the experimental stage (Purcell *et al.*, 2002; Purcell 2004; Uthicke and Purcell 2004; see also Loneragan and Abraham, 2011). As far as we know, no release studies have been carried out with *I. fuscus* to date. Putative problems associated with the use of captive-bred sea cucumbers to replenish or build up wild stocks include the cost of production and low survival rates, making for a potentially costly management tool. The specter of genetic diversity loss has also been evoked together with the unknown outcomes of interaction between wild and introduced stocks. Similar preoccupations relate to the impact of escapees (at the larval or juvenile stages) on the genetic profile and diversity of sea cucumbers over small and large geographical scales.

14.7 Acknowledgements

We would like to thank the many scientists and technicians from around the world who work on sea cucumber aquaculture for the many publications and discussions that made this review possible. We also thank I. Eeckhaut and R. Espinosa for sharing photographs of their installations with us.

14.8 References

AGUDO N (2007) *Sandfish hatchery techniques.* The WorldFish Center, Secretariat of the Pacific Community and Australian Centre for International Agricultural Research, Nouméa.

ANDERSON S C, FLEMMING J M, WATSON R and LOTZE H K (2010) Serial exploitation of global sea cucumber fisheries. *Fish and Fisheries*, 12, 317–339.

BATTAGLENE S C (1999) Culture of tropical sea cucumbers for stock restoration and enhancement. *Naga*, 22, 4–11.

BATTAGLENE S C and SEYMOUR J E (1998) Detachment and grading of the tropical sea cucumber sandfish, *Holothuria scabra*, juveniles from settlement substrates. *Aquaculture*, 159, 263–274.

BATTAGLENE S C, SEYMOUR J E, RAMOFAFIA C and LANE I (2002) Spawning induction of three tropical sea cucumbers, *Holothuria scabra*, *H. fuscogilva* and *Actinopyga mauritiana*. *Aquaculture*, 207, 29–47.

© Woodhead Publishing Limited, 2013

BECKER P, GILLAN D, LANTERBECQ D, JANGOUX M, RASOLOFONIRINA R, RAKOTOVAO J and EECKHAUT I (2004) The skin ulceration disease in cultivated juveniles of *Holothuria scabra* (Holothuroidea, Echinodermata). *Aquaculture*, 242, 13–30.

BECKER P, EECKHAUT I, YCAZA R H, MERCIER A and HAMEL J-F (2009) Protozoan disease in larval culture of the edible sea cucumber *Isostichopus fuscus*, in Harris L G, Bottger S A, Walker C W and Lesser M P (eds), *Echinoderms: Durham*. London: CRC Press, 571–573.

BELL J and NASH W (2004) When should restocking and stock enhancement be used to manage sea cucumber fisheries? in Lovatelli A, Conand C, Purcell S, Uthicke S, Hamel J-F and Mercier A (eds), *Advances in Sea Cucumber Aquaculture and Management,* Fisheries Technical Paper No. 463. Rome: FAO, pp. 173–179.

BELL J D, LEBER K M, BLANKENSHIP H L, LONERAGAN N R and MASUDA R (2008) A new era for restocking, stock enhancement and sea ranching of coastal fisheries resources. *Reviews in Fisheries Science*, 16, 1–9.

BRUCKNER A W (2005) The recent status of sea cucumber fisheries in the continental United States of America. *SPC Beche-de-mer Information Bulletin*, 22, 39–46.

CHEN J (2003) Overview of sea cucumber farming and sea ranching practices in China. *SPC Beche-de-mer Information Bulletin*, 18, 18–23.

CONAND C (2004) Present status of world sea cucumber resources and utilisation: An international overview, in Lovatelli A, Conand C, Purcell S, Uthicke S, Hamel J-F and Mercier A (eds), *Advances in Sea Cucumber Aquaculture and Management*, Fisheries Technical Paper. No. 463. Rome: FAO, 13–23.

EECKHAUT I, PARMENTIER E, BECKER P, SILVA S G D and JANGOUX M (2004) Parasites and biotic diseases in field and cultivated sea cucumbers, in Lovatelli A, Conand C, Purcell S, Uthicke S, Hamel J-F and Mercier A (eds), *Advances in Sea Cucumber Aquaculture and Management*, Fisheries Technical Paper No. 463. Rome: FAO, 311–326.

HAIR C, KAURE T, SOUTHGATE P and PICKERING T (2011) Potential breakthrough in hatchery culture of sandfish *Holothuria scabra* by using algal concentrate as food. *SPC Beche-de-mer Information Bulletin*, 31, 60–61.

HAMEL J-F and MERCIER A (1996) Early development, settlement, growth, and spatial distribution of the sea cucumber *Cucumaria frondosa* (Echinodermata: Holothuroidea). *Canadian Journal of Fisheries and Aquatic Sciences*, 53, 253–271.

HAMEL J-F and MERCIER A (2008) Population status, fisheries and trade of sea cucumbers in temperate areas of the northern hemisphere, in Toral-Granda V, Lovatelli A and Vasconcellos M (eds), *Sea Cucumbers. A Global Review of Fisheries and Trade*. Fisheries and Aquaculture Technical Paper No. 516. Rome: FAO, 257–292.

HAMEL J-F, CONAND C, PAWSON D L and MERCIER A (2001) The sea cucumber *Holothuria scabra* (Holothuroidea: Echinodermata): its biology and exploitation as beche-de-mer. *Advances in Marine Biology*, 41, 129–223.

HUILING S, MENGQING L, JINGPING Y and BIJUAN C (2004) Nutrient requirements and growth of the sea cucumber, *Apostichopus japonicus*, in Lovatelli A, Conand C, Purcell S, Uthicke S, Hamel J-F and Mercier A (eds), *Advances in Sea Cucumber Aquaculture and Management*, Fisheries Technical Paper No. 463. Rome: FAO, 327–332.

ITO S and KITAMURA H (1998) Technical development in seed production of the Japanese sea cucumber, *Stichopus japonicus*. *SPC Beche-de-mer Information Bulletin*, 10, 24–28.

KANG K H, KWON J Y and KIM Y M (2003) A beneficial coculture: charm abalone *Haliotis discus hannai* and sea cucumber *Stichopus japonicus*. *Aquaculture*, 216, 87–93.

KATO S, TSURUMARU S, TAGA M, YAMANE T, SHIBATA Y, OHNO K, FUJIWARA A, YAMANO K and YOSHIKUNI M (2009) Neuronal peptides induce oocyte maturation and gamete

© Woodhead Publishing Limited, 2013

spawning of sea cucumber, *Apostichopus japonicus*. *Developmental Biology*, 326, 169–176.

KISHIMOTO T and KANATANI H (1980) Induction of oocyte maturation by disulfide-reducing agent in the sea cucumber, *Stichopus japonicus*. *Development Growth & Differentiation*, 22, 163–167.

LAVITRA T, RASOLOFONIRINA R, JANGOUX M and EECKHAUT I (2009) Problems related to the farming of *Holothuria scabra* (Jaeger, 1833). *SPC Beche-de-mer Information Bulletin*, 29, 20–30.

LONERAGAN N and ABRAHAM I (2011) *4th International Symposium on Stock Enhancement and Sea Ranching – Book of abstracts*. Perth: Murdoch University.

LOVATELLI A, CONAND C, PURCELL S, UTHICKE S, HAMEL J-F and MERCIER A (2004) *Advances in Sea Cucumber Aquaculture and Management*, Fisheries Technical Paper No. 463. Rome: FAO.

MADEALI M I, TANGKO A M, PANTAI D E R and MAROS B P P B (1993) Polyculture of sea cucumber, *Holothuria scabra* and seaweed, *Eucheuma cottoni* in Battoa waters, Polmas Regency, South Sulawesi. *Prosiding Seminar Hasil Penelitian*, 11, 105–109.

MERCIER A, BATTAGLENE S C and HAMEL J-F (1999) Daily burrowing cycle and feeding activity of juvenile sea cucumbers *Holothuria scabra* in response to environmental factors. *Journal of Experimental Marine Biology and Ecology*, 239, 125–156.

MERCIER A, BATTAGLENE S C and HAMEL J-F (2000a) Periodic movement, recruitment and size-related distribution of the sea cucumber *Holothuria scabra* in Solomon Islands. *Hydrobiologia*, 440, 81–100.

MERCIER A, BATTAGLENE S C and HAMEL J-F (2000b) Settlement preferences and early migration of the tropical sea cucumber *Holothuria scabra*. *Journal of Experimental Marine Biology and Ecology*, 249, 89–110.

MERCIER A, YCAZA R H and HAMEL J-F (2004) Aquaculture of the Galapagos sea cucumber, *Isostichopus fuscus*, in Lovatelli A, Conand C, Purcell S, Uthicke S, Hamel J-F and Mercier A (eds), *Advances in Sea Cucumber Aquaculture and Management*, Fisheries Technical Paper No. 463. Rome: FAO, 347–358.

MERCIER A, YCAZA R H and HAMEL J-F (2007) Long-term study of gamete release in a broadcast-spawning holothurian: predictable lunar and diel periodicities. *Marine Ecology Progress Series*, 329, 179–189.

MERCIER A, YCAZA R H, ESPINOZA R, HARO V M A and HAMEL J-F (2012) Hatchery experience and useful lessons from *Isostichopus fuscus* in Ecuador and Mexico. *ACIAR-SPC Asia-Pacific tropical sea cucumber aquaculture symposium*, Vol. 136, 79–90.

MORGAN A D (1999) *Husbandry and spawning of the sea cucumber* Holothuria scabra *(Echinodermata: Holothuroidea)*, MSc thesis, University of Queensland.

MORGAN A D (2001) The effect of food availability on early growth, development and survival of the sea cucumber *Holothuria scabra* (Echinodermata: Holothuroidea). *SPC Beche-de-mer Information Bulletin*, 14, 6–12.

MORGAN A D, SCHOPPE S, SAMYN Y, LAMBETH L, LEVIN V S, GUDIMOVA E N and BATTAGLENE S (2001) Aspects of sea cucumber broodstock management (Echinodermata: Holothuroidea). *SPC Beche-de-mer Information Bulletin*, 13, 2–8.

OLAVIDES R D D, RODRIGUEZ B D R and JUINIO-MEÑEZ M A (2011) Simultaneous mass spawning of *Holothuria scabra* in sea ranching sites in Bolinao and Anda municipalities, Philippines. *SPC Beche-de-mer Information Bulletin*, 31, 23–24.

PASCAL B and ROBINSON G (2011) *Handbook for sandfish farming*. Mauritius: Regional Coastal Management Programme of the Indian Ocean Countries (ReCoMaP).

PITT R and DUY N D Q (2004) Breeding and rearing of the sea cucumber *Holothuria scabra* in VietNam, in Lovatelli A, Conand C, Purcell S, Uthicke S, Hamel J-F and Mercier A (eds), *Advances in Sea Cucumber Aquaculture and Management*, Fisheries Technical Paper No. 463. Rome: FAO, 333–346.

© Woodhead Publishing Limited, 2013

PURCELL S W (2004) Criteria for release strategies and evaluating the restocking of sea cucumbers, in Lovatelli A, Conand C, Purcell S, Uthicke S, Hamel J-F and Mercier A (eds), *Advances in Sea Cucumber Aquaculture and Management*, Fisheries Technical Paper No. 463. Rome: FAO, 181–191.

PURCELL S W (2010) Diel burying by the tropical sea cucumber *Holothuria scabra*: effects of environmental stimuli, handling and ontogeny. *Marine Biology*, 157, 663–671.

PURCELL S, GARDNER D and BELL J (2002) Developing optimal strategies for restocking sandfish: a collaborative project in New Caledonia. *SPC Beche-de-mer Information Bulletin*, 16, 2–4.

PURCELL S W, PATROIS J and FRAISSE N (2006) Experimental evaluation of co culture of juvenile sea cucumbers, *Holothuria scabra* (Jaeger), with juvenile blue shrimp, *Litopenaeus stylirostris* (Stimpson). *Aquaculture Research*, 37, 515–522.

PURCELL S W, MERCIER A, CONAND C, HAMEL J-F, TORAL-GRANDA V, LOVATELLI A and UTHICKE S (in press) Sea cucumber fisheries: global review of stock status, management measures and drivers of overfishing. *Fish and Fisheries*, DOI: 10.1111/j.1467-2979.2011.00443.x.

RACHMANSYAH MADEALI M I, TANGKO A M, TONNEK S and ISMAIL D A (1992) Polyculture of sea cucumber, *Holothuria scabra* and seaweed, *Eucheuma* sp. in pen culture at Parepare Bay, South Sulawesi. *Journal Penelitian Budidaya Pantai*, 8, 63–70.

RASOLOFONIRINA R, LÉONET A, JANGOUX M and EECKHAUT I (2009) A new method to induce oocyte maturation in holothuroids (Echinodermata). *Invertebrate Reproduction and Development*, 53, 13–21.

RENBO W and YUAN C (2004) Breeding and culture of the sea cucumber, *Apostichopus japonicus*, Liao, in Lovatelli A, Conand C, Purcell S, Uthicke S, Hamel J-F and Mercier A (eds), *Advances in Sea Cucumber Aquaculture and Management*, Fisheries Technical Paper No. 463. Rome: FAO, 277–286.

SEO J-Y and LEE S-M (2011) Optimum dietary protein and lipid levels for growth of juvenile sea cucumber *Apostichopus japonicus*. *Aquaculture Nutrition*, 17, 56–61.

STEPHENS P A, SUTHERLAND W J and FRECKLETON R P (1999) What Is the Allee Effect? *Oikos*, 87, 185–190.

SUI X (1990) Seed production and cultivation of sea cucumber. *Agriculture Press of China*, 107–153.

TANGKO A M, RACHMANSYAH MADEALI M I, TONNEK S and ISMAIL A (1993) Polyculture of sea cucumber, *Holothuria scabra* and seaweed, *Eucheuma* sp. in Sanisani Bay waters, Kolaka Regency, Southeast Sulawesi. *Prosiding Seminar Hasil Penelitian*, 11, 85–89.

TORAL-GRANDA V (2008) Population status, fisheries and trade of sea cucumbers in Latin America and the Caribbean, in Toral-Granda V, Lovatelli A and Vasconcellos M (eds), *Sea Cucumbers. A Global Review of Fisheries and Trade,* Fisheries and Aquaculture Technical Paper No. 516. Rome: FAO, 211–229.

TORAL-GRANDA V, LOVATELLI A and VASCONCELLOS M (2008) *Sea Cucumbers. A Global Review of Fisheries and Trade*, Fisheries and Aquaculture Technical Paper No. 516. Rome: FAO.

UTHICKE S and BENZIE J A H (2000) Allozyme electrophoresis indicates high gene flow between populations of *Holothuria* (*Microthele*) *nobilis* (Holothuroidea: Aspidochirotida) on the Great Barrier Reef. *Marine Biology*, 137, 819–825.

UTHICKE S and PURCELL S (2004) Preservation of genetic diversity in restocking of the sea cucumber *Holothuria scabra* investigated by allozyme electrophoresis. *Canadian Journal of Fisheries and Aquatic Sciences*, 61, 519–528.

UTHICKE S, WELCH D and BENZIE J A H (2004) Slow growth and lack of recovery in overfished holothurians on the Great Barrier Reef: evidence from DNA fingerprints and repeated large-scale surveys. *Conservation Biology*, 18, 1395–1404.

© Woodhead Publishing Limited, 2013

UTHICKE S, SCHAFFELKE B and BYRNE M (2009) A boom-bust phylum? Ecological and evolutionary consequences of density variations in echinoderms. *Ecological Monographs,* 79, 3–24.

XILIN S (2004) The progress and prospects of studies on artificial propagation and culture of the sea cucumber, *Apostichopus japonicus*, in Lovatelli A, Conand C, Purcell S, Uthicke S, Hamel J-F and Mercier A (eds), *Advances in Sea Cucumber Aquaculture and Management*, Fisheries Technical Paper No. 463. Rome: FAO, 273–276.

XIYIN L, GUANGHUI Z, QIANG Z, LIANG W and BENXUE G (2004) Studies on hatchery techniques of the sea cucumber, *Apostichopus japonicus*, in Lovatelli A, Conand C, Purcell S, Uthicke S, Hamel J-F and Mercier A (eds), *Advances in Sea Cucumber Aquaculture and Management*, Fisheries Technical Paper No. 463. Rome: FAO, 287–295.

YANG H, YUAN X, ZHOU Y, MAO Y, ZHANG T and LIU Y (2005) Effects of body size and water temperature on food consumption and growth in the sea cucumber *Apostichopus japonicus* (Selenka) with special reference to aestivation. *Aquaculture Research*, 36, 1085–1092.

YAQING C, CHANGQING Y and SONGXIN (2004) Pond culture of sea cucumbers, *Apostichopus japonicus*, in Dalian, in Lovatelli A, Conand C, Purcell S, Uthicke S, Hamel J-F and Mercier A (eds), *Advances in Sea Cucumber Aquaculture and Management*, Fisheries Technical Paper No. 463. Rome: FAO, 269–272.

YIN-GENG W, CHUN-YUN Z, XIAO-JUN R, JIE-JUN C and CHENG-YIN S (2004) Diseases of cultured sea cucumber, *Apostichopus japonicus*, in China, in Lovatelli A, Conand C, Purcell S, Uthicke S, Hamel J-F and Mercier A (eds), *Advances in Sea Cucumber Aquaculture and Management*, Fisheries Technical Paper No. 463. Rome: FAO, 297–310.

YOSHIDA S, SHIMADA Y, KONDOH D, KOUZUMA Y, GHOSH A K, JACOBS-LORENA M and SINDEN R E (2007) Hemolytic C-type lectin CEL-III from sea cucumber expressed in transgenic mosquitoes impairs malaria parasite development. *PLoS Pathogens*, 3, 1962–1970.

ZHOU Y, YANG H, LIU S, YUAN X, MAO Y, LIU Y, XU X and ZHANG F (2006) Feeding and growth on bivalve biodeposits by the deposit feeder *Stichopus japonicus* Selenka (Echinodermata: Holothuroidea) co-cultured in lantern nets. *Aquaculture*, 256, 510–520.

© Woodhead Publishing Limited, 2013

Part III

Closing the life-cycle and overcoming challenges in hatchery production for selected fish species

15

Closed-cycle hatchery production of tuna

G. J. Partridge, Australian Centre for Applied Aquaculture Research, Australia

DOI: 10.1533/9780857097460.3.457

Abstract: The reliance on wild-caught juvenile tuna is seen as the Achilles heel of tuna ranching industries worldwide and it is widely recognised that closed-cycle hatchery production is essential to sustain the demand for tuna and reduce pressure on wild stocks. The many features that make tuna such a unique group of fish also give rise to the many challenges associated with their culture and despite some forty years of effort, commercial scale hatchery production has yet to be fully realised. This chapter summarises the history of closed-cycle tuna production and discusses the recent and significant achievements that have been made towards the achievement of this elusive goal.

Key words: tuna, broodstock, larvae, Thunnini, nutrition, disease, maturation, hormone induction, spawning, nursery production, hatchery environment.

15.1 Introduction

Members of the Scombrid family and the Thunnini tribe, tuna are found throughout the world's oceans from the sub-polar regions to the tropics. Regarded by many as 'the king of fish', tuna are the ultimate apex predator with a biological sophistication far exceeding that of most teleost species. They are 'fast, powerful, streamlined, and equipped with specializations that enable them to perform their duties better than any other fish in the ocean' (Ellis, 2008b). Their biological sophistication includes being endothermic and having very high metabolic rates and cardiac outputs, ability for trans-oceanic migrations and a homing nature (Block and Stevens, 2001).

There are 17 tuna species within the Thunnini tribe, seven of which form the basis of the world's most valuable fishery totalling approximately 6–9 million tonnes worth approximately US$ 6 billion (Mylonas *et al.*, 2007). The contribution of the various species within this total catch varies greatly

© Woodhead Publishing Limited, 2013

in terms of both volume and value. Skipjack and yellowfin tuna, for example, contribute over 80 % of the total tuna catch by weight, but are low in value. The three species of bluefin tuna on the other hand, represent less than 3 % of the total tuna catch but are at the other end of the price spectrum. Once considered inedible and fit only for pet food, the popularity of bluefin has grown exponentially over the last 50 years to become the most sought after fish in the world. The practice of eating raw tuna became commonplace in Japan only after the Second World War and the increasing demand for tuna since that time is exemplified by the almost 10-fold increase in catch since the 1950s (Miyake *et al.*, 2004). Today sashimi and sushi are enjoyed the world over, however, the Japanese continue to dominate the market, particularly for bluefin tuna, where over 75 % of the world bluefin catch is consumed. Although bluefin is the tuna of choice for sashimi, the popularity of yellowfin and bigeye is increasing, particularly in the face of declining bluefin stocks and in countries other than Japan. The record price paid for a single bluefin tuna was set at the Tsukiji Market in Tokyo early in 2011 at ¥32.5 million ($396 000); considerably higher than the previous record of ¥16.3 million set in 2010. The major incentive for farming bluefin tuna is obvious, but tuna have many other characteristics that make them good aquaculture candidates, including a rapid growth rate, high fillet recovery, a wide temperature tolerance and an apparent hardiness against diseases and parasites (Anon., 2007). There are currently six tuna species under investigation for hatchery production (Table 15.1)

Table 15.1 Maximum size and size and age at maturation of the six tuna species currently under investigation for hatchery production

Species	Common name	Maximum size (kg)	Size at maturity (kg)	Age at maturity (years)	References
Thunnus orientalis	Pacific bluefin	600	60	3–5	Masuma *et al.* (2008)
Thunnus thynnus	Atlantic bluefin	680	20	3–5[a]	Rooker *et al.* (2007)
			135	8–12[b]	Rooker *et al.* (2007)
Thunnus maccoyi	Southern bluefin	160	100	10–14	Schaefer (2001)
Thunnus obesus	Bigeye	200	35	3	Calkins (1980) in Schaefer (2001)
Thunnus albacares	Yellowfin	175	20	2	Margulies *et al.* (2007b)
Thunnus atlanticus	Blackfin	20	2	2	Vieira *et al.* (2005)

[a] Eastern stocks; [b] Western stocks.

© Woodhead Publishing Limited, 2013

Despite these many positive aquaculture characteristics, closed cycle tuna aquaculture is not yet a commercial reality and the global demand for tuna is currently satisfied predominantly from wild caught fish, either directly or via ranching. Ranching typically involves the capture and transfer of adult or sub-adult fish into pens where they are fed over a short period, primarily to increase their percentage body fat and subsequently their value, but also to realise some overall weight gain and to regulate the supply of fish to the market (Mylonas *et al.*, 2010). The exceptions to this generalisation are in Japan and Croatia where much smaller fish are captured and reared for longer periods. This technique is generally termed farming, rather than ranching (Mylonas *et al.*, 2010). In Japan yearling tuna as small as 300 g are caught and transferred to pens for growout to market size of 30–70 kg in three to four years (Ikeda, 2002) and approximately 300 000–400 000 juveniles are caught annually in Japan for this purpose (Normile, 2009). In Croatia, 8–20 kg sub-adults are caught and grown to a market size of 30–50 kg over approximately 2.5 years (Mladineo and Miletic, 2008, Mylonas *et al.*, 2010).

Although dating back to Canada in the late 1960s (Miyake *et al.*, 2003), modern tuna ranching has its origins in South Australia where, in the early 1990s, new techniques were developed for ranching southern bluefin tuna. These techniques transformed southern bluefin tuna from a low value fish sold to canneries to the heights of bluefin extravagance. Not surprisingly, many other countries quickly followed the Australians' lead and ranching of bluefin soon began in almost every country where bluefin are found, including most countries which border the Mediterranean Sea, Mexico and Korea.

A consequence of their very high value is that the stocks of most bluefin tuna are now under serious threat. Worldwide, regional fisheries management organisations (RFMO) are implementing reductions in total allowable catches to protect threatened stocks. In 2006, the organisation responsible for the conservation of tuna in the Atlantic Ocean, ICCAT (the International Commission for the Conservation of Atlantic Tunas), decreased the total allowable catch (TAC) of the eastern stock of Atlantic bluefin from 32 000 tonnes to 25 500 tonnes, despite advice suggesting the new TAC should have been ca 15 000 tonnes. Wide criticism followed and the Convention on International Trade in Endangered Species (CITES) attempted unsuccessfully to cease the international trade of the bluefin tuna. In 2009, ICCAT set a new TAC of 13 500 tonnes and made the provision to completely ban fishing if new scientific evidence suggests an imminent stock collapse (Nomura and Watanabe, 2010). A similar story exists for the southern Bluefin tuna, whose RFMO is the Council for Conservation of the Southern Bluefin Tuna (CCSBT). The catch of southern bluefin tuna has declined from approximately 75 000 tonnes in the 1970s to a current TAC of 9449 tonnes (Stehr, 2010). The Western and Central Pacific Fisheries Commission is the RFMO responsible for managing Pacific

© Woodhead Publishing Limited, 2013

bluefin. With Japan being the largest fishing nation for this species and with the natural spawning grounds primarily surrounding this country, Japan is taking the leadership in managing this stock and in 2010 published the first policy on their management (Koya, 2010). Unlike the bluefin species, stocks of tropical tunas are more resistant to overfishing due to their early age at maturity, extended spawning seasons and rapid growth (Fonteneau and Fromentin, 2002).

The reliance on wild-caught seed stock is seen as the Achilles heel of tuna ranching industries worldwide, and it is widely recognised that in order to sustain the demand for tuna the move must be made away from wild caught seed to hatchery produced juveniles (Aranda *et al.*, 2011). The extremely high value of the bluefin tuna coupled with the status of their wild stocks has seen the majority of the research effort into hatchery production focused on these species and bluefin tuna propagation programmes exist in most countries or regions in which the species are ranched. Important research is also occurring into other tuna species such as yellowfin, bigeye and blackfin tuna. Such species are seen as both important aquaculture candidates in their own right and useful surrogates for gaining further insights into their larger bluefin cousins.

The incentives for closed-cycle hatchery production of tuna are many, yet the features that make tuna so unique also give rise to the many challenges associated with their culture. Despite many significant achievements over some 40 years of effort, commercial-scale hatchery production has yet to be fully realised. This chapter summarises these achievements and the challenges that still remain to achieve this goal.

15.2 Broodstock systems and management

There are many unique challenges associated with maintaining and manipulating tuna broodstock. Many of these stem from the fishes' very large size and the practical challenges this creates in simply housing and handling such fish; yet other factors that revolve around their biological sophistication also contribute. These include their fast and continuous swimming behaviour and obligate ram ventilation, difficulties with anaesthesia and susceptibility to physical damage. Another very important consideration is the fact that tuna broodstock are often highly valuable; particularly those species that mature at a large size and which have been held in captivity for a long period of time prior to reaching maturity (Table 15.1). These various factors conspire to complicate the processes of obtaining the key reproductive indicators such as sex and stage of maturity that are required to effectively manage a broodstock population. These factors create particular problems for the large bluefin species, which are not only the most valuable and slowest to mature but which also do not exhibit reliable

© Woodhead Publishing Limited, 2013

natural spawning in captivity and for which obtaining these key reproductive indicators is therefore so important.

Broodstock programmes for bluefin tuna exist in nearly all regions in which these species are being ranched including Japan, Korea, Europe and Australia. Currently nine facilities are involved in broodstock programmes with Pacific bluefin tuna in Japan, involving government, universities and industry (Masuma *et al.*, 2011). The same species is farmed in Korea and Mexico and a number of land-based broodstock facilities have recently been constructed in Korea. Although the species' natural spawning grounds are limited to the western Pacific Ocean near Japan and Korea (Zertuche-González and Rodriguez, 2008), ranched fish are being ongrown in cages as potential future broodstock in Mexico (Anon., 2011). In Europe, a collaborative effort between industry participants and research partners from several countries has been undertaken under the banner of two 'DOTT' (domestication of *Thunnus thynnus*) projects aimed at gaining a greater understanding of the reproductive physiology of the eastern stock of Atlantic bluefin and subsequently developing methods to control their reproduction in captivity. The first of these projects, REPRO-DOTT (reproduction of the bluefin tuna in captivity – feasibility study for the domestication of *Thunnus thynnus*), ran from 2003 until 2005 and the subsequent project, SELFDOTT (from capture based to self-sustained aquaculture and domestication of bluefin tuna, *Thunnus thynnus*), from 2008 to 2011. A regional Italian research consortium (ALLOTUNA) collaborates with and has similar goals to the DOTT programmes (de la Gándara *et al.*, 2009; De Metrio *et al.*, 2009). There are no projects involving the western stock of Atlantic bluefin due to the considerably larger size at which it matures (at least eight years and >135 kg) compared with the eastern stock (three to five years and 25 kg) (Rooker *et al.*, 2007) (Table 15.1). Efforts in Australia for southern bluefin tuna are being driven by private company Clean Seas Tuna with collaboration and investment from the Australian government and a number of government research partners (Stehr, 2009).

In addition to bluefin tuna, broodstock programmes exist (or are currently being developed) throughout the world for other species including yellowfin tuna in Panama, Indonesia, Peru and the USA and blackfin tuna and bigeye in the USA. These smaller tuna species as well as other scombrids such as bonito are often investigated under the guise of 'surrogates' or working models for their more valuable and larger bluefin cousins (Anon., 2008, 2009). The small size and comparatively young age of maturity of these three tuna species (approximately two to three years for each species, Table 15.1) has clear advantages in terms housing and handling stock. A further advantage of yellowfin tuna is that it spawns naturally in captivity (in both tanks and seacages), providing extended access to larvae for the experimentation necessary to overcome the larviculture constraints described in greater detail below. Although the study of these smaller tuna

© Woodhead Publishing Limited, 2013

species will certainly contribute to the knowledge of bluefin hatchery production and is a valid argument for the study of such species, it should also be recognised that these species also have merit as aquaculture candidates in their own right.

Given the many challenges of handling, manipulating and hormonally inducing bluefin tuna broodstock, a multi-disciplinary approach involving the development of new tools and techniques is being adopted under the different broodstock programs that exist throughout the world. The most novel approach of overcoming the constraints involved in housing, handling and manipulating such large and valuable broodstock is the concept of using true surrogate broodstock (as opposed to the aforementioned working models that are often termed surrogates), whereby smaller and more 'conventional' fish species are manipulated to produce tuna gametes. Although great progress has been made towards achieving this ambitious goal, including successful surrogacies in other species (Okutsu *et al.*, 2007), it is yet to be achieved in tuna. Another novel approach being investigated is to induce the early onset of puberty in late maturing species by manipulating the KISS system, the 'gatekeeper' of puberty (Elizur *et al.*, 2009; Masuma, 2010). Whilst such novel techniques have great potential for improving control over reproduction in tuna, the refinement and adaptation of more conventional approaches are continuing in parallel with these new technologies.

15.2.1 Broodstock systems

Seacage systems

Seacages represent the cheapest and most convenient method for housing large tuna broodstock, particularly for operations already using such infrastructure for tuna ranching. Seacages for tuna broodstock are used widely in Japan and Europe for Pacific and Atlantic bluefin, respectively. Southern bluefin broodstock in Australia are also housed in seacages when young, but are transferred via helicopter into a land-based facility as they approach maturity. Early investigations with yellowfin tuna occurred in cages in various locations in southern Japan from the early 1970s through to the early 1990s (Harada *et al.*, 1980) but, as is detailed below, all yellowfin tuna broodstock programmes are now land-based.

Bluefin tuna broodstock cages are larger than those used for ranching in order to minimise any stress that may impact on maturation and spawning, with stocking densities typically less than 1 kg/m^3 (Masuma *et al.*, 2011). A significant mortality of tuna broodstock in cages is associated with collisions or contact with the net following burst swimming and such encounters are therefore reduced in larger cages. In Japan, most net pens are oval in shape with dimensions of 20–80 m × 20–50 m and 10–15 m deep (Masuma *et al.*, 2008), whilst those used in Europe typically consist of round 'polar circle' type cages ranging in size from 25 to 50 m in diameter and with depths up to 20 m (Mylonas *et al.*, 2007; Caggiano *et al.*, 2009). A very unique 'sea cage'

© Woodhead Publishing Limited, 2013

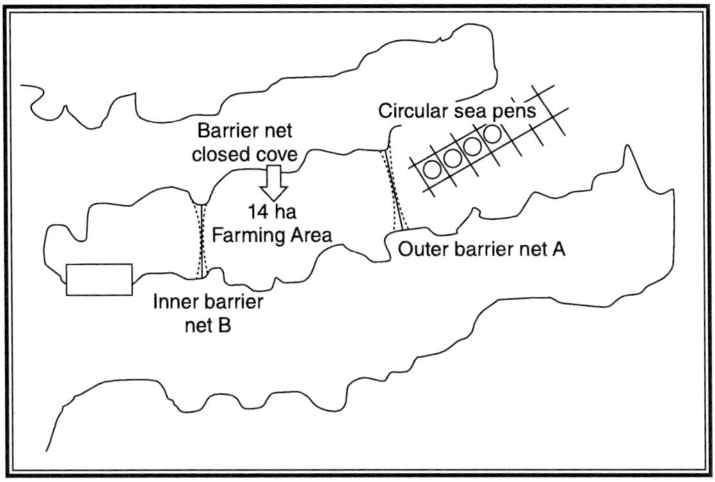

Fig. 15.1 The 14 ha broodstock cove, barrier-netted from the ocean for PBT broodstock at Amami Station, NCSE, FRA, Japan (Masuma *et al.*, 2011). Reprinted with permission from Elsevier © 2011.

system for Pacific bluefin broodstock was established at the Amami Station of the Fisheries Research Agency in southern Japan in 1995. Broodstock are held within a 14 ha, 20 m deep cove which is barrier netted off from the ocean (Fig. 15.1). This facility has the largest broodstock (weighing up to 400 kg) and the widest range in natural water temperatures and has subsequently provided the most stable and reliable spawnings of all the facilities in Japan. In addition to these very large broodstock, younger broodstock are also held in conventional 40 m diameter cages outside of this enclosure.

One of the challenges associated with maintaining broodstock in cages is the effective collection of spawned eggs. In Japan, such eggs are collected with dip nets or purse seines or, in the case of the aforementioned closed-bay in Amami, by following spawning fish in a small boat with a plankton net. All of these methods, however, are acknowledged as being inefficient due to the difficulties imposed by factors including weather and predators (Mylonas *et al.*, 2007; Masuma *et al.*, 2008). A study on the effect of salinity on the buoyancy of Atlantic bluefin tuna eggs conducted under the SELF-DOTT programme determined that at ambient salinity, eggs spawned at 10 m of depth take 38 min to rise to the surface and are therefore highly susceptible of being removed from the cage by water currents (Anon., 2009). Several in-cage egg collection nets have been designed and tested to resolve this problem, the challenge being to develop a sufficient sized egg collector that does not impact on the drag of the overall cage. The most effective method was a barrier curtain hung around the perimeter of the cage with inbuilt funnels and collection tubes (Mylonas *et al.*, 2007; De

© Woodhead Publishing Limited, 2013

Metrio *et al.*, 2010). A further complicating factor in the collection of eggs from seacages is the spawning of other wild fishes in the nearby vicinity. Due to the very similar appearance of all marine fish eggs, other eggs within the cages can be confused for tuna eggs and DNA analysis has been utilised to confirm the parentage of such eggs.

A further disadvantage of utilising seacages for broodstock is the inability to manipulate environmental conditions to control or extend the natural spawning season. Despite being spread over the world's oceans, all tuna species spawn only in sea surface temperatures greater than 23–24 °C (Schaefer, 2001; Rooker *et al.*, 2007; Masuma *et al.*, 2011). For the various bluefin species, significant trans-oceanic migrations occur from their extensive, cool-water foraging grounds and into their smaller, localised tropical spawning grounds (Fonteneau and Fromentin, 2002). In the Mediterranean and most of Japan, this corresponds to a very narrow spawning window, typically between June and July, although in the warmer, southern waters of Japan spawning has been reported to extend to as late as November (Corriero *et al.*, 2007; Rooker *et al.*, 2007; Masuma *et al.*, 2011). The disadvantage of such short spawning seasons is the limited window of time each year into which investigations into overcoming the challenges associated with larval rearing can occur. Given that Block and Stevens (2001) suggest that this temporary migration into warm-water spawning grounds is highly physiologically demanding, it could be that maintaining such fish in cages within these tropical conditions throughout the year may impose stress that would negatively impact on their maturation.

Land-based systems

As is the case for all marine fish species, land-based systems offer advantages for the management of tuna broodstock, particularly the ability to control and manipulate those environmental conditions required to induce maturation and spawning. The disadvantage of land-based broodstock systems for tuna is that the need to maintain low stocking densities combined with the fishes' large size means that the systems themselves must be very large and therefore expensive to construct. The use of land-based systems has therefore been more common for smaller tuna species such as yellowfin; however, this tropical species spawns naturally without photoperiod or temperature control and therefore does not benefit from the inherent advantage that land-based systems offer. Despite this historical tendency, more systems are being built (or proposed) for bluefin tuna as these species will benefit more from the ability to extend and control the environment (Mylonas *et al.*, 2007). Table 15.2 summarises the land-based systems either currently in use or under development throughout the world.

The first and most successful land-based system for true a Thunnid species was constructed at the Inter-American Tropical Tuna Commission's Achotines facility in Los Santos, Panama in 1996. This dedicated yellowfin tuna facility was constructed under a collaborative agreement with the

© Woodhead Publishing Limited, 2013

Table 15.2 Land-based tuna broodstock systems either currently in use or under development throughout the world

Species	Country	Organisation	Tank volume (m^3)	Year constructed
Southern bluefin	Australia	Clean Seas Tuna	3000	2006
Pacific bluefin	Japan	Tokyo Sea Life Park	2200	1989
	South Korea	National Fisheries Research and Development Institute	N/A	N/A
Atlantic bluefin	Spain	Futuna Blue	2 × 4000	2011
Yellowfin	Panama	Inter-American Tropical Tuna Commission	1300	1996
	Indonesia	Gondol Research Institute for Mariculture	1500	2003
	USA	Hawaii Oceanic Technology	300	2011
		Pacific Planktonics	300	2011
Blackfin	USA	University of Miami	80	2009
Bigeye	USA	Hawaii Oceanic Technology	300	2011
		Pacific Planktonics	300	2011

Japanese Overseas Fishery Cooperation Foundation (OFCF) and the Panamanian government. The system is described in detail by Wexler *et al.* (2003) and Scholey *et al.* (2001). In brief, it comprises six concrete tanks housed under a 1300 m^2 roofed shed with open walls. A translucent panel directly above each tank provides natural light. The two largest tanks are used for tuna. The acclimation tank (170 m^3; 8.5 m diameter × 3 m deep) houses newly caught sub-adult fish which are used to replenish the main broodstock population as mortalities occur. The largest tank (1360 m^3; 17 m diameter × 6 m deep) houses the main broodstock population. Both tanks are built partially into the ground to insulate from noise and vibration and each has its own independent recirculating systems described in detail by Wexler *et al.* (2003). Further detailed information relating to filtration systems appropriate for captive tuna can be found in Farwell (2001).

In the main tank, the water is turned over seven or eight times each day. Pumping water into the tank in a clockwise direction takes advantage of the Coriolis effect and creates a 0.5–1 knot current at the edge of tank and a large vortex in the centre. New water is filtered to 16 μm and equates to a flow rate of 1–5 % of the system volume per day. To meet the high oxygen demands of the tuna, a large aeration tower aerates and degasses a large percentage of the recirculating water. Parasites have never been an issue with tuna broodstock in this system and this may be partially attributable to the very high flow rates. The smaller acclimation tank has a lower water

© Woodhead Publishing Limited, 2013

turnover rate of two or three times per day, but a similar exchange rate of new water. As previously described, there is no temperature or photoperiod manipulation in this system and fish spawn naturally, and almost daily, for 10–11 months per year when water temperature is above 23.3 °C (Margulies *et al.*, 2007b, 2010). Successful spawning has occurred with fish numbers ranging from five to 55 yet, despite such a wide range in fish numbers, an overall stocking density is always maintained at a maximum of 0.75 kg/m^3 (Wexler *et al.*, 2003).

Another land-based system for yellowfin tuna exists at the Gondol Research Institute for Mariculture in Bali, Indonesia. This system was also developed under a collaborative project with the Japanese OFCF and the Ministry of Marine Affairs and Fisheries (Republic of Indonesia). It was built more recently than that in Panama in 2003, but is similar in many respects. Slight differences include the main tank being slightly larger (1500 m^3; 18 m diameter × 6 m deep) and there being two acclimation tanks (135 m^3; 8 m diameter × 3 m deep). Each system also operates as an independent recirculating system without photoperiod or temperature control. A more detailed description of the system can be found in Hutchison *et al.* (2011). Yellowfin tuna spawned naturally in this system from 2004 to 2008. When the OFCF project finished in 2006, broodstock replacement ceased and numbers slowly declined to a level where spawning ceased. With investment from the Australian Federal Government's Australian Centre for International Agricultural Research (ACIAR), broodstock numbers have been replenished and spawning recommenced in 2011.

Although yellowfin tuna broodstock have survived in both facilities for up to five years and reached a size of 120 kg, the average survival time is approximately two years (Margulies *et al.*, 2007a; Hutapea pers. com.). Most deaths are caused by collisions with the tank wall as a result of the fish being startled. Wall strikes usually occur in the dark and in the early morning hours and are more common when stocking densities exceed ca 0.5 kg/m^3 and in larger fish, perhaps due to the fact that they are less able to make the rapid, sharp turns required to avoid the collision. In an effort to minimise wall strikes, the walls of all tuna tanks are painted with vertical black stripes to improve contrast and help the tuna see the wall (Wexler *et al.*, 2003). At the Tuna Research and Conservation Centre in California, tuna holding tanks are also fitted with pliable plastic curtains (also with vertical black stripes) to soften such collisions (Farwell, 2001). Avoiding stray lighting at night is another key management practice used to prevent the startle reflex and subsequent collisions. In both Panama and Indonesia, approximately 30 % of the population is lost annually to wall strikes, demonstrating the importance of locating such facilities close to a source of wild fish for constant replenishment. Although the Tuna Research and Conservation Centre has developed a very successful method of transporting young yellowfin and Pacific bluefin tuna over long distances (Farwell, 2002; Partridge, 2009), such methods are expensive and local collection is preferable.

© Woodhead Publishing Limited, 2013

A land-based system for blackfin tuna has recently been established at the University of Miami's Experimental Marine Fish Hatchery (UMEH) (Benetti *et al.*, 2009). This species is the smallest of the true Thunnid tuna species and with a relatively limited natural geographical range compared with most tuna (Schaefer, 2001). Given the small size at which the species matures (ca 2 kg), broodstock are being housed in comparatively small tanks of 80 m^3. Although the tanks are small, sufficient numbers of mature fish can be held within such tanks without exceeding the aforementioned maximum stocking density of 0.75 kg/m^3 as recommended for yellowfin tuna in land-based systems. Fish ranging in size from 2 to 14 kg are caught from local waters and quickly transferred directly into broodstock tanks. The practice of firstly quarantining fish in 12–15 m^3 tanks was abandoned after experiencing a higher incidence of 'puffy snout syndrome' (Kaya *et al.*, 1984) in these fish (Stieglitz pers. com.). UMEH have been employing similar recirculating systems and management practices to those previously described for the successful facility in Panama in an effort to maximise the chances of spawning success. No volitional spawning has occurred during the first two years of the project despite the presence of mature fish in the tanks, as evidenced by post-mortem of wall-strike mortalities. It is possible that despite their small size, blackfin tuna may be more similar in their reproductive physiology to bluefin tuna than yellowfin and therefore more difficult to spawn in captivity. As such, preliminary trials have commenced using the same slow release hormone implants designed for bluefin tuna and described in further detail below. Plans are also in place to commission a larger broodstock tank.

Clean Seas Tuna in Port Lincoln, South Australia have been pioneering the hatchery production of southern bluefin tuna and were the first company in the world to build a land-based system for any bluefin species. In this circumstance, a land-based system was the only option as the temperate waters off South Australia are over 4500 km from the species' natural, tropical spawning grounds between northern Australia and Indonesia. The age at which southern bluefin are caught for ranching in South Australia is two to three years, but they do not mature until approximately 10–14 years (Schaefer, 2001). In 1999, Clean Seas Tuna retained approximately 500 fish from their ranching quota for ongrowing as broodstock. In 2006, they commissioned a $AUD6 million recirculating broodstock system from Uni-Aqua, Denmark. The state-of-the-art facility comprises a 3000 m^3 tank (25 m in diameter and 6 m deep), with a sophisticated recirculating system capable of mimicking all the natural conditions of the Java Sea including temperature, day length and water currents (Ellis, 2008a). The first broodstock weighing in excess of 100 kg were transferred into the facility from nearby seacages via helicopter in 2006 (Ellis, 2008a; Stehr, 2010). Using hormonal manipulation techniques developed under the REPRO-DOTT programme (Elizur *et al.*, 2009) and described in further detail below, male southern bluefin tuna began releasing milt in the tank in 2007. In 2008, a

© Woodhead Publishing Limited, 2013

small number of fertilised eggs were collected, and every year since large numbers of eggs have been collected and used for developing larval rearing protocols.

Although there are no dedicated land-based broodstock facilities for Pacific bluefin tuna in Japan, successful spawning has been reported from within a 2200 m^3 public exhibit at the Tokyo Sea Life Park. The 'deformed' doughnut shaped tank has an inner diameter of 20 m and outer diameter of 30 m (Mimori *et al.*, 2008). The aquarium has a high internal water turnover rate of one exchange per hour and an exchange rate of new water of approximately 3 % per day. No spawning occurred in the 10 years following the opening of the exhibit when photoperiod from artificial lighting was maintained at the constant rate of 14 hours of light and 10 hours of darkness and with an annual temperature variation of 18–23.5 °C. In March of 1999, photoperiod manipulation began by decreasing the amount of daylight during the period of decreasing water temperature. In July, photoperiod and temperature had reached their minima of 11 hours and 19 °C, respectively. By August, when photoperiod and temperature had increased to 13 hours and 23 °C, natural spawning began (Mimori, pers. com.). Although these data point to the importance of photoperiod as an important cue for the spawning of bluefin tuna, it is noteworthy that only one of many females within the tank responded to this stimulus and was involved in these spawning events. This female ceased spawning one month later, shortly after new fish were introduced to the tank. No further spawnings occurred for another seven years, despite on-going temperature and photoperiod manipulation and the release of milt into the water by males and despite females reaching maturity (as evidenced by the high gonadosomatic index of females which died) (Mimori *et al.*, 2008). These data are consistent with the unreliable spawnings and failure of females to undergo final oocyte maturation in Japanese seacages described below.

The first land-based system for Atlantic bluefin has recently been constructed in Cadiz, Spain by private company Futuna Blue. This is the largest bluefin facility in the world, with a floor space of 30 000 m^2 and two broodstock tanks each of 4000 m^3. Like that previously described for Clean Seas Tuna, these tanks operate under sophisticated recirculating systems capable of completely controlling all aspects of the environment (Urup pers. com.).

15.2.2 Broodstock nutrition and feeding

Providing adequate nutrition to broodstock is critical for ensuring eggs are of the highest possible quality. As outlined by Mourente and Tocher (2002), little is known about the exact nutritional requirements of tuna broodstock and, in the absence of such data, broodstock are typically fed on diets that closely mimic what they feed on in the wild (either directly or on locally available substitutes) and usually supplemented with a general vitamin and mineral premix (Wexler *et al.*, 2003; Hutapea and Permana, 2007). An early

© Woodhead Publishing Limited, 2013

project under the REPRO-DOTT programme tested the effect of supplementing broodstock diets with paprika (which contains the antioxidant carotenoids capsorbin and capxanthin). This supplementation resulted in increased levels of carotenoids in the gonads, which it was believed would improve the reproduction performance of broodstock (Anon., 2007). Under the more recent SELFDOTT programme, a detailed proximate analysis of the gonads of mature wild bluefin tuna was undertaken as an indicator of the fish's dietary requirements during maturation (Anon., 2008). These data and those from Mourente and Tocher (2002) demonstrated that wild bluefin have a much higher DHA concentration and DHA:EPA ratio in their gonads than other marine fish. Based on these data and those from the aforementioned REPRO-DOTT study, a modified broodstock diet consisting of a 1:1 ratio of squid and mackerel supplemented with 2 % paprika powder was fed to broodstock three months prior to the beginning of the spawning season. This diet resulted in significant improvements in all reproductive indices in both male and female Atlantic bluefin relative to those fed on a control diet of mackerel and herring. Both sexes had higher concentration of sex steroids within their plasma, higher gonado-somatic indices (GSI) and improved condition factors. In addition, females had greater oocyte size and a lower incidence of apoptotic vitellogenic oocytes, whilst males had improved sperm concentration and motility (Anon., 2009). The inclusion of squid is considered essential to obtaining high quality eggs as it is very high in phospholipids and cholesterol which are important components of vitellegenin, as well having a superior protein content (Pousis *et al.*, 2011).

Broodstock are typically fed to satiety with intake rates ranging from 2 to 11 % of their body weight per day on a wet weight basis, depending on water temperature, the species and their age and size (Farwell *et al.*, 1997; Wexler *et al.*, 2003; Mimori *et al.*, 2008; De Metrio *et al.*, 2010; Masuma *et al.*, 2011). An alternative approach described by Farwell (2001) is to feed a fixed calorific intake rather than a percentage of body weight. For example, yellowfin tuna held at 20 °C at the Tuna Research and Conservation Centre in California are fed 20 Kcals per kilogram of body weight per day, whilst Pacific bluefin, which have a higher metabolic rate, are fed 30 Kcals/kg/day. Feeding in excess of these intakes was found to result in the deposition of excessive fat within the muscle (Farwell *et al.*, 1997). Margulies *et al.* (2007a) reported that short-term increases in daily ration of up to 33 % resulted in increased egg production up to 234 %.

15.2.3 Broodstock handling and anaesthesia

Perhaps surprisingly, tuna are relatively easy to handle when unanaesthetised due the phenomenon of 'tonic immobility' whereby they enter a natural state of paralysis when turned upside down (Brill, 2002). Indeed, invasive procedures such as the surgical implantation of archival tags are

© Woodhead Publishing Limited, 2013

routinely carried out on unanaesthetised tuna with a very high rate of success (Gunn and Block, 2001; Brill, 2002). Such handling techniques work well on small captive tuna that are easy to corral and capture using soft barrier nets and slings (Farwell, 2002). Yellowfin tuna broodstock, for example, are captured as immature fish weighing less than 5 kg and preferably between 2 and 3 kg (Wexler *et al.*, 2003; Hutapea *et al.*, 2009). At this size, the fish are relatively easy to handle without anaesthesia. Once they are transferred into the main broodstock tank they spawn naturally and therefore require no further handling. Handling large mature bluefin tuna by this method, however, is not possible without stress and physical damage.

The typical method of anaesthetising marine fish broodstock involves immersion in an anaesthetic bath. The difficulties associated with capturing very large fish without stress and damage makes this type of anaesthesia inappropriate for tuna. Several reports have also described other problems associated with bath anaesthesia that are independent of the challenges associated with firstly getting the fish into the bath. Brill (2002) described that maintaining an acceptable plane of anaesthesia in a bath is difficult, yet lethal overdose is a constant threat. Further problems associated with such baths include the fact that recovering fish are highly susceptible to physical damage and collisions with tank walls prior to regaining full consciousness. The fact that tuna are also negatively buoyant and obligate ram ventilators creates additional problems during this recovery period as they are susceptible to suffocation when not swimming (Brill, 2002).

An alternative to bath immersion is an intramuscular injection of anaesthetic. This method has the advantage of being able to be administered with self-discharging dart guns, eliminating the need to firstly capture the fish. Difficulties associated with such injections include the lack of a subcutaneous space that does not exist in fish (but that does exist in mammals), which leads to leakage of the anaesthetic from the injection site and therefore creates difficulties in delivering accurate dosages (Anon., 2008). A number of studies have been conducted on various injectable anaesthetics for tuna and other scombrids. Williams *et al.* (2004) tested the effects of an intramuscular combination of ketamine and medetomidine on related scombrid species bonito and mackerel and found this combination to be effective and reversible with atipamezole. Under the REPRO-DOTT programme, however, ketamine and detomidine (alone or in combination) were tested with 100 kg Atlantic bluefin, but the times required for the animals to be restrained were unacceptable and mortalities followed (Anon., 2007). In 2008, investigators in the SELFDOTT programme conducted further work on these same anaesthetics and one other (xylazin). The results were again unacceptable, with only one of 12 fish losing equilibrium. The poor results were attributed mainly to the method of application rather than the drugs themselves and further trials are planned to improve delivery, including different pneumatic darts, multiport needles and intraperitoneal rather than intramuscular injections (Anon., 2008).

© Woodhead Publishing Limited, 2013

15.2.4 Maturation and hormone induction

The lack of reliability and control over reproduction in bluefin tuna is considered a major bottleneck to their successful culture. Although tuna do spawn naturally in cages in Japan and Europe, spawning is unreliable. In Japan, for example, successful natural spawning was first achieved in 1979 with five year old caged Pacific bluefin tuna. These fish spawned again in 1980 and 1982 but, for reasons unknown, spawning then ceased. Successful spawning did not occur again in Japan until 1993, and even today the percentage of female Pacific bluefin tuna maturing in cages only ranges from 10 to 20 %, despite all fish being of sufficient age and size (Masuma, 2010; Masuma *et al.*, 2011). Croatian-based private company Kali Tuna have the world's largest holding of bluefin tuna broodstock comprising approximately 1000 fish in cages in both Croatia (Atlantic bluefin) and Mexico (Pacific bluefin). The company has achieved natural spawning in their Croatian cages over the past three years but from only a small number of these fish. Similarly, spawning of Pacific bluefin tuna at Tokyo Sea Life Park has been inconsistent over the past 20 years, despite providing phototherm manipulation for 10 of these years.

This unreliability in natural spawning in bluefin tuna highlights the need for a greater understanding of tuna reproductive physiology, the cues that bluefin tuna require for successful spawning and methods to control these processes. The failure to undergo final oocyte maturation (FOM) occurs in many captive marine fish and is caused by the lack of appropriate environmental cues and/or the effects of chronic stress imposed by the captive environment (Zohar and Mylonas, 2001). The development of successive generations of fully domesticated fish has been demonstrated in other marine fish species to improve the reliability of natural spawning and alleviate the need for hormonal manipulation. This is presumably due to the lack of chronic stressors on domesticated fish, which are accustomed to the captive environment (Mylonas *et al.*, 2007). The only bluefin species for which the lifecycle has been closed and for which first-generation broodstock have spawned naturally is the Pacific bluefin tuna where in Japan, six to seven year old broodstock ranging in weight from 30 to 150 kg spawned in 2002, producing viable eggs that were successfully reared to market size (Sawada *et al.*, 2005). Unfortunately, however, the natural spawning obtained from these fish, at least in their first spawning season, was still unreliable in terms of periodicity and the numbers of egg produced (Sawada *et al.*, 2005).

That wild yellowfin tuna spawn naturally in captivity demonstrates that no such cues are missing for this species and suggests that perhaps the missing cues in bluefin may be related to the long migrations that are associated with their spawning behaviour and/or the deep diving that occurs in nature and that cannot be mimicked in captivity (Rooker *et al.*, 2007).

The ability to assess the sex and reproductive status of live tuna broodstock in a stress-free manner is the first critical step required to understand their reproductive physiology (Pousis *et al.*, 2011). Although there is some

© Woodhead Publishing Limited, 2013

evidence of hermaphroditism in skipjack tuna, all Thunnid tunas are gonochoristic (Schaefer, 2001) and, apart from during spawning events when male tuna display vertical black bars, there are no external features differentiating the sexes. The methods used for sexing and determining maturation status on more 'conventional' aquaculture species involves invasive gonadal biopsies on anaesthetised fish. Given the aforementioned difficulties associated with handling and anaesthetising large tuna, these methods are unsuitable and alternate methods have been developed.

In Japan and Europe, methods have been developed for obtaining tissue samples and determining sex without the need to handle fish. Although obtaining direct gonadal samples under such techniques would be extremely difficult, biochemical and molecular techniques on other tissues including plasma and muscle have been used successfully for the same purpose. In 1995, Japanese researchers developed a specialised device attached to a speargun for tagging and collecting muscle biopsies from caged yellowfin tuna. From this homogenised muscle, vitellogenin was measured and, although only two samples were obtained, the results suggested that female fish could be positively identified (Takemura and Oka, 1998). Under the REPRO-DOTT programme, a similar 'jab stick' was developed for obtaining a muscle biopsy whilst simultaneously implanting both electronic and visual identification tags and data loggers into Atlantic bluefin broodstock (Fig. 15.2). Extraction and quantification of various sex steroids as well as the proteins vitellogenin and zona radiata from these muscle samples demonstrated that sex could be determined with a relatively high degree of accuracy when the results of all of these tests were taken into account (Anon., 2007). Although determining sex from muscle samples was not as

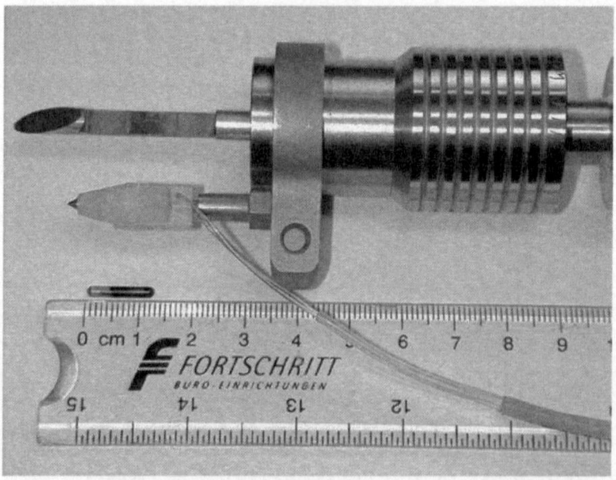

Fig. 15.2 The combination adapter for muscle biopsy tissue sampling and electronic and visual tag application. With permission from C.R.Bridges (Univ. Düsseldorf).

© Woodhead Publishing Limited, 2013

accurate as from plasma, obtaining plasma samples without handling and stress is not possible. Preliminary evidence from the Gondol Research Institute for Mariculture suggests that measuring 11-ketotestosterone in mucus extracted from fin clips can be used for determining the sex of immature yellowfin tuna. High levels of this steroid were measured in fish as small as 6 kg (despite them not maturing until approximately 20 kg) and such fish were confirmed to be males once they had matured and died (Hutchinson *et al.*, 2011). This is a valuable tool in the sexing of yellowfin tuna, which are handled only briefly after capture as immature fish and fin clips can be taken during this transfer without stress.

Recent studies into the reproductive endocrinology of bluefin tuna have contributed a great deal towards the goal of achieving reliable spawning (Corriero *et al.*, 2007, 2009; Pousis *et al.*, 2011). The development of bluefin specific assays for gonadotropin releasing hormone (GnRH), gonadotropins (GtHs) and vitellogenin A and B under the European DOTT programmes were used to determine that luteinising hormone (LH) accumulates in the pituitary of captive bluefin tuna at the same rate as wild fish and that vitellogenesis occurs, demonstrating that the endocrine system is functioning normally in preparation for the natural spawning season (Corriero *et al.*, 2007; Pousis *et al.*, 2011). The high pituitary LH concentration, together with low GSI values, low plasma LH levels and low numbers of vitellogenic oocytes (compared with wild fish), suggests that the failure of captive bluefin tuna to undergo final oocyte maturation and ovulation is due to the absence of LH release from the pituitary (Corriero *et al.*, 2007). These findings also suggested that those similar GnRH-based therapies used successfully to trigger the natural release of gonadotropins in other marine species should also be effective for bluefin tuna. In these more 'conventional' species, an agonist of GnRH is applied as a simple liquid injection to anaesthetised broodstock at the appropriate stage of maturation. Two features complicate this process for tuna broodstock. Firstly, due to the difficulties associated with obtaining gonadal biopsies, it is difficult to confirm that the broodstock are at the appropriate maturational stage for this hormone induction to be successful. Although the aforementioned methods of measuring reproductive hormones and proteins from muscle are appropriate for sexing fish, they are not sufficiently sensitive to accurately determine the stage of maturation. The solution to this problem has been to simply wait until water temperature is above the 23 °C threshold and assume that the broodstock are adequately mature (Corriero *et al.*, 2007).

The second complicating factor is that multiple injections of GnRHa are often required to sustain sufficient levels of gonadotropins in the plasma, which further complicates this technique for tuna. An effective alternative to the use of multiple injections is to administer a slow-release implant that can maintain elevated plasma levels of gonadotropins for long periods (Mylonas, 2002; Mylonas *et al.*, 2007). In 2004 and 2005, an agonist of GnRH prepared within a matrix of poly[Ethylene-Vinyl Acetate] (EVAc) was

© Woodhead Publishing Limited, 2013

administered to nine and 26 Atlantic bluefin tuna broodstock of both sexes, respectively, at rates ranging from 40–100 μg/kg (Mylonas *et al.*, 2007; Corriero *et al.*, 2009; Aranda *et al.*, 2011). The implants were administered on a custom-designed arrowhead (Fig. 15.3) via a speargun into the muscle above the lateral line. The implant was attached to the arrowhead along

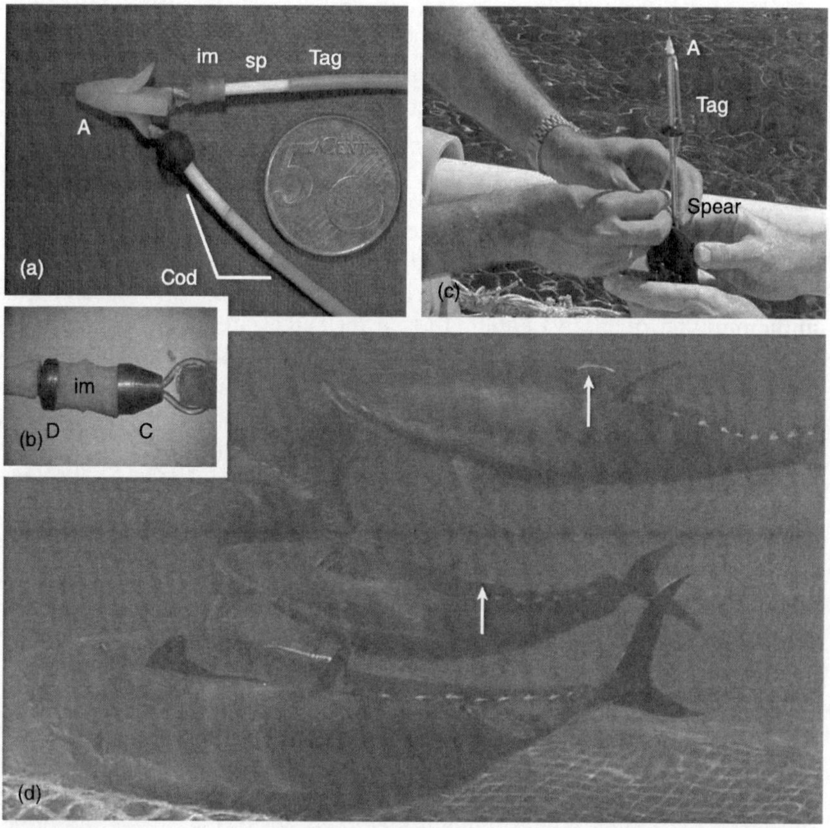

Fig. 15.3 Preparation of GnRHa implants and implantation of captive-reared bluefin tuna. (a) The two GnRHa implants (im) were attached to a polyethylene arrowhead (A) via a 0.5 mm nylon monofilament, followed by 10 mm reporter-spacer (sp), a 10–15 cm pink tag (tag), and a three colour individual identification code (cod). (b) In the second year experiment (2005), a modification was made to the implant assembly with the addition of a metal, conical leader (C) and disc stopper (D) around the implants (im), in order to prevent them from rubbing against the skin and muscle block of the fish during insertion, and being pushed over the tubing. (c) The arrowhead was fitted to a specially prepared spear and was administered to the fish using a spear gun. (d) The implant was administered at the posterior part of the tail (arrows) and the success of the implantation procedure was first evaluated by the diver, after visually examining if the reporter-spacer was inside the fish (Mylonas *et al.*, 2007). Reprinted with permission from Taylor and Francis © 2007.

© Woodhead Publishing Limited, 2013

with an indicator to assess penetration depth and a colour coded tag to identify individual fish (Corriero *et al.*, 2007; Mylonas *et al.*, 2007; De Metrio *et al.*, 2010). Two to eight days later, all fish were sacrificed to assess their reproductive status. The number of males expressing milt was higher (26 %) in the implanted group compared with the control group (12 %) (Mylonas *et al.*, 2007, 2010); however, it was suggested by Corriero *et al.* (2009) that longer term hormone treatment is probably required to yield improvements in testicular maturation and sperm production. Of the females, induction was found to stimulate oocyte maturation and ovulation and prevent atresia; however, it was found that batch fecundity was not increased by such implantation (Corriero *et al.*, 2007, 2010; Aranda *et al.*, 2011). Although no naturally spawned eggs were collected in the days following implantation, the presence of post-ovulatory follicles in sacrificed fish suggested that spawning did occur. In addition, two female fish sacrificed three days following implantation in 2005 were found to have ovulated eggs in their oviduct. These eggs were successfully fertilised by sperm collected from implanted and sacrificed males (Mylonas *et al.*, 2010). In 2008, induction of fish in Italy under the ALLOTUNA project using these same methods resulted in spawnings over four consecutive days and the collection of over 20 million eggs (De Metrio *et al.*, 2010). In 2009, induction in Spain and Italy was successful with 140 million eggs collected over 18 days (Anon., 2009) and 37 million eggs collected over two weeks, respectively (Caggiano *et al.*, 2009). In 2010, successful spawning was again obtained from broodstock held in Spain with over 60 million eggs collected over a period of one month (de la Gándara *et al.*, 2010), and approximately 130 million eggs were collected in Italy under the ALLOTUNA project (Caggiano *et al.*, 2011). Using similar methods, Clean Seas Tuna in Australia collected over 50 million southern bluefin eggs over a 35-day spawning period in 2009 and a similar number in 2010 over a 92-day spawning period (Thomson *et al.*, 2010).

15.2.5 Spawning

Tuna are highly fecund, with egg production estimates ranges from >90 oocytes per gram for the Atlantic and Pacific bluefin species, down to 30 oocytes per gram for bigeye tuna (Schaefer, 2001; Mylonas *et al.*, 2010). Southern bluefin and yellowfin tuna have intermediate fecundities of 57 and 67 oocytes per gram, respectively (Schaefer, 2001). In addition, tuna spawn very frequently within their spawning seasons. Bigeye tuna, for example, spawn the most frequently, on average every 1.05 days, whilst southern bluefin tuna have the least frequent spawning, but still spawn on average every 1.6 days (Rooker *et al.*, 2007). It has been estimated that each female bluefin tuna spawns for three weeks per season (Mylonas *et al.*, 2010).

As previously described, spawning in all tuna species occurs at sea surface temperatures above approximately 23 °C. For tropical tuna such as

© Woodhead Publishing Limited, 2013

yellowfin, there appears to be no correlation between photoperiod or lunar phase and spawning (Margulies *et al.*, 2007b); however, there is evidence to suggest that photoperiod is important for bluefin (Mimori *et al.*, 2008). Margulies *et al.* (2007b) also described the sensitivity of yellowfin tuna to temperature change, with spawning ceasing within one day of water temperatures decreasing by only 0.1–0.2 °C.

The natural spawning of tuna in dedicated broodstock tanks and public aquaria has allowed detailed observations to be made of the spawning processes, which appear similar between species. At the Tokyo Sea Life Park, pre-spawning male Pacific bluefin tuna change colour to black with a pattern of vertical bands (Mimori *et al.*, 2008). Pre-spawning behaviour lasts several hours and involves chasing and paired swimming (Margulies *et al.*, 2007b). Between one and three males pursue a single female, who then swim together in tightening circles before the female spawns in a tail up position near the water surface. Margulies *et al.* (2007b) described that males are not monogamous to single spawning groups, but move from one spawning female to another. Spawning begins in the early evening and continues into the night (Schaefer, 2001). From detailed egg collection records from the IATTC facility in Panama, it has been determined that yellowfin tuna adjust the time of day in which they spawn depending on water temperature, such that their eggs hatch the following day between 3 and 9 pm (Margulies *et al.*, 2007b).

Yellowfin tuna eggs remain positively buoyant until 2–4 h prior to hatching when they begin to sink, independent of salinity within the range of 26–36 ppt (Margulies *et al.*, 2007b). Yellowfin tuna eggs typically measure 900–970 μm in diameter, with a single oil globule ranging in size from 150–280 μm (Margulies *et al.*, 2007b; Hutapea *et al.*, 2009). Pacific bluefin eggs are typically slightly larger at 1010 μm in diameter with an oil globule of 270 μm (Sawada *et al.*, 2005), whilst southern bluefin eggs have not been measured over 1000 μm (Thomson *et al.*, 2010). Results in Europe also demonstrate a declining egg diameter during the spawning season from a peak of 1060 μm down to 1010 μm at the end of the season (Anon., 2009); however, average egg diameter from this species has also been measured as small as 843 μm (De Metrio *et al.*, 2010). In addition to size, egg numbers and hatch rate also declined as the season progressed (Anon., 2009). Margulies *et al.* (2007b) also determined that egg diameter was inversely proportional to water temperature and directly proportional to the size of the spawning female.

Like all fish species, incubation time is temperature dependent. For yellowfin tuna, hatching occurs 20 h post fertilisation at 30 °C and in 28 h at 23 °C (Margulies *et al.*, 2007b). Hutapea (2008) suggested that the optimum incubation temperature for yellowfin was 28 °C and found that below 24 °C and above 32 °C a high percentage of embryos developed abnormally. For Atlantic bluefin, incubation time ranges from 26 h at 27 °C to 44 h at 21 °C (Anon., 2009; De Metrio *et al.*, 2010). At hatch, all species measure

© Woodhead Publishing Limited, 2013

2.6–2.9 mm in length (Miyashita *et al.*, 2001; Anon., 2007; Margulies *et al.*, 2007b).

15.3 Larval rearing and nursery production

The first attempts at rearing larval tuna were with yellowfin in Japan in 1970, and of these larvae the oldest survived to 20 days post hatch (dph) (Mori *et al.*, 1971). Rearing efforts in the late 1970s and early 1980s with Pacific bluefin saw improvements over the first yellowfin attempts, with the oldest larvae surviving to 57 dph (Sawada *et al.*, 2005). Considering the techniques for rearing marine finfish larvae were still in their infancy at these times and given the specific difficulties associated with rearing tuna larvae, these early attempts were positive. Progress has continued since that time and, when natural spawning of Pacific bluefin resumed in Japan in 1994, over 13 000 juveniles were produced over the following three years. Survival rates of these larvae to the end of the nursery period (40–54 dph) ranged from 0.07 to 0.40 % (Sawada *et al.*, 2005). During the transfer of these fish from the land-based nursery tanks into seacages, over 80 % mortality occurred over the following month; however, sufficient numbers survived to become the aforementioned first-generation broodstock that have since spawned naturally in captivity. The mortality that occurred during these initial transfers has now been largely overcome and further improvements in larval rearing techniques have continued. In 2007 Kinki University successfully transferred 1500 juvenile Pacific bluefin tuna to seacages for ongrowing by private companies and in 2009 this number had increased to 32 000 (Anon., 2010; Ishibashi, 2010).

Despite a much shorter history, considerable progress in larval rearing has also been made in recent years in both Europe and Australia for Atlantic and southern bluefin tuna, respectively. Through the development of the aforementioned exogenous hormone therapies, reliable spawnings are now being achieved in these species and the eggs obtained have been used for the development of larval rearing protocols. The approach taken under the SELFDOTT programs in Europe has been to distribute eggs amongst as many collaborating research institutes and commercial hatcheries as possible in order to test a diverse range of larval rearing protocols under different operational circumstances. The first larval rearing attempts made under this programme in 2008 saw larvae surviving only to 14 dph, yet in 2011 approximately 3000 juveniles were produced at the Instituto Español de Oceanografia in Spain, many of which were transferred to sea (de la Gándara pers. com.). Likewise, steady progress has been made at Clean Seas Tuna in Australia with southern bluefin tuna. From their first successful spawning in 2008 where larvae were reared for only seven days, three years later (2011) saw 150 juveniles transferred to seacages for the first time.

© Woodhead Publishing Limited, 2013

The majority of research on yellowfin tuna at the Inter-American Tropical Tuna Commission in Panama has been focused on gaining a greater understanding of their wild ecology and, as such, commercial quantities of juveniles have not been produced for ongrowing. However, juveniles have been grown up to 100 dph (Scholey *et al.*, 2001; Margulies *et al.*, 2009). Although the focus of the Gondol Research Institute for Mariculture in Indonesia is on commercial culture, larval rearing attempts have been hindered by the occurrence of an egg parasite described in further detail below and the oldest juveniles have been reared to 53 dph (Hutapea, pers. com.).

Despite some successes, the survival of tuna larvae is highly variable and generally considered too low for efficient commercial production. Current survival rates to weaned juveniles (35–50 mm) range from 0.01 to 4.5 % (Masuma *et al.*, 2008). The Japanese describe three critical periods of mortality in Pacific bluefin tuna larvae, which interestingly span the majority of the rearing cycle: the first occurs during the first 10 dph, the second between 14 and 30 dph and the third from 30 to 60 dph (Takashi *et al.*, 2006; Masuma *et al.*, 2011). The causative factors of mortality at each stage include physical and nutritional factors during the first stage, cannibalism and nutritional factors in the second stage and collision mortality in the third. Each of these factors is described in further detail in the relevant sections below.

15.3.1 Larval development

A representation of typical tuna development is shown in Fig. 15.4. For most tuna species, flexion begins at approximately 10 dph (Kaji *et al.*, 1996; Kawamura *et al.*, 2003; Partridge *et al.*, 2011) and, prior to this, larval development of tuna is similar to most other marine fish larvae and growth is relatively slow (Kaji *et al.*, 1996; Kawamura *et al.*, 2003; Thomson *et al.*, 2010). At first feeding, most tuna species measure between 3.5 and 4.0 mm in length (Kaji *et al.*, 1996; Kawamura *et al.*, 2003; Margulies *et al.*, 2007a; Hutchinson, 2009). Kawamura *et al.* (2003) and de la Gándara (2009) suggest that swimbladder inflation begins a few days after first feeding at between 5 and 6 dph; however, Takashi *et al.* (2006) reported that inflation begins at 2 dph and that 80 % inflation has occurred by 3 dph. These differences may be due to the time at which appropriate conditions for inflation were presented to the larvae. For example, in the SELFDOTT trials, surface skimming did not begin until 4 dph (Anon., 2009). One feature that appears somewhat unique to tuna during this pre-flexion stage is a high body density and the inability of the inflated swimbladder to compensate for this density until approximately 9 dph (Takashi *et al.*, 2006). This phenomenon and its implications are described in further detail below.

Early tuna larvae have a very high number of free neuromasts on their body surface; even higher than most pelagic species. These mechanoreceptors make tuna larvae particularly sensitive to stress due to handling and transfer and may also result in stress caused by noise and vibration within

© Woodhead Publishing Limited, 2013

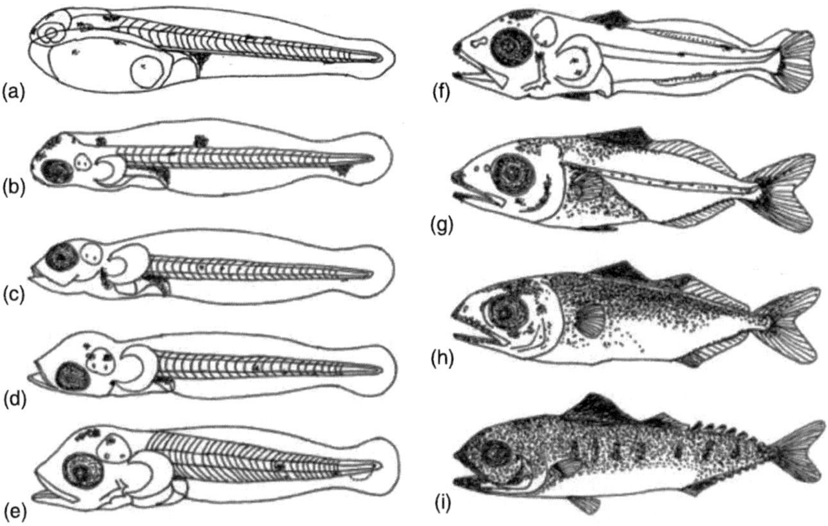

Fig. 15.4 Bluefin tuna larvae and typical features of morphological development; (a) newly hatched yolk-sac larvae; (b) larvae 2 days old with pectoral fins formed, yolk and oil globule nearly resorbed, and mouth and anus open; (c) larvae 3 days old, eyes well pigmented, small amounts of yolk and oil globule still present, and bent and swollen gut with food; (d) larvae 5 days old with first inflated swimbladder; (e) larvae 9 days old with rudiments of the hypural plate and anal fin base, four teeth on the upper jaw, and three teeth on the lower jaw; (f) larvae 12 days old with fully flexed notochord and forked caudal fin; (g) larvae 16 days old with complete rays and spines in all fins except pectorals; (h) larvae 19 days old with almost fusiform body; (i) juvenile 33 days old with scale formation in progress (Kawamura *et al.*, 2003).

the rearing tanks themselves (Kawamura *et al.*, 2003). Similar findings were also recently reported for Atlantic bluefin (Anon., 2009).

Flexion in all species is typically complete by 18 dph when larvae measure between 6.5 and 8.5 mm (Kaji *et al.*, 1996; Fukuda *et al.*, 2010; Thomson *et al.*, 2010). Following flexion, the larvae take on morphological and physiological characteristics that are more typical of all scombrid larvae and less typical of most cultured marine fish larvae (Miyashita *et al.*, 2001). The head and also the eyes and mouth increase greatly in size relative to the rest of the body and the development of the digestive system begins. The onset of acidic digestive capacity, as indicated by the development of the gastric gland and differentiation of pyloric caeca, occurs much earlier in tuna than in other species. In Pacific bluefin, for example, these features are present at 11 and 14 dph, respectively. This compares with 15 and 18 dph in yellowtail kingfish, a pelagic marine fish larvae also considered to have very early digestive tract development (Chen *et al.*, 2006). More typical times for the appearance of the gastric gland in other cultured marine fish larvae are between 22 and 36 days (Chen *et al.*, 2006). These features of precocious

© Woodhead Publishing Limited, 2013

digestive development are characteristic of tuna's natural early piscivorous behaviour. Indeed, the problem of cannibalism begins at the same time that these digestive changes occur, concurrent with the development of the teeth and jaws and earlier than in most other species (Margulies *et al.*, 2007a; Sabate *et al.*, 2010). Post-flexion tuna also exhibit a very high level of growth hormones and a high RNA/DNA ratio (Kaji, 2002). These various morphological and physiological characteristics are responsible for the large scope for growth and, indeed, once piscovory begins larval growth rate increases from 20–30 %/day prior to flexion to 30–50 % per day (Margulies *et al.*, 2007a). Metamorphosis is typically complete by 30 dph, at which time most tuna species measure approximately 15–25 mm in length (Kaji *et al.*, 1996; Scholey *et al.*, 2001; Fukuda *et al.*, 2010).

15.3.2 The physical rearing environment

There have been and remain many different approaches to rearing tuna larvae, demonstrating that the best suite of rearing conditions is yet to be determined. Larvae are reared in both indoor intensive recirculating hatcheries under artificial light and outdoors under enclosures with natural diffused lighting. In Japan, large and deep tanks, either round or octagonal and ranging in size from 20 to 50 m^3, are typical (Sawada *et al.*, 2005; Masuma *et al.*, 2011). Such large tanks are, however, not essential for achieving good survival. At the IATTC in Panama, for example, most larval rearing of yellowfin tuna occurs in small tanks with volumes of 700–1200 L and similar survival rates are achieved compared with bluefin in Japan (Margulies *et al.*, 2007a; Partridge *et al.*, 2011). Similarly, under the ALLOTUNA and SELFDOTT programmes in Europe, tank sizes ranging from 500 L to 40 m^3 have been used (Anon., 2009; Caggiano *et al.*, 2009). There appear to be no documented studies on optimal tank colour for tuna larvae and a wide range of colours including black, dark green, light blue and yellow have been used. Likewise, a diverse range of rearing strategies from extensive outdoor ponds to intensive clear water systems have been considered or trialled in Europe and Australia (Anon., 2009; Fielder *et al.*, 2009). Under the SELFDOTT programme it has, however, been demonstrated that clear water techniques give inferior results compared with green water techniques. Larvae reared in clear water survived to only 20 dph, had very poor swimbladder inflation, poor growth and higher stress levels compared with the green water techniques, which survived up to 73 dph (Anon., 2009). Very similar results were also obtained in bonito, a surrogate scombrid (Anon., 2009). In both Japan and Panama, green water systems are used routinely and concentrations of *Nannochloropsis* (either alone or in combination with a brown algae such as T.Iso) ranging from 0.5×10^6 to 1×10^6 are typically maintained (Margulies *et al.*, 2001; Nakagawa *et al.*, 2007; Partridge *et al.*, 2011). Under these green water techniques, it is common for the cultures to remain static (i.e. without flow) for

© Woodhead Publishing Limited, 2013

the first few days of culture before gradually increasing water exchange rates to 300–500 % per day.

Although some attempts have been made to rear tuna larvae at high densities (Caggiano *et al.*, 2009; Hutchinson, 2009), most rearing occurs at very low densities, typically less than 10 larvae per litre (Kaji *et al.*, 1996). Several experiments investigating the effects of larval density on the growth of yellowfin tuna have been conducted at the IATTC in Panama. Increasing larval density from ca 2/L to ca 18/L resulted in growth reductions of up to 35 % (Margulies *et al.*, 2007a). Although Hutchinson (2009) found no difference in growth of southern bluefin tuna larvae reared at 20, 40 or 60 larvae/L, growth in all three treatments was very slow and perhaps the result of these high larval densities. In contrast, Atlantic bluefin tuna larvae from the same cohort grew faster (8.1 mm by 20 dph) when stocked at 30–60/L in 500 L intensive green water tanks than at 5–10/L in a 40 m^3 semi-intensive mesocosm (7.1 mm at 20 dph); however, many factors other than stocking density could have contributed to this difference and again growth rates in both systems were relatively slow (Anon., 2009).

As described above, very high levels of mortality occur in tuna larvae during the first 10 days and this mortality has been largely attributed to physical characteristics of the rearing environment. Japanese researchers describe two phenomena somewhat unique to tuna known as 'floating death' and 'sinking death' (Masuma *et al.*, 2011). The phenomenon of floating death (also known as surface tension related death) occurs between 0 and 4 dph and is caused by the larvae becoming trapped in the water surface tension (Munday *et al.*, 2003). This mortality is reduced by placing a thin layer of oil on the water surface to reduce this surface tension (Sawada *et al.*, 2005); however, this can interfere with swimbladder inflation. Sinking death is more problematic than floating death and is caused by the fact that tuna larvae between the ages of 4 and 9 dph are more dense than sea water. Even an inflated swimbladder during this period is insufficient to offset the high body density and larvae sink at night (Takashi *et al.*, 2006). Failure of larvae to inflate their swimbladder further intensifies sinking death (Ishibashi, 2010). Various strategies have been trialled to prevent such sinking. The most common has been to employ improved hydrodynamics and 'upwelling' systems with air and/or water. Kato *et al.* (2008) measured the effects of different turbulence levels on first feeding Pacific bluefin larvae. 'Medium' to 'semi-high' turbulence levels equating to upwelling water flow rates of 20–30 L/min in a 500 L tank resulted in the highest survival and greatest ingestion rates of rotifers. The use of water pumps has also been recently trialled in the same species in large-scale (50 m^3) rearing tanks to create upward and anticlockwise rotational currents (Masuma *et al.*, 2011). This approach significantly improved survival to 10 dph; however, the authors pointed out that, despite achieving survival rates of up to 60 % to this age, results remain highly variable between batches and work is continuing to obtain more stable production by further

© Woodhead Publishing Limited, 2013

manipulating both water current and lighting (Masuma *et al.*, 2011). Similar 'medium' rates of turbulence have also been demonstrated to be optimal for growth and survival of yellowfin tuna (Kimura *et al.*, 2004; Scholey *et al.*, 2004; Margulies *et al.*, 2007a).

An alternative approach to upwelling to prevent night-time sinking was recently trialled with yellowfin tuna larvae and involved the provision of a continuous photoperiod to prevent the larvae from sinking at night (Partridge *et al.*, 2011). When a decreased night-time light intensity (10 μmoles m^{-2} s^{-1} = 30 % of the daytime intensity) was provided within a continuous photoperiod regime, survival and growth were not significantly improved over the control photoperiod of 12L:12D. However; when the light intensity was maintained at a constant 30 μmoles m^{-2} s^{-1} for the entire 24 h photoperiod, survival increased by approximately 10 fold. These larvae were also significantly larger and more developmentally advanced than those reared under the control photoperiod, presumably due to the fact that tuna larvae are visual feeders (Margulies, 1997) and an increased photoperiod therefore provides a greater foraging time and subsequent food intake. A disadvantage of maintaining a 24 h period is that it appears to inhibit swimbladder inflation by repelling the larvae from the surface (Anon., 2009; Partridge *et al.*, 2011). A decrease in light intensity from 30–60 μmoles m^{-2} s^{-1} to 1–3 μmoles m^{-2} s^{-1} during 6–7 dph resulted in improved swimbladder inflation in Atlantic bluefin tuna larvae; however, the effects on survival during the two days of low light were not reported (Anon., 2009). Although continuous photoperiods have been trialled with many other marine fish larvae, the improvements in survival have not been as great as those reported in tuna larvae, further demonstrating that mortality caused by night-time sinking is a greater issue for tuna than for other marine fish.

15.3.3 Feeding and nutrition

A typical feeding strategy for tuna larvae is shown in Fig. 15.5. Although this feeding strategy is for larval Pacific bluefin, it is very similar to that used for all other tuna species (Margulies *et al.*, 2007a; Caggiano *et al.*, 2009; Thomson *et al.*, 2010). This schedule demonstrates that the early feeding regime for tuna larvae is similar to that used for other marine fish, with enriched rotifers offered as the first feed through until around 19–25 dph (Sawada *et al.*, 2005; Masuma *et al.*, 2008; Anon., 2009). In addition to the aforementioned physical factors causing early mortality, Masuma *et al.* (2008) suggested that nutrition-related issues such as unsuccessful initial feeding, unsuitable food or insufficient nutrition may also contribute to this early high mortality. That only 20 % of Atlantic bluefin tuna larvae offered rotifers had a full stomach on the second day of feeding, compared with yellowfin tuna offered copepod nauplii, which had a 100 % feeding incidence by the second day supports this theory (Margulies *et al.*, 2001; Anon., 2009).

© Woodhead Publishing Limited, 2013

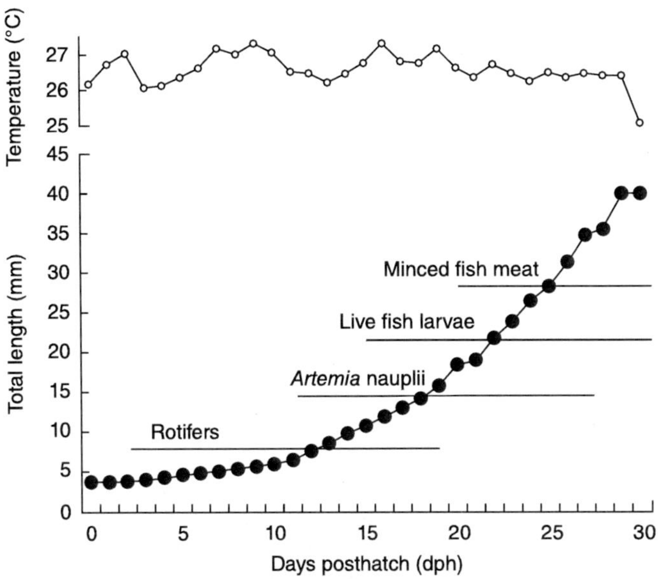

Fig. 15.5 The representative feeding regime used in the larviculture of the Pacific bluefin tuna, *Thunnus orientalis* (Temminck et Schlegel), in 1998 (Sawada *et al.*, 2005). Reprinted with permission from John Wiley and Sons © 2005.

Historically, feeding on *Artemia* nauplii has commenced between 10 and 13 dph and is continued until 25–28 dph (Kaji *et al.*, 1996; Sawada *et al.*, 2005; Masuma *et al.*, 2008). However, as is described in greater detail below, there has been a recent shift away from such a prolonged use of *Artemia* based on aspects of their nutritional profile which appear inappropriate for tuna and which cannot be overcome through the typical enrichment processes.

The main difference between the feeding schedule of tuna larvae and other marine fish begins around the post-flexion stage when the larvae are offered yolk sac larvae, rather than being weaned directly onto artificial diets. These yolk sac larvae may be of their own species (Margulies *et al.*, 2010) or any other available marine fish larvae (Masuma *et al.*, 2008). Often the feeding of these larvae commences at the same time as *Artemia* (ca 10 dph) or slightly later (15 dph). The rationale for this approach is that this diet matches the piscivorous feeding behaviour of wild tuna larvae and results in superior growth and survival. The feeding of such larvae typically continues to approximately ~32 dph (Sawada *et al.*, 2005) by which time the larvae have transitioned onto a diet of chopped fresh fish.

Although the use of *Artemia* in marine finfish larval rearing is almost ubiquitous, they are not a natural food source for any species and lack certain key nutrients required for the growth and development of marine fish larvae. The deficiency of *Artemia* in essential fatty acids has been long

© Woodhead Publishing Limited, 2013

recognised (Sargent *et al.*, 1999) and protocols and products have subsequently been developed for increasing these concentrations to levels that make them suitable for most marine finfish larvae. Despite this, enriched *Artemia* appear unsuitable for tuna larvae as their use results in 'growth failure', or stunting, compared with those reared on yolk sac larvae (Biswas *et al.*, 2006), even when the DHA level of the *Artemia* is matched to that of yolk sac larvae. Seoka *et al.* (2007), for example, demonstrated that 15 dph Pacific bluefin tuna larvae fed on yolk sac fry of Japanese parrot fish for nine days had a final body weight more than three times greater than those fed on *Artemia* enriched with an equivalent DHA concentration.

There are a number of factors that likely contribute to *Artemia* being an inferior food source for tuna larvae. The majority of DHA within yolk sac larvae is found within the polar lipid fraction whilst in enriched *Artemia*, DHA is stored within the neutral lipid fraction (Sargent *et al.*, 1999). The polar lipids (consisting predominantly of the phospholipids) are of greater importance than neutral lipids for developing larvae, as they are key structural components of cellular membranes, particularly in the eyes and brain (Sargent *et al.*, 1999), whilst neutral lipids are used primarily for energy. Other disadvantages of *Artemia* are their rapid retroconversion of DHA into EPA and a naturally high concentration of 18:3n-3 in their polar lipid fraction. Natural tuna larval diets have very high levels of DHA and relatively low levels of EPA, resulting in a high DHA to EPA ratio. Seoka *et al.* (2007) demonstrated that when *Artemia* were enriched using only DHA to achieve the same DHA concentration as knife jaw yolk sac larvae, this process of retroconversion resulted in an EPA concentration three times higher than the knife jaw larvae, and subsequently a much lower DHA to EPA ratio (1.1 vs 3.3).

The importance of polar phospholipids to tuna larvae was further demonstrated by Seoka *et al.* (2008), who compared the performance of 18 dph tuna larvae fed artificial diets containing isoproteic and isolipidic diets that differed only in their lipid class for 10 days against those fed on enriched *Artemia* or yolk sac larvae. Those larvae fed on the artificial diet containing polar lipids grew significantly faster than those on enriched *Artemia* and those fed the artificial diet containing only neutral lipids. Interestingly, in a similar experiment with juvenile Pacific bluefin, those fed the artificial diet high in polar lipids not only had improved growth, but suffered significantly fewer mortalities associated with wall strikes. These findings may serve to reinforce the importance of polar lipids in the development of neural and visual systems and/or that the stress-induced startle behaviour that leads to wall strikes is lessened in juveniles fed a diet high in polar lipids.

Whilst the importance of dietary inclusion of polar lipids is not unique to tuna and their benefits also widely demonstrated for many other species of marine fish larvae (Navarro, 1997; Cahu *et al.*, 2003; Hadas, 2003; Gisbert,

© Woodhead Publishing Limited, 2013

2005), none has been shown to undergo the same growth failure experienced by tuna larvae when fed on enriched *Artemia*. The physiological mechanisms responsible for these differences have not been elucidated, but these findings reinforce that important differences exist between tuna larvae and those of other marine fish species.

Based on these findings, the aforementioned tuna larval feeding regimes under which *Artemia* are fed for periods of up to 10–15 days (Fig. 15.5) are now being reviewed. It is now more typical for yolk sac fry to be used as the main live feed following rotifers and the use of *Artemia* restricted to just a three to four day co-feeding period – primarily to reduce size variation between larvae (Seoka *et al.*, 2008). The provision of yolk sac larvae together with *Artemia* (compared with *Artemia* alone) has also been demonstrated to reduce cannibalism (Ishibashi, 2010).

Given the challenges associated with rearing tuna larvae, it is somewhat surprising that there are no documented studies on the use of copepods in their culture. Copepods are a natural food source for all marine fish larvae, including tuna (Uotani *et al.*, 1980) and have been repeatedly demonstrated to deliver superior rates of survival and growth in 'difficult' marine fish larvae (Schipp *et al.*, 1999; Støttrup, 2000; Payne *et al.*, 2001; McKinnon *et al.*, 2003; Kraul, 2006). In an ecological study on the food preferences of larval yellowfin tuna fed zooplankton assemblages comprising 19 taxa during the first four days of feeding, Margulies *et al.* (2001) found that copepods accounted for up to 96 % of all prey consumed. Rotifers, on the other hand, were highly avoided, accounting for only 0.4–2.1 % of the total number of prey consumed, despite representing 31–40 % of the available prey items. Feeding incidence in this study was also very high, with 95 % of yellowfin tuna larvae consuming copepods on the first day of feeding and 100 % the following day (Margulies *et al.*, 2001). This contrasts with a study under the SELFDOTT programme in which only 20 % of Atlantic bluefin larvae had rotifers in their gut on the second day of feeding (Anon., 2009). Given their high metabolic rate and limited visual acuity, early stage tuna larvae are highly susceptible to starvation (Margulies, 1997; Margulies *et al.*, 2007a) and initiating feeding quickly is therefore critical to their survival. The low first feeding incidence of rotifer-fed tuna larvae may therefore be an important contributing factor to the high rates of early mortality. It is also noteworthy that rotifers share some of the aforementioned negative attributes of *Artemia* regarding their lipid metabolism, such as the predominant incorporation of lipid enrichments within the body as neutral lipids (Rainuzzo *et al.*, 1994; Fernández-Reiriz and Labarta, 1996).

In light of the previous discussion on the lipid composition of both *Artemia* and rotifers, another major benefit of copepods as a food item for tuna larvae is their lipid profile. Like yolk sac larvae, copepods are dominated by polar lipids (80–90 % of total lipid) and have both a high DHA concentration and high DHA to EPA ratio (McKinnon *et al.*, 2003). The

© Woodhead Publishing Limited, 2013

major limiting factor in the use of copepods as a prey item for cultured marine fish larvae is difficulties associated with their mass production, yet, given that tuna are harvested at a market size of one to two orders of magnitude heavier than most other marine fish, significantly fewer juveniles must be produced from a hatchery to obtain the same final marketable biomass. This, coupled with their very high value, suggests that providing copepods to tuna larvae could be economically viable. Private company Futuna Blue concur with this premise and the general philosophy that copepods will be of great benefit in the mass production of tuna juveniles, and their new multi-species hatchery in Cadiz, Spain has been designed specifically around this concept. The facility has six large (3200 m^3) outdoor tanks dedicated to the semi-intensive production of copepods. Due to the recent completion of this facility, the company do not yet have their own broodstock. However, in their first year of operation (2011) they sourced a very small number of fertilised eggs for a commissioning run in which to test their larval rearing systems. During this preliminary run, excellent survival (>10 %) was achieved to 35 dph on a diet of only copepods and some of these fish remain in their land-based nursery (Urup pers. com. 2011). Futuna Blue plan to produce between 500 000 and 1 000 000 bluefin tuna juveniles per year by 2014.

Also in Europe, SELFDOTT have recently reported a collaborative project with the SINTEF institute in Norway, in which trials will be undertaken feeding copepod nauplii to Atlantic bluefin larvae.

15.3.4 Disease

Although tuna are considered very robust in grow-out and have very few reported disease issues, some diseases and parasite issues have been reported in larviculture. Viruses including viral nervous necrosis (VNN) and iridovirus have been reported in bluefin tuna and incoming hatchery water is often sterilised with UV or ozone to prevent the transmission of such diseases (Munday *et al.*, 2003; Sawada *et al.*, 2005; Nishioka *et al.*, 2010).

The egg endoparasite *Ichthyodinium chabelardi* has caused mass mortality of yolk sac larvae of yellowfin tuna in Indonesia. A free-swimming stage of the parasite infects eggs horizontally (i.e. via the rearing water) within 2 h of spawning. Once inside the egg, the parasite rapidly multiplies, giving the yolk a mottled appearance. On hatching, the parasites are released where they can survive up to three days before re-infecting other eggs (Hutapea and Permana, 2007). The very frequent spawnings of yellowfin tuna in captivity therefore make it difficult to break this cycle. Bathing in formalin within 1 h of spawning or the rapid removal of the eggs into sterile water are effective means of preventing infection but are difficult to administer in practice. Disinfection of incoming sea water or complete removal of all eggs from the broodstock tanks prior to hatch may be more effective means of preventing infection (Hutchinson *et al.*, 2011).

© Woodhead Publishing Limited, 2013

15.3.5 Weaning and nursery culture

As previously described, tuna larvae are transitioned from a diet of yolk sac larvae and onto one of chopped fish. It has been acknowledged that the development of appropriate weaning diets is an urgent priority for the development of commercial-scale tuna hatchery production (Masuma *et al.*, 2008; Margulies *et al.*, 2010). Seoka *et al.* (2008) noted that, although Pacific bluefin larvae do ingest commercially available weaning diets, their exclusive use results in rapid emaciation of the juvenile tuna. The very early development of the digestive system in larval tuna suggests that this failure to thrive on artificial diets is not due to a lack of capacity to digest it. Indeed, Seoka *et al.* (2008) obtained significantly faster growth of 18 dph Pacific bluefin larvae on an experimental artificial diet compared with those fed enriched *Artemia*, confirming that such diets can be well digested and achieve rapid growth. Survival of larvae on this diet, however, was inferior to those fed *Artemia*. The lipid content of this experimental diet comprised solely polar lipids, and these authors therefore suggested that the poor performance of bluefin larvae on commercially available weaning diets may be due to the fact they are prepared with conventional fish oils high in neutral lipids. Further research has continued to demonstrate that artificial diets can deliver satisfactory results for juvenile tuna (Ji *et al.*, 2008; Biswas *et al.*, 2009; Takii *et al.*, 2010) and anecdotal evidence suggests that dedicated juvenile tuna diets are now being produced in Japan by feed manufacturer Hayashikane Sangyo (http://www.minato-tsukiji.com/news_detail_3185.html).

Although cannibalism begins in the post-flexion stage, it continues into the nursery stage and can account for up to 50 % of mortality within a cohort (Masuma *et al.*, 2011). Schooling behaviour does not develop until 25–27 dph (Fukuda *et al.*, 2010; Sabate *et al.*, 2010), at which time cannibalism becomes less problematic (Margulies *et al.*, 2009). Although cannibalism in other marine fish species has been linked to social hierarchy, in tuna it is believed to be linked only to hunger (Sabate *et al.*, 2010) and can therefore be minimised by adequate food provision as well as size grading (Sawada *et al.*, 2005). Grading tuna larvae is, however, more difficult than in most marine fish due to their sensitivity to handling, and even very gentle hand selection grading has resulted in heavy mortalities (Caggiano *et al.*, 2009). These mortalities were minimised in 20–25 dph Atlantic bluefin larvae by anaesthesia with 50–100 ppm of phenoxy-ethanol and the use of soft nets (Anon., 2009).

A new problem that arises in the post-metamorphic, nursery stage is the collision of the juveniles with tank or net walls. This is due to the fish's remarkable propulsive power created by a fully formed caudal fin, whilst lacking the adult braking and steering capability due to the pectoral and abdominal fins not being as well developed (Normile, 2009). Such collisions may be instantly fatal due to a broken vertebral column, or caused by secondary bacterial infections due to the skin abrasions. Many of these

© Woodhead Publishing Limited, 2013

collisions are caused by the fish being startled by rapid changes in light intensity or by loud noises, and avoiding such stimuli is required to minimise such mortalities (Masuma *et al.*, 2011). As previously described for broodstock yellowfin tuna, it has been suggested that collisions can be reduced by providing a visual contrast on the tank or net wall, such as vertical stripes, polka dots or a lattice pattern (Ishibashi, 2010).

The transfer of juveniles out of tanks and into much larger nursery cages helps minimise encounters with such surfaces; however, the process of transfer itself has created challenges. When such transfers were first attempted in 1994, nearly 50 % mortality occurred the following day and 95 % mortality within one month. Slight improvements occurred over the next few years; however, 40 % mortality over the three to five days post-transfer remained common (Ishibashi *et al.*, 2009). It has since been determined that juvenile Pacific bluefin tuna have a very poor scotopic visual threshold, poor light sensitivity and temporal resolution compared with other juvenile marine fish (Ishibashi *et al.*, 2009; Ishibashi, 2010). These data suggest that collisions are due to the fact that the juveniles cannot easily perceive the net walls and that this perception is further inhibited under low light conditions. By providing night-time lighting with an intensity of 200–3000 lux over nursery cages, survival increased from 12 % in the un-illuminated control cages to 73 % in the 23 days post transfer.

15.4 Conclusions and future trends

The challenges associated with producing commercial quantities of juvenile tuna in a hatchery environment have long been recognised and, given that such attempts have a 40-year history, doubts have arisen whether this elusive goal would ever be achieved. In the last 10 years, however, significant progress has been made on many fronts and the future is now looking very positive.

Obtaining reliable spawning from bluefin tuna has been a major constraint to their commercial culture. Through the development of slow release hormone implants and effective techniques for their administration, reliable spawnings are now being obtained in several bluefin species. This advancement has enabled large numbers of fertilised tuna eggs to be obtained, thereby facilitating further studies into overcoming larval rearing bottlenecks. Although the very short natural spawning period of bluefin tuna still constrains egg availability, there is some evidence that land-based broodstock facilities may enable this season to be extended and, indeed, the use of such systems for bluefin tuna is increasing. That the life-cycle has now been closed for Pacific bluefin tuna is another recent and significant achievement and will hopefully yield more reliable natural spawnings. Hatchery reared juvenile Atlantic bluefin are also now being ongrown for use as broodstock.

© Woodhead Publishing Limited, 2013

Excellent progress has also been made towards identifying and overcoming causes of larval mortality in each of the three critical phases. For example significant progress has been made in minimising sinking death through the manipulation of tank hydrodynamics and lighting regimes. Causative factors for the growth failure seen under an *Artemia* feeding regime have been suggested and, although these nutritional constraints remain, management approaches to minimise these negative effects are now being tested. Perhaps novel techniques to modify lipid storage and control retroconversion within rotifers and *Artemia* can be developed to overcome these constraints. Alternatively, providing natural live foods such as copepods will almost certainly yield benefits in terms of growth and survival over rotifers, and *Artemia* and their provision may be cost-effective for tuna, given their large size at harvest and high market value.

The inability to wean tuna larvae onto artificial diets has also been seen as a significant constraint to tuna larval culture; however, recent research has demonstrated that good growth can be obtained on artificial diets. The development of tuna-specific weaning diets high in polar lipids is underway and such diets may also become viable alternatives to *Artemia* and enable early weaning directly from rotifers.

In 2009, approximately 32 000 juvenile Pacific bluefin tuna were transferred to seacages in Japan for ongrowing to market. This was a 20-fold increase from 2007 and represents approximately 10 % of the total number of wild juvenile fish caught annually for ranching in Japan. Similar exponential increases have also been achieved in very recent years in Europe with Atlantic bluefin tuna. This demonstrates that the multi-disciplinary research efforts into tuna hatchery production are yielding commercial outcomes. If similar gains continue to be made each year, it will not be long until the reliance on wild-caught juvenile tuna for farming is eliminated and the goal of fully closed-cycle tuna aquaculture finally achieved.

15.5 Acknowledgements

The author acknowledges and thanks the following people for their input and assistance: Bent Urup, Syd Kraul, Daniel Benetti, John Stieglitz, John Hutapea, Fernando de la Gándara, Chris Bridges, Ryosuke Mimori and Chuck Farwell.

15.6 References

ANON (2007) *Reproduction of the Bluefin Tuna in Captivity – feasability study for the domestication of Thunnus thynnus*. Final Report of the REPRO-DOTT Project. Mazarrón: Instituto Español de Oceanografía.

ANON (2008) *From capture based to SELF-Sustained Aquaculture and Domestication of Bluefin Tuna, Thunnus thynnus*, Periodic Report of the SELFDOTT Project. Mazarrón: Instituto Español de Oceanografía.

© Woodhead Publishing Limited, 2013

ANON (2009) *From capture based to SELF-Sustained Aquaculture and Domestication of Bluefin Tuna, Thunnus thynnus*. Periodic Report of the SELFDOTT Project. Mazarrón: Instituto Español de Oceanografía.

ANON (2010) Future of bluefin tuna in doubt, researchers try to boost numbers. *The Nikkei Weekly*, 24 May, 17.

ANON (2011) Umami Sustainable Seafood Confirms Natural Bluefin Tuna Spawning at its Farm in Croatia, press release, available at: http://www.umamiseafood.com/news/umami-sustainable-seafood-confirms-natural-bluefin-tuna-spawning-its-farm-croatia-3 (accessed September 2012).

ARANDA G, ARAGÓN L, CORRIERO A, MYLONAS C C, LA GÁNDARA F D, BELMONTE A and MEDINA A (2011) GnRHa-induced spawning in cage-reared Atlantic bluefin tuna: An evaluation using stereological quantification of ovarian post-ovulatory follicles. *Aquaculture*, 317, 255–259.

BENETTI D, STIEGLITZ J, HOENIG R, WELCH A, BROWN P, SARDENBERG B and MIRALAO S (2009) Developments in Blackfin tuna *Thunnus atlanticus* aquaculture, in Allan G, Booth M, Mair G, Clarke S and Biswas A (eds). *The 2nd Global COE Program Symposium of Kinki University*. Osaka: Kinki University Press, 12–14.

BISWAS A K, NOZAKI J, KURATA M, TAKII K, KUMAI H and SEOKA M (2006) Effect of *Artemia* enrichment on the growth and survival of Pacific bluefin tuna *Thunnus orientalis* (Temminck & Schlegel) larvae. *Aquaculture Research*, 37, 1662–1670.

BISWAS A K, JI S-C, BISWAS A K, SEOKA M, KIM Y-S, KAWASAKI K-I and TAKII K (2009) Dietary protein and lipid requirements for the Pacific bluefin tuna *Thunnus orientalis* juvenile. *Aquaculture*, 288, 114–119.

BLOCK B A and STEVENS E D (2001) Preface, in Block B A and Stevens E D (eds), *Tuna: Physiology, Ecology and Evolution.* San Diego, CA: Academic Press, xi–xiii.

BRILL R W (2002) Handling and manipulating Tunas in captivity: a physiologist's perspective, in Bridges C, Garcia-Gomez A and Gordin H (eds), *Domestication of Thunnus thynnus – DOTT. Proceedings of the First International Symposium*, 3–8 February, University of Cartagena, Spain, 30–32.

CAGGIANO M, CAMPANA M, MOSCATO M, CORRIERO A, DELFORIO M, GRILLI G, INTINI A, VALENZA M and DE METRIO G (2009) Recent developments in larval and juvenile rearing of Atlantic bluefin tuna *Thunnus thynnus*, in Allan G, Booth M, Mair G, Clarke S and Biswas A (eds), *The 2nd Global COE Program Symposium of Kinki University*. Osaka: Kinki University Press, 25–30.

CAGGIANO M, CAMPANA M, MOSCATO M, BRIDGES C, MYLONAS C, DELFORIO M, SANTAMARIA N, ZUPA R, POUSIS C, GRILLI G, INTINI A, VALENZA M, DE METRIO G and CORRIERO A (2011) Atlantic bluefin tuna *Thunnus thynnus* larval and juvenile rearing; three consecutive years of experiments, *Proceedings of the World Aquaculture Conference 2011: Aquaculture for a Changing World*. Baton Rouge, CA: World Aquaculture Society, 195.

CAHU C L, INFANTE J L Z and BARBOSA V (2003) Effect of dietary phospholipid level and phospholipid: neutral lipid value on the development of seabass (*Dicentrarchus labrax*) larvae fed a compound diet. *The British Journal of Nutrition*, 90, 21–28.

CHEN B N, QIN J G, KUMAR M S, HUTCHINSON W and CLARKE S (2006) Ontogenetic development of the digestive system in yellowtail kingfish *Seriola lalandi* larvae. *Aquaculture*, 256, 489–501.

CORRIERO A, MEDINA A, MYLONAS C C, ABASCAL F J, DEFLORIO M, ARAGÓN L, BRIDGES C R, SANTAMARIA N, HEINISCH G, VASSALLO-AGIUS R, BELMONTE A, FAUVEL C, GARCIA A, GORDIN H and DE METRIO G (2007) Histological study of the effects of treatment with gonadotropin-releasing hormone agonist (GnRHa) on the reproductive maturation of captive-reared Atlantic bluefin tuna (*Thunnus thynnus* L.). *Aquaculture*, 272, 675–686.

© Woodhead Publishing Limited, 2013

CORRIERO A, MEDINA A, MYLONAS C C, BRIDGES C R, SANTAMARIA N, DEFLORIO M, LOSURDO M, ZUPA R, GORDIN H, DE LA GANDARA F, BELMONTE A, POUSIS C and DE METRIO G (2009) Proliferation and apoptosis of male germ cells in captive Atlantic bluefin tuna (*Thunnus thynnus* L.) treated with gonadotropin-releasing hormone agonist (GnRHa). *Animal Reproduction Science*, 116, 346–357.

DE LA GÁNDARA F, MYLONAS C, COVÈS D, BRIDGES C, BELMONTE RÍOS A, VASSALLO-AGIUS R, ROSENFELD H, MEDINA A, DEMETRIO G, FALCON J, SVEINSVOLL K, GHYSEN A and DEGUARA W (2009) The challenge of domestication of bluefin tuna *Thunnus thynnus* – highlights of the SELFDOTT Project from 2008–2009, in Allan G, Booth M, Mair G, Clarke S and Biswas A (eds), *The 2nd Global COE Program Symposium of Kinki University*, Adelaide. Kinki University Press, 70–72.

DE LA GÁNDARA F, MYLONAS C, COVÈS D, ORTEGA A, BRIDGES C R, BELMONTE RÍOS A, VASSALLO-AGIUS R, PAPANDROULAKIS N, ROSENFELD H, TANDLER A, MEDINA A, DE METRIO G, CORRIERO A, FAUVEL C, FALCON J, SVEINSVOLL K, GHYSEN A, DEGUARA S and GORDIN H (2010) Seedling production of Atlantic bluefin tuna (ABFT) *Thunnus thynnus*. The SELFDOTT Project, in Miyashita S, Takii K, Sakamoto W and Biswas A, (eds), *Joint International Symposium of Kinki University and Setouchi Town on The 40th Anniversary of Pacific Bluefin Tuna Aquaculture*. Osaka: Kinki University Press, 45–52.

DE METRIO G, DEFLORIO M, MYLONAS C, BRIDGES C, VASSALLO-AGIUS R, GORDIN H, CAGGIANO M, CAPRIOLI R, SANTAMARIA N, ZUPA R, POUSIS C, DI GIOIA T, LOSURDO M, SPEDICATO D and CORRIERO A (2009) The Atlantic bluefin tuna (*Thunnus thynnus*) spawning in captivity, in Allan G, Booth M, Mair G, Clarke S and Biswas A, (eds), *The 2nd Global COE Program Symposium of Kinki University*. Osaka: Kinki University Press, 73–75.

DE METRIO G, BRIDGES C R, MYLONAS C C, CAGGIANO M, DEFLORIO M, SANTAMARIA N, ZUPA R, POUSIS C, VASSALLO-AGIUS R, GORDIN H and CORRIERO A (2010) Spawning induction and large-scale collection of fertilized eggs in captive Atlantic bluefin tuna (*Thunnus thynnus* L.) and the first larval rearing efforts. *Journal of Applied Ichthyology*, 26, 596–599.

ELIZUR A, DIECHMANN M, WISE M, ZOHAR Y, NOCILLADO J, LEE Y, ABRAHAM L, BRIDGES C, MYLONAS C, BIRAN J, SIVAN B, KNIBB W, STOKOE R, BUBNER E, BROOKS P, SULLIVAN B, YOSHIZAKI G, TAKEUCHI Y, THOMAS P and FOSTER C (2009) Strategies to control reproduction in southern bluefin tuna (*Thunnus maccoyii*) in South Australia, in Allan G, Booth M, Mair G, Clarke S and Biswas A, (eds), *Proceedings of the 2nd Global COE Program Symposium of Kinki University*. Osaka: Kinki University Press, 31–33.

ELLIS R (2008a) The Bluefin in Peril. *Scientific American*, March, 70–77.

ELLIS R (2008b) *Tuna: A Love Story*. New York: Alfred A. Knopf.

FARWELL C J (2001) Tunas in captivity, in Block B A and Stevens E D (eds), *Tuna: Physiology, Ecology and Evolution*. San Diego, CA: Academic Press, 391–410.

FARWELL C (2002) Management of captive tuna: collection and transportation, holding facilities, nutrition, growth, and water quality, in Bridges C, Garcia-Gomez A and Gordin H, (eds), *Domestication of Thunnus thynnus – DOTT. Proceedings of the First International Symposium*, 3–8 February, University of Cartagena, Spain, 54–57.

FARWELL C, DARROW C and BLOCK B A (1997) Yellowfin tuna husbandary at the Monterey Bay Aquarium. *Proceedings of the 4th International Aquarium Congress*, 23–27 June, Tokyo, 63–65.

FERNÁNDEZ-REIRIZ M J and LABARTA U (1996) Lipid classes and fatty acid composition of rotifers (*Brachionus plicatilis*) fed two algal diets. *Hydrobiologia*, 330, 73–79.

FIELDER D G, ALLAN G, GO J, THOMSON M and HARDY-SMITH P (2009) Biosecurity protocols for translocation of southern bluefin tuna (*Thunnus maccoyii*) eggs from South Australia to New South Wales, Port Stephens Fisheries Institute, in Allan

© Woodhead Publishing Limited, 2013

G, Booth M, Mair G, Clarke S and Biswas A (eds), *The 2nd Global COE Program Symposium of Kinki University*, Osaka: Kinki University Press, 79–80.

FONTENEAU A and FROMENTIN J-M (2002) The Atlantic bluefin tuna: a global perspective, in Bridges C, Garcia-Gomez A and Gordin H (eds), *Domestication of* Thunnus thynnus – *DOTT. Proceedings of the First International Symposium*, 3–8 February, University of Cartagena, Spain, 61–64.

FUKUDA H, TORISAWA S, SAWADA Y and TAKAGI T (2010) Ontogenetic changes in schooling behaviour during larval and early juvenile stages of Pacific bluefin tuna *Thunnus orientalis*. *Journal of Fish Biology*, 76, 1841–1847.

GISBERT E, VILLENUEVE L, ZAMBONINO-INFANTE J.L, QUAZUGUEL P, CAHU C.L (2005) Dietary phospholipids are more efficient than neutral lipids for long chain polyunsaturated fatty acid supply in european sea bass *Dicentrarchus labrax* larval development. *Lipids*, 40, 609–618.

GUNN J and BLOCK B A (2001) Advances in acoustic, archival and satellite tagging of tunas, in Block B A and Stevens E D (eds), *Tuna: Physiology, Ecology and Evolution.* San Diego, CA: Academic Press, 167–224.

HADAS E, KOVEN W, SKLAN D, TANDLER A (2003) The effect of dietary phosphatidylcholine on the assimilation and distribution of ingested free oleic acid (18:1*n*-9) in gilthead seabream (*Sparus aurata*) larvae. *Aquaculture*, 217, 577–588.

HARADA T, MURATA O and ODA S (1980) Rearing of and morphological changes in larvae and juveniles of yellowfin tuna. *Memoirs of the Faculty of Agriculture of Kinki University*, 13, 33–36.

HUTAPEA J (2008) Interaksi suhu dan salinitas media inkubasi terhadap laju pemanfaatan nutrisi endogenous embrio ikan tuna sirip kuning, *Thunnus albacares*. *Prosiding Seminar Nasional Biodiversiti II. Buku II*, Universitas Erlangga, 91–95.

HUTAPEA J H and PERMANA I G N (2007) Life cycle of endoparasite, *Ichthyodinium chabelardi* which infect marine fish eggs. *3rd Marine and Fisheries National Seminar Proceedings*, Surabaya Indonesia, 68–72 (in Indonesian).

HUTAPEA J H, PERMANA I G N and GIRI I N A (2009) Achievements and bottlenecks for yellowfin tuna, *Thunnus albacares,* propagation at the Gondol Research Institute for Mariculture, Bali, Indonesia, in Allan G, Booth M, Mair G, Clarke S and Biswas A (eds), *The 2nd Global COE Program Symposium of Kinki University*, Osaka: Kinki University Press, 34–37.

HUTCHINSON W G (2009) Southern bluefin tuna (*Thunnus maccoyii*) larval rearing advances at the South Australian Research and Development Institute and collaborating institutions, in Allan G, Booth M, Mair G, Clarke S and Biswas A (eds), *Proceedings of the 2nd Global COE Program Symposium of Kinki University*. Osaka: Kinki University Press, 38–42.

HUTCHINSON W, PARTRIDGE G J and HUTAPEA J (2011) *Achieving consistent spawning of captive yellowfin tuna (*Thunnus albacares*) broodstock at Gondol Research Institute for Mariculture, Bali, Indonesia*. Canberra: Australian Centre for International Agricultural Research.

IKEDA S (2002) Market and domestic production of cultured tuna in Japan – cultured tuna in the Japanese market, in Bridges C, Garcia-Gomez A and Gordin H (eds), *Domestication of* Thunnus thynnus – *DOTT. Proceedings of the First International Symposium*, 3–8 February, University of Cartagena, Spain. 83–84.

ISHIBASHI Y (2010) Seedling production of the Pacific bluefin tuna, *Thunnus orientalis* at Kinki University, in Miyashita S, Takii K, Sakamoto W and Biswas A (eds), *Joint International Symposium of Kinki University and Setouchi Town on The 40th Anniversary of Pacific Bluefin Tuna Aquaculture*. Osaka: Kinki University Press, 64–70.

ISHIBASHI Y, HONRYO T, SAIDA K, HAGIWARA A, MIYASHITA S, SAWADA Y, OKADA T and KURATA M (2009) Artificial lighting prevents high night-time mortality of juvenile

© Woodhead Publishing Limited, 2013

Pacific bluefin tuna, *Thunnus orientalis*, caused by poor scotopic vision. *Aquaculture*, 293, 157–163.

JI S-C, TAKAOKA O, BISWAS A K, SEOKA M, OZAKI K, KOHBARA J, UKAWA M, SHIMENO S, HOSOKAWA H and TAKII K (2008) Dietary utility of enzyme-treated fish meal for juvenile Pacific bluefin tuna *Thunnus orientalis*. *Fisheries Science*, 74, 54–61.

KAJI T (2002) Bluefin tuna larval rearing and developments in the state of the art, in Bridges C, Garcia-Gomez A and Gordin H (eds), *Domestication of* Thunnus thynnus – *DOTT. Proceedings of the First International Symposium*, 3–8 February, University of Cartagena, Spain, 74–78.

KAJI T, TANAKA M, TAKAHASHI Y, OKA M and ISHIBASHI N (1996) Preliminary observations on development of Pacific bluefin tuna *Thunnus thynnus* (Scombridae) larvae reared in the laboratory, with special reference to the digestive system. *Marine and Freshwater Research*, 47, 261–269.

KATO Y, TAKEBE T, MASUMA S, KITAGAWA T and KIMURA S (2008) Turbulence effect on survival and feeding of Pacific bluefin tuna *Thunnus orientalis* larvae, on the basis of a rearing experiment. *Fisheries Science*, 74, 48–53.

KAWAMURA G, MASUMA S, TEZUKA N, KOISO M, JINBO T and NAMBA K (2003) Morphogenesis of sense organs in the bluefin tuna *Thunnus orientalis*, in Browman H I and Skiftesvik A B (eds), *The Big Fish Bang. Proceedings of the 26th Annual Larval Fish Conference*. Bergen: Institute of Marine Research, 124–135.

KAYA C M, QUEENTH M K K and DIZON A E (1984) Capturing and restraining technique for experimental work on small tuna in large laboratory holding tanks. *The Progressive Fish Culturist*, 46, 288–290.

KIMURA S, NAKATA H, MARGULIES D, SUTER J and HUNT S (2004) Effect of oceanic turbulence on the survival of yellowfin tuna larvae. *Nippon Suisan Gakkaishi*, 70, 175–178.

KOYA T (2010) Actions to be introduced by government of Japan toward effective conservation and management for Pacific bluefin tuna, in, Miyashita S, Takii K, Sakamoto W and Biswas A (eds), *Joint International Symposium of Kinki University and Setouchi Town on The 40th Anniversary of Pacific Bluefin Tuna Aquaculture*, Osaka: Kinki University Press, 12–15.

KRAUL S (2006) Live food for marine fish larvae, in Cruz Suarez L E, Ricque Marie D, Tapia Salazar M, Nieto Lopez M G, Villarreal Cavazos D, Puello Cruz A C and Garcia Ortega A (eds), *VIII Simposium Internacional de Nutricion Acuicola*, Nuevo Leon, Universidad Autonoma de Nuevo Leon, 55–61.

MARGULIES D (1997) Development of the visual system and inferred performance capabilities of larval and early juvenile scombrids. *Marine and Freshwater Behaviour and Physiology*, 30, 75–98.

MARGULIES D, WEXLER JB , BENTLER K T, SUTER J M, MASUMA S, TEZUKA N, TERUYA K, OKA M, KANEMATSU M and NIKAIDO H (2001) Food selection of yellowfin tuna, *Thunnus albacares*, larvae reared in the laboratory. *Inter-American Tropical Tuna Commission Bulletin*, 22, 9–51.

MARGULIES D, SCHOLEY V P, WEXLER J B, OLSON R J, SUTER J M and HUNT D M (2007a) *A Review of IATTC Research on the Early Life History and Reproductive Biology of Scombrids Conducted at the Achotines Laboratory from 1985 to 2005*. La Jolla, CA: Inter-American Tropical Tuna Commission.

MARGULIES D, SUTER J M, HUNT S L, OLSON R J, SCHOLEY V P, WEXLER J B and NAKAZAWA A (2007b) Spawning and early development of captive yellowfin tuna (*Thunnus albacares*). *Fishery Bulletin*, 105, 249–265.

MARGULIES D, SCHOLEY V P, WEXLER J B and SANTIAGO M C (2009) Research on the reproductive biology and rearing of larvae and juveniles of yellowfin tuna (*Thunnus albacares*) at the IATTC's Achotines laboratory, Republic of Panama, in Allan G, Booth M, Mair G, Clarke S and Biswas A (eds), *The 2nd Global COE Program Symposium of Kinki University*. Osaka: Kinki University Press, 43–45.

© Woodhead Publishing Limited, 2013

MARGULIES D, SCHOLEY V P, WEXLER J B and SANTIAGO C B (2010) Research on the reproductive biology and early life history of yellowfin tuna (*Thunnus albacares*) at the IATTC's Achotines Laboratory, Republic of Panama, in Miyashita S, Takii K, Sakamoto W and Biswas A (eds), *Joint International Symposium of Kinki University and Setouchi Town on The 40th Anniversary of Pacific Bluefin Tuna Aquaculture*. Osaka: Kinki University Press, 59–63.

MASUMA S (2010) Seedling Production of Pacific Blefin Tuna–1: National Activities, in Miyashita S, Takii K, Sakamoto W and Biswas A (eds), *Joint International Symposium of Kinki University and Setouchi Town on The 40th Anniversary of Pacific Bluefin Tuna Aquaculture*. Osaka: Kinki University Press, 64–70.

MASUMA S, MIYASHITA S, YAMAMOTO H and KUMAI H (2008) Status of bluefin tuna farming, broodstock management, breeding and fingerling production in Japan. *Reviews in Fisheries Science*, 16, 385–390.

MASUMA S, TAKEBE T and SAKAKURA Y (2011) A review of the broodstock management and larviculture of the Pacific northern bluefin tuna in Japan. *Aquaculture*, 315, 2–8.

MCKINNON A D, DUGGAN S, NICHOLS P D, RIMMER M A, SEMMENS G and ROBINO B (2003) The potential of tropical paracalanid copepods as live feeds in aquaculture. *Aquaculture*, 223, 89–106.

MIMORI R, TADA S and ARAI A (2008) Overview of husbandry and spawning of Bluefin tuna in the aquarium at Tokyo Sea Life Park. *7th International Aquarium Congress*, 19–24 October, Shanghai, C130-C136.

MIYAKE P M, DE LA SERNA J M, DI NATALE A, FARRUGIA A, KATAVIC I, MIYABE N and TICINA V (2003) General review of bluefin tuna farming in the Mediterranean area. *Collective volume of scientific papers. International Commission for the Conservation of Atlantic Tuna*, 55, 114–124.

MIYAKE M P, MIYABE N and NAKANO H (2004) *Historical trends of tuna catches in the world*. Rome: FAO, 74.

MIYASHITA S, SAWADA Y, OKADA T, MURATA O and KUMAI H (2001) Morphological development and growth of laboratory-reared larval and juvenile *Thunnus thynnus* (Pisces: Scombridae). *Fishery Bulletin*, 99, 601–616.

MLADINEO I and MILETIC I (2008) Rearing and disease of Northern Atlantic bluefin tuna (*Thunnus thynnus*) in the Adriatic Sea. *World Aquaculture*, 3a, 26–29.

MORI K, UEYANAGI S and NISHIKAWA Y (1971) The development of artificially fertilised and reared larvae of the yellowfin tuna, *Thunnus albacares*. *Bulletin of The Far Seas Fisheries Research Laboratory*, 5, 219–232.

MOURENTE G and TOCHER D R (2002) An approach to study the nutritional requirements of the bluefin tuna (*Thunnus thynnus thynnus*, L.), in Bridges C, Garcia-Gomez A and Gordin H (eds), *Domestication of* Thunnus thynnus – *DOTT. Proceedings of the First International Symposium*, 3–8 February, University of Cartagena, Spain, 125–131.

MUNDAY B L, SAWADA Y, CRIBB T and HAYWARD C J (2003) Diseases of tunas, *Thunnus* spp. *Journal of Fish Diseases*, 26, 187–206.

MYLONAS C (2002) Hormonal induction of spawning with reference to the bluefin tuna, in Bridges C, Garcia-Gomez A and Gordin H (eds), *Domestication of* Thunnus thynnus – *DOTT. Proceedings of the First International Symposium*, 3–8 February, University of Cartagena, Spain, 132–133.

MYLONAS C C, BRIDGES C R, GORDIN H, BELMONTE A, GARCIA-GOMEZ A, DE LA GÁNDARA F, FAUVEL C, SUQUET M, MEDINA A, PAPADAKI M, HEINISCH G, DE METRIO G, CORRIERO A, VASSALLO-AGIUS R, GUZMAN J M, MAÑANOS E and ZOHAR Y (2007) Preparation and administration of gonadotropin-releasing hormone agonist (GnRHa) implants for the artificial control of reproductive maturation in captive-reared Atlantic bluefin tuna (*Thunnus thynnus thynnus*). *Reviews in Fisheries Science*, 15, 183–210.

© Woodhead Publishing Limited, 2013

MYLONAS C C, DE LA GÁNDARA F, CORRIERO A and BELMONTE RIOS A (2010) Atlantic bluefin tuna (*Thunnus thynnus*) farming and fattening in the Mediterranean Sea. *Reviews in Fisheries Science*, 18, 266–280.

NAKAGAWA Y, EGUCHI M and MIYASHITA S (2007) Pacific bluefin tuna, *Thunnus orientalis*, larvae utilize energy and nutrients of microbial loop. *Aquaculture*, 267, 83–93.

NAVARRO J C, MCEVOY L A, BELL M V, AMAT F, HONTORIA F and SARGENT J R (1997) Effect of different dietary levels of docohexaenoic acid (DHA) on the DHA composition of lipid classes in sea bass larval eyes. *Aquaculture International*, 5, 509–516.

NISHIOKA T, MORI K-I, SUGAYA T, TEZUKA N, TAKEBE T, IMAIZUMI H, KUMON K, MASUMA S and NAKAI T (2010) Involvement of viral nervous necrosis in larval mortality of hatchery-reared Pacific bluefin tuna *Thunnus orientalis*. *Fish Pathology*, 45, 69–72.

NOMURA I and WATANABE H (2010) Current status and problems of fisheries and farming of Atlantic bluefin tuna, in Miyashita S, Takii K, Sakamoto W and Biswas A (eds), *Joint International Symposium of Kinki University and Setouchi Town on The 40th Anniversary of Pacific Bluefin Tuna Aquaculture*, Setouchi Town, Japan. Kinki University Press, 16–25.

NORMILE D (2009) Persevering researchers make a splash with farm-bred tuna. *Science*, 324, 1260–1261.

OKUTSU T, SHIKINA S, KANNO M, TAKEUCHI Y and YOSHIZAKI G (2007) Production of trout offspring from triploid salmon parents. *Science*, 317, 1517.

PARTRIDGE G J (2009) *Hatchery Production of Yellowfin Tuna*. Melbourne: International Specialised Skills Institute.

PARTRIDGE G J, BENETTI D D, STIEGLITZ J D, HUTAPEA J, MCINTYRE A, CHEN B, HUTCHINSON W and SCHOLEY V P (2011) The effect of a 24-hour photoperiod on the survival, growth and swim bladder inflation of pre-flexion yellowfin tuna (*Thunnus albacares*) larvae. *Aquaculture*, 318, 471–474.

PAYNE M F, RIPPINGALE R J and CLEARY J J (2001) Cultured copepods as food for West Australian dhufish (*Glaucosoma hebraicum*) and pink snapper (*Pagrus auratus*) larvae. *Aquaculture*, 194, 137–150.

POUSIS C, DE GIORGI C, MYLONAS C C, BRIDGES C R, ZUPA R, VASSALLO-AGIUS R, DE LA GÁNDARA F, DILEO C, DE METRIO G and CORRIERO A (2011) Comparative study of liver vitellogenin gene expression and oocyte yolk accumulation in wild and captive Atlantic bluefin tuna (*Thunnus thynnus* L.). *Animal Reproduction Science*, 123, 98–105.

RAINUZZO J R, REITAN K I and OLSEN Y (1994) Effect of short- and long-term lipid enrichment on total lipids, lipid class and fatty acid composition in rotifers. *Aquaculture International*, 2, 19–32.

ROOKER J R, BREMER J R A, BLOCK B A, DEWAR H, DE METRIO G, CORRIERO A, KRAUS R T, PRINCE E D, RODRIGUEZ MARIN E and SECOR D H (2007) Life history and stock structure of Atlantic bluefin tuna (*Thunnus thynnus*). *Reviews in Fisheries Science*, 15, 265–310.

SABATE F D L S, SAKAKURA Y, TANAKA Y, KUMON K, NIKAIDO H, EBA T, NISHI A, SHIOZAWA S, HAGIWARA A and MASUMA S (2010) Onset and development of cannibalistic and schooling behavior in the early life stages of Pacific bluefin tuna *Thunnus orientalis*. *Aquaculture*, 301, 16–21.

SARGENT J, BELL G, MCEVOY L, TOCHER D and ESTEVEZ A (1999) Recent developments in the essential fatty acid nutrition of fish. *Aquaculture,* 177, 191–199.

SAWADA Y, OKADA T, MIYASHITA S, MURATA O and KUMAI H (2005) Completion of the Pacific bluefin tuna *Thunnus orientalis* (Temminck and Schlegel) life cycle. *Aquaculture Research*, 36, 413–421.

SCHAEFER K M (2001) Reproductive biology of tunas, in Block B A and Stevens E D (eds), *Tuna: Physiology, Ecology and Evolution.* San Diego, CA: Academic Press, 225–270.

© Woodhead Publishing Limited, 2013

SCHIPP G R, BOSMANS J M P and MARSHALL A J (1999) A method for hatchery culture of tropical calanoid copepods, *Acartia* spp. *Aquaculture*, 174, 81–88.

SCHOLEY V P, MARGULIES D, OLSON R J, WEXLER J B, SUTER J and HUNT S (2001) Lab culture and reproduction of yellowfin tuna in Panama. *Global Aquaculture Advocate*, 4, 17–18.

SCHOLEY V P, MARGULIES D, WEXLER J B and HUNT S (2004) Larval tuna research mimics ocean conditions in lab. *Global Aquaculture Advocate*, 7, 38.

SEOKA M, KURATA M and KUMAI H (2007) Effect of docosahexaenoic acid enrichment in *Artemia* on growth of Pacific bluefin tuna *Thunnus orientalis* larvae. *Aquaculture*, 270, 193–199.

SEOKA M, KURATA M, TAMAGAWA R, BISWAS A K, BISWAS B K, SEOK KIAN YONG A, KIM Y-S, JI S-C, TAKII K and KUMAI H (2008) Dietary supplementation of salmon roe phospholipid enhances the growth and survival of Pacific bluefin tuna *Thunnus orientalis* larvae and juveniles. *Aquaculture,* 275, 225–234.

STEHR H (2009) Preface, in Allan G, Booth M, Mair G, Clarke S and Biswas A (eds), *The 2nd Global COE Program Symposium of Kinki University*. Osaka: Kinki University Press, 7.

STEHR H (2010) Building a sustainable Southern bluefin tuna aquaculture industry in South Australia – a progress report, in Miyashita S, Takii K, Sakamoto W and Biswas A (eds), *Joint International Symposium of Kinki University and Setouchi Town on The 40th Anniversary of Pacific Bluefin Tuna Aquaculture*. Osaka: Kinki University Press, 30–35.

STØTTRUP J G (2000) The elusive copepods: their production and suitability in marine aquaculture. *Aquaculture Research*, 31, 703–711.

TAKASHI T, HIROTOSHI KOHNO, WATARU SAKAMOTO, SHIGERU MIYASHITA, OSAMU MURATA and YOSHIFUMI SAWADA (2006) Diel and ontogenetic body density change in Pacific bluefin tuna, *Thunnus orientalis* (Temminck and Schlegel), larvae. *Aquaculture Research*, 37, 1172–1179.

TAKEMURA A and OKA M (1998) Immunochemical sexing of living yellowfin tuna, *Thunnus albacares* (Bonnaterre), using a vitellogenin like protein. *Aquaculture Research*, 29, 245–249.

TAKII K, BISWAS B K, YOSHIDA A, SAKOGUCHI K, IKEUE Y and BISWAS A (2010) Needs of taurine, feeding stimulant and lecithin in enzyme treated fish-meal diet for juvenile Pacific bluefin tuna. *Aquaculture 2010*. Baton Rouge, CA: World Aquaculture Society, 386.

THOMSON M, DEICHMANN M, CYYPIONKA K, CZYPIONKA A, CRAWFORD J, MILLER A, HUTCHINSON W and CHEN B (2010) Recent developments in Southern bluefin tuna larval and juvenile rearing, in Miyashita S, Takii K, Sakamoto W and Biswas A (eds), *Joint International Symposium of Kinki University and Setouchi Town on The 40th Anniversary of Pacific Bluefin Tuna Aquaculture*. Osaka: Kinki University Press, 53–58.

UOTANI I, SAITO T, HIRANUMA K and NISHIKAWA Y (1980) Feeding habit of bluefin tuna *Thunnus thynnus* larvae in the western north Pacific Ocean. *Nippon Suisan Gakkuishi*, 56, 713–717.

VIEIRA K R, OLIVEIRA J, BARBALHO M and ALDATZ J (2005) Aspects of the dynamic populations of blackfin tuna (*Thunnus atlanticus* – Lesson, 1831) caught in the Northeast Brazil. *Collective volume of scientific papers. International Commission for the conservation of Atlantic Tuna*, 58(5), 1623-1628.

WEXLER J B, SCHOLEY V P, OLSON R J, MARGULIES D, NAKAZAWA A and SUTER J M (2003) Tank culture of yellowfin tuna, *Thunnus albacares*: developing a spawning population for research purposes. *Aquaculture*, 220, 327–353.

WILLIAMS T D, ROLLINS M and BLOCK B A (2004) Intramuscular anesthesia of bonito and Pacific mackerel with ketamine and medetomidine and reversal of anesthesia

© Woodhead Publishing Limited, 2013

with atipamezole. *Journal of the American Veterinary Medical Association*, 225, 417–421.

ZERTUCHE-GONZÁLEZ J A, SOSA-NISHIZAKI O, VACA RODRIGUEZ J G, DEL MORAL SIMANEK R, YARISH C and COSTA-PIERCE BA (2008) *Marine Science Assessment of Capture-Based Tuna (*Thunnus orientalis*) Aquaculture in the Ensenada Region of Northern Baja California, Mexico.* Final Report to The David & Lucile Packard Foundation, 300 Second Street, Los Altos, CA, 95.

ZOHAR Y and MYLONAS C (2001) Endocrine manipulations of spawning in cultured fish: From hormones to genes. *Aquaculture*, 197, 99–136.

© Woodhead Publishing Limited, 2013

16

Developments in hatchery technology for striped catfish (*Pangasianodon hypophthalmus*)

P. T. Nguyen, T. M. Bui and T. A. Nguyen, Can Tho University, Vietnam and S. De Silva, Network of Aquaculture Centres in Asia and Pacific (NACA), Thailand and Deakin University, Australia

DOI: 10.1533/9780857097460.3.498

Abstract: Striped catfish (*Pangasianodon hypophthalmus*) farming in the Mekong Delta, Vietnam, is considered as a major, aquaculture development both in Vietnam and globally. One of the main drivers responsible for the explosive growth of the sector is considered to be the development and commercialisation of techniques for artificial propagation of the species. This chapter looks first at the life-cycle of the striped catfish and historical developments in hatchery technology before going on to discuss induced breeding of catfish in hatcheries together with larval and fry nursing. Finally, harvesting and transportation are described and possible future directions in the sector.

Key words: striped catfish, hatchery, fingerling, fry, spawning.

16.1 Introduction

The Mekong Delta in the Southern part of Vietnam (8°33′–10°55′N; 104°30′–106°50′E) is renowned for catfish farming. There are two genera of catfish in Vietnam; the genus *Pangasius* comprising 10 species and two species of the genus *Pangasianodon*. Of these, the striped catfish (*Pangasianodon hypophthalmus*) (Fig. 16.1) is the most important and has been farmed for decades. For several decades, this species was farmed in small ponds using wild-caught seed (Nguyen, 2009); larger-scale commercial culture followed in cages, pens and ponds commencing with the development of artificial mass seed production in the early part of the last decade (Tuan *et al.* 2003; Phan *et al.*, 2009; Bui *et al.*, 2011). The total production of the striped catfish reached 1.2 million tonnes in 2011 (Fig. 16.2) (Fisheries

© Woodhead Publishing Limited, 2013

Fig. 16.1 Striped catfish *Pangasianodon hypophthalmus* (Sauvage, 1878).

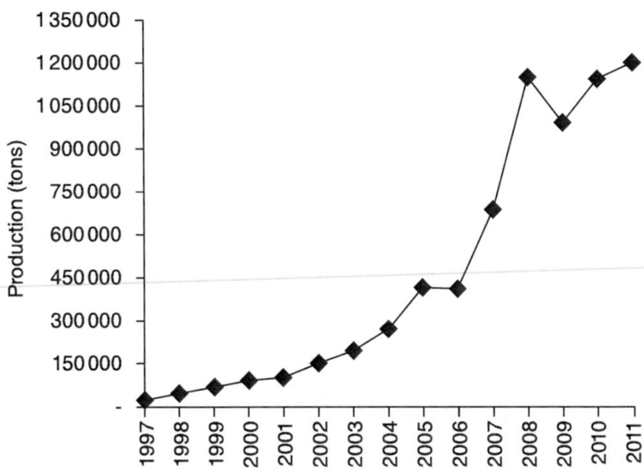

Fig. 16.2 Growth of striped catfish production in Vietnam, 1997–2011.

Directorate, 2011). This fish species is now known worldwide due to its products being exported to 136 countries and territories in 2010 (De Silva and Phuong, 2011). The striped catfish has become the 'Princess in Vietnamese aquaculture' in recent years (Phuong and Oanh, 2010). One of the key drivers for the fast growth of the striped catfish farming sector is the successes of seed production and associated development of hatchery techniques, including the uptake and successful adoption of the techniques by the farming community.

16.1.1 Life-cycle of the striped catfish

The striped catfish (*P. hypophthalmus*) is a migratory riverine species that undertakes long-distance migrations of more than several hundred kilometres between its upstream refuges and spawning habitats and its downstream feeding and nursery habitats (Van Zalinge *et al.*, 2002; Baran, 2006). The life-cycle of the striped catfish is intimately tied to the annual monsoon

© Woodhead Publishing Limited, 2013

flood cycle, with spawning in May–June at the start of the monsoon season (FAO, 2010–2012). The spawning ground of the striped catfish is generally known to be upstream of the Mekong River Delta, more specifically, below the Khone Falls on the Laos–Cambodia border (Van Zalinge *et al.*, 2002). The fish spawns at the beginning of the rainy season and the adhesive eggs are deposited on roots of aquatic macrophytes and other substrates. The newly hatched larvae drift downstream and are swept into floodplain areas in southern part of Cambodia and the Mekong Delta of Vietnam. The striped catfish is a facultative air-breather (Lefevre *et al.*, 2011) the air-breathing organ of this fish consisting of tiny blood vessels located around the palate which allow the fish to withstand low levels of dissolved oxygen.

16.1.2 Demand for striped catfish seeds

The success of the artificial seed production is considered to be one of the key drivers for the explosive growth of the striped catfish in Vietnam (Phuong and Oanh, 2009). The production of hatchery-reared seed has increased rapidly during the past years; larvae and fingerling production increased 18-fold and 26-fold, respectively, between 2002 and 2011 (Fig. 16.3). The seed demand is mostly for fingerling size for grow-out stocking.

16.1.3 Historical developments in striped catfish hatchery technology

Initially, the seed stock for striped catfish farming was wild-caught, primarily from Cambodian waters at the confluence of the Mekong, Bassac and Tonle Sap Rivers, the main nursery grounds of this species (Nguyen, 2009). However, the Cambodian authorities banned the capture of wild stocks in

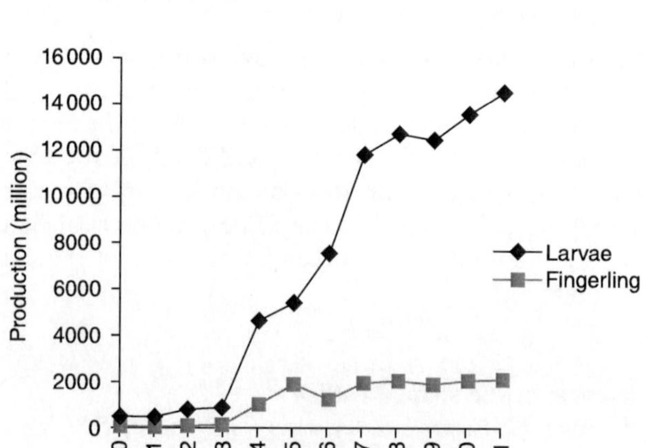

Fig. 16.3 Growth of striped catfish larvae and fingerlings in Vietnam (compiled by Phuong, 2011).

© Woodhead Publishing Limited, 2013

1994 (Ngor, 1999; Nguyen, 2009; Phuong and Oanh, 2010) and this ban led to a hiatus in the expansion of striped catfish farming. However, it also led to a concerted effort to study and develop artificial propagation techniques.

Artificial propagation of the striped catfish was first done in late 1978 (Xuan, 1994). However, the results were not sufficiently reliable for mass seed production and the research activities were discontinued. Then research on induced spawning of striped catfish re-commenced in 1995 under an EU funded project, with the involvement of four partner organisations from France and Vietnam, which was led by Can Tho University (Phuong and Oanh, 2010). This research led to the development of techniques for the induced spawning of the striped catfish in 1996, and transferred to hatchery operators in 2000 (Cacot, 1999; Cacot *et al.,* 2002). This initial development was followed by further improvements in the hypophysation technique on striped catfish, thereby consolidating the processes (Legendre *et al.*, 2000; Manosroi *et al.,* 2004). Since then, seed production of the striped catfish has increased significantly, currently fulfilling industry demand.

16.2 Striped catfish seed production: induced breeding in hatcheries

Striped catfish seed production in Vietnam is structured within two main sectors-hatchery and nursery (Bui *et al.*, 2010) (Fig. 16.4). Hatcheries produce large numbers of larvae which are mostly sold to nursery farms (94 %), while the nursery sector grows fry and fingerling for sale to the grow-out farms.

16.2.1 Hatchery design

The size of hatchery depends on the target production of larvae, fry and fingerling. Based on the total larvae produced, the hatcheries are divided into three groups: ≤ 300 million fry/year (about 36.4 %); 300–500 million larvae/year (27.3 %); and ≥ 500 million larvae/year (36.4 %) (Le and Le, 2010). Bui *et al.* (2010) reported that the size of hatcheries varied from 0.2 to 15 ha (average 2.5 ha), with 0.05–10 ha (average 1.59 ha) under water. The area of hatchery houses varied from 120 to 500 m^2. The larval production of hatcheries ranged from 10 to 3500 million, of which approximately 94 % were sold at larval stage (prior to commencement of feeding) to nursery farms (Fig. 16.5).

Hatcheries are generally designed with four main components: broodstock tanks, hatching jars (Zoug jar and Weiss shaped incubators) (Fig. 16.6), larval handling tanks and broodstock ponds (including potential and conditioning ponds). The hatching jars have a volume of 6–200 L (average 40 L). The total hatching jar volume of the hatchery reflects the production

© Woodhead Publishing Limited, 2013

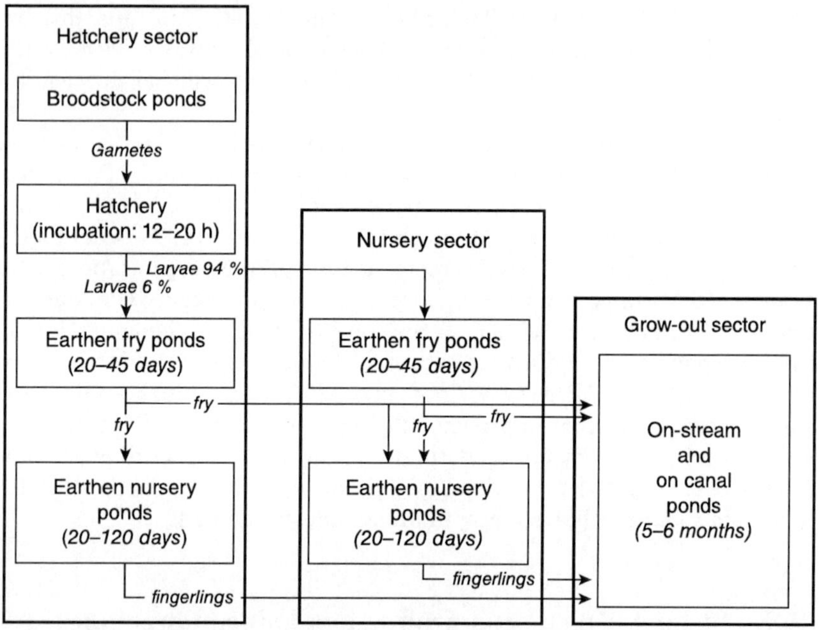

Fig. 16.4 Structure of the hatchery and nursery sectors of the striped catfish seed production in the Mekong delta and the movement of stock between each sector (from Bui *et al.*, 2010).

Fig. 16.5 Hatchery and broodstock ponds. (Photo DN Long)

© Woodhead Publishing Limited, 2013

Fig. 16.6 Weiss shaped incubators. (Photo DN Long)

capacity. According to Bui *et al.* (2010), the production of larvae varied from 1.8 to 3.8 million/L incubator/year.

It is estimated that there is a total of 172 hatcheries and 5775 nurseries operating in the Mekong Delta, including 'backyard' hatcheries. The latter are simple and generally cater to needs of a single farmer with an integrated hatchery, fry to fingerling and grow-out facilities. However, the great bulk of seed production occurs in facilities that are dedicated for this purpose.

Most hatcheries maintain large number of potential brood fish (varying from 350 to 29200), but only a small proportion of this stock is used for breeding in a year. Therefore, many hatcheries have large areas of potential and conditioning broodstock ponds. The need to maintain such large numbers of potential broodstock, at a relatively high maintenance cost, which has been the tradition, has been questioned. The recently developed guidelines on 'Better Management Practices' for the catfish farming sector recommend that for an average hatchery operation to be successful the number of potential broodstock maintained could be around 200 (De Silva *et al.*, 2011).

16.2.2 Broodstock sourcing

Broodstock sources include pond (domesticated) and wild collected. Le and Le (2010) found that 78.1 % of hatcheries were using domesticated broodstock (collected from grow-out ponds), 6.3 % using wild-collected broodstock and 15.6 % using both sources. There is a clear trend towards in the increased use of pond reared broodstock. Belton *et al.* (2010) reported in detail on broodstock procurement and concluded that broodstock could be sourced from extensive grow-out, export-orientated grow-out, capture

© Woodhead Publishing Limited, 2013

fisheries, own hatcheries or other hatcheries, in that order of importance. Bui *et al.* (2010) reported that male and female broodstock (including potential broodstock) should be from 0.5–8 kg and 0.5–12 kg in weight (average 3–6 kg), respectively. The age of fish at breeding should be over three years (normally five to six years).

Broodstock are stocked at a rate of 5 t/ha (varying from 4 to 6 t/ha). The best stocking density is considered to be 4–5 t/ha. Males and females can be maintained at the ratio of three to four females to six to seven males, either separately or together. Broodstock selected for induced spawning must weigh at least 1.75 kg for females and 1.5 kg for males. Broodstock are generally discarded when they reach 10 kg or when the relative productivity (number of viable eggs produced) is less than 5 % of the female weight. Hatcheries normally recruit new broodstock on a regular basis of every two to three years; new broodstock are obtained both from grow-out farms and the wild. The replacement and procurement of fresh broodstock are currently done on an *ad hoc* basis and based entirely on the farmers' experience/intuition. This is an area that needs much scientific input to ensure genetic diversity is maintained.

16.2.3 Broodstock conditioning and maturation culture

Broodstock culture systems

Broodstock are cultured in earthen ponds. Normally, hatcheries have many potential broodstock ponds and a smaller number of conditioning and maturation culture ponds (Fig. 16.7). On average, about eight (range 3–25) ponds are used for maintaining the large number of potential broodfish. The ponds are around 0.16 ha (from 0.02–3.0 ha) and 3–4 m deep.

Fig. 16.7 Broodstock potential and conditioning ponds. (Photo DN Long)

© Woodhead Publishing Limited, 2013

Broodstock ponds are carefully treated before use, usually by drying, sludge removal, liming (quick lime) and salting (common salt). Ponds are re-treated yearly or biannually. Most hatcheries use a river or canal as the main water source. Less than 50 % of farms screen the inlet water, but most farms treat the water once ponds are filled (Bui *et al.*, 2010). The pond water is exchanged at around 10 % during the days of high tide (six to eight days/month) during the conditioning culture period, and is exchanged daily up to 20–30 % during the maturation culture period.

Feed and feeding

Both manufactured pelleted feed and farm-made feed are used for striped catfish broodstock, singly or in combination, the latter being used more as a conditioning diet. The formulation of farm-made feed is relatively simple, often comprising a combination of locally available ingredients such as sun-dried trash fish (20 %), fresh trash fish (40 %), rice bran (30 %) and vegetable (10 %) (Bui *et al.*, 2010; Phuong, 2012).

The overall feeding rates for broodstock range from 0.2–10 % body weight/day (average 2.8 %) and 0.2–25 % body weight/day (average 5.4 %) for manufactured pelleted feed and farm-made feeds, respectively (Bui *et al.*, 2010). Phuong (2012) reported that the feeding rates for broodstock vary according to the culture period; fish are fed 4–5 % body of weight for the preparatory period, and 1.5–2 % for the maturation and spawning periods.

16.2.4 Maturity and spawning season

The striped catfish can reach full maturation in captive conditions and can be induced to spawn (CAB International, 2006). The fish spawns throughout the year, but the peak breeding period is in May–July (as in nature), which corresponds to the onset of the rainy season (Bui *et al.*, 2010). Good quality broodstock from potential broodstock ponds are normally selected, based almost entirely on farmer experience, and transferred to maturation ponds for induced spawning about two to three months prior to spawning.

16.2.5 Hormone treatment and gametes collection

Broodstock selection

Principally, the selected females and males must be healthy, without visible injury or abnormal signs. The females should have a big belly, thin abdominal skin, swollen, reddish genitals and well-developed ovarian follicles (Fig. 16.8). However, sexual dimorphism is not clearly evident externally; therefore monitoring oocyte development using an intra-ovarian biopsy with a flexible catheter is often employed to evaluate the maturity state (Fig. 16.9). Well-matured broodstock should have an oocyte diameter of 1.0–1.1 mm,

© Woodhead Publishing Limited, 2013

Fig. 16.8 Selection of broodstock for induced spawning basing on external appearance.

Fig. 16.9 Checking oocyte of broodstock using a flexible catheter. (Photo DNLong)

and the males should discharge milt on application of gentle pressure on the abdomen.

Hormone injection
There are several hormones and stimulating agents used to induce the spawning in striped catfish, e.g. hCG (human chorionic gonadotropin), ovaprim and pituitary gland. However, hCG is most commonly used because of its proven high efficacy. hCG is injected in females at doses of 200–6500 IU/kg at a time and may be injected up to four or five times before

© Woodhead Publishing Limited, 2013

the fish are finally induced (Bui *et al.*, 2010). However, the injection can be two or three times in the peak spawning season and three or four times in the offseason. In practice, the total doses of hCG vary from 5500 to 6500 IU/kg. The males receive only one injection of 1000 IU/kg, coinciding with the time of the third injection for the females (Fig. 16.10 and Table 16.1).

16.2.6 Egg and sperm collection, fertilisation and incubation

Ovulation occurs around 10 hours after the last injection. Eggs are collected as soon as possible by stripping (Fig. 16.11), because the survival time of eggs is short. According to Campet *et al.* (1999), the proportion of deformed larvae increased (24 %) and the hatching rate declined (35 %) if

Fig. 16.10 Hormone injection for broodstock.

Table 16.1 Hormone (hCG) dose rates and timing of hormone injection used to induced spawning in the striped catfish

Female			Male		
Injection No.	Time (hr)	Dose (UI/kg)	Injection No.	Time (hr)	Dose (UI/kg)
1st	0	200–2000 (542)			
2nd	8–24	200–2000 (597)			
3rd	16–18	200–2000 (893)	1st	20–48	167–3500 (1060)
4th	24–72	800–6500 (3442)			
Combined		3000–8150 (5400)			

Note: Values in parentheses are the mean.
Source: Modified after Bui *et al.*, 2010.

© Woodhead Publishing Limited, 2013

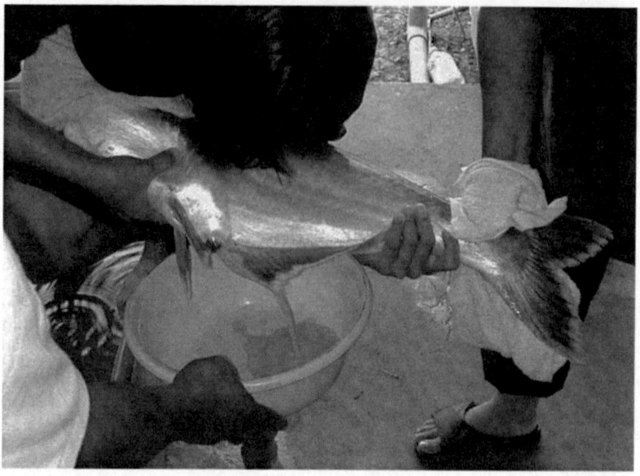

Fig. 16.11 Stripping eggs.

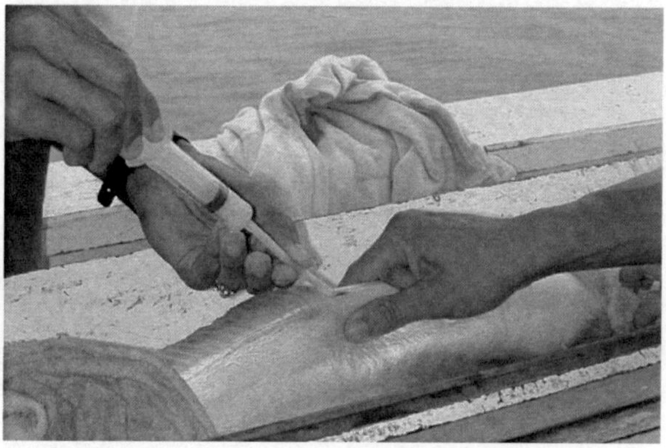

Fig. 16.12 Collection of sperm by syringe. (Photo Cacot)

the eggs were collected 3 h after ovulation. Derivaz *et al.* (2000) also recognised that 4 h after ovulation, the proportion of normal larvae was reduced by half. The average relative fecundity of striped catfish is 150 000 eggs/kg female.

Milt is collected into an immobilisation solution (containing 10 g Tris-HCL in 1 L of 9 ppt water or physiological solution) container by pressing gently on the abdomen of the fish or using a syringe (Fig. 16.12). The milt is diluted five times in immobilisation solution for direct use or temporary storage at 4–5 °C for 24 h.

A dry fertilisation method is normally used for the striped catfish, when eggs and milt are mixed gently. Fertilisation solution (containing 3 g urea and 4 g salt in 1 L of water) is added to the mixture of eggs and milt to

© Woodhead Publishing Limited, 2013

trigger fertilisation after 5 min. The fertilised eggs are then treated with tannic acid solution of 1‰ with a ratio of 1:1 (one volume of eggs and one volume of tannic acid solution) for 5–10 s. The fertilised eggs are then transferred into Zoug Jars or a Weiss incubation system. The incubators have a volume of 6 L or 40 L and are stocked at 0.23 kg eggs/L (varying 0.02–1.5 kg eggs/L).

Fertilisation and hatching rates vary between the peak and off-season production periods. The fertilisation rates vary from 10–99 % (averaging 86 %) and 28–95 % (averaging 71 %); and the hatching rates are 60–100 % (averaging 88 %) and 50–100 % (averaging 77 %) during the peak and off-season production periods, respectively (Bui *et al.,* 2010).

16.3 Striped catfish seed production: larval and fry nursing

The larval rearing can be a part of the hatcheries but, in most cases, larvae are nursed to fry and fingerling stages by a nursery sector, which is separated from the hatchery activities. The commercial nursing of larvae-to-fry and fry-to-fingerling is done in earthen ponds to avoid mass mortality, because of the cannibalistic nature of the fish during the first week post-hatching.

16.3.1 Nursery pond construction

The nursery ponds are located at on-stream and on-canal sites, to facilitate water exchange and ease of transportation of fry and fingerlings. The most popular pond shape is rectangular with a length to width ratio of three:four, an area of 1000–5000 m^2 and a water depth of 1.5–2 m. The inlet and outlet usually have a diameter of 20–40 cm depending on the pond size and are located at the opposite sides of the ponds.

16.3.2 Pond preparation

Nursery ponds for larval to fry rearing are prepared about a week before stocking to encourage the growth of live food. The pond preparation includes the removal of bottom sludge, liming 10–15 kg/100 m^2, drying the pond bottom for three to five days and killing all unwanted organisms. In cases where the ponds cannot be completely drained, the use 0.5–1 kg/100 m^2 derris root (*Derris elliptica*) containing rotenone or saponin products of 1 kg for 300–500 m^3 should be effective to kill all unwanted organisms.

The water supplied into the pond must be of high quality (such as pH from 6.4–8.5, dissolved oxygen ≥ 3 mg/L, free of toxicants). The water is screened by a fine mesh to prevent the entrance of eggs and larvae of other undesirable organisms. The water in the pond is levelled up to 1 m and

© Woodhead Publishing Limited, 2013

commonly treated with chlorine (1 kg/1000 m^3) or formalin (25 mg/L). However, chlorine is most commonly used because of the lower cost.

16.3.3 Larvae to fry nursing

Larvae of the striped catfish are cannibalistic, usually causing low survival rate during the first days of the nursing period. Low stocking density and creation of natural food in rearing water are very important to reduce this mortality.

Pond fertilisation

The nursing pond must be prepared well to permit the growth of natural food by adding fertiliser 24 h after treatment. This is important for the fish larvae in the first days of post-stocking. Fish powder (or low value fish meal) (2–3 kg), soybean meal (2–3 kg) or other products (such as Zeofish 4 kg + 1 kg blood powder DP92, or super benthos 6–8 kg) can be added into 1000 m^2 of pond. In addition, supposedly beneficial bacterial products (or microbial-products) can be added into the pond at a rate of 300 g/1000 m^3, together with 1–2 kg of live food (such as *Moina*). It should be pointed out that explicit scientific evidence is not available at present to indicate the beneficial effects, if any, of the addition of commercial products as such Zeofish, etc., which are readily available in the market and very aggresively marketed. Nevertheless, most farmers tend to use such products, incurring high costs in spite of the lack of scientific evidence of the claimed benefits.

Stocking

The larvae of the striped catfish have to be transferred to rearing tanks or ponds within 24 h after hatching and fed live food (Fig. 16.13). Good quality

Fig. 16.13 Collection of larvae after hatching.

© Woodhead Publishing Limited, 2013

larvae usually have no abnormal signs, are of uniform size, swim actively and respond to external stimuli quickly. The stocking density of larvae in ponds varies from 500 to 800 larvae/m^2. However, Bui *et al.* (2010) reported that the stocking density of larvae was highly variable among farmers and ranged from 250 to 2000 larvae/m^2 (average 863 larvae/m^2). Larvae are transported from hatcheries to nursing ponds in oxygenated bags (5000–8000 larvae/L of water) in the early morning (7:00–10:00 am) or late afternoon to avoid direct exposure to sunlight. The larvae are acclimated to rearing pond water by keeping the bags in the pond for 15–30 min before releasing.

First feeding

The striped catfish larvae commence exogenous feeding two days after hatching (or 48 h) even when the yolk is not completely absorbed and the digestive tract is still not fully functional; at this time, the larvae require live food organisms for optimal growth and development (CAB International, 2006). In tank conditions, *Artemia* nauplii, *Moina* and *Tubifex* are usually fed to the larvae at a high feeding rate combined with slight aeration. However, in pond conditions, pond fertilisation to stimulate the growth of natural food together with the additional stocking of zooplankton and zoobenthos species (such as *Moina, Artemia, Tubifex*) are important to enhance the survival rate of the larvae. Hung *et al.* (2002) and Jacques *et al.* (1999) reported that *Artemia* is an excellent starter food for striped catfish, and gives the best growth performance. The feeding schedule for larvae to fry nursing is given in Table 16.2.

Pond management

Water quality and larval behaviour need monitoring/checking early every morning. Water colour should be maintained green (similar to banana leaf colour). Presence of predators such as snakes, frogs and carnivorous fish, insects etc. should be regularly checked and all precautions taken to prevent/minimise their entry into ponds. The use of a light at the pond surface in the evening to gather harmful insects (such as *Notonecta* and dragonfly larvae), a net fence to prevent entry of frogs and scooping out tadpoles from the water surface must be regularly carried out during the nursing period. Overfeeding should be avoided to prevent deterioration of water quality. However, microbial-products such as EM, Bio-Tab®, Zeofish, yucca can be administered weekly to enhance water quality. The application of lime (such as dolomite or $CaMg(CO_3)_2$, $CaCO_3$) at the rate of 3–5 kg/100 m^2 pond is required after heavy rain.

The larvae metamorphose to fry (3000–4000 fry/kg) 20–45 days post-stocking. Fry can be graded and transferred to other ponds for nursing to fingerling size (Fig. 16.4). The survival rate of the larvae to fry varies from 30 to 50 %.

© Woodhead Publishing Limited, 2013

Table 16.2 Feeding table for the striped catfish larvae

Descriptions	Week 1st (*)	Week 2nd (*)	Week 3rd & later
Feeds	Farm-made feed, which is formulated from soybean meal or fish meal (250 g), fish milk product (250 g), protein yeast (NuPro®) (50 g) and Bio-Mos (1 g)	Concentrated powder (40 % protein): 0.5 kg and nutritional products (50 g Nupro + 1 g Bio-Mos) for each feeding	Commercial pellet (30–35 % protein) with instruction for feed size according to larval age with addition of Bio-Mos 2 kg/ton of feed and vitamin C (1–2 kg/ton of feed)
Feeding rate	The mentioned amount is for 1 feeding	The mentioned amount is for 1 feeding and is increased 10–15 % daily	According to the instruction of feed manufacturer
Feeding frequencies	Five feedings: at 7 h, 10 h, 13 h, 16 h and 19 h	Four feedings: at 7 h, 11 h, 15 h and 19 h	Three to four feedings a day
Feeding methods	Mixing mixture with water and spraying over the pond surface	Mixing mixture with water and spraying over the pond surface	Soaking Bio-Mos, NuPro in water for 15 min then spraying onto commercial pellet before feeding the fish

*Calculated for 1 million stocked larvae.

16.3.4 Fry to fingerling nursing

Nursing of fry to fingerling is conducted in earthen ponds. Normally, nursery farms have ponds for both nursing fry and fingerlings, which have similar characteristics. Fry of 20–45 days are harvested and transferred to fingerling nursing ponds within a farm, or sold to other nursery farms.

Stocking density

The stocking density of fry varies from 200 to 300 ind./m^2; fry should be in good health, indicated by active swimming and no signs of disease and/or injury, and of uniform size. The survival rate of the fry to fingerlings ranges from 40 to 50 %.

Feed and feeding

Fry are fed manufactured pelleted feeds containing 30–45 % crude protein according to size. The feeding rates vary from 6 to 8 % body weight with two or three feedings daily (Phuong, 2012). Nutrient supplements such as vitamin C (1–2 kg/t of feed) and Bio-Mos® (2 kg/t of feed) are regularly used during the nursing period.

© Woodhead Publishing Limited, 2013

16.3.5 Larvae to fingerling nursing

Nursing of larvae to fingerling is also conducted in earthern ponds. The nursing procedures are similar to those used for nursing larvae to fry and fry to fingerling. However, the average stocking density is 724 ind./m^2. The nursing period is 2.72 months (varying from 2.5 to 3 months). The average survival rate is around 16.6 % depending on season, 15–20 % in the peak season (March–May) but only 5–7 % in the remaining months .

16.3.6 Economic aspects of fry and fingerling production

In general, and similar to many cultured species, there is very little published data on the economics of fry and fingerling production. However, Le and Le (2010) studied the economic aspects of both larvae production and larvae to fry and fingerling rearing for catfish in the Mekong delta. Tables 16.3 and 16.4, respectively, give details on each of the above. The information in these tables confirms that there are many cost factors in these operations, and also provides further insights into each of the operations. For example, it is evident from Table 16.3 that most females are spawned more than once in a year with the best mean net income obtained when females were spawned five or six times. Similarly, in larvae to fingerling rearing the best net income was achieved when water was exchanged every five days with stocking densities of 500–700 m^2 (Table 16.4).

Table 16.3 Factors affecting the yield (fry production) and net income for catfish hatcheries

Variable	Fish yield	Net income
	Mean ± SD	Mean ± SD
Unit	Million fry/L	VND million/L/yr
Total volume of Weiss tank		
<100 L	3.1 ± 1.6	6.3 ± 3.0
100–200 L	3.0 ± 3.3	7.7 ± 4.2
200–300 L	3.8 ± 3.1	8.2 ± 4.6
>300 L	1.8 ± 1.0	3.8 ± 3.3
Number of times of spawning per brooder/year		
≤2 times	2.6 ± 0.9	5.2 ± 1.8
3–4 times	1.9 ± 1.3	5.7 ± 4.1
5–6 times	5.3 ± 3.6	9.3 ± 5.1
≥7 times	4.2 ± 3.5	7.3 ± 4.5

Note: 18 000 VND = 1 US$
Source: Modified from Le and Le, 2010.

© Woodhead Publishing Limited, 2013

Table 16.4 Factors affecting the yield, production costs and net income of nursery rearing of striped catfish

Variable	Fish yield	Production costs	Net income
	Mean ± SD	Mean ± SD	Mean ± SD
Unit	1000 mil. fingerlings/1000 m^2 pond/cycle	VND mil./1000 m^2 of pond area/cycle	VND mil./1000 m^2 of pond area/cycle
Water depth of the nursery pond			
≤1.5 m	75.0 ± 44.5	10.4 ± 7.6	9.9 ± 12.2
1.5–2.0 m	119 ± 76.1	16.5 ± 10.9	41.1 ± 63.0
2.0–2.5 m	128 ± 65.3	11.6 ± 8.4	49.6 ± 48.8
≥2.5	156 ± 97.3	11.9 ± 6.0	18.3 ± 15.6
Frequency of water exchange			
Daily (1 day/time)	133 ± 78.5	13.4 ± 9.20	32.9 ± 41.3
3 days/time	87.6 ± 43.3	13.5 ± 8.90	37.1 ± 60.4
5 days/time	119 ± 63.4	11.8 ± 6.70	45.9 ± 64.3
7 days/time	137 ± 99.0	13.8 ± 11.2	22.3 ± 22.7
Stocking density of hatchlings			
≤250/m^2	63.9 ± 44.1	13.4 ± 10.5	34.5 ± 75.8
250–500/m^2	111 ± 55.1	15.8 ± 9.9	39.4 ± 50.0
500–700/m^2	116 ± 52.6	11.8 ± 4.1	45.5 ± 48.3
≥750/m^2	184 ± 103	7.5 ± 2.8	17.0 ± 13.8
Size of harvested fingerling (height)			
≤1.5 cm	140 ± 96.5	10.2 ± 4.6	20.1 ± 43.6
1.5–2.0 cm	119 ± 69.5	14.3 ± 10.7	35.5 ± 44.1
≥2.0 cm	103 ± 42.7	15.8 ± 10.4	51.9 ± 53.7

Note: 18000 VND = 1 US$
Source: Modified from Le and Le, 2010.

16.4 Harvesting and transportation

The harvest of fry and fingerling in ponds is done by seining. The ponds are normally disturbed in order to train the fish to adapt to transport conditions about three or four days before harvest. The seined fish are sorted in pond, with groups of uniform size fish kept in hapas (Fig. 16.14) before transferring to transportation facilities (bags or boats).

The transportation of small fry is conducted in oxygenated plastic bags, while larger fry are transported in composite tanks with aeration (open transportation). The transport density of fry depends on size and duration, but it can range from 100 to 40000 ind./L of water (average 7314) (Bui *et al.*, 2010). The fish could be in transit for 6–25 h, and are normally treated with common salt prior to or during transport.

The transportation of fingerlings to grow-out ponds is by boat with aeration (Fig. 16.15). The boats usually have a capacity of 20–30 t. Fingerlings

© Woodhead Publishing Limited, 2013

Fig. 16.14 Conditioning fry in hapa before transportation. (Photo DN Long)

Fig. 16.15 Boat used to transport fry and fingerling.

are stocked at a density of three fish/L for fingerlings of 30–33 g, and 6–6.5 fish/L for fingerlings of 14–16 g. The water is exchanged 20–30 % every 6 h during transport.

16.5 Future trends

The demand for striped catfish seeds will increase in the coming years to keep pace with the increase in nationally planned production. The induced spawning technique of the striped catfish is relatively well developed and

© Woodhead Publishing Limited, 2013

is being gradually improved in order to obtain a higher productivity and a better larval, fry and fingerling quality. However, each of the stages in the cycle needs improvement, mortality needs to be reduced at each stage and costs need to be rationalised. Hence further research is needed.

Research is being conducted on broodstock source selection and management, broodstock feeds, hormone and stimulating agents to induce spawning and genetic improvement for enhancing growth performance. Research is needed in the coming years to develop genetic improvement for specific disease resistance and saline water tolerance. Research on nursery techniques has also been planned to improve the quality and survival rate at different nursing stages (larvae to fry and fry to fingerling) by improving compound feeds and live feed generation in ponds. As pointed out previously, most farmers are encouraged to use substances and compounds purporting to enhance/improve production through reduced mortality, disease occurrence, etc. These treatments and substances impose substantial costs to farmers, yet the inefficacy has not been proven, and research in this regard is urgently needed, especially so the catfish farming sector can maintain long-term economic viability.

16.6 References

BARAN E (2006) *Fish migration triggers in the Lower Mekong Basin and other tropical freshwater systems*, MRC Technical Paper No. 14. Vientiane: Mekong River Commission.

BELTON B, LITTLE D C and LE X SINH (2010) Pangasius catfish seed quality in Vietnam. Part 1. User and producer perceptions on broodstock and hatchery production. *Aquaculture Asia*, March–April, 36–39.

BUI TAM M, PHAN LAM T, INGRAM B A, NGUYEN THUY T T, GOOLEY G J, NGUYEN HAO V, NGUYEN PHUONG V and DE SILVA S S (2010) Seed production practices of striped catfish, *Pangasianodon hypophthalmus* in the Mekong Delta region, Vietnam. *Aquaculture*, 306, 92–100

CAB INTERNATIONAL (2006) *Case study: Artificial seed production of Pangasiid catfish in the Mekong Delta, Vietnam*, Aquaculture Compendium 2006. Wallingford: CABI.

CACOT P (1999) Description of the sexual cycle related to the environment and set up of the artificial propagation in *Pangasius bocourti* (Sauvage, 1880) and *P. hypophthalmus* (Sauvage, 1878) reared in floating cages and in ponds in the Mekong Delta, in Legendre M and Pariselle A (eds), *The Biological Diversity and Aquaculture of Clariid and Pangasiid Catfishes in South East Asia. Proceedings of the mid-term workshop of the 'Catfish Asia Project' Cantho, Vietnam, 11–15 May 1998.* Toulouse: PARAGRAPHIC, 71–89.

CACOT P, LEGENDRE M, DAN T Q, TUNG L T, LIEM P T, MAROJOULS C and LAZARD J (2002) Induced ovulation of *Pangasius bocourti* (Sauvage, 1880) with a progressive hCG treatment. *Aquaculture*, 213, 199–206.

CAMPET M, CACOT P, LAZARD J, DAN T Q, MUON D T and LIEM P T (1999) Egg quality of an Asian catfish of the Mekong River (*Pangasius hypophthalmus*) during the process of maturation induced by HCG injections, in Legendre M and Pariselle A (eds), *The Biological Diversity and Aquaculture of Clariid and Pangasiid Catfishes in South East Asia. Proceedings of the mid-term workshop of the 'Catfish*

© Woodhead Publishing Limited, 2013

Asia Project' Cantho, Vietnam, 11–15 May 1998. Toulouse: PARAGRAPHIC, 113–117.

DE SILVA S S and PHUONG N T (2011) Striped catfish farming in the Mekong Delta, Vietnam: A tumultuous path to a global success. *Reviews in Aquaculture,* 3, 45–73.

DE SILVA S S, NGUYEN V H, NGUYEN P T, BUI M T, NGUYEN N, VO M S, PHAN L T, DUONG N L, NGUYEN T T T, GOOLEY G J and INGRA B A (2011) *Better Management Practices for Striped Catfish (tra) Farming in the Mekong Delta, Viet Nam,* Version 3. Agricultural Publishing house, Ho Chin Minh City.

DERIVAZ M, CHAU L H, VU N H, CAMPET M, CACOT P and LAZARD J (2000) Survival of the ova in vivo: compared study of *Pangasius bocourti, P. hypophthalmus* and their hybrid (female *P. hypophthalmus* x male *P. bocourti*). *Final Workshop of the Catfish Asia Project,* May 15–20, Bogor, Indonesia.

FAO (2010–2012) Cultured Aquatic Species Information Programme. *Pangasius hypophthalmus.* Text by D Griffiths, P Van Khanh and T Q Trong. In: FAO Fisheries and Aquaculture Department [online]. Rome. Updated 14 January 2010. (Accessed September 2012) (available at: http://www.fao.org/fishery/culturedspecies/Pangasius_hypophthalmus/en).

FISHERIES DIRECTORATE (2011) Review of the 2011 fisheries plan; and proposed responsibility and key solutions for the 2012 fisheries plan (in Vietnamese).

HUNG L T, TUAN N A, CACOT P and LAZARD J (2002) Larval rearing of the Asian catfish, *Pangasius bocourti* (Siluroidei, Pangasiidae): alternative feeds and weaning time. *Aquaculture,* 212, 115–127.

JACQUES S, HUNG L T, SUBAGJA J and LEGENDRE M (1999) Effects of type of prey, feeding level, prey accessibility and water aeration on growth and survival of *Pangasius hypophthalmus* larvae (Siluroidae, Pangasidae), in Legendre M and Pariselle A (eds), *The Biological Diversity and Aquaculture of Clariid and Pangasiid Catfishes in South East Asia. Proceedings of the mid-term workshop of the 'Catfish Asia Project' Cantho, Vietnam, 11–15 May 1998.* Toulouse: PARAGRAPHIC, 137–145.

LE X S and LE L H (2010). Supply and use of catfish (*Pangasianodon hypophthalmus*) seed in the Mekong delta of Vietnam. *Aquaculture Asia,* January–March, 27–32.

LEFEVRE S, HUONG D T T, WANG T, PHUONG N T and MARK B (2011) Hypoxia tolerance and partitioning of bimodal respiration in the striped catfish (*Pangasianodon hypophthalmus*). *Comparative Biochemistry and Physiology, Part A,* 158, 207–214.

LEGENDRE M, SLEMBROUCK J, SUBAGJA J and KRISTANTO A H (2000) Ovulation rate, latency period and ova viability after GnRH- or hCG-induced breeding in the Asian catfish *Pangasius hypophthalmus* (Siluriformes, Pangasiidae). *Aquatic Living Resources,* 13, 145–151.

MANOSROI J, MENG-UMPHAN K and MANOSROI A (2004) Maturation induction of *Pangasius hypophthalmus* using gonadotropin releasing hormone analogue (GnRHa) in combination with domperidone, in oil suspension dosage. *Asian Fisheries Science,* 17, 39–42.

NGOR P B (1999) Catfish fry collection in the Mekong River of Kandal and Phnom Penh, in: Van Zalinge N P, Nao T and Deap L (eds), *Present Status of Cambodia's Freshwater Capture Fisheries and Management Implications.* Nine presentations given at the Annual Meeting of the Department of Fisheries, Phnom Penh, 19–21 January. Phnom Penh: Mekong River Commission Secretariat and Department of Fisheries, 116–134.

NGUYEN T T T (2009) Patterns of use and exchange of genetic resources of the striped catfish *Pangasianodon hypophthalmus* (Sauvage 1878). Reviews in Aquaculture, 1, 222–229.

PHAN L T, BUI T M, NGUYEN T T T, GOOLEY G J, INGRAM B A, NGUYEN H V, NGUYEN P T and DE SILVA SENA S (2009) Current status of farming practices of striped catfish,

© Woodhead Publishing Limited, 2013

Pangasianodon hypophthalmus in the Mekong Delta, Vietnam. *Aquaculture*, 296, 227–236.

PHUONG N T (2012) On-farm feed management practices for striped catfish (*Pangasianodon hypophthalmus*) in Mekong River Delta, Viet Nam, in Hasan M R and Shipton T (eds) *On-farm feeding and feed management in aquaculture*, Fisheries and Aquaculture Technical Paper. Rome: FAO (in press).

PHUONG N T and OANH D T H (2010) Striped catfish (*Pangasianodon hypophthalmus*) aquaculture in Viet Nam: An unprecedented development within a decade, in De Silva S S and Davy F B (eds) *Success Stories in Asian Aquaculture*. Amsterdam: Springer-IDRC-NACA, 133–150.

TUAN N A, PHUONG N T, LIEM P T and THUONG N V (2003) Results of the study on Pangasius catfishes and their future development. *Journal of Mekong Fisheries*. pp. 129–134. Agricultural publishing house. Ho Chi Minh city. Viet Nam (in Vietnamese).

VAN ZALINGE N, SOPHA L, BUN N P, KONG H and JORGENSEN J V (2002) *Status of the Mekong Pangasianodon hypophthalmus resources, with special reference to the stock shared between Cambodia and Viet Nam*, MRC Technical Paper No. 1. Phnom Penh: Mekong River Commission.

XUAN T T (1994) Some biological characteristics and artificial reproduction of river catfish (*Pangasus micronemus* Bleeker) in the South Vietnam, *International Workshop on the Biological Bases for Aquaculture of Siluriformes*, May 24–27, Montpellier.

© Woodhead Publishing Limited, 2013

17

Aquaculture production of meagre (*Argyrosomus regius*): hatchery techniques, ongrowing and market

N. J. Duncan and A. Estévez, IRTA, Spain, H. Fernández-Palacios, Universidad de las Palmas de Gran Canaria, Spain, I. Gairin, IRTA, Spain, C. M. Hernández-Cruz, J. Roo and D. Schuchardt, Universidad de las Palmas de Gran Canaria, Spain and R. Vallés, IRTA, Spain

DOI: 10.1533/9780857097460.3.519

Abstract: Meagre (*Argyrosomus regius*) have good potential for aquaculture. Culture characteristics include controlled spawning in captivity, relatively easy larval rearing, fast growth, good feed conversion ratios and no maturation during ongrowing. Whilst characteristics of the final product include good processing yield, low fat content, good taste and firm texture. This chapter details the present knowledge on the biology of the species, the culture protocols being used through the entire production cycle (broodstock management, larval rearing, pre-ongrowing and ongrowing), product quality, market and economics of production. Finally future prospects and challenges to achieve mass production are examined.

Key words: meagre, *Argyrosomus regius*, aquaculture, broodstock management, larval rearing, ongrowing, market, economics.

17.1 Introduction

The family *Sciaenidae* includes 70 genus and 270 species (Nelson, 1994) which are distributed in the Atlantic, Indian and Pacific oceans. Sciaenids are commonly referred to as drums or croakers for their ability to produce drumming sounds during the spawning period (Lagardère and Mariani, 2006) with a special sonic muscle in close association with the swimbladder. Two sciaenid species can be considered to have an established aquaculture production, the red drum (*Sciaenops ocellatus*) (FAO, 2005–2011a) and the yellow croaker (*Larimichthys crocea*) (Chen *et al.*, 2003). In 2009, production of red drum was 51 476 t and yellow croaker was 66 021 t (FAO, 2011).

© Woodhead Publishing Limited, 2013

Several sciaenids are new to aquaculture and have good aquaculture potential: *Argyrosomus japonicus* in Australia, southern Africa and Taiwan, *Sciaena umbra* in Greece and Turkey, *S. ocellatus* in China, USA, Israel, Mexico and Taiwan, *Umbrina cirrosa* in Cyprus, Spain, Greece, Italy and Turkey and *A. regius* in Spain, Egypt, France, Italy, Morocco and Turkey. Characteristics that give meagre high aquaculture potential and which have made other Sciaenids such as red drum successful aquaculture specie are: relatively easy broodstock management to obtain good quality eggs; relatively easy larvae rearing with standard industry live feeds and formulated diets; and juveniles that do not present reproductive maturation during ongrowing, grow fast and that can achieve good feed conversion ratios (Mañanós *et al.*, 2009).

The meagre, *A. regius* (Fig. 17.1), is a sciaenid distributed in the Mediterranean and Black Sea and along the Atlantic coasts of Europe and Africa (Poli *et al.*, 2003). The fish can grow up to 2 m and reach 50 kg in the wild (FAO, 2005–2010b). In spring, coincident with periods of freshwater discharge (Hall, 1984), meagre enter the mouths of estuaries in groups for spawning, which takes place until late summer. Areas that remain important for spawning are the estuary of the river Gironde, Bay of Biscay, France (Quéméner *et al.*, 2002), river Tagus, Portugal and river Guadiana, Southern Spain and Portugal (Gonzalez-Quiros *et al.*, 2011). During winter, meagre return to deeper waters possibly of the west coast of Africa to feed (Quéméner *et al.*, 2002). Juveniles leave estuaries at the end of summer and spend two to three years in coastal waters before migrating to offshore feeding areas. Water temperature seems to be the most important factor determining the migrations and reproduction of meagre (FAO, 2005–2010b).

Global fisheries production for all *Sciaenidae* species ranges from 5000 to 10 000 t per annum. In Europe, meagre fisheries production is lower, ranging from a few hundreds to 1500 t (FAO, 2011). The combination of

Fig. 17.1 Meagre (*Argyrosomus regius*). (Photo by Neil Duncan).

© Woodhead Publishing Limited, 2013

low catches principally in the areas of spawning has resulted in lucrative niche markets in Portugal, Southern Spain and parts of France and Italy. Aquaculture of meagre began in the late 1990s in France with an annual production of 30 t in 1997. In 2009, production reached 2200 t in Egypt, 1348 t in Spain, 418 t in France, 102 t in Italy and 44 t in Portugal (FAO, 2011). Ongrowing techniques are similar to those used for bass (European seabass, *Dicentrarchus labrax*) and bream (gilthead seabream, *Sparus aurata*), principally large circular (25 m diameter) sea surface cages. Meagre is a promising aquaculture species for its high growth rate (around 1 kg per year) and good feed conversion ratio (Jiménez *et al.*, 2005). Meagre flesh quality is very well considered by the consumer due to its low muscle fat content (Poli *et al.*, 2003; Piccolo *et al.*, 2008; Monfort 2010; Grigorakis *et al.*, 2011). The meagre is also interesting for recreational purposes (aquariums), considering its high adaptation to captivity and the wide range of temperatures and salinities tolerated.

Until recently, all juvenile production of meagre was from a few hatcheries (FAO, 2005–2010b); the hatchery techniques were not generally available with little published information. This situation has at times restricted the availability of juveniles to the industry and, therefore, potentially restricted the growth of aquaculture of meagre. This chapter aims to summarise and provide the available information on the culture methods used for meagre.

17.2 Broodstock management

The basis of a spawning protocol is an understanding of the species reproductive strategies and maturational cycle, i.e. information on sexual differentiation, size at first maturity, reproductive dietary requirements, maturational development in relation to environmental changes, reproductive endocrinology, spawning behaviour and egg parameters that enable the holding of broodstock in conditions that allow maturation to advance to late stages of gametogenesis or spawning. Critical points, selection of the broodstock, holding conditions, broodstock nutrition and the captive environment during gametogenesis and spawning must be addressed to ensure the spawning of good quality eggs.

The *Sciaenidae* are gonochoristic with fixed sex after sexual differentiation. The period of sexual differentiation has been described in meagre to be completed by 10–12 months (Schiavone *et al.*, 2012). First maturation or puberty is generally at a large size in *Sciaenidea* and minimum size of wild meagre females in vitellogenesis was 8 kg in the fishery of the coast of the Algarve, Portugal (Dr Nuno Prista, University of Lisboa, Faculty of Sciences, Instituto de Oceanografia) or 70–110 cm in southern Spain (González-Quirós *et al.*, 2011). However, meagre reared from the egg in captivity were observed to mature at 5.4+ kg. A stock of 140 meagre that were reared in

© Woodhead Publishing Limited, 2013

seacages off the coast of the Canary Islands were transferred to 10 m^3 flowthrough tanks at 1.5 kg and ongrown to 9.8 ± 1.3 kg using a diet of Vitalis REPRO (Skretting) and under natural conditions of photoperiod and temperature that ranged from summer highs of 24 °C to winter lows of 18 °C (Schuchardt *et al.*, 2007). At the age of three years, all females over the size of 5.4 kg and all males over the size of 4.2 kg had either oocytes in advanced stages of vitellogenesis or flowing sperm. Schiavone *et al.* (2012) reported a considerably lower size at first maturity, two years old and 0.92 ± 0.08 kg in males and three years old and 1.61 ± 0.09 kg in females. The males at this size had flowing sperm and histology showed that the females had post-ovulatory follicles. However, neither the quality nor collection of eggs was reported from these small broodstock.

Therefore, broodstock should be formed from fish over 6 kg (captivity-reared fish) or over 8 kg (wild fish) and larger fish 8+ kg gave improved fecundities. Andree *et al.* (2010) have described microsatellites that have been used to characterise both wild-caught and captivity-reared broodstock. When forming a broodstock, the genetic variability should be examined to have a starting point from which the genetic variability can be maintained while selecting for desired traits such as growth and flesh quality. The genetic variability of three broodstock were tested (unpublished results Duncan *et al.*): two broodstock of 12 and 32 wild fish caught on the Algarve coast, Portugal that were set up on different dates in IRTA, Sant Carles de la Rapita, Tarragona, Spain and a third broodstock that was set up in ICCM, Las Palmas, Canaries, Spain with 140 fish that were hatched and reared in captivity (see above). The two wild broodstock exhibited a heterogeneous genetic variability as would be expected from a wild population and the captive broodstock (tested $n = 19$) that were originally obtained from a single hatchery in France and ongrown in the Canaries exhibited a reduced genetic variability compared to the wild fish, suggesting that the group originated from a small broodstock pool. This situation may be unavoidable in a captivity-reared broodstock as all the initial production (previous to 2008) of meagre juveniles originated from few hatcheries in France (FAO, 2005–2011b). It is, therefore, important to ensure that the genetic variability is not further eroded by crossing within one family and plans should be made to add fish to the stock from different genetic origins to increase the genetic variability.

Once the origin of the broodstock has been selected, the holding conditions and nutrition are two critical factors that must be optimal to ensure that the broodstock reach late stages of maturity from which spawns can be obtained. Holding tanks for fish of 6+ kg need to give sufficient space for the migratory swimming behaviour and spawning behaviour. Wild broodstock with an average weight of 11 kg that were grown to 15 kg (20 fish, i.e. biomass of 220→300 kg) were successfully acclimatised and matured to advanced stages of vitellogenesis and spermiation during each spawning season in a 50 m^3 D-ended raceway with a 1 m depth. Captivity-reared

© Woodhead Publishing Limited, 2013

broodstock were grown from an average weight of 1.5 kg to 10 kg in rectangular 10 m^3 tanks (3 m × 3 m × 1.5 m depth) maintaining the biomass below 80 kg per tank and the fish matured to late stages of vitellogenesis and spermiation during each spawning season and no stock holding problems were encountered. Meagre, particularly wild meagre, will jump from a tank or within a tank resulting in impact damage, such as bruising, loss of scales, wounds and often eye damage, i.e. exophthalmia. Therefore, tanks should be covered with a strong top net (multi-monofilament knotless nylon netting with 30 mm mesh suitable for seine fishing, Badinotti, Milan, Italy). Problems were encountered holding both wild and captive broodstock in an 18 m^3 rectangular tank (3 × 6 m) with a depth of 80 cm. The wild fish (two groups of six fish with a mean of 20 kg) presented a significant mortality in these holding conditions. The wild fish were successfully acclimatised to the 18 m^3 tanks, but during the spawning season husbandry procedures associated with maturity sampling and/or a change in the behaviour of the fish resulted in the eventual mortality of all the fish, three fish died due to jumping from the tank during husbandry procedures and six from over-inflation of the swimbladder. No parasitic or bacterial infection was found in relation to the swimbladder problem and it was concluded that the holding conditions, husbandry procedures and behavioural changes associated with spawning caused increased stress and activity which resulted in impact damage and over-inflation of the swimbladder. Ideally, an adequate holding tank should have densities below 5 $kg.m^{-3}$ with sufficient space and depth for the fish to avoid each other during spawning activity.

The newly acquired broodstock need to be treated for parasites and to be encouraged to feed. The only parasites that were found to cause problems and mortality during acclimatisation were monogenean worms and skin copepods (sea lice) that infect the skin and gills. These parasites have killed meagre due to infestation of the gills and excess of mucus. The feeding of newly caught wild broodstock can also be problematic. The following protocol was used to acclimatise wild broodstock caught in October: The temperature and photoperiod decreased naturally, but temperature was maintained above a minimum of 16 °C. During the first week in captivity, the fish were left with no disturbance and no feed was offered. During the following three weeks, the broodstock were given a weekly (Monday) 1 h bath of 100 ppm of formalin and offered sardines and squid Tuesday–Friday. After 13 days in captivity, the first feeding response was obtained when a fish ate squid and by day 21 all the fish appeared to be feeding. Once the fish were feeding well, praziquantel was fed (placed inside pieces of squid) at approximately 5 mg kg^{-1}. Observations of feeding behaviour have identified that temperature and parasitic load influence the appetite of the meagre both during acclimation and after the fish have been acclimatised to captivity. Temperature has been observed to be an important factor and, if winter temperatures pass below 14 °C or summer temperatures above 25 °C, the broodstock have been observed not to eat food that is offered. A

© Woodhead Publishing Limited, 2013

wild broodstock ($n = 32$) that was caught in November and held at temperatures below 12.3 °C during December, January and February did not feed during the period December–March and consequently did not mature during March–May. Re-occurrences of monogenean worms and sea lice have resulted in respiratory stress where the meagre did not feed and repeatedly raised the head out of the water. Treatments of formalin (100 ppm at weekly intervals) and praziquantel have controlled the monogenean worms and sea lice. When the broodstock were not feeding, praziquantel was administered suspended in saline solution (9 %) with a canula directly into the stomach of anesthetised fish (5 mg/kg). Meagre were anaesthetised with 0.3 mL L^{-1} of 2-phenoxyethanol or 67.5 mg L^{-1} of MS222 (Tricaine methanesulfonate) (Serezli *et al.*, 2012; personal observation). An amyloodinium infection resulted in mortalities and the meagre became sedentary, resting on the bottom of the tank and not feeding. The infection was controlled with two baths of 200 ppm H_2O_2 (1 h duration) applied a week apart.

When meagre broodstock have been set up with heterogeneous genetic variability, in adequate holding conditions and are feeding well, reproductive dietary requirements must be supplied to ensure that the broodstock have the nutritional components for gametogenesis. The wild broodstock have successfully matured and spawned with a diet of sardine, squid and Vitalis REPRO CAL (Skretting, Burgos, Spain). The broodstock were fed close to satiation so that feed was not left uneaten. Maximum appetite was observed during spring and autumn when the broodstock were fed Monday to Friday, the percentage fed per week of the total biomass was 1.6 % REPRO Cal Vitalis, 1 % squid and 0.5 % sardine. As temperatures increased and decreased feeding was reduced in relation to the appetite, feeding was reduced by feeding fewer days per week rather than many days with small amounts of feed. The cultured broodstock (ICCM) have successfully matured and spawned with a diet of Vitalis REPRO (Skretting) two times a week and fresh diet, mussels and cuttlefish, once a week.

Gametogenesis appears to begin in December/January and the spawning period started during March–April and finished during the period June–September. Gametogenesis was indicated to have started as levels of oestradiol (females) and testosterone (males) have been observed to rise in studies on the two captivity-reared broodstock held in ICCM, Canaries and IRTA, Sant Carles de la Rapita, Tarragona (Vallés *et al.*, 2011). Both stock were in advanced stages of vitellogenesis and spermiating in late March and the earliest spawns were obtained on the 29th March in IRTA, Sant Carles de la Rapita and 16th March in the ICCM, Canarias. The last spawns of the spawning season were obtained on the 1st June in IRTA, Sant Carles de la Rapita and 27th September in the ICCM, Canarias. Both stocks were held under natural conditions with a natural photoperiod and temperature profile. In the Canaries, photoperiod ranges from the longest day in summer of 13:55 hours to 10:22 hours in the winter (from official sunset

© Woodhead Publishing Limited, 2013

to sunrise, i.e. not including twilight periods) and temperatures range from winter low of 18 °C to summer highs of 24 °C. In the IRTA, Sant Carles de Rapita photoperiod ranges from the longest day in summer of 15 h to 9 h in the winter (official sunset to sunrise) and temperatures were controlled to give a winter low of 16 °C and summer highs of 25 °C. In the ICCM, Canarias the temperature range of the spawning period within which spawns were successfully induced ranged from 19 to 23 °C and in IRTA, Sant Carles de la Rapita from 16 to 21 °C. Sampling in June that coincided with temperatures of 23.5 °C found that females were in gonadal regression with either atretic oocytes or immature oocytes and males had no sperm or poor quality sperm, indicating that the photoperiod and temperature profile had entrained the end of the spawning period.

The development of the oocytes appeared to be group-synchronous as females with vitellogenic oocytes ready for induction (> 580 μm) exhibited a modal distribution of oocyte size. Ovarian biopsies were taken from 10 broodstock and 1468 oocytes (118–150 oocytes per fish) were randomly selected in wet mounts and the diameter measured. Two modes in the frequency distribution were observed (Fig. 17.2). A bimodal distribution of oocytes is typical of group-synchronous oocyte development (Wallace and Selman, 1981; Tyler and Sumpter, 1996). At maximum oocyte sizes of less than 400 μm, the distribution of oocyte size was skewed to larger oocytes.

To date, broodstock held in IRTA or ICCM have not spawned spontaneously without hormone induction. Hormone therapies have been developed for both wild and cultured stocks using a single intramuscular injection of GnRHa (des-Gly10, [D-Ala6]-gonadotropin releasing hormone ethylamide, Sigma, Spain, which sold the hormone as LHRHa, des-Gly10, [D-Ala6]-Lutienizing hormone releasing hormone with product code L4513). The

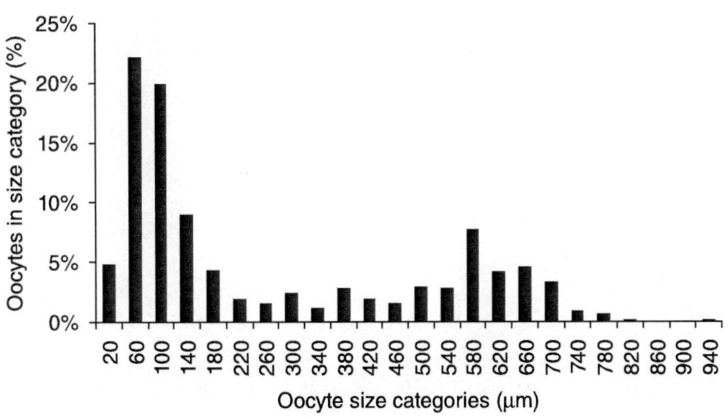

Fig. 17.2 Percentage of oocytes in each category of oocyte diameter for ovarian samples taken from 10 female meagre (*Argyrosoumus regius*) before hormonally inducing spawning, 118–150 oocytes were measured from each fish, total oocyte $n = 1468$.

© Woodhead Publishing Limited, 2013

dose of the injection and stage of ovarian development, i.e. oocyte size, are the bases of a hormone therapy to induce spawning (Ibarra-Castro and Duncan, 2007; Mañanós *et al.*, 2009). Studies in the ICCM with the stock of 8–10 kg cultured broostock have identified that a single injection of a dose of 15 μg kg^{-1} applied to females with an oocyte size > 500 μm successfully induced spawning (Fernández-Palacios *et al.*, 2009a,b, unpublished results). During two spawning seasons (May–June 2009 and 2010) at weekly intervals for each of the following doses, 0, 1, 5, 10, 15, 20, 25, 30, 40 and 50 μg/kg, a female (oocytes > 500 μm) and a male meagre were given a single injection GnRHa (Sigma, Spain). Each dose was tested on a minimum of three females and three times per female. The GnRHa was dissolved in 9 % saline solution. The spawning variables of fecundity (number of eggs per kg) and egg quality (percentage fertilisation, hatching and larvae with fully absorbed yolk) were determined and all the variables including number of larvae with fully absorbed yolk (Fig. 17.3) indicated that a dose of 15 μg. kg^{-1} was optimal. The optimal dose of 15 μg kg^{-1} gave an average of 2 ± 0 spawns per hormone administration, a total number of eggs per spawn of 127 571 ± 58 848 eggs kg^{-1} and respective average percentage of fertilisation, hatching and larvae with yolk sac of 99.3 ± 1.0 %, 95.1 ± 3.6 % and 92.3 ± 6.3 %.

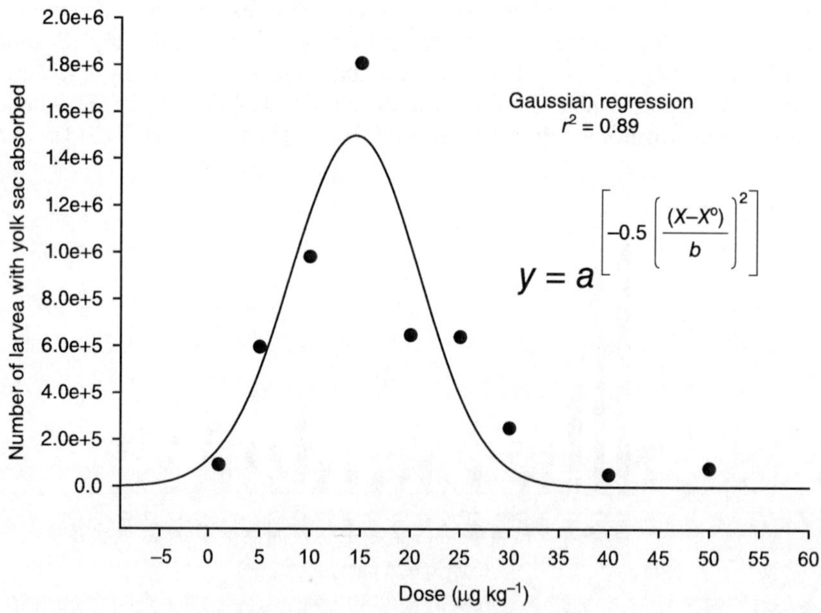

Fig. 17.3 Gaussian regression analysis for the dependent variable total number of larvae with yolk sac against the independent variable GnRHa dose (μg kg^{-1} BW) used to induce spawning of meagre (*Argyrosomus regius*).

© Woodhead Publishing Limited, 2013

In IRTA, the wild broodstock of 10–30 kg have been induced to spawn successfully with an injected dose of 20 μg kg^{-1} applied to females with vitellogenic oocytes larger than 580 μm. Males were treated with a half dose of 10 μg kg^{-1}. Of females ($n = 18$) with oocytes > 580 μm 89 % spawned and egg quality was good with an average hatching success of 84 ± 25 %. The first spawn was obtained 30–72 h after GnRHa was applied. At temperatures of 11.9–14.9 °C, recently spawned egg were collected 72 h after treatment and at temperatures from 15–18.1 °C eggs in blastrula to gastrula were collected 48 h after treatment. A female spawned from one to three times during the week after treatment and the second spawn was obtained the day after the first spawn. Generally, higher temperatures resulted in fewer spawns of higher fecundity. No further spawning was obtained from injected broodstock after the initial week of spawning. The average (± SD) relative fecundity per hormone induction for the wild broodstock was 130 900 ± 94 200 eggs kg^{-1}. Implants were also applied to six (three females and three males) wild broodstock and a dose of 50 μg/kg (Evac implants) applied to females with oocytes greater than 550 μm and males implanted with a half dose of 25 μg kg^{-1} spawned successfully (Duncan *et al.*, 2007, 2008, 2012). The implants induced similar quality and quantity as the injected broodstock during the week after hormone application. After the initial week of spawning, implanted broodstock continued to spawn small (>1 000 000 eggs per spawn) amounts of eggs as late as 20 days after hormone application and this gave the implanted treatment a higher quantity of eggs per hormone treatment. Mylonas *et al.* (2011) obtained daily spawning for periods for 5–19 days from meagre implanted with 50 μg kg^{-1} (Evac implants) with 70 % of the eggs being spawned during the first four spawns after hormone application. Both in IRTA and the ICCM, broodstock have been hormonally induced to spawn more than once during the spawning season. In IRTA, six broodstock were induced (20 μg kg^{-1}) to spawn a second time after three to four weeks, resulting in an average (± SD) relative fecundity per female for the spawning season of 359 000 ± 108 500 eggs kg^{-1}. In ICCM, individual broodstock were hormonally treated (15 μg kg^{-1}) every three, four, five, six and seven days. All broodstock responded with multiple spawns during the trial. The highest fecundities and egg quality were obtained from fish treated every five to seven days (Fernández-Palacios *et al.*, 2011).

Eggs are pelagic and can be collected in egg collectors that receive the spawning tanks' surface water in a mess basket. Egg diameter was 0.99 ± 0.02 mm from wild broodstock and the initial morphology was transparent, with a single large yolk cell, a smaller group of dividing cells and usually a single oil droplet, but it was also common to observe more than one oil droplet. The embryo was formed during the first 12 h and at 17–18 °C the eggs would hatch after 48 h. Eggs can be incubated in separate incubation tanks in the same way as bream and bass eggs, conical bottomed tanks of 100–250 L, stocked with 10 000–15 000 eggs/L with gentle aeration and water exchange (Moretti *et al.*, 1999). Water exchange is increased before

© Woodhead Publishing Limited, 2013

and during hatching. Debris from the hatching is purged before the larvae are transferred by gravity and water flow to the larval rearing tanks.

17.3 Larviculture

The most important features of meagre larvae ontogeny have been reported in detail by Jiménez *et al.* (2007) and Pastor and Cárdenas (2007). In addition, a complete morphometric approach from hatching to 30 days after hatching (dah) was reported by Fernández-Palacios *et al.* (2007). These studies were also complemented with the histological description performed by Hernandez-Cruz *et al.* (2007) and Abreu *et al.* (2009), whereas the development of skeletal structures has been recently described by Cardeira *et al.*, 2012. Thus, at hatching, meagre larvae has an average standard length (SL) of 2.82 ± 0.37 mm presenting a very simple structure and development. At rearing temperatures of 20 ± 1 °C, mouth and anus were open at 3 dah (SL = 3.25 mm); at this time, yolk sac was completely depleted, the digestive system merely composed of midgut, hindgut and foregut, and the oil droplet was maintained until 7 dah. From 3 dah, eosinophilic affinity was detected in the hepatocytes suggesting glycogen accumulation; moreover, zymogen granules and endocrine pancreas were observed at the same time as an indication of initial digestive capacity and exogenous feeding. Similarly, a pigmented retina and the presence of precursor cells of the swimbladder were observed. Swimbladder inflation starts in part of the larval population at 4 dah, and by day 8 all the larvae show a functional swimbladder. Growth in total and standard length was very similar until 19 dah when notochord flexion and urostile formation begins. First gastric glands were described at around 6.00 mm SL while glandular stomach was identified when larvae reach 8.50 mm SL. Meagre larval growth can reach 16–19 mm total length by day 30 after hatching at 20–21 °C; however, this parameter is clearly influenced by rearing conditions and temperature. Thus, Rodríguez-Rúa *et al.* (2007) cited 15.1 and 11.66 mm SL of larvae at 30 dah reared at 20–23 °C or 20.5–24.8 °C, respectively. Fernández-Palacios *et al.* (2007), Vallés and Estévez (2009) and Cardeira *et al.* (2012) reported a range between 6–8 mm SL at 20–21 °C and 6.06–8.45 mm at 19–20 °C for the same age, respectively (Fig. 17.4).

Culture conditions and protocols for meagre larval rearing differ amongst different research stations and private companies, although all of them generally include the use of filtered and UV sterilised sea water and natural salinity of 32–40 ppt (Cárdenas, 2010), with dissolved oxygen of 5.61 ± 0.14 mg/L, and temperature of 18–25 °C (Cárdenas, 2010). Green water technique is usually applied adding live and/or dry or frozen phytoplankton (*Nannochloropsis* sp., *Isochrysis galbana*) to maintain a concentration of 250 000 cells ml^{-1} in the rearing tanks. Larval culture is generally conducted under 12/12 (dark/light) photoperiod by means of artificial light or mixed

© Woodhead Publishing Limited, 2013

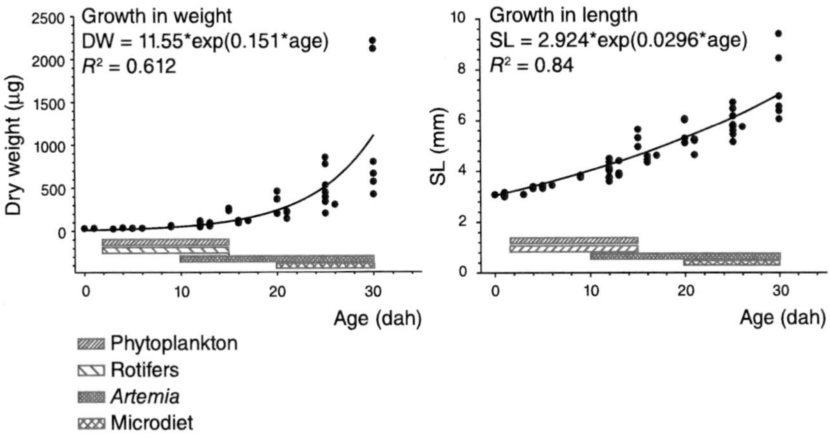

Fig. 17.4 Growth in dry weight and standard length and feeding sequence used for meagre (*Argyrosomus regius*) larval rearing in IRTA.

with natural light, reaching a light intensity at the water surface ranging between 1000 and 3500 lux (Estévez *et al.*, 2007; Rodríguez-Rúa *et al.*, 2007; Roo *et al.*, 2007, 2009, 2010; Cárdenas, 2010). However, both photoperiod and light intensity have a clear effect on larval growth and survival, affecting behavior (24hL:0hD increases cannibalism) and larval quality (24hL:0hD and 1000 lux induced hyperinflation of the swimbladder – (Vallés and Estévez, 2009, 2012; Cardeira *et al.*, 2012)). A photoperiod of 12hL:12hD and 500 lux light intensity is highly recommended.

Limited detailed information about feeding schedules for *A. regius* is available. In general, rotifers (*Brachionus* sp.) are used at 5–10 rotifers ml^{-1}, either enriched with commercial products or with microalgae and fed twice a day from first feeding (2 dah) to 15–20 days after hatching (Fig. 17.4). Similarly, *Artemia* introduction may vary according to larval rearing protocols. Roo *et al.* (2007, 2010) have shown that early introduction of *Artemia* was reflected in a lower survival rate and higher larval length as big larvae quickly adapt to *Artemia* feeding, whereas small and weaker larvae were unable to feed on *Artemia*, dying shortly after rotifer removal. Similarly, the delay in *Artemia* introduction until day 15 was reported to reduce larval length and survival, probably as a consequence of a high energy cost/benefit related to the small size of the prey. Thus, a protocol including *Artemia* metanaupli from day 12 might be considered the most suitable for meagre larval rearing. Other authors (Duran *et al.*, 2008) have also found very good results of growth and survival using only rotifers (days 2 to 20) and artificial diets (day 18 onwards). A positive correlation between larval survival and early *Artemia* replacement with microdiets has also been found by Hernández-Cruz *et al.* (2007). Thus, microdiets can be introduced as early as from 15 dah, with a complete weaning at 20 dah and 40–45 % survival rate. A co-feeding period of *Artemia* and microdiets of 10, 15 and 20 days

© Woodhead Publishing Limited, 2013

produced a significantly higher larval growth than shorter periods of zero and five days (Fernandez-Palacios *et al.*, 2009b) without affecting larval survival after an air resistance activity test or the biochemical composition of the larvae. From these studies, it is possible to conclude that meagre larvae can be directly weaned from rotifer to an inert feed from 15 dah with a modest survival rate (7.7–13.1 %). However, a co-feeding period with *Artemia* implies higher survival rate, higher growth of the larvae and better resistance to activity tests.

One factor that highly affected the results in growth and survival was larval density. Several studies carried out by different authors (Estévez *et al.*, 2007; Roo *et al.* 2007, 2009, 2010) indicate that meagre larvae grow (four times higher dry weight at 25 than at 100 larvae L^{-1}) and resist the activity test better if reared using less than 50 larvae L^{-1}. Even larval survival rate was better in low than in high larval densities being as high as 53.2 ± 12.0 % (Roo *et al.*, 2007, 2010).

In a study comparing two different rearing systems, intensive (75 larvae L^{-1}, 2 m^3 tanks) and semi-extensive (7.5 larvae L^{-1}, 40 m^3 tanks), Roo *et al.* (2009) showed that at low densities growth in length (16.0 ± 1.5 mm vs 19.1 ± 2.3 mm) and weight (6.5 ± 0.5 mg vs 13.1 ± 2.4 mg), as well as survival after an activity test at 30 dah, were higher when semi-extensive conditions were used. Production cost is relatively lower (0.015 vs 0.02 €) under intensive culture vs semi-extensive, considering the cost of microalgae, rotifer, *Artemia* and microdiet used for weaning (Roo *et al.*, 2009, Table 17.1). Under these conditions, the survival rate under intensive conditions was similar to that reported previously (Estévez *et al.*, 2007; Rodriguez-Rua *et al.*, 2007) and higher when semi-extensive culture was used (25.7 %). However, survival rate can be improved if the cannibalistic behaviour observed in 20 dah larvae is minimised either by an early grading of the fish or taking care that the metamorphosing juveniles are fed to satiation.

First attempts to evaluate meagre larval quality in terms of skeletal deformities were reported by Estévez *et al.* (2007) and Roo *et al.* (2009). These authors showed that there was no significant interaction between the

Table 17.1 Comparison of feed quantities (per fry produced) and feeding costs according to the rearing system after weaning (35 dah) for meagre (*Argyrosomus regius*)

Treatment	Microalgae (ml)	Rotifers (millions)	*Artemia* (millions)	Dry feed (g)	Cost (Euro)
Intensive	26.28[b]	21.4	8.2[b]	0.025[b]	0.015[b]
Semi-extensive	37.07[a]	18.2	5.1[a]	0.051[a]	0.020[a]

Note: Different letters in superscript in the same column indicate a significant difference ($p < 0.05$).

© Woodhead Publishing Limited, 2013

larval density or the rearing system used and the number of deformed fish ($\chi^2 = 1.970; p = 0.160$) and only 4.3 % of fry produced showed some skeletal deformities, with fusion affecting vertebrae 10 to 15 the most frequent anomaly that was observed.

17.4 Ongrowing and harvest

Generally the technologies and practices used for ongrowing meagre are the same as those used for bream and bass, but the meagre presents superior growth rates and can achieve superior feed conversion ratios (FCR). In Europe, Spain has the highest production of meagre, producing 1348 t in 2009 (FAO, 2011). Over 95 % of the Spanish production was produced from seacages on the Mediterranean coast. The production cycle is pre-ongrowing, juvenile transport, stocking, feeding and harvest.

Pre-ongrowing is practised principally in land-based facilities and consists of growing the weaned meagre to 3–15 g. The pre-ongrowing period is approximately two to four months, depending on temperature and desired juvenile size for transfer. At rearing temperatures of 20 °C, the juveniles will approximately double in weight every two weeks in the size range of 3–15 g. The principal problem during pre-ongrowing is cannibalism, and frequent feeding and approximately bi-monthly size grading is required to maintain populations with a low variation in the size frequency distribution to avoid cannibalism of smaller fish by larger members of the cohort. Flexi-bacteria and, to a lesser extent, vibrios can be problematic during pre-ongrowing, particularly attacking fin edges, and should be controlled with formalin, H_2O_2 baths and antibiotics. It is particularly important that juveniles close to sea transfer are free from flexi-bacteria as in the seacages the flexi-bacteria can cause mortalities and the treatment of flexi-bacteria in the cages is more complicated than in land-based tanks.

Transport from the hatchery of 10 g meagre is identical to bream and bass in trucks fitted with 1–2 m^3 cubic tanks supplied with air and oxygen and with a large discharge valve to discharge by gravity through a 15 cm diameter flexible pipe. The transport density used is maximum 30 kg m^{-3} and oxygen is maintained between 120 and 160 % saturation – higher saturation can stress the meagre juveniles. On arrival at the site, the water in the transport tanks is changed with water from the site. The site water should have 100 % oxygen saturation and be free from suspended material and contaminants such as heavy metals, pesticides and hydrocarbons. Once acclimatised, the juveniles are passed to a boat fitted with transport tanks and taken to be stocked into the cages. The meagre juveniles are more sensitive to handling and particularly scale loss compared to bream. Counting or handling the juveniles more than is necessary during the stocking procedure is not recommended. The cages most commonly used are 25 m diameter (depth 15 m) and each cage is stocked with 100 000 juveniles

© Woodhead Publishing Limited, 2013

(approx. 14 juveniles per m^3). The stock is maintained in the cage until harvest with net changes made when necessary.

Commercial diets are available for meagre and bream diets have been successfully used. Meagre, like bream and bass, are carnivorous fish and initially diets were recommended with 45–48 % protein and 20–24 % lipid (FAO, 2005–2011b). Chatzifotis *et al.* (2012) confirmed that highest growth rates were obtained with 40–50 % protein. However, meagre appear to have lower lipid requirements than used in the initial diets and higher growth rates were obtained with a dietary lipid level of 17 % compared to levels of 21 % and 13 % (Chatzifotis *et al.*, 2010, 2012). A commercial diet used by Estévez *et al.* (2010) had a crude fat content of 11.5 %. Plant protein is being used in commercial diets and up to 315 g kg^{-1} of plant protein (or 76.2 % of total protein content) was included without any reduction in growth rates (Estévez *et al.*, 2011). During the initial period after transfer, cannibalism can be a problem (up to 15 g) and feeding frequently, two to four times a day, to excess is recommended to aid the transfer and avoid cannibalism and feeding problems. As the meagre grow, feeding frequency can be reduced to once a day. The meagre feed low in the water column and take time to rise to the surface to feed. Meagre require approximately double the ration used for bream and can be fed 1–2 % body weight in agreement with manufacture's feeding tables. FCR of 1:1 has been achieved, but can be variable between different juvenile batches, environments and producers. For ongrowing, FCRs of 1.7 (FAO, 2005–2011b), 0.9–1.2 (Monfort, 2010) and 1.8 (personal communication) have been reported. On the Southern Mediterranean coast of Spain, growth from 5–1100 g was achieved in the first 12 months and to 2500 g in 24 months (Fig. 17.5), with seawater temperatures in the range of 14–26 °C.

There are few reports of disease problems during ongrowing, leading to the suggestion that meagre are resistant to diseases compared to other

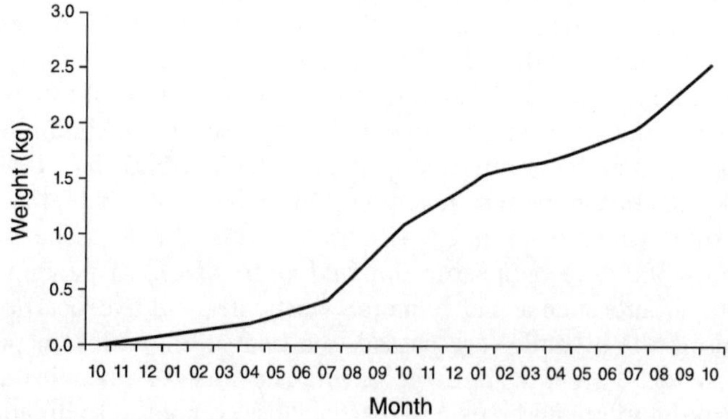

Fig. 17.5 Growth, wet weight of meagre (*Argyrosomus regius*) reared in cages on the Mediterranean coast of Spain with sea water temperatures in the range of 14–26 °C.

© Woodhead Publishing Limited, 2013

marine fish species (FAO, 2005–2011b). However, species new to aquaculture often do not present disease problems because the transfer of pathogens amongst the small population is limited. Problems with potential to seriously affect production have been bacterial, particularly flexi-bacteria when transferred with the fish from the hatchery, and parasite infections similar to those reported above for wild-caught broodstock. Infections of monogenean parasites, *Benedenia sciaenae* (Toksen *et al.*, 2007) and *Sciaenacotyle panceri* (Merella *et al.*, 2009; Ternengo *et al.*, 2010), have been reported for meagre ongrown in cages. *B. sciaenae* caused skin lesions and loss of scales (Toksen *et al.*, 2007) and *S. panceri* that appeared to originate from contact with wild fish populations infected the gills and caused mortalities (Merella *et al.*, 2009; Ternengo *et al.*, 2010). Merella *et al.* (2009) reported mortality levels of 5–10 % in large fish (780–1500 g) and no significant mortality (<2 %) in smaller fish (170–457 g). The infection in large fish was reduced and mortality controlled by reducing stocking density through harvesting. It is probable that a feed-based parasite control and/or careful management reducing contact with wild fish and separation of year classes will be required to avoid monogenea infections. However, to date these problems have been relatively isolated cases and when 10 g juveniles were stocked, survival during ongrowing has commonly been 90 %.

Harvesting is similar to bass and care must be taken not to damage the meagre during crowding, which can cause scale loss. Fish damaged during crowding and not harvested develop lesions and a proportion die in the following days and weeks. Meagre has been slaughtered with cold shock and suffocation in harvest bins with an iced slurry of sea water. EFSA (2009) scientific opinion for the European Parliament was that slaughter on ice was not recommended. Electrical stunning was a recommended alternative for European seabass (Lambooij *et al.*, 2008) that is being used for commercial slaughter and should be considered for the slaughter of meagre. For large (2+ kg) meagre percussion methods (mechanical blow to the head) commercially used for the slaughter of salmon (RSPCA, 2007) should also be considered.

17.5 Product, market and economic aspects

17.5.1 Product and market

The fishery production and the market of the meagre has historically been quite small with a mean annual catch 3183 ± 1448 t (±SD) during the period 1950–2008 (maximum of 7228 t in 1978 and minimum of 800 t in 1970) (FAO, 2011). It would appear that the combination of small catches landed close to the spawning areas and the good marketing qualities such as attractive fish shape, good processing yield, good nutritional values, low fat content, excellent taste and firm texture have given rise to good niche markets in the areas that meagre were traditionally landed, such as

© Woodhead Publishing Limited, 2013

Portugal, Southern, Spain and parts of France and Italy (Monfort 2010). The European meagre fishery has declined to low levels (less than 500 t per year during 1984–2002) and the catch has been sustained from the fishery in north and western Africa (FAO, 2011) and more recently from aquaculture. The decline of the European fishery has contributed to creating lucrative European niche markets in which meagre commands prices in the middle range of 6–12 €/kg, which can be compared to the market for cod, salmon, haddock and hake (Monfort, 2010). Aquaculture of meagre started in 1997 and European production in 2009 was 1912 t. Monfort (2010) reviewed the European markets where cultured meagre is being sold. Farmed meagre is mainly sold fresh, whole or as fillets into the niche European markets. Over 50 % of the fish are sold at 1–2 kg, 30 % above 2 kg, and the rest below 1 kg. Only small volumes are sold frozen. Meagre has many good marketing qualities and is appreciated within the niche markets. However, meagre is not known outside of the niche markets. The niche markets do not have the capacity to absorb a high aquaculture production of meagre. Monfort (2010) predicts a production of meagre of approximately 10 000 in 2010 based on juveniles stocked during 2008. However, the global financial crisis has contributed to a reduced demand and stocking of juveniles in 2009 and 2010. It is unclear if the market price can be sustained with the predicted productions for 2010–2012, but Monfort (2010) concludes that marketing is required to expand the meagre market and maintain the medium price level. If production rises without a constructive marketing policy, the oversupply of meagre will drive the market price down possibly to a low price market (<€5/kg), where pangasius and Nile perch can be found (Monfort, 2010).

17.5.2 Economics

At present, contacts in the industry report a cost of production for meagre that is similar to the cost of production of bream and bass. The cost of production was estimated to be approximately 3.9 € kg^{-1} (Table 17.2). However, the production cost was reported to be variable with different batches of meagre juveniles presenting very different costs, principally due to variations in feeding, feed conversion ratios and growth. In comparison, production costs of bream are more stable and predictable with profitable companies producing for 3.5–3.8 € kg^{-1}, similar to values reported in 2005 (Merinero *et al.*, 2005; FAO, 2005–2011c). In addition to the similarity of total production cost, the breakdown of production cost of meagre and bream are also similar (Table 17.2). Monfort (2010) reports a market price for cultured meagre of 6 € kg^{-1} making meagre production economically attractive. However, as mentioned above, the niche market for meagre is small, the product is unknown in other markets and without marketing increased aquaculture production is expected to reduce the meagre market price. There would appear to be room to reduce meagre production costs

© Woodhead Publishing Limited, 2013

Table 17.2 Comparison of production costs per kilo broken down into different concepts, juvenile cost, feed, personnel, processing and other costs, for the ongrowing of meagre (*Argyrosomus regius*) and gilthead seabream (*Sparus auratus*)

	Meagre[a]		Bream[b]	
Concept	Cost kg^{-1}	% of cost	Cost kg^{-1}	% of cost
Juvenile	0.8 €	21 %	0.83 €	21 %
Feed	1.6 € (FCR 1.8)	41 %	1.51 €	39 %
Personnel	0.5 €	13 %	0.87 €	22 %
Processing	0.7 €	18 %	0.36 €	9 %
Others (medication etc)	0.3 €	8 %	0.34 €	9 %
Total	3.9 €	100 %	3.91 €	100 %

[a] Costs estimated with the help of the authors' contacts within the industry
[b] Merinero *et al.*, 2005.

and sustain a profitable production with a lower market price. Monfort (2010) gives a production cost of 3 € kg^{-1} and suggests this could be reduced to 2.5 € kg^{-1}. In particular, juvenile cost (21 % of cost) and feed costs (41 % of cost) could be reduced. Meagre juveniles cost approximately 1 € per juvenile with all production pre 2008 coming from few producers (FAO, 2005–2011b; Monfort, 2010). The entrance into the market of new meagre juvenile producers and fully horizontally integrated producers (controlling the whole life-cycle, broodstock, larval rearing and ongrowing) should reduce the cost of juveniles. Meagre larval rearing is similar to bream, which cost 0.1–0.28 € each depending on size (2–5 g) and rearing system (FAO, 2005–2011c). Monfort (2010) suggests meagre juvenile price could drop to 0.3 € per juvenile. Feed costs can be lower than the figure used here, meagre grow faster than bream and can have FCR of 0.9–1.2 (Monfort, 2010). However, technologies, feed formulation and juvenile quality need to be improved to consistently obtain FCRs close to these values.

17.6 Future trends

The principal and immediate problem for meagre culture is the marketing. Monfort (2010) indicates that meagre are presently positioned as a medium-priced product in a small niche market. Outside of the niche market (Portugal and parts of Spain, France and Italy) the product is not known. For the success of meagre culture, this market needs to be expanded safeguarding a profitable market price. Rapid expansion of the market without adequate marketing may cause the price of meagre to fall into low-priced market and, thereby, threaten the profitability of the production. Production in 2010 onwards is expected to pass the niche market size and the future market and potential profitability of the production will be

© Woodhead Publishing Limited, 2013

determined in the coming years (Monfort, 2010). The situation is such that a larger production is required to enable consumers to know the product, but a larger production supplied to a public that do not know the product will probably lower the market price. If cyclic variations in market price are to be avoided, a marketing strategy that informs consumers as production increases is required (Monfort, 2010).

The marketing situation can be aided by improving production technologies to reduce the cost of production. Meagre is a species new for aquaculture in south Atlantic and Mediterranean countries and, as the producers' knowledge of the species increases, improved production can be expected. Genetic selection for organisms adapted to culture and for traits such as growth and product quality will reduce variability between growth and FCR of batches of juveniles. The newly established producers of juveniles are often using captive-bred broodstock from a small genetic pool (possibly one family) and these companies need use genetic tools (Andree *et al.*, 2010) to evaluate stocks, programme the introduction of new genetic material and set up genetic improvement programmes. Future studies are needed on the nutritional requirements and possible implications of nutrition on cannibalism and variable FCRs. Hatchery and juvenile production techniques need to be improved to optimise culture productivity and larval quality, especially to avoid dispersion in larval growth that has severe consequences in the cannibalistic behaviour of 20 dah larvae (6 mm, Roo *et al.*, 2009) and in the hyperinflation of the swimbladder. Swimbladder hyperinflation has been observed in meagre larvae in absence of gas super-saturation of the rearing water and as a consequence of early *Artemia* feeding and high light intensity or long photoperiods. Several causes have been pointed out for this abnormality: bacterial load of the nauplii and stress related to larval density and light conditions (Roo *et al.*, 2007; Cardeira *et al.*, 2012). Broodstock management needs to be improved to define the environment required to achieve spontaneous spawning. Out-of-season egg production methods should be applied to ensure a continuous market supply of juveniles and processed meagre. Studies in IRTA and the ease of use of photoperiod controlled out-of-season egg production in red drum and other Sciaenids (Thomas *et al.*, 1995) would suggest that this can be achieved without difficulty. The emergence of disease problems in ongrowing can be expected, and initially the provision of in-feed products to control skin and gill parasites should be developed. Lastly, harvesting and processing should be improved. Alternatives to harvesting on ice will be needed in the future and Monfort (2010) suggests that marketing in packaged fillets can be attractive to consumers.

In conclusion, as an aquaculture species the meagre appears to be an excellent candidate with many of the attributes that both producers and consumers look for in a cultured species. It has been stated that meagre can be the salmon of southern Europe. The principal hurdle to this metaphor is that consumers do not know the product and experience shows that selling an unknown product in the competitive fish market is at best

© Woodhead Publishing Limited, 2013

difficult. At the lower end of the market, pangasius has been introduced very successfully. It remains to be seen if meagre can achieve a similar success, while maintaining a profitable market price.

17.7 Acknowledgements

Thank you to the many people that have helped with information and the research that has made this chapter possible, especially industry contacts and technicians in IRTA Sant Carles de la Rapita and ICCM, Canary Islands. Most of the research reported here for meagre culture protocols was funded by 1) Spanish Ministerio de Medio Ambiente y Medio Rural y Marino, Secretaria General del Mar, JACUMAR, project Spanish national plan for meagre culture, coordinated nationally by Salvador Cardenas, in Catalonia by Alicia Estévez and in Las Canaries by Hipólito Fernández-Palacios and 2) INIA-FEDER project RTA2008–00107 coordinated by Neil Duncan.

17.8 References

ABREU N, SOCORRO J A, BETANCOR M, CABALLERO M J, FERNÁNDEZ-PALACIOS H, HERNÁNDEZ-CRUZ C M *et al.* (2009) Nuevas aportaciones al estudio de la organogénesis en larvas de corvina (*Argyrosomus regius*, Asso 1801), in Beaz Paleo, D, Villarroel Robinson, M and Cardénas Rojas S (eds), *Libro de Resúmenes, XII Congreso Nacional de Acuicultura*, 24-26 Noviembre, Madrid. Madrid: CICE-GRAF Artes Graficas, 510–511 (in Spanish, abstract in English).

ANDREE K, AXTNER J, BAGLEY M J, BARLOW E J, BEEBEE T J C *et al.* (2010) Permanent Genetic Resources Note. Permanent Genetic Resources added to Molecular Ecology. Resources Database 1 April 2010 – 31 May 2010. *Molecular Ecology Resources* 10: 1098–1105.

CARDEIRA J, VALLÉS R, DIONISIO G, ESTÉVEZ A, GISBERT E, POUSAO-FERREIRA J, CANCELA M L and GAVAIA J P (2012) Osteology of the axial and appendicular skeletons of the meagre *Argyrosomus regius* (Sciaenidae) and early skeletal development at two rearing facilities. *Journal of Applied Ichthyology*, 28 (3): 464–470.

CÁRDENAS S (2010) *Crianza de la corvina (Argyrosomus regius)*, Cuadernos de Acuicultura n° 3, Madrid: Fundación OESA.

CHATZIFOTIS S, PANAGIOTIDOU M, PAPAIOANNOU N, PAVLIDIS M, NENGAS I and MYLONAS C C (2010) Effect of dietary lipid levels on growth, feed utilization, body composition and serum metabolites of meagre (*Argyrosomus regius*) juveniles. *Aquaculture* 307: 65–70.

CHATZIFOTIS S, PANAGIOTIDOU M and DIVANACH P (2012) Effect of protein and lipid dietary levels on the growth of juvenile meagre (*Argyrosomus regius*). *Aquaculture International* 20: 91–98.

CHEN X H, LIN K B and WANG X W (2003) Outbreaks of an iridovirus disease in maricultured large yellow croaker, *Larimichthys crocea* (Richardson), in China. *Journal of Fish Diseases* 26: 615–619.

DUNCAN N J, ESTÉVEZ A and MYLONAS C C (2007) Efecto de la inducción hormonal mediante implante o inyección de GnRHa en la cantidad y calidad de puestas de corvina (*Argyrosomus regius*), in Cerviño A, Guerra A and Pérez C (eds), *Libro*

© Woodhead Publishing Limited, 2013

de Actas Toma II XI Congreso Nacional de Acuicultura, 24–28 Septiembre, Vigo. Pontevedra: Graficas Salnes SL, 731–734 (in Spanish, abstract in English).

DUNCAN N, ESTÉVEZ A, PADROS F, AGUILERA C, MONTERO F E, NORAMBUENA F *et al.* (2008) Acclimation to captivity and GnRHa-induced-spawning of meagre (*Agyrosomus regius*). *CYBIUM International Journal of Ichtyology* 32: 332–333.

DUNCAN N, ESTÉVEZ A, PORTA J, CARAZO I, NORAMBUENA F, AGUILERA C, GAIRIN I, BUCCI F, VALLES R and MYLONAS C C (2012) Reproductive development, GnRHa-induced spawning and egg quality of wild meagre (*Argyrosomus regius*) acclimatised to captivity. *Fish Physiology and Biochemistry*, DOI 10.1007/s10695-012-9615-3

DURAN J, PASTOR E, GRAU A, MASSUTI-PASCUAL E, VALENCIA J M and GIL M M (2009) Total replacing of *Artemia* by an artificial diet in larval rearing feeding protocol of meagre (*Argyrosomus regius*, Asso 1801). *Aquaculture Europe 2009*, European Aquaculture Society, 14–17 August, Trondheim.

EFSA (2009) Scientific Opinion of the Panel on Animal Health and Welfare on a request from the European Commission on welfare aspect of the main systems of stunning and killing of farmed seabass and seabream. *The EFSA Journal* 1010: 1–52.

ESTÉVEZ A, TREVIÑO L and GISBERT E (2007) La densidad larvaria inicial afecta al crecimiento pero no a la supervivencia de las larvas de corvina (*Argyrosomus regius*) en cultivo, in Cerviño A, Guerra A and Pérez C (eds), *Actas XI Congreso Nacional de Acuicultura*, 24–28 Septiembre, Vigo. Pontevedra: Graficas Salnes SL, 747–750.

ESTÉVEZ A, TREVINO L, KOTZAMANIS Y, KARACOSTAS I, TORT L and GISBERT E (2011) Effects of different levels of plant proteins on the ongrowing of meagre (*Argyrosomus regius*) juveniles at low temperatures. *Aquaculture Nutrition*, 17: e572–e582.

FAO (2011) FAO Fisheries Department, Fishery Information, Data and Statistics Unit. FISHSTAT Plus: Universal software for fishery statistical time series. Version 2.3. 2000. Data sets: Aquaculture production: quantities and values 1950–2009; Capture production 1950–2009.

FAO © 2005–2011a, Cultured Aquatic Species Information Programme. *Sciaenops ocellatus.* Cultured Aquatic Species Information Programme. Text by Cynthia K Faulk, A In: FAO Fisheries and Aquaculture Department [online]. Rome. Updated 9 February 2005. [Cited 20 September 2012]. http://www.fao.org/fishery/culturedspecies/Sciaenops_ocellatus/en.

FAO © 2005–2011b, Cultured Aquatic Species Information Programme. *Argyrosomus regius*. Cultured Aquatic Species Information Programme. Text by Stipa, P; Angelini, M In: FAO Fisheries and Aquaculture Department [online]. Rome. Updated 10 February 2005. [Cited 20 September 2012]. http://www.fao.org/fishery/culturedspecies/Argyrosomus_regius/en.

FAO © 2005–2011c, Cultured Aquatic Species Information Programme. *Sparus aurata*. Cultured Aquatic Species Information Programme. Text by Colloca, F; Cerasi, S In: FAO Fisheries and Aquaculture Department [online]. Rome. Updated 8 February 2005. [Cited 20 September 2012]. http://www.fao.org/fishery/culturedspecies/Sparus_aurata/en.

FERNÁNDEZ-PALACIOS H, SCHUCHARDT D, ROO J, BORRERO C, HERNÁNDEZ-CRUZ C M and SOCORRO J (2007) Estudio morfométrico de la corvina (*Argyrosomus regius*, Asso 1801) durante el primer mes de vida, in Cerviño A, Guerra A and Pérez C (eds), *Actas XI Congreso Nacional de Acuicultura*, 24–28 Septiembre, Vigo. Pontevedra: Graficas Salnes SL, 755–758.

FERNÁNDEZ-PALACIOS H, SCHUCHARDT D, ROO J, HERNÁNDEZ-CRUZ C M and DUNCAN N (2009a) Efecto de distintas dosis de GnRHa sobre la calidad de la puesta de corvina (*Argyrosomus regius*), in Beaz Paleo D, Villarroel Robinson M and Cardénas Rojas S (eds), *Libro de Resúmenes, XII Congreso Nacional de Acuicultura*,

© Woodhead Publishing Limited, 2013

24–26 Noviembre, Madrid. Madrid: CICEGRAF Artes Graficas, 554–555 (in Spanish, abstract in English).

FERNÁNDEZ-PALACIOS H, SCHUCHARDT D, ROO J, HERNÁNDEZ-CRUZ C M and DUNCAN N (2009b) Eficacia de la inducción hormonal con distintas dosis de GnRHa en corvina (*Argyrosomus regius*), in Beaz Paleo D, Villarroel Robinson M and Cardénas Rojas S (eds), *Libro de Resúmenes, XII Congreso Nacional de Acuicultura*, 24–26 Noviembre, Madrid. Madrid: CICEGRAF Artes Graficas, 556–557 (in Spanish, abstract in English).

FERNÁNDEZ-PALACIOS H, SCHUCHARDT D, ROO J, HERNÁNDEZ-CRUZ C M, SABATER C and DUNCAN N (2011) Efecto de diferentes intervalos de tiempo entre inyecciones con GnRHa, sobre las puestas de reproductores de corvina (*Argyrosomus regius*), *Libro de Resúmenes, XIII Congreso Nacional Acuicultura*, 21–24 de Noviembre, Barcelona (in Spanish, abstract in English).

GONZÁLEZ-QUIRÓS R, DEL ÁRBOLA J, GARCÍA-PACHECOA M M, SILVA-GARCÍA A J, NARANJOA J M and MORALES-NINC B (2011) Life-history of the meagre *Argyrosomus regius* in the Gulf of Cádiz (SW Iberian Peninsula). *Fisheries Research* 109: 140–149.

GRIGORAKIS K, FOUNTOULAKI E, VASILAKI A, MITTAKOS I and NATHANAILIDES C (2011) Lipid quality and filleting yield of reared meagre (*Argyrosomus regius*). *International Journal of Food Science and Technology* 46: 711–716.

HALL D A (1984) The Coorong: Biology of the major fish species and fluctuations in catch rates 1976–1983. *SAFIC* 8: 3–17.

HERNÁNDEZ-CRUZ C M, SCHUCHARDT D, ROO J, BORRERO C and FERNÁNDEZ-PALACIOS H (2007) Optimización del protocolo de destete de corvina (*Argyrosomus regius* Asso, 1801), in Cerviño A, Guerra A and Pérez C (eds), *Actas XI Congreso Nacional de Acuicultura*, 24–28 Septiembre, Vigo. Pontevedra: Graficas Salnes SL, 751–754.

IBARRA-CASTRO L and DUNCAN N J (2007) GnRHa-induced spawning of wild-caught spotted rose snapper *Lutjanus guttatus. Aquaculture* 272: 737–746.

JIMÉNEZ MT, PASTOR E, GRAU A, ALCONCHEL JI and CÁRDENAS S (2005) Revisión sobre el cultivo de esciénidos en el mundo y presentación del Plan nacional de Cría de corvina (*Argyrosomus regius*), in *Actas del X Congreso Nacional de Acuicultura*, 17–21 Octubre, Valencia, 396–397.

JIMÉNEZ M T, RODRÍGUEZ DE LA RÚA A, SÁNCHEZ R, CÁRDENAS S (2007) Atlas de desarrollo de la corvina *Argyrosomus regius* (Pisces: Sciaenidae) durante su primer mes de vida. *REDVET Revista electrónica de Veterinaria* (http://www.veterinaria.org/revistas/redvet) Vol. VIII, n° 1.

LAGARDERE J P and MARIANI A (2006) Spawning sounds in meagre *Argyrosomus regius* recorded in the Gironde estuary, France. *Journal of Fish Biology* 69: 1697–1708.

LAMBOOIJ B, GERRITZEN M A, REIMERT H, BURGGRAAF D, ANDRE G and VAN DE VIS H (2008) Evaluation of electrical stunning of sea bass (*Dicentrarchus labrax*) in seawater and killing by chilling: welfare aspects, product quality and possibilities for implementation. *Aquaculture Research*, 39: 50–58.

MAÑANÓS E, DUNCAN N and MYLONAS C C (2009) Reproduction and control of ovulation, spermiation and spawning in cultured fish, in Cabrita E, Robles V and Harraez P (eds), *Methods in Reproductive Aquaculture*. Boca Raton, FL: CRC Press Taylor and Francis Group, 3–80.

MERELLA P, CHERCHI S, GARIPPA G, FIORAVANTI M L, GUSTINELLI A, SALATI F (2009) Outbreak of *Sciaenacotyle panceri* (Monogenea) on cage-reared meagre *Argyrosomus regius* (Osteichthyes) from the western Mediterranean Sea. *Diseases of Aquatic Organisms* 86: 169–173.

MERINERO S, MARTÍNEZ S, TOMÁS A and JOVER M (2005) Análisis económico de alternativas de producción de Dorada en jaulas marinas en el litoral Mediterráneo español. *Revista AquaTIC* 23: 1–19 (in Spanish).

© Woodhead Publishing Limited, 2013

MONFORT M C (2010) *Present market situation and prospects of meagre (Argyrosomus regius), as an emerging species in Mediterranean aquaculture.* Studies and Reviews, General Fisheries Commission for the Mediterranean No. 89. Roma: FAO.

MORETTI A, PEDINI FERNANDEZ-CRIADO M, CITTOLIN G and GUIDASTRI R (1999) *Manual on hatchery production of seabass and gilthead seabream*, Volume 1. Rome, FAO.

MYLONAS C C, MITRIZAKIS N, SIGELAKI I, and PAPADAKI M (2011) Spawning kinetics of individual female meagre (*argyrosomus regius*) after treatment with GnRHa implants. *Indian Journal of Science and Technology*, 9th ISRPF Issue, 4: No. S8 p. 232–233.

NELSON J S (1994) *Fishes of the world* (3rd edi). New York: Wiley.

PASTOR E and CÁRDENAS S (2007) Cultivo larvario de la corvina *Argyrosomus regius* (Asso, 1801), in Cerviño A, Guerra A and Pérez C (eds), *Actas XI Congreso Nacional de Acuicultura*, 24–28 Septiembre, Vigo. Pontevedra: Graficas Salnes SL, 739–742.

PICCOLO G, BOVERA F, DE RIU N, MARONO S, SALATI F, CAPPUCCINELLI R and MONIELLO G (2008) Effect of two different protein/fat ratios of the diet on meagre (*Argyrosomus regius*) traits. *Italian Journal of Animal Science*, 7: 363–371.

POLI B M, PARISI G, ZAMPACAVALLO G, IURZAN F, MECATTI M, LUPI P and BONELLI A (2003) Preliminary results on quality and quality changes in reared meagre (*Argyrosomus regius*): body and fillet traits and freshness changes in refrigerated commercial-size fish. *Aquaculture International*, 11: 301–311

QUÉMÉNER L, SUQUET M, MERO D and GAIGNON J-L (2002) Selection method of new candidates for finfish aquaculture: the case of the French Atlantic, the Channel and the North Sea coasts. *Aquatic Living Resources*, 15: 293–302.

RODRÍGUEZ-RÚA A, GRAU A, JIMÉNEZ M T, VALENCIA J M, ROSANO M, DURÁN J *et al.* (2007) Cultivo larvario de la corvina *Argyrosomus regius* (Asso, 1801), in Cerviño A, Guerra A and Pérez C (eds), *Actas XI Congreso Nacional de Acuicultura*, 24–28 Septiembre, Vigo. Pontevedra: Graficas Salnes SL, 739–742.

ROO J, HERNÁNDEZ-CRUZ C M, BORRERO C, FERNÁNDEZ-PALACIOS H, SHUCHARDT D (2007) Efecto de la densidad larvaria y secuencia alimentaria en el cultivo larvario de corvina (*Argyrosomus regius* Asso, 1801) durante el primer mes de vida, in Cerviño A, Guerra A and Pérez C (eds), *Actas XI Congreso Nacional de Acuicultura*, 24–28 Septiembre, Vigo. Pontevedra: Graficas Salnes SL, 743–746.

ROO F J, HERNÁNDEZ-CRUZ C M, FERNÁNDEZ-PALACIOS H, SCHUCHARDT D and IZQUIERDO M S (2009) Effect of rearing system intensiveness on biological features, culture performance and larval quality of meagre (*Argyrosomus regius* Asso, 1801) larvae, in Hendry C I, Van Stappen G, Wille M and Sorgeloos P (eds), *LARVI' 09 – 5th Fish & Shellfish Larviculture Symposium*. Europ. Aquacul. Soc. Spec. Publ 38, p. 371–374.

ROO J, HERNÁNDEZ-CRUZ C M, BORRERO C, SCHUCHARDT D, FERNÁNDEZ-PALACIOS H (2010) Effect of larval density and feeding sequence on meagre (*Argyrosomus regius*; Asso, 1801) larval rearing. *Aquaculture*, 302, 82–88.

RSPCA (2007) *RSPCA Welfare standards for farmed Atlantic salmon.*

SCHUCHARDT D, FERNÁNDEZ-PALACIOS H, ROO J, HERNÁNDEZ-CRUZ C M (2007) Estabulación y mantenimiento de un stock de reproductores de corvina (*Argyrosomus regius*, Asso, 1801), en Canarias, in *Libro de Actas Toma II XI Congreso Nacional de Acuicultura*, 24–28 Septiembre, Vigo. Pontevedra: Graficas Salnes SL, 727–730 (in Spanish, abstract in English).

SCHIAVONE R, ZILLI L, STORELLI C and VILELLA S (2012) Changes in hormonal profile, gonads and sperm quality of *Argyrosomus regius* (Pisces, Scianidae) during the first sexual differentiation and maturation. *Theriogenology*, 77: 888–898.

SEREZLI R, BASARAN F, GUNGOR MUHTAROGLU C and KAYMAKCI BASARAN A (2012) Effects of 2-phenoxyethanol anaesthesia on juvenile meagre (*Argyrosomus regius*). *Journal of Applied Ichthyolgy*, 28: 87–90.

© Woodhead Publishing Limited, 2013

TERNENGO S, AGOSTINI S, QUILICHINI Y, EUZET L and MARCHAND B (2010) Intensive infestations of *Sciaenocotyle pancerii* (Monogenea, Microcotylidae) on *Argyrosomus regius* (Asso) under fish-farming conditions. *Journal of Fish Diseases* 33: 89–92.

THOMAS P, ARNOLD C R and HOLT G J (1995) Red drum and other sciaenids, in Bromage NR and Roberts RJ (eds), *Broodstock Management and Egg and Larval Quality*. Oxford: Blackwell Science, 118–137.

TOKSEN E, BUCHMANN K and BRESCIANI J (2007) Occurrence of *Benedenia sciaenae* van Beneden, 1856 (Monogenea: Capsalidae) in cultured meagre (*Argyrosomus regius* Asso, 1801) (Teleost: Sciaenidae) from western Turkey. *Bulletin of the European Association of Fish Pathologists*, 27(6): 250.

TYLER C R and SUMPTER J P (1996) Oocyte growth and development in teleosts. *Reviews in Fish Biology and Fisheries* 6: 287–318.

VALLÉS R and ESTÉVEZ A (2009) Efecto del fotoperiodo en el crecimiento y supervivencia de larvas de corvina (*Argyrosomus regius*) en cultivo intensivo, in Beaz Paleo D, Villarroel Robinson M and Cardénas Rojas S (eds), *Libro de Resúmenes, XII Congreso Nacional de Acuicultura*, 24–26 Noviembre, Madrid: CICEGRAF Artes Graficas, 614–615 (in Spanish, abstract in English).

VALLÉS R and ESTÉVEZ A (2012) Light conditions during larval rearing of meagre (*Argyrosomus regius*). *Aquaculture International* (submitted)

VALLÉS R, BAYARRI M J, MAÑANÓS E and DUNCAN N (2011) Plasma sex steroid profiles in meagre (*Argyrosomus regius*). *Indian Journal of Science and Technology*, 9th ISRPF Issue, Vol. 4 No. S8 p. 108

WALLACE R A and SELMAN K (1981) Cellular and dynamic aspects of oocyte growth in teleosts. *American Zoologist* 21: 325–343.

© Woodhead Publishing Limited, 2013

18

Hatchery production of yellowtail kingfish (*Seriola lalandi*)

D. Stewart Fielder, Port Stephens Fisheries Institute, Australia

DOI: 10.1533/9780857097460.3.542

Abstract: Yellowtail kingfish (*Seriola lalandi*; YTK) hatchery technology has been largely developed in New Zealand and Australia where a burgeoning seacage grow-out industry exists. Wild-caught broodstock can be maintained in land-based tanks and induced to spawn using hormone therapy or more commonly will spawn spontaneously within one to two breeding seasons of domestication using phototherm manipulation. The chapter looks at seriola aquaculture, detailing broodstock management and larviculture. Two peaks of mortality have been found to occur – one at 3–4 dph when larvae commence feeding, and a second caused by cannibalism which occurs when larvae are 6–10 mm total length – and the chapter discusses strategies for mitigating these problems. The major bottlenecks to YTK juvenile production have included low survival and high rates of deformity, and the chapter finishes by looking at how commercial and government research institutes are addressing these.

Key words: yellowtail kingfish, *Seriola lalandi*, aquaculture, broodstock management, larval culture.

18.1 Introduction

Yellowtail kingfish (*Seriola lalandi*; YTK), also known as gold striped amberjack, is a member of the family Carangidae, commonly referred to as jacks and pompanos. YTK are found circum-globally mainly in high salinity (marine) waters but prefer temperate and sub-tropical waters (18–24 °C) (Kailola *et al.*, 1993). Populations are disjunct, occurring in the Indo-Pacific (South Africa, Walter Shoals, Amsterdam Island, Japan, Australia, New Zealand, New Caledonia, Hawaii, Rapa, Pitcairn Island and Easter Island) and the Eastern Pacific (British Columbia, Canada to Chile. Eastern Atlantic: St Helena, South Africa) (Eschmeyer *et al.*, 1983; Smith-Vaniz, 1990).

Schools of juveniles commonly comprising hundreds of fish up to 7 kg are generally found close to the coast. YTK occasionally enter brackish

© Woodhead Publishing Limited, 2013

estuarine waters in pursuit of prey comprising small fish, squid and crustaceans, while larger fish are more common around deep reefs and offshore islands out to the edge of the continental shelf. Schools are commonly associated with floating debris or weed that provides a focus for gathering. Tagging programmes have shown widespread movements of YTK. These include trans-Tasman (from Australia to New Zealand and vice versa) crossings and many large-scale (>500 km) movements along the coast of New South Wales (Gillanders *et al.*, 2001).

YTK have a maximum recorded length, weight and age of 250 cm, 96.8 kg and 21 years, respectively. In Australia, the largest recorded YTK was about 200 cm in length and 70 kg in weight but, as indicated by size and age frequency catch data for New South Wales, fish considered large are commonly around 100 cm and 10–15 kg and 10–12 years old (Stewart *et al.*, 2004). New South Wales stocks of YTK grow to a mean length of about 450 mm in their first year. Thereafter, to an age of about 11 years annual growth increments are essentially constant, progressively diminishing to about 90 mm in year 6. Growth of New South Wales stocks of YTK under relatively warm temperature regimes is considerably faster, but terminal size is considerably smaller than counterparts in colder New Zealand and US waters. Annual growth rates of 144 mm y^{-1} for 500 mm total length (TL) fish in New South Wales compare with 98 mm y^{-1} in New Zealand and a range of 34–10 mm y^{-1} in the USA for similar size fish (Gillanders *et al.*, 1999; Stewart *et al.*, 2004).

Predation by YTK on small schooling pelagic fish (e.g. sardines, anchovies, jack mackerel and Pacific mackerel) and cephalopods has been reported off California, Australia, New Zealand and in the Gulf of Mexico. There have been fewer studies on the diet of YTK food in the south-western Atlantic, with reported diet comprising juvenile Argentine anchovy (81 %), jack mackerel (7 %) and chub mackerel (1.9 %). In Argentina, YTK tend to be restricted to rocky reefs, which have a scarce and patchy distribution in the region, and this affinity to reef habitats may be related to their feeding behaviour (Vergani *et al.*, 2008).

Aquaculture of *Seriola* spp. is well established, especially in Japan where approximately 57 % of all cultured marine finfish production (139 000 t per annum) is from *S. quinqueradiata*, *S.dumerili* and *S. lalandi* at a value of US$1.1 billion per annum (Portenaar *et al.*, 2003; Ohara *et al.*, 2005). Interestingly, the Japanese industry has developed without the need for hatchery production of seed stock for on-growing. Juvenile *Seriola* spp. are captured with round haul nets from the natural aggregations which occur around drifting seaweeds (Sakakura and Tsukamoto, 1996; Kolkovski and Sakakura, 2004). Aquaculture of *S. lalandi* (YTK) in New Zealand and Australia is a new industry having commenced in 2001 (Miller *et al.*, 2011). Pilot projects for YTK culture have been done in New South Wales and Western Australia, but the industry is predominantly located in South Australia which currently produces 3000–4000 t per annum (Miller *et al.*, 2011). In contrast

© Woodhead Publishing Limited, 2013

to Japan, all YTK aquaculture in Australia and New Zealand is reliant on hatchery culture of juveniles. Basic hatchery technology for production of YTK juveniles in Australia has developed rapidly and mostly by a small number of private hatcheries using techniques already developed for other species including Australian snapper (*Pagrus auratus*) and mulloway (*Argyrosomus japonicus*) (Fielder and Heasman, 2011). More recently, government research organisations in Australia and New Zealand have been assisting industry to improve and develop hatchery technology for YTK. The aim of this chapter is to provide a summary of available information on the methods available for hatchery culture of juvenile YTK.

18.2 Broodstock management

The estimated mean size and age of onset of sexual maturity in YTK varies with gender and geographic location. In New South Wales, the mean size and age at maturity recorded for female YTK was 834 mm and 3+ years, respectively and for males 471 mm and 0.9 years, respectively (Gillanders *et al.*, 1999). Corresponding sizes and ages reported for YTK stocks in New Zealand are 944 mm and 7–8 years for females and 812 mm and 4 years for males. The large differences in size and age of sexual maturity between New South Wales and New Zealand populations may be due to different growing conditions, e.g. warmer water temperatures in NSW, or behavioural and physiological differences between populations. Although no fixed genetic differences have been identified between NZ and NSW populations (Miller *et al.*, 2011), large-scale movements between these populations are uncommon (Poortenaar *et al.*, 2001).

As with many other fish, viable gametes can be obtained from either wild, captive or cultured YTK broodstock. However, wild YTK broodstock can suffer severe stress at capture and offshore spawning sites are relatively inaccessible. In addition, breeding condition regresses while the wild-caught fish are held in interim quarantine holding facilities. Consequently, land-based breeding programmes of captive stock are mostly used by hatchery operators. Breeding tanks are typically 20–70 m^3 or occasionally larger, and at least 2 m deep with reported stocking rates of 5–14 kg/m^3.

YTK are serial spawners, breeding under ambient conditions beginning in late spring or early summer and continuing to autumn over temperatures in the range 17–24 °C (Gillanders *et al.*, 1999; Poortenaar *et al.*, 2001) depending on seasonal sea temperature ranges of particular locations. Photoperiod at the start of the natural spawning season (November) is about 13.5 h light:10.5 h dark, and increases to a maximum day length in mid-December (14 h light:10 h dark) before decreasing to 11.5 h light:12.5 h dark at the end of the spawning season (April). In the southern hemisphere, when held under constant temperature 20 ± 1 °C and natural photoperiods, YTK spawning and larviculture occurs from November until the end of

© Woodhead Publishing Limited, 2013

February, after which time egg production and quality tend to decrease to a degree that makes it unviable for commercial purposes.

Spawning occurs either just prior to dawn during the first half of the spawning season and 1 h either side of dusk in the latter half. Courtship behaviour involves one male and female, and consists of a high-speed pursuit punctuated by stalling, nipping and touching. This lasts for approximately 0.5–1.5 h until, immediately prior to spawning, males nip at the female gonoduct, presumably to induce spawning. An additional male becomes involved at this stage in 50 % of spawns. Release of gametes involves frenzied circling behaviour near the bottom of the tank and lasts 20–25 s (Moran *et al.*, 2007). In captive YTK, no mass spawning events (those involving more than one female) were recorded, although two or three individual females spawned within an hour of each other on several occasions. This suggests that female YTK may spawn close together, but not necessarily at the same time (Moran *et al.*, 2007).

In commercial hatchery operations, it is often desirable to extend the natural spawning season or stimulate out-of season spawning to increase production. This is achieved with YTK by controlling the temperature and photoperiod into abbreviated seasonal cycles called photothermS. At the Port Stephens Fisheries Institute, year-round, on-demand spawning of captive YTK broodfish has been achieved from a single tank of fish containing seven pairs of wild collected fish. After initial exposure to a truncated phototherm (Table 18.1), fish have been held at a constant 10 h:14 h light:dark regime and at 16 °C. Spontaneous spawning can be induced to occur three to four days after the water temperature is increased from 16 to 22 °C within 24–48 h. Spawning events then continue every two to four days. Spawning is stopped when water temperature is decreased to 16 °C.

Wild-caught YTK broodstock spawn spontaneously in captivity within one or two breeding seasons of domestication. If in the shorter term, hormone treatments are required to stimulate maturation and spawning, the timing of treatments needs to coincide with critical stages of reproductive development. Females with oocytes less than 650 μm in diameter do not spawn, females with oocytes around 700 μm spawn, but subsequent fertilisation and hatch rates are poor, whereas females with oocytes greater than 800 μm in diameter produced vast quantities of good quality eggs. Samples of oocytes can be collected from anaesthetised broodstock by inserting a biopsy tube (catheter) into the genital opening. Three different hormonal treatments have been successfully used to induce oocyte maturation and ovulation in kingfish, including a single injection of hCG at 500 $IUkg^{-1}$, priming injections of hCG one day after the single hCG, and injection and single implantation of LHRHa at 220–400 $\mu g kg^{-1}$ in a cholesterol pellet. While the latter (LHRH implantation) is superior in terms of egg quality and yields of eggs, a single injection of hCG is a much cheaper procedure.

© Woodhead Publishing Limited, 2013

Table 18.1 Compressed seasonal photoperiod and temperature regime (phototherm) used in at the NSW Department of Primary Industries, Port Stephens Fisheries Institute

Date	Day length	Light ON	Light OFF	Temperature (°C)	Day
19-Dec	10.5	6 45	17 15	16	1
24-Dec	11.25	6 15	17 30	16.5	5
29-Dec	11.75	6 00	17 45	16.5	10
3-Jan	12.5	5 30	18 00	16.7	15
8-Jan	13	5 15	18 15	16.95	20
13-Jan	13.5	5 00	18 30	17.2	25
18-Jan	13.5	5 00	18 30	17.65	30
23-Jan	14.25	4 45	19 00	18.1	35
28-Jan	14.25	4 45	19 00	18.95	40
2-Feb	14.25	4 45	19 00	19.8	45
7-Feb	14.25	4 45	19 00	20.4	50
12-Feb	13.25	5 15	18 30	21	55
17-Feb	12.75	5 30	18 15	21.4	60
22-Feb	12.25	5 45	18 00	21.8	65
27-Feb	11.75	6 00	17 45	22	70
4-Mar	11.25	6 15	17 30	22.2	75
9-Mar	10.75	6 30	17 15	21.85	80
14-Mar	10.25	6 45	17 00	21.5	85
19-Mar	10.25	6 45	17 00	20.5	90
24-Mar	10	7 00	17 00	19.5	95
29-Mar	10	7 00	17 00	18.85	100
3-Apr	10	7 00	17 00	18.2	105
8-Apr	10	7 00	17 00	16	110
13-Apr	10	7 00	17 00	16	115
18-Apr	10.5	6 45	17 15	16	120

Note: Compressed (15 min) regime.

Although captive broodstock are usually fed fresh or frozen premium quality diets, e.g. chopped fish, squid, it is recommended that they are weaned onto a semi-moist or dry pellet (50.5 % protein, 24 % lipid) with vitamin and mineral supplements. The reason for the latter is that in kingfish (*Seriola* species), a diet of soft-dry pellets may produce two to five times more eggs and larvae of superior quality (higher fertilisation rates and fingerling yields) (Verakunpiriya *et al.*, 1997). While the majority of research on broodstock nutrition in marine fish concerns the levels of essential fatty acids (EFAs) required to support normal larval development, the optimal amounts of various EFAs for *Seriola* broodstock are not known. It does, however, appear that the ratio of EFAs is likely to be more important than the absolute quantities of the individual EFAs. Several studies on the broodstock nutrition of congeners of YTK (*S. quinqeradiata*) have shown that certain carotenoids, in particular astaxanthin, improve egg and larval quality. In one study by Verakunpiriya *et al.* (1997), astaxanthin improved egg buoyancy, fertilisation and hatching rate and prolonged the period of

© Woodhead Publishing Limited, 2013

egg production. However, not all carotenoids appear beneficial to broodstock. For example β-carotene is very poorly absorbed and high levels of β-carotene in broodstock diets have little effect on the subsequent egg and larval quality. Feeds are administered at 1–3 % or 10 % of total fish weight daily for pellet and wet diets, respectively.

18.3 Larviculture

Spawned eggs from well-nurtured YTK broodstock are positively buoyant, range from 1.33–1.50 mm in diameter and have a single oil droplet 0.30–0.33 mm in diameter and can have a high fertilisation rate (90–99 %). Mean ± s.d. egg viability within the floating fraction over a complete spring/summer reproductive season has been reported as 74 ± 17 % (approximate range 50–90 %) (Moran *et al.*, 2007). Indistinct cell margins and asymmetrical cleavage during blastomere formation are the most common deformities observed in eggs. Egg and oil droplet volume may decrease by 15–20 % over the spawning season, although no associated fall in egg and larval viability has been reported. For instance, while larvae hatch at a smaller length with a larger yolk sac and oil droplet at warmer incubation temperatures, there is no substantial difference in the maximum larval length reached at the onset of first feeding across incubation temperatures in the range 16–24 °C (Moran *et al.*, 2007). Incubation (time to hatch) is 120–40 h depending on water temperature in the range 16 °C and 24 °C, respectively.

In the case of natural spawning, fertilised eggs are collected from the surface of broodstock tanks with nets or screens or using automatic systems. Collected eggs are rinsed and treated with 100 mg L^{-1} formalin or preferably with ozone (CT = 1 mg L^{-1} min) to disinfect them of bacteria, fungi and viruses. Disinfected fertilised eggs are placed in sloping-bottom tanks typically ranging in size from 200 to 2000 L and maintained under darkness until hatching.

Kingfish (*Seriola* spp) larvae grow faster than many other marine fish. The first feeding larvae (three to four days post hatching; dph) are relatively large, averaging about 4.7 mm standard length (SL), therefore typical marine fish hatchery equipment and rearing protocols can be followed (Table 18.2) (Fielder and Heasman, 2011).

YTK larvae are cultured in rearing vessels commonly in the range 1–10 m^3 and stocked at densities ranging from 20 to 100 larvae L^{-1} into water conditioned with microalgae. Aeration or upwelling is used to maintain the larvae in a gentle rolling suspension to help reduce the incidence of nocturnal larval sinking and subsequent deformities and early mortalities. As for most other physoclistous larvae, surface skimmers are fitted to larval rearing tanks to ensure normal swimbladder inflation (Woolley and Qin, 2010) which commences at 2 dph (Fielder, 2012, unpublished data).

© Woodhead Publishing Limited, 2013

Table 18.2 Optimal rearing parameters and feeding schedule for Yellowtail kingfish (*Seriola lalandi*) larvae used at the NSW Department of Primary Industries, Port Stephens Fisheries Institute

Parameter	Target	dph	Adjustment
pH	7.6–8.2	0+	
Dissolved oxygen (mg L^{-1})	> 6.00	0+	Use compressed oxygen diffuser to maintain saturation level
Temperature (°C)	24.5	0+	Increase post SB inflation
Salinity (ppt)	25–35	0+	
Water exchange (%/day)	100–200	0+	Increase exchange as larvae develop
Surface skimmer (h/day)	24	2+	Monitor skimmer to ensure larvae at water surface are not affected
Photoperiod (L:D)	(12:12) (18:06)	(0+) (6+)	Increase post SB inflation
Lux	14500–35000	0+	
Green water (cells mL^{-1})	3–5 × 10^5	0+	Pro-Aqua* concentrate 57 × 10^9 per ml
Rotifer (R mL^{-1})	20.0–5.0	3+	Initial 20/mL until feeding and then increase frequency of reduced concentration (e.g. 4 × 5/mL/d)
Artemia (A mL^{-1})	0.2–2.0	12+	0.2/mL until weaned, then increase concentration and frequency
Weaning diet size (μm)	200–800	18+	Commence weaning at 10 mm TL

* Algae concentrate used Rotifer Diet-3600 (*Nannocholoropsis/Tetraselmis* blend) from Reed Mariculture Instant Algae, imported via Proaqua Australia. http://www.proaqua.net.au

YTK larvae begin feeding 3–4 dph once the yolk sac has been absorbed and jaw development is completed.

Feeding protocols are essentially the same as those described for other temperate marine fish larvae including mulloway (*Argyrosomus japonicus*) (Fielder and Heasman, 2011). Hatchery-reared YTK larvae are initially fed small or large strain rotifers (*Brachionus* spp.) enriched with commercial products at 5–20 rotifers mL^{-1}. Enriched *Artemia* meta-nauplii are subsequently added to the diet 10–14 days after hatch. Metamorphosis occurs approximately 20 days after hatching and 10 mm SL, is coincident with weaning onto inert formulated foods and is usually completed 40–50 days after hatching.

Feeding performance of YTK larvae increases with age and light intensity under both clearwater and greenwater rearing conditions, demonstrating that visual feeding proficiency increases with larval stage (Carton, 2005).

© Woodhead Publishing Limited, 2013

Feeding intensity remains low over the first three days of feeding regardless of light intensity. On days 6 and 7 after hatch, larvae show considerably higher feeding intensity particularly at light intensities in the range 8 and 17 $\mu mol\,sec^{-1}\,m^{-2}$ (≈1600–3400 lux). Recent studies have shown that growth, swimbladder inflation and survival of early-stage YTK larvae is increased as light intensity is increased up to 14000 lux (Stuart and Drawbridge, 2011) and 35000 lux (Fielder, 2012, unpublished data). This improvement indicates an ontogenetic shift in sensory acuity and/or swimming competence. First-feeding larvae perform equally well in clear water and green water up to algal cell densities of 8×10^4 mL^{-1}, although at a low light intensity of 0.1 $\mu mol\,sec^{-1}\,m^{-2}$ (≈20 lux), feeding performance is significantly constrained. The ability of YTK larvae to capture and consume free-swimming prey during the first-feeding window up to 5 dph is also impeded under high algae cell densities above 16×10^4 cells mL^{-1}. After this period, however, green water with an algal cell concentration of $3\text{–}5 \times 10^5$ cells mL^{-1} is necessary to optimise survival and growth of YTL larvae by improving feeding response (Stuart and Drawbridge, 2011).

Water temperature is typically maintained at the higher end of the optimum range (20–28 °C), although recent data suggest that 24.5 °C may be optimal for survival and low incidence of larval deformities (Fielder, 2012, unpublished data). The water exchange rate is gradually increased from 4 L/min (1–2 × exchanges per day) at the time of stocking, up to 20 L min^{-1} (4–8 × exchanges per day) immediately prior to weaning.

As in other marine finfish larvae, first feeding in YTK larvae is a major hurdle and adequate nutrition is critical to the success of this phase. Also in common with larvae of most other marine finfish, essential fatty acids (EFAs), in particular docosahexaenoic acid (DHA), are critical for normal development (Masuda *et al.*, 1998, 1999). DHA is accumulated in the central nervous system of YTK larvae and is essential not only for activity and vigour but also for the development of schooling behaviour in juveniles. Studies on the effect of the different EFAs on the growth and survival of kingfish larvae such as those of *S. quinqueradiata* have shown that the growth and survival rate of those fed DHA-enriched *Artemia* at 2.1–2.5 % dry wt day^{-1} is up to ten times better (88 %) than larvae fed *Artemia* enriched with other EFAs including eicosapentaenoic acid (EPA), arachidonic acid (AA) or oleic acid (OA).

Generally there are two peaks of mortality in rearing kingfish (*Seriola* spp) larvae (Kolkovski and Sakakura, 2004). The first is the so-called 'critical period' of high mortality from hatch to first feeding especially during the mouth-opening phase/first feeding stage (3 or 4 dph), during which larvae sink to the bottom of rearing tanks. This phenomenon can be mitigated by imposing strong upwelling currents that are also of substantial benefit for retaining the inert particles longer in the water column especially if live feeds are partially or totally substituted by inert microparticulate diets. The second mortality peak is caused by cannibalism.

© Woodhead Publishing Limited, 2013

The first obvious signs of aggressive interactions becomes evident as early as 12 days after hatch (6–7 mm total length), with both the large and medium size individuals displaying threatening (aiming) behaviour at smaller siblings (Moran, 2007). The onset of cannibalism does not occur in small larvae but coincides with increased growth rate and size heterogeneity and the onset of metamorphosis 18–22 dph at a mean size of 10 mm TL, cannibals being able to successfully prey on fish up to half of their own body size.

Fortunately cannibalism is a relatively fleeting phenomenon in YTK, progressively waning from around 30 dph (> 12 mm TL) as schooling behaviour takes over. A clear dominance hierarchy exists within schools of post-larvae, although it is likely that the ranking order changes with time. The aggressive behaviour of larvae is affected by temperature and light intensity, stocking density, feeding level, starvation and the size difference between fish (heterogeneity). Aggression in general, and cannibalism in particular, is exacerbated by increasing temperature over the range 15–30 °C, and by starvation (food deprivation) periods exceeding 12 h and is highest at medium light intensity. Numbers of aggressive encounters decreases as density is increased, probably due to the inability of predators to single out and attack individual smaller prey at very high densities. Aggression in dominant fish also increases at higher densities and, based on practical experience with other highly aggressive carnivorous fish species, it is likely that the overall level of cannibalism at intermediate densities will increase with increasing density. Although the above factors exaggerate the level of aggression in larvae, aggression still persists under low density, well-fed conditions amongst individuals of the same size. In order to reduce the level of cannibalism in cultured conditions, the recommended optimum stocking density is three fish L^{-1}. Fish should also be size-graded regularly (e.g. weekly) beginning at about 10 mm TL. At night time, juveniles cease swimming and drift at the surface in dense aggregations and are therefore easy to handle and grade. Indeed, night-time grading, has shown to increase survival yields by 1.5 to three times and is therefore recommended (Moran, 2007).

While relatively low rates of survival are to be expected with YTK larvae as with those of other species of kingfish, survival rates to metamorphosis and weaning have nevertheless been progressively improved from initial rates of 0.1–2 % up to 5–10 % currently with adoption of the measures discussed above.

An important lingering problem with hatchery production of YTK and other kingfish is the level of deformities. This problem is common to fish cultured in different places, including Japan, Australia and New Zealand. These deformities range from fused vertebrae and scoliosis, bent and/or shortened lower jaws, incomplete or absent gill covers (opercula) and compacted body and tails (Cobcroft *et al.*, 2004). In some instances it has been necessary to cull up to 50 % of a batch of hatchery-reared YTK due to

© Woodhead Publishing Limited, 2013

excessive level of juvenile fish deformity. This is expensive and increases the cost of fingerling production as detection of deformed fish under commercial conditions is not possible until the fish are advanced (e.g. 1–2 g; 40–50 mm TL) and sorting of deformed fish requires high levels of skilled labour. Although the full array of factors causing deformities is yet to be determined and negated, significant progress in mitigating these deformities has been made through improved larval nutrition and a better understanding of optimal abiotic larval rearing conditions. For example, feeding *Artemia* metanauplii (but not rotifers) enriched with 'mega' doses of vitamins E and C to YTK larvae is effective in significantly lowering the incidence of these deformities (Kolkovski and Sakakura, 2004).

Based on scales of deformity used by commercial operators to determine whether fish are retained or culled, the percentage of fish with excessive deformity and therefore culled from batches of YTK produced using current hatchery techniques is reliably less than 5 %. However, 20–40 % of the population can still have some degree of deformity (Fielder 2012, unpublished data).

18.4 Future trends

The major challenge for development of sustainable, profitable hatchery production of juvenile YTK remains the development of techniques to increase survival and quality of fingerlings. Although advances in technology have been made, particularly over the last few years, hatchery survival remains relatively low at approximately 10 %. This fact coupled with on occasion a high level of deformities in hatchery progeny and their subsequent culling can have a major impact on the cost per fingerling. In Australia, these issues are being addressed through coordinated research involving industry and research partners. Areas of research include understanding the influences that management of broodstock and abiotic and biotic larval rearing parameters have on production of juvenile YTK.

18.5 References

CARTON A G (2005) The impact of light intensity and algal-induced turbidity on first feeding *Seriola lalandi* larvae. *Aquac. Res.* 36, 1588–1594.

COBCROFT J M, PANKHURST P M, POORTENAAR C, HICKMAN B and TAIT M (2004) Jaw malformation in cultured yellowtail kingfish (*Seriola lalandi*) larvae. *N.Z. J. Mar. Freshwater Res.*, 38, 67–71.

ESCHMEYER W N, HERALD E S and HAMMANN H (1983) *A Field Guide to Pacific Coast Fishes of North America*. Boston, MA: Houghton Mifflin.

FIELDER D S and HEASMAN M P (2011) *Hatchery Manual for the Production of Australian Bass, Mulloway and Yellowtail Kingfish*. Orange, NSW: Industry and Investment NSW.

© Woodhead Publishing Limited, 2013

GILLANDERS B M, FERRELL D J and ANDREW N L (1999) Size at maturity and seasonal changes in gonad activity of yellowtail kingfish (*Seriola lalandi:* Carangidae) in New South Wales, Australia. *N.Z. J. Mar. Freshwater Res.*, 33, 457–468.

GILLANDERS B, FERRELL M J F and ANDREW N L (2001), Estimates of movement and life-history parameters of yellowtail kingfish (*Seriola lalandi*): how useful are data from a cooperative tagging program? *Mar. Freshwater Res.*, 52, 179–192.

KAILOLA P J, WILLIAMS M J, STEWART P C, REICHELT R E, MCNEE A and GRIEVE C (1993) *Australian Fisheries Resource*. Bureau of Resource Sciences, Canberra.

KOLKOVSKI S and SAKAKURA Y (2004) Yellowtail kingfish from larvae to mature fish – problems and opportunities, in Cruz Suárez L E, Ricque Marie D, Nieto López M G, Villarreal D, Scholz U, and González M (eds), *Avances en Nutrición Acuicola VII. Memorias del VII Simposium Internacional de Nutrición Acuicolae*, 16–19 Noviembre. Hermosillo, Sonora, México, 109–125.

MASUDA R, TAKEUCHI T, TSUKAMOTO K, ISHIZAKI Y, KANEMATSU M and IMAIZUMI K (1998) Critical involvement of dietary docosahexaenoic acid in the ontogeny of schooling behaviour in the yellowtail, *J. Fish Biol.*, 53, 471–484.

MASUDA R, TAKEUCHI T, TSUKAMOTO K, SATO S, SHIMIZU K and IMAIZUMI K (1999) Incorporation of dietary docosahexaenoic acid into the central nervous system in the yellowtail *Seriola quinqueradiata*, *Brain Behav. Evol.*, 53, 173–179.

MILLER P A, FITCH A J, GARDNER M, HUTSON, K and MAIR G (2011) Genetic population structure of Yellowtail Kingfish (*Seriola lalandi*) in temperate Australasian waters inferred from microsatellite markers and mitochondrial DNA, *Aquaculture*, 319, 328–336.

MORAN D (2007) Size heterogeneity, growth potential and aggression in juvenile yellowtail kingfish (*Seriola lalandi* Valenciennes), *Aquac. Res.*, 38, 1254–1264.

MORAN D, SMITH C K, GARA B G and POORTENAAR C W (2007) Reproductive behaviour and egg development in yellowtail kingfish (*Seriola lalandi* Valenciennes 1833), *Aquaculture*, 262, 95–104.

OHARA E, NISHIMURA T, NAGAKURA Y, SAKAMOTO T, MUSHIAKE K and OKAMOTO N (2005) Genetic linkage maps of two yellowtails (*Seriola quinqueradiata* and *Seriola lalandi*), *Aquaculture*, 244, 41–48.

POORTENAAR C W, HOOKER S H and SHARP N (2001) Assessment of yellowtail kingfish (*Seriola lalandi lalandi*) reproductive physiology, as a basis for aquaculture development, *Aquaculture*, 201, 271–286.

POORTENAAR C, JEFFS A, HEATH P and HOOKER S (2003) *Commercial Opportunities for Kingfish Aquaculture in Northland.* Auckland: Enterprise Northland Trust, National Institute of Water and Atmospheric Research.

SAKAKURA Y and TSUKAMOTO K (1996) Onset and development of cannibalistic behaviour in early life stages of yellowtail, *J. Fish Biol.*, 48, 16–29.

SMITH-VANIZ W F, QUÉRO J C, and DESOUTTER M (1990) Carangidae, in Quéro JC, Hureau JC, Karrer C, Post A and Saldanha A (eds), *Check-list of the Fishes of the Eastern Tropical Atlantic (CLOFETA)*. Lisbon: JNICT; Paris: SEI and UNESCO, Vol. 2, 729–755.

STEWART J, FERRELL D J and VAN DER WALT B (2004) Sizes and ages in commercial landings with estimates of growth, mortality, and yield per recruit of yellowtail kingfish (*Seriola lalandi*) from New South Wales, Australia, *Mar. Freshwater Res.*, 55, 489–497.

STUART K R and DRAWBRIDGE M (2011) The effect of light intensity and green water on survival and growth of cultured larval Californian yellowtail (*Seriola lalandi*), *Aquaculture*, 321, 152–156.

VERAKUNPIRIYA V, MUSHIAKE K, KAWANO K and WATANABE T (1997) Supplemental effect of astaxanthin in broodstock diets on the quality of yellowtail eggs, *Fish. Sci.*, 63, 816–823.

© Woodhead Publishing Limited, 2013

VERGANI M, ACHA E M, DIAZ DE ASTARLOA J M and GIBERTO D (2008) Food of the yellowtail amberjack *Seriola lalandi* from the south-west Atlantic, *J. Mar. Biol. Assoc. UK*, 88, 851–852.

WOOLLEY L D and QIN J G (2010) Swimbladder inflation and its implication to the culture of marine finfish larvae, *Rev. Aquac.*, 2, 181–190.

© Woodhead Publishing Limited, 2013

Part IV

Aquaculture hatcheries for conservation and education

19

Hatchery production for conservation and stock enhancement: the case of Australian freshwater fish

S. J. Rowland, NSW Fisheries, Australia

DOI: 10.1533/9780857097460.4.557

Abstract: Freshwater fishes are one of the great natural resources of the world, but anthropogenic activities have adversely affected aquatic ecosystems and many species are threatened with extinction. Throughout the 1900s, the distribution and abundance of native freshwater fishes in inland and coastal drainages in south-eastern Australia were significantly reduced, and there were extirpations of some populations. This chapter reviews these declines and the use of captive breeding and stocking programmes since the late 1970s to aid the recovery of threatened species, establish and maintain large recreational fisheries, and restore lost biodiversity. The successes of these programmes have been due to the development of best practices for broodfish management, artificial breeding, larval rearing, fingerling production and management of infectious diseases, and guidelines which minimise the effects of domestication, maintain genetic identity of stocked populations, and maximise genetic variation providing a basis for fitness and evolutionary potential of each species.

Key words: biodiversity in freshwater ecosystems, conservation of threatened species, stock enhancement, hatchery production, fish stocking.

19.1 Introduction

Freshwater fishes are one of the great natural resources of the world, and play important cultural, social and economic roles in many countries. Anthropogenic activities have adversely affected freshwater aquatic ecosystems, biodiversity is declining rapidly, and many fishes are now threatened with extinction on all continents (e.g. Warren and Burr, 1994; Jelks *et al.*, 2008). Freshwater fishes are one of the most threatened groups of vertebrates, and the loss of biodiversity in fresh water exceeds that observed

© Woodhead Publishing Limited, 2013

in both terrestrial and marine environments (Bruton, 1995; Ricciardi and Rasmussen, 1999).

Australia is the driest inhabited continent on earth and its freshwater ecosystems and fishes are threatened (Lake, 1971). Australia has a depauperate and unique freshwater fish fauna compared to other countries and continents due primarily to its isolation over the past 40 million years and the geological and climatic conditions over that time (Merrick, 2006). The inland Murray–Darling River System (MDRS) (Fig. 19.1) is Australia's largest drainage covering over 1 million km^2 or 14 % of the land mass, and is characterised by low rainfall and runoff, and relatively low discharge (Lake, 1971; Lintermans, 2007). Coastal drainages such as the Clarence River System, the Richmond River System and the Mary River System are relatively short, with high flows and rapid discharge, and a fish fauna containing catadromous and anadromous, as well as potamodromous species (Lake, 1971; Merrick and Schmida, 1984; Allen *et al.*, 2002).

Freshwater fish play important roles in Australian society. Native species are prominent in aboriginal culture and mythology, and inland fish were exploited and used as fresh food by explorers, pioneers and early settlers

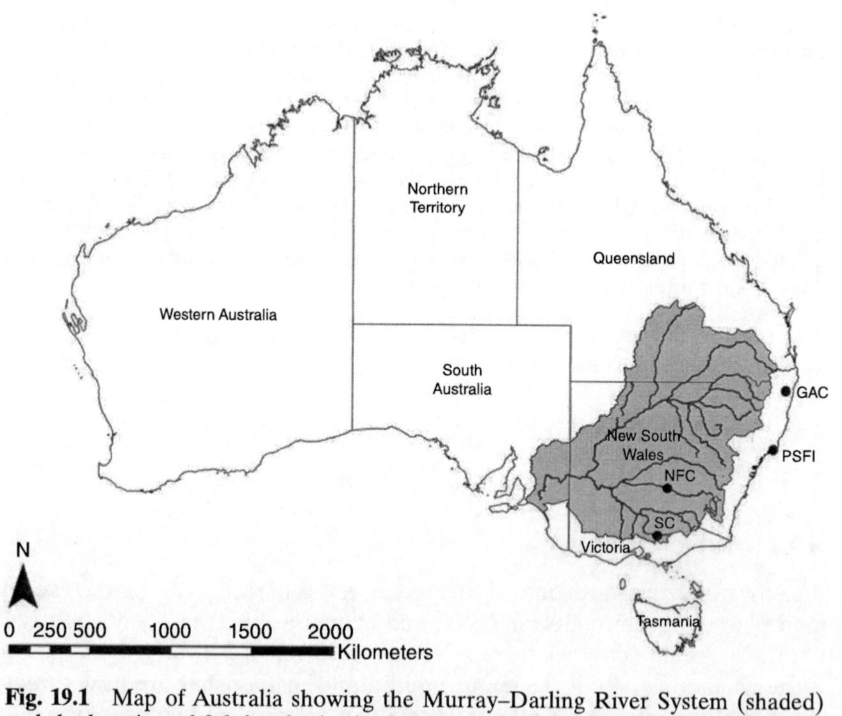

Fig. 19.1 Map of Australia showing the Murray–Darling River System (shaded) and the location of fish hatcheries in the states of New South Wales and Victoria: Narrandera Fisheries Centre (NFC), Grafton Aquaculture Centre (GAC), the Port Stephens Fisheries Institute (PSFI) and the Snobs Creek Hatchery (SC).

© Woodhead Publishing Limited, 2013

(Rowland, 1989, 1993). Many species are highly regarded for their edible and sporting qualities, and are the basis of popular recreational fisheries and developing commercial aquaculture industries (Harris and Rowland, 1996; Ingram and De Silva, 2004; Rowland, 2005, 2009). Australia's native fishes have remarkable adaptations to the extreme conditions and regimes of floods and droughts that are characteristic of many of its drainages (Lake, 1971). However, changes including significant degradation to aquatic environments and river regulation in both inland and coastal river drainages in south-eastern Australia have provided numerous and widespread threats to native fish, and the perilous state of many freshwater fishes was evident by the 1970s (Lake, 1971; Pollard *et al.*, 1980; Merrick and Schmida, 1984; Rowland, 1989). In response to the decline of freshwater fishes, government hatcheries in New South Wales (NSW) and Victoria began producing native fish fingerlings in the late 1970s to re-establish some species, to increase fish abundance and to enhance recreational fishing opportunities (Rowland *et al.*, 1983; Cadwallader and Gooley, 1985). In the 1980s, conservation of threatened fishes became a major focus for fisheries agencies (Merrick and Schmida, 1984; Harris, 1987; Rowland and Barlow, 1988).

Fish stocking has been an important and successful conservation and management tool in Australia since the late 1970s. In this chapter, the decline of Australian native freshwater fishes, their conservation status, the development of captive breeding and stocking programmes and the management of potentially detrimental factors such as domestication, population genetics and infectious diseases are reviewed.

19.1.1 Decline of freshwater fishes

Most of the cods and perches of Australia were originally named by early naturalists and explorers, and are taxonomically different to those of other continents and oceans (Berra, 2001). There was concern about the declining number of fish in the MDRS as early as 1880 (Macleay *et al.*, 1880), and by the 1930s the commercial fishery for Murray cod (*Maccullochella peelii* (Mitchell)) was not profitable for large-scale operators (Dakin and Kesteven, 1938). During the mid-1900s, there were further reductions in the catch of Murray cod and other species, and catches remained low until closure of the commercial fishery in 2001 (Rowland, 1989; Reid *et al.*, 1997; Gilligan, 2005a). In addition, trout cod (*M. macquariensis* (Cuvier)), Macquarie perch (*Macquaria australasica* (Cuvier)), golden perch (*M.* (Richardson)), and silver perch (*Bidyanus bidyanus* (Mitchell)) were in decline and disappeared completely from some areas (Lake, 1971; Cadwallader, 1977, 1981; Ingram and Douglas, 1995; Ingram *et al.*, 2000; Koehn, 2001). Lake (1971) considered that one third of Australia's freshwater fish fauna was in jeopardy, and the findings of Harris and Gehrke (1997) emphasised the

© Woodhead Publishing Limited, 2013

imperiled status of native fish in the MDRS. There were also significant declines of freshwater fishes in coastal drainages in south-eastern Australia, and by the mid-1900s populations of eastern freshwater cod (*M. ikei* Rowland) and Mary River cod (*M. mariensis* Rowland) were extirpated (Rowland, 1985a, 1993, 1996a; Simpson and Jackson 1996) and the distribution of Australian bass (*M. novemaculeata* (Steindachner)) was severely reduced (Harris, 1984). The cumulative effect of these declines, extirpations and reduced genetic variation in some species (Keenan *et al.*, 1995; Bearlin and Tikel, 2003) constituted a significant loss of biodiversity in inland and coastal drainages of south-eastern Australia.

Many Australian freshwater fishes breed during spring and summer in response to various environmental cues, including increasing water temperatures, lengthening photoperiod and rising water levels associated with rainfall, snow-thaw in southern areas, and monsoonal rains in northern areas (Lake, 1967a, b; Mackay, 1973; Rowland, 1983a, 1998; Merrick and Schmida, 1984; Battaglene, 1991; Ingram and Douglas, 1995; Ingram *et al.*, 2000; Butler and Rowland, 2009). The catadromous Australian bass is an exception and migrates downstream to estuaries to spawn during autumn and winter (Harris and Rowland 1996). Recruitment of larvae and juveniles has been linked to the occurrence of increasing water levels, high rivers, and/or floods in rivers and impoundments (e.g. Mackay, 1973; Reynolds, 1976; Rowland, 1983a, 1998; Geddes and Puckridge, 1989; Battaglene, 1991; Mallen-Cooper and Stuart, 2003; Ye and Zampatti, 2007; King *et al.*, 2009).

Causes of the decline of Australian native fishes are multi-factorial and primarily linked to urban and rural development. During the late 1800s and early 1900s, key factors were overfishing, natural and man-made pollution, exotic species and habitat degradation (Lake, 1971; Cadwallader, 1977; Merrick and Schmida, 1984; Mackay and Shafron, 1989; Rowland 1989, 1993, 2005; Richardson, 1994). By the mid-1950s, river regulation associated with the construction of dams, high-level weirs and levee banks had dramatically altered flow, flooding and temperature regimes and reduced access to and inundation of the floodplains in the MDRS; these dramatic changes adversely affected spawning and/or recruitment of native fishes (Lake, 1971; Reynolds, 1976; Walker, 1980; Rowland, 1989; Richardson, 1994; Gehrke *et al.*, 1995; Humphries and Lake, 2000; Koehn, 2001). Besides affecting reproduction, weirs and dams are barriers to species known to move or migrate to breed, including Murray cod, golden perch, silver perch and Australian bass (Reynolds, 1983; Harris and Rowland 1996; Koehn *et al.*, 2009). Many large-scale changes caused by anthropogenic activities are difficult or impossible to reverse and, while some factors such as barriers to movement and migration can be ameliorated through the construction of fish ladders in weirs (Barrett, 2008; Stuart *et al.*, 2008), the altered flow and temperature regimes, restricted inundation of the floodplain, exotic

© Woodhead Publishing Limited, 2013

fishes and extensive agricultural development remain key environmental characteristics throughout much of the MDRS.

19.1.2 Conservation classifications and actions

Conservation classifications of Critically Endangered, Endangered, Vulnerable and Potentially Threatened relate to the threat of extinction, e.g. Endangered species have undergone a very large reduction in abundance, geographic distribution or genetic diversity, are affected by threatening processes, and face a very high risk of extinction in the near future. International, national and state bodies, as well as specialist organisations list threatened species, and Moore *et al.* (2010) provides a summary of conservation and fisheries acts at national and state levels in Australia. For example, native fish in NSW are listed under the *NSW Fisheries Management Act 1994*, the Commonwealth *Environment Protection and Biodiversity Conservation Act 1999 (EPBC)*, by the International Union for Conservation of Nature and Natural Resources (IUCN) and the Australian Society for Fish Biology (ASFB). The classifications can vary between the listing agencies, e.g. Murray cod is not listed or protected in NSW and Queensland, but is listed in Victoria and by the Commonwealth (Lintermans *et al.*, 2005). The conservation status and summary of stocking programmes of some threatened fishes are summarised in Table 19.1.

Conservation of threatened fishes became a high priority in Australia in the 1980s. For example, in NSW there was a rapid response to the imperiled status of the eastern freshwater cod after it was identified as a separate species to Murray cod; in 1984, it was listed as Totally Protected under section 19 of the *NSW Fisheries and Oyster Farms Act* (Rowland, 1985a; Talbot *et al.*, 2004). Subsequently, construction of the Eastern Freshwater Fish Research Hatchery (later named the Grafton Aquaculture Centre (GAC)) commenced in 1984, and a conservation breeding and stocking programme was established (Rowland, 1985b, 1988a; Talbot *et al.*, 2004). In addition, trout cod was protected in both NSW and Victoria, and hatchery programmes began at the Narrandera Fisheries Centre (NFC) and Snobs Creek (Rimmer, 1987; Ingram and Rimmer, 1992). In 1985, the ASFB held a threatened fishes conference in Melbourne and eight species were listed as Endangered or Vulnerable (Harris, 1987). In 1988, the IUCN listed 12 Australian fishes, including eastern freshwater cod and trout cod as Endangered or Vulnerable (IUCN, 1988).

Recovery Plans and Threat Abatement Plans have been subsequently developed for threatened species; trout cod (Douglas *et al.*, 1994), Mary River cod (Simpson and Jackson, 1996), silver perch (Clunie and Koehn, 2001; Anon., 2006) and eastern freshwater cod (NSW Fisheries, 2002; Talbot *et al.*, 2004). There have also been a number of published reviews and case

© Woodhead Publishing Limited, 2013

Table 19.1 Conservation status and stocking programmes of some threatened Australian freshwater fishes

Species	Conservation status	Stocking programmes	Comments
Eastern freshwater cod *Maccullochella ikei*	E	NSW 1988 and 1989 1997–2002	Increased distribution and abundance; now in two river systems
Mary River cod *Maccullochella mariensis*	CE	Qld Since 1983	Increased distribution and abundance; now in four river systems
Trout cod *Maccullochella macquariensis*	E – NSW, ACT CE – Victoria	NSW, Victoria Since 1987	Increased distribution and abundance; now four or five stocks, with three self-maintaining
Murray cod *Maccullochella peelii*	V – *EPBC*, South Australia E – Vic. Not listed – NSW, Qld	NSW, Victoria, Qld Since 1978	Increased distribution; significant increase in abundance in many NSW waters
Silver perch *Bidyanus bidyanus*	V	NSW Since 1978 Impoundments only	Abundant in some impoundments; rare or uncommon in most rivers
Macquarie perch *Macquaria australasica*	E	Victoria, NSW No large-scale hatchery techniques; low numbers stocked	No recovery

CE = Critically Endangered; E = Endangered; V = Vulnerable; *EPBC* = *Environment Protection and Biodiversity Conservation Act 1999*

studies of threatened species in Australia, e.g. Pollard *et al.* (1990), Ingram *et al.* (1990) and Morris *et al.* (2001).

19.2 Captive breeding and stocking

19.2.1 Justification

The stocking of hatchery-bred fish into the wild in freshwater, estuarine and marine environments is used worldwide to maintain or enhance fish stocks for sport or recreation, to provide food and income, and to conserve threatened species (Cowx, 1994; Welcomme and Bartley, 1998). Many endangered species of animals require captive breeding to save them from extinction and a range of species have been preserved in captivity (Frankham *et al.*, 2002).

© Woodhead Publishing Limited, 2013

In Australia, the establishment of native fish hatcheries was first recommended in 1936, but early attempts to produce Murray cod for restocking were unsuccessful (Dakin and Kesteven, 1938). By the 1970s, following research by the late John Lake (Lake, 1967a, b), it was apparent that habitat degradation and river regulation had removed or reduced the incidence of the key environmental stimuli necessary for breeding and recruitment, and the maintenance of self-maintaining populations of native fishes in many parts of the MDRS. In response to the long-term decline of native fish, the imperiled status of some species, and the lack of recreational fishing opportunities, NSW Fisheries commenced a major research project in 1972 to develop hatchery techniques for Murray cod, golden perch and silver perch (Rowland *et al.*, 1983). In addition, Victorian Fisheries commenced research into the hatchery production of Murray cod at Lake Charlegrark and later at Snobs Creek (Cadwallader and Gooley, 1985). By 1982, techniques enabling large-scale production of fingerlings of those species were developed, and in 1982/83 the technology was extended to a commercial industry (Rowland, 1983a, b, c, d; Cadwallader and Gooley, 1985; Rowland and Tully, 2004). These techniques also provided the basis for research and production of other species in following years including trout cod, eastern freshwater cod and Macquarie perch as part of conservation programmes (Gooley, 1986; Rimmer, 1987; Gooley and McDonald, 1988; Rowland, 1990; Ingram and Rimmer, 1992; Ingram *et al.*, 1994) and Australian bass for stock enhancement (Battaglene *et al.*, 1989; Fielder and Heasman, 2011). The native fish stocking programme commenced in NSW in 1976/77 (Rowland *et al.*, 1983), and in Victoria in 1978/79 (Cadwallader and Gooley, 1985).

19.2.2 Concerns and criticisms of stocking

The practice of fish stocking is contentious and has received strong criticism based on proven and perceived negative impacts of stocked hatchery-reared fish, particularly on genetics, population structure, reproduction and fitness of wild stocks, and ecosystems (e.g. Allendorf, 1991; Lynch and O'Hely, 2001; Utter, 2004; Eby *et al.*, 2006). In Australia, potential threats have been highlighted at workshops and, in an Environmental Impact Statement (Moore and Hughes, 2000; New South Wales Fisheries, 2003; Phillips, 2003; Lintermans and Phillips, 2005; Moore *et al.*, 2010), stocking was listed as a threat to Murray cod by Koehn (2005b), and Lintermans *et al.* (2005), and Gillanders *et al.* (2006) identified numerous potential adverse effects of stocking fish in the MDRS. There is no stocking of native fish in South Australia (Gillanders and Ye, 2011). The main concerns with stocking are associated with genetics, competition, predation, diseases and ecosystem effects.

While some of the criticisms of fish stocking in Australia are justified, many have been made on a theoretical basis, influenced by overseas

© Woodhead Publishing Limited, 2013

findings, with little or no empirical evidence, and without taking into account the imperiled status of freshwater fishes in the 1970s, the loss of genetic diversity and biodiversity during the 1900s, the urgent need for rapid action to increase abundance and distribution of many species, and the difficulties and long timeframes required to ameliorate environmental degradation.

19.3 Actions to address concerns in Australia

The potential detrimental effects of stocking have long been recognised by fisheries authorities in Australia, and significant effort has been made to eliminate or minimise such effects and to ensure best hatchery and stocking practices and genetic management (Brown, 1985, 1987; Rowland and Barlow, 1988; O' Connor, 1989; Rowland, 1995; New South Wales Fisheries, 2003; Sanger and Talbot, 2003; Rowland and Tully, 2004). Aquaculture research in the 1970s and early 1980s led to successful, efficient techniques for the large-scale production of fry and fingerlings, as well as management of broodfish, water quality and fish diseases (e.g. Rowland, 1983a, b, 1986; Rowland and Ingram, 1991). On-going research since the mid-1980s has improved fish breeding and larval rearing techniques, and management of broodfish, genetics, infectious diseases and fish health (e.g. Rowland and Ingram, 1991; Rowland and Bryant, 1995; Rowland, 1996b; Rowland and Tully, 2004; Ingram *et al.*, 2005; Ingram and De Silva, 2007; Read *et al.*, 2007; Rowland *et al.*, 2007; Ingram, 2009).

19.3.1 Genetics

Genetics is of fundamental importance in conservation, as well as stock enhancement programmes where stocked fish may mix with fish in wild, outbreeding populations. Genetic changes caused by genetic drift, inbreeding and outbreeding depression, as well as characteristics selected for in captivity are overwhelmingly disadvantageous in the natural environment, and can lead to reduced 'fitness' in the wild (Frankham, 1999, 2008; Lynch and O'Hely, 2001; Frankham *et al.*, 2002). In many hatchery programmes throughout the world, small numbers of 'founding' fish are used to establish hatchery breeding stocks, and then fish are selected from each following generation for use as future broodfish. This potentially creates a genetic bottleneck, leading to a loss of genetic variation and domestication of broodfish (Brown 1987). Subsequent hatchery releases have the potential to reduce genetic variation and effective population size in wild populations; the latter is called the 'Ryman–Laikre effect' (Ryman and Laikre, 1991). However, while such practice is common in northern hemisphere species, particularly salmonids, only wild broodfish are used to produce fingerlings for conservation and stock enhancement in Australian native fish hatcheries (Brown 1987; Rowland and Barlow, 1988; Rowland and

© Woodhead Publishing Limited, 2013

Tully, 2004). Minimising generations in captivity and mimicking natural environmental conditions can reduce genetic adaption to captivity (Lynch and O'Hely, 2001; Frankham, 2008).

The management of genetics in threatened species and small populations was addressed by workshops in Victoria and NSW in 1985 (Brown, 1985; Harris, 1987; Rowland and Barlow 1988). Brown (1985) discussed the principles of genetic variation within and between populations in relation to the establishment of captive breeding populations, and Brown (1987) recommended broodfish management goals to maximise genetic variation and minimise domestication. Fisheries scientists and conservation geneticists discussed hatchery techniques, genetics of wild and hatchery-reared fish, and genetics of threatened species and small populations, and established the following guidelines: (i) use of wild fish only, and subsequent replacement after three to four years; (ii) 100 broodfish per generation (20–25 per year); (iii) use of hormone-induction techniques where possible to avoid 'same-matings'; (iv) stocking progeny from at least five pairs to minimise loss of genetic variation (Rowland and Barlow, 1988). These guidelines were subsequently used in captive breeding and stocking programmes at government and commercial hatcheries, and have been updated as new information became available (e.g. Douglas *et al.* 1994; Gilligan, 2000; Moore, 2000; Thurstan, 2000; Frankham *et al.*, 2002; Hallerman, 2003; Rowland and Tully, 2004; Rourke *et al.* 2011; Ingram *et al.* 2011).

Genetically-responsible stocking is dependent on a good knowledge the taxonomy and population genetics of fish. Although there are still some uncertainties, the taxonomy and genetic structure of threatened and other native freshwater fishes are reasonably well known, e.g. eastern freshwater cod and Mary River cod (Rowland, 1985a, 1993; Nock *et al.*, 2010, 2011); Murray cod (MacDonald, 1978; Bearlin and Tikel, 2003; Rourke *et al.*, 2009, 2010, 2011); trout cod (MacDonald, 1978; Douglas *et al.*, 1995; Bearlin and Tikel, 2003); silver perch (Keenan *et al.*, 1995; Bearlin and Tikel, 2003); Macquarie perch (MacDonald, 1978; Dufty, 1986; Faulks *et al.*, 2010a, 2011), purple-spotted gudgeon (Faulks *et al.*, 2008; Hammer, 2008), golden perch (MacDonald, 1978; Musyl and Keenan, 1992; Keenan *et al.*, 1995; Faulks *et al.*, 2010b) and Australian bass (Jerry, 1997). Recently, there has been a review of the management of genetic resources for fish and crustaceans in the Murray–Darling Basin (Moore *et al.*, 2010), and the development of a comprehensive set of DNA barcodes that distinguishes between 58 native and introduced species of freshwater fish in the Murray–Darling Basin will enhance capacity in conservation biology, including captive breeding and restocking programmes (Hardy *et al.*, 2011).

19.3.2 Domestication

Domestication results in adaptation to the captive environment, and can include changes in behaviour, performance, morphology and predator

© Woodhead Publishing Limited, 2013

avoidance (Álvarez and Nicieza, 2003; Hallerman, 2003; Mayer *et al.*, 2011). Differential selection in captivity is caused by factors such as crowding, change of food, absence of predators and inadvertent selection for tameness, and can lead to poor performance of fish stocked in natural environments (Frankham and Loebel, 1992; Hallerman, 2003). Examples of changes associated with captivity in Australian fishes include faster growth and earlier sexual maturation of domesticated silver perch (Rowland, 2004), and changes in swimming, schooling and feeding behaviour of weaned Murray cod and silver perch (Rowland *et al.*, 1994; Ingram and De Silva, 2004).

Methods to reduce effects of domestication include making the captive environment similar to wild environments, use of wild broodfish of both sexes and equalising family size (Frankham and Loebel, 1992; Allendorf, 1993; Harada *et al.*, 1998). The following management and rearing practices are commonly used on Australian native fish hatcheries to minimise the effects of domestication: (i) use of only wild fish, and turnover of broodfish each three to five years; (ii) stocking broodfish in earthen ponds (Macquarie perch and golden perch are held in tanks on some hatcheries); (iii) use of natural feed for broodfish such as live fish and crustaceans; (iv) holding larvae under natural temperature and photoperiod regimes; (v) rearing larvae in a semi-natural environment of fertilised earthen ponds where they feed on zooplankton and insects, and are subject to natural photoperiod and water temperatures, predation by birds and aquatic insects, and in some species cannibalism; and (vi) minimising the period in captivity by harvesting fingerlings (30–50 mm) after 5–10 weeks.

19.3.3 Infectious diseases and health management

Infectious diseases are relatively common in aquaculture, and have also been reported in Australian native fish in the wild (Langdon *et al.*, 1985; Puckridge *et al.*, 1989; Rowland and Ingram, 1991; Boys *et al.*, 2012). The introduction of pathogens via stocked fish is a significant problem throughout the world (Paperna, 1991) and so a good understanding of diseases and implementation of appropriate health management strategies are essential to minimise the risks.

Diseases of Australian freshwater fishes are well known. Parasitic protozoans, gill flukes and the fungus *Saprolegnia parasitica* are common pathogens, whereas the incidence of bacterial and viral diseases is relatively low (Rowland and Ingram, 1991; Callinan and Rowland, 1995; Lancaster *et al.*, 2003; Ingram *et al.*, 2005; Rowland *et al.*, 2007; Fielder and Heasman 2011). Australian freshwater cod are particularly susceptible to infestations of the ciliates *Ichthyophthirius multifiliis* and *Chilodonella hexasticha*, and infections of *S. parasitica* both under culture conditions and in the wild (Rowland and Ingram, 1991). Although these pathogens can cause high mortalities, techniques for diagnosis, treatment and prevention have been established,

© Woodhead Publishing Limited, 2013

diagnostic manuals are available and health management strategies have been developed (Rowland and Ingram, 1991; Ingram *et al.*, 2005; Read *et al.*, 2007; Rowland *et al.*, 2007). The recent report of the disease epizootic ulcerative syndrome on native fish in the MDRS and its possible introduction on infected fish from farms in coastal drainages (Boys *et al.*, 2012) demonstrates the risks associated with movements of hatchery fish and the need for appropriate hatchery practices and policies. Good hatchery and health management practices and adherence to criteria in the hatchery quality assurance programme (HQAP; see Section 19.3.5 below) will eliminate or minimise the transfer of pathogens from native fish hatcheries (Rowland and Tully, 2004; Fielder and Heasman 2011).

19.3.4 Translocation

Translocation of fish is a major problem worldwide with the potential to cause significant ecological problems that are extremely difficult to manage. It is a potential problem associated with fish stocking in the MDRS (Arthington, 1991; Rowland, 1995; Phillips, 2003). Translocation involves not only the target fish species, but non-target fish and other aquatic organisms that may be moved in water. Fish culture has been implicated in the production and release of the Boolarra strain of common carp (*Cyprinus carpio*) in the MDRS (Rowland and Tully, 2004) and the banded grunter (*Amniataba percoides*) into impoundments in south-eastern Queensland and the Clarence River System in northern NSW (Rowland, 2001; Rowland and Tully, 2004).

19.3.5 Hatchery quality assurance programme

In response to the need for a hatchery accreditation scheme across the Murray–Darling Basin (Murray–Darling Basin Commission, 2003; Phillips, 2003), a hatchery quality assurance programme (HQAP) was developed by NSW Fisheries (Rowland and Tully, 2004). The HQAP identified best hatchery practices, determined effective population sizes for conservation stockings ($N_e = 100$) and stock enhancement ($N_e = 50$), and designated essential criteria for site selection, design and operation, broodfish management, breeding programs, fish health and water quality management, and harvest and stocking procedures. The HQAP is administered through a Hatchery Quality Assurance Scheme, and all hatcheries in NSW that produce fish for stocking programmes must be accredited and undergo regular audits. Breeding and stocking programmes must be closely linked, and are designed to maintain the genetic identity of wild populations, maximise genetic variation in stocked fish, prevent swamping by siblings and minimise domestication. Essential criteria in the HQAP also aim to prevent the dispatch of diseased fish and trash fish from hatcheries.

© Woodhead Publishing Limited, 2013

19.3.6 Stocking policy and guidelines

In NSW, management of native fish stocking has been based on fisheries regulations and policy that ensure fish from government hatcheries are only stocked into waters within their natural or historical ranges, and only into waters that do not contain significant breeding populations of that species (O'Connor, 1989; Rowland, 1995). An Environmental Impact Statement on freshwater fish stocking was prepared, and a Fishery Management Strategy for Fish Stocking was developed (New South Wales Fisheries, 2003). All stockings from both government and commercial hatcheries require a permit, and threatened species and endangered ecological communities are considered before any permit is issued (New South Wales Fisheries, 2003; Sanger and Talbot, 2003). There are also guidelines for stocking public waters in other Australian states (see Moore *et al.*, 2010).

19.4 Australian government hatcheries

The following government hatcheries have been involved in conservation and stock enhancement.

19.4.1 Narrandera Fisheries Centre (formally Inland Fisheries Research Station)

The Narrandera Fisheries Centre (NFC) is located adjacent to the Murrumbidgee River at Narrandera in southern NSW (Fig. 19.1). NFC is a pond-based facility (Fig. 19.2a) consisting of 45 earthen ponds (0.05–0.5 ha surface area), a reservoir, effluent dams, a hatchery complex, and offices, workshops and associated equipment. The main water supplies are the Murrumbidgee River and groundwater. Established in 1960–61 (Lake, 1967c), NFC was the first facility built in Australia for research into inland native fishes.

19.4.2 Grafton Aquaculture Centre (formally Eastern Freshwater Fish Research Hatchery)

The Grafton Aquaculture Centre (GAC) is located near Grafton, on the NSW North Coast (Fig. 19.1). It is a pond-based facility (Fig. 19.2b) consisting of 19 earthen ponds (0.1–0.35 ha surface area), two reservoirs (8.5 and 9 ML capacity), an effluent/settlement dam (43 ML), a hatchery/office/laboratory complex and a large workshop/storage shed. The main water supply is the Clarence River, and all effluent water is stored, settled and either reused for fish culture or used for irrigation. The facility was established in the mid-1980s for the conservation of the eastern freshwater cod (Rowland, 1988a, 1990) and subsequently, between 1990 and 2009, GAC

© Woodhead Publishing Limited, 2013

Fig. 19.2 NSW Government hatcheries used in the conservation of threatened fishes: (a) the Narrandera Fisheries Centre (photo taken 1978); and (b) the Grafton Aquaculture Centre (1996).

was used for the hatchery production and aquaculture research and development of silver perch (Rowland, 2009).

19.4.3 Port Stephens Fisheries Institute

This is a multipurpose marine hatchery located on the coast of NSW (Fig. 19.1) that enables the use of brackish and freshwater for breeding Australian bass. The facility has a combination of outdoor ponds, tanks and indoor culture facilities for algae, rotifers, *Artemia* and fish. Detailed descriptions are found in Fielder and Heasman (2011).

19.4.4 Snobs Creek Hatchery

Snobs Creek Hatchery is located near the Goulburn River, Victoria (Fig. 19.1). It has a long history of salmonid culture and, since the 1980s, native

© Woodhead Publishing Limited, 2013

warmwater fish, including threatened species. The facility consists of 28 earthen ponds (0.09–0.39 ha) for broodfish and larval rearing, inlet channels which supply gravity-fed water from Snobs Creek or pumped water from the Goulburn River, effluent-settlement ponds and a hatchery complex which includes recirculation facility to maintain relatively high water temperatures for breeding warmwater species. The facilities are used for both research and fish production. A number of species have been produced for stocking, including Murray cod, trout cod, Macquarie perch and golden perch.

19.4.5 Hatchery technology

The following is a brief, general overview of hatchery techniques; further information, including fish husbandry, pond management and fish health management are available in Rowland (1983c, d, 1986), Thurstan and Rowland (1995), Ingram and De Silva (2004), Rowland and Tully (2004), Ingram *et al.* (2005), Rowland *et al.* (2007) and Fielder and Heasman (2011).

19.4.6 Broodfish management

Only fish from the wild are used in breeding programmes; there are no line-bred or domesticated fish used in any conservation or stock enhancement programme. Sexually mature fish are collected from the wild, quarantined and then stocked in earthen ponds, tanks or cages; most are held in ponds to limit potential effects of domestication. Fish with abnormalities or deformities are rejected. All fish are tagged, and males and females are usually stocked together at ratios near 1:1. In NSW, hatcheries must have sufficient broodfish from each target species and genetic population to meet the breeding requirements of the HQAP. In the breeding season, sexually mature broodfish are selected randomly with respect to size and appearance, and mated according to guidelines in the HQAP. Broodfish are replaced with fish from the wild after three to five years in captivity to meet requirements of the breeding programmes and because the reproductive performance of some species declines after this period in captivity (Rowland and Tully, 2004). Ripe Australian bass broodfish are captured directly from rivers during the breeding season (Fielder and Heasman, 2011). Historically, Macquarie perch broodfish were collected from the wild immediately prior to hormone-induce breeding, but in recent years pond-held broodfish have been used (Brett Ingram and Matthew McLellan, pers. comm.). Anaesthetics are used to facilitate broodfish handling (see Rowland *et al.*, 2007; Fielder and Heasman, 2011).

Earthen ponds at most hatcheries are artificially aerated to ensure good water quality, especially dissolved oxygen concentrations, and the variables DO, pH, temperature and ammonia are regularly monitored (Rowland and Bryant, 1995; Rowland *et al.*, 2007). Broodfish are not sampled from ponds or tanks for disease monitoring because handling is known to induce atresia and resorption of oocytes in some species, particularly *Maccullochella* spp.

© Woodhead Publishing Limited, 2013

(Rowland, 1988b). Ponds are treated prophylactically with formalin in winter and spring to prevent infestations of *C. hexasticha* and other ectoparasites (Rowland and Ingram, 1991).

19.4.7 Induced spawning and incubation of eggs

Initial research into Murray cod, golden perch and silver perch identified breeding seasons, and established optimal hormone dosages for induced breeding and techniques for broodfish management, egg incubation, pond preparation and the rearing of larvae and production of fingerlings in ponds (e.g. Rowland, 1983a, c, d, 1984, 1986, 1988b, 1992, 1996b; Rowland *et al.*, 1983; Cadwallader and Gooley, 1985). Broodfish undergo normal gonadal development in captivity, and in spring and summer final oocyte maturation, ovulation and spawning are induced using species-specific dosages of human chorionic gonadotrophin (HCG). HCG is a uniform, measureable source of gonadotrophin that is readily available, easily stored and a successful inducing agent in Australian freshwater fishes (Rowland, 1983b). In addition to HCG, other agents such as gonadotrophin releasing hormone (GnRH) and GnRH analogues such as salmon gonadotrophin releasing hormone analogue (sGnRHa) and Ovaprim® (which contains SGnRHa) have been used to control reproduction in trout cod, Macquarie perch, eastern freshwater cod and Australian bass (Rimmer, 1987; Rowland, 1990; Ingram and Rimmer, 1992; Ingram *et al.*, 1994; Ingram and Gooley, 1996; Battaglene and Selosse, 1996; Ingram and Thurstan, 2008). Australian bass are spawned in autumn and winter, and eggs are incubated in saline water mimicking natural conditions (Fielder and Heasman, 2011).

Although eggs can be stripped from Murray cod after the induction of final oocyte maturation and ovulation, this technique is labour intensive and the quality of eggs may be variable (Rowland, 1988b). Murray cod broodfish prepare nests, undertake complex reproductive behaviour and spawn readily on hard surfaces in earthen ponds (Rowland, 1983a; Cadwallader and Gooley, 1985), and production on most government and commercial hatcheries is based on collection of naturally-fertilised eggs from artificial spawning sites placed in ponds in spring (Thurstan, 2000; Ingram and De Silva, 2004). Although most Murray cod spawnings are monogamous, polygamous and polyandrous spawnings and repeat spawnings of some fish can occur in a single breeding season (Rourke *et al.*, 2009). Consequently, an unknown number of broodfish may contribute to the production of Murray cod fingerlings each season (Rourke *et al.*, 2009; Ingram *et al.*, 2011).

Eggs of all species are incubated in glass aquaria, fibreglass tanks or troughs under controlled conditions of photoperiod, water temperature and water quality. Water is filtered and aerated, and supply may be flowthrough or recirculation. Eggs are counted (usually volumetrically or by weight) and fertilisation rates and hatch rates determined. Hatch rates can vary according to different factors (e.g. history and age of broodfish, egg quality) and

© Woodhead Publishing Limited, 2013

are used to estimate the number of larvae. Larvae use endogenous nutrients and energy from the yolk sac before exogenous feeding commences.

19.4.8 Rearing of larvae and production of fingerlings

Larval rearing has long been a bottleneck in hatchery production of many species, especially those with small, poorly developed larvae like silver perch, golden perch and Australian bass. There has been significant research to identify key factors affecting survival of native fish larvae, and to develop management practices that optimise environmental conditions and maximise survival (e.g. Arumugam 1986; Arumugam and Geddes, 1987; Rowland, 1992, 1996b; Thurstan and Rowland, 1995; Ingram and De Silva, 2007; Ingram, 2009). To reduce potential effects of domestication, first-feeding larvae, except Australian bass, are stocked directly into earthen ponds where they feed on natural zooplankton (rotifers, copepods, cladocerans) and aquatic insects, especially chironomid larvae; the fish larvae and fingerlings are not weaned onto artificial feeds. Ponds are left dry during winter to desiccate pathogens, to provide suitable conditions for dormant stages of zooplankton, and to facilitate good water quality. Ponds provide a semi-natural environment with natural photoperiod and light intensities, diurnal fluctuations in water quality, and exposure to predators, including aquatic organisms and birds (Thurstan and Rowland, 1995; Ingram, 2009). Australian bass larvae are reared primarily in tanks where they are fed intensively-cultured rotifers (*Brachionus* spp.) and brine shrimp (*Artemia*) (see Fielder and Heasman, 2011).

Water quality and fish health are regularly monitored during the larval rearing phase. Post-larvae and fingerlings are susceptible to infectious diseases, particularly infestations of the ciliate protozoans *I. multifiliis* and *C. hexasticha* that can cause high mortalities unless diagnosed and treated promptly (Rowland and Ingram, 1991; Ingram *et al.*, 2005; Rowland *et al.*, 2007). Survival of larvae can be extremely variable, and is largely dependent on the availability of adequate densities of suitable-sized zooplankton in the first few weeks of stocking and feeding (e.g. Rowland, 1992, 1996b; Thurstan and Rowland, 1995; Ingram, 2009). Swimbladder inflation and nutritional deficiencies are of particular importance in the culture of Australian bass, and optimal conditions are described by Fielder and Heasman (2011). Survival rates in species with small larvae, such as golden perch, silver perch and Australian bass, are typically 30–50 % (Thurstan and Rowland, 1995; Rowland, 1996b; Fielder and Heasman, 2011), but usually over 60 % in Murray cod and trout cod (Rowland, 1992; Ingram, 2009). After 6–10 weeks, fry (~25 mm) or fingerlings (to 60 mm) are harvested, quarantined for several days and then transported and stocked. The high fecundity of silver perch, golden perch and Australian bass enables production of large numbers of fingerlings (Rowland *et al.*, 1983; Thurstan and Rowland, 1995; Rowland, 1996b, 2004; Fielder and Heasman, 2011).

© Woodhead Publishing Limited, 2013

Fingerlings are used for conservation, stock enhancement, stocking farm dams, and commercial aquaculture. Production varies from hatchery to hatchery and year to year, and 5–10 million native fish are stocked annually (Rowland and Tully, 2004; Ingram *et al.*, 2011). Murray cod, golden perch, silver perch and Australian bass from government hatcheries are stocked only into impoundments for stock enhancement and conservation, and threatened species are stocked into rivers and impoundments. Fish from commercial hatcheries are sold for stocking rivers, lakes and dams. In NSW and Victoria, revenue from recreational fishing licenses is used to fully or partially fund stocking programmes.

19.5 Threatened species: decline, stocking and recovery

19.5.1 Eastern freshwater cod

Eastern freshwater cod is a large fish (see Fig. 19.3a) endemic to the Clarence River System, a coastal drainage in north-eastern NSW; freshwater cod were also found in the adjacent Richmond River System (Fig. 19.4) (Rowland, 1985a, 1993). Cod were abundant in all freshwater reaches of these systems, from the upper estuaries through to the headwaters of most streams and rivers in both catchments (Rowland, 1993; Butler, 2009). In the 1900s, both the distribution and abundance declined significantly. Cod disappeared from much of the Clarence River System after a number of massive fish kills during the late 1920s and the 1930s; long dry periods, followed by extensive bush fires and then heavy summer rains preceded some of the kills (Rowland, 1993). In addition, pollution from tin and gold mines, and dynamiting of the rivers during the construction of the north coast railway contributed to the demise of cod in the Clarence and Richmond systems (Rowland, 1993). Between 1930 and the 1980s, habitat degradation and overfishing probably prevented any recovery of cod (Rowland, 1993, 1996a; Butler, 2009). By the 1970s, cod were extinct in the Richmond River System, and eastern freshwater cod remained only in isolated parts of the Clarence River System (Fig. 19.4) (Rowland, 1993). Currently, the species is fragmented in the Clarence River System with four genetic populations: (i) upper Mann and Nymboida rivers; (ii) lower Mann and Nymboida rivers; (iii) Guy Fawkes and Sara rivers; and (iv) Washpool Creek (Nock *et al.*, 2011).

Conservation of eastern freshwater cod was the first such programme in Australia. A key priority was to increase the distribution in the Clarence River System and to establish a self-maintaining population in the Richmond River System (Rowland, 1985b, 1988a; Ingram *et al.*, 1990). In 1988 and 1989, a total of 29 000 fish produced at the Eastern Freshwater Fish Research Hatchery was stocked at selected sites in both the Clarence and Richmond systems (Rowland, 1990). Subsequently, between 1997 and 2002, approximately 300 000 fingerlings, produced at a commercial hatchery

© Woodhead Publishing Limited, 2013

Fig. 19.3 Threatened freshwater fishes of Australia: (a) an adult eastern freshwater cod, *Maccullochella ikei*, sampled from the Clarence River System; (b) silver perch, *Bidyanus bidyanus*, a popular angling fish that also has significant potential for commercial aquaculture.

licensed by NSW Fisheries, were released into the two river systems (Talbot *et al.*, 2004; Butler, 2009). In both programmes, wild-caught, captive broodfish were used to produce fingerlings.

The stockings, in combination with a fishing prohibition, have resulted in a significant increase in distribution and abundance (Fig. 19.4). There has been a dramatic recovery of the remnant populations of eastern freshwater cod, in particular the population in the Nymboida and Mann rivers. Cod in other parts of the Clarence River System are patchily distributed (Pollard and Wooden, 2002; Butler, 2009). Eastern freshwater cod has been recorded at most stocking sites in the Richmond River System, and juveniles are regularly angled from parts of the Richmond River and tributaries such as Grady's Creek (Kevin Clark, pers. comm.), suggesting there is some breeding and recruitment. However, cod are not common and the lack of suitable spawning sites and poor quality habitat are thought to be limiting this stocked population (Butler, 2009; Butler and Rowland, 2009).

© Woodhead Publishing Limited, 2013

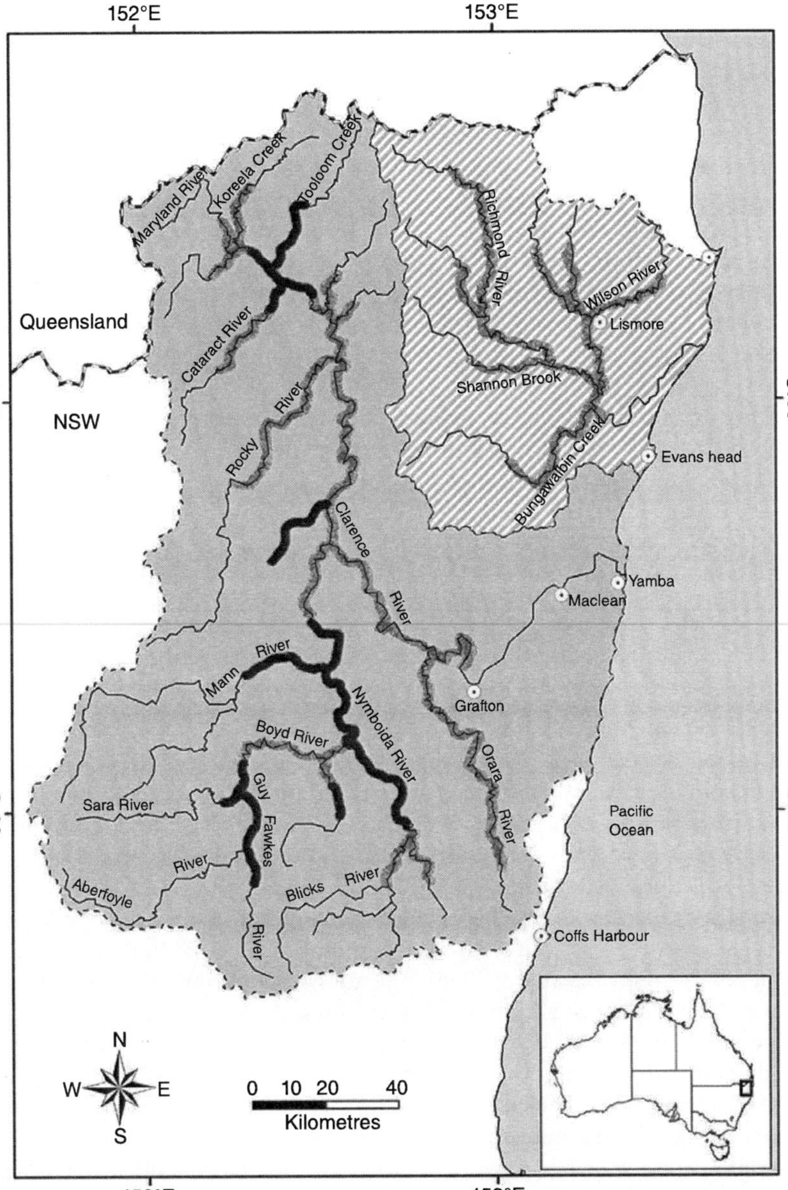

Fig. 19.4 Distribution of the eastern freshwater cod, *Maccullochella ikei*: remnant populations (black) and after conservation stocking programmes in 1987–1989 and 1997–2002 (grey). Clarence River System shaded and the Richmond River System striped. From Butler (2009).

© Woodhead Publishing Limited, 2013

19.5.2 Mary River cod

Mary River cod is endemic to the coastal Mary River System of south-eastern Queensland and was very abundant in the late 1800s and early 1900s before declining dramatically (Rowland, 1993). The known natural distribution of Mary River cod extends over about 170 km stream length compared to an historical distribution of at least 700 km, and in 1996, the total number of cod was estimated to be less than 1000 individuals (Simpson and Jackson, 1996; Pusey *et al.*, 2004). It is also possible that this species existed in other river systems, including the Brisbane, Logan–Albert and Coomera rivers where freshwater cod were common (Pusey *et al.*, 2004). If so, Mary River cod became extirpated from a number of these coastal river systems, leaving only the small remnant population in a restricted area of the Mary River System. Causes of the decline are thought to include over-fishing, habitat degradation and loss, and barriers to movement (Simpson and Jackson, 1996; Simpson and Mapleston, 2002).

Since 1983, hatchery-reared fingerlings have been restocked into the Mary River System, as well as other drainages in south-eastern Queensland. Distribution and abundance have increased markedly; the species is now in four river systems and numerous impoundments and, while the major goal of stocking is conservation, there is also a recreational fishery for this threatened species (Peter Kind, pers. comm.).

19.5.3 Trout cod

Trout cod is endemic to the southern tributaries of the MDRS. Historically, trout cod was common and widespread, and in some locations more abundant than Murray cod (Lake, 1971; Cadwallader, 1977). The distribution of trout cod was extended by translocation in the early 1900s into Cataract Dam, a Sydney water supply dam, Sevens Creek, a tributary of the Goulburn River, and Lake Sambell; however, trout cod hybridise with Murray cod in Cataract Dam, and so this population is not considered pure (Wajon, 1983; Harris and Rowland, 1996). The range and abundance of trout cod have declined markedly since European settlement (Cadwallader and Gooley, 1984). Lake (1971) considered trout cod as seriously threatened, and it was extirpated in the Murrumbidgee River by the 1970s (Llewellyn, 1983). The rapid decline lead to only two self-maintaining, remnant populations by the 1970s; one in the Murray River between Yarrawonga and Barmah State Forest, and the translocated population in Seven Creeks (Ingram *et al.*, 1990).

Captive breeding programmes for trout cod were established at the NFC and Snobs Creek to produce and release juveniles into selected sites with the aim of re-establishing breeding stocks (Rimmer, 1987; Ingram *et al.*, 1990). Native Fish Australia also runs a small, independent breeding programme in Victoria (Ingram and Thurstan, 2008). Between 1986 and 2011, approximately 1.5 million fingerlings have been released (Brett Ingram,

© Woodhead Publishing Limited, 2013

pers. comm.), with good survival at many locations in rivers in both NSW and Victoria (Ingram and Douglas, 1995; Gilligan, 2005b; Lyon *et al.*, 2007). Douglas and Brown (2000) reported natural recruitment in a population of trout cod stocked into Loombah Weir in Victoria, and naturally-spawned juveniles were captured in 2001 and 2002 confirming success in establishing a self-maintaining population of trout cod in the lower Murrumbidgee River (Gilligan, 2005b). Besides the lower Murrumbidgee, natural recruitment has been confirmed in the Goulburn River and is suspected in at least three other rivers (Koster *et al.*, 2004; Ingram and Thurstan, 2008).

19.5.4 Murray cod

Murray cod is Australia's largest freshwater fish growing to 113.5 kg, and is regarded as the icon of the MDRS, forming the basis of a very popular recreational fishery (Rowland, 2005). Originally, it was widely distributed throughout the MDRS, including the upper reaches of many rivers, and because of its great popularity had been translocated into many locations outside its natural range (Rowland, 1989). Since the late 1800s, the range and abundance of Murray cod declined due to series of anthropogenic factors including overfishing, pollution, predation by and competition with redfin (*Perca fluviatilis*), reduced survival and recruitment of larvae and juveniles following river regulation (Rowland, 1989, 2005), and habitat loss, barriers to movement, changes to water quality, cold water pollution, desnagging, loss to irrigation systems, low population numbers, deviations from sustainable population structures, and diseases (Koehn, 2001, 2005b). Murray cod became uncommon in many areas, particularly above dams, and was rare or extirpated from some rivers and creeks, including those on the Northern Tablelands of NSW (Lake, 1971; Rowland, 2005). Its abundance has remained low in many Victoria waters, and in South Australia (Rowland, 2005; Ye and Zampatti, 2007). Although Kearney and Kildea (2001) considered that the persistence of Murray cod was not of immediate concern, the decline of cod led to its listing as Vulnerable under the *EPBC Act 1999*, threatened in Victoria (Lintermans *et al.*, 2005) and Critically Endangered by the IUCN (www.iucnredlist.org). Murray cod is not listed or protected in NSW (Lintermans *et al.*, 2005) because it is widespread and relatively abundant, and there is evidence of a significant recovery in many rivers and impoundments since the early 1990s (Rowland, 2005). There are four known genetic populations of Murray cod: the Murray, Murrumbidgee and lower Darling; the Lachlan River; the Macquarie River; and the northern rivers of the MDRS (Rourke *et al.*, 2011).

Hatchery production commenced in the late 1970s (Rowland, 1983a, 1986; Cadwallader and Gooley, 1985), and stockings were undertaken into impoundments and other locations where the species was rare or extirpated (Rowland, 2005). Since 1976, approximately 16.1 million Murray cod have been produced at government and private commercial hatcheries for stock

© Woodhead Publishing Limited, 2013

enhancement and conservation (Rowland and Tully, 2004; Ingram *et al.*, 2011; Brett Ingram, pers. comm.). Stocking has reintroduced Murray cod to many areas, including northern NSW, and has contributed to the significant recovery of stocks in many other NSW waters since the early 1990s (Rowland, 2005). Analysis of data collected from 28 long-term monitoring sites between 1994 and 2011 indicate that Murray cod abundance has increased by 740 % in the last 17 years (Dean Gilligan, unpublished data). Despite this recovery in NSW, there are concerns about the population structure, absence of small age/size classes and poor recruitment since the 1970s in the Murray River in South Australia (Ye and Zampatti, 2007). If there is a decrease in reproductive performance of old fish as suggested by Rowland (2005), the Murray cod stock in the lower Murray River may be at risk of extirpation.

19.5.5 Silver perch

Silver perch (see Fig. 19.3b) was once widespread throughout the MDRS with the exception of higher altitudes (Merrick and Schmida, 1984). It is an excellent sports fish and highly regarded for its edible qualities, and it has significant potential for large-scale, commercial aquaculture (Lake, 1967c; Rowland, 2009). There was a rapid decline of silver perch between 1960 and 1985. In the early 1960s, silver perch was considered the species least affected by dams and weirs (Lake, 1964); however, the commercial catch of silver perch peaked in 1958/59 before collapsing in 1984/85 (Reid *et al.*, 1997). Silver perch has remained uncommon or rare in most rivers, with the exception of the middle reaches of the Murray River (Merrick, 1996; Clunie and Koehn, 2001; Lintermans, 2007). It is virtually absent from the Darling and Lachlan rivers, and some riverine populations may be extirpated. There are good stocks in impoundments that are maintained by stocking (Rowland *et al.*, 1983; Hamlyn and Brooks, 1992; Rowland, 1995). Factors causing the decline of silver perch are thought to be river regulation, barriers to spawning migrations of adults and juveniles, thermal pollution and interaction with carp and redfin (Lake, 1971; Merrick, 1996; Mallen-Cooper and Stuart, 2003; Lintermans, 2007).

Techniques for the large-scale hatchery production of silver perch were developed in the early 1980s, and fingerlings are produced annually at government and commercial hatcheries for stocking impoundments and farm dams, and for commercial aquaculture (Rowland *et al.*, 1983; Rowland and Tully, 2004). In NSW, approximately 12 million fingerlings have been stocked since 1976. All stockings have been restricted to impounded waters, with no silver perch stocked into rivers. Silver perch is now common in many impoundments and is part of large, mixed-species recreational fisheries in NSW and Queensland (Hamlyn and Brooks, 1992; Rowland, 1995; Simpson *et al.*, 2002). However, silver perch remains rare or uncommon in most rivers (Harris and Gehrke, 1997; Gilligan, 2005a, b; Davies *et al.*, 2008), and only

© Woodhead Publishing Limited, 2013

20 silver perch were recorded in a survey involving 351 randomly selected sites covering 16 river valleys in the MDRS (Lintermans, 2007).

19.5.6 Macquarie perch

Macquarie perch was originally widespread throughout the southern tributaries of the western, inland MDRS, as well as the Hawkesbury–Nepean and Shoalhaven River Systems in the eastern drainage of NSW (Cadwallader, 1981; Dufty, 1986; Harris and Rowland, 1996). The eastern populations consist of both endemic and translocated stocks, and Macquarie perch was also translocated to the eastern Yarra River in Victoria (Cadwallader, 1981; Harris and Rowland, 1996; Lintermans, 2007; Faulks *et al.*, 2010a). It was a popular angling fish that was easily caught and highly regarded for its edible qualities (Will Trueman, pers. comm.). In the early 1900s, large numbers were recorded passing through Euston Weir in the lower Murray River (Cadwallader, 1977), and in the 1960s it was considered to be reasonably abundant in the cooler, higher reaches in and above impoundments, particularly Wyangala and Burrinjuck dams (Lake, 1967c, 1971). However, Lake (1971) had identified Macquarie perch as seriously threatened and, since then, its numbers have continued to decline, stocks and recreational fisheries have been lost, and there is now substantial fragmentation and high levels of diversity and divergence (Ingram *et al.*, 1990, 2000; Lintermans, 2007; Faulks *et al.*, 2010a).

Macquarie perch is considered extinct from the Macquarie River after which it is named (Will Trueman, pers. comm.), and in the last 13 years a population in the Kangaroo River (Shoalhaven River System) is thought to have been extirpated (Faulks *et al.*, 2010a). Factors contributing to the decline are overfishing, river regulation, pollution, habitat degradation including sedimentation, barriers to dispersal, introduced species including redfin and salmonids, and possibly the infectious disease epizootic haematopoietic necrosis (Ingram *et al.*, 1990, 2000; Harris and Rowland, 1996). Macquarie perch has specific habitat preferences, and habitat fragmentation has played a key role in its decline and in determining connectivity and genetic diversity in the species (Gilligan *et al.*, 2010; Faulks *et al.*, 2011). Only small remnant populations remain and there are serious concerns for the long-term conservation of this species.

Research to develop large-scale hatchery techniques commenced in the 1980s, and captive breeding programmes for the 'western' Macquarie perch have been undertaken at government hatcheries in Victoria and NSW (Gooley, 1986; Gooley and McDonald, 1988; Ingram *et al.*, 1994; Ingram and Gooley, 1996), as well as by Native Fish Australia (Victoria). Unfortunately, there have been substantial difficulties in establishing reliable techniques (Ingram *et al.*, 1994, 2000). Although early success was achieved using mature broodfish captured from Dartmouth Dam, the success of hormone induction of pond-held fish was limited (Gooley and McDonald, 1988;

© Woodhead Publishing Limited, 2013

Ingram *et al.*, 1994). Between 1986 and 1997 approximately 456 000 juveniles were released into nine sites, but survival was low (Ingram *et al.*, 2000). Although techniques for larval rearing are reliable (Ingram, 2009), the decreasing size of fish from Dartmouth Dam, and decline in the performance of broodfish are of concern (Ingram and Gooley, 1996). In recent years, there has been limited success with both tank-held and pond-held broodfish being induced to spawn (Brett Ingram and Dean Gilligan, pers. comm.), but hatchery production of Macquarie perch remains very low compared to other species.

19.6 Stock enhancement: golden perch and Australian bass

Golden perch and Australian bass are highly regarded for their edible and angling qualities, and in recent decades they have become very popular sports fish. Reproduction and recruitment in both species have been adversely affected by river regulation and barriers to migration (Harris and Rowland, 1996). Golden perch undertakes long, upstream migrations associated with breeding, and spawns in response to rising water levels and floods (Lake, 1967a, b; Mackay, 1973; Reynolds, 1983). Consequently, suitable conditions for breeding usually do not occur in impounded waters. The distribution of golden perch was restricted by its inability to breed successfully in dams and by barriers to migration (Harris and Rowland, 1996). By the mid-1900s, it was uncommon in most impoundments with the exception of Lake Keepit, and was absent from waters above Yarrawonga Weir on the Murray River (Lake, 1971; Cadwallader, 1977; Battaglene, 1991).

Australian bass is a catadromous species found in coastal drainages from the Mary River in southern Queensland to Wilson's Promontory, Victoria (Harris and Rowland, 1996). Adult bass migrate from fresh water reaches to estuarine waters to spawn in autumn and winter, and then both adults and juveniles return to fresh water (Harris 1986). Man-made barriers have significantly impeded or obstructed migrations in up to half the natural range of Australian bass (Harris, 1984). The range and abundance of bass have subsequently declined, and there are no natural fisheries in impoundments.

Techniques for the hormone-induced spawning and larval rearing of golden perch and Australian bass were developed in the 1980s and large-scale hatchery production and stock enhancement of both species has been on-going since then (Rowland *et al.*, 1983; Battaglene *et al.*, 1989; Battaglene and Selosse, 1996; Fielder and Heasman, 2011). Over 33 million golden perch have been stocked in NSW since 1976. Stockings have been highly successful and golden perch and Australian bass are key species in large, economically important recreational fisheries in impoundments in south-eastern Australia (Hamlyn and Brooks, 1992; Rowland, 1995; Simpson *et al.*, 2002; Rolfe *et al.*, 2005; Fielder and Heasman, 2011). Because neither species

© Woodhead Publishing Limited, 2013

can breed successfully in impoundments, these fisheries are dependent on regular stocking. Australian bass has also been stocked into some rivers with varying degrees of success (Cameron and Baumgartner, 2012).

19.7 Hatchery production and stocking: a success story in Australia

Success of initial stockings of native fish was apparent by the early 1980s, with golden perch and silver perch being sampled by research staff and caught by anglers from areas between Yarrawonga and Hume weirs, the Ovens River, as well as Wyangala, Copeton, Pindari dams and Lake Wyangan (Rowland *et al.*, 1983). Over the following decade these initial successes were being repeated throughout NSW, Queensland and Victoria with expanding distributions, increasing abundance, and new recreational fisheries in impoundments (Hamlyn and Brooks, 1992; Rowland, 1995). There is evidence that stocking has also increased the abundance of golden perch in rivers such as the Murrumbidgee River (Gilligan, 2005b). Recently, chemical marking techniques were used to show that stocking contributes significantly to golden perch populations in both lakes (47–90 %) and rivers (18–100 %) in southern parts of the MDRS, particularly where natural recruitment is low (Crook *et al.*, 2010; Hunt *et al.*, 2010). Similarly, the development of hatchery techniques and stock enhancement of the catadromous barramundi (*Lates calcarifer*) has created large, economically successful fisheries in impoundments where the species is unable to breed, as well as augmenting stocks in rivers in northern Queensland (MacKinnon and Cooper, 1987; Rutledge *et al.*, 1990; Rimmer and Russell, 1998; Hogan, 2000).

Reintroduction of eastern freshwater cod, Mary River cod, trout cod and Murray cod to areas where populations were extirpated during the early and mid-1900s has fully or partly restored the distribution of each of these species. There are now two or more stocks of each species, significantly reducing the likelihood of extinction. For species where conservation breeding and stocking programmes have not been completely successful (e.g. Macquarie perch) or where stocking is restricted to impoundments (e.g. silver perch), the recovery of wild, riverine stocks has been very limited. The success with *Maccullochella* spp. but lack of recovery of Macquarie perch and silver perch demonstrates the value of conservation stockings in assisting the recovery of threatened species. Macquarie perch may benefit from increased efforts to develop large-scale hatchery techniques, and silver perch from a strategic stocking programme in selected rivers. Similarly, recovery of Murray cod in the Murray River in South Australia may require stocking, particularly if there is continued recruitment failure.

Although habitat protection and rehabilitation are of major importance for the recovery of native fish and key ecosystem processes within the

© Woodhead Publishing Limited, 2013

MDRS (Murray-Darling Basin Commission, 2003; Barrett, 2004; Koehn, 2005a), the adverse effects of some environmental changes are difficult or impossible to ameliorate. It is highly unlikely that the rapid recovery of the threatened species of *Maccullochella* could have been achieved by the elimination or reduction of key threatening processes and habitat rehabilitation alone. The low abundance and extirpation of populations made the recovery of these threatened species dependent, at least initially, on captive breeding and stocking in combination with protection provided by fisheries regulations including fishing prohibitions. Where species have limited dispersal abilities or are absent, or where levels of connectivity are low, habitat restoration alone may not be sufficient for re-establishment (Hughes, 2007). Despite significant effort and vast monetary resources on endangered fish recovery and ecosystem restoration in the USA, there has only been very limited success, partially due to a focus on degraded habitats on small spatial scales (Williams *et al.*, 2011). While captive breeding and stocking programmes have focused on key large-bodied, high profile species in Australia, they have also provided reference points and are powerful drivers of change such as improved environmental flows, protection of key habitat areas, and increased community awareness of fish conservation, biodiversity and ecosystem processes (Talbot *et al.*, 2004). For example, recent efforts to conserve the purple spotted gudgeon (*Mogurnda adspersa*) and other small-bodied species in the lower Murray River are based on captive breeding and stocking into restored wetland habitats (Hammer *et al.*, 2009; Hall *et al.*, 2010).

The success of conservation and stock enhancement of Australian freshwater fishes has been based on the establishment and continual improvement of hatchery practices and genetic guidelines. The production of high quality fingerlings with appropriate genetics is fundamental for the continued evolutionary potential of each species. Results of recent studies suggest that effects of stocking on the genetics of receiving populations are dependent on the quality of breeding and stocking programmes. To date, there have been no major temporal changes in genetic diversity, heterozygosity, allelic richness and effective population size of pre- and post-stocking Murray cod (1949–1954 and 1994–2005) in the southern parts of the MDRS (Rourke *et al.*, 2010). The authors suggested that the use and regular replacement of wild-caught broodfish, and mixing progeny from different spawnings before stocking contributed to the lack of genetic change. Ingram *et al.* (2011) modelled various hatchery and broodfish scenarios and suggested that large, well-managed breeding and stocking programmes could help maintain genetic diversity of Murray cod across the Murray–Darling Basin and potentially increase genetic diversity within a few generations in stocks where the effective population size is very small. It is also predicted that stocking would significantly increase total recruitment and population sustainability of Murray cod where natural recruitment was reduced,

© Woodhead Publishing Limited, 2013

presuming the reproductive efficiency of hatchery-origin fish was not compromised (Rogers *et al.*, 2010).

There is no genetic evidence for survival of stocked Murray cod in the Lachlan River catchment, suggesting that recovery of this population in recent decades has been due to natural recruitment alone (Rowland, 2005; Rourke *et al.*, 2011). There is some mixing of genetic stocks in the northern parts of the MDRS (Rourke *et al.*, 2011), possibly due to inadvertent stocking in the 1980s and 1990s before the population structure of Murray cod was determined by Rourke (2007). There has been a significant loss of heterozygosity and allelic richness in eastern freshwater cod since stocking (Nock *et al.*, 2011). This loss of genetic diversity may be due to the use of insufficient broodfish and low hatchery N_e in 1988 and 1989 (Rowland, 1990), and/or swamping of the lower Mann–Nymboida population by hatchery-reared fingerlings following stockings in 1997–2002 (Nock *et al.*, 2011). These findings demonstrate the need for good genetic data and strict adherence to the breeding and stocking guidelines in the HQAP. Poor hatchery and stocking practices have the potential to cause irreversible biological damage (Rowland and Tully, 2004). Biodiversity and the genetic diversity within species and populations are the basis of evolution (Frankham *et al.*, 2002), and the maintenance of genetic diversity is critical to species and ecosystem resilience in the MDRS (Moore *et al.*, 2010).

19.8 Conclusions

Captive breeding and stocking programmes enabled a rapid response to the imperiled status of freshwater fishes in Australia, and have played an important and continuing role in the recovery of native freshwater fishes since the 1970s and the restoration of some of the lost biodiversity in the MDRS and coastal drainages.

While on-going captive breeding and stocking programmes are necessary to maintain stocks in impoundments, in altered and severely degraded habitats, and where species are absent or rare, they should not be seen as surrogates for poor habitat and natural breeding and recruitment. Continuing efforts to rehabilitate fish habitats, improve flow management, modify temperature regimes below impoundments, provide fish passage, minimise pollution and reduce or eliminate exotic fishes are essential for the long-term survival and continuing evolution of Australia's freshwater fishes. There is a close link between hatcheries and the environment because all conservation and stock enhancement breeding programmes in Australia are dependent on the availability of wild broodfish. Hatcheries should operate according to sound scientific principles, use best practices, respond to new information on biology, culture techniques and monitoring, and have

© Woodhead Publishing Limited, 2013

well-defined breeding and stocking programmes that are not open-ended, but have clearly defined objectives and timetables.

19.9 Acknowledgements

I acknowledge the late John Lake and Dr Hamar Midgley who pioneered the conservation and stocking of native fish in Australia, and Dr Brett Ingram, Dr Mike Rimmer, Dr Phil Cadwallader, Dr Stephen Battaglene, Geoff Gooley, the late Gerald Cook and Ray Mepham for their contributions to the early development of artificial breeding and hatchery techniques. I thank Cameron Westaway and Dr Ben Doolan for information and data on fish stocking in NSW, Ian Wooden and Tim McGarry for preparing Fig. 19.1, and Drs Gavin Butler, Brett Ingram and Geoff Allan for their comments on a draft of the chapter.

19.10 References

ALLEN G R, MIDGLEY S H and ALLEN M (2002) *Field guide to the freshwater fishes of Australia*. Perth: Western Australian Museum.

ALLENDORF F W (1991) Ecological and genetic effects of fish introductions: synthesis and recommendations, *Canadian Journal of Fisheries and Aquatic Sciences*, 48 (Suppl. 1), 178–181.

ALLENDORF F W (1993) Delay of adaptation to captive breeding by equalising family size, *Conservation Biology*, 7, 416–419.

ÁLVAREZ D and NICIEZA A G (2003) Predator avoidance behaviour in wild and hatchery-reared brown trout: the role of experience and domestication, *Journal of Fish Biology*, 63, 1565–1577.

ARTHINGTON A H (1991) Ecological and genetic impacts of introduced and translocated freshwater fishes in Australia, *Canadian Journal of Fisheries and Aquatic Sciences*, 48, 33–43.

ARUMUGAM P T (1986) *An experimental approach to golden perch (Macquaria ambigua) fry-zooplankton interaction in fry rearing ponds, south-eastern Australia*, PhD Thesis, Adelaide, University of Adelaide.

ARUMUGAM P T and GEDDES M C (1987) Feeding and growth of golden perch larvae and fry (*Macquaria ambigua* Richardson), *Transactions of the Royal Society of South Australia*, 111, 59–65.

ANON. (2006) *Silver perch NSW recovery plan*. Port Stephens: NSW Department of Primary Industries.

BARRETT J (2004) Introducing the Murray–Darling Basin Native Fish Strategy and initial steps towards demonstrations reaches, *Ecological Management and Restoration*, 5, 15–23.

BARRETT J (2008) *The sea to Hume Dam: restoring native fish passage in the Murray River*. Canberra: Murray-Darling Basin Commission.

BATTAGLENE S C (1991) *The golden perch, Macquaria ambigua (Pisces: Percichthyidae) of Lake Keepit, NSW*, MSC Thesis, Sydney, University of New South Wales.

BATTAGLENE S C and SELOSSE P M (1996) Hormone-induced ovulation and spawning of captive and wild broodfish of the catadromous Australian bass, *Macquaria*

© Woodhead Publishing Limited, 2013

novemaculeata (Steindachner), (Percichthyidae), *Aquaculture Research*, 27, 191–204.

BATTAGLENE S C, TALBOT R B and BEAVERS P J (1989) Australian bass culture – recent advances, *Australian Fisheries*, 49, 28–39.

BEARLIN A R and TIKEL D (2003) Conservation genetics of Murray-Darling Basin fish; silver perch (*Bidyanus bidyanus*), Murray cod (*Maccullochella peelii*), and trout cod (*M. macquariensis*), in Phillips B (ed.), *Managing fish translocation and stocking in the Murray-Darling Basin workshop, Canberra, 25–26 September 2002: statement, recommendations and supporting papers*. Sydney: World Wildlife Fund, 59–83.

BERRA T M (2001) *Freshwater fish distribution*. Sydney: Academic Press.

BOYS C A, ROWLAND S J, GABOR M, GABOR L, MARSH I B, HUM S and CALLINAN R B (2012) Emergence of epizootic ulcerative syndrome in native fish of the Murray–Darling River System, Australia: hosts, distribution and possible vectors, *PLoS ONE*, 7(4), e35568.

BROWN A M (1985) Intraspecific variation in animal populations, in Brown A M (ed.), *Proceedings of the state conservation strategy workshop on genetic diversity*, Technical Report Series No. 11. East Melbourne: Fisheries and Wildlife Service, 39–43.

BROWN A M (1987) Genetic aspects of the propagation and dispersion of fishes, in Harris J H (ed.), *Proceedings of the conference on Australian threatened fishes*. Sydney: Division of Fisheries, Department of Agriculture of New South Wales, 45–54.

BRUTON M N (1995) Have fishes had their chips? The dilemma of threatened fishes, *Environmental Biology of Fishes*, 43, 1–27.

BUTLER G (2009) *Review of the biology and conservation of the endangered eastern freshwater cod Maccullochella ikei, with particular reference to defining environmental water requirements*. Sydney: NSW Department of Water & Energy.

BUTLER G L and ROWLAND S J (2009) Using underwater cameras to describe the reproductive behaviour of the endangered eastern freshwater cod *Maccullochella ikei*, *Ecology of Freshwater Fish*, 18, 337–349.

CADWALLADER P L (1977) *J.O. Langtry's 1949–50 Murray River investigations*, Fisheries and Wildlife Paper Victoria No. 13. East Melbourne: Fisheries and Wildlife Division.

CADWALLADER P L (1981) Past and present distributions and translocations of Macquarie perch (Pisces: Percichthyidae), with particular reference to Victoria, *Proceedings of the Royal Society of Victoria*, 93, 23–30.

CADWALLADER P L and GOOLEY G J (1984) Past and present distributions and translocations of Murray cod *Maccullochella peeli* and trout cod *M. macquariensis* (Pisces: Percichthyidae) in Victoria, *Royal Society of Victoria Proceedings*, 96, 33–43.

CADWALLADER P L and GOOLEY G (1985) *Propagation and rearing of Murray cod Maccullochella peeli at the warmwater fisheries station pilot project Lake Charlegrark*. Melbourne: Government Printer.

CALLINAN R B and ROWLAND S J (1995) Diseases of silver perch, in Rowland S J and Bryant C (eds), *Silver perch culture: proceedings of silver perch aquaculture workshops, Grafton and Narrandera, April, 1994*. Sandy Bay: Austasia Aquaculture, 67–75.

CAMERON L and BAUMGARTNER L (2012) *Assessment of Australian bass restocking in the Snowy River*, Fisheries Final Report Series. Cronulla: NSW Department of Primary Industries.

CLUNIE P and KOEHN J (2001) *Silver perch: a recovery plan*, Final Report for Natural Resource Management Strategy Project R7002. Canberra: Murray–Darling Basin Commission.

© Woodhead Publishing Limited, 2013

COWX I G (1994) Stocking strategies, *Fisheries Management and Ecology*, 6, 21–34.

CROOK D A, GILLANDERS B M, SANGER A C, MUNRO A R, O'MAHONY D J, WOODCOCK S H, THURSTAN S and BAUMGARTNER L J (2010) *Methods for discriminating hatchery fish and outcomes of stocking in the Murray–Darling Basin*, Final Report, Murray–Darling Basin Authority, Native Fish Strategy Project MD741. Heidelberg: Arthur Rylah Institute for Environmental Research.

DAKIN W J and KESTEVEN G L (1938) The Murray cod [*Maccullochella macquariensis* (Cuv. et Val.)], *N.S.W. State Fisheries Research Bulletin*, 1, 1–18.

DAVIES P E, HARRIS J H, HILLMAN T J and WALKER K F (2008) *Sustainable rivers audit report 1: a report on the ecological health of rivers in the Murray-Darling Basin, 2004–2007*. Canberra: Murray–Darling Basin Commission.

DOUGLAS J W and BROWN P (2000) Notes on successful spawnings and recruitment of a stocked population of the endangered Australian freshwater fish, trout cod, *Maccullochella macquariensis* (Cuvier) (Percichthyidae), *Proceedings of the Linnean Society of New South Wales*, 122, 143–147.

DOUGLAS J W, GOOLEY G J and INGRAM B A (1994) *Trout cod. Maccullochella macquariensis (Cuvier) (Pisces: Percichthyidae), resource handbook and research and recovery plan*. Alexandra: Conservation and Natural Resources.

DOUGLAS J W, GOOLEY G J, INGRAM B A, MURRAY N D and BROWN L D (1995) Natural hybridization between Murray cod, *Maccullochella peelii peelii* (Mitchell), and trout cod, *Maccullochella macquariensis* (Cuvier) (Percichthyidae), in the Murray River, Australia, *Marine and Freshwater Research*, 46, 729–734.

DUFTY S (1986) *Genetic and morphological divergence between populations of Macquarie perch (Macquaria australasica) east and west of the Great Dividing Range*, Honours Thesis, Sydney, University of New South Wales.

EBY L A, ROACH J, CROWDER L B and STANFORD J A (2006) Effects of stocking-up freshwater food webs, *Trends in Ecology and Evolution*, 21, 576–584.

FAULKS L K, GILLIGAN D M and BEHEREGARAY L B (2008) Phylogeography of a threatened freshwater fish (*Mogurnda adspersa*) in eastern Australia: conservation implications, *Marine and Freshwater Research*, 59, 89–96.

FAULKS L K, GILLIGAN D M and BEHEREGARAY L B (2010a) Evolution and maintenance of divergent lineages in an endangered freshwater fish, *Macquaria australasica*, *Conservation Genetics*, 11, 921–934.

FAULKS L K, GILLIGAN D M and BEHEREGARAY L B (2010b) Clarifying an ambiguous evolutionary history: range-wide phylogeography of an Australian freshwater fish, the golden perch (*Macquaria ambigua*), *Journal of Biogeography*, 37(7), 1329–1340.

FAULKS L K, GILLIGAN D M and BEHEREGARAY L B (2011) The role of anthropogenic vs. natural in-stream structures in determining connectivity and genetic diversity in an endangered freshwater fish, Macquarie perch (*Macquaria australasica*), *Evolutionary Applications*, 4(4), 589–601.

FIELDER D S and HEASMAN M P (2011) *Hatchery manual for the production of Australian bass, mulloway and yellowtail kingfish*. Orange: Industry and Investment NSW.

FRANKHAM R (1999) Quantitative genetics in conservation biology, *Genetic Research*, 74, 237–244.

FRANKHAM R (2008) Genetic adaptation to captivity in species conservation programs, *Molecular Ecology*, 17, 325–333.

FRANKHAM R and LOEBEL D A (1992) Modelling problems in conservation genetics using captive *Drosophila* populations: rapid genetic response to captivity, *Zoo Biology*, 11, 333–342.

FRANKHAM R, BALLOU J D and BRISCOE D A (2002) *Introduction to conservation genetics*. Cambridge: Cambridge University Press.

© Woodhead Publishing Limited, 2013

GEDDES M C and PUCKRIDGE J T (1989) Survival and growth of larval and juvenile native fish the importance of the floodplain, in *Proceedings of the workshop on native fish management, Canberra 16–17 June 1988*. Canberra: Murray–Darling Basin Commission, 101–114.

GEHRKE P C, BROWN P, SCHILLER C B, MOFFAT D B and BRUCE A M (1995) River regulation and fish communities in the Murray–Darling River System, *Regulated Rivers: Research & Management*, 11, 363–375.

GILLANDERS B M and YE Q (2011) *Ecological risk assessment of stocking Murray cod in South Australia*. Adelaide: University of Adelaide and South Australian Research and Development Institute.

GILLANDERS B M, ELSON T S and MUNRO A R (2006) *Impacts of native fish stocking on fish in the Murray–Darling Basin*. Adelaide: University of Adelaide.

GILLIGAN D M (2000) Conservation genetics and stocking for species recovery programs, in Moore A and Hughes R (eds), *Proceedings of a workshop on stock enhancement of marine and freshwater fisheries*. Lismore: Australian Society for Fish Biology and Southern Cross University, 39–45.

GILLIGAN D (2005a) *Fish communities of the Murrumbidgee catchment: status and trends*, Fisheries Final Report Series No. 75. Cronulla: NSW Department of Primary Industries.

GILLIGAN D (2005b) *Fish communities of the Lower Murray-Darling catchment: status and trends.* Fisheries Final Report Series No. 83. Cronulla: NSW Department of Primary Industries.

GILLIGAN D, MCGARRY T and CARTER S (2010) *A scientific approach to developing habitat rehabilitation strategies in aquatic environments: A case study on the endangered Macquarie perch (Macquaria australasica) in the Lachlan catchment*, Fisheries Final Report Series No. 128. Cronulla: Industry & Investment NSW.

GOOLEY G J (1986) Culture methods for Macquarie perch at Dartmouth, Victoria, *Australian Fisheries*, 45, 18–20.

GOOLEY G J and MCDONALD G L (1988) *Preliminary study on the hormone-induced spawning of Macquarie perch, Macquaria australasica (Cuvier) (Percichthyidae), from Lake Dartmouth, Victoria*, Technical Report Series No. 80. Heidelberg: Arthur Rylah Institute for Environmental Research.

HALL A, HAMMER M and FREARS A (2010) Rescue, recovery and restoration: a South Australian fish tale, in *Native Fish Forum 2010, Australian National Museum*. Canberra: Murray–Darling Basin Authority, 92–96.

HALLERMAN E M (2003) *Population genetics: principles and applications for fisheries scientists*. Bethesda, MD: American Fisheries Society.

HAMLYN A and BROOKS S (1992) *Recreational fishing programme Lake Cania review of monitoring data 1986/87 to 1990/91*. Brisbane: Department of Primary Industries Fisheries.

HAMMER M (2008) *A molecular genetic appraisal of biodiversity and conservation units in freshwater fishes from Southern Australia*, PhD Thesis, Adelaide, University of Adelaide.

HAMMER M, BARNES T, PILLER L and SORTINO D (2009) *Reintroduction plan for the purple-spotted gudgeon in the Southern Murray–Darling Basin*. Canberra: Murray–Darling Basin Authority.

HARADA Y, YOKOTA M and IIZUKA M (1998) Genetic risk of domestication in artificial fish stocking and its possible reduction, *Researches in Population Ecology*, 40, 311–324.

HARDY C M, ADAMS M, JERRY D R, COURT L N, MORGAN M J and HARTLEY D M (2011) DNA barcoding to support conservation: species identification, genetic structure and biogeography of fishes in the Murray–Darling Basin, Australia, *Marine and Freshwater Research*, 62, 887–901.

© Woodhead Publishing Limited, 2013

HARRIS J H (1984) Impoundment of coastal drainages of south-eastern Australia, and a review of its relevance to fish migrations, *Australian Zoologist*, 21, 235–250.

HARRIS J H (1986) Reproduction of the Australian bass, *Macquaria novemaculeata* (Perciformes; Percichthyidae) in the Sydney basin, *Australian Journal of Marine and Freshwater Research*, 37, 209–235.

HARRIS J H (1987) *Proceedings of the conference on Australian threatened fishes*. Sydney: Division of Fisheries, Department of Agriculture of New South Wales.

HARRIS J H and GEHRKE P C (1997) *Fish and rivers in stress, the NSW rivers survey*. Cronulla: NSW Fisheries and CRC for Freshwater Ecology.

HARRIS J H and ROWLAND S J (1996) Family Percichthyidae, Australian freshwater cods and basses, in McDowall R M (ed.), *Freshwater fishes of south-eastern Australia*. Sydney: Reed Books, 150–163.

HOGAN A (2000) The Tinaroo barra fishery – from infamy to the Holy Grail, in Moore A and Hughes R (eds), *Proceedings of a workshop on stock enhancement of marine and freshwater fisheries*. Lismore: Australian Society for Fish Biology and Southern Cross University, 6–10.

HUGHES J M (2007) Constraints on recovery using molecular methods to study connectivity of aquatic biota in rivers and streams, *Freshwater Biology*, 52, 616–631.

HUMPHRIES P and LAKE P S (2000) Fish larvae and the management of regulated rivers, *Regulated Rivers: Research & Management*, 16, 421–432.

HUNT T L, ALLEN M S, DOUGLAS J and GASON A (2010) Evaluation of a sport fish stocking program in lakes of the southern Murray-Darling Basin, Australia, *North American Journal of Fisheries Management*, 30, 805–811.

INGRAM B A (2009) Culture of juvenile Murray cod, trout cod and Macquarie perch (Percichthyidae) in fertilised earthen ponds, *Aquaculture* 287, 98–106.

INGRAM B A and DOUGLAS J W (1995) Threatened fishes of the world: *Maccullochella macquariensis* (Cuvier, 1829) (Percichthyidae), *Environmental Biology of Fishes*, 43, 38.

INGRAM B A and DE SILVA S S (2004) *Development of intensive commercial aquaculture production technology for Murray cod*. Snobs Creek: Primary Industries Research Victoria.

INGRAM B A and DE SILVA S S (2007) Diet composition and preference of juvenile Murray cod, trout cod and Macquarie perch (Percichthyidae) reared in fertilised earthen ponds, *Aquaculture*, 271, 260–270.

INGRAM B A and GOOLEY G (1996) Hormone-induced spawning of the threatened Macquarie perch (*Macquaria australasica*): an Australian native freshwater fish, in Hancock D A and Beumer J P (eds), *Developing and sustaining world fisheries resources: The state of science and management*. Brisbane: Second World Fisheries Congress, 97.

INGRAM B A and RIMMER M A (1992) Induced breeding and larval rearing of the endangered Australian freshwater fish trout cod, *Maccullochella macquariensis* (Cuvier) (Percichthyidae), *Aquaculture and Fisheries Management*, 24, 7–17.

INGRAM B A and THURSTAN S (2008) Re-introduction of trout cod into the Murray-Darling River Basin, Australia, in Soorae P S (ed.), *Global re-introduction perspectives*. Abu Dhabi: IUCN/SSC Re-introduction Specialist Group, 37–41.

INGRAM B A, BARLOW C G, BURCHMORE J J, GOOLEY G J, ROWLAND S J and SANGER A C (1990) Threatened native freshwater fishes in Australia – some case histories, *Journal of Fish Biology*, 37A, 175–182.

INGRAM B A, RIMMER M A and ROWLAND S J (1994) Induced spawning trials with captive Macquarie perch, *Macquaria australasica* (Percichthyidae), *Proceedings of the Linnean Society of NSW*, 114, 109–116.

INGRAM B A, DOUGLAS J W and LINTERMANS M (2000) Threatened fishes of the world: *Macquaria australasica* Cuvier, 1830 (Percichthyidae), *Environmental Biology of Fishes*, 59, 68.

© Woodhead Publishing Limited, 2013

INGRAM B A, GAVINE F and LAWSON P (2005) *Fish health management guidelines for farmed Murray cod*, Fisheries Victoria Research Report Series No. 32. Snobs Creek: Department of Primary Industry.

INGRAM B A, HAYES B and ROURKE M L (2011) Impacts of stock enhancement strategies on the effective population size of Murray cod, *Maccullochella peelii*, a threatened Australian fish, *Fisheries Management and Ecology*, 18(6), 467–481.

IUCN (1988) *1988 IUCN Red list of threatened animals*. Cambridge: IUCN.

JELKS H J, WALSH S J, BURKHEAD N M, CONTRERAS-BALDERAS S, DÍAZ-PARDO E, HENDRICKSON D A, LYONS J, MANDRAK N E, MCCORMICK F, NELSON J S, PLATANIA S P, PORTER B A, RENAUD C B, SCHMITTER-SOTS J J, TAYLOR E B and WARREN M L (2008) Conservation status of imperiled North American freshwater and diadromous fishes, *Fisheries*, 33, 372–405.

JERRY D R (1997) Population genetic structure of the catadromous Australian bass from throughout its range, *Journal of Fish Biology*, 51, 909–920.

KEARNEY R E and KILDEA M A (2001) *The status of Murray cod in the Murray–Darling Basin*. Canberra: Environment Australia and National Heritage Trust.

KEENAN C P, WATTS R J and SERAFINI L G (1995) *Population genetics of golden perch (Macquaria ambigua), silver perch (Bidyanus bidyanus) and eel-tailed catfish (Tandanus tandanus) within the Murray–Darling Basin*, Final Report to the Natural Resources Management Strategy, Project No. M262. Canberra: Murray–Darling Basin Commission.

KING A J, TONKIN Z and MAHONEY J (2009) Environmental flow enhances native fish spawning and recruitment in the Murray River, Australia, *River Research and Applications*, 25, 1205–1218.

KOEHN J D (2001) Ecological impacts of cold water releases on fish and ecosystem processes, in Phillips B (ed.), *Thermal pollution of the Murray–Darling Basin waterways: workshop held at Lake Hume 18–19 June 2001*. Sydney: Inland Rivers Network and World Wildlife Fund, 7–11.

KOEHN J D (2005a) Rehabilitation of fish habitats in the Murray-Darling Basin where have we been and where are we going? in *Native fish rehabilitation and management in the Murray–Darling Basin*. Albury: Murray–Darling Basin Commission and Cooperative Research Centre for Freshwater Ecology, 9–17.

KOEHN J D (2005b) Threats to Murray cod, in Lintermans M and Phillips B (eds), *Management of Murray cod in the Murray–Darling Basin workshop: statement, recommendations and supporting papers, proceedings of a workshop held in Canberra ACT, 3–4 June 2004*. Canberra: Murray–Darling Basin Commission and Cooperative Research Centre for Freshwater Ecology, 30–37.

KOEHN J D, MCKENZIE J A, O'MAHONY D J, NICOL S J, O'CONNOR J P and O'CONNOR W G (2009) Movements of Murray cod (*Maccullochella peelii peelii*) in a large Australian lowland river, *Ecology of Freshwater Fish*, 18, 594–602.

KOSTER W, CROOK D and FAIRBROTHER P (2004) *Surveys of fish communities in the lower Goulburn River*. Heidelberg: Arthur Rylah Institute for Environmental Research.

LAKE J S (1964) Silver perch is widespread, *Journal of the Federation of Riverina Inland Anglers' Clubs*, 1, 36.

LAKE J S (1967a) Rearing experiments with five species of Australian freshwater fishes. I. Inducement to spawning, *Australian Journal of Marine and Freshwater Research*, 18, 137–153.

LAKE J S (1967b) Rearing experiments with five species of Australian freshwater fishes. II. Morphogenesis and ontogeny, *Australian Journal of Marine and Freshwater Research*, 18, 155–173.

LAKE J S (1967c) *Freshwater fishes of the Murray-Darling River System*, New South Wales State Fisheries Research Bulletin, 7. Sydney: Chief Secretary's Department.

© Woodhead Publishing Limited, 2013

LAKE J S (1971) *Freshwater fishes and rivers of Australia*. Sydney: Thomas Nelson (Australia).

LANCASTER M J, WILLIAMSON M M and SCHROEN C J (2003) Iridovirus-associated mortality in farmed Murray cod (*Maccullochella peelii peelii*), *Australian Veterinary Journal*, 81, 633–634.

LANGDON J S, GUDKOVS N, HUMPHREY J D and SAXON E C (1985) Deaths in Australian freshwater fishes associated with *Chilodonella hexasticha* infection, *Australian Veterinary Journal*, 62, 409–412.

LINTERMANS M (2007) *Fishes of the Murray–Darling Basin: an introductory guide*. Canberra: Murray–Darling Basin Commission.

LINTERMANS M and PHILLIPS B (2005) *Management of Murray cod in the Murray–Darling Basin: statement, recommendations and supporting papers, proceedings of a workshop held in Canberra ACT, 3–4 June 2004*. Canberra: Murray–Darling Basin Commission, and Cooperative Research Centre for Freshwater Ecology.

LINTERMANS M, ROWLAND S, KOEHN J, BUTLER G, SIMPSON B and WOODEN I (2005) The status, threats and management of freshwater cod species *Maccullochella* spp. in Australia, in Lintermans M and Phillips B (eds), *Management of Murray cod in the Murray-Darling Basin workshop: statement, recommendations and supporting papers, proceedings of a workshop held in Canberra ACT, 3–4 June 2004*. Canberra: Murray–Darling Basin Commission and Cooperative Research Centre for Freshwater Ecology, 15–29.

LLEWELLYN L L (1983) *The distribution of fish in New South Wales*, Special Publication No. 7. Australian Society for Limnology.

LYNCH M and O'HELY M (2001) Captive breeding and the genetic fitness of natural populations, *Conservation Genetics*, 2, 363–378.

LYON J, MCDONALD A, DEASON B and NICOL S (2007) *Implementation of recovery plan for the management and monitoring of trout cod Maccullochella macquariensis in the Ovens River, Victoria*. Melbourne: Arthur Rylah Institute for Environmental Research.

MACDONALD M C (1978) Morphological and biochemical systematics of Australian freshwater and estuarine percichthyid fishes, *Australian Journal of Marine and Freshwater Research*, 29, 667–698.

MACKAY N J (1973) Histological changes in the ovaries of the golden perch, *Plectroplites ambiguus*, associated with the reproductive cycle, *Australian Journal of Marine and Freshwater Research*, 24, 95–101.

MACKAY N J and SHAFRON M (1989) Water quality, in *Proceedings of the workshop on native fish management, Canberra 16–17 June 1988*. Canberra: Murray–Darling Basin Commission, 137–147.

MACKINNON M R and COOPER P R (1987) Reservoir stocking of barramundi for enhancement of the recreational fishery, *Australian Fisheries*, 46, 34–37.

MACLEAY W, COX J C, DALLEY W B, DANGAR H C, DRIVER R, FARNELL J S, HILL R, HIXSON F, HOLT T, OLIVER A, RAMSAY E P, SKARRATT C C, THORNTON G and WANT G F (1880) *Fisheries enquiry commission. Report of the Royal Commission, to enquire into and report upon the actual state and prospect of the fisheries of this colony*. Sydney: Government Printer.

MALLEN-COOPER M and STUART I G (2003) Age, growth and non-flood recruitment of two potamodromous fishes in a large semi-arid/temperate river system, *River Research and Applications*, 19, 697–719.

MAYER O, MEAGER J, SKJÆRAASEN J E, RODEWALD P, SVERDRUP G and FERNÖ A (2011) Domestication causes rapid changes in heart and brain morphology in Atlantic salmon (*Gadus morhua*), *Environmental Biology of Fishes*, 92, 181–186.

MERRICK J R (1996) Family Terapontidae freshwater grunters or perches, in McDowall R (ed.), *Freshwater fishes of south-eastern Australia*. Sydney: Reed Books, 164–167.

© Woodhead Publishing Limited, 2013

MERRICK J R (2006) Australasian freshwater fish fauna: diversity, interrelations, radiations and conservation, in Merrick J R, Archer M, Hickey G M and Lee M S Y (eds), *Evolution and biogeography of Australasian vertebrates*. Oatlands: Auscipub Pty Ltd, 195–224.

MERRICK J R and SCHMIDA G E (1984) *Australian freshwater fishes biology and management*. North Ryde: John R. Merrick.

MOORE A (2000) The genetic impact of stock enhancement programs using captively bred fish, in Moore A and Hughes R (eds), *Proceedings of a workshop on stock enhancement of marine and freshwater fisheries, 7–12 August 2000, Albury*. Lismore: Australian Society for Fish Biology and Southern Cross University, 33–38.

MOORE A and HUGHES R (2000) *Proceedings of a workshop on stock enhancement of marine and freshwater fisheries 7–12 August 2000, Albury*. Lismore: Australian Society for Fish Biology and Southern Cross University.

MOORE A, INGRAM B, FRIEND S, KING HO H, ROBINSON N, MCCORMACK R, COUGHRAN J and HAYES B (2010) *Management of genetic resources for fish and crustaceans in the Murray-Darling Basin*. Canberra: Bureau of Rural Sciences.

MORRIS S A, POLLARD D A, GEHRKE P C and POGONOSKI J J (2001) *Threatened and potentially threatened freshwater fishes of coastal New South Wales and the Murray-Darling Basin*, NSW Fisheries Final Report Series No. 33. Sydney: NSW Fisheries.

Murray–Darling Basin Commission (2003) *Native fish strategy for the Murray–Darling Basin 2003–2013*. Canberra: Murray–Darling Basin Commission.

MUSYL M K and KEENAN C P (1992) Population genetics and zoogeography of Australian freshwater golden perch, *Macquaria ambigua* (Richardson 1845) (Teleostei: Percichthyidae), and electrophoretic identification of a new species from the Lake Eyre basin, *Australian Journal of Marine and Freshwater Research*, 43, 1585–1601.

New South Wales Fisheries (2003) *Environmental impact statement of freshwater fish stocking in NSW*. Cronulla: NSW Fisheries.

NOCK C J, ELPHINSTONE M S, ROWLAND S J and BAVERSTOCK P R (2010) Phylogeography and revised taxonomy of the Australian freshwater cod genus, *Maccullochella* (Percichthyidae), *Marine and Freshwater Research*, 61, 980–991.

NOCK C J, OVENDEN J R, BUTLER G L, WOODEN I, MOORE A and BAVERSTOCK P R (2011) Population structure, effective population size and adverse effects of stocking in the endangered Australian eastern freshwater cod *Maccullochella ikei*, *Journal of Fish Biology*, 78, 303–321.

NSW Fisheries (2002) *Eastern (freshwater) cod (Maccullochella ikei): recovery plan*. Port Stephens: NSW Fisheries.

O'CONNOR P F (1989) Fisheries management in inland NSW, in *Proceedings of the workshop on native fish management, Canberra 16–17 June 1988*. Canberra: Murray–Darling Basin Commission, 19–23.

PAPERNA I (1991) Diseases caused by parasites in the aquaculture of warm water fish, *Annual Review of Fish Disease*, 1991, 155–194.

PHILLIPS B (2003) *Managing fish translocation and stocking in the Murray–Darling Basin workshop, Canberra, 25–26 September 2002: statement, recommendations and supporting papers*. Sydney: World Wildlife Fund.

POLLARD D A and WOODEN I (2002) *Interim results of surveys in the Clarence and Richmond river systems*. Cronulla: NSW Fisheries.

POLLARD D A, LLEWELLYN L C and TILZEY R D J (1980) Management of freshwater fish and fisheries, in Williams W D (ed.), *An ecological basis for water resource management*. Canberra: Australian National University Press, 227–270.

POLLARD D A, INGRAM B A, HARRIS J H and REYNOLDS L F (1990) Threatened fishes in Australia – an overview, *Journal of Fish Biology*, 37A, 67–78.

© Woodhead Publishing Limited, 2013

PUCKRIDGE J T, WALKER K F, LANGDON J S, DALEY C and BEAKES G W (1989) Mycotic dermatitis in a freshwater gizzard shad, the bony bream, *Nematalosa erebi* (Gunther), in the River Murray, South Australia, *Journal of Fish Diseases*, 12, 205–221.

PUSEY B, KENNARD M and ARTHINGTON A (2004) *Freshwater fishes of north-eastern Australia*. Collingwood: CSIRO Publishing.

READ P, LANDOS M, ROWLAND S J and MIFSUD C (2007) *Diagnosis, treatment and prevention of the diseases of the Australian freshwater fish silver perch (Bidyanus bidyanus)*. Sydney: NSW Department of Primary Industries.

REID D, HARRIS J H and CHAPMAN D J (1997) NSW *Inland commercial fishery data analysis*. Sydney: NSW Fisheries.

REYNOLDS L F (1976) Decline of the native fish species in the River Murray, *South Australian Fisheries Industries Council*, 8, 19–24.

REYNOLDS L F (1983) Migration patterns of five fish species in the Murray–Darling River System, *Australian Journal of Marine and Freshwater Research*, 34, 857–871.

RICCIARDI A and RASMUSSEN J B (1999) Extinction rates of North American freshwater fauna, *Conservation Biology*, 13, 1220–1222.

RICHARDSON B A (1994) The human impacts on the ecology of freshwater fish in Western New South Wales, in Lunney D, Hand S, Reed P and Butcher D (eds), *Future of the fauna of western New South Wales*. Mosman: The Royal Zoological Society of New South Wales, 169–176.

RIMMER M A (1987) Trout cod bred for first time at Narrandera, *Australian Fisheries*, 46, 33–34.

RIMMER M A and RUSSELL D J (1998) Survival of stocked barramundi, *Lates calcarifer* (Bloch), in a coastal river system in far northern Queensland, Australia, *Bulletin of Marine Science*, 62, 325–335.

ROGERS M A, ALLEN M S, BROWN P, HUNT T L, FULTON W and INGRAM B A (2010) A simulation model to explore the relative value of stock enhancement versus harvest regulations for fishery management, *Ecological Modelling*, 221, 919–926.

ROLFE J, PRAYAGA P, LONG P and CHEETHAM R (2005) *Estimating the value of freshwater recreational fishing in three Queensland dams*. Rockhampton: Central Queensland University.

ROURKE M L (2007) *Population genetic structure of Murray cod (Maccullochella peelii peelii) and impacts of stocking in the Murray-Darling Basin*, PhD Thesis, Clayton, Monash University.

ROURKE M L, MCPARTLAN H C, INGRAM B A and TAYLOR A C (2009) Polygamy and low effective population size in a captive Murray cod (*Maccullochella peelii peelii*) population: genetic implication for wild restocking programs, *Marine and Freshwater Research*, 60, 873–883.

ROURKE M L, MCPARTLAN H C, INGRAM B A and TAYLOR A C (2010) Biogeography and life history ameliorate the potentially negative effects of stocking Murray cod (*Maccullochella peelii peelii*), *Marine and Freshwater Research*, 61, 918–927.

ROURKE M L, MCPARTLAN H C, INGRAM B A and TAYLOR A C (2011) Variable stocking effect and endemic population genetic structure in Murray cod *Maccullochella peelii*, *Journal of Fish Biology*, 79(1), 155–177.

ROWLAND S J (1983a) Spawning of the Australian freshwater fish Murray cod *Maccullochella peeli* (Mitchell), in earthen ponds, *Journal of Fish Biology*, 23, 525–534.

ROWLAND S J (1983b) The hormone-induced ovulation and spawning of the Australian freshwater fish golden perch, *Macquaria ambigua* (Richardson) (Percichthyidae), *Aquaculture*, 35, 221–238.

ROWLAND S J (1983c) The hormone-induced spawning and larval rearing of Australian native fish, with particular emphasis on the golden perch, *Macquaria*

© Woodhead Publishing Limited, 2013

ambigua, in *Proceedings of the first freshwater aquaculture workshop, Narrandera, February 1983*. Department of Agriculture New South Wales, 23–32.

ROWLAND S J (1983d) Design and operation of an extensive aquaculture system for breeding warmwater fishes, in *Proceedings of the first freshwater aquaculture workshop, Narrandera, February 1983*. Department of Agriculture New South Wales, 121–144.

ROWLAND S J (1984) The hormone-induced spawning of silver perch, *Bidyanus bidyanus* (Mitchell) (Teraponidae), *Aquaculture*, 42, 83–86.

ROWLAND S J (1985a) *Aspects of the biology and artificial breeding of the Murray cod, Maccullochella peeli, and the eastern freshwater cod, M. ikei sp. nov. (Pisces: Percichthyidae)*, PhD Thesis, North Ryde, Macquarie University.

ROWLAND S J (1985b) Identification and conservation of the Eastern Freshwater Cod, in *Biennial report, July 1983–June 1985, Agricultural Research and Advisory Station, Grafton*. Grafton: Department of Agriculture New South Wales, 16–17.

ROWLAND S J (1986) Hatchery production of native warmwater fishes in New South Wales, in Pyne R (ed.), *Advances in aquaculture; the proceedings of a workshop, Darwin, 1986*, Technical Report No. 3. Northern Territory, Fisheries Division, Department of Primary Industry & Fisheries, 79–92.

ROWLAND S J (1988a) Eastern freshwater cod (*Maccullochella ikei*); identification, status, conservation and potential threats, in Rowland S J and Barlow R (eds), *Proceedings of a fish genetics workshop, Cronulla, N.S.W. 1985*. Grafton: NSW Agriculture & Fisheries, 18–23.

ROWLAND S J (1988b) Hormone-induced spawning of the Australian freshwater fish Murray cod, *Maccullochella peeli* (Mitchell) (Percichthyidae), *Aquaculture*, 70, 371–389.

ROWLAND S J (1989) Aspects of the history and fishery of the Murray cod, *Maccullochella peeli* (Mitchell) (Percichthyidae), *Proceedings of the Linnean Society of NSW*, 113(3), 201–213.

ROWLAND S J (1990) Conservation of the eastern freshwater cod, in *Biennial research report, July 1987–June 1989, Agricultural Research and Advisory Station, Grafton*. Grafton: NSW Agriculture & Fisheries, 8–9.

ROWLAND S J (1992) Diet and feeding of Murray cod (*Maccullochella peeli*) larvae, *Proceedings of the Linnean Society of NSW*, 113(4), 193–201.

ROWLAND S J (1993) *Maccullochella ikei*, an endangered species of freshwater cod (Pisces: Percichthyidae) from the Clarence river system, NSW, and *M. peelii mariensis* a new subspecies from the Mary River, Qld, *Records of the Australian Museum*, 45, 121–145.

ROWLAND S J (1995) Stocking of freshwater fishes and policy in New South Wales, in Prokop F B (ed.), *Translocation issues in Western Australia*, Fisheries Management Paper No. 83. Perth: Fisheries Department of Western Australia, 50–61.

ROWLAND S J (1996a) Threatened fishes of the world: *Maccullochella ikei* Rowland, 1985 (Percichthyidae), *Environmental Biology of Fishes*, 46, 350.

ROWLAND S J (1996b) Development of techniques for the large-scale rearing of the larvae of the Australian freshwater fish golden perch, *Macquaria ambigua* (Richardson, 1845), *Marine and Freshwater Research*, 47, 233–242.

ROWLAND S J (1998) Aspects of the reproductive biology of Murray cod, *Maccullochella peelii peelii*, *Proceedings of the Linnean Society of NSW*, 120, 147–162.

ROWLAND S J (2001) Record of the banded grunter *Amniataba percoides* (Teraponidae) from the Clarence River, New South Wales, *Australian Zoologist*, 31 (4), 603–607.

ROWLAND S J (2004) Domestication of silver perch (*Bidyanus bidyanus*) broodfish, *Journal of Applied Aquaculture*, 15, 75–84.

ROWLAND S J (2005) Overview of the history, fishery, biology and aquaculture of Murray cod (*Maccullochella peelii peelii*), in Lintermans M and Phillips B (eds),

© Woodhead Publishing Limited, 2013

Management of Murray cod in the Murray–Darling Basin workshop: statement, recommendations and supporting papers, proceedings of a workshop held in Canberra ACT, 3–4 June 2004. Canberra: Murray–Darling Basin Commission and Cooperative Research Centre for Freshwater Ecology, 38–61.

ROWLAND S J (2009) Review of aquaculture research and development of the Australian freshwater fish silver perch, *Bidyanus bidyanus*, *Journal of the World Aquaculture Society*, 40(3), 291–324.

ROWLAND S J and BARLOW R (1988) *Proceedings of a fish genetics workshop, Cronulla, NSW, 1985*. Sydney: NSW Agriculture & Fisheries.

ROWLAND S J and BRYANT C (1995) *Silver perch culture: proceedings of silver perch aquaculture workshops, Grafton and Narrandera, April 1994*. Sandy Bay: Austasia Aquaculture.

ROWLAND S J and INGRAM B A (1991) *Diseases of Australian native freshwater fishes, with particular emphasis on the ectoparasitic and fungal diseases of Murray cod (Maccullochella peeli), golden perch (Macquaria ambigua) and silver perch (Bidyanus bidyanus)*, Fisheries Bulletin No. 4. Sydney: NSW Agriculture & Fisheries.

ROWLAND S J and TULLY P (2004) *Hatchery quality assurance program for Murray cod (Maccullochella peelii peelii), golden perch (Macquaria ambigua) and silver perch (Bidyanus bidyanus)*. Cronulla: NSW Department of Primary Industries.

ROWLAND S, DIROU J and SELOSSE P (1983) Production and stocking of golden and silver perch in NSW, *Australian Fisheries*, 42(9), 24–28.

ROWLAND S J, ALLAN G L, CLARK K, HOLLIS M and PONTIFEX T (1994) Production of fingerling *Bidyanus bidyanus* (Teraponidae) at two densities in earthen ponds in Australia, *The Progressive Fish-Culturist*, 56, 296–298.

ROWLAND S J, LANDOS M, CALLINAN R B, ALLAN G L, READ P, MIFSUD C, NIXON M, BOYD P and TULLY P (2007) *Development of a health management strategy for the silver perch aquaculture industry*, Final Report to the Fisheries Research Development Corporation on Projects 2000/067 and 2004/089, Fisheries Final Report Series No. 93. Cronulla: NSW Department of Primary Industries.

RUTLEDGE W P, RIMMER M A, RUSSELL D J, GARRETT R and BARLOW C (1990) Cost benefit of hatchery-reared barramundi, *Lates calcarifer* (Bloch), in Queensland, *Aquaculture and Fisheries Management*, 21, 443–448.

RYMAN N and LAIKRE L (1991) Effects of supportive breeding on the genetically effective population size, *Conservation Biology*, 5, 325–329.

SANGER A and TALBOT B (2003) Management of fish stocking in New South Wales, in Phillips B (ed.), *Managing fish translocation and stocking in the Murray–Darling Basin workshop, Canberra, 25–26 September 2002: statement, recommendations and supporting papers*. Sydney: World Wildlife Fund, 88–93.

SIMPSON R R and JACKSON P D (1996) *The Mary River cod research and recovery plan*. Canberra: Australian Nature Conservation Agency Endangered Species Program.

SIMPSON R R and MAPLESTON A J (2002) Movements and habitat use by the endangered Australian freshwater Mary River cod, *Maccullochella peelii mariensis*, *Environmental Biology of Fishes*, 65, 401–410.

SIMPSON B, HUTCHINSON M, GALLAGHER T and CHILCOTT K (2002) *Fish stocking in impoundments: a best practice manual for eastern and northern Australia*. Deception Bay: Department of Primary Industries.

STUART I G, ZAMPATTI B P and BAUMGARTNER L J (2008) Can a low-gradient vertical-slot fishway provide passage for a lowland river fish community? *Marine and Freshwater Research*, 59, 332–346.

TALBOT B, MOLLOY S, CHAPMAN R and RICHES M (2004) The rivers to the sea: experience with two endangered aquatic fish species, in Hutchings P, Lunney D and Dickman C (eds), *Threatened species legislation: is it just an Act?* Mosman: Royal Zoological Society, 125–134.

© Woodhead Publishing Limited, 2013

THURSTAN S (2000) Practical management for genetic stock management of native fish hatcheries, in Moore A and Hughes R (eds), *Proceedings of a workshop on stock enhancement of marine and freshwater fisheries*. Lismore: Australian Society for Fish Biology and Southern Cross University, 46–48.

THURSTAN S and ROWLAND S J (1995) Techniques for the hatchery production of silver perch, in Rowland S J and Bryant C (eds), *Silver perch culture: proceedings of silver perch aquaculture workshops, Grafton and Narrandera, April, 1994*. Sandy Bay: Austasia Aquaculture, 29–39.

UTTER F (2004) Population genetics, conservation and evolution in salmonids and other widely cultured fishes: some perspectives over six decades, *Reviews in Fish Biology and Fisheries*, 14, 125–144.

WAJON S (1983) *Hybridization between Murray cod and trout cod in Cataract Dam, NSW*, Honours Thesis, Sydney, University of New South Wales.

WALKER K F (1980) The downstream influence of Lake Hume on the River Murray, in Williams W D (ed.), *An ecological basis for water resource management*. Canberra: Australian National University Press, 182–191.

WARREN M L and BURR B M (1994) Status of freshwater fishes of the United States: overview of an imperilled fauna, *Fisheries*, 19, 6–18.

WELCOMME R L and BARTLEY D M (1998) Current approaches to the enhancement of fisheries, *Fisheries Management and Ecology*, 5, 351–382.

WILLIAMS J E, WILLIAMS R N, THUROW R F, ELWELL L, PHILIPP D P, HARRIS F A, KERSHNER J L, MARTINEZ P J, MILLER D, REEVE G H, FRISSELL C A and SEDELL J R (2011) Native fish conservation areas: a vision for large-scale conservation of native fish communities, *Fisheries*, 36, 267–288.

YE Q and ZAMPATTI B (2007) *Murray cod stock status: the lower River Murray, South Australia*. Adelaide: South Australian Research and Development Institute.

© Woodhead Publishing Limited, 2013

20

Developing educational programs in partnership with aquaculture hatchery facilities

D. Meritt and D. Webster, University of Maryland, USA

DOI: 10.1533/9780857097460.4.596

Abstract: The hands-on nature of the hatchery setting is a natural supplement to curriculum units for students of all ages and experience levels. This chapter addresses the benefits of experiential education and provides descriptions of its use for a variety of aquaculture-based learning experiences pertinent to a number of subject areas. Recommendations for implementing such programs are presented for each grade level and discipline. The chapter provides a number of examples and resources to aid the reader in developing or enhancing instruction in the hatchery or classroom.

Key words: aquaculture, hatchery, education, curriculum, experiential, hands-on.

20.1 Introduction

In aquaculture, a hatchery is typically where animals are brought to reproductive potential and spawned, with resulting young raised through the larval stage to the point where they enter growing operations. The hands-on (experiential) nature of the hatchery setting is a natural supplement to curriculum units for students of all age and experience levels. Aquaculture hatcheries provide a variety of learning experiences pertinent to a number of subject areas. This chapter addresses the value of experiential education and demonstrates how educational programs can be developed in partnership with aquaculture facilities. The reader will find examples of how a hatchery-based program can be tailored to meet the learning needs of students from elementary school to adulthood. Recommendations for and examples of hatchery-based educational programs are presented for the following sectors:

© Woodhead Publishing Limited, 2013

- primary school
- secondary school
- undergraduate
- graduate/post-graduate
- technical/vocational
- teacher training
- extension, outreach, and technology transfer

The spectrum of disciplines that are part of hatchery design, construction and management illustrate the many areas that can be used for education. Existing programs that provide *in situ* and hands-on experience have demonstrated strong reinforcement in learning, as will be discussed later. Among the disciplines inherent to hatchery education are:

- *biology* – demonstrations of the life-cycles of animals and in particular their early life history, as well as aspects of nutrition, reproduction and genetics.
- *chemistry* – studies of the analysis and quality of water used in culture systems and their manipulation for the maintenance of proper quality.
- *engineering* – experience with hatchery design and operation, development of water supply systems and aeration, site selection, material handling, heating, cooling and current trends in automation/computerization.
- *economics and business management* – exposure to the operational considerations for public and private hatcheries with an emphasis on profitable business functions.
- *mathematics* – development of proficiency at daily tasks such as calculating volumes, concentrations, serial dilutions, means, ranges, standard deviations, and other mathematical skills for successful hatchery operations.
- *social studies* – addressing the environmental, social, and political issues affecting aquaculture, such as the acquisition of permits and the process of gaining support from the local community.

We have provided a number of resources to guide the reader in creating or enhancing either a hatchery educational program or a curriculum unit supplemented by hands-on aquaculture experiences. Examples of successful programs conducted at both public and private hatcheries yield a perspective of how hatchery education projects are and can be used for instructional purposes.

20.1.1 Theories of learning

The great difficulty in education is to get experience out of ideas. George Santayana, philosopher, 1905

© Woodhead Publishing Limited, 2013

While didactic, or rote, learning has often been standard practice in education, and is useful in certain instances, it can have drawbacks when one is trying to enthuse students and enrich the learning process. Interest in a learning activity naturally fosters the intrinsic motivation to succeed. Experts would agree that learning takes place most effectively when the student is willingly engaged in the process (deCharms, 1976; Thomas, 1980) and applies skills in order to conceptualize the experience (CDSL, 2000). Learning is further enhanced when the pupil is given the opportunity to analyze and problem-solve, both independently and cooperatively (Barrows and Tamblyn, 1980; Johnson *et al.*, 2000). Knowledge is continuously gained through repeated personal and environmental interactions, but should be underpinned through active involvement, reflection, and problem-solving exercises. Such exercises give meaning to a lesson, which allows students to process information at a deeper level (Forgus and Schwartz, 1957; Craik and Lockhart, 1972). In this way, the learner may derive the greatest understanding and retention of the subject matter.

One of the most well-known leaders in the field of pedagogy is John Dewey, a reformer who championed new educational philosophies during the first half of the twentieth century. Dewey (1916, 1938) advocated balancing the delivery of knowledge with the experiences and interests of students He became a staunch proponent of hands-on learning, or experiential education, and advocated that good teaching is accomplished by the teacher acting as an educational guide, or facilitator. In this role, the teacher forms a learning partnership with the students, who are encouraged to independently discover meaning within an area of investigation.

While Dewey was an innovator in his concept of experiential education (Dewey, 1938), he was certainly not the only one. The approach of 'learning by doing' has been practiced since the time of Aristotle. Additional models of learning beyond the traditional lecturer/student paradigm became popularized by other educational theorists. Among these are experiential learning (Kolb and Fry, 1975), cooperative learning (Johnson and Johnson, 1994), action learning (Ravens, 1980), active learning (Hamer, 2000), free choice learning (Oregon Sea Grant, 2012), and service learning (Gordon, 2007). We reference papers that provide useful applications of these concepts. All of these forms of learning spring from participatory involvement in the learning process by the students. Direct experience provides a more significant contribution to learning than listening to a lecture alone (Dale, 1969; Hamer, 2000; Farkas, 2003). Evidence of this is demonstrated in Fig. 20.1. It is the visual and hands-on process of acquiring knowledge that is thought to stay with students far longer than simply listening to lectures (Kvam, 2000; Beers and Bowden, 2005).

One of the most apparent applications of experiential education exists within the study of environmental sciences. Here, studies have shown that being outdoors and observing natural processes in a group setting is more effective at engaging students than attending lectures (Tanner 1980).

© Woodhead Publishing Limited, 2013

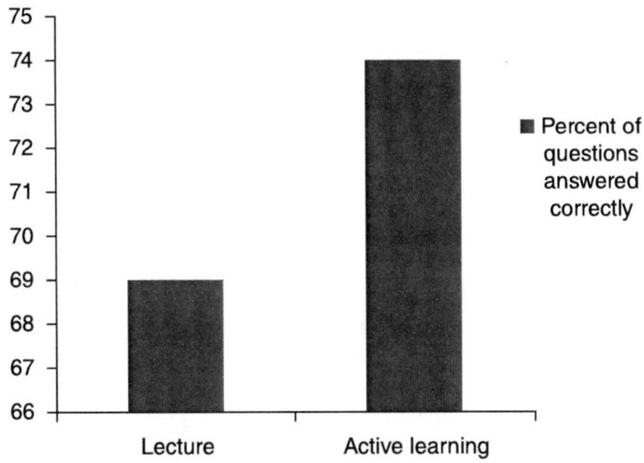

Fig. 20.1 Effect of class format on student performance (adapted from Hamer, 2000).

Hatcheries provide fertile ground for this type of education, as they expose students to an environment in which multiple disciplines have been integrated; furthermore, they require students to apply their new skills to real-world situations.

20.1.2 Overview of aquaculture hatcheries

Aquaculture hatcheries can be classified by the species they produce, namely finfish or shellfish. They are further delineated by the intended use of the product: consumption, ornamental, baitfish, or habitat restoration. (Within the field of aquaculture there are also producers of aquatic plants, but this is not within the scope of this chapter.)

Hatcheries can also be categorized by application and production. Commercial hatcheries are typically in the private sector and produce animals for sale. They may be further described as either stand-alone businesses that produce juvenile animals for sale to others to grow or they may be one component of a vertically integrated business that produces animals for ultimate consumer sale.

Public sector hatcheries may be designated for research and/or production. The former are often affiliated with academic institutions or state and federal agencies where they are used to answer scientific queries about a species and to acquire information about its life-cycle and nutritional requirements. Public production hatcheries may also provide animals for restoration or enhancement stocking. The practice of using aquaculture to produce food has been around for 3000 years; by comparison, the production of aquatic animals for use as a means to restore depleted populations or, more recently, ecosystems is a relatively new field. Enhancement production is typified by governmental fish hatcheries that raise and distribute

© Woodhead Publishing Limited, 2013

species popular with recreational fishermen. This production is normally supported by funds paid through license or tag fees by those who fish for the animals.

Many fish hatcheries not only produce animals for restoration or commercial purposes, but they serve as a valuable educational resource for the conservation community. This is especially true for many government hatcheries. However, often these facilities do not have the resources to adequately fund the educational component of the operation. Many were built before the environmental movement took hold in the 1970s and were not designed to conduct educational programs. Therefore, some of these hatcheries work cooperatively with other local environmental and educational groups to address the educational components. If properly organized, this can be a benefit to both the educational community and the hatchery.

As an alternative, a number of hatcheries have created support organizations dedicated to providing aquaculture-based educational programs. An example can be found at the Decorah Fish Hatchery in Iowa. This facility was constructed in the 1930s as part of the Civilian Conservation Corps program. A major renovation in this facility was completed in 1989, which not only improved the actual culture facilities but modified them to better accommodate visitors and provide an educational experience. A non-profit organization called the Friends of the Decorah Fish Hatchery was formed to 'support and improve the services delivered to the public by the Decorah Fish hatchery, primarily through educating the public on the conservation, enhancement, and protection of natural resources that is vital to the effort of the hatchery and through conducting fundraising efforts to support these activities' (http://www.decorahfishhatchery.org). This organization has a website and a Facebook page where they offer merchandise for sale and promote their activities.

There are many examples of this type of cooperative relationship between hatcheries and the environmental or educational community. Some organizations like the Pennsylvania Steelhead Association, a non-profit organization dedicated to the growth, enhancement and protection of Lake Erie cold water fisheries, operate a cooperative hatchery facility in conjunction with the S.O.N.S. (Save Our Native Species) of Lake Erie and the Pennsylvania Fish and Boat Commission. They conduct programs similar to that described above. Some hatchery programs have taken the educational component even further by developing a web-based interactive game that uses a typical day in the hatchery to teach basic math and science skills and to demonstrate how those skills are important to fish hatchery operation (Colorado Division of Wildlife, 2011).

Regardless of the species they produce, the final application of their product, or whether they are operated by the public or private sector, hatcheries can be used to develop quality educational programs that can provide students with a broad range of knowledge and skills, and deepen their understanding. This can lead to career paths in commercial

© Woodhead Publishing Limited, 2013

aquaculture production, restoration and enhancement aquaculture, or development of ancillary industries. Education should be considered an important part of the hatchery operation when assessing the skills required of those employed to operate them.

20.2 The hatchery as a multi-disciplinary educational tool

While there have been educational programs taught at hatcheries for many years, these are now expanding as more are developed and the information about them is shared through organized means. One has only to look at the schedules of national and international aquaculture conferences and meetings to see that many, if not most, of these now include sections devoted to education. Those involved in aquaculture education have begun organizing to create more formal lines of communication, many through the internet. This has provided the opportunity for broad dissemination of ideas about organizing, managing, and evaluating programs, while highlighting the publications and other media efforts and activities created for their support.

20.2.1 Disciplines of study within the hatchery

The design, construction, and operation of fish and shellfish hatcheries are multi-disciplinary enterprises. Some of the disciplines are quite apparent to those who operate hatcheries, such as the need for biological knowledge about the animals being produced, but others may only become apparent when the hatchery is viewed as a total operation. In the private sector, for instance, business practices relating to the production, marketing, and sale of the product are important and affect the overall success of the business as an operating enterprise.

Biology

Biology includes the examination of the structure and function of an animal, as well as its origin, evolution, growth, distribution, and taxonomy. Becoming familiar with the life-cycle of an animal is considered a precursor to aquaculture production. Because hatchery operations involve early life phases of animals they are of great use in teaching developmental biology and embryology. The transition of organisms from fertilization through the progressive development of later stages provides sequential learning about the totality of life processes. For successful hatchery operations, this will also include its nutritional requirements, reproductive needs and sequence, genetics, and, to some degree, ethology.

Microbiology is the science of studying microbes or unicellular and cell-cluster organisms. A great deal of this field is concerned with the study of the immune system which is extremely important in the health-related research of cultured aquatic animals. Microbiologists have also advanced

© Woodhead Publishing Limited, 2013

the study of bacteria, which degrade or transform potentially toxic substances and render them safe in culture systems. Treatments are typically referred to as 'pro-biotic' and can be helpful in hatchery situations. Microbe culture is critical in the success of recirculating culture systems; these organisms convert ammonia and nitrite to less toxic nitrate and ensure the health of the culture animals.

Phycology, the study of algae, is a sub-field of biology and of great importance to aquaculture, particularly shellfish hatcheries where it is essential to the proper nutrition of the animals. Being filter-feeders, shellfish require an abundant supply of algae to thrive. Understanding the nutritional contribution of various algal species and knowing how to culture them is essential to the success of a shellfish hatchery. The culture of algae requires proper management of the factors affecting survival and growth of these microscopic plants. The culture process must include the proper balance of light, nutrients, and carbon dioxide (CO_2) in order to provide the plants with the conditions supporting photosynthesis. Automated systems that measure algal concentrations in culture vessels and provide them to the hatchery in computer-controlled programs have helped boost shellfish production in recent years.

Chemistry

Chemistry is the study of the makeup of matter and its changes. Subdivisions of this field include inorganic and organic chemistry and biochemistry. In hatcheries, it includes concern for the ever-changing makeup of the culture water which affects the health of the animals being cultured. Thus, for students, its mastery contributes to the success or failure of the hatchery operation. Other factors include pH and alkalinity, salinity, dissolved oxygen, and carbon dioxide levels. Hatchery operators often require skills to mix and apply chemical treatments to culture water, including the addition of therapeutic agents ranging from salt to antibiotics. The ability to determine proper concentrations when applying those agents is a task that any hatchery operator will need to master. This provides educators with another teachable moment, one that results in a direct feedback mechanism for students. No two water sources are equal as far as water quality is concerned and students need to learn the critical components of water quality as they pertain to the species under culture.

Ecology

Ecology is an inter-disciplinary branch of biology that provides study of relationships between organisms and their environment. Perhaps one of the most interesting aspects of this comes in the area of water reuse or recirculating aquaculture systems. These systems have become attractive because of their minimal discharge of waste material and use the same water many times over, adding more only periodically. They are often used when water must be heated or cooled. In these, the waste products from the animals

© Woodhead Publishing Limited, 2013

must be transformed from their toxic state to benign or non-toxic compounds that will not lead to degraded water quality that could kill the animals. Microbes are used in the decomposition process, with multiple species utilized to transform products such as ammonia to nitrite and then nitrate. In doing so, the culturist is actually caring for multiple species since the micro-organisms that are used in the process must be provided with conditions that allow them to survive and thrive. An interesting area of aquaculture is that of aquaponics, which combines the culture of fish and plants in systems that use each species in relationships that benefit each and lead to two or more products to sell.

Engineering

Engineering encompasses a range of hatchery design and operation skills, including water supply and aeration systems, filtration, site selection, material handling, automation and computerization, as well as heating and cooling, and so its study provides students with a variety of interesting tasks. Developing programs or projects that lead to the design and maintenance of systems can be very rewarding for students, and can encourage them to work in teams, which is a key element in many aspects of experiential education. Furthermore, examining the engineering challenges of designing hatchery systems requires students to draw upon other skill sets, such as mathematic fluency and computer literacy, and highlights the multidisciplinary nature of the field.

Economics and business management

An aquaculture hatchery also provides many opportunities for lessons in economics and business management, whether in the public or private sector. Private sector hatcheries, by their very nature, require a profit for their long-term financial viability, but public hatcheries also must meet the goal of staying within budget targets. The management of the operation is an area that readily lends itself to the development of educational programs and can be an important building block in the educational background of students. Understanding how finance, costs and returns, and profitability relate to hatchery operations requires that students apply math skills, create and interpret figures, and practice decision-making skills. With public sector hatcheries, understanding the budget process and how to plan for long-term operation (and possibly expansion) are areas that can be of great value to the student.

Agricultural economics is a sub-field that involves the supply and demand of cultivated food and fiber; aquacultural economics deals specifically with the production of aquatic plants and animals. Many academic institutions include agricultural economics in departments with resource economics, the latter of which deals with economic issues as they pertain to natural resources. These have become more important in recent decades due to their role in policy-making by governments at all levels. This illustrates how

© Woodhead Publishing Limited, 2013

the study of hatchery economics provides a natural transition into the study of policy-making.

Social studies

According to the National Curriculum Standards set forth by the US government, the subject area of social studies includes, but is not limited to, a study of culture, government, and civic ideals (NCSS, 2012). It is often very closely related to the aforementioned economic elements of production, distribution, and consumption as well. Working with the director of a hatchery allows students to witness the importance of these issues within the field of aquaculture. As is the case with oysters, there is often a historically and culturally significant component surrounding the fishery. The hatchery must positively engage local oyster harvesters for the wellbeing of the program and all parties involved. Furthermore, administrators in public hatcheries tend to have a vast understanding of politics, both at the local and state level. Working with government officials can help to ensure funding for the operation, as can becoming an active participant in the community.

Mathematics

It goes without saying that students will have ample opportunities to refine their mathematical skills when involved in a hatchery-based education program. From measuring volumes to calculating food quantities based on tank density to determining the appropriate concentration of a chemical treatment, math is an everyday part of a hatchery operation.

20.3 Levels of hatchery education

20.3.1 Primary school

Students in early studies are challenged to learn basic concepts that they can build upon as they mature. Even young children understand a great deal about the principles of biology (CDSL, 2000). They are naturally curious at this age and their enthusiasm can be harnessed to enhance their understanding of the subject matter and instil a love of learning. Demonstrations can be particularly effective with early childhood students, who may lack the self-control or motor skills necessary to conduct an experiment themselves. The following example of such a demonstration illustrates the environmental benefit of oysters. A simple setup consisting of two aquaria – one with and one without oysters, but both with added algae – is all that is required to show students how oysters can clean the water (Fig. 20.2). After less than an hour, the tank with the bivalves will be clear and the other will remain turbid. This lesson not only provides information about the benefits of bivalves, but also introduces students to the concept of experimental design: how to set up a simple experiment to test

© Woodhead Publishing Limited, 2013

Fig. 20.2 Demonstration of the oysters' filtering ability using hatchery-cultured algae. Top: aquaria at the beginning of demonstration, Bottom: aquaria after oysters have filtered the algae from the water.

a treatment against a control. Children can be encouraged to make hypotheses before the demonstration and discuss or draw conclusions upon its completion.

Young students can also learn about the habitat value of oysters by examining an aquarium containing a mock oyster reef, complete with invertebrates and fish. Youngsters can be challenged to 'seek and find' animals that are listed on a handout as dwelling in an oyster reef. They will get much enjoyment from using flashlights and magnifying glasses to find and examine the organisms in the tank. Instructors can enhance the lesson by placing an anemone-covered oyster shell in water under a dissecting scope. A video camera streaming to a large monitor will allow students to clearly see small organisms (Fig. 20.3). Following the activity with a discussion about how people's homes are similar to and different from those of the animals will bring real meaning to the lesson. These types of activities have a great deal more interest to children than simply hearing about how oysters filter and who lives in their reefs.

As children proceed through the ages of 12–14, they should expand on knowledge acquired from the aforementioned lessons and begin to develop their own investigations. Activities at this stage should still remain fairly simple, with the objective being to encourage the development of proper inquiry and experimental design skills. Collaborative activities are often well-suited to meet these goals. Shared or team education is a basic concept

© Woodhead Publishing Limited, 2013

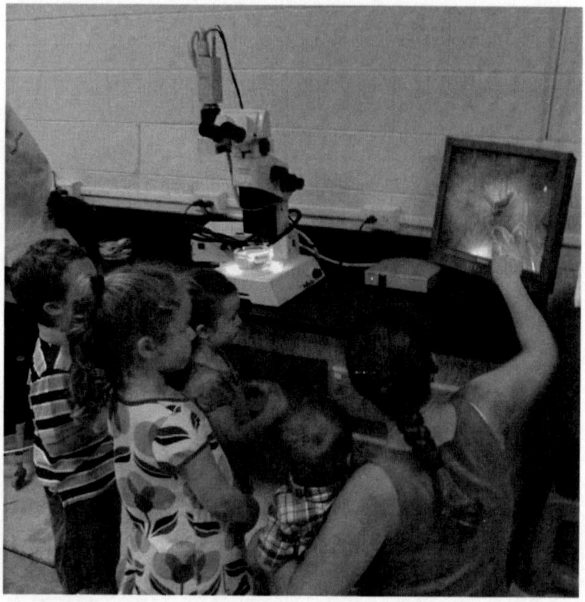

Fig. 20.3 Microscopes can be effective learning tools for young students. Here they learn about some of the organisms typically found living on oysters.

that runs through many variations of experiential education (Association for Experiential Education, www.aee.org). Students can work together to create investigations during which water quality variables are manipulated, for example. They are usually mature enough to be successful at utilizing technology (e.g. digital thermometer, microscope, or basic spreadsheet software) to assist them in executing the study. It is the ability to develop shared goals and to participate in all aspects of the project that makes this type of learning successful (Johnson *et al.*, 2000). The teachers or instructors should gradually develop the role of facilitator or mentor as students become more comfortable about their part in thinking through problems, developing hypotheses, designing experiments to test them, and analyzing results (Smith and MacGregor, 1992).

A field trip to an aquaculture facility is a lesson in collaboration, as hatcheries require teamwork in order to function efficiently. To emphasize this important point, hatchery personnel can engage students in group projects. These types of programs start to become most effective for those of 12–14 years. They are also an opportunity for students to apply knowledge that has been acquired in the classroom, thus reinforcing it as valuable when problem-solving. Implementing such projects often requires repeated visits to the facility. The first visit will often be a tour, an explanation of the hatchery, and a review of background material, while the project would commence on the second visit. There should be background material

© Woodhead Publishing Limited, 2013

provided and the visits should be integrated into a thematic unit by the classroom teacher. As an example, middle school students can work together to evaluate legislation that might affect a fishery or hatchery, develop a stance regarding the regulations, and defend it to the group. Bringing aquaria into the classroom and providing demonstrations of the life-cycle of fish can be effective as well. These 'mini-hatcheries' can be set up to allow fish to reproduce while students tend to the tanks. They can work together to measure water quality, calculate the carrying capacity of tanks with fish, and provide the management for fish that are simple to tend.

A European network has been initiated to support the sustainable development of aquatic resources (http://www.aquatt.ie/index.php/1/home). The network links many aspects of aquaculture and includes strong educational support to student programs with a goal of linking environmental and commercial sectors. In doing so, educational material can be provided to teach students about the beneficial aspects of aquaculture while respecting other users. The PlanetAqua project provides a link that moves into the holistic nature of water and its critical importance in the life of the Earth and its inhabitants.

20.3.2 Secondary schools

By the time the students enter secondary school, they are well on their way to more advanced subject matter. This is a particularly effective time to incorporate educational programs into hatchery activities. At this stage, high school students are working with advanced mathematics and sciences. The ability to teach them about data gathering and analysis is strong, as is the opportunity to interest them in various aspects of technology and engineering. Creating a team of students and hatchery employees that work together to design a project can be especially rewarding. For example, having a group design a small hatchery would involve looking at product flow, finding equipment and prices, instituting a working unit, and perhaps including a simple business plan that would project expenditures and revenue to project profitability. The team members would have many components to examine, further honing research and collaborative skills amongst the students. Engaging collaborative learning provides students with peer support that can yield strong results. This form of cooperation includes positive interdependence, in which students must take responsibility for a task or role and must participate fully within the group (Johnson *et al.*, 1998). Each member helps to promote the success of others by explaining what they have learned and by assisting each other in understanding the parts as they relate to the whole. Students involved with projects of this type therefore develop accountability while building social skills in leadership, communication, decision-making, and conflict management. Periodically, the group must evaluate its progress and effectiveness and refine it for improvement.

© Woodhead Publishing Limited, 2013

Participating in the care and spawning of animals can also be an effective educational activity for students above the age of 14 and there are many resources that show how to develop and manage aquaculture systems in a classroom setting. However, as previously stated, care must be taken to ensure that the requirements of the target species are capable of being provided by the classroom facility or the exercise is likely to fail. One program in the state of Maryland in the USA has partnered hatcheries with secondary schools where fish whose populations are undergoing restoration are distributed to classrooms. The students raise them for eventual release into the wild while learning about the species and their culture techniques. This is an extension of a project that was suggested earlier for younger students (Aquaculture In Action, 2012). The schools provide field trips to the hatcheries where the students are exposed to the entire operation. Fingerlings are then provided to the schools post spawning and the students raise the fish over a period of time. Students learn care of the animals as well as how to track growth and mortality. This provides a natural opportunity to learn and exercise proper data collection and record keeping. The fish are later returned to the hatchery for release to rebuild the population; often students can participate in this last step. Documenting the release through organized publicity can help provide visibility for the project and aid in attracting public or private funds for its continued support. Classroom teachers can assess students' performance by evaluating data logs, formal reports, and oral presentations.

Because secondary school students are larger and stronger than their younger counterparts, they are also able to be involved in some of the more physically demanding hatchery tasks – work that is often carried out in wet or muddy conditions (Fig. 20.4). Helping to handle fish and shellfish, moving bags or containers of cultch, and performing other such jobs may seem onerous to those who have done them many times in the course of their employment, but students can find fun and humor in tasks like these. An important part of these types of activities is that the students feel that they are accomplishing positive goals and contributing to the operation.

Secondary school-aged students may even find part-time or summer employment at hatcheries. Often formal programs exist that may be advertised in educational institutions. Schools may have a work–study program where students are allowed to work in the community during school hours. If the school is in proximity to a hatchery, this can provide an excellent opportunity for the students to interact with the hatchery. Students will gain valuable experience working in the hatchery, while the facility will likely receive a valuable addition to their labor pool. However, it should be noted that any hatchery contemplating this sort of arrangement should outline the duties expected of the student and then work with him or her to schedule these at times mutually beneficial to both the student and the hatchery. Treating the student as simply an additional labor input will not result in a productive relationship or a positive educational experience. Proper

© Woodhead Publishing Limited, 2013

Fig. 20.4 Students can help supplement a hatcheries labor force. Here students help unload an oyster setting tank.

planning and scheduling can result in a beneficial learning experience for the student and, potentially, a valued addition to the hatchery work force. While many institutional hatcheries and some government hatcheries may provide summer internships, private hatcheries should be encouraged to employ young people as a way of interesting them in the potential for ultimate employment. Many worthwhile employees of companies started with summer jobs.

A consideration that must be undertaken by businesses offering to host students is that of insurance and indemnity. The business will need to check with its insurance carrier to ensure coverage of young people. This is especially critical due to the wet and slippery environment in hatcheries and other potential hazards such as service on vessels. At the least, any sending organization should be asked to provide a release or otherwise indemnify the business in the event of accident. Government health and safety agencies should likewise be consulted to discern any legal restrictions on youths in or around hatcheries.

20.3.3 Undergraduate

Students entering college often have a rough notion of what interests them most academically. Others seek to answer that question by attending the plethora of courses offered by an undergraduate institution. Aquaculture

© Woodhead Publishing Limited, 2013

hatcheries provide students with an immersive experience that helps them define their interests or hone already acquired skills. Some universities have aquaculture programs on campus and these usually include hatchery operations as a part of their work (e.g. University of California at Davis, USA and King Mongkut's University of Technology Thonburi, Bangkuntien, Bangkok, Thailand). These may be related to general aquaculture or more specific studies such as nutrition or genetic improvement of stocks. Those with aquaculture engineering departments may be designing and testing various components for system development and instrumentation. In all, students may have access to a variety of academic disciplines while working on the various aspects of hatchery technology.

Undergraduate opportunities for learning about hatchery technology also become wider in areas where public and private sector operations exist. Many of these operations specifically target students for summer employment as a way of finding potential full-time future employees. Hatcheries can attract students by posting information on college campuses. Students can seek opportunities by contacting state and federal hatcheries and searching the internet. If the applicant possesses background knowledge about the science of the operation as well as enthusiasm for the job, he or she will likely be successful at gaining employment. Students desiring to work in a hatchery should be proactive and contact hatcheries to determine summer internship availability. Often if one hatchery is not hiring, others in the region could be and neighboring hatcheries are often aware of other openings.

The student who is interested in hatchery work and is fortunate enough to work on a campus where facilities exist should also look into the possibility of performing an independent project. This could be developed with an instructor and conducted for credit. Usually these types of projects are semester long and require a presentation and written report at their conclusion. This allows the student the opportunity to organize and conduct a study that can be of value to the hatchery. For those considering pursuing a graduate degree, such a project will not only provide scientific experience, but also can be useful as a point of consideration in graduate school applications.

Internships placing students at businesses to gain practical work experience are also offered in some institutions. The University College Cork, Ireland has an educational module which allows a student to be accredited for at least six weeks' work in an appropriate enterprise. The Asian Institute of Technology has placed many students in projects in over 30 countries since its inception in 1981. These students have engaged in projects on commercial farms in Thailand, as well as institutional facilities through the region and are collaborating with the NGO Aquaculture Without Frontiers to expand the program to other areas.

The undergraduate experience should aid students in determining their career paths as well as deciding whether or not to pursue a graduate

© Woodhead Publishing Limited, 2013

education. In many instances, the experiences gained by working on applied projects, such as those found in the operation of hatcheries, will lead students to pursue further education.

Aqua-tnet is an EU funded project which encourages industry involvement in student practicums and the development of appropriate curricula for relevant experiences. The goal is to develop partnerships between academia and the aquaculture industry that provide well trained and motivated students to become employees or, possibly, new entrepreneurs,

20.3.4 Graduate/post-graduate

At the graduate and post-graduate level, students interested in hatchery technology focus specifically on studies that can be applied to that field. Students enrolled in a graduate program not only take higher level courses to enrich their knowledge base, but most often are required to design and implement an original research project (Fig. 20.5). Some of these programs are specifically geared towards aquaculture, while others award a degree in a broader field, such as chemistry, biology, engineering, or ecology.

Those interested in chemistry may work on a range of topics, from water quality measurement and manipulation to biochemical treatments applied

Fig. 20.5 Graduate students often rely upon hatcheries for their research. Here a grad student extracts blood from an Atlantic sturgeon.

© Woodhead Publishing Limited, 2013

to stock species. Biochemistry is an area with particular application to hatcheries and aquaculture since it deals with the chemical processes in living organisms. Much of biochemistry involves cellular components that include proteins, carbohydrates, and lipids.

Biology students may choose to work in areas such microbiology or genetics. Students involved in the former field work with hatcheries to identify diseases of the animals and develop methods of controlling epizootics through the development of antibiotics. Those who are interested in the latter might work with a hatchery that participates in a breeding program to develop new or enhanced lines of animals.

Aquaculture genetics is a growing field; a number of public and private hatcheries cooperate with universities on research programs designed to enhance the performance of hatchery-produced animals used in grow-out or stocking operations. These partnerships may also pursue work on nutrition in order to maximize growth, survival, and other desired characteristics which are the hallmarks of cutting edge aquaculture production. Becoming knowledgeable about aquaculture genetics and nutrition is critical for success in aquaculture, as is keeping hatchery operators familiar with new advances in animal husbandry and hatchery practices.

Engineering studies at this level may include the development of hatchery equipment and techniques to enhance production or survival. There are many opportunities for engineering graduate students to observe and research such methodology. The use of recirculating or 'closed system' technology is a particularly attractive field that has had major advances in recent decades. Various components have been developed to make it easier to separate solids from culture water and to enhance biological filtration for increased survival and growth of the animals. The automation and computerization of hatcheries and culture systems is a growing field of investigation, with commercial equipment being created or adapted to the unique needs of aquaculture. For example, recent application of computerized monitoring and distribution systems in oyster hatcheries has made it possible to keep a determined number of phytoplankton cells available to feed larvae, the results of which increase their growth and survival through metamorphosis (Fig. 20.6).

20.3.5 Technical school/training programs

For those who are interested in hatcheries but do not wish to attend college, or for those who have some skills and are seeking to upgrade them, it would be advisable to consider technical schools and training programs. These are organized to provide very specific skill sets to trainees, skills that can be immediately applied in vocational settings. There are many examples of these throughout the world such as the North Coast Institute in Australia, which offers a certification course in Fish Hatchery Technology (NSW TAFE, 2012), and the US Fish and Wildlife Service's National Conservation

© Woodhead Publishing Limited, 2013

Fig. 20.6 Hatchery algologist using an automated monitoring and feeding system to feed larvae in a production-scale oyster hatchery.

Training Center (http://training.fws.gov), which offers a wide range of courses that include warm and cold water hatchery management.

Training programs may also be developed to offer skills to potential workers. These may be organized by either private or public sector groups and, while they do not provide degrees, they may issue certificates of completion. Participants gain specific knowledge, skills, and competencies to aid them in entering and advancing in industry. For hatcheries, training programs are often organized around specific aspects of the business, such as fish hatchery management or fish health diagnosis and control.

Many technical schools are two-year colleges. These institutions specialize in providing employment skills for trained labor. Normally the courses that they offer are based on local industry demand so that the potential for post graduation employment is enhanced. These will be in areas where aquaculture is well established and the skills learned can be put to immediate use by employees. Because training courses in these programs are specific to the industries they support and provide concentrated learning experiences, students frequently gain far more knowledge through them than they would from on-the-job training during similar time periods. Frequently, the advisory committees that develop and monitor these programs include prominent members of local industry who are very competent at commenting on the skills required of new employees. Students who attend

© Woodhead Publishing Limited, 2013

these technical schools may also become interested enough to decide to pursue undergraduate studies at a four-year institution.

20.3.6 Teacher training

Teacher workshops, usually at the state or regional level, are useful in creating information networks among education professionals. Some states provide this information in a formal program while others rely on teachers to share their own information. That has become a bit easier with the advent of social media. For those with formal programs, there are annual training sessions that are sometimes carried out at hatchery sites so that teachers can participate in the many activities that it takes to operate a facility and develop their curriculum units accordingly.

Networks, such as the National Aquaculture Educators Network (NAEN) in the USA, provide ready sources of information that can be used by educators in the classroom. One of their expressed purposes is to 'help educators motivate students using aquaculture and water-related topics in science as well as math, language arts, social studies and the fine arts' (NAEN, 2012). There are groups in other countries and in many states and regions that support aquaculture information and often there are ties to hatchery components.

20.3.7 Extension, outreach and technology transfer

An important area of education is that which takes place off campus and generally does not result in a degree or certification process. This is the essence of 'lifelong learning' since it brings information to those who require it for problem solutions in a 'real world' setting. While related, there are significant differences between these categories. Extension is the process of identifying needs of individuals or groups and organizing a series of educational activities designed to provide research-based solutions to them, with a final evaluation of the results (Seevers *et al.*, 1997). Outreach is a simpler and often informal dissemination of information to audiences through various means, targeting people or groups thought to be able to use or apply it in their area of interest (UMass Amherst, http://ag.umass.edu/index.php/outreach 2011). Technology transfer refers to the process of moving knowledge about scientific developments – often as equipment or production processes – to those most likely to be able to put it into application. Universities are now frequently stressing the need for both scientific research and technology commercialization (Owen-Smith, 2003) and one frequently finds references to the process of moving innovation to application by proponents of this form of knowledge advancement.

Extension in the USA is typified by the Land Grant College system, which includes teaching, research, and extension as components in supporting agriculture and aquaculture industries (Seevers *et al.*, 1997).

© Woodhead Publishing Limited, 2013

Research-generated knowledge is taught to students in campus educational programs while extension programs bring that information directly to producers through organized means. These methods include demonstration plots, lectures, and other educational programs, as well as print and electronic information.

There have been many definitions of extension over the years, but all involve expanding or extending the work of universities beyond the boundaries of the campus by providing information to those who can put it into application. Many nations have instituted extension services in an effort to better people's lives through educational programs; the international agencies have seen these as beneficial, although operationally they have undergone several distinct periods. In developing nations, a system of Training & Visitation (T&V) was used during the 1970s and 1980s (van den Ban, 1996). This was a top-down system that sought to bring the benefits of knowledge developed elsewhere to rural communities via experts. As such, it was paternalistic and required persuasion on the part of the teacher to get locals to adopt the practices. In some cases, these systems did not produce sustained or desired results (van den Ban and Hawkins, 1996). The method fell out of favor in preference to new methods that invoke empowerment of locals and participatory methods, such as farmer-to-farmer programs with participants encouraged to make their own decisions. Extension education is conducted at all levels of aquaculture, from small-scale to large-scale, in many nations and is recognized as an effective tool for turning scientific information from research into development (Pillay, 1947).

Extension educators often work with hatchery operators, helping them identify problems and assessing their need for information or decision-making tools. They organize programs designed to address these problems and use a variety of educational tools to provide information to the operators. Rather than supplying what the educator thinks is the answer, he or she will provide the operators with a range of options and work with them to select, implement, and evaluate their chosen path of action. The concept of evaluation is inherent to extension programs in which personnel must document the measurement of actual progress through the application of knowledge and its role in modifying behavior. It is this adherence to measuring the 'outcomes' of applied change rather than the 'outputs' of counting how many attend a particular program that is necessary to judge success.

One extension component in the USA that may have particular application to future development in hatchery training is the 4-H Program. This formal part of the Land Grant system has involved the training of youth for many decades. Participants may enroll in a local club that combines students with adult leaders who help them to choose and develop projects in many agricultural fields. While 4-H provides skills to youngsters through the establishment of formal projects, it also encourages leadership development to provide them with skills in areas such as public

© Woodhead Publishing Limited, 2013

speaking. The goal has been to develop future generations of leaders through education.

While there have been very few aquaculture 4-H and youth programs so far, this should be considered an important area for future development. One club in the USA has been formed to actually manage grounds, and grow and sell oysters in cooperation with a commercial enterprise (4-H Big Quil Enterprises, 2012). A few others have implemented oyster growing as a means of interesting children in the process of aquaculture. Others have used small-scale fish production as an educational project area. However, an important link could be developed between hatcheries and youth by integrating young people into the production process and allowing them to learn and understand it first-hand. Creating formal programs that would include print and electronic media could provide education across a wide geographic area and find integration into other nations.

Outreach education provides information to those who may need it but may not use a formal process of needs assessment. It also differs from extension by not including a broad program of education but rather focusing on dissemination to a limited audience; for example, a researcher who publishes the results of his or her investigations in a journal or industry periodical is targeting the select readers of that literature. The information may be found and used by those who have a need for it, but there is no formal series of programs that are created to disseminate the results. Outreach can also include scientific papers delivered at a conference or single presentations at a meeting organized around a particular issue. In any event, these are generally limited in their distribution and without the formal evaluation process of applied results that extension provides or without a well planned and sustained series of educational events or without formal evaluation of the results.

In creating outreach education, researchers would be well advised to search publications that are used by hatchery operators and to target articles and press releases to them. This will provide the opportunity for information to reach the required audience where it may be put into effect if the operator believes it will be beneficial or provide a prospective solution to a problem. With the current level of electronic technology available to many in the hatchery field, there is a strong demand for this type of information and those seeking to deliver it should thoroughly investigate the use of the internet and social media for its distribution. Along with written material, thought should be given to voiceover PowerPoint® presentations and video clips to further illustrate the points or concepts that are being disseminated.

Technology transfer (which is sometimes known as 'transfer of technology', 'tech transfer', 'T-square' or 'Tech Xfer') is utilized for getting the results of research into the hands of industry and is simply the process of transferring knowledge and skills to others. Technology transfer is used by businesses, government, and academic research institutions to ensure that

© Woodhead Publishing Limited, 2013

technological developments are made accessible to users to further exploit these new products. In some instances extension programs have been supplanted by governmental technology transfer programs.

Those involved in developing concepts and technology frequently seek legal protection by obtaining patents or copyrights. This allows them to license the results of their research when it is applied and used for processes in industry. Other business forms may be developed in order to apply these results to specific industries through the development of 'spin-offs'. Government institutions such as universities now include offices that patent the technology created by their research faculty in order to license it to industry. This raises additional funds for more research. In this regard, the universities' roles include: preserving intellectual property rights; facilitating partnerships to support further research or product development; and protecting academic research enterprises (AUTM, 2012).

Education is a key component of technology transfer because of the need for those transferring the technology to provide background on their products to those at the user level. Often the developer will need to work with the industry on the actual application of the technology to the particular problem or process that the user seeks to apply it to. In recent years there has been an increase in technology brokers and those who are intermediaries in the transfer of technology.

20.4 Important considerations for educational programs that utilize aquaculture

Education programs that utilize aquaculture or hatcheries must recognize the needs and constraints of the species under culture. Some aspects of animal husbandry require specialized culture systems, specific water quality parameters and controls, or other unique factors that are not easily duplicated in a normal school classroom. For example, bivalve spawning can be a very visual and hands-on educational tool but typically requires the ability to control the temperature and salinity of sea water for spawning to occur. It is unlikely that many school systems will have the infrastructure to support this activity. It may also be challenging for schools to obtain shellfish in a ripened state ready for spawning. For this reason, it is more appropriate to work with an operational hatchery to arrange a demonstration at their facility. Similarly, activities such as larval rearing may be too complicated for most students to successfully accomplish in the classroom. The school would need a consistent supply of sea water at the correct temperature, the ability to filter that water, cultured phytoplankton to feed the larvae, and someone to care for them seven days a week. Again, unless a local hatchery will agree to provide larvae and perhaps cultured phytoplankton during the normal school year, this activity may be problematic. Care should be taken when preparing a classroom activity to make sure

© Woodhead Publishing Limited, 2013

that the required facilities are available for the lesson, that a sufficient water supply is available, and that the correct life stages of the organism are obtainable. Inability to meet these requirements will usually yield a failed experiment that will be of little educational value to the students.

In order for students to derive the greatest value from a hatchery-based learning experience, it is advisable for hatchery personnel to understand where the lesson will fit in to the overall classroom syllabus. This can be done by consulting the teacher or professor in charge of the group or reviewing the appropriate curriculum requirements. Such requirements can often be found by visiting national or state education department websites which should provide a great deal of information about the content and skills that students need to master at different stages. Considering these objectives during the development of a hatchery-based learning program will help ensure that lessons are pertinent and effective.

Curriculum units that incorporate hatcheries into their planned activities can be exciting for students. Seeing science and mathematics in action gives meaning to classroom lessons, broadens the learner's base of knowledge, and allows students to practice skills that can aid them throughout their lives or perhaps inspire them to pursue career paths in aquaculture. The experience of working in a hatchery or of having hatchery components brought to the school provides practical, hands-on projects for learners. Even demonstrations by hatchery personnel contribute to the success of a curriculum unit when properly designed and implemented in cooperation with the classroom teacher. It should be noted that most hatcheries are located near the site where the actual grow-out or restoration activity is likely to occur. This usually means that citizens in that region have some understanding of the species being cultured. Often, there is much misinformation in the general community about the species. Well-planned and coordinated educational programs aimed at providing accurate information to people will go a long way towards correcting this misinformation.

Creating a partnership with a school, or even a local non-profit organization, benefits the educational and environmental communities as well as the hatchery facility and staff. The educational community has enhanced access to the hatchery facilities and the hatchery staff receive assistance with educational activities and/or volunteers to help with the operation. The more the public is involved, the better they understand the facts that surround a species and the more likely they are to support appropriate management activities. Furthermore, creating a group of individuals that understand the positive benefits of hatcheries can help to generate political support for hatchery operations. Having this sort of program in place at a hatchery may mean the difference during tough budget times when agencies are looking for facilities where they may cut budgets or close as part of fiscal reduction programs. The more important a facility is to the educational and environmental communities the less likely it will be to suffer severe budget reductions.

© Woodhead Publishing Limited, 2013

The relationship of public involvement to continued funding is particularly important for public sector hatcheries, most of which often understand the importance of education and include outreach in their scheduled activities. However, private sector hatcheries can also find that their participation in educational projects yields beneficial publicity and gains support for their industry by diffusing the knowledge about their operations and the resulting products. This can help them find investors as their business or overall industry expands and can provide a continuing source of new employees from the students who participate in projects or programs.

20.4.1 Education: A continuing effort

We have come to recognize that education is a lifelong experience. It makes sense to assume that those who continue to learn are those who generally have the greatest chance at success. In the field of hatcheries, education has many facets and applications, as we have shown. There is the education that occurs prior to when one decides to go into that field as well as the continuing education that is required to build skills and apply new technology to solve problems.

Education is truly a collaborative process in many aspects of hatchery operation, with those in the field educating each other and outsiders about their operations, problems, and needs. Researchers provide the required technology that can be applied to these issues, while developing educational programs that are effective and have a positive impact.

We can foster interest and learning early in a student's life by providing basic education to youngsters during field trips, open houses at hatcheries, and classroom activities. In later years, as their knowledge and skills increase, we can show them how to apply those in a practical setting by interesting them in hatcheries as places of learning. Throughout the college years we should actively promote the inclusion of students in hatchery work and research to support the students in nourishing their careers.

Those who are already operating hatcheries know that a skilled and interested workforce is necessary for effective production. Continuing employee education through attendance at meetings, conferences, and field days should be a part of any hatchery's plan for management. Creating a workplace where education is a valued prize can contribute to the overall success of its mission or profitability.

20.5 Future trends

The expansion of both public and private hatcheries to support public and private aquaculture is occurring in many nations and offers the

© Woodhead Publishing Limited, 2013

opportunity to create effective educational programs. Such programs offer exciting educational experiences for training new generations of technically competent students. The sharing of knowledge in the creation of these educational programs has become an integral part of the organized programs in many national and international aquaculture conferences. This exchange has allowed others to learn about innovative and successful efforts and to adopt concepts and adapt teaching methods and goals to meet local species and needs.

Perhaps the greatest boon to information sharing has been the development and expansion of electronic technology such as the internet. This has placed a world of information into the hands of people across the world that otherwise never would have contact with each other. Text, photos, video, and other presentation formats have made the expansion and diffusion of information easier with each passing year. But while the increase in information technology has been innovative, it has also brought with it the need to be cautious about the source and to verify claims and material that are placed in the public domain.

Some private hatcheries practice restricted or limited entry policies in which access to their facilities is not allowed for non-employees. Often this is caused by insurance stipulations or legal concerns. Biosecurity is often of great concern in hatcheries due to the fragility of early life forms or the potential for those from outside to transfer microbes that could be injurious to the animals. Sometimes these facilities may have quarantine areas where non-native or other animals of concern are kept. While instances where proprietary activities, equipment, or procedures may require limiting access, future hatchery operators will likely need to develop relationships with the many government agencies that are scrutinizing their operational practices. Additionally, hatchery operators should strive to publicize the environmentally friendly nature of their businesses. An excellent method is to establish a close working relationship with local schools, environmental groups, and regulatory agencies. An open door policy can aid in establishing strong, positive relationships.

Proprietary practices, equipment, and techniques that need to be protected should be accommodated, but, in general, as one Pacific Northwest Coast hatchery owner stated, 'what makes a hatchery operation successful typically cannot be photographed'. An added benefit from an open door policy is the development of a network of operators who provide a valuable resource when problems arise.

The electronic information age is important, and will become more so, but greater impact can be gained by witnessing animals being cultured. Hands-on activities become a hatchery operator's ally in educating groups or individuals about the economic and environmental benefits that their facility brings to the community. The educational component of any hatchery should not be looked upon merely as an option; rather it is an important

© Woodhead Publishing Limited, 2013

piece of the overall operation of the facility that, when given proper emphasis, enhances production and/or profit.

20.6 Sources of further information

Below is a listing of websites that demonstrate the global variety of aquaculture-based educational programs. These online resources may aid hatchery operators in developing or enhancing such programs. They also provide valuable information for teachers who are looking to incorporate aquaculture into their curriculum. Regardless of the species addressed on these websites, the content may spur ideas from hatchery personnel or educators and can often be adapted to pertain to a relevant animal.

- Dr Gavin Partridge, International ISS Institute/DEEWR Trades Fellowship, Australia:
  http://issinstitute.org.au/wp-content/media/2011/04/ISS-FEL-REPORT-G-PARTRIDGE-low-res.pdf

This outreach publication describes the emerging field of the hatchery production of yellowfin tuna and is the product of a fellowship supported by the Australian Government's Department of Education, Employment, and Workplace Relations.

- Seymour Salmonid Society, Canada:
  http://www.seymoursalmon.com/gds.php

This is a description of 'Gently Down the Seymour', a field trip program designed to give elementary school students hands-on experience both in the field and in a salmon hatchery.

- The Moyola Hatchery, Ireland:
  http://www.salmon-ireland.com/salmon-fishing-articles/hatching-a-plan-for-the-moyola/hatching-a-plan-for-the-moyola-micheal-mcglade.jsp

Here is an example of how a trout and salmon hatchery use the internet to educate readers about its hatchery operations in detail.

- National Trout Centre, New Zealand:
  http://www.greatlaketaupo.com/new-zealand/Nationaltroutcentre/

This website provides a description of the educational offerings at this hatchery in New Zealand.

- SEAFDEC Aquaculture Department, Philippines:
  http://www.seafdec.org.ph/training_schedule.html

Provided as an example of the technical training courses available, this website has a course schedule pertaining to a variety of species and topics.

© Woodhead Publishing Limited, 2013

- King Mongkut's University of Technology, Thailand: http://www.pdti.kmutt.ac.th/en/

The information on this website describes the variety of aquaculture-based research programs at this university in Thailand.

- Education at the National Lobster Hatchery in Padstow Cornwall, UK: http://www.nationallobsterhatchery.co.uk/education/index.htm

An overview of the educational experience at this hatchery in Great Britain, as well as information about lobster biology and activities, is provided on this website.

- Aquaculture Information Center – DOC/NOAA, USA: http://www.lib.noaa.gov/retiredsites/docaqua/k12.html

This website, maintained by the Department of Commerce and the National Oceanic and Atmospheric Association, gives a listing of K-12 educational sites related to aquaculture.

- National Geographic Chesapeake Classrooms Community, USA: http://www.ngsednet.org/community/resources_category.cfm?community_id=151&category_id=228

Here there are a number of lesson plans and teaching resources about the Chesapeake Bay for grades K-12.

- Chesapeake Bay Trust, USA: http://www.cbtrust.org/site/c.miJPKXPCJnH/b.5457695/k.E7B8/Education.htm

This website provides educational resources, as well as a listing of 'Aquaculture in the Classroom' programs for schools in Maryland.

- Aquaculture in Action, USA: http://www.mdsg.umd.edu/programs/education/AinA/

This program, developed by the Maryland Sea Grant College, partners aquaculture educators with local school systems. Teacher training workshops, a web-based component for data entry, and technical support throughout the school year are key components.

- University of Maryland College of Agriculture and Natural Resources, USA: http://www.nrac.umd.edu/EducationalResources/K-12%20education.cfm

This source has links to presentations, videos, and print publications regarding the advantage of and best approach to bringing aquaculture into the classroom

- Southold Project in Aquaculture Training, USA: http://ccesuffolk.org/spat/

© Woodhead Publishing Limited, 2013

This program involves the community by enlisting volunteers to work in the hatchery and grow their own shellfish gardens.

- Department of Agriculture Farm Service Agency, USA: http://www.fsa.usda.gov/FSA/webapp?area=home&subject=oued&topic=landing

Here you will find information about seminars, grants, and technical support for farmers and hatchery operators.

- Friends of the Decorah Fish Hatchery, USA: http://www.facebook.com/pages/Friends-of-the-Decorah-Fish-Hatchery/136921346344860?sk=wall

This is an example of how a nonprofit organization, created to help fund and organize the education programs at a fish hatchery, uses social media to promote its cause.

- University of California Davis, Biological and Agricultural Engineering Department, USA: http://bae.engineering.ucdavis.edu/pages/Research/aquaculture.html:

This is another example of an undergraduate/graduate research program in the field of aquaculture engineering.

20.7 References

4-H BIG QUIL ENTERPRISES (2012) http://bigquil.blogspot.com/search/label/Big%20Quil%20Enterprises (accessed September 2012).

AQUACULTURE IN ACTION (2012): http://www.mdsg.umd.edu/programs/education/AinA/(accessed September 2012).

AUTM (2012) *University technology transfer: 'Why we do what we do'*. Deerfield, IL, Association of University Technology Managers, available at: http://www.autm.net/AM/Template.cfm?Section=White_Papers&Template=/CM/ContentDisplay.cfm&ContentID=1895 (accessed September 2012).

BARROWS H S and TAMBLYN R M (1980) *Problem-based learning: An approach to medical education*. New York: Springer.

BEERS G W and BOWDEN S (2005) The effect of teaching method on long-term knowledge retention, *Journal of Nursing Education*, 44, 511–514.

CDSL (2000) *How people learn: brain, mind, experience, and school*. Washington DC: National Academies Press.

COLORADO DIVISION OF WILDLIFE (2011) The hatchery game: An interactive math and science resource, available at: http://wildlife.state.co.us/Education/TeacherResources/FishHatcheryGame/Pages/HatcheryManager.aspx (accessed September 2012).

CRAIK F and LOCKHART R (1972) Levels of processing: A framework for memory research, *Journal of Verbal Learning and Verbal Behavior*, 11, 671–684.

DALE E (1969) *Audio-visual methods in teaching* (3rd edn). New York: Holt, Rinehart and Winston.

© Woodhead Publishing Limited, 2013

DECHARMS R (1976) *Enhancing motivation: Change in the classroom*. New York: Irvington.

DEWEY J (1916) *Democracy and education: An introduction to the philosophy of education*. New York: WLC Books.

DEWEY J (1938) *Experience and education*. New York: Collier Books.

FARKAS R (2003) Effects of traditional versus learning-styles instructional methods on middle school students, *The Journal of Educational Research*, 97, 42–51.

FORGUS R and SCHWARTZ R (1957) Efficient retention and transfer as affected by learning method, *Journal of Psychology: Interdisciplinary and Applied*, 43, 135–139.

GORDON L M (2007) Service learning and teacher education in reading, *Academic Exchange Quarterly*, 11, 23–27.

HAMER L O (2000) The additive effects of semi-structured classroom activities on student learning: An application of classroom-based experiential learning techniques, *Journal of Marketing Education*, 22, 25–34.

JOHNSON D and JOHNSON R (1994) *Learning together and alone, cooperative, competitive and individualistic learning*. Needham Heights, MA: Prentice-Hall.

JOHNSON D, JOHNSON R and HOLUBECK E (1998) *Cooperation in the classroom*. Boston, MA: Allyn and Bacon.

JOHNSON D, JOHNSON R and STANNE M B (2000) Cooperative learning methods: A meta-analysis, available at: http://www.tablelearning.com/uploads/File/EXHIBIT-B.pdf (accessed September 2012).

KOLB D A and FRY R (1975) Toward an applied theory of experiential learning, in Cooper C (ed). *Theories of Group Process*. London: John Wiley, 33–58.

KVAM P (2000) The effect of active learning methods on student retention in engineering statistics. *The American Statistician*, 54, 136–140.

NAEN (2012) National Aquaculture Educators Network, http://www.aces.edu/dept/fisheries/education/NationalAquacultureEdNetwork.php (accessed September 2012).

NCSS (2011) *National curriculum standards for social studies: introduction*. Silver Spring, MD: National Council for Social Studies, available at: http://www.socialstudies.org/standards/introduction (accessed September 2012).

NSW TAFE (2012) *Hatchery operations: Statement of attainment in seafood industry studies*. Grafton, NSW: New South Wales Technical and Further Education Commission, available at: http://northcoast.tafensw.edu.au/natfish/Documents/Hatchery%20SOA%202012.pdf (accessed September 2012).

OREGON SEA GRANT (2012) *Free choice learning*. Corvalis, OR: Oregon Sea Grant, available at: http://seagrant.oregonstate.edu/education/free-choice-learning (accessed September 2012).

OWEN-SMITH J (2003) From separate systems to a hybrid order: Accumulative advantage across public and private science at Research One universities, *Research Policy*, 32(6), 1081–1104.

PILLAY T V R (1947) *Planning of Aquaculture Development – an introductory guide*. Rome: UN Fish and Agriculture Organization (FAO).

RAVENS R W (1980) *Action learning: New techniques for management*. London: Blond & Briggs, Ltd.

SEEVERS B, GRAHAM D L and GAMON J A (1997) *Education Through Cooperative Extension*. Albany, NY: Delmar.

SMITH B and MACGREGOR J (1992) What is collaborative learning? in Goodsell A, Maher M, Tinto V, Smith B and MacGregor J (eds), *Collaborative Learning: A Sourcebook for Higher Education*. University Park, PA: Pennsylvania State University. available at: http://www.clinicalresearchlearning.org/index.php?option=com_content&view=article&id=13&Itemid=46 (accessed September 2012).

© Woodhead Publishing Limited, 2013

TANNER T (1980) Significant Life Experiences: A New Research Area in Environmental Education, *Journal of Environmental Education*, 11, 20–24.
THOMAS J W (1980) Agency and achievement: Self-management and self-regard, *Review of Educational Research*, 50, 213–240.
VAN DEN BAN A W and HAWKINS H S (1996) *Agricultural Extension* (2nd edn). London: Blackwell Science Ltd.

© Woodhead Publishing Limited, 2013

Index

© Woodhead Publishing Limited, 2013

© Woodhead Publishing Limited, 2013

© Woodhead Publishing Limited, 2013

© Woodhead Publishing Limited, 2013

© Woodhead Publishing Limited, 2013

© Woodhead Publishing Limited, 2013

© Woodhead Publishing Limited, 2013

© Woodhead Publishing Limited, 2013

© Woodhead Publishing Limited, 2013

© Woodhead Publishing Limited, 2013

© Woodhead Publishing Limited, 2013

© Woodhead Publishing Limited, 2013

© Woodhead Publishing Limited, 2013

© Woodhead Publishing Limited, 2013

© Woodhead Publishing Limited, 2013

© Woodhead Publishing Limited, 2013

© Woodhead Publishing Limited, 2013

© Woodhead Publishing Limited, 2013